ADVANCES IN CERAMICS – ELECTRIC AND MAGNETIC CERAMICS, BIOCERAMICS, CERAMICS AND ENVIRONMENT

Edited by **Costas Sikalidis**

Advances in Ceramics – Electric and Magnetic Ceramics, Bioceramics, Ceramics and Environment
Edited by Costas Sikalidis

Published by InTech
Janeza Trdine 9, 51000 Rijeka, Croatia

Publishing Process Manager Niksa Mandic

Technical Editor Teodora Smiljanic

Cover Designer InTech Design Team

Additional hard copies can be obtained from orders@intechopen.com

Advances in Ceramics – Electric and Magnetic Ceramics, Bioceramics,
Ceramics and Environment, Edited by Costas Sikalidis
p. cm.
ISBN 978-953-307-350-7

For EU product safety concerns:
IN TECH d.o.o., Prolaz Marije Krucifikse Kozulić 3, 51000 Rijeka, Croatia,
info@intechopen.com or visit our website at intechopen.com.

Meet the editor

Dr. Costas Sikalidis, born in Thessaloniki Greece in 1948, matriculated in the Chemistry Department at the Aristotle University of Thessaloniki in 1966. After earning his BSc, he served for three years as an officer in the Hellenic Army. He pursued postgraduate studies in Ceramic Technology at the University of Northstaffordshire, UK and completed training & professional development in the British Ceramic Industry. He worked as a production manager at Philkeram-Johnson, subsidiary of Johnson Tile manufacturers. In 1981 he joined the Department of Chemical Engineering as a faculty member holding a PhD in Chemical Engineering. An Associate Professor and Head of the Laboratory of Industrial Inorganic Raw Materials and Industrial Ceramics, since 2008 he has published more than 50 papers in peer-reviewed scientific journals, presented his research in more than 80 International Conferences, received several patents and has been a reviewer in 8 journals. Dr. Sikalidis has mentored more than 18 MSc graduates and produced 2 PhD graduates, whereas he has earned funding via a number of competitive European research grants. Both an academic and a person originating from the Industry he has been maintaining ties, collaborations and outreach with the Ceramic and Building Materials Industry. A proud father of two sons, married to Irene a Medical Doctor and a Chemist.

Contents

Preface

Ceramic materials research is currently increasing as ceramics entail a quickly expanding field due to the vast range of both traditional and special applications in accordance to their characteristics and properties. Electric and magnetic ceramics, bioceramics and ceramics related to the improvement of environmental parameters, consist important areas of research demonstrating high potential and particularly great interest. Research in these fields requires combined knowledge from several scientific fields of study (engineering, physical sciences, biology, chemistry, medicine) rendering them highly interdisciplinary. Consequently, for optimal research progress and results, close communication and collaboration of various differently trained researchers such as medical doctors, bioscientists, chemists, physicists and engineers (chemical, mechanical, electrical) is vital.

Some of today's most interesting research topics in the electric and magnetic ceramics, included in this volume, are covered by discussing studies on: $Ba(Zr_{0.2}Ti_{0.8})O_3$ and $BaTiO_3$ ceramics prepared by spark plasma sintering targeting to the fabrication of lead-free piezoelectrics; the influence of Sr and Zr content on the dielectric properties of $(Ba_{1-x}Sr_x)(Ti_{0.5}Zr_{0.5})O_3$ ceramics; the influence of various dopants and manufacturing parameters on properties; integrated piezoceramics as a base of intelligent actuators; the ceramic based intelligent piezoelectric energy harvesting devices and of ultrasonic piezoelectric ceramic power transducers. Additionally several other studies are discussed like, the studies of advanced PMN-PT ferroelectric systems and of modern types of hexaferrites in connection to their applications in agriculture, medicine, computer engineering, telecommunication and television. Also, the studies of relaxor dielectrics, of the dielectics anomalies at high temperatures, of the properties of polymer based nanodielectrics and those of glass ceramics with para, anti or ferroelectric active phases, are developed in this book. The studies on structural, morphological, magneto-transport and thermal properties of antimony substituted $(La,Pr)_{2/3}Ba_{1/3}Mn_{1-x}Sb_xO_3$ perovskite manganites and of metal oxide ZnO-based varistor ceramics used to protect circuits against excessive voltages are also discussed.

Ceramics are now commonly used in the medical fields giving rise to the category of ceramics termed "bioceramics". Bioceramics and bioglasses are ceramic materials that are biocompatible, non-toxic and non-inflammation causing. These materials must encompass certain properties and to be used in certain applications it is required to be, bioinert (not interactive with biological systems), bioactive (can undergo interfacial interactions with surrounding tissues), biodegradable, soluble or resorbable (eventually replaced or incorporated into tissue). Much work has been carried out the last years on bioceramics since there is a great demand on the market of biomedical applications. Ambitious goals appeared in the field including developing sensing devices made with ceramics and organic matter (i.e., collagen), developing ceramics with piezoelectric properties, manufacturing electromagnetic wave-sensing ceramics or composites for transfer of energy in the eye or the ear as well as for nerve signal transduction. Also, the combination of nanotechnology and biomaterials seems to have unlimited potential for future applications.

Some very interesting topics from today's research in bioceramics, included in this volume, are: The studies on biocompatibility; studies on the Role of Aluminum Ceramics in Total Hip Arthroplasty assisting the development of new ceramic components that will be beneficial for patients; the biological, mechanical and optical properties of zirconia dental ceramics which have emerged as versatile and promising materials; the studies of silicon nitride ceramics and the rationale for their use in biomedical applications; the marine-based carbon and silicon carbide scaffolds with patterned surface for tissue engineering applications; the use of bioglass additive in zirconia for reducing the latter's sintering temperature and production cost while maintaining the biocompatibility of the final product; the comparative studies in patients after implantation of apatite-wollastonite glass-ceramic and b-tricalcium phosphate used to fill the bones after curettage of bone cycts

Furthermore, the role of ceramics in the field of environmental protection has been increasing drastically over the last years. Environmental cleaning/decontamination, stabilizing toxic and hazardous elements, waste treatment and recycling, environmental conservation, ecological building, protection from electromagnetic fields and radiation are some of the today's fields of research in the brad area that relates ceramics to the environment.

Rather interesting research, included in this volume, is focused on water cleaning from toxic elements, through the development of techniques for obtaining potassium polytitanate nanoabsorbent ceramics for the removal of lead ions from water. Other interesting included is research on the developments in stabilization mechanisms, incorporation efficiencies, metal leaching properties of product phases aiming to stabilize metal bearing toxic wastes by incorporation into the matrix of ceramic products. Ceramics' production result to thousand of tones of wastes and very useful research is underway on the utilization of these wastes in ecoefficient concrete and precast concrete products.

The current book consists of twenty-four chapters divided into three sections.

Section I includes fourteen chapters in electric and magnetic ceramics which are dealing with modern specific research on dielectrics and their applications, on nanodielectrics, on piezoceramics, on glass ceramics with para-, anti- or ferro-electric active phases, of varistors ceramics and magnetic ceramics.

Section II includes seven chapters in bioceramics. These chapters include review information and research results/data on biocompatibility, on medical applications of alumina, zirconia, silicon nitride, ZrO_2, bioglass, apatite-wollastonite glass ceramic and b-tri-calcium phosphate.

Section III includes three chapters in applications of ceramics in environmental improvement and protection, in water cleaning, in metal bearing wastes stabilization and in utilization of wastes from ceramic industry in concrete and concrete products.

Constantinos A. Sikalidis
Department of Chemical Engineering
Aristotle University of Thessaloniki, GR

Part 1

Topics in Electric and Magnetic Ceramics

1

Preparation and Properties of $BaTiO_3$ and $Ba(Zr,Ti)O_3$ Ceramics by Spark Plasma Sintering

Hiroshi Maiwa
Shonan Institute of Technology
Japan

1. Introduction

In this chapter, the dielectric and electromechanical properties of $Ba(Zr_{0.2}Ti_{0.8})O_3$ and $BaTiO_3$ ceramics prepared by spark plasma sintering (SPS) are reported. Those of ceramics prepared by conventional sintering are also reported for comparison with the SPS-prepared ceramics. The obtained information is helpful for possible application to the fabrication of lead-free piezoelectrics.

1.1 $BaTiO_3$ and $Ba(Zr,Ti)O_3$

Barium titanate has recently attracted attention due to the demand for lead-free piezoelectrics. Barium titanate ceramics prepared by microwave sintering (Takahashi et al., 2006, 2008) or two-step sintering (Karaki et al., 2007) with fine grains approximately 1 μm in size show excellent piezoelectric properties. These high piezoelectric properties are considered to be due to the small grain size. It is well known that the suppression of grain growth results in low-density samples by conventional sintering. Therefore, spark plasma sintering (SPS) was applied in the present study.

Zr-doped $BaTiO_3$ (BZT) ceramics are interesting materials that exhibit linear field-induced strain for actuator applications. We have previously reported the microstructure and the dielectric and electromechanical properties of these materials in thin film form (Maiwa et al., 2010); however, a characterization of the ceramic BZT with fine grains has not yet been carried out.

1.2 Spark plasma sintering

SPS is a process that uses electrical discharge between particles under pressure of several megapascals. SPS enables a compact powder to be sintered to a high density at a relatively low temperature and with a shorter sintering period.(Munir et al., 2006, 2011) In addition, SPS has an advantage over conventional sintering in that it suppresses exaggerated grain growth. Thus far, SPS has been applied to fabricate various piezoelectric ceramics, the reported results indicate that SPS is a powerful technique and opens the possibility of processing ceramics with controlled sub-micron grain sizes.(Hungria et al., 2009) Lead titanate (Kakegawa, 2004), $NaNbO_3$ (Wada et al., 2003), and $Na_xK_{1-x}NbO_3$ (Zhang, 2006) ceramics have been prepared by SPS. SPS has also applied to preprare $BaTiO_3$ ceramics and exhibited high dielectric constant of 10000; however, electromechanical properties has not been reportred(Takeuchi et al., 1999).

2. Experimental

2.1 Sample preparation

The starting powder used were commercial $Ba(Zr_{0.1}Ti_{0.9})O_3$ and $Ba(Zr_{0.2}Ti_{0.8})O_3$ ceramic powder (Sakai Chemicals, Japan) and $BaTiO_3$ ceramic powder (Toda Kogyo, Japan) . These powders were prepared by the hydrothermal method. The purities of the sample powder were more than 99%. In the case of conventional sintering, the powder was supplemented with 1% polyvinyl alcohol (PVA) binder, pressed in a die at a pressure of 80 MPa and sintered in air for 2 h from 1100 to 1450°C. In the case of SPS, no binder was added to avoid residual organics. Since the pellet is pressed during SPS, a binder is not required. For SPS, SPS-511S (SPS Syntex Inc., Japan) was used; raw powder was placed in a graphite die (10 mm diameter), and sintering was carried out in air atmosphere at a pressure of 60 MPa. The temperature was increased to 900-1100°C within 11 minutes and maintained at that temperature for 5 minutes, after which the pressure was released and the sample was cooled to room temperature. Since the pellet as-sintered by SPS at 1100°C is black and conductive, the pellet was annealed at 900-1400°C for 12 h in air.

2.2 Characterization

The surface of the sintered ceramics was observed by scanning electron microscopy (SEM, Hitachi S-2100A). The sintered samples were polished and then produced electrodes using a silver paste. Measurements of the electric field-induced displacement and polarization in BZT ceramics were performed using displacement sensor (Mahr GmbH, Millimar Nr. 1301, Germany) and a charge-amplifier circuit (Kitamoto Electronics, POEL-101, Japan). An alternating electric field of 0.1 Hz was used in these measurements. Prior to the small-signal measurements, including resonant-antiresonant methods and d_{33} measurements with the d_{33} meter, the ceramic specimens were polarized for 20 min in a silicone bath under a DC field of 20 kV/cm at room temperature. The resonant-antiresonant methods were carried out using an impedance analyzer (HP 4192A) for an additional 24 h after the polarization. The d_{33} meter (Chinese Academy of Science ZJ-3B, China) was used for the d_{33} measurements.

3. Structure and properties

3.1 $Ba(Zr,Ti)O_3$

Dielectric properties were evaluated in both $Ba(Zr_{0.1}Ti_{0.9})O_3$ and $Ba(Zr_{0.2}Ti_{0.8})O_3$ ceramic. Structure and electromechanical and piezielectric properties were evaluated mainly in $Ba(Zr_{0.2}Ti_{0.8})O_3$(BZT20) ceramic.

3.1.1 Density and microstructure

$Ba(Zr_xTi_{1-x})O_3$ (BZT, x=0.1, and 0.2) ceramics are prepared by SPS and conventional sintering. By application of SPS, the $Ba(Zr,Ti)O_3$ ceramics with more than 96% relative densities could be obtained by the sintering at 1100˚C for 5 minutes in air atmosphere. The grain growth is suppressed in the ceramics prepared by SPS, the average grain sizes were less than 1micron. Carbon contents of SPS prepared BZT ceramics and the conventionally sintered BZT are 0.15% and 0.024%, respectively.(Maiwa, 2008b) It should be noted that the SPS prepared BZT ceramics examined carbon content contained organic binder intentionally for comparison. Since the organic binder is not added to SPS prepared ceramics usually, carbon contents of the SPS prepared ceramics would be less than 0.15%.

The SPS-BZT20 ceramics prepared by SPS at 1100°C and then annealed at 1100°C, 1200°C, and 1300°C were 5.89, 5.87, and 5.83 g/cm³, respectively. These ceramics were almost fully sintered. SEM images of the BZT20 ceramics prepared by SPS and normal sintered are shown in Figs. 1 and 2, respectively. The SPS-BZT20 ceramics annealed at 1200°C were found to have very small grains less than 1 μm in diameter. In the SPS-BZT20 ceramics annealed at 1300°C, small grains less than 1 μm in diameter and relatively large grains several tens of microns in diameter coexisted. Since SPS provided rapid sintering at 1100°C within 5 minutes, grain growth was suppressed. As described later, grain growth is limited by normal sintering at 1300°C and lower, and the fine grains of as-SPS ceramics are taken over after a lower post-annealing temperature of 1200 and 1300°C. The grains of the SPS-BZT20 ceramics annealed at 1400°C were all relatively large, more than several tens of microns in diameter.

In conventional sintering, the relative density was found to increase and the grains grew with increases in the sintering temperature. The relative densities of the ceramics sintered at 1300, 1350, 1400, and 1450°C were 4.67, 5.03, 5.68, and 5.77 g/cm³, respectively. These values were lower than those of the SPS-BZT20 ceramics. It should be noted that the normally sintered ceramics contained pores, as show in Fig. 2. The average grain sizes were approximately 1 μm for the samples annealed at 1300-1400°C, with the size increasing slightly with temperature. Grain growth occurred over the range from 1400-1450°C.

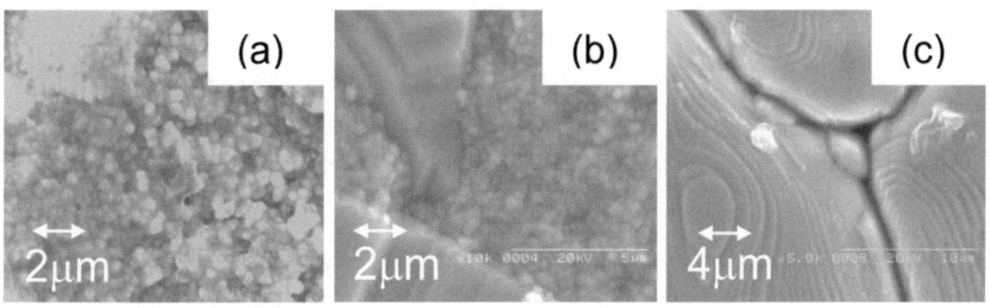

Fig. 1. SEM images of the $Ba(Zr_{0.2}Ti_{0.8})O_3$ ceramics SPS-prepared at 1000°C and annealed at (a) 1200, (b)1300°C, and (d)1400 °C.

Fig. 2. SEM images of the $Ba(Zr_{0.2}Ti_{0.8})O_3$ ceramics normally sintered at (a) 1300, (b) 1350, (c) 1400, and (d)1450.

3.1.2 Dielectric properties

Figure 3 and 4 show the temperature dependence of the dielectric constant and loss tangent of the $Ba(Zr_{0.1}Ti_{0.9})O_3$ and $Ba(Zr_{0.2}Ti_{0.8})O_3$ ceramics in the temperature range of 30 - 150°C, respectively. The dielectric anomalies in these $Ba(Zr_{0.1}Ti_{0.9})O_3$ and $Ba(Zr_{0.2}Ti_{0.8})O_3$ ceramics occur at approximately 90 and 40°C, respectively. These coincide well with reported values.

With increasing sintering temperature in normal sintering, the jumps accompanying the dielectric anomaly become clear. The temperature dependence of the SPS-prepared BZT ceramics is relatively mild, and the nominal value is higher than that of the sample sintered at relatively lower temperatures of 1200 and 1250°C. The mild temperature dependence of the SPS BZT ceramics is considered to be due to small grains. The relatively high dielectric constant is attributed to the high density.

It was reported that by Takeuchi et al. $BaTiO_3$ SPS-prepared at 1000°C and with grains less than 1 μm does not show a broadening transition but a sharp transition. (Takeuchi et al, 1999) Kinoshita et al. reported that fine-grained $BaTiO_3$ ceramics with 1.1 μm grains exhibit sharp transition.(Kinoshita & Yamaji 1976) In this study, the broad transitions are observed in $Ba(Zr_{0.1}Ti_{0.9})O_3$ ceramics with 0.56 μm grains and $Ba(Zr_{0.2}Ti_{0.8})O_3$ ceramics with 0.86 μm and 0.53 μm grains. At present, the reason for the difference in transition between pure $BaTiO_3$ and zirconium-containing $BaTiO_3$ system is not clear. It is considered that the relaxor nature of the zirconium-containing $BaTiO_3$ system induces a broadening of the transition in ceramics with larger grains, than pure $BaTiO_3$ ceramics. In the $BaTiO_3$-$BaZrO_3$ system, dielectric relaxation is induced by the addition of nonpolar $BaZrO_3$, and disorder is considered to be expanded by separating a large number of small grains. The transition is further broadened in small-grained samples of SPS $Ba(Zr_{0.2}Ti_{0.8})O_3$, supporting the above speculations.

There is another point to be compared with pure $BaTiO_3$. It has been reported that the dielectric constant of pure $BaTiO_3$ at room temperature has been determined for ceramics of approximately 1 μm grain size. In this study, no marked increase in dielectric constant at room temperature is observed in $Ba(Zr_{0.1}Ti_{0.9})O_3$ and $Ba(Zr_{0.2}Ti_{0.8})O_3$ ceramics. Compared with the conventionally sintered ceramics of the same grain size, the SPS-prepared $Ba(Zr_{0.1}Ti_{0.9})O_3$ and $Ba(Zr_{0.2}Ti_{0.8})O_3$ ceramics exhibit relatively larger dielectric constants at room temperature; however, this is mainly due to the increase in their densities.

The dielectric properties of the SPS-prepared ceramics are understood to be dependent on the enlargement of the small grains of the ceramics sintered at low temperature, as shown in Fig. 1. It is reasonable to say that no marked increase in the dielectric properties of the materials occurs in the SPS-prepared samples; however, the elimination of pores plays a major role in increasing dielectric constant. It is difficult to discuss the effect of density on dielectric constant quantitatively. Conductivity affects dielectric constant through capacitance; however, is difficult to take conductance into account in the calculation of dielectric constant. Moreover, the formularization of pore distribution is difficult. By assuming a model of a series of capacitors consisting of air and dielectrics materials, the thicknesses ratio of air to the dielectrics of 1/1000, and a dielectric constant of 5000 in the dielectrics, the measured dielectric constant becomes 17% of the pure dielectrics. Although this model is too simple; however, it shows that the elimination of pores enhances the dielectric constant, and it can roughly explain the high dielectric constant of the dense ceramics prepared by SPS in the entire throughout measured temperature range. In other words, the dielectric constants of the small-grained $Ba(Zr_{0.1}Ti_{0.9})O_3$ and $Ba(Zr_{0.2}Ti_{0.8})O_3$ ceramics are weakly temperature-dependent basically. The low dielectric constants of the small-grained ceramics normally sintered are due to the low density. While, SPS-prepared ceramics are dense and composed of small grains, their dielectric constant is high and weakly temperature-dependent.

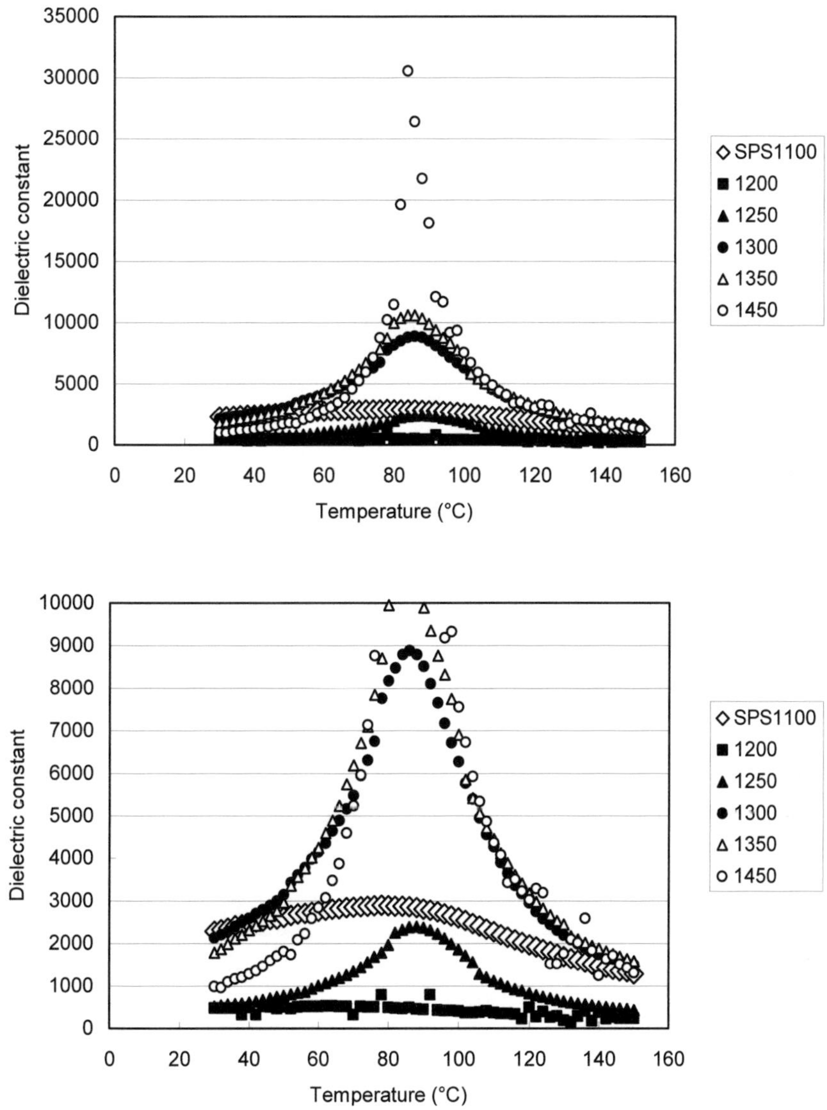

Fig. 3. Temperature dependences of (a) dielectric constant, (b) dielectric constant (expanded), of $Ba(Zr_{0.1}Ti_{0.9})O_3$ ceramics.

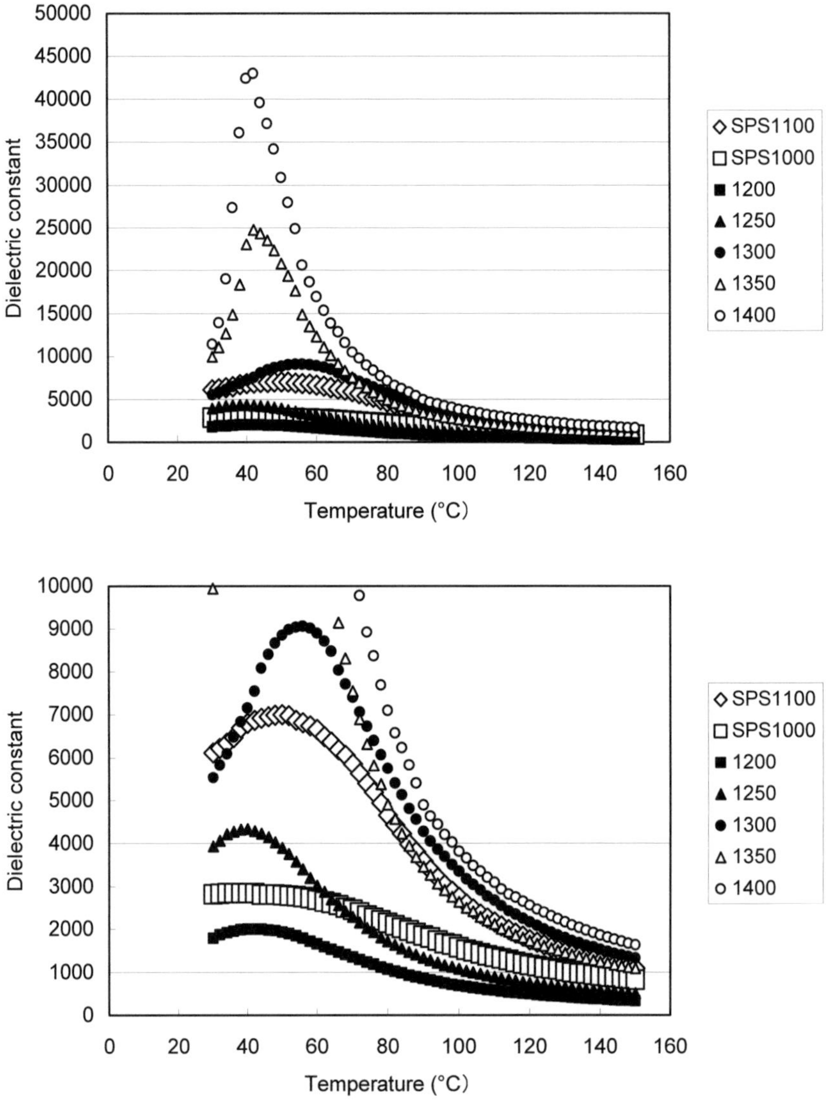

Fig. 4. Temperature dependence of (a) dielectric constant, (b) dielectric constant (expanded), of $Ba(Zr_{0.2}Ti_{0.8})O_3$ ceramics.

The ceramics with low density exhibit a relatively high loss tangent, probably due to a large numbers of defects. SPS and normally sintered ceramics with high densities exhibit smaller loss tangent, generally less than 2%. An increase in loss tangent accompanying dielectric anomaly is more clearly seen in the $Ba(Zr_{0.2}Ti_{0.8})O_3$ ceramics, probably due to the greater diffusivity of the transition.

3.1.3 Electromechanical properties

The Strain/Field in the SPS $Ba(Zr_{0.1}Ti_{0.9})O_3$ and $Ba(Zr_{0.2}Ti_{0.8})O_3$ are 76pm/V under 24.5kV/cm and 252pm/V under 13.9kV/cm, respectively. Figures 5 and 6 show the field-induced strain of SPS-BZT20 ceramics annealed at 1200 and 1300°C and the BZT20 ceramics normally sintered at 1300 - 1450 °C, respectively. The SPS-BZT20 ceramics annealed at 1200°C exhibited a strain loop with less shrinkage and lower displacement. Since the strain hysteresis behavior accompanying shrinkage is derived from ferroelectric domain switching, ferroelectric domain activities are suppressed, probably due to the residual stress or small grains, or both. The SPS-BZT20 annealed at 1300°C exhibited a strain loop with large shrinkage and larger displacement. The SPS-BZT annealed at 1400°C was too leaky to measure the dynamic strain loop under application of a DC field of 10kV/cm and higher.

In the case of normal sintering, the BZT20 ceramics sintered at 1350°C exhibited the highest strain among the samples measured. The sample sintered at 1300°C exhibited smaller strain due to the small grains, low density, or both. The samples sintered at 1400 and 1450°C exhibited smaller displacement than that sintered at 1350°C. The grain sizes of these samples were considered to be larger than appropriate for this material. The strain loops became more hysteretic with increasing sintering temperature. In the case of pure $BaTiO_3$, the ceramics with a grain size of 0.61-0.74 μm exhibited the largest field-induced strain, and the ceramics with smaller and larger grains exhibited lower strain. The results obtained here for BZT followed the grain size dependencies seen in pure $BaTiO_3$.

The unipolar field-induced strains of these samples were also measured. The general tendencies were the same as those observed with the bipolar strain loops. The dynamic strain/field at 20 kV/cm of SPS-BZT annealed at 1300°C and the BZT20 ceramics normally sintered at 1350 °C were 290 and 280 pm/V, respectively. These two values are comparable; however, the SPS-BZT20 yielded a more linear strain curve compared with the sample normally sintered at 1350°C. This difference can be shown clearly in the dependence of electric field on dynamic d_{33}, which is calculated from the strain/field. The results are shown in Fig. 7. Considering the relatively low dielectric constant of 1204, the reason for the linear strain of SPS-BZT20 is considered to be due to the suppressed polarization rotation.(Maiwa 2008a) The lower hysteretic strain with good linearity for the SPS-BZT20 ceramics is unique and might be desirable for actuator applications that require analogue operations.

3.1.4 Static piezoelectric properties

The clear resonances were observed only for the SPS-BZT annealed at 1300°C and the BZT20 ceramics normally sintered at 1400 °C, as shown in Fig. 6. The piezoelectric properties calculated by the resonance method are included in Table 1. In the case of the SPS-BT, the Qm and kp values are 44-62 and 16.2-17.5, respectively, which are smaller than the Qm and kp values of 325 and 25.1 (%) for the SPS-BZT20 annealed at 1300°C. The reason for this difference is not clear, but the unique microstructure composed of submicron and coarse grains might play a role in producing the reasonably high Qm and kp values. These ceramics exhibit a small loss tangent, generally less than 2%.

The d_{33} values measured with the d_{33} meter are shown in Fig. 7. This measurement method is more sensitive than dynamic measurement. The low values are derived from the insufficient polarization due to the grains being too small or high conductivity of the samples. The d_{33} values for the SPS-BZT20 annealed 1300°C and the BZT20 ceramics normally sintered at 1400 °C are of 126 and 138pC/N, respectively. Yu et al. have reported d_{33} values of 130pC/N measured in $Ba(Zr_{0.08}Ti_{0.92})O_3$ ceramics by resonance-antiresonance measurements (Yu et al., 2002), and the values obtained in this study are reasonable in comparison with the values.

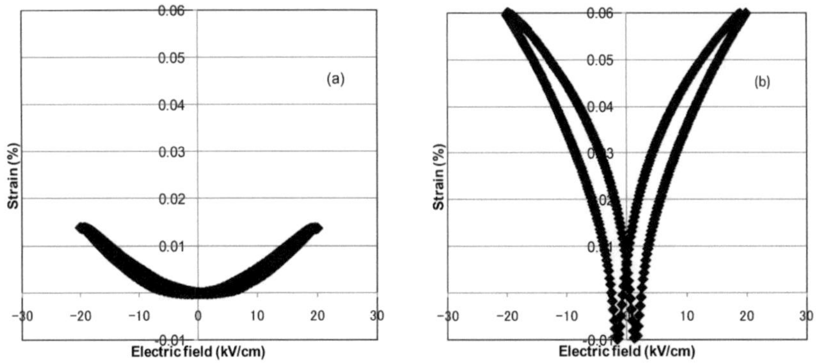

Fig. 5. Field-induced strain of the $Ba(Zr_{0.2}Ti_{0.8})O_3$ ceramics SPS-prepared at 1100°C and then annealed at (a)1200 and (b)1300°C.

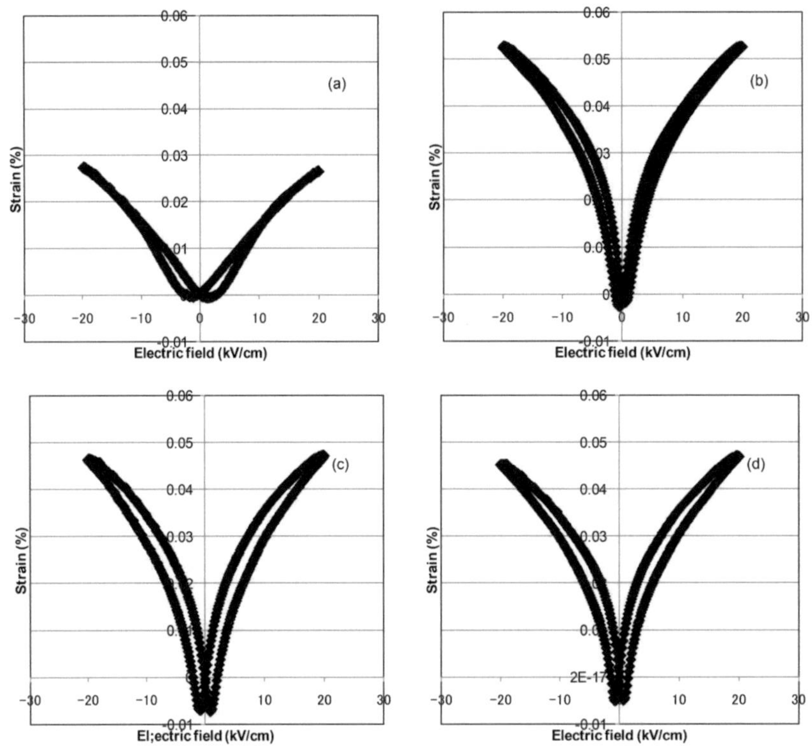

Fig. 6. Field-induced strain of the $Ba(Zr_{0.2}Ti_{0.8})O_3$ ceramics normally sintered at (a)1300, (b)1350, (c)1400, and (d)1450 °C.

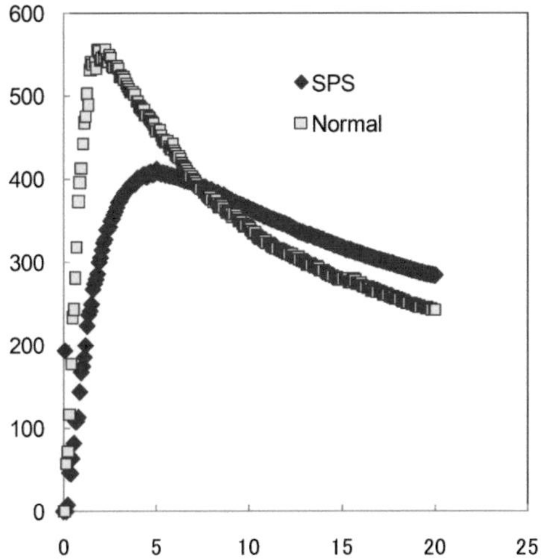

Fig. 7. Field dependence of the dynamic d_{33} calculated from the unipolar field-induced strain of the $Ba(Zr_{0.2}Ti_{0.8})O_3$ ceramics SPS-prepared at 1100°C(SPS) and then annealed at 1300°C and normally sintered at 1400°C(Normal).

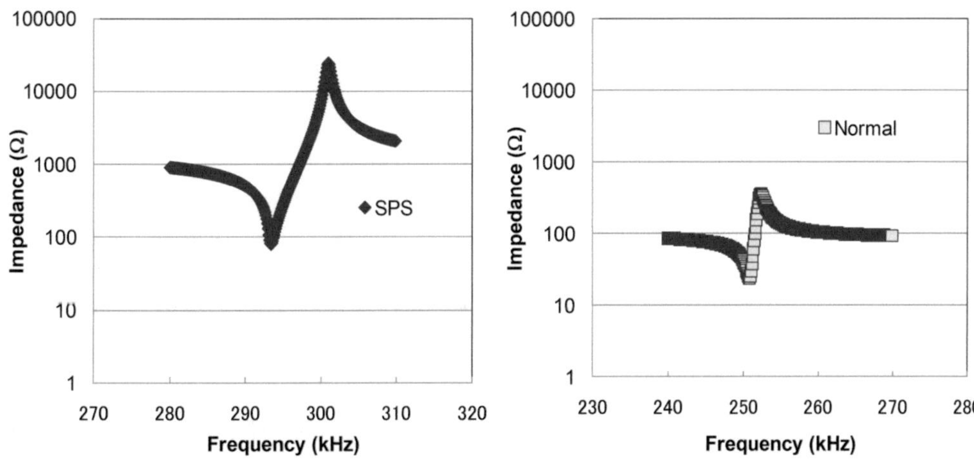

Fig. 8. Resonance-antiresonance measurement of the $Ba(Zr_{0.2}Ti_{0.8})O_3$ ceramics SPS-prepared at 1100°C and then annealed at 1300°C(SPS) and normally sintered at 1400°C(Normal).

Sample	Qm	kp (%)	Dielectric constant	$d_{31}(10^{-12}C/N)$
SPS1300	312	25	1204	43
Normal 1400	119	21	7869	91

Table 1. Piezoelectric properties of the $Ba(Zr_{0.2}Ti_{0.8})O_3$ ceramics SPS-prepared at 1100°C and then annealed at 1300°C (SPS1300) and normally sintered at 1400°C(Normal 1400).

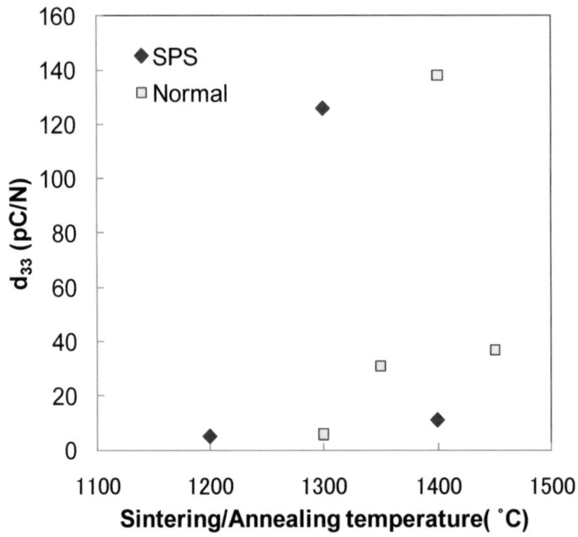

Fig. 9. The d_{33} values measured by d_{33} meter.

3.2 $BaTiO_3$

Density, dielectric and preliminary electromechanical properties were evaluated in the $BaTiO_3$ creramics SPS-prearred at 900°C. Electromechanical and piezielectric properties were evaluated mainly in the $BaTiO_3$ creramics SPS-prearred at 950°C.

3.2.1 Density and microstructure

The as-sintered pellet prepared by SPS at 900-1100 °C was black and conductive. Although SPS was carried out in air atmosphere, the samples were deoxidized by heating the carbon die. By post-annealing at 900-1200 °C for 12 h in air, the pellet was oxidized and became white and insulating. These features are similar to those of the $Ba(Zr,Ti)O_3$. The relative densities of the BT ceramics prepared by SPS at 900-1100 °C were 5.84-5.97 g/cm^3. These ceramics are almost fully sintered. Compared with $Ba(Zr_{0.2}Ti_{0.8})O_3$ ceramics, the BT ceramics can be sintered at lower temperatures. This is due to the smaller particle size or the nature of the pure BT. In conventional sintering, the relative densities of the samples increase with sintering temperature, as shown in Fig. 10. SEM images of the BT ceramics prepared by conventional sintering and by SPS are shown in Figs. 11 and 12, respectively. The average grain sizes of samples of these ceramics are shown in Fig. 13. It is noted that grain growth is promoted by high-temperature annealing and suppressed by SPS. The average grain sizes of the BT ceramics prepared by SPS at 900-1200 °C, which increase with annealing temperature, are below 1 μm. These values are almost equivalent to those of ceramics of the same composition conventionally sintered at 1100-1200 °C. Figures 14 and 15 show the X-ray diffraction patterns of the BT ceramics prepared by normal sintering and by SPS, respectively. The diffraction peaks of the starting powder are broad and no splits due to the distortion from cubic structure are observed. The cubic structure of fine BT powder has been frequently reported. The peaks of the as-SPS ceramics, the ceramics prepared by SPS without annealing, are broad and shifted to a lower angle. The lattice elongation is caused by deoxidization. The X-ray diffraction patterns of the ceramics conventionally sintered at

1100 °C and the ceramics prepared by SPS at 900 °C and then annealed at 900-1100 °C were different in terms of the normal splitting peaks, with 1:2 intensity ratio of tetragonal (002) and (200) observed in the ceramics conventionally sintered at 1200 and 1300 °C. This is due to the structural change derived from the stress in the small grain below 1 μm.

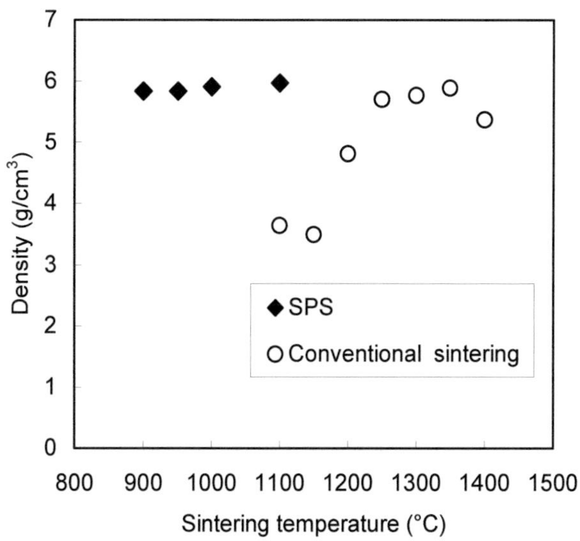

Fig. 10. Relative densities of $BaTiO_3$ samples.

Fig. 11. SEM images of the $BaTiO_3$ ceramics normally sintered at (a) 1100, (b) 1200, (c) 1300, and (d) 1400 °C.

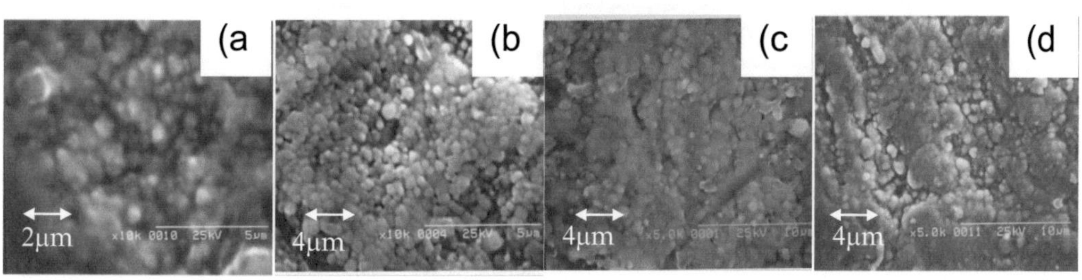

Fig. 12. SEM images of the $BaTiO_3$ ceramics SPS-prepared at 900 °C and then sintered at (a) 900, (b)1000, (c) 1100, and (d)1200 °C.

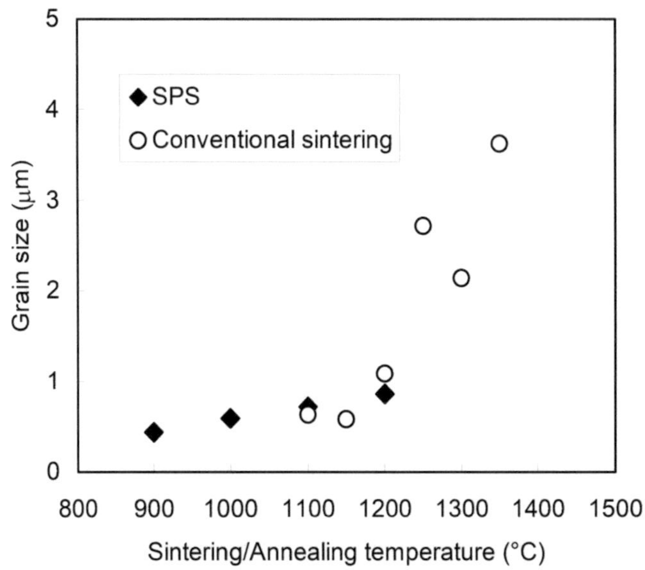

Fig. 13. Average grain sizes of $BaTiO_3$ samples.

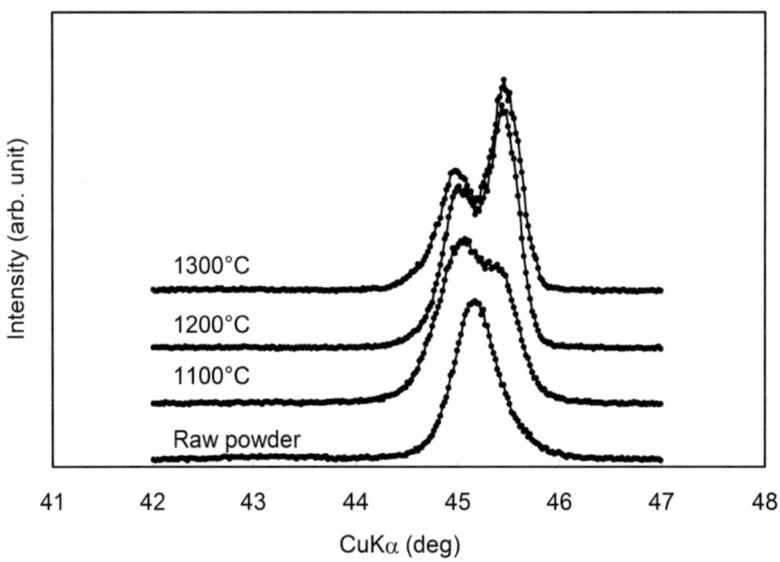

Fig. 14. X-ray diffraction patterns of starting powder and the $BaTiO_3$ ceramics conventionally sintered at 1100, 1200, and 1300 °C.

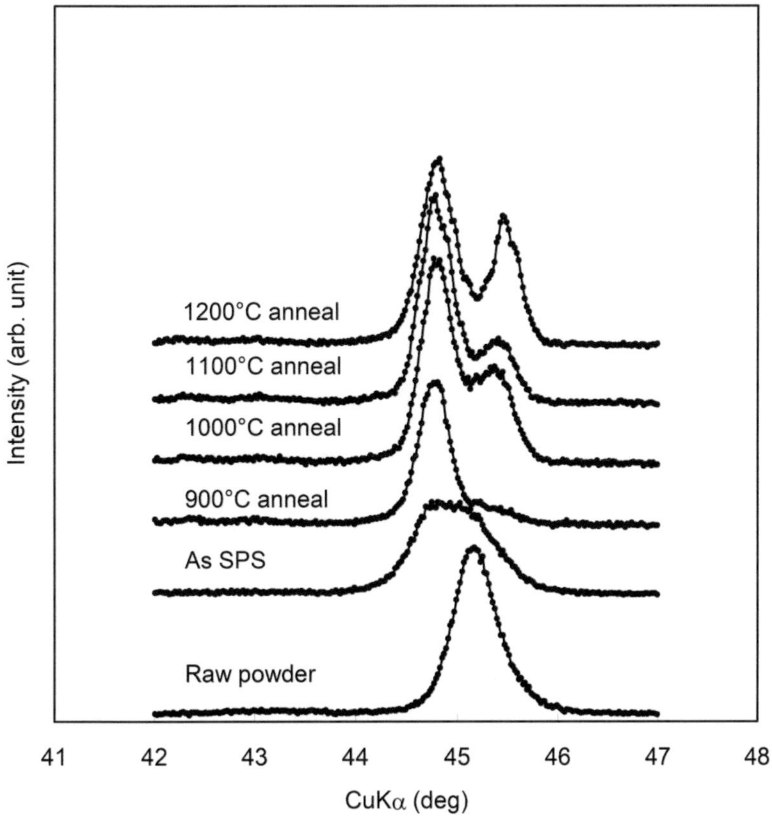

Fig. 15. X-ray diffraction patterns of starting powder and the $BaTiO_3$ ceramics SPS-prepared at 900°C and then sintered at (a) 900, (b)1000, (c) 1100 and (d)1200 °C.

3.2.2 Dielectric properties

Figure 16 shows the dielectric constant of the BT ceramics at room temperature. Figure 17 shows the temperature dependence of the dielectric constant of the BT ceramics prepared by normal sintering and by SPS in the temperature range of 30-150 °C. The dielectric constants at room temperature of the BT ceramics prepared by SPS and then annealed at 1000 and 1100 ˚C, whose grains are 0.61-0.74 μm in size, are highest among the samples prepared in this study. This result agrees well with the reported values. It was reported by Takeuchi et al. that BT ceramics SPS-prepared at 1000 °C with grains of less than 1 μm size, showed a dielectric constant of 10000. (Takeuchi et al., 1999) Kinoshita and Yamaji reported that the fine-grained BT ceramics with 1.1 μm grains exhibited a high dielectric constant of approximately 5000.(Kinoshita & Yamaji, 1976) Arlt et al. reported that the fine-grained BT ceramics with 0.7 μm grains exhibited a high dielectric constant of approximately 5000 at room temperature.(Arlt et al., 1985)

The transition temperature is another point to be compared with previous reports on pure BT. The dielectric anomalies in the BT ceramics occur at approximately 120 °C. In this study, a lowering dielectric anomaly with annealing temperature was observed, as shown in Fig. 17. Kinoshita and Yamaji reported that the fine-grained $BaTiO_3$ ceramics with 1.1 μm grains exhibited a negligible shift compared with the ceramics with 53 μm grains.(Kinoshita & Yamaji, 1976) Takeuchi et al. reported that the BT ceramics SPS-prepared at 1000 °C and with grains of less than 1 μm size showed a shift in the Curie temperature.(Takeuchi et al., 1999) It has been reported that the Curie temperature of the pure BT ceramics with fine grains shifts to a lower temperature. (Line & Glass, 1977) (Xu et al., 1989) The decrease in the Curie temperature is due to the effects of the fine grains.

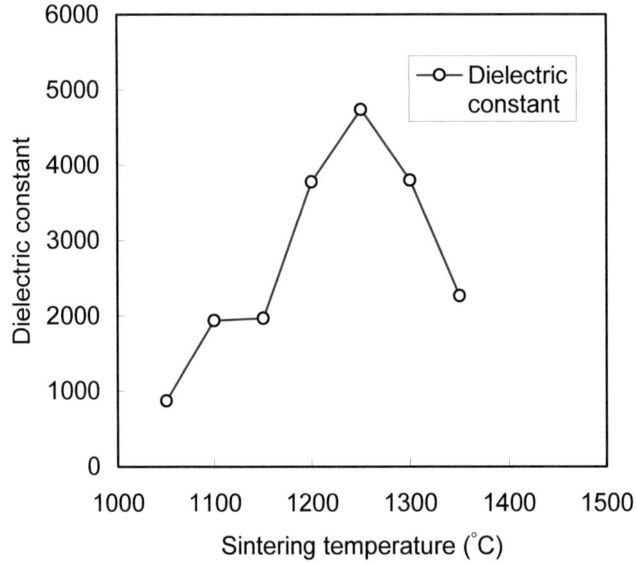

Fig. 16. Dielectric constant of the conventionally sintered $BaTiO_3$ ceramics.

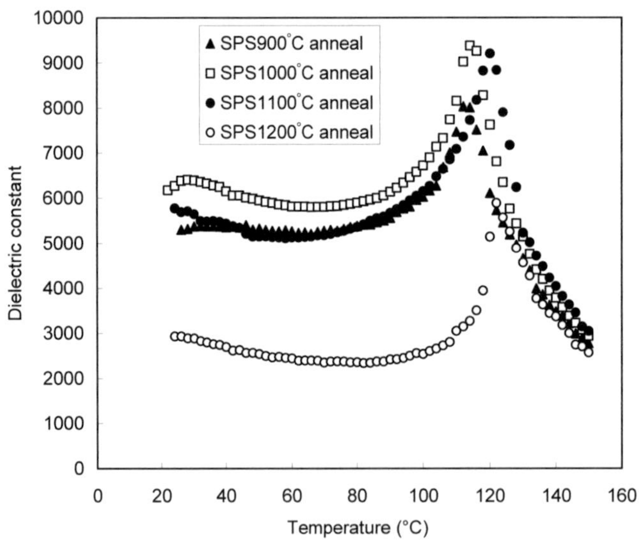

Fig. 17. Temperature dependence of dielectric constant of the $BaTiO_3$ ceramics SPS-prepared at 900°C and then sintered at (a) 900, (b)1000, (c) 1100 and (d)1200 °C.

3.2.3 Electroimechanical properties

The field-induced displacement of the BT ceramics prepared by SPS at 900 °C were measured. The bipolar field-induced strain loops of the BT ceramics prepared by SPS and then annealed at 900-1200 °C are shown in Fig. 18. With increasing annealing temperature, the strain loops became slim. This is due to the ease of the domain motion in larger grains. A larger displacement of the BT ceramics prepared by SPS was observed in the samples annealed at 1000 and 1100 °C, which have larger dielectric constants than the samples annealed at 900 and 1200 °C. The unipolar field-induced strain loops and calculated dynamic d_{33} of the BT ceramics prepared by SPS and then annealed at 1000 and 1100 °C are meaured.(Maiwa 2008a) The strain/field values at 15 kV/cm of the BT ceramics prepared by SPS and then annealed at 1000 and 1100 °C are 540 and 530 pm/V, respectively. These values are comparable to the reported high remanent d_{33} values of the $BaTiO_3$ ceramics prepared by microwave sintering or two-step sintering.

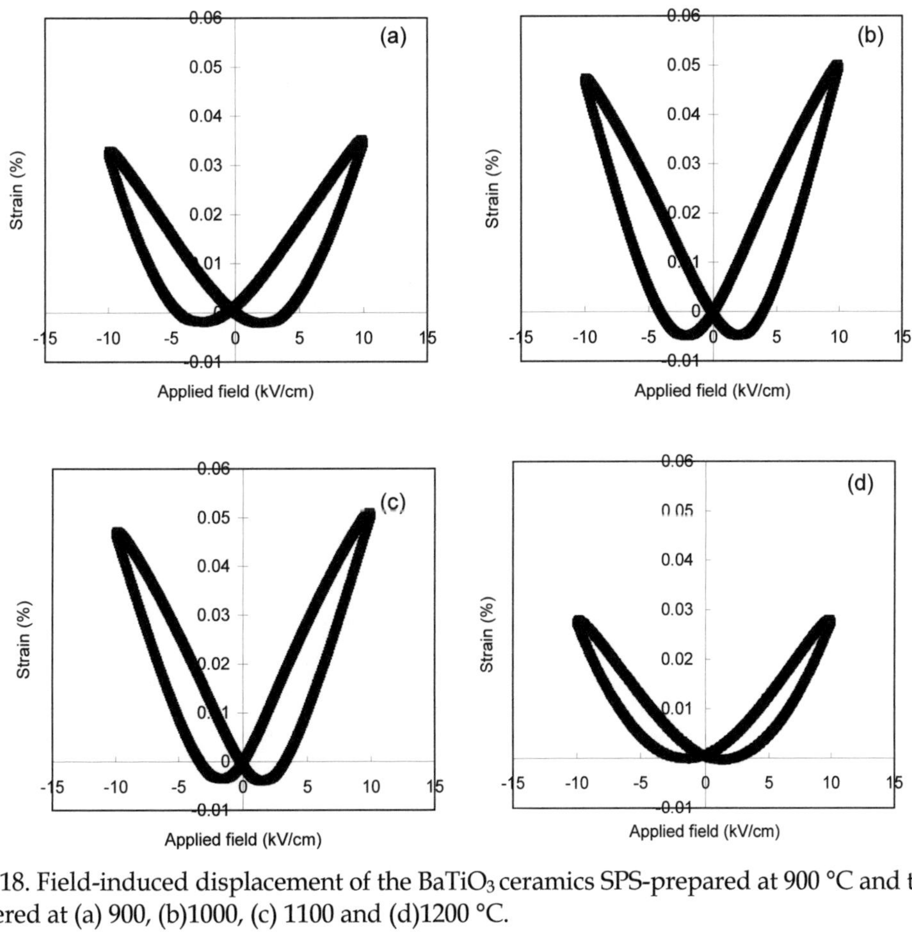

Fig. 18. Field-induced displacement of the $BaTiO_3$ ceramics SPS-prepared at 900 °C and then sintered at (a) 900, (b)1000, (c) 1100 and (d)1200 °C.

The bipolar polarization and field-induced strain loops of the SPS-BT, conventionally sintered, and two-step-sintered BT ceramics are shown in Figs.19-21, respectively. Note that the SPS-BT ceramics exhibit relatively thin polarization loops. Large hysteretic strain loops are obtained in the ceramics sintered at a high temperature of 1400 ˚C. The large hysteresis is due to the effects of polarization rotation in larger grains. The unipolar strain and derived dynamic d_{33} values of these BT ceramics are measured.(Maiwa, 2009) SPS-BT ceramics exhibit relatively large strains regardless of their fine grains. Compared with the fine-grained BT ceramics fabricated by other methods, such as the BT ceramics conventionally sintered at 1300 °C and the BT ceramics two-step-sintered at 1300 °C, the SPS-BT ceramics exhibit high d_{31}. The field-induced strain loops of the SPS-BT ceramics are linear; this corresponds to the flat dynamic d_{33} behavior under high field. Large hysteretic strain loops and high calculated d_{33} values are observed in the ceramics sintered at high temperature. The calculated d_{33} values of these samples decrease markedly under high field. These phenomena can be explained as follows. The samples exhibit large induced strains due to polarization rotation under low field, and the induced strains decrease together with the completion of polarization rotation. In the case of the SPS-BT ceramics, the observed linear strain behavior is considered to be due to the suppressed polarization rotation and electrostrictive strain reflected by the high dielectric constant, or both mechanisms.

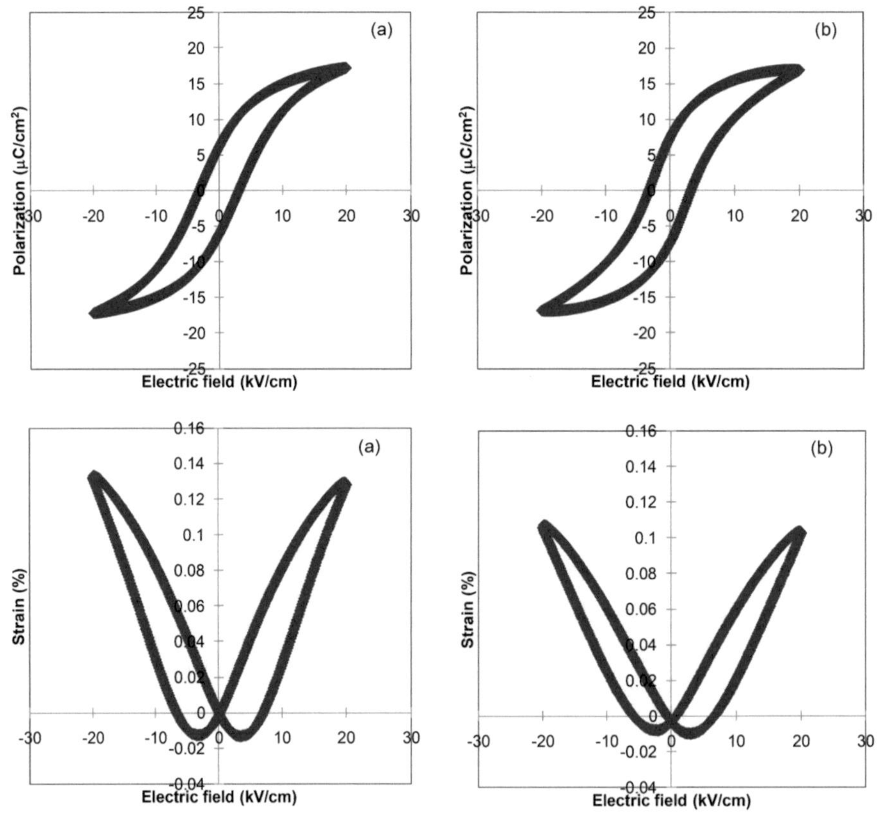

Fig. 19. Polarizations and field-induced strains of the BT ceramics SPS-prepared at 950 °C and then annealed at (a)1000 and (b)1200 °C.

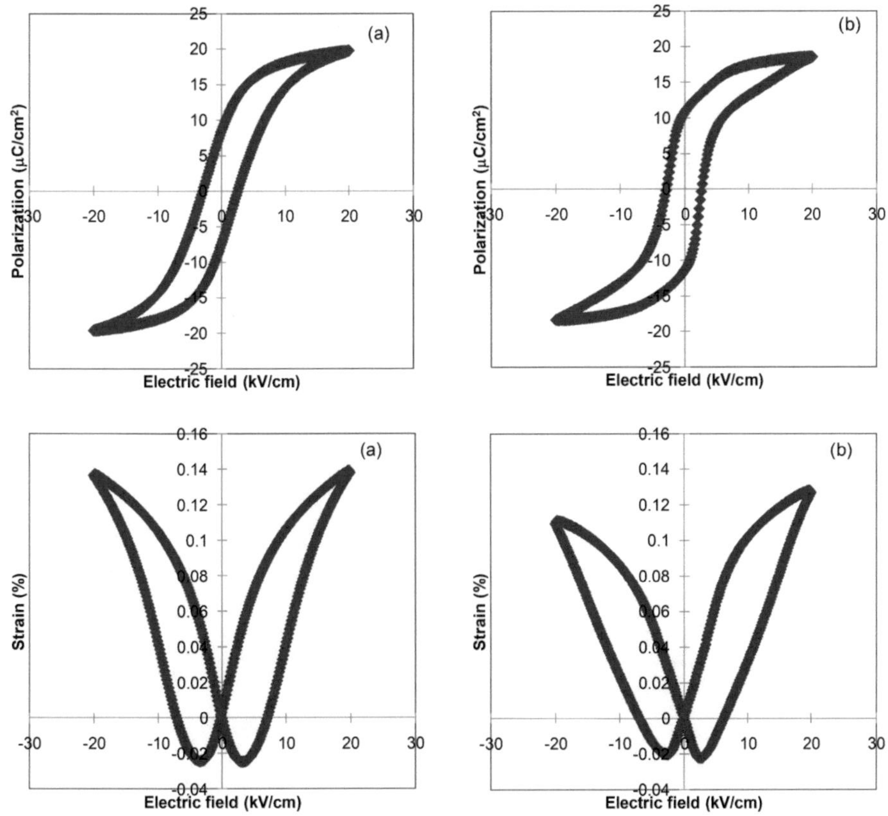

Fig. 20. Polarizations and field-induced strains of the BT ceramics conventionally sintered at (a) 1300 and (b) 1400 °C.

3.1.4 Static piezoelectric properties

The piezoelectric properties calculated by resonance methods are included in Table 2. The SPS-BT ceramics are characterized by high d_{31} and low Qm. Compared with the fine-grained BT ceramics fabricated by other methods, the SPS-BT ceramics exhibit high d_{31}. Relatively high d_{31} values correspond to large field-induced strains. Here, I discuss the origin of the large difference between d_{31} obtained by the resonance methods and d_{33} calculated from the slope of the dynamic-field-induced strain measurement. The major differences of these measurements lie in the driving frequency and field amplitude. The frequencies and electric field amplitudes in the resonance methods and the dynamic measurement are 200-300 kHz and 14-15 V/cm, and 0.1 Hz and 20-30 kV/cm, respectively.

A slow and high-field dynamic measurement detects the displacement including polarization rotation that requires large energy. A rapid and low-field resonance method eliminates the displacement due to polarization rotation. The dynamic d_{33} values calculated from the linear part of the slope under a high field of more than 20 kV/cm are 200-350 pm/V generally. These values roughly correspond to twice the d_{31} value calculated by the resonance method, indicating that the explanations above are reasonable. Qm is related to the internal stress. The low Qm of the SPS-BT ceramics suggests the presence of high

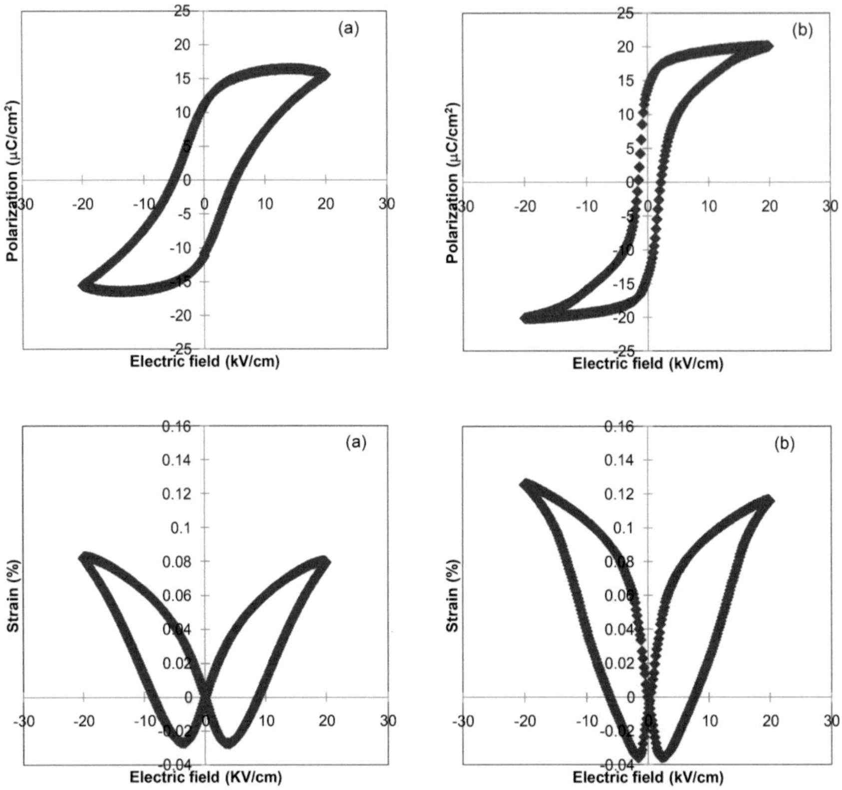

Fig. 21. Polarizations and field-induced strains of the BT ceramics two-step-sintered at (a) 1300 and (b) 1400 °C.

Sample	Density (g/cm^3)	Grain size (μm)	Qm	kp (%)	Dielectric constant	d_{31} (10^{-12} C/N)	Loss tangent	Np (Hz·m)
SPS1000	5.98	0.7	62	16.8	5444	66	0.02	2753
SPS1100	6.00	1.3	44	17.5	5491	73	0.024	2653
SPS1200	5.95	2.5	51	16.2	3406	51	0.039	2743
Normal1300	5.87	3.6	169	14	3356	44	0.014	2796
Normal1400	5.78	26	936	38.7	2333	99	0.0068	2942
two-step1300	5.15	1.2	216	8.9	3850	39	0.015	2408
two-step1350	5.55	3.1	243	17.1	3360	56	0.013	2806
two-step1400	5.82	7.5	530	38	2236	87	0.0075	2978

Table 2. Piezoelectric properties of the $BaTiO_3$ ceramics.

internal stress in the SPS-BT samples. By considering the distorted X-ray diffraction peaks and decrease in transition temperature, it is speculated that the high internal stress still remains in the SPS-BT ceramics even after 12 h postannealing. Regardless of the fabrication methods, the kp of the fine-grained BT ceramics is low generally. This is probably due to the insufficient poling treatment of the fine-grained samples.

4. Conclusion

By application of SPS, dense $Ba(Zr,Ti)O_3$ and $BaTiO_3$ ceramics with fine grains can be obtained. The properties are quite unique, which cannot be easily obtained by other methods. And this method offers applicability to other dielectric and piezoelectric materials and may yields unique properties similar with the ones observed in $BaTiO_3$ based ceramics.

5. Acknowledgment

I would like to thank Mr. Masakazu Kawahara of SPS Syntex Inc. for the spark plasma sintering of the BZT and BT powders. This study was supported in part by a grant-in-aid from the ministry of education (No. 22560671), sports, culture, and technology, Japan and the Iketani science and technology foundation. Some figures and some parts of the sentences are used under the permission of the Japanese Journal of Applied Physics.

6. References

Arlt, G.; Hennings, D. & De With, G. (1985) Dielectric properties of fine-grained barium titanate ceramics, *Jounal of Applied Physics,* Vol. 58, No. 4 (1985) pp. 1619-1925, ISSN 00218979

Hennings, D.; Schnell, A. & Simon, G. (1982) Diffuse ferroelectric phase transitions in $Ba(Ti_{1-y}zr_y)O_3$ ceramics. *Journal of the American Ceramic Society* Vol. 65, No. 11, (November 1982) pp. 539-544, ISSN 0002-7820

Hungria, T.; Galy, J. & Castro, A. (2009) Spark plasma sintering as a useful technique to the nanostructuration of piezo-ferroelectric materials, *Advanced Engineering Materials,* Vol. 11, No. 8, (August 2009) pp. 615-631, ISSN14381695

Kakegawa, K. et al., (2004) Sintering of lead titanate using a spark-plasma-sintering technique, *Journal of the American Ceramic Society* Vol. 87, No. 4, (April 2004) pp. 541-545, ISSN 0002782

Karaki, T. et al., (2007). Lead-free piezoelectric ceramics with large dielectric and piezoelectric constants manufactured from $BaTiO_3$ nano-powder. *Japanese Journal of Applied Physics,* Vol.46,No. 4-7 (February 2007), pp.L97-L98, ISSN 00214922

Kinoshita, K. & Yamaji, A. (1976) Grain-size effects on dielectric properties in barium titanate ceramics. *Journal of Applied Physics,* Vol. 47, No.1 (1976), pp. 371-373, ISSN 00218979.

Lines, M. E. & Glass, A. M. (1977) *Principle and Applications of Ferroelectrics and Related Materials.* (Oxford University Press, Oxford, U.K., 1977) p. 532.

Maiwa, H. (2007) Dielectric and electromechanical properties of $Ba(Zr_xTi_{1-x})O_3$ (x = 0.1 and 0.2) ceramics prepared by spark plasma sintering *Japanese Journal of Applied Physics,* Vol.46, No. 10b (October 2007), pp. 7013-7017, ISSN 00214922

Maiwa, H. (2008a) Preparation and properties of $BaTiO_3$ ceramics by spark plasma sintering *Japanese Journal of Applied Physics,* Vol.47, No.9-2 (September 2008), pp. 7646-7649, ISSN 00214922

Maiwa, H. (2008b) Structure and properties of $Ba(Zr_{0.2}Ti_{0.8})O_3$ ceramics prepared by spark plasma sintering *Journal of Materials Sciences,* Vol.43, No. 19 (October 2008), pp. 6385-6390, ISSN 00222461

Maiwa, H. (2009) Electromechanical properties of $BaTiO_3$ ceramics prepared by spark plasma sintering and other methods. *Japanese Journal of Applied Physics,* Vol.48, No. 9-2 (2009), pp. 09KD041-09KD044, ISSN 00214922

Maiwa, H.; Ohashi, K. & Hayashi, T. (2010). Temperature dependence of the electrical and electromechanical properties of $Ba(Zr_{0.2}Ti_{0.8})O_3$ thin films prepared by chemical solution deposition, *Journal of the Ceramic Society of Japan,* Vol. 118, Issue 1380, (August 2010), pp. 735-737, ISSN 18820743

Munir, Z. A.; Anselmi-Tamburini, U. & Ohyanagi, M. (2006) The effect of electric field and pressure on the synthesis and consolidation of materials: A review of the spark plasma sintering method, *Journal of Materials Science* Vol. 41, No. 3, (February 2006), pp. 763-777, ISBN 00222461

Munir, Z. A.; Quach, D. V. & Ohyanagi, M. (2011) Electric current activation of sintering: A review of the pulsed electric current sintering process, *Journal of the American Ceramic Society,* Vol. 94, No.1, (January 2011), pp. 1-19, ISBN 0027820

Neirman, S. M. (1988). The Curie point temperature of $Ba(Zr_{1-x}Zr_x)O_3$ solid solutions, *Journal of Materials Science,* Vol. 23, No. 11 (November 1988), pp. 3973-3980, ISBN 00222461

Takahashi, H. et al., (2006). Piezoelectric properties of $BaTiO_3$ ceramics with high performance fabricated by microwave sintering. *Japanese Journal of Applied Physics,* Vol. 45, No. 9B (September 2006), pp. 7405-7408, ISSN 00214922

Takahashi, H. et al., (2008). Considerations for $BaTiO_3$ ceramics with high piezoelectric properties fabricated by microwave sintering method. *Japanese Journal of Applied Physics,* Vol.47, Issue 11, (November 2008), pp. 8468-8471, ISSN 00214922

Takeuchi, T. et al., (1999) Preparation of dense $BaTiO_3$ ceramics with submicrometer grains by spark plasma sintering, *Journal of the American Ceramic Society,* Vol. 82, No. 4, (1999), pp. 939-943, ISSN 00027820

Wada, T. (2003) Ferroelectric $NaNbO_3$ ceramics fabricated by spark plasma sintering, *Japanese Journal of Applied Physics,* Vol. 42, No. 9B, (September 2003), pp. 6110-6114, ISSN 00214922

Xu, J.; Shaikh, A. S. & Vest, R. W. (1989). High K $BaTiO_3$ films from metalloorganic precursors. IEEE Transactions on Ultrasonics, Ferroelectrics, and Frequency Control, Vol. 36, No. 3, (May 1989), pp. 307-312, ISSN 08853010.

Yu, Z. et al., (2002) Piezoelectric and strain behavior of $Ba(Zr_{1-x}Zr_x)O_3$ ceramics, *Journal of Applied Physics,* Vol. 92, (August 2002) pp. 1489-1493, ISSN 00218979

Zhang, B. P. (2006) Compositional dependence of piezoelectric properties in $Na_xK_{1-x}NbO_3$ lead-free ceramics prepared by spark plasma sintering, *Journal of the American Ceramic Society,* Vol. 89, No. 5, (May 2006) pp. 1605-1609, ISSN 00027820

2

Integrated Piezoceramics as a Base of Intelligent Actuators

Frank Bärecke, Muhammed Abed Al Wahab and Roland Kasper
Otto-von-Guericke University Magdeburg
Germany

1. Introduction

Embedded functionality is one main focus of technology in the upcoming century. The embedded structures are able to work as sensors and/or actuators at the same time, making a passive structure to a mechatronic device. In this chapter we focus on structures comprehending piezoceramics as actuators. Pierre and Jacques Curie discovered the piezoelectric effect in 1880. With this effect a material itself can effect a small displacement by voltage application. Only for resonance applications like speaker the reachable displacement is adequate. For most technical drive systems the reachable displacement of a piezoelectric actuator seems insufficient. This changes with the utilization of displacement amplification systems. A displacement amplification system transforms the high force but small displacement of the piezoactuator to a moderate force with useful displacement.
These displacement amplification systems are composed of passive structures of steel, polymer or ceramics and the piezoactuator itself. Basic amplification systems are bending actuators and flat leverage actuators. Bending actuators or bimorph actuators are composed of a thin piezoceramics actuator and a passive steel or ceramic plate opposite. The shrinking of the piezoceramic actuator's length by voltage application results in a bending movement allowing only very small forces. The flat leverage actuator uses the piezoactuator, the leverages and joints. Critical for these systems are the joints and the stiffness. If there is minimal backlash in one joint it will absorb the displacement of the piezo. With rising stiffness of the system the applicable force rises, but the reachable displacement will shrink.
In this chapter we give an introduction to piezoceramics, conventional piezoactuators and a new manufacturing technology. Then we focus on displacement amplification systems developed by the authors in recent years. We optimise the structures for high amplification, maximal stiffness and maximum speed of the actuator depending on the field of application. Afterwards our focus changes to two applications for amplification systems developed by the authors. Finally we discuss the capabilities of piezoceramics from the technical point of view.

2. Piezoceramics

The piezoelectric effect means the linear electromechanical interaction between the mechanical and the electrical state in crystalline materials with no inversion symmetry. An overview for sensor applications gives (Gautschi 2002) for actuator applications one can use (Janocha 2010). The direct piezo effect discovered in 1880 by Pierre and Jacques Currie means that a structure disposed to physical loads generates a proportional charge or voltage. The direct piezoelectric effect is used for different types of sensors transforming physical loads to an electric signal. The inverse piezoelectric effect means to deform a structure by charge or voltage application. Using the inverse piezo effect the ceramics reaches a maximal elongation change of about 0.02 to 0.1 per cent of the initial length depending on the used material by an applied electric field of about 2000 V/mm. The piezoelectric (3,1)-effect or transversal effect means the extension in the direction of the field. The (3,1)-effect or longitudinal effect means a contraction orthographic to the field. The actuator has to be polarised before use in field direction. It is important not to excess the materials Curie temperature to avoid the loss of polarisation. The actual field of application of piezoceramics comprises ink-jet printers, loud speakers, valves and diesel fuel injection systems (Janocha 2010).

Piezoactuator basics and applications

With (Ruschmeyer 1995) one can find a basic book about piezoceramics. Constitutive equations describe the piezoelectric effect of all materials using a linear model. The constitutive equations (1) and (2) represent the coupled behaviour of mechanical and electrical properties of an element, neglecting thermal or other coupling effects. On the electrical side displacement D and on the mechanical side strain S is composed of one part resulting from the electrical field E and a second part resulting from mechanical stress T

$$D = d \cdot T + \varepsilon^T \cdot E \tag{1}$$

$$S = s^E \cdot T + d \cdot E \tag{2}$$

Material data is given by electric permittivity ε, piezoelectric load constant d and mechanical compliance s. Equations (1) and (2) give constitutive equations in scalar formulation, where the electrical field and strain have the same direction, which corresponds to the (3,3)-piezoelectric effect, or are perpendicular as in the case of the (3,1)-effect. Stress and strain always have the same direction.
The constitutive equations applied to cuboids (Figure 1) with basic area A and length L, whose electrodes are supplied with a voltage $U = E \cdot L$, deliver the charge $Q = D \cdot A$ stored on that element

$$Q = d \cdot F + C \cdot U \tag{3}$$

and the change of length ΔL

$$\Delta L = \frac{F}{c_m} + d \cdot U \tag{4}$$

Here

$$F = T \cdot A \tag{5}$$

describes the force acting on the surface A, $c_m = \frac{A}{s^E \cdot L}$ specifies the spring constant and $C = \varepsilon \cdot \frac{A}{L}$ is the capacity of the element. Eq. (5) shows how the actuator's force can be influenced directly by adapting its area A. But eq. (4) indicates that the elongation ΔL will be small, due to the restrictions of supply voltage and the piezoelectric constant of available materials.

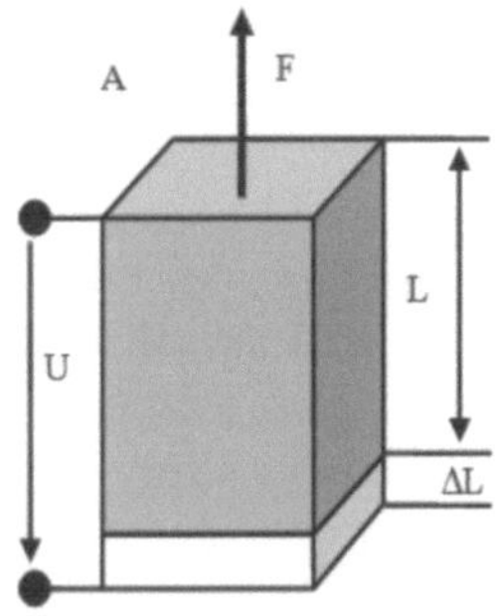

Fig. 1. Piezoelectric element

To overcome these restrictions and to generate larger motion, several construction principles of piezoactuators are common. The bender type uses the (3,1)-piezo-effect in relative long thin strips. At least two of these strips are connected together in assembly comparable to a bimetal, to get the required bending. Benders generate large motion in the mm range with the disadvantage of decreasing forces, which is the limiting factor for applications. Stack actuators use the (3,3)-piezo-effect and add the elongations of a large number of n thin elements to generate a larger motion

$$\Delta L_A = n \cdot \Delta L = \frac{F}{c_m / n} + d \cdot L_A \cdot E \qquad (6)$$

As effective area A as well as the acting stress T remains unchanged, eq. (5) shows that very high forces are possible. Simply adding stacks allows only limited motions, related to the total length $L_A \approx n \cdot L$ of the actuator and suffers from a reduced stiffness of the actuator expressed by the reduced spring constant in eq. (6). Thus, in the case of stack actuators, range of motion is the limiting factor for applications. There are further actuator construction types like crossbow actuators etc. that promise a better compromise between force and motion but with the need for a more complex assembly and much higher cost.

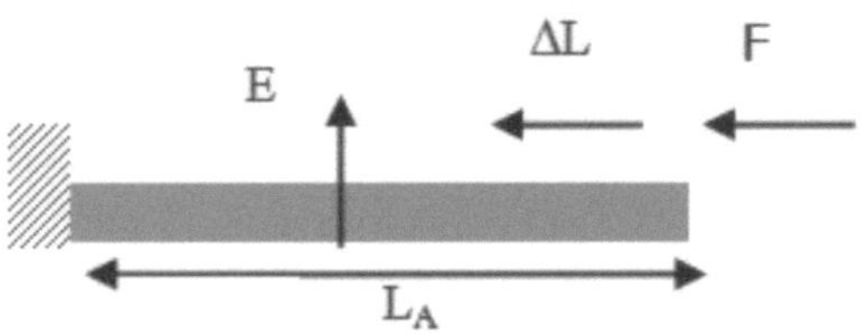

Fig. 2. Planar piezoactuator

To adapt actuator characteristics to the requirements of applications, motion amplification systems (MAS) can be used (Kasper, Heinemann und Wagner 1998), (Kasper 2002). In this paper the focus is set on mechanical solutions, as they offer a better way of integration into ceramic structures. Mechanical MAS in many cases utilize implementations of the law of lever

$$F_1 \cdot l_1 = F_2 \cdot l_2 \quad \text{und} \quad \frac{x_1}{l_1} = \frac{x_2}{l_2} \tag{7}$$

where F is force, x is motion, l is the effective lever length and index 1 means the primary, index 2 the secondary side of the lever. They offer a simple mechanical construction, but have problems with friction, backlash and stiffness. To avoid these drawbacks and to ease handling, frame constructions are used, which integrate several tasks like motion amplification, pre-stress generation and safe mounting. Unfortunately all types of MAS increase size and moved masses of the actuator. This is a contradiction to an optimized actuator system, fine tuning all its components, which is required for today's applications.

2.1 Structured actuators

Analyzing the structure of typical piezo-driven systems in order to improve integration of separated functional elements, the actuator itself and the MAS are good candidates for doing a first step. Both elements are connected by a critical interface, where very small motions have to be transmitted together with very high forces. As mentioned above, piezoelectric actuators use two principles to generate force and motion as required by an application. Benders use the (3,1)-effect to generate motion in the mm range. Due to bending the complete actuator with a very large lever arm, only forces up to 1 N can be generated. Stack actuators on the other hand use the (3,3)-effect to generate much higher forces in the range of several kN. But range of motion is restricted to 50 to 100 µm.
An alternate approach is to use the (3,1)-effect like a bender actuator to generate a large motion due to the large effective length L_A of a planar actuator. Figure 3 shows how the size of motion ΔL is determined by the actuator's length L_A and strength of the electrical field E. The necessary supply voltage therefore is given by the actuator's thickness. To reduce it, the actuator can be produced from several layers. From a practical point of view, only a small number of layers will be needed. Integrating a MAE into this planar actuator can be done in several ways. An important distinguishing feature of an MAE is the integration in the same plane as the actuator or generating motion perpendicular to this plane.

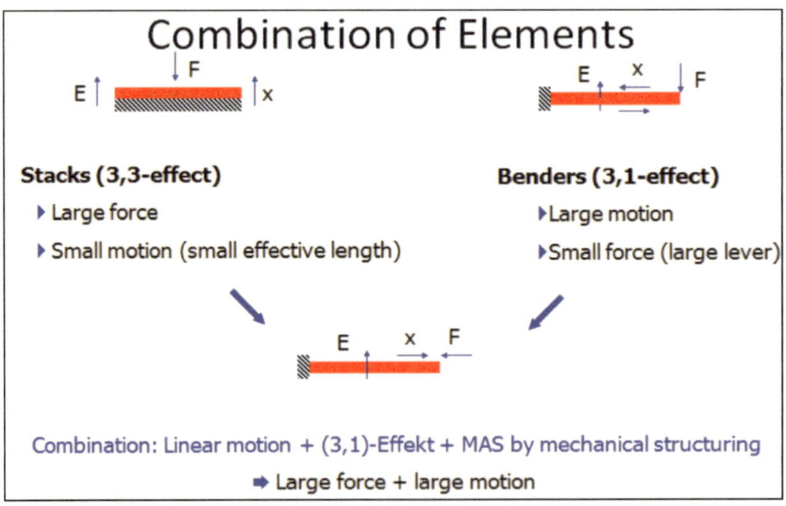

Fig. 3. Combination of Elements

Planar systems offer the advantage that all components are located in the same physical plane. As a consequence they can be produced from piezoceramic plates or discs within one structuring process. Figure 4 shows two variants of planar systems. Variant a) is driven by the length reduction of a central plate. Two levers, one on each side, amplify this motion up to 2x88 µm, while delivering a force of 4 N. Variant b) utilizes another method of motion amplification by adding respectively subtracting the motion generated by the bars of a meander structure.

Planar Systems		Spatial Systems			
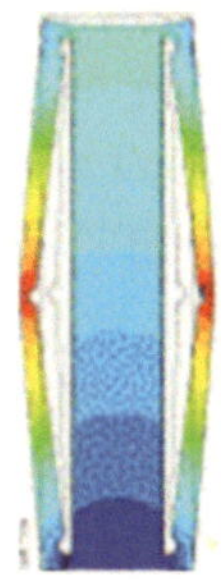		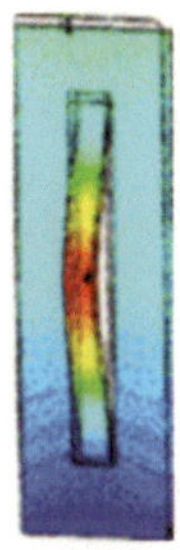	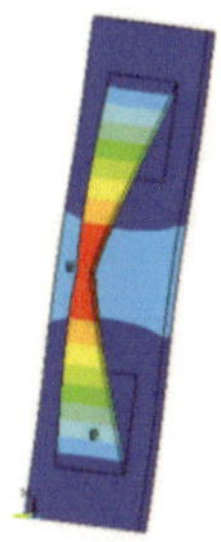		
a) Two-sided Lever system	b) Symmetric Two-sided Meander System	c) One-sided Lever system	d) Hybrid One-sided Lever system	e) Hybrid One-sided Disc system	
2x88	2x47	64	281	511	Motion [µm]
4	0.8	1	20	40	Force [N]
70x25	70x25	70x25	70x25	$p \cdot 35^2$	Area [mm^2]

Fig. 4. Actuator prototypes using several construction principles

Variant c) uses a tongue in the middle of a base plate. If the base plate shrinks, due to an electric field, the passive tongue acts as a kind of two-side supported lever system and bends perpendicular to the base plate. Optimizing the tension distribution and designing solid joints with the aid of a FE-simulation, a motion of 64 µm can be achieved. Unfortunately forces are down on a level of 1 N. To avoid these limitations that are closely related to the solid joints used in these designs, a hybrid construction can be chosen. Variant d), which is constructed very similar to variant c), but replaces the ceramic lever by one made of spring steel, gets much better results of 281 µm motion and 20 N force. The lever will be fixed in outbreaks of the ceramic plate using special mounting elements. The strongest advantage of the hybrid type is the flexibility to adapt geometric and material data at critical points in a wider range than this can be done for a pure ceramic actuator. Variant e) shows the best results until now. More than 500 µm motion and a force of 40 N can be generated by combining a ceramic base disc with star-like MAE made of spring steel. Activating the piezoceramic disc will shrink its diameter. This radial motion is amplified and transformed to a movement of the top of the MAE perpendicular to the disc.

2.2 Synthesis of tense piezoceramic structures

Here is presented a methodology developed for synthesis of closed structures for micro- and nano-applications, utilizing the advantages of structured piezoceramics, tense piezoactuators and closed robot kinematics structures. The synthesis of closed kinematic structure with piezoceramic actuators is investigated for three case studies:

- Synthesis for parallel structures in which the basic links are connected, only by means of driving chains of the piezoceramic actuators.
- Synthesis for parallel structures in which the basic links are connected in a serial chain. The driving chains of the piezoceramic actuators are attached parallel to the links of the basic serial chain.
- Synthesis for parallel structures in which the basic links are connected in a parallel chain. The driving chains of the piezoceramic actuators are attached parallel to the links of these chains.

A synthesis of kinematics schemes with definite degrees of freedom based on the synthesised structures is developed in the paper. The class of the kinematic joints of the links and the immovable link are selected. Examples and graphic interpretation of the solutions are presented in the paper.

Synthesis of closed structures based on piezo structured ceramics.

In order to be tensed, piezoceramic structures must be composed with parallel or closed topology. To achieve tension in closed piezoceramic structures it is possible to use predefined deformation in elastic joints or antagonistic interaction of the redundant actuators (Chakarov, et al. 2007).

Here the synthesis of basic closed structures for micro- & nano-manipulation tasks will be presented. According to the Mechanism and Machine Theory the basic closed kinematic structures include links with a number of 3, 4 or more kinematic joints, which we describe with n_j ($j \geq 3$). The following relation (Chakarov und Parushev 1994) among the number of basic links in the closed kinematic structure is used:

$$2 \cdot i = \sum_{j=3}^{i+3} (j-2) \cdot n_j \tag{8}$$

Here

$$i = p - n \tag{9}$$

is the difference between the number of the all kinematic joints p and the number of all links n in the closed kinematic chain:

$$n = \sum_{j=2}^{i+3} n_j \, . \tag{10}$$

The possible variants of the basic links of the closed kinematic structures are derived using (8), which are presented in Table 1. In the table there are limitations on the difference (9) $i \leq 4$, and on the number n^0 of the basic links in the closed structure:

$$n^0 = \sum_{j=3}^{i+3} n_j \leq 3 \, . \tag{11}$$

The variants of the basic links shown in Table 1 allow closed structures synthesis by means of link coupling directly among them, or by means of sequential kinematics chains including one or more links of type n_2.

n^0	i=1	i=2	i=3	i=4
1	$n_4=1$			
2	$n_3=2$	$n_4=2$ $n_3=1, n_5=1$	$n_4=1, n_6=1$ $n_5=2$	$n_5=1, n_7=1$ $n_6=2$
3		$n_3=2, n_4=1$	$n_3=2, n_6=1$ $n_3=1, n_4=1, n_5=1$ $n_4=3$	$n_3=1, n_5=1, n_6=1$ $n_3=1, n_4=1, n_7=1$ $n_4=2, n_6=1$ $n_4=1, n_5=2$

Table 1. Possible variants of basic links of closed kinematic structures

In Table 2 are shown the possible closed structures received from Table 1, included only basic links. In Table 2 the links n_3, n_4 and so on are presented by the symbols 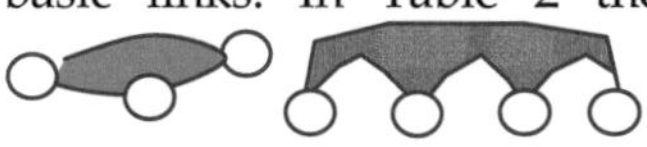 , representing the number of their joints by bubbles.

n^0	i=1	i=2		i=3				
1	4 							
2	3 – 3 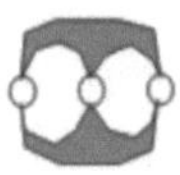	4 – 4 	3 -5 	5 – 5 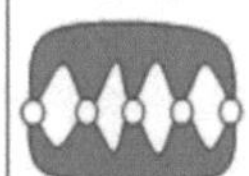	4 - 6 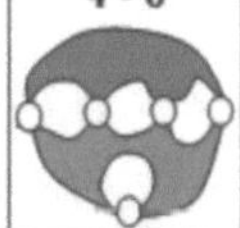			
3		3 – 3 – 4 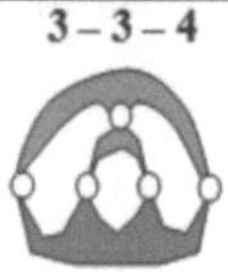	3 – 4 – 3 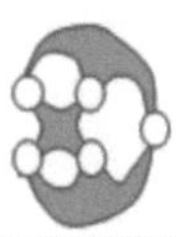	3 – 6 – 3 	3 – 4 – 5 	3 – 5 – 4 	4 – 3 – 5 	4 – 4 – 4 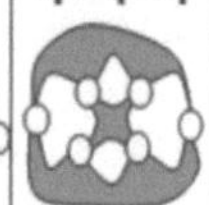

Table 2. Closed kinematic structures obtained by direct coupling of basic links

The variants of the basic links shown in Table 1 allow closed structures synthesis of link coupling by means of sequential driven kinematic chains based on piezoceramics. Piezoceramic structures can be assumed according to the Mechanism and Machine Theory as a combination of rigid links and polarised ceramic elements. The polarised ceramic elements can be estimated as actuators for linear motion, which can be modelled by the kinematic chain shown in Figure 5. It includes two links 1 and 2 and a joint T with linear motion, which can perform drive functions and two rotational joints R by which the actuator is attached to the driven links. The rotational joints can be created by introducing elastic areas in the ceramics or by additional elastic joints.

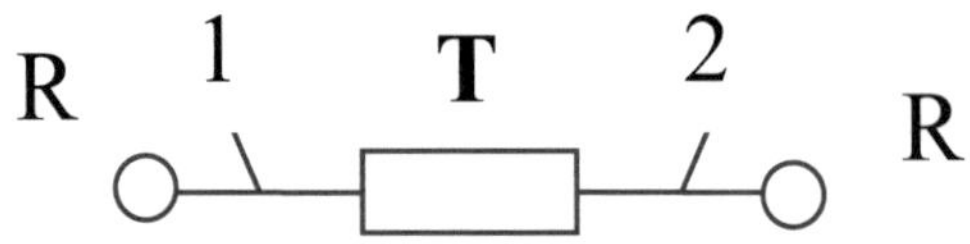

Fig. 5. Kinematic chain of linear actuator.

Three case studies of closed kinematic structure synthesis for micro- and nano-manipulation tasks

The closed kinematic structure synthesis for micro- and nano- manipulation tasks can be investigated for three case studies:

n0	i =0	i =1	i =2	i=3	i =4
2	2 - 2 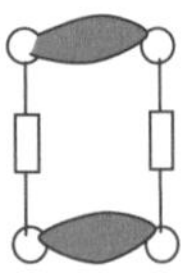	3 - 3 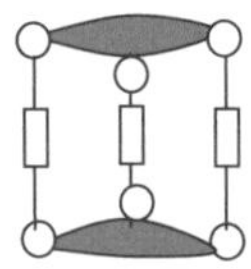	4 - 4 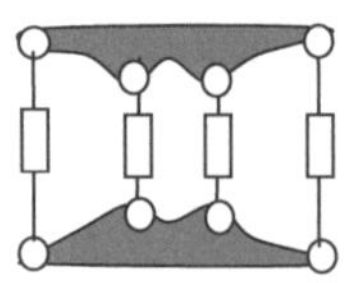	5 - 5 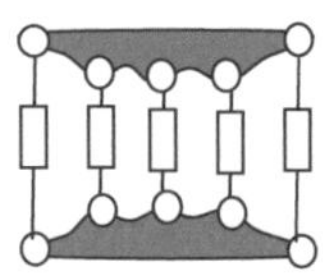	6 - 6 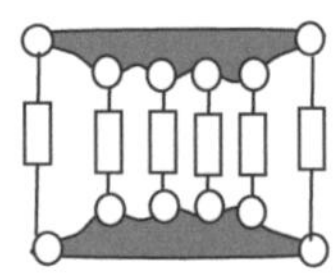

Table 3. Synthesised closed kinematic structures for case A)

A) Synthesis for parallel structures in which the links n_j, ($j \geq 3$) (Table 2) are connected in between, only by means of sequential kinematics chains like as driving chains clearly seen in Figure 5. The driving kinematics chains, presented in Figure 5 are sequential chains including two links from type n_2 and three joints.
The possible kinematics structures for cases $i \leq 4$ and $n^0 = 2$ defined as a result of the synthesis are shown graphically in Table 3, where the driving chains are shown by the symbol in Figure 5. The simplest closed structure 2-2 ($n_2 = 2$) can be found in Table 3, too. The number of driving chains m depends on the number of the basic links (9) and on the difference (11) as follows:

$$m = i + n^0. \tag{12}$$

B) Serial - parallel structures synthesis including the basic links n_j, ($j \geq 3$) in Table 1 connected in a serial chain that can be included and binary links n_2. The total number n_B^0 of the links of basic serial chain includes links n_j, ($j \geq 3$) shown in Table 1 and binary links n_2. The driving chains, presented in Figure 5 connect the free joints of the links n_B^0 of the basic serial chain in between them. A restriction is imposed for structures synthesis only including number m of the driving chains higher than the number of the joints connecting the links of the serial chain. The reason for this restriction is the fact that all the joints of the basic serial chain are driven by means of parallel driving chains. The kinematics structures defined as a result of synthesis for $n^0=1, 2$ according to Table 1 are shown graphically in Table 4 on the left side. In Table 4 the possible kinematic structures are shown for case $i \leq 2$ and the number of the links of basic serial chain $n_B^0 = 2, 3$. The number of the driving chains m in this case is defined by i in equation (9):

$$m = i + 1. \quad (13)$$

Here, the same graphical symbols as in Table 3 are used for the link and the driving chain definition. The difference in the arrangement of the links in Table 2 is also assessed. The simplest closed structure $n_2 = 2$ is shown in Table 4, too.

C) Parallel structures synthesis including the basic links n_j, ($j \geq 3$) in Table 1 connected in between in a parallel chain that can be included and binary links n_2. The total number n_B^0 of the links of the basic parallel chain includes links n_j, ($j \geq 3$), shown in Table 1, and binary links n_2.

The driving chains (Figure 5) connect the free joints of the links n_B^0 of the basic parallel chain in between them. The number of the driving chains in this case is defined by the difference (9) as: m = i.

The kinematic structures defined as a result of synthesis for $n^0 = 2$, 3 and i =1, 2 according Table 1 are shown graphically in Table 4 on the left. On the right of Table 4 the possible kinematic structures are shown for case i =1, 2 and the number of the links of basic parallel chain $n_B^0 = 3$.

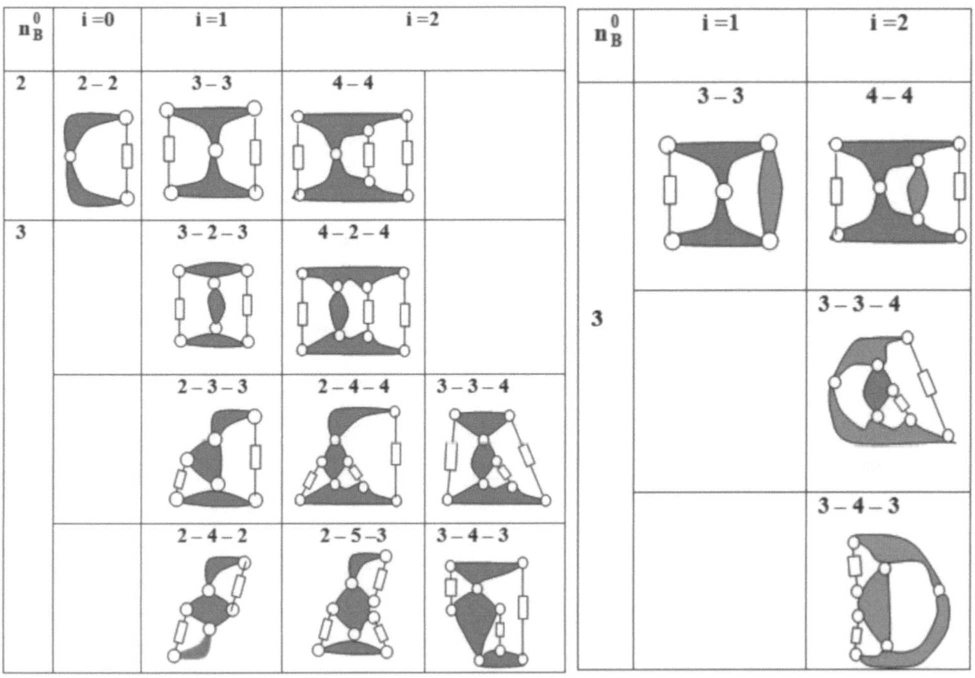

Table 4. Synthesised closed kinematic structures for case B) and for case C)

Synthesis of kinematics schemes for micro- & nano-manipulation tasks based on the synthesised structures.

A synthesis of kinematics schemes with definite degrees of freedom h based on the synthesised structures can be developed. To create the kinematics schemes with definite degrees of freedom, the class j of the kinematics joints p_j of the links and the immovable link are selected. The synthesis is performed separately for case A), B) and C) about structures of Tables 3 and 4.

A) For parallel structures of Table 3 the number of the degrees of freedom usually is equal to the number of the actuators h = m. If the number of the actuators m is larger than that of the degrees of freedom m≥h we have actuation redundancy. This can be used for achieving the tensity, improving the dynamic parameters and removing any backlash in the device. The redundancy is presented by the difference:

$$m - h = r. \tag{14}$$

To create the kinematics schemes with definite degrees of freedom, the class j of the kinematic joints p_j of the links has to be selected. The number and the class of the kinematics joints are defined by the actuators. Each actuator according to Figure 5 possesses one linear motor joint (T) and two revolution joints (R). The linear joints apply 5 restrictions to the movements, thus the joint possess one degree of freedom. Their number p_5^l is equal to the number of the actuators:

$$p_5^l = m \,. \tag{15}$$

The revolution joints of the parallel structures of Table 3 can apply different number of restriction j = 1...5, defining the degrees of freedom of the device using the well known equation:

$$h = 6 \cdot n - \sum_{j=1}^{5} j \cdot P_j \tag{16}$$

where n is the number of all mobile links and P_j is the number of all kinematics joints of class j.
As any actuator has two movable links and one of the basic links n^0 is immovable, the number of all mobile links is:

$$n = n^0 - 1 + 2 \cdot m. \tag{17}$$

As the number of all kinematics joints of class j=5 includes the number of linear joints p_5^l and the number of revolution joints p_5^r of class 5 according (15) it leads to:

$$P_5 = p_5^l + p_5^r = m + p_5^r\,. \tag{18}$$

According (17) and (18) equation (16) assumes the form:

$$h = 6 \cdot (n^0 - 1) + 7 \cdot m - \sum_{j=1}^{5} j \cdot p_j^r\,. \tag{19}$$

As any actuator according Figure 5 has two revolution joints, the total number of revolution joints is defined by the number of actuators m:

$$\sum_{j=1}^{5} p_j^r = 2 \cdot m \,. \tag{20}$$

For any structural scheme of Table 3 defined by parameters n^0 and i ($i=m+n^0$), above two equations (19) and (20) allow to determine one or more variants of allocation of the different class joints and synthesis of kinematics schemes with desired degrees of freedom $h = m - r$.
If we use elastic revolution joints (geometric closed), the class of joints is limited j = 3, 4. 5.
The possible revolution joints p_j^r according equations (19) and (20) for structures of Table 3 ($n^0=2$) with h=1,2,3 and i=0,…,4 are shown in Table 5.

h	i=0, m=2	i=1, m=3	i=2, m=4	i=3, m=5	i=4, m=6
1	$p_4=1, p_5=3$	$p_4=4, p_5=2$ $p_3=1, p_4=2, p_5=3$ $p_3=2, \quad p_5=4$	$p_4=7, p_5=1$ $p_3=1, p_4=5, p_5=2$ $p_3=2, p_4=3, p_5=3$ $p_3=3, p_4=1, p_5=4$	$p_4=10$ $p_3=1, p_4=8, p_5=1$ $p_3=2, p_4=6, p_5=2$ $p_3=3, p_4=4, p_5=3$ $p_3=4, p_4=2, p_5=4$ $p_3=5, \quad , p_5=5$	$p_3=1, p_4=11$ $p_3=2, p_4=9, p_5=1$ $p_3=3, p_4=7, p_5=2$ $p_3=4, p_4=5, p_5=3$ $p_3=5, p_4=3, p_5=4$ $p_3=6, p_4=1, p_5=5$
2	$p_3=1, p_5=3$	$p_4=5, p_5=1$ $p_3=1, p_4=3, p_5=2$ $p_3=2, p_4=1, p_5=3$	$p_4=8$ $p_3=1, p_4=6, p_5=1$ $p_3=2, p_4=4, p_5=2$ $p_3=3, p_4=2, p_5=3$ $p_3=4, \quad p_5=4$	$p_3=1, p_4=9, \quad p_3=2,$ $p_4=7, p_5=1$ $p_3=3, p_4=5, p_5=2$ $p_3=4, p_4=3, p_5=3$ $p_3=5, p_4=1, p_5=4$	$p_3=2, p_4=10$ $p_3=3, p_4=8, p_5=1$ $p_3=4, p_4=6, p_5=2$ $p_3=5, p_4=4, p_5=3$ $p_3=6, p_4=2, p_5=4$ $p_3=7, \quad , p_5=5$
3	X	$p_4=6$ $p_3=1, p_4=4, p_5=1$ $p_3=2, p_4=2, p_5=2$	$p_3=1, p_4=7$ $p_3=2, p_4=5, p_5=1$ $p_3=3, p_4=3, p_5=2$ $p_3=4, p_4=1, p_5=3$	$p_3=2, p_4=8, \quad p_3=3,$ $p_4=6, p_5=1$ $p_3=4, p_4=4, p_5=2$ $p_3=5, p_4=2, p_5=3$ $p_3=6, \quad , p_5=4$	$p_3=3, p_4=9$ $p_3=4, p_4=7, p_5=1$ $p_3=5, p_4=5, p_5=2$ $p_3=6, p_4=3, p_5=3$ $p_3=7, p_4=1, p_5=4$

Table 5. The possible revolution joints for case A)

B) Serial - parallel structures of Table 4 (left side) allow the basic serial chain to determine the degrees of freedom of the device h. The joints of this chain can be elastic and can bear the necessary motion restrictions. The actuators are treated as identical structural groups with zero degrees of freedom (Figure 6). The class of the rotation joints of the actuators are chosen in such a way according to (16), do not change the degrees of freedom of the basic chain. The class of the rotation joints for the spatial (3D) case are shown in Figure 6.

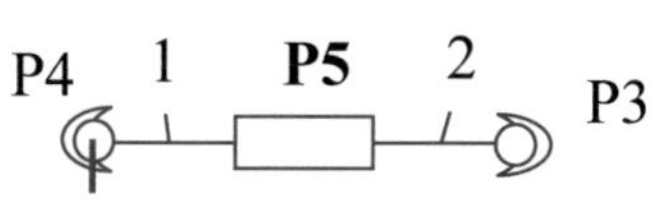

Fig. 6. Class of the actuator joints for the spatial (3D) case

The number of the actuators usually is equal to the number of the degrees of freedom m=h. If the number of the actuators m is larger than that of the degrees of freedom m≥h we have actuation redundancy presented through difference (14). This can be used for achieving the tensity, improving the dynamic parameters and removing the backlashes in the device.
The class of the joints of the basic serial chain is determined according equation (16) presented as:

$$h = 6 \cdot (n_B^0 - 1) - \sum_{j=1}^{5} j \cdot p_j^0 . \tag{21}$$

where n_B^0 and p_j^0 are the number of the links and the joints of the basic serial chain. The joints number of the basic serial chain is determined according the number of its movable links using equality:

$$n_B^0 - 1 = \sum_{j=1}^{5} p_j^0 . \tag{22}$$

For each structure of Table 4 with desired degrees of freedom h, above two equations (21) and (22) allow determining the class of basic chain joints and their number. If the class of revolution joints is limited j = 3, 4. 5, the possible joints p_j of the basic serial chain with h=1, 2,3,4,5 degrees of freedom and n_B^0 =2, 3 are shown in Table 6.

H	1	2	3	4	5
n_B^0 =2	p5=1	p4=1	p3=1	X	X
n_B^0 =3	X	p5=2	p4=1, p5=1	p4=2	p3=1, p4=1

Table 6. The possible joints p_j of the basic serial chain

Each basic joint of class j must be included in (6-j) closed cycles with the actuators of structure to get the drivability. The joints with definite class of the basic chain can be either revolution or linear, as well.

C) In the parallel structures of Table 4 (right side) the actuators are treated too as identical structural groups with a zero degrees of freedom (Figure 6). The basic parallel chain determines the degrees of freedom of the device h. The number of the actuators is equal to the number of the degrees of freedom m = h, or it is larger of them m > h. The class of joints p_j^0 of the basic parallel chain is defined by equation (16) as:

$$h = 6 \cdot (n_B^0 - 1) - \sum_{j=1}^{5} j \cdot p_j^0 . \tag{23}$$

where n_B^0 and p_j^0 are the number of the links and the joints of the basic parallel chain. The total number of the joints of the basic parallel chain is equal to the number of its links:

$$n_B^0 = \sum_{j=1}^{5} p_j^0 . \tag{24}$$

Above two equations allow for each structure of Table 5 (right side) included nB0 links, to specify the class of the joints of the basic parallel chain with h degrees of freedom. If the class of elastic revolution joints is limited j = 3, 4, 5, the possible solutions for structures with h=1, 2,3,4,5 degrees of freedom and nB0=3, 4 are shown in Table 7.

h	1	2	3	4	5
n_B^0 =3	p3=1, p4=2	p3=2, p4=1	p3=3	X	X
	p3=2, p5=1				
n_B^0 =4	p3=1, p4=1, p5=2	p3=1, p4=2, p5=1	p3=2, p4=1, p5=1	X	X

Table 7. The possible joints p_j of the basic parallel chain

3. Integrated displacement amplification system examples

The first and the second example show two displacement amplification systems for very precise positioning in 3D. The aim was to synthesize a mechatronic handling device with 3 degrees of freedom - two degrees affect the orientation the third affects the translation. The amplification system is appropriate for cell manipulation. For this purpose two prototypes were experimentally investigated utilizing the synthesized kinematic structures based on cases A) and B).
The development of an adaptive Gas-Spring-Damper System (GSD) resulted in a need for a very fast high flow pneumatic fluid restrictor. The appropriate application area of an adaptive Gas-Spring-Damper is a passenger cars suspension strut. Other possible application areas are seat or cabin dampers. GSDs comprises mostly of two or three air chambers with different working directions. A fluid restrictor connects these air chambers. The GSD behaviour as a spring and damper system depends on geometries of the air chambers and the air flow controlled by the fluid restrictor. For the development of an adaptive GSD one requires a variable fluid restrictor for controlling the damping mode of it. The second example presents the fluid restrictor and the integrated piezoelectric displacement amplification system developed for controlling the GSDs gas flow. The displacement amplification system comprises a piezoceramic ring, where the piezoelectric (3,1)-effect is used, and spokes which connect the ring to an inner bearing. To improve the amplification we designed several structures which differ in spoke thickness, material and position. Alteration of the piezo design does not seem promising. Plungers extend the spokes for sealing the long holes embedded in the housing, enabling building up a fluid restrictor from the displacement amplification system. The third example shows the integration of the displacement amplification in a pneumatic fluid restrictor.

3.1 Mechatronic handling device with three piezoelectric actuators

The mechatronic handling device is based on the kinematic structure with two links of the basic serial chain $n_B^0=2$ (structure type of case B)) and three degrees of freedom h=3. The number of the actuators is equal to the number degrees of freedom of the device m = h = 3. According to equation (13) and Table 4 (left side), structure 4–4 corresponds to the chosen parameters. The structure is symmetric and each basic link can be chosen as an immovable one as it is shown in Figure 7 a). Following equations (21), (22) and results in Table 6, the possible joints distribution of the basic serial chain is $p_3=1$. On the above it can be build up a kinematics scheme of a device for micro- and nano-manipulation tasks as shown in Figure

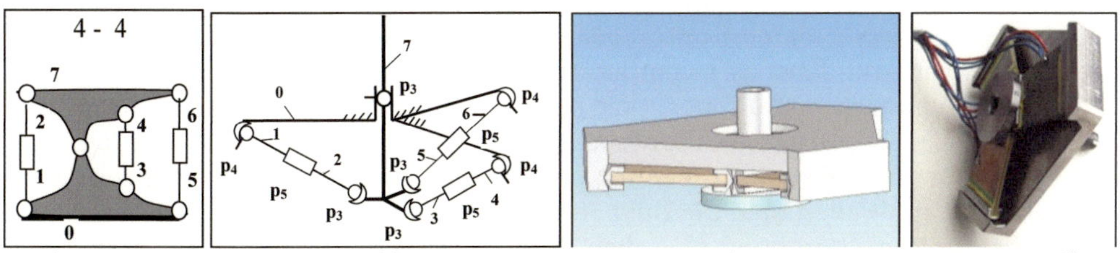

a) b) c) d)

Fig. 7. Development of the handling device with structured piezoceramics (type B) with $n_B^0=2$ and h=m=3

7b). Actuator chains in the scheme are chosen according Figure 6. The basic chain includes two links marked as 0 and 7, connected by means of one elastic joint p_3, allowing one linear and two rotational movements. Link 0 is chosen as immovable one. Based on this scheme a simulation model (Figure 7 c) and a device prototype (Figure 7.d) with structured piezoceramics are developed. (Kasper, Al Wahab, et al. 2006)

3.2 Mechatronic handling device with six piezoelectric actuators

It is based on the kinematic structures with two links of the basic serial chain n_0=2 (structure type A) and three degrees of freedom h=3. The number of the actuators is 3 and therefore bigger than the number of degrees of freedom of the device, as follows: m = h+r =3+3 = 6. According equation (12) and Table 4 (left side) structure 6-6 corresponds to the chosen parameters. The structure is symmetric and each basic link can be chosen as immovable one as it is shown in Figure 7 a). Based on the equations (21) and (22) and results in Table 6, the possible revolution joints distributions are the following:

p_3=3, p_4=9; p_3=4, p_4=7,p_5=1; p_3=5, p_4=5, p_5=2; p_3=6, p_4=3, p_5=3 and p_3=7, p_4=1, p_5=4.

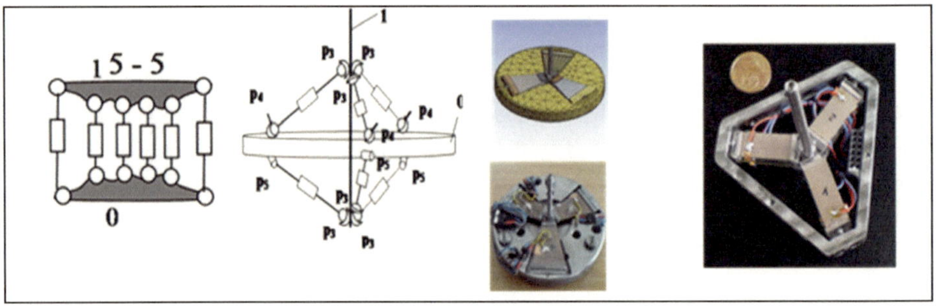

a) b) c) d)

Fig. 8. Phases of design during development of the handling device with structured piezoceramics (type A) with n_0=2 and h=3, m=6.

In this case a kinematics scheme with joints p_3=6, p_4=3, p_5=3, is chosen as it is shown on Figure 7 b). Actuator chains in the scheme are chosen according to Figure 6. The basic two links are indicated as 0 and 1, while link 0 is chosen as the immovable one. A simulation model and a device prototype with structured piezoceramics built up that scheme are shown on Figure 8 c) and d).

3.3 Air flow restrictor

The requirements of the fluid restrictor are speed and high flow rate to be able controlling the GSD damping mode. In most operational mode the pressure difference between the air chambers is small and depends in particular on the stimulation of the road. It can vary between a few Pa and some bar for extreme road unevenness at high speed. The operational pressure for a GSD application lies between 5 and 20 bars. Restrictions are here the overall size, which rises with decreased pressures, and the technology to provide the compressed air. Due to the application area the fluid restrictor must resist all shocks caused by road unevenness (Bärecke, Kasper und Al Wahab 2008) and (Kasper, Bärecke, et al. 2008).

Displacement amplification system

The displacement amplification system comprises a piezoceramic ring (Figure 9 and Figure 10). Voltage application on the piezoelectric actuator causes it to enlarge in thickness.

The piezoelectric (3,1)-effect causes a contraction of the ring and coincidental reduces its diameter. Six spokes connect an inner bearing to the piezoceramic ring. They can't contract like the piezoceramic ring due to their high longitudinal stiffness. This leads them to bend. This results in an angular movement of the actuator and a rotation on the outside of the piezoceramic ring. Important influencing factors for the displacement amplification are the adjustment, and thickness of the spokes. Thicker spokes have a higher bending stiffness, so they are working against the rotation, therefore decreasing the maximum amplification. For maximum displacement amplification one can reduce the eccentricity, but on the other hand this reduces the operational torque.

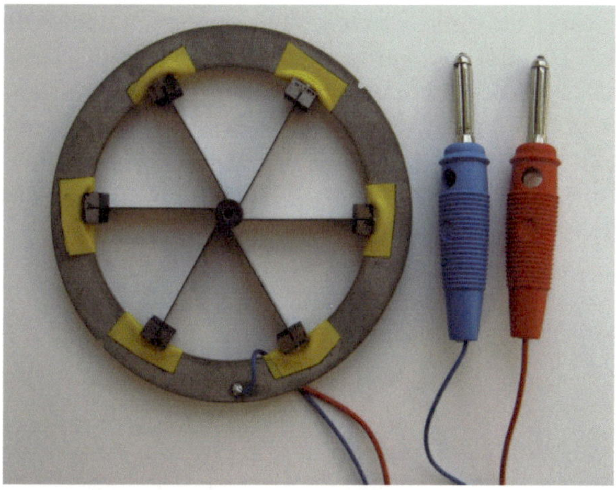

Fig. 9. Discrete displacement amplification prototype

Figure 9 shows an early prototype for the displacement amplification system. The piezoelectric ring is cut out of a quadratic plate by jet cutting. Bearings and spring steel spokes are eroded from a steel block. This amplification system is manually assembled. The amplification system has a diameter of about 80 mm. The piezoelectric actuator is 1.5 mm, the spokes are 0.4 mm thick and have an eccentricity of 0.8 mm. To avoid electrical breakdown between the piezoceramics ring and the bearing resides adhesive tape. The measured displacement of this prototype is 0.45 mm.

The left side of Figure 10 shows a prototype manufactured with the technology of insert moulding. A carbon fibre polymer is mould over the piezoceramic ring. This prototype reaches a displacement of about 0.3 mm. The right part of that Figure shows a FEM simulation for this amplification system. The results of the simulation and the measured values are about the same size. This could be aroused by geometry meanderings like draft angle due to the moulding process. The insert moulded spokes have a trapezoid format which is need for the demoulding process.

Further development of the displacement amplification system

Calculations showed that the needed amplification wasn't reached by these first prototypes. An improved amplification system has to be developed. To reduce the restoring force of the bended spokes thinner spokes have to be used. It is possible to use spring steel inlays with a thickness of 0.3 mm or even thinner dimension. Insert moulded spokes have a boundary for spokes thickness at about 0.5 mm. Thinner spokes could clog the capillary of the moulding

form which is handled by an increased moulding pressure. This increased moulding pressure degrades the piezoceramics so we can't use it. In summary with spring steel inlays we could decrease injection pressure and rejection rate due to broken piezoceramics.
The left of Figure 11 shows a displacement amplification system with spring steel inlays. The spring steel spokes are connected by insert moulding to the inner bearing. Outer bearings connect the spring steel inlays to the piezoceramic ring. The spring steel inlays ends are rounded by brazing to prevent them cutting the carbon polymer. The moulding process has to fill seven independent cavities. The right of Figure 11 shows the mounting position of the displacement amplification system. It is arranged in a housing containing six outlet notches of 1x10 mm. In the closed position discs at the end of the displacement amplification system cover the outlet notches. With an applied voltage the ring rotates and a voltage proportional air flow crossing area is uncovered. The valve additionally incorporates six areas of air flow with an approximate crossing section of 0.7x10 mm. Very important for the use of the fluid restrictor in the GSD is the leakage of the closed restrictor. The circumferential line surrounding all outlet notches is proportional to the leakage. The additional length of this parameter is 130 mm. It was calculated, that a leakage minimization is possible with a production accuracy of about 10 μm.

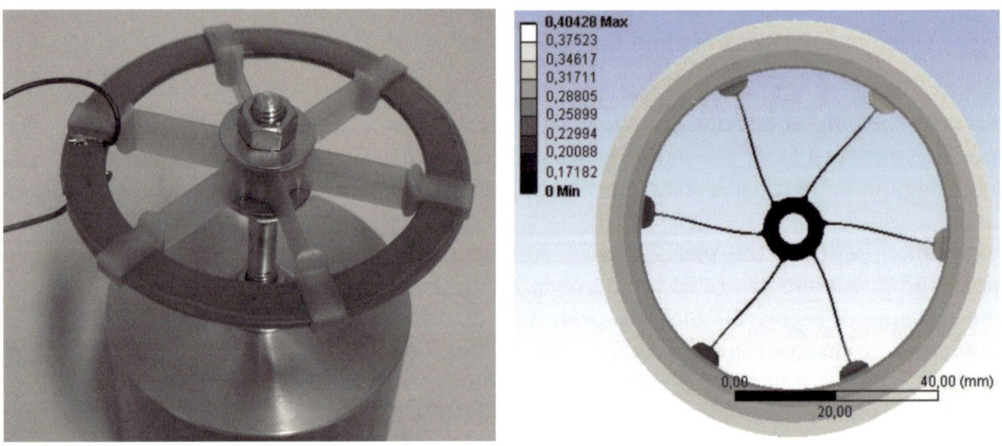

Fig. 10. Displacement amplification system made by insert moulding (left), displacement amplification system FEM simulation results (right)

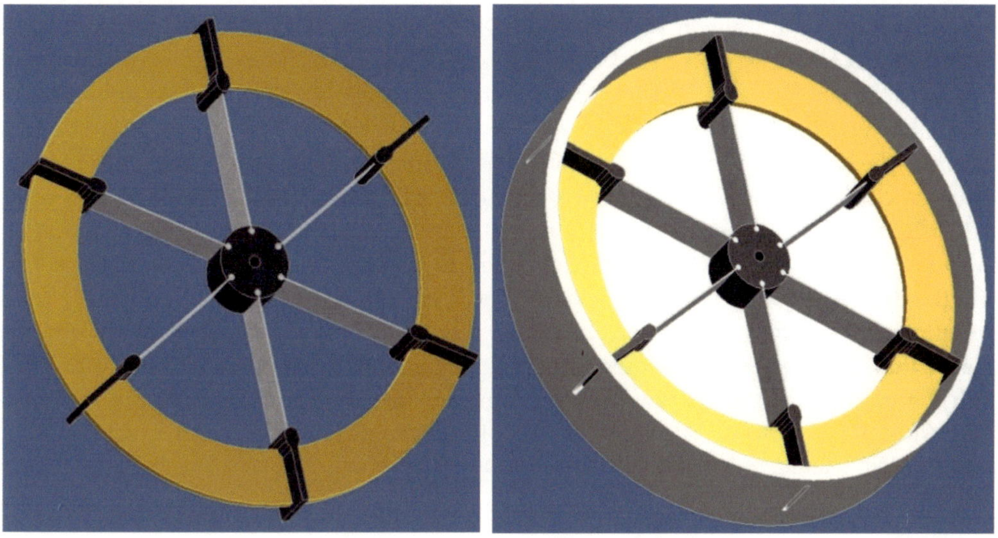

Fig. 11. Improved displacement amplification System using steel spokes (left), fluid restrictor with improved displacement amplification system (right)

Actuator	Displacement µm	Maximal moment Nm	First eigen frequency Hz
Discrete	410	6	248
Insert moulding	400	0,18	121
New 1.0	470	0,54	197
New 0.8	570	0,44	155
New 0.7	630	0,38	144
New 0.6	710	0,33	129

Table 8. Displacement amplification for miscellaneous actuators

Table 8 illustrates a comparison of different types of displacement amplification systems: The discrete amplification system shown in Figure 9, the first insert moulding prototype shown on the left side of Figure 10 and the first improved amplification system (New 1.0) shown in Figure 11 are similar with respect to the maximum displacement. This results from the constant eccentricity of the spokes of one mm. From these the discrete prototype using only steel as bearing and spokes provides the maximal torque. The new amplification system has a tripled force compared with the first insert moulding prototype and also improves the amplification. Varying the spokes eccentricity down to 0.6 mm with the new displacement amplification system allowed us improving the displacement further up to over 700 µm. With a restoring force of the previous actuator it had to overbear this was not possible. This force is proportional to the second moment of area of the spokes which rises with the third order of the spokes thickness. The Figure shows the first natural frequency of the actuator in addition. This should be higher than 100Hz for our application; all actuators achieve that. The higher eigenfrequencies of the amplification systems are all above 1000 Hz. With the example of the air flow restrictor one can see the benefits of the integration of different materials in one displacement amplification system. Therefore a sophisticated manufacturing technology like piezoceramic injection moulding and piezoceramic insert moulding is needed.

Piezoceramic manufacturing technology

The main common shaping process of piezoceramics is isostatic dry pressing, where powder granulate is pressed into the form. The slip casting is used for producing thin foils with controllable tolerances. These technologies have an issue with complex geometries. These complex geometries are needed for the integrated amplification systems. Therefore a new shaping technology for piezoceramics is needed.

A new shaping process of piezoceramics is the Ceramic Injection Moulding (CIM), which is a manufacturing technology that allows generating very complex structures of piezoceramics. For this process the PZT-powder is mixed with a thermoplastic binder system to a feedstock, which can be moulded at a temperature of 160°C. The required moulding tool that includes the three-dimensional cavity for the actuator has to be designed in CAD considering the materials shrinking behaviour on later process steps. To avoid incomplete filling one simulates the filling process with the CAD-model of the filling cavity under consideration of material properties. In general melt fronts due to different injection points for one cavity have to be avoided in the CIM process, because of the appearance of entrapments when the part gets sintered. After the forming, the binder system must be removed by thermal treatment. To compact the piezo grain structure one sinters the piezoceramic actuator at 1300°C. Next process steps were the metallization, which is done by sputtering and polarization with 4 kV per one millimetre thickness at 90°C. While the thermal disposal of the binder and the sintering process the actuator shrinks. This allowance must be reminded while designing the cavity for the actuator. The ceramic injection moulding process can be used for all of the actuators in the displacement amplification systems getting geometric optimized actuators.

Ceramic injection moulding supersedes finishing treatment of the actuator preparing it for its use in the amplification system. For early prototypes one can use water jet cutting for two dimensional processing. Here the ceramic plate is glued with thermal glue on a metal plate, preventing it from breaking while water jet cutting. Afterwards it is heated and separated

from the metal plate. One has to observe using thermal glue with lower melting point than the curie temperature of the piezoceramic actuator.
Another possible finishing technology is laser cutting. A high energy laser is needed which gives only a very short energy impulse avoiding depolarisation. Ultrasonic manufacturing is also possible with moderate costs. With ultrasonic manufacturing the ceramic could break due to the finishing process. For a large scaled amount of actuators these finishing processes are too expensive.

5. Conclusion

This chapter gave a short introduction to piezoceramics. Starting from flat and bending actuators with the need for moderate force and displacement at the same time, structured actuators where introduced. The structured amplification systems started from flat ones, went on with lever systems and finished with three examples of complex amplification systems. While the first amplification system is only from piezoceramics, the later ones integrate steel gaining more force and displacement amplification. The last amplification system integrates additional carbon fibre polymer. This allows an integrated manufacturing process and increases displacement amplification.

6. Acknowledgment

The research work of the authors is supported by the European Regional Development Fund (EFRE).

7. References

Bärecke, F, R Kasper, und M Al Wahab. „A structured piezo ceramic valve for an adaptive car gas damping system." 5th International Symposion on Mechatronics and its Applications ISMA. Amman, Jordan: IEEE, 2008.

Chakarov, Dimitar, Muhammed Al Wahab, Roland Kasper, und Kostadin Kostadinov. „Synthesis of tense piezo structures for local micro- and nano manipulations." 8. Magdeburg Maschinenbautage & 7. Mahreg Innovationsforum. Magdeburg: Universität Magdeburg, 2007. 173-180.

Gautschi, G. Piezoelectric Sensorics. Berlin Heidelberg: Springer Verlag, 2002.

Janocha, H. Unkonventionelle Aktoren. München: Oldenburg, 2010.

Kasper, R. „Mechatronisches Design integrierter Piezoaktoren." ISOM 2002 "Innovative Antriebstechnik". Chemnitz, 2002. 379-392.

Kasper, R, F Bärecke, M Al Wahab, M Harmann, und Matthias Hartmann. „High flow piezo ceramic valve for an adaptive vehicle gas spring damper." Actuator 2008. Bremen, 2008. 927-930.

Kasper, R, M Al Wahab, W Heinemann, K Kostadinov, und D Chakarov. „Mechatronic handling device based on piezo ceramic structures for micro and nano applications." Actuator 2006. Bremes, 2006. 154-158.

Kasper, R, W Heinemann, und A Wagner. „Modelling and Control for Piezoelectric Actuators for high speed applications." Movic. Zürich, 1998. 231-236.

Ruschmeyer, Karl. Piezokeramik: Grundlagen, Werkstoffe, Applikationen. Renningen-Malmsheim: expert-Verlag, 1995.

3

Recent Advances in Processing, Structural and Dielectric Properties of PMN-PT Ferroelectric Ceramics at Compositions Around the MPB

Eudes Borges Araújo
Universidade Estadual Paulista, Departamento de Física e Química Ilha Solteira, SP
Brazil

1. Introduction

Ferroelectric (1-x)[$Pb(Mg_{1/3}Nb_{2/3})O_3$]-x$PbTiO_3$ (PMN-PT) solid solutions are known for their exceptional electromechanical properties, sometimes one order of magnitude larger than classical $PbZr_{1-x}Ti_xO_3$ (PZT) ceramics. Prepared with a suitable x composition, PMN-PT is technologically important for fabricating some of the most important solid state devices such as a piezoelectric transducer, actuator, FERAM, etc. PMN-PT ceramics, thin films or single crystal forms can be prepared with high piezoelectric coefficients, a high dielectric constant and a low dielectric loss. Some compositions of PMN-PT single crystals exhibit a very high piezoelectric coefficient (d_{33}) and electromechanical coupling coefficients (k_{33}) (d_{33} ~ 1240 pC/N and k_{33} ~ 0.923), a high dielectric constant (ε ~ 3100) with low dielectric loss (tanδ ~ 0.014) compared to those of polycrystalline ceramics (d_{33} ~ 690 pC/N and k_{33} ~ 0.73) (Park & Shrout, 1997; Viehland et al., 2001). Recently, researchers have also reported that PMN-PT single crystals have high remanent polarization (P_r ~ 35 μC/cm^2) at a low coercive field (E_c ~ 3.4 kV/cm), a high dielectric constant (ε ~ 2500), low loss tangent (tanδ ~ 0.031), the highest piezoelectric coefficient (d_{33} ~ 1500 pC/N) and a high electrochemical coupling coefficient (k_{33} ~ 0.82) for <112> grain-oriented PMN-PT ceramics (Sun et al., 2004). The piezoelectric coefficient dij determines the stress levels induced by a given electric field and thus is the parameter most frequently used to describe the performance of an actuator.

PMN-PT solid solutions present a perovskite ABO_3 structure, where the A site is occupied by the Pb^{2+} ion, while the B site is randomly occupied by Mg^{2+}, Nb^{5+} and Ti^{4+} ions. Different compositions of the PMN-PT present distinct physical properties. The complex perovskite $Pb(Mg_{1/3}Nb_{2/3})O_3$ (x = 0) is a typical relaxor ferroelectric, characterized by a diffuse maximum of the dielectric constant associated with considerable frequency dispersion, that exhibits a non-polar paraelectric phase at high temperatures, similar in many aspects to normal ferroelectrics (Bokov & Ye, 2006). After cooling, a transformation occurs from the paraelectric phase to the ergodic relaxor state, characterized by the presence of polar nanoregions randomly distributed by the specimen, at the Burns temperature (T_B). This transformation is not accompanied by changes in the crystal structure on the macroscopic or mesoscopic scale and therefore cannot be considered a structural phase transition. In general, the state of a relaxor crystal at $T < T_B$ is frequently considered a new phase different from the paraelectric phase, since the polar nanoregions substantially affect the behavior of the crystal properties. Cooling the crystal hundreds of degrees below T_B, the polar nanoregions become frozen in a nonergodic state at around the freezing temperature (T_f) which can be irreversibly transformed into a ferroelectric state below T_f. The substitution of Ti^{4+} ions for the $(Mg_{1/3}Nb_{2/3})^{4+}$ ions in the B site of PMN-PT results in a long-range breaking leading to complex domain structures (Ye & Dong, 2000). The fascinating relaxor effect tends to disappear with this substitution, and the so-called morphotropic phase boundary (MPB) emerges in the composition-temperature (x-T) phase diagram of the PMN-PT system located at $x \approx 0.35$ (Noblanc et al., 1996). In the past decade, intensive research work was undertaken on the scientific understanding of the MPB nature in ferroelectric solid solutions (Noheda, 2002). This interest was mainly motivated by understanding the mechanisms responsible for the high piezoelectric response of ceramic compositions around the MPB. Finally, at the other extreme of the x-T phase diagram (x = 1), the lead titanate $PbTiO_3$ shows a typical ferroelectric-paraelectric phase transition presenting a high Curie temperature (T_C = 490 °C).

One of the most common problems related to PMN–PT emerges during synthesis. Indeed, obtaining high-density ceramics is very difficult without rigorous control of several steps in the processing, which makes using these materials for practical applications difficult. In the processing of PMN-PT solid solutions, the formation of an undesirable pyrochlore phase has long been recognized as a major problem for widespread use of these materials in the technology. In the last few years, several attempts, including modifications of the columbite method and other solid state reaction routes by using ultrafine powders, have been made to develop a processing technique in which formation of the pyrochlore phase is suppressed. Based on these processes, high-density ceramics could be obtained, and consequently, some properties could be improved. The purpose of this chapter is to provide an overview of the current understanding of some issues of the PMN-PT ferroelectric system at compositions around the MPB, including processing, structure and dielectric properties.

2. Processing and synthesis of PMN-PT

A major problem concerning the scientific studies and applications of PMN and PMN-PT electroceramics is the difficulty in producing a single-phase material with a perovskite structure. The main obstacle in PMN-PT synthesis is forming a lead niobate-based pyrochlore phase, frequently formed in the initial stage of the reactions processes for different methods. The presence of such a pyrochlore phase in the PMN-PT system is commonly believed to significantly degrade their dielectric properties, and therefore, the resulting material is inadequate for technological applications or a systematic scientific study (Mergen & Lee, 1997; Swartz & Shrout, 1982; Shrout & Swartz, 1983). Thus, choosing the synthesis method is fundamental to prepare pyrochlore-free PMN-PT of good quality. In general, several complex problems in materials science, such as the correct determination of the structure of the PZT solid solution in the MPB, remain unsolved in part due to the difficulty comparing results obtained from different samples. Some differences observed in peak widths of the neutron diffraction pattern suggest that the precise structural arrangements in these materials depend on the method of preparation (Yokota et al., 2009).

Compounds of the general formula $A_2B_2O_7$ (A and B metals) represent a family of phases isostructural to the mineral pyrochlore $(NaCa)(NbTa)O_6F/(OH)$ (Subramanian, M.A. et al., 1983). Although $A_2B_2O_7$ compounds exhibit a wide variety of interesting physical properties, in high-performance ferroelectrics such as PMN-PT these compounds should be avoided due to their paraelectric characteristic (Kamba et al., 2007). The compound $Pb_{1.83}Mg_{0.29}Nb_{1.71}O_{6.39}$ (PMN) exhibits a pyrochlore structure (space group Fd3m), frequently growing in the perovskite PMN ceramics, where the dielectric constant was found to increase as the temperature decreased, presenting an anomalous peak near 20 K, attributed to relaxation phenomena (Shrout & Swartz, 1983).

Chemical methods used to obtain complex precursor powders are normally less expensive and enable better stoichiometric control of complex oxides. Among several chemical methods, the approaches for obtaining mixed-cation oxide powders may be frequently grouped into two categories: the sol-gel process that uses alkoxide compounds as starting precursors (Johnson, 1985) and the polymeric precursor method that uses chelating agents such as citric acid (Lessing, 1989). In the sol-gel process, the high surface area of the dried gels results in high reactivity that in turn permits low temperature processing or even the formation of nonequilibrium phases. This process can be categorized in polymerized alkoxides and colloidal sols. The main advantages in using the polymerized alkoxides are low temperature, densification and ease doping compared to the disadvantage of slow drying. However,, colloidal sols present rapid drying and the low cost of the materials as advantages and doping difficulty as a disadvantage. In the popular polymeric precursor method, the synthesis of a cross-linked polymer resin provides some benefits such as homogeneous mixing of the cations and less tendency for segregation during calcinations. The main feature of the polymeric precursor method is that it allows a very homogeneous dispersion of the cations along the polymer. Thus, choosing a specific method to prepare mixed-cation oxide powders depends on several factors, including the kind of desired ceramic and the final objectives of each study.

Suppressing the pyrochlore phase is extremely difficult and depends on the synthesis method. To prepare pyrochlore-free PMN-PT ceramics or single crystals, several methods have been developed over the past three decades. The columbite process was known as a classical solid-state reaction method for synthesizing PMN (Swartz & Shrout, 1982). This method for solving the perovskite-pyrochlore reaction problem was proposed to bypass the formation of the intermediate pyrochlore phase reaction. By using the conventional mixed oxide technique and commercially starting materials, the following reaction sequences were proposed: $MgO + Nb_2O_5 \rightarrow MgNb_2O_6$ and $MgNb_2O_6 + 3PbO \rightarrow 3(PMN)$. This procedure was adopted keeping in mind that the kinetics for liberating Nb_2O_5 from the $MgNb_2O_6$ columbite phase may be slow enough to prevent the pyrochlore phase from forming and that structure of the $MgNb_20_6$ is a structure similar to the perovskite structure. At 900 °C, the reaction above is complete with small residuals of the pyrochlore phase. To remove this small amount of the pyrochlore phase, an MgO excess is introduced before the $MgNb_20_6$ calcination step. Thus, powders with perovskite structure have been successfully synthesized with the two-step calcination technique described, where the columbite ($MgNb_2O_6$) phase is initially synthesized at 1000 °C and then reacts with PbO and TiO_2 at 900 °C to form the pyrochlore-free PMN-PT phase. The effects of the MgO excess on the microstructure and dielectric properties of PMN and PMN-PT ceramics prepared by the columbite revealed that excess MgO completely inhibits formation of the pyrochlore phase and increases the grain size without substantially changing the density (Swartz et al., 1984). A significant increase in the dielectric constant values for PMN and PMN-PT, with $x = 0.1$, was observed with the addition of excess MgO and the increase in grain size and sintering temperature, reaching 18000 and 31000 for PMN and PMN-PT, respectively. Combining these factors to improve the dielectric constant in PMN-PT ceramics with the columbite

method leads to a decrease at the Curie temperature (T_C) as the grain size increases. These facts clearly indicate that the processing method plays an important role in the investigation of the phenomenology of ferroelectric ceramics based on the dielectric properties, since variations such as the T_C shift are attributed to the clamping of domains walls or internal stresses generated at the grain boundaries (Randall et al., 1998). This method was also used to prepare PMN-PT ceramics compositions around the MPB, with a dense microstructure, grain size around 2-3 μm and excellent electromechanical properties (Kelly et al., 1997). In addition to the columbite method, routes such as coprecipitation (Sekar & Halliyal, 1998), sol-gel (Yoon et al., 1995), modified sol-gel (Babooran et al., 2004), combustion synthesis (Cruz et al., 2002), semi-wet chemical route (Panda & Sahoo, 2005) and others have been proposed to prepare pyrochlore-free PMN powders, which can be also extended to PMN-PT synthesis. Alternatively, methods that require only one calcination step to produce single-phase perovskite PMN–PT and PMN powders were also proposed (Gu et al., 2003). In this case, the Nb_2O_5 powder is coated with $Mg(OH)_2$ in the first step, and in the second step, the $Mg(OH)_2$-coated Nb_2O_5 powder is mixed with PbO for calcinations. The pyrochlore-free perovskite powders obtained could be sintered to almost full density at 1150 °C. The main advantage of this method over the columbite method is the single calcination step.
Variants of the columbite method were proposed to optimize their electrophysical properties or to be used as alternative routes to obtain fine powders. Often, oxide precursor powders with specific characteristics, with controlled particle size as an example, are not available commercially. An alternative for each research laboratory is producing its own power with the desired characteristic. Thus, variants of the processes already established have arisen. A variant is the polymerized complex method based on the Pechini-type reaction route (Pechini, 1967). The general idea of this approach is to distribute the metallic ions homogeneously throughout the polymeric resin, in which chelate is formed between dissolved ions and a hydroxycarboxylic acid (citric acid is the usual). Heating the resin in air causes a breakdown of the polymer, and subsequent calcinations at 500 to 900 °C are necessary to form the desired mixed oxides (Lessing, 1989). This alternative was also applied to produce $MgNb_2O_6$ powders, which were used as precursors to produce the PMN phase, obtained by the solid-state reaction between $MgNb_2O_6$ and PbO (Guarany, 2006; Guerra et al., 2009).
The polymerized based on the Pechini method combined with the columbite method was also applied to obtain good-quality PMN-PT ceramics at compositions around the MPB with a tendency to <111>-orientation (Araújo et al., 2007). The texture tendency observed in these ceramics was attributed to shear rates introduced during the ceramic pressing when a fine powder was used. Similar <112> and <001> grain-oriented PMN-PT ceramics, at x = 0.30 and 0.32, respectively. The first one was prepared with a directional solidification method of the compressed powders prepared by the columbite method (Sun et al., 2004), and the second ceramic was produced with the templated grain growth method (Sabolsky et al., 2001). The latter method, responsible for synthesizing PMN-PT ceramics with good-quality texture, is a processing routine that uses a tabular (001)-oriented $SrTiO_3$ as a template (Kwon et al., 2005), at which the volume fraction of the oriented material increases as the thermal treatment continues at the expense of the finer grain matrix (Kwon et al., 2005). Despite the different methods proposed in the literature, studies of the phenomenology behind the MPB often have used ceramics prepared with the columbite classic method or some variation thereof. Table 1 summarizes the physical, dielectric and piezoelectric properties of PMN and PMN-PT ceramics for different compositions (x) around the MPB.

	PMN ceramics		
References	K_{max} (at 1 kHz)	T_C (°C)	Comments
Swartz et al. (1984)	16200	-10	Columbite method
Wang & Schulze (1990)	13700	-10	Modified columbite method
Kong et al. (2001)	13979	-	High-energy ball milling
Panda & Sahoo (2005)	10335	-	Semi-wet chemical route
Paula et al. (2008)	19000	-6	Modified columbite method

	PMN-PT ceramics				
	x	d_{33} (pC/N)	k_{33}	k_t	
Sun et al. (2004)	0.30	~1500	0.82	0.51	<112> Grain-oriented ceramics
Sabolsky et al. (2001)	0.32	~1200	0.755	-	<001> Textured ceramics
	0.30	450	-	0.41	
Kelly et al. (1997)	0.33	640	-	0.43	Random-oriented ceramics
	0.35	700	-	0.44	
Park et al. (1997)	0.30	670	-	-	Poled and unpoled ceramics
	0.33	690	0.73	-	

Table 1. Summary of the maximum dielectric constant and Curie temperature (T_C) of PMN ceramics prepared from different methods, and piezoelectric properties of PMN-PT ceramics for different compositions (x) around the MPB.

The preparation method also has a strong influence on the microstructure and dielectric properties of ferroelectric ceramics. In general, chemical methods that produce fine powders are more favorable for producing PMN-PT ceramics with a dense microstructure and larger grain size, but the density and grain size both depend on the sintering temperature. For ceramics prepared with the columbite method, the density decreases as the sintering temperature increases while the grain size increases (Swartz et al., 1984; Kong et al., 2002). An increase in the dielectric constant as the sintering temperature increases has also been observed for different compositions of PMN-PT (Baek et al., 1997). Grain-size dependence of the dielectric constant of PMN-PT ceramics was observed in the ferroelectric and paraelectric region, and this observation indicates that it is not a domain or stress effect (Swartz et al., 1984).

3. Structure of PMN-PT compositions at the MPB

The highest piezoelectric and electromechanical coupling coefficients of the PMN-PT (Park & Shrout, 1997) are found for compositions in the MPB at $0.27 < x < 0.37$, between the tetragonal and rhombohedral regions of the x-T phase diagram (Noblanc et al., 1996). A typical example of the MPB between the tetragonal and rhombohedral perovskite phases can be found in the x-T phase diagram of the PZT system at around $x \approx 0.52$ (Jaffe et al., 1971), where the highest piezoelectric coefficients are also observed. Since the 1970s, the MPB has been interpreted in terms of different models that attempt to explain the high piezoelectric response in the PZT system. The assumption that the morphotropic phase change is a first-order transition that defines a finite region where both tetragonal and rhombohedral phases coexists was one of the most discussed models (Arigur & Benguigui, 1975) until the mid-1990s (Mishra et al., 1996). The assumption of a metastable character for one of the two phases within the finite coexistence region (Isupov, 1975) and the propose of a true boundary explained with compositional fluctuations (Kakegawa & Mohri, 1977) were other models considered to understand the nature of the MPB in the PZT system. As often happens in science, complicated models reflect the degree of difficulty in understanding a problem, and new paradigms must be considered. These models began to decline when a study published in 1998 (Du et al., 1998), based on the phenomenological approach, revealed a great enhancement of piezoelectric coefficient d_{33} of rhombohedral PZT compositions around the MPB oriented along the tetragonal polar direction (001). This result was unexpected since the expectative for the largest piezoelectric deformations was to occur along the polar direction, with (001) for the tetragonal phase and (111) for the rhombohedral phase. Although this result has been obtained with thin films, since due to the lack of single crystals of PZT preferential orientations of PZT thin films can be adequately obtained by choosing an appropriate substrate, deposition method or drying temperature (Chen & Chen, 1994), it can be considered a catalyst for changing our perception of piezoelectric materials.

The discovery of a stable monoclinic phase in the PZT system in 1999, at compositions in the MPB (Noheda et al., 1999, 2000, 2001), provides a new perspective for explaining the high piezoelectric response of PZT ceramics (Guo et al. 2000, Singh et al. 2008, Bellaiche et al. 2000), attributed before to the coexistence of tetragonal and rhombohedral phases. After the initial discovery of a monoclinic phase with space group *Cm* (M_A type) on the PZT system, other new monoclinic phases with *Pm* and *Cc* space groups were also discovered (Singh & Pandey, 2001; Hatch et al., 2002; Ranjan et al., 2005). In addition, phenomenological and Raman scattering studies were performed on different PZT compositions around the MPB to confirm the presence of the monoclinic *Cm* and *Cc* phases (Souza et al., 2000; Souza et al., 2002). Consequently, a new x-T phase diagram of the PZT system was proposed (Pandey & Ragini, 2003), and more recently (Woodward et al., 2005; Pandey et al., 2008), a revised phase diagram was suggested including a region of monoclinic phases with *Cm* and *Cc* symmetries at compositions corresponding to the MPB. In addition to this context, more recent studies (Singh et al., 2007) suggest the presence of a monoclinic phase on the PZT system with the *Cm* space group instead of the rhombohedral phase with the *R3m* space group for Zr-rich compositions ($0.40 \leq x \leq 0.475$). Thus, the limits for the monoclinic phase existence in the x-T phase diagram of the PZT system are still open for discussion. Two years after the discovery of the monoclinic phase in the PZT system, a similar monoclinic phase with the space group *Pm* was also observed in the PMN-PT system (Singh & Pandey, 2001). Other work followed these studies demonstrating the existence of another low-temperature monoclinic phase in the PMN-PT system, where the monoclinic phase with the *Cm* space group transforms into the another monoclinic phase with the *Cc* space group (Ranjan et al., 2002). This *Cm*-to-*Cc* phase transition in PMN-PT could not be observed in the x-ray diffraction patterns because this transition leads to superlattice reflections, observed only in the electron and neutron diffraction patterns.

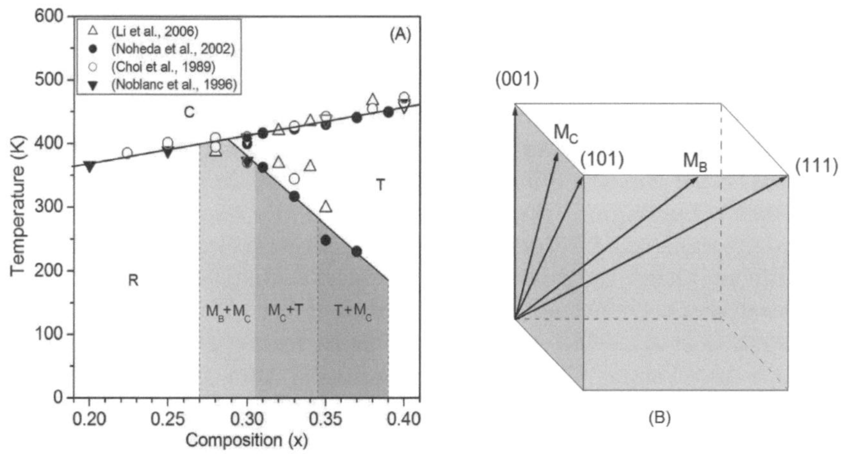

Fig. 1. A) Modified x-T phase diagram of PMN-PT around the MPB, adapted from Noheda et al. (Noheda et al., 2002a), Noblanc et al. (Noblanc et al., 1996), Choi et al. (Choi et al., 1989) and Li et al. (Li et al., 2006). The MPB includes the phase coexistences M_B+M_C (*Cm*+*Pm*), M_C+T (*Pm*+*P4mm*) and T+M_C (*P4mm*+*Pm*) based on Singh and Pandey's results (Singh & Pandey, 2003). B) Direction of the polarization in the tetragonal phase, M_C monoclinic phase, M_B monoclinic phase and rohombohedral phase according to Vanderbilt and Cohen's theory (Vanderbilt & Cohen, 2001).

Ferroelectric solid solutions exhibit high piezoelectric and electromechanical coupling coefficients and high dielectric permittivity at compositions in the MPB. But, what is the real importance of the monoclinic phases on the physical properties of these systems? Since the discovery of a monoclinic phase in PZT and PMN-PT solid solutions, several works have been published to understand the mechanisms responsible for the high piezoelectric response of these materials. The discovery of the monoclinic phases in these systems has catalyzed several exciting experimental (Guo et al., 2000; Bokov & Ye, 2004; Viehland et al., 2004) and theoretical developments (Fu & Cohen, 2000; Vanderbilt & Cohen, 2001). Interesting field-induced phase transitions were discovered in PMN PT (Chen et al., 2002; Zhao et al., 2003), highlighting the important role of the intermediate monoclinic phase for the large electromechanical response of this material. Theoretical first principles calculations and phenomenological theory considerations have not only confirmed the existence of monoclinic phases in the MPB region but also revealed the dominant role of the rotational instabilities of the polarization vector across the MPB compositions (Fu & Cohen, 2000; Vanderbilt & Cohen, 2001). Further, phenomenological theory considerations have shown that the Devonshire free energy expansion up to the eighth-order term is needed to stabilize the monoclinic phases, indicating the highly anharmonic nature of these materials (Vanderbilt & Cohen, 2001). All these studies have also led to significant modifications of the x-T phase diagrams of these solid solutions (Noheda et al., 2002), including the phase diagram of the PMN-PT system, previously established by Noblanc et al. (Noblanc et al., 1996) and Choi et al. (Choi et al., 1989). Thus, based on these results, a new x-T phase diagram for PMN-PT near the MPB could be constructed, as shown in Fig. 1-(A) with the shaded area.

The monoclinic phase with the *Cm* space group in the PMN-PT system was first proposed by Ye et al. (Ye et al., 2001) based on high-resolution synchrotron x-ray diffraction studies

on single crystals for x = 0.35. Almost simultaneously, Singh and Pandey (Singh & Pandey, 2001) observed a new monoclinic phase with the space group *Pm*, instead of *Cm*, as Ye et al. proposed, in a study with Rietveld refinements of the x-ray diffraction data from sintered powders with different compositions in the MPB. These results showed that the intrinsic width of the MPB is an order of magnitude smaller than previously believed and that the structure changes abruptly around x = 0.34-0.35, from monoclinic (*Pm*) to tetragonal (*P4mm*). Subsequent studies revealed that the structure of the PMN-PT is rhombohedral (*R3m*) for $x < 0.27$, M_B+M_C (*Cm+Pm*) phase coexistence for $0.27 < x < 0.30$, M_C+T (*Pm+P4mm*) phase coexistence for $0.30 < x < 0.34$, T+M_C (*P4mm+Pm*) phase coexistence for $0.34 < x < 0.39$ and tetragonal (*P4mm*) for $x > 0.39$ (Singh & Pandey, 2003; Singh et al., 2002; Li et al., 2006). A study of the molar fractions of different phases in the MPB by Singh and Pandey (Singh & Pandey, 2003) shows that for M_B+M_C phase coexistence the amount of M_B is majority (80-100%), the M_C phase is dominant (70-100%) for M_C+T phase coexistence, while for the T+M_C phase coexistence region the amount of tetragonal one increases from 40% to 70% when x increases from 0.35 to 0.39. Finally, for $x > 0.39$ the phase is purely tetragonal, as indicated in Fig. 1-A. The MPB region shown in Fig. 1-A was constructed based on room temperature dielectric measurements and Rietveld analysis of powder x-ray diffraction data obtained by Singh and Pandey (Singh & Pandey, 2003). Although these results were obtained for room temperature, the x-T phase diagram delineated in Fig. 1-A gives an idea of the complexity of the phase transitions involved in the MPB. Obviously, the structure of the PMN-PT system in the MPB as a function of temperature was much more complicated.
In the above paragraphs, the monoclinic phases M_A, M_B and M_C types observed in the PZT and PMN-PT systems were mentioned. The notation can be understood by considering the arguments of stability for the *Pm* and *Cm* phases described elsewhere (Vanderbilt & Cohen, 2001). The analysis of the phenomenology of the phase transitions in ferroelectrics is based on the order parameter **P**. In the space group *Pm3m* cubic phase, $P_X=P_Y=P_Z=0$. If the polarization P is constrained to a symmetry axis along (001) or (111), the resulting phase becomes tetragonal (space group *P4mm*) and rhombohedral (space group *R3m*), respectively. A similar analysis can be applied to the M_A (space group *Cm*), M_B (space group *Cm*) and M_C (space group *Pm*) monoclinic phases. For the M_A phase, the magnitudes of the polarization to the pseudocubic cell are $P_X=P_Y\neq P_Z$ with $P_Z>P_X$, for the M_B phase $P_X=P_Y\neq P_Z$ with $P_X>P_Z$ and for the M_C phase $P_X=0$ and $P_Y\neq P_Z$ (Vanderbilt & Cohen, 2001). The directions of the polarization in the tetragonal, monoclinic (M_B and M_C) and rhombohedral phases in the PMN-PT system are shown in Fig. 1-(B).
The observed phases and phase transition in barium titanate, $BaTiO_3$ and ferroelectrics in general were explained by Devonshire (Devonshire, Phil Mag 40, 1040 1949) in terms of the phenomenological Landau theory, based on a sixth-order expansion of the free energy in terms of order parameter **P**. However, the sixth-order Devonshire expansion fails to describe the existence of the monoclinic phase M_A observed initially in the PZT system. A work published in 2000 (Souza et al., 2000) confirms that the Devonshire phenomenological theory is not adequate for predicting the observed monoclinic phase. Then, Vanderbilt and Cohen (Vanderbilt & Cohen, 2001) showed that the simplest extension of the Devonshire theory, including an eighth-order term in the ferroelectric order parameter, admits the monoclinic phases M_A, M_B and M_C. An extension of the model to a twelfth-order term also predicts a triclinic ferroelectric phase.

3.1 Structural phase transition near the MPB: temperature-dependence

In addition to the complexity interpretation of the PMN-PT transformations for different compositions, the temperature dependence of the structural phase transitions in the MPB is another challenge. Several efforts have been made in the past few years to solve the phase transition problem for the PMN-PT system near the MPB, including structural studies conducted with high-resolution X-ray diffraction, spectroscopic studies (Araújo et al., 2008; Hlinka et al., 2006a) and dielectric analysis (Li et al., 2006; Hlinka et al., 2006b).

Spectroscopic techniques, such as Raman and infrared spectroscopy (IR), are useful tools for investigating phase transitions in polycrystalline materials and single crystals. Infrared spectroscopy can be used to study ferroelectric phase transitions since infrared vibrational frequencies, and consequently the interatomic forces, are affected by the onset of the ferroelectric state due to the temperature phase transitions. In the infrared spectra of the PMN-PT at composition x = 0.35, a broad band is observed from 472 cm^{-1} to 870 cm^{-1} (Araújo et al., 2008). This band is associated with ν_1-(Nb-O), ν_1-(Ti-O) and ν_1-(Mg-O) stretching modes in the PMN-PT structure. The IR curve shape changes slightly when the temperature increases, and these discrete changes as a function of the temperature may be attributed to the phase transitions involved in this system, introduced by distortions in BO_6 octahedra. Considering that an observed infrared spectrum for PMN-PT is a composition of the stretching modes in the BO_6 octahedron, each IR spectrum, recorded for different temperatures, can be fitted by considering three lorentzian functions:

$$I = \sum_i I_0 \frac{AW_i}{(\omega - \omega_f)_i^2 + W_i^2}, \qquad (1)$$

where I_0 is the constant, ω is the experimental frequency, ω_f is the fitting frequency, A is the area under curve (associated with the number of the oscillators) and W is the half width. The number of lorentzian functions used to fit each spectrum is assigned based on possible vibration modes in PMN-PT. The PMN-PT system presents a general ABO_3 perovskite structure and different phases depending on the PMN/PT ratio. The infrared vibrations for this perovskite family may be explained based on vibrations of the BO_6 octahedron (B = Mg, Nb and Ti), similarly to classical perovskites such as the $BaTiO_3$, $SrTiO_3$, $PbTiO_3$ and $PbZrO_3$ octahedron (Last, 1957; Spitzer et al., 1962; Perry et al., 1964). In these structures, the BO_6 octahedron presents four distinct vibration modes: ν_1-*stretching* at higher-frequency and lower-frequency ν_2-*torsion*, ν_3-*bending* and ν_4-cation-(BO_3) vibrations. These arguments justify the use of three lorentzian functions for analyzing ν_1-*stretching* within the 400-1000 cm^{-1}, since the ν_2, ν_3 and ν_4 vibrations are located in the far infrared region. The ν_1-*stretching* vibration mode of the BO_6 octahedra in the perovskite structure is schematically illustrated in Fig. 2-A.

The shift for lower frequencies of ν_1-*stretching* modes, as a function of increasing temperature (nonlinear behavior), can be explained by the inverse relationship between atomic separation and vibrational frequency if there is no structural phase transition in the observed temperature range. If we consider that the vibrational amplitude is proportional to the increasing temperature ($A \propto T$), then the nonlinear case described by the restoring force can be written as follows:

$$F(T) = -kT + \varepsilon(T), \qquad (2)$$

where k is the corresponding force constant and $\varepsilon(T)$ represents the departure from linearity. Expanding $\varepsilon(T)$ as a power series in equation (2) and solving the differential equation with the successive approximations method, the wavenumber $\bar{\nu}(T)$ can be expressed as:

$$\bar{\nu}(T) = \sqrt{C_1 - C_2 T^2}\ , \tag{3}$$

in which C_1 and C_2 are constants. This equation describes the behavior of the ν_1-*stretching* modes as a function of increasing temperature.

The phase transitions for the PMN-PT system by using IR spectroscopy can be understood from the point of view of the group representation, considering the possible symmetries involving the BO_6 octahedron. The cubic symmetry, point group O_h, presents an F_{1u} species and a single triply degenerate band structure is expected, since three equivalent axes exist in the case of the cubic lattice. When the cubic phase transforms into the tetragonal phase, point group C_{4v}, an E and A_1 species appear with the cubic triple degeneracy partially removed. Consequently, a double band structure is expected for the tetragonal phase. Finally, the E and A_1 transform into A′ and A″ species when the symmetry changes from tetragonal to monoclinic. Thus, a double partially degenerate band structure is also expected for the monoclinic phase.

Under the above considerations, the monoclinic and tetragonal phase coexistence in PMN-PT must be interpreted as a superposing of two double partially degenerate band structures. Considering that these bands occur at very close frequencies, using three lorentzian functions is sufficient to fit each IR spectrum by using Equation (1). To illustrate this discussion, Fig. 2 shows the behavior of the ν_1-(Ti-O) mode as a function of temperature, obtained from the PMN-PT single crystal for composition x = 0.35 (Araújo et al., 2008). The singularities observed at around 230 K, 300 K and 422 K were interpreted as the result of structural phase transitions induced by the temperature increase while the curves in this figure represent the theoretical fitting of the experimental data by using Equation (3). The anomalous behavior observed for the ν_1-(Ti-O) mode between 230 K and 300 K, where $\bar{\nu}(T)$ increases rather than decreases, as predicted by Equation (3), suggests a possible explanation if the coexistence of the monoclinic and tetragonal phases is considered. Thus, based on the previous discussion and supported by the results in the literature (Noheda et al., 2002; Li et al., 2006), the occurrence of the following *monoclinic* (M_C) → *monoclinic* (M_C) + *tetragonal* (T) → *tetragonal* (*T*) → *cubic* (C) phase transition can be assumed, as indicated in Fig. 2. The behavior of the ν_1-(Ti-O) mode as a function of temperature observed in Fig. 2 by itself does not allow assessment of the sequence of the above-proposed phase transition without considering the results in the literature obtained from other techniques. However, the linear increase in the ν_1-(Ti-O) mode between 230 K and 300 K is strong evidence in favor of the phase coexistence hypothesis within this temperature interval.

The stability of the monoclinic phase at low temperatures and the phase coexistence in the MPB of the PMN-PT system has also been frequently discussed in the literature. The phase coexistences in this system on zero-field cooling were considered in recent experimental works by using neutron diffraction, high-resolution X-ray diffraction and dielectric analysis (Singh & Pandey, 2006; Li et al., 2006). These studies have discussed the possibility of

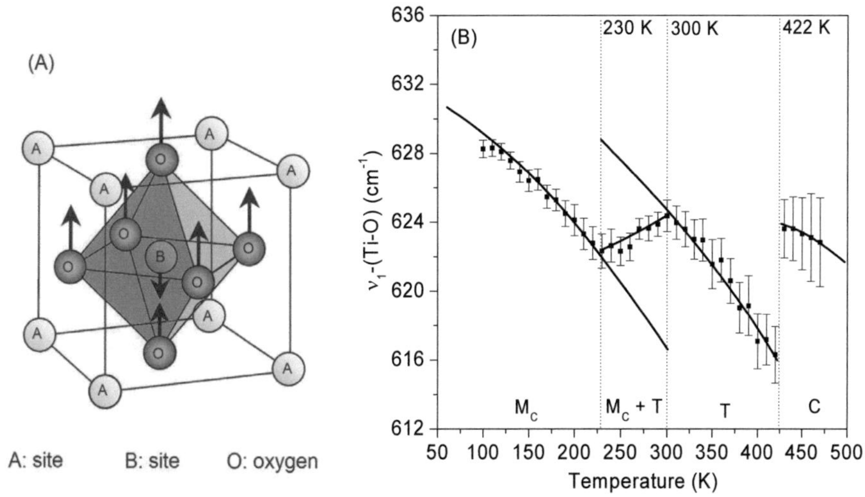

Fig. 2. (A) Schematic infrared-normal active higher-frequency ν_1-*stretching* vibration of the BO_6 octahedra in the perovskite structure. (B) Behavior of the *ν_1-(Ti-O)-stretching* mode in the BO_6 octahedron in ABO_3 structure of the PMN-PT as a function of the temperature. (Adapted from Araújo et al., 2008).

tetragonal-monoclinic (T-M_C) and tetragonal-monoclinic (T-M_B) phase coexistence in unpoled PMN-PT ceramics in a wide temperature range. From these studies, the metastable phase diagram for this system in the MPB (see Fig. 7 in Li et al., 2006) was constructed based on the arguments that the (T-M_B) phase coexistence is a non-equilibrium state and is the result of a local stress field and clampdown and blocking effect. A more recent study on the heterophase structures and the domain state-interface in the PMN-PT system, including the intermediate monoclinic phases (M_B or M_C), provides insight into understanding the stress-relief conditions at the phase coexistence near the MPB (Topolov, 2009). A correlation between the domain structures and the unit-cell parameters of adjacent M_B and M_C phases in certain temperature ranges could be suggested as the possible origin of the different sequences of phase transitions in solid solutions near the MPB.

3.2 The origin of the morphotropic phase boundary

To explain the large piezoelectric and electromechanical coupling coefficients observed in ferroelectric systems that exhibit an MPB, a polarization rotation model was proposed (Fu & Cohen, 2000; Guo et al., 2000). A giant piezoelectric response has been observed for $Pb(Zn_{1/3}Nb_{2/3})O_3$-$PbTiO_3$ (PZN-PT) and PMN-PT single crystals in the rhombohedral phase, which have a polarization along the [111] direction. The crystals respond mechanically with a remarkable increase in strain if a small electric field is applied along the (001) direction. However, when the magnitude of the electric field is increased, a very flat strain-versus-field slope is observed (Fu & Cohen, 2000; Park & Shrout, 1997). Thus, the giant piezoelectric response could be explained by the reorientation of the polar domains, which depends on the mesoscopic structure or ordering, or by the polarization rotation mechanism at the atomic level (Fu & Cohen, 2000). An intuitive explanation for this mechanism can be delineated based on fact that $PbTiO_3$ is a ferroelectric, which does not present a rhombohedral phase, with a large strain level of 6%. However, if there were a rhombohedral phase in $PbTiO_3$, a colossal mechanical strain would be induced by applying an electric field. The model was proposed based on first-principles calculations on the classical $BaTiO_3$ ferroelectric perovskite, by selecting different polarization directions and calculating their internal energies, provided explanations for the experimental observations (Park & Shrout, 1997; Park et al., 1999). Bellaiche et al. (Bellaiche et al., 2000) have succeeded in deriving the monoclinic phase of the PZT system at compositions around the MPB from first-principles calculations, showing that the predicted structural data were in good agreement with the measurements and consistent with the monoclinic phase as an intermediate phase between the tetragonal and rhombohedral phases.

The proposed polarization vector rotation for monoclinic phases in the PMN-PT system is schematically illustrated in Fig. 1-B. As the monoclinic phase represents the structural bridge between the tetragonal (space group *P4mm*) and rhombohedral (space group *R3m*) phases, the symmetry constrains the polarization vector to remain in the monoclinic plane, but is free to rotate within it (Noheda, 2002). The polarization vector of the M_C phase can rotate along the *ac* plane between the (001) and (101) directions, while the M_B phase can rotate along the *bc* plane between the (101) and (111) directions, configuring the polarization rotation model. The rotation of the polarization vector along these planes is responsible for the high values observed for the piezoelectric coefficients d_{33} in the PZT and PMN-PT solid solutions, as calculated previously (Bellaiche et al., 2000; Fu & Cohen, 2000). Interesting reviews of advances in understanding the role of the monoclinic phases in structure and high piezoelectricity in lead oxide solid solutions can be found in several references (Noheda, 2002; Noheda & Cox, 2006). Although several authors have reported the existence of the M_B phase for the PMN-PT system, this phase has not yet been effectively observed.

The revision of results and traditional notions acquired about ferroelectrics in the past few decades has led to new ideas and interpretations, demanding new experiments and models to explain the nature of the MPB (Kreisel et al., 2009; Khachaturyan, 2010). Some partial conclusions indicate that the MPB is considered more of a region than a boundary, characterized by two competing coexisting phases. However, some points still remain unclear to explain the giant electromechanical response of ferroelectric solid solutions near the MPB. The apparent violation of the Landau theory of the second-order phase transition, which does not permit a gradual rotation of the polarization vector under an applied electric field in the monoclinic phase, since the transition between different monoclinic phases is expected to occur, has resulted in alternative propositions to explain the phenomenology in the MPB (Khachaturyan, 2010). If the monoclinic phase is considered an *adaptive phase*, however, an adaptive state can be assumed in the MPB, and thus, gradual polarization rotation is now possible because of the easy rearrangement of nanodomains under the reduced polar anisotropy. The theory of an adaptive phase was developed to predict the ferroelectric transitions in solid solutions near the MPB and is based on stress-accommodating nanotwins domains (Jin et al., 2003; Bhattacharyya et al., 2008). The adaptive concepts have been extended to explain the origin of X-ray and neutron diffraction patterns of ferroelectric solid solutions systems near the MPB, but the existence of the monoclinic phase cannot be completely ruled out at low temperatures (Khachaturyan, 2010). These results demonstrate the necessity of additional experimental studies on traditional systems such as PZT and PMN-PT. The giant electromechanical response studied in PMN-PT single crystals was interpreted as a critical phenomenon such as in the electric-field–temperature–composition (*E-T-x*) phase diagram of PMN-PT a first-order paraelectric–

ferroelectric phase transition terminates in a line of critical points where the piezoelectric coefficient is the maximum (Kutnjak et al., 2006). In addition, the large piezoelectric response observed at PMN-PT relaxors could be also related to the competition between ordered states at nanoscale in the presence of disorder and to the fact that any weak stress or electric field strongly affects such competition (Dkhil et al., 2009). All this work has provided valuable contributions to understanding the phenomenology behind the MPB.

Exploring theoretical predictions and experimental results for the similar MPB in the pure lead titanate $PbTiO_3$, which appear under high hydrostatic pressure and cryogenics temperatures leading to tetragonal → monoclinic → rhombohedral phase transition, a more general idea of the morphotropic phase boundary could be introduced with perspectives of developing new high-performance electromechanical materials based on the chemical pressure idea (Ahart et al., 2008). In addition to these very attractive possibilities, observation of a low-symmetry monoclinic phase in the $PbTiO_3$ opened very recent discussions on the arguments that justify the origin of morphotropic phase boundaries in ferroelectrics (Ahart et al., 2008) and arguments against the validity of the rotation polarization model to describe the stability of the phases and the sequence of phase transformation in the MPB in pure $PbTiO_3$ (Frantti et al., 2008). This controversy is very interesting because the discussions illustrate how different ideas promote the advancement of science, showing the richness behind the research in phase transitions of systems that exhibit the MPB and that this subject is still open to new contributions.

4. Conclusion

Most modern piezoelectric materials used for technological applications are solid solutions that display a transition region in their composition-temperature phase diagrams, known as the morphotropic phase boundary, where the structure changes abruptly and the electromechanical properties exhibit maximal values. Some composition of PMN-PT ceramics, which are simpler to prepare than single crystals, is frequently used for technological applications due to their very high piezoelectric coefficient and electromechanical coupling coefficients. A major problem concerning the application of these electroceramics is the difficulty in producing a pyrochlore-free perovskite structure. Therefore, to prepare high-quality pyrochlore-free PMN-PT ceramics, several methods have been developed over the past three decades. In addition to the technological interest, the scientific comprehension of the origin of the morphotropic phase boundary in these ferroelectric solid solutions was a great challenge in past decades. The discovery of a stable monoclinic phase in the PZT system provides a new perspective to explain the high piezoelectric response of ferroelectric ceramics. The subsequent experimental and theoretical studies provided valuable contributions to understanding the phenomenology behind the MPB and promoted a significant advancement in the area. However, recent new ideas about and interpretations of the structural phase transition in high-performance ferroelectric systems and the nature of the MPB demonstrated that this subject is still open for discussion.

5. Acknowledgment

The author is grateful to Brazilian agencies FAPESP, CNPq and CAPES for the financial support for the correlated research. My special thanks to Pró-Reitoria de Pós-Graduação da

UNESP (PROPG) and Fundação para o Desenvolvimento da UNESP (FUNDUNESP) for their financial support for this work.

6. References

Ahart, M. et al. (2008). Origin of morphotropic phase boundaries in ferroelectrics. *Nature* 451, 545-549. ISSN 0028-0836

Araújo, E.B. et al. (2007). Synthesis of slightly <111>-oriented $0.65Pb(Mg_{1/3}Nb_{2/3})O_3$–$0.35PbTiO_3$ ceramic prepared from fine powders. *Materials Chemistry and Physics* 104, 40-43. ISSN 0254-0584

Araújo, E.B. et al. (2008). Structural phase transition studies on PMN-0.35PT using Infrared Spectroscopy. *Ferroelectrics* 369, 35-42. ISSN 1563-5112

Arigur, P. & Benguigui, L. (1975). Direct determination of coexistence region in solid-solutions $Pb(Zr_xTi_{1-x})O_3$. *Journal of Physics D: Applied Physics* 8, 1856-1862. ISSN 1361-6463

Babooram, K. et al. (2004). Phase formation and dielectric properties of $0.90Pb(Mg_{1/3}Nb_{2/3})O_3$–$0.10PbTiO_3$ ceramics prepared by a new sol–gel method. *Ceramics International* 30, 1411-1417. ISSN 0272-8842

Baek, J.G. et al. (1997). Synthesis of pyrochlore-free $0.9Pb(Mg_{1/3}Nb_{2/3})O_3$–$0.1PbTiO_3$ ceramics via soft mechanochemical route. *Journal of the American Ceramic Society* 80, 973-981. ISSN 1551-2916

Bellaiche, L. et al. (2000). Finite-Temperature properties of $Pb(Zr_{1-x}Ti_x)O_3$ alloys from first principles. *Physical Review Letters* 84, 5427-5430. ISSN 1079-7114

Bhattacharyya, S. et al. (2008). Direct high-resolution transmission electron microscopy observation of tetragonal nanotwins within the monoclinic M_C phase of $Pb(Mg_{1/3}Nb_{2/3})O_3$-$0.35PbTiO_3$ crystals. *Applied Physics Letters* 92, 1429041-1429043. ISSN 1077-3118

Bokov, A.A. & Ye, Z.G. (2004). Domain structure in the monoclinic Pm phase of $Pb(Mg_{1/3}Nb_{2/3})O_3$-$PbTiO_3$ single crystals. *Journal of Applied Physics* 95, 6347-6359. ISSN 1089-7550

Bokov, A.A. & Ye, Z.G. (2006). Recent progress in relaxor ferroelectrics with perovskite structure. *Journal of Materials Science* 41, 31-52. ISSN 1573-4803

Chen, K.P. et al. (2002). Electric-field-induced phase transition in <001>-oriented $Pb(Mg_{1/3}Nb_{2/3})O_3$-$PbTiO_3$ single crystals. *Journal of Physics: Condensed Matter* 14, L571-L576. ISSN 1361-648X

Chen, S.Y. & Chen, I.W. (1994). Temperature-time texture transition of $Pb(Zr_{1-x}Ti_x)O_3$ thin films: II, Heat treatment and compositional effects. *Journal of the American Ceramic Society* 77, 2337-2344. ISSN 1551-2916

Choi, S.W. et al. (1989). Morphotropic phase boundary in $Pb(Mg_{1/3}Nb_{2/3})O_3$-$PbTiO_3$ system. *Materials Letters* 8, 253-255. ISSN 0167-577X

Cruz, L.P. et al. (2002). An easy way to $Pb(Mg_{1/3}Nb_{2/3})O_3$ synthesis. *Materials Research Bulletin* 37, 1163-1173. ISSN 0025-5408

Dkhil, B. et al. (2009). Intermediate temperature scale T^* in lead-based relaxor systems. *Physical Review B* 80, 064103-064109. ISSN 1550-235x

Du, X.J. et al. (1998). Crystal orientation dependence of piezoelectric properties of lead zirconate titanate near the morphotropic phase boundary. *Applied Physics Letters* 72, 2421-2423. ISSN: 1077-3118

Frantti, J. et al (2008). Evidence against the polarization rotation model of piezoelectric perovskites at the morphotropic phase boundary. *Journal of Physics: Condensed Matter* 20, 472203-472203. ISSN 1361-648X

Fu, H.X. & Cohen, R.E. (2000). Polarization rotation mechanism for ultrahigh electromechanical response in single-crystal piezoelectrics. *Nature* 403, 281-283. ISSN 0028-0836

Guerra, J.D.S. et al. (2009). Investigation on dielectric response of PMN ceramics around paraelectric–ferroelectric diffuse phase transition. *Materials Science and Technology* 25, 1316-1320. ISSN 1743-2847

Gu, H. et al. (2003). Single-calcination synthesis of pyrochlore-free $0.9Pb(Mg_{1/3}Nb_{2/3})O_3$–$0.1PbTiO_3$ and $Pb(Mg_{1/3}Nb_{2/3})O_3$ ceramics using a coating method. *Journal of the American Ceramic Society* 86, 217-221. ISSN 1551-2916

Guo, R. et al. (2000). Origin of the high piezoelectric response in $PbZr_{1-x}Ti_xO_3$. *Physical Review Letters* 84, 5423-5426. ISSN 1079-7114

Hatch, D.M. et al. (2002). Antiferrodistortive phase transition in $Pb(Ti_{0.48}Zr_{0.52})O_3$: Space group of the lowest temperature monoclinic phase. Physical Review B 65, 212101-212103. ISSN 1550-235x

Hlinka, J. et al. (2006). Infrared dielectric response of relaxor ferroelectrics. *Phase Transitions* 79, 41-78. ISSN 1029-0338

Hlinka, J. et al. (2006). Anisotropic dielectric function in polar nanoregions of relaxor ferroelectrics. *Physical Review Letters* 96, 027601-027604. ISSN 1079-7114

Johnson, D.W. (1985). Sol-gel processing of ceramics and glass. *American Ceramic Society Bulletin* 64, 1597-1602. ISSN 1551-2916

Jin, Y.M. et al. (2003). Adaptive ferroelectric states in systems with low domain wall energy: Tetragonal microdomains. *Journal of Applied Physics* 94, 3629-3640. ISSN 1089-7550

Kamba, S. et al. (2007). Quantum paraelectric behavior of pyrochlore $Pb_{1.83}Mg_{0.29}Nb_{1.71}O_{6.39}$. *Physical Review B* 76, 054125-054131. ISSN 1550-235x

Kakegawa, K. & Mohri, J. (1977). A compositional fluctuation and properties of $Pb(Zr,Ti)O_3$. *Solid State Communications* 24, 769-772. ISSN 0038-1098

Kelly, J. et al. (1997). Effect of composition on the electromechanical properties of $(1-x)Pb(Mg_{1/3}Nb_{2/3})O_3$-$xPbTiO_3$ ceramics. *Journal of the American Ceramic Society* 80, 957-964. ISSN 1551-2916

Khachaturyan, A.G. (2010). Ferroelectric solid solutions with morphotropic boundary: Rotational instability of polarization, metastable coexistence of phases and nanodomain adaptive states. *Philosophical Magazine* 90, 37–60. ISSN 1478-6443

Kong, L.B. et al. (2001). Preparation of PMN powders and ceramics via a high-energy ball milling process. *Journal of Materials Science Letters* 20, 1241-1243. ISSN 1573-4811

Kreisel, J. et al. (2009). Phase transitions and ferroelectrics: revival and the future in the field. *Phase Transitions* 82, 633–661. ISSN 1029-0338

Kutnjak, Z. et al. (2006). The giant electromechanical response in ferroelectric relaxors as a critical phenomenon. *Nature* 441, 956-959. ISSN 0028-0836

Kwon, S. et al. (2005). High strain, <001> textured $0.675Pb(Mg_{1/3}Nb_{2/3})O_3$–$0.325PbTiO_3$ ceramics: Templated grain growth and piezoelectric properties. *Journal of the American Ceramic Society* 88, 312-317. ISSN 1551-2916

Last, J.T. (1957). Infrared-absorption studies on barium titanate and related materials. *Physical Review* 105, 1740-1750. ISSN 1536-6065

Li, J.B. et al. (2006). Structural transition in unpoled (1-x)PMN-xPT ceramics near the morphotropic boundary. *Journal of Alloys and Compounds* 425, 373-378. ISSN 0925-8388

Mergen, A. & Lee, W.E. (1997). Fabrication, characterisation and formation mechanism of $Pb_{1.83}Mg_{0.29}Nb_{1.71}O_{6.39}$ pyrochlore. *Journal of the European Ceramic Society* 17, 1033-1047. ISSN 0955-2219

Mishra, S.K. et al. (1996). Effect of phase coexistence at morphotropic phase boundary on the properties of $Pb(Zr_xTi_{1-x})O_3$ ceramics. *Applied Physics Letters* 69, 1707-1709. ISSN 1077-3118

Noblanc, O. et al. (1996). Structural and dielectric studies of $Pb(Mg_{1/3}Nb_{2/3})O_3$-$PbTiO_3$ ferroelectric solid solutions around the morphotropic boundary. *Journal of Applied Physics* 79, 4291-4297. ISSN 1089-7550

Noheda, B. et al. (1999). A monoclinic ferroelectric phase in the $Pb(Zr_{1-x}Ti_x)O_3$ solid solution *Applied Physics Letters* 74, 2059-2062. ISSN 1077-3118

Noheda, B. et al. (2001). Tetragonal-to-monoclinic phase transition in a ferroelectric perovskite: The structure of $PbZr_{0.52}Ti_{0.48}O_3$. *Physical Review* B 61, 8687-8695. ISSN 1550-235x

Noheda, B. et al. (2001). Stability of the monoclinic phase in the ferroelectric perovskite $PbZr_{1-x}Ti_xO_3$. *Physical Review* B 63, 014103-014112. ISSN 1550-235x

Noheda, B. et al. (2002). Phase diagram of the ferroelectric relaxor (1-x)$PbMg_{1/3}Nb_{2/3}O_3$-x$PbTiO_3$. *Physical Review B* 66, 054104-054114. ISSN 1550-235x

Noheda, B. (2002). Structure and high-piezoelectricity in lead oxide solid solutions. *Current opinion in solid state & materials science* 6, 27-34. ISSN 1359-0286

Noheda, B. & Cox, D.E. (2006). Bridging phases at the morphotropic boundaries of lead oxide solid solutions. *Phase Transitions* 79, 5-20. ISSN 1029-0338

Panda, P.K. & Sahoo, B. (2005). Preparation of pyrochlore-free PMN powder by semi-wet chemical route. *Materials Chemistry and Physics* 93, 231-236. ISSN 0254-0584

Pandey, D. & Ragini (2003). On the discovery of new low temperature monoclinic phases with Cm and Cc space groups in $Pb(Zr_{0.52}Ti_{0.48})O_3$: an overview. *Zeitschrift Fur Kristallographie* 218, 1-7. ISSN 0044-2968

Pandey, D. et al. (2008). Stability of ferroic phases in the highly piezoelectric $Pb(Zr,Ti)O_3$ ceramics. *Acta Crystallographica Section A* 64, 192-203. ISSN 0108-7673

Park, S.E. & Shrout, T.R. (1997). Characteristics of relaxor-based piezoelectric single crystals for ultrasonic transducers. *IEEE Transactions on Ultrasonics, Ferroelectrics and Frequency Control* 44, 1140-1147. ISSN 0885-3010

Pechini, M. (1967). Method of preparing lead and Al-alkaline-earth titanates and niobates and coating method using the same to form a capacitor. U.S. Pat. N° 3 330 697.

Perry, C.H. et al. (1964). Infrared studies of perovskite titanates. *Physical Review* 135, A408-A412. ISSN 1536-6065

Randall, C.A. et al. (1998). Intrinsic and extrinsic size effects in fine-grained morphotropic-phase-boundary lead zirconate titanate ceramics. *Journal of the American Ceramic Society* 81, 677-688. ISSN 1551-2916

Ranjan, R. et al. (2002). Antiferrodistortive phase transition in $Pb(Ti_{0.48}Zr_{0.52})O_3$: A powder neutron diffraction study. *Physical Review B* 65, 060102-060106. ISSN 1550-235x

Ranjan, R. et al. (2005). Comparison of the Cc and R3c space groups for the superlattice phase of $Pb(Zr_{0.52}Ti_{0.48})O_3$. *Physical Review B* 71, 092101-092104 (2005). ISSN 1550-235x

Sabolsky, E.M. et al. (2001). Piezoelectric properties of <001> textured $Pb(Mg_{1/3}Nb_{2/3})O_3$–$PbTiO_3$ ceramics. *Applied Physics Letters* 78, 2551-2553. ISSN 1077-3118

Sekar, M.M.A. & Halliyal, A. (1998). Low-temperature synthesis, characterization, and properties of lead-based ferroelectric niobates. *Journal of the American Ceramic Society* 81, 380-388. ISSN 1551-2916

Shrout, T.R. & Swartz, S.L. (1983). Dielectric-properties of pyrochlore lead magnesium niobate. *Materials Research Bulletin* 18, 663-667. ISSN 0025-5408

Singh, A.K. & Pandey, D. (2001). Structure and the location of the morphotropic phase boundary region in $(1-x)[Pb(Mg_{1/3}Nb_{2/3})O_3]-xPbTiO_3$. *Journal of Physics: Condensed Matter* 13, L931-L936. ISSN 1361-648X

Singh, A.K. et al. (2002). Powder neutron diffraction study of phase transitions in and a phase diagram of $(1-x)[Pb(Mg_{1/3}Nb_{2/3})O_3]-xPbTiO_3$. *Physical Review* B 66, 0541011-0241019. ISSN 1550-235x

Singh, A.K. & Pandey, D. (2003). Evidence for M_B and M_C phases in the morphotropic phase boundary region of $(1-x)[Pb(Mg_{1/3}Nb_{2/3})O_3]-xPbTiO_3$: A Rietveld study. *Physical Review B* 67, 064102-064114. ISSN 1550-235x

Singh, A.K. & Pandey, D. (2006). Powder neutron diffraction study of phase transitions in and a phase diagram of $(1-x)[Pb(Mg_{1/3}Nb_{2/3})O_3]-xPbTiO_3$. *Physical Review B* 74, 024101-024119. ISSN 1550-235x

Singh, A.K. et al. (2007). High-resolution synchrotron x-ray diffraction study of Zr-rich compositions of $Pb(Zr_xTi_{1-x})O_3$ ($0.525 \leq x \leq 0.60$): Evidence for the absence of the rhombohedral phase. *Applied Physics Letters* 91, 192904-192907. ISSN 1077-3118

Singh, A.K. et al. (2008). Origin of high piezoelectric response of $Pb(Zr_xTi_{1-x})O_3$ at the morphotropic phase boundary: Role of elastic instability. *Applied Physics Letters* 92, 022910-022913. ISSN 1077-3118

Souza, A.G. et al. (2000). Monoclinic phase of $PbZr_{0.52}Ti_{0.48}O_3$ ceramics: Raman and phenomenological thermodynamic studies. *Physical Review B* 61, 14283-14286. ISSN 1550-235x

Souza, A.G. et al. (2002). Raman scattering study of the $PbZr_{1-x}Ti_xO_3$ system: Rhombohedral-monoclinic-tetragonal phase transitions. *Physical Review B* 66, 132107-132111. ISSN 1550-235x

Subramanian, M.A. et al. (1983). Oxide pyrochlores - A review. *Progress in Solid State Chemistry* 15, 55-143. ISSN 0079-6786

Sun, S. et al. (2004). Fabrication and electrical properties of grain-oriented $0.7Pb(Mg_{1/3}Nb_{2/3})O_3$-$0.3PbTiO_3$ ceramics. *Applied Physics Letters* 84, 574-576. ISSN 1077-3118

Swartz, S.L. & Shrout, T.R. (1982). Fabrication of perovskite lead magnesium niobate. *Materials Research Bulletin* 17, 1245-1250. ISSN 0025-5408

Swartz, S.L. et al. (1984). Dielectric properties of lead-magnesium niobate ceramics. *Journal of the American Ceramic Society* 67, 311-315. ISSN 1551-2916

Spitzer, W.G. et al. (1962). Far infrared dielectric dispersion in $BaTiO_3$, $SrTiO_3$ and TiO_2. *Physical Review* 126, 1710-1962. ISSN 1536-6065

Topolov, V.Yu. (2009). Monoclinic phases and stress-relief conditions in (1-x)$Pb(Mg_{1/3}Nb_{2/3})TiO_3$–x$PbTiO_3$ solid solutions. *Journal of Alloys and Compounds* 480, 568-574. ISSN 0925-8388

Isupov, V.A. (1975). Comments on the paper "x-ray study of the pzt solid solutions near the morphotropic phase transition". *Solid State Communications* 17, 1331-1333. ISSN 0038-1098

Viehland, D. et al. (2001). Piezoelectric instability in <011>-oriented $Pb(B_{1/3}{}^{I}B_{2/3}{}^{II})O_3$–$PbTiO_3$ crystals. *Applied Physics Letters* 79, 1006-1009. ISSN 1077-3118

Viehland, D. et al. (2004). Domain structure changes in (1-x)$Pb(Mg_{1/3}Nb_{2/3})O_3$-x$PbTiO_3$ with composition, dc bias, and ac field. *Journal of Applied Physics* 96, 3379-3381. ISSN 1089-7550

Vanderbilt, D. & Cohen, M.H. (2001). Monoclinic and triclinic phases in higher-order Devonshire theory. *Physical Review B* 63, 094108-094117. ISSN 1550-235x

Wang, H.C. & Schulze, W.A. (1990). The role of excess magnesium oxide or lead oxide in determining the microstructure and properties of lead magnesium niobate. *Journal of the American Ceramic Society* 73, 825-832. ISSN 1551-2916

Woodward, D.I. et al. (2005). Review of crystal and domain structures in the $PbZr_xTi_{1-x}O_3$ solid solution. *Physical Review B* 72, 104110-04118. ISSN 1550-235x

Ye, Z.G. & Dong, M. (2000). Morphotropic domain structures and phase transitions in relaxor-based piezo-/ferroelectric (1-x)$Pb(Mg_{1/3}Nb_{2/3})O_3$-x$PbTiO_3$ single crystals. *Journal of Applied Physics* 87, 2312-2320. ISSN 1089-7550

Ye, Z.G. et al. (2001). Monoclinic phase in the relaxor-based piezoelectric/ferroelectric $Pb(Mg_{1/3}Nb_{2/3})O_3$-$PbTiO_3$ system. *Physical Review B* 64, 184114-184119. ISSN 1550-235x

Yokota, H. et al. (2009). Crystal structure of the rhombohedral phase of $PbZr_{1-x}Ti_xO_3$ ceramics at room temperature. *Physical Review B* 80, 104109-104121. ISSN 1550-235x

Yoon, K.H. et al. (1995). Characteristics of lead magnesium niobate thin-film prepared by sol-gel processing using a complexing agent. *Journal of the American Ceramic Society* 78, 2267-2270. ISSN 1551-2916

Zhao, X.Y. et al. (2003). Effect of a bias field on the dielectric properties of 0.69$Pb(Mg_{1/3}Nb_{2/3})O_3$-0.31$PbTiO_3$ single crystals with different orientations. *Journal of Physics: Condensed Matter* 15, 6899-6908. ISSN 1361-648X

4

Advances in Engineering and Applications of Hexagonal Ferrites in Russia

Marina Y. Koledintseva[1], Alexey E. Khanamirov[2], and Alexander A. Kitaitsev[2]

[1]Missouri University of Science and Technology, Missouri
[2]Moscow Power Engineering Institute (Technological University), Moscow
[1]U.S.A.
[2]Russia

1. Introduction

Richard Feynman (Feynman, 2005) once stated that ferrites were one of the most difficult areas for theoretical description, but the most interesting for studies and practical applications. These words are especially true when dealing with a special type of ferrites, which have a hexagonal crystallographic structure - hexagonal ferrites, or hexaferrites.

The world's first permanent magnets based on ferroxdure - hexagonal barium ferrite $BaFe_{12}O_{19}$ (equivalent to $BaO \cdot 6(Fe_2O_3)$, also called BaM) appeared in 1951 (Rathenau et al., 1952). The main engineering problem that was solved at that time was the replacement of cumbersome metallic (Ni- and Co-alloy) magnets by comparatively compact and light-weight permanent magnetic systems. The systematic study and applications of gyromagnetic properties of hexaferrites started in 1955 (Weiss & Anderson, 1995; Weiss, 1955; Sixtus et al., 1956). Currently, in the world, enormous progress in fundamental theoretical and experimental laboratory studies of various properties of hexaferrites, their synthesis, and engineering of a wide range of microwave and mm-wave coatings and devices on their basis has been achieved - see, for example, papers (Harris et al., 2006, 2009) and references therein.

Hexaferrites as the materials for extermely high-frequency (EHF) range, *Ka* (27-40 GHz), *U* (40-60 GHz), *V* (60-80 GHz), *W* (80-100 GHz) bands, and higher, have been also studied and applied in Russia since middle 1950s. Authors of this paper, being apprentices and followers of the outstanding Russian scientists, V.A. Kotelnikov, L.K. Mikhailovsky, and K.M. Polivanov. V.A. Kotelnikov named the millimeter waveband "a nut in a hard shell", deeply believe that the practical development of this waveband is possible only when using hexaferrites. Herein, the summary of achievements in engineering hexagonal ferrites and various devices of on their basis in Russia for the past over 50 years is presented.

In 1955-1956, a then young scientist from Radio Engineering Department of Moscow Power Engineering Institute, L.K. Mikhailovsky, studied microwave ferrites and developed devices operating at the ferrimagnetic resonance (FMR) with new functional possibilities, such as a magnetic detector and a gyromagnetic cross-multiplier. For mm-wave applications, Mikhailovsky proposed to use instead of huge bias magnets just the internal field of crystallographic magnetic anisotropy inside a ferrite. According to Kittel's formula (Kittel, 1948, 1949), the frequency of the FMR in a ferrite resonator magnetized to the saturation along the $\hat{Z}$ direction is

$$f_{res} = \gamma / \sqrt{H_x \cdot H_y} \text{ with}$$

$$H_x = H_0 + H_A - (N_z - N_x) \cdot 4\pi M_S \text{ and } H_y = H_0 + H_A - (N_z - N_y) \cdot 4\pi M_S, \quad (1)$$

where $\gamma = 2.8 \frac{\text{MHz}}{\text{Oe}}$ is the gyromagnetic ratio; H_A is the crystallographic anisotropy field; H_0 is the external bias magnetic field; $4\pi M_S$ is the saturation magnetization (G); N_x, N_y, N_z are the demagnetization shape (form) factors of the ferrite sample. The sample may be of an ellipsoidal shape (spheroid, sphere, elongated cylinder, disk), or it may be a flat slab. In (1), the vectors $H_0, H_A, \hat{Z}$ are collinear.

However, in 1950s, no ferrites with significant internal magnetic fields were available in the USSR. In 1958-1962, Mikhailovsky initiated the pioneering work on the study of electromagnetic energy absorption by magneto-uniaxial ferrites. The very first magneto-uniaxial ferrite Ba-Zn (Zn_2W) was synthesized by S.A. Medvedev, who had previously worked in France and got some experience in making ferrites with high crystallographic anisotropy (but not magneto-uniaxial). The results of this first research were published in 1960 (Polivanov et al, 1960), and it was concluded that the absorbed energy at the natural ferrimagnetic resonance (NFMR), i.e., without any bias magnetic field, is significantly higher than the resistive or pure dielectric polarization loss. Thus the NFMR differs from the FMR only by the significantly lower magnetic field needed for the resonance operation of microwave (mm-wave) devices.

In 1962, the Industrial Ferrite Laboratory („OPLF") with Moscow Power Engineering Institute (MPEI) was founded. The OPLF from the very beginning united three working groups from three different Departments of the MPEI: Radio Engineering (with Mikhailovsky as the Head), Electrical Engineering (lead by Polivanov), and Electro-Mechanical Technology (lead by Medvedev). Also, in late 1950-ies, the State Research and Development Institute of Magneto-Dielectrics („NIIMD") was founded in Leningrad in order to engineer and manufacture new types of ferrites. The main goal of the OPLF as a research laboratory within a university was to intensively collaborate with and conduct R&D projects determined by this leading enterprise of the USSR electronics industry. Thus since early 1960s, the OPLF has been one of the world leaders on synthesis, theoretical and experimental research of ferrites, including hexagonal ferrites, and development of new unique microwave and mm-wave designs on their basis.

As a result of the research activity of the OPLF during nearly 50 years of existence, the new school of thought in gyromagnetism was founded. This school continues the best traditions of such Russian physicists as Lebedev, Arkadiev, Polivanov, and Kotelnikov. In 1980s, Mikhailovsky founded, and till now has been leading, the scientific field of „currentless" spin electronics and gyrovector electrodynamics. „Gŷros" in Greek means „revolution". Gyromagnetism in classical phenomenological representation arises from the relation between the angular momentum and the magnetization vector of a magnetic medium. The motion of the magnetization vector in magnetic (ferrite, ferri-, ferro-and antiferromagnetic) media at the magnetic resonance is associated with spin moment rotation of magnetic atoms, and is represented as the precession around the static bias magnetic field direction (Landau & Lifshitz, 1935, 1960). Mikhailovsky has developed a novel theory, which he called „the gyrovector formalism“, or „the gyrovector algebra“. This theory explains the mechanism of absorption of electromagnetic energy at ferrimagnetic resonance by microwave ferrites, including hexaferrites (Mikhailovsky, 2002). The background of this theory is Maxwell's hydrodynamic model (Maxwell, 1856) and Dirac's quantum spinor electrodynamics (Dirac, 1975). Thus the gyrovector formalism mathematically unites classical and quantum physics approaches, and explains a local quantum (energy) interaction of electromagnetic field with centers of absorption and radiation of a gyromagnetic medium.

This theory lays the basis for many engineering applications, including the hexaferrite radioabsorbing materials with electrical conductivity close to zero; omnidirectionally matched with free space protecting coatings; and devices for spectral analysis and frequency-selective measurements of microwave and mm-wave power. Also, as soon as a new class of ferrite materials, magneto-uniaxial hexagonal ferrites with high internal fields of crystallographic anisotropy, were synthesized, it has become possible to develop gyromagnetic resonance devices operating without external bias magnetization, or with low bias magnetization needed for ferrite saturation and tuning of resonance frequency.

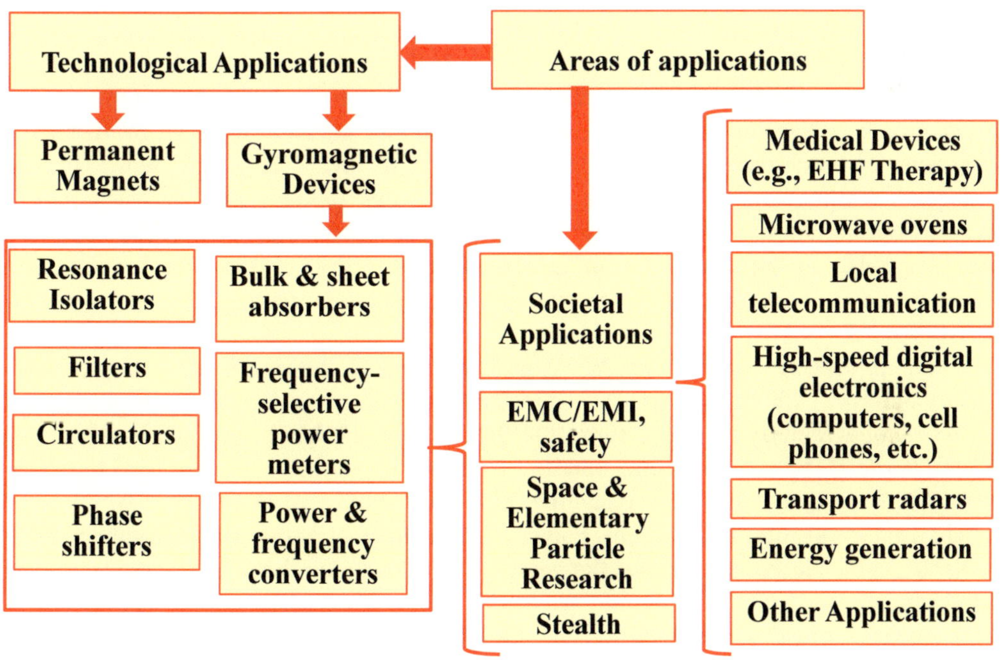

Fig. 1. Fields of application of hexagonal ferrites

Unfortunately, these achievements could not be published in open literature with wide international access for many decades. Very limited number of papers on this topic were published, mainly in Russian. The objective of the present Chapter is to cover this gap, and allow readers to get acquainted with these works not only from retrospective point of view. They contain the present-day novelty, and can be useful for engineers designing electronic equipment operating in a wide frequency range from about 2 GHz to 300 GHz, and potentially even higher. An application of hexagonal ferrites is proven and remains very perspective for solving numerous problems related to microwave engineering, radar engineering, electromagnetic compatibility (EMC), electromagnetic immunity (EMI), and signal integrity (SI). Hexaferrites can be used for detection and suppression of unwanted radiation and coupling paths; for frequency-selective measurements of signal parameters; and for providing proper non-reciprocal isolation in channels of generation, transmission, and reception over the selected frequency bands within the wide range up to a few hundred GHz.

In this Chapter, the review of the engineered modern types of hexagonal ferrites for SHF and EHF frequency bands will given, as well as an overview of research and design experience for various hexagonal ferrite devices gained during multi-year collaboration between MPEI (TU) and Russian industry, in which the co-authors have been directly involved. Different engineering, societal, and other applications of hexagonal ferrites will be also discussed, include agricultural and medical applications, computer engineering, telecommunication, and television. Fig. 1 shows some application fields of hexaferrites.

2. Hexagonal ferrites as advanced ceramic materials for microwave and millimeter wave engineering

Hexaferrites are known to be magneto-dielectric, specifically ferrimagnetic materials with hexagonal magnetoplumbite-type crystallographic structure (Smit & Wijn, 1959). Ferrimagnetic magnetoplumbite has the general chemical formula $MeO{\cdot}6Fe_2O_3$, in which Me may be Ba^{2+}, Sr^{2+}, or Pb^{2+}. The ferric ions can be also partially replaced by Al^{3+}, Ga^{3+}, Cr^{3+}, Sc^{3+}, or combinations of ions, for example, Co^{2+} with Ti^{4+}, Zn^{2+} with Ti^{4+}, etc. Hexagonal ferrites, unlike the other groups of ferrites (spinels and garnets), have a pronounced internal effective magnetic field H_A, associated with the magnetic crystallographic anisotropy. From a crystallographic point of view, a hexaferrite is characterized by the hexagonal basis plane and the axis of symmetry that is orthogonal to the basis plane.

The scanning electron microscopy (SEM) picture in Fig. 2 shows the microstructure of a hexagonal ferrite containing hexagonal shaped flakes. If the direction of easy magnetization is the axis of symmetry of the hexagonal structure, then the ferrite is called a *magnetically uniaxial ferrite*. If the easy magnetization direction belongs to the basis plane, this is a *planar ferrite*. Monocrystalline and polycrystalline magnetically uniaxial hexaferrites are the most widely used in practical applications. Polycrystalline uniaxial hexaferrites are commercially available. As for planar hexaferrites, the possibilities of studying them are limited by the low Curie temperatures.

The concept of a field of magnetic crystallographic anisotropy, or briefly called "anisotropy field", is widely used for phenomenological description of hexaferrite behavior. It is calculated approximately as (Gurevich & Melkov, 1996)

$$H_A \approx 2\,|K_1|\,/M_s, \tag{2}$$

where M_s is the saturation magnetization, and K_1 is the first constant of anisotropy, such that $K_1 > 0$ for uniaxial ferrites, and $K_1 < 0$ for planar ferrites. The dependence of crystallographic magnetic anisotropy energy of hexagonal ferrites (Gurevich & Melkov, 1996)

$$U_A = K_1 sin^2\theta + K_2 sin^4\theta + \frac{1}{2}\mu_0(N - N_C)(M_S sin\theta)^2. \tag{3}$$

upon the angle θ between the equilibrium magnetization vector $\vec{M}_0$ and crystallographic axis *c* for uniaxial and planar ferrites is shown in Fig. 3. The second constant of anisotropy for hexagonal ferrites is much smaller than the first constant of anisotropy $|K_2| << |K_1|$.

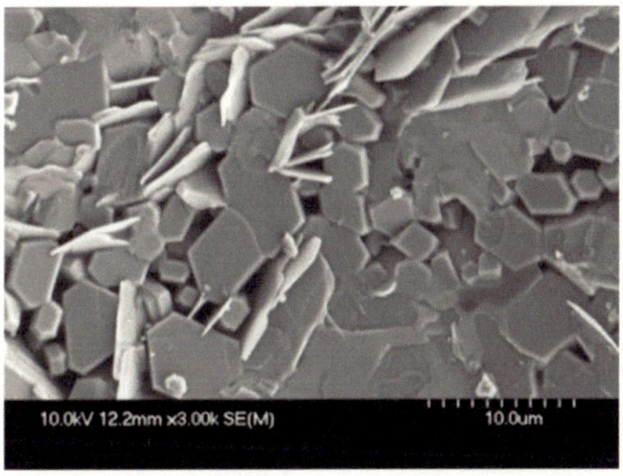

Fig. 2. Microstructure of a Ba-SrM polycrystalline ferrite obtained by SEM

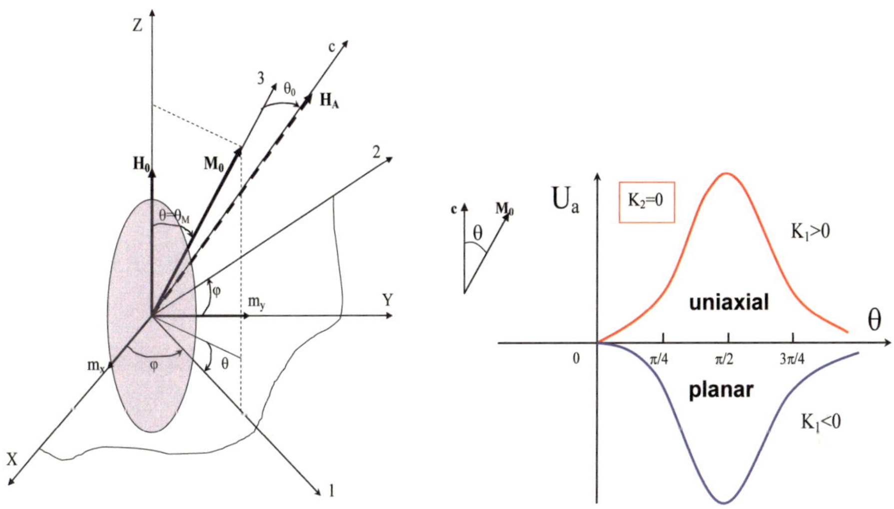

Fig. 3. Dependence of anisotropy energy of hexagonal ferrites U_a upon the angle θ between the equilibrium magnetization vector $\vec{M}_0$ and crystallographic axis c.

The crystallographic magnetic anisotropy field determines the conditions for *ferrimagnetic (gyromagnetic) resonance* (FMR) in hexagonal ferrites. The resonance frequency of a magneto-uniaxial ferrite is related to the magnetization field and orientation of a equilibrium magnetic moment with respect to the constant bias magnetic field. Boris P. Pollak, a scientist from MPEI, in 1964 theoretically and experimentally obtained the curves for H_A field of magneto-uniaxial monocrystalline hexaferrites in negative bias fields, extending Weiss's curves (Weiss, 1955). The dependences shown in Fig. 4 are known as „Weiss-Pollak curves" (Polivanov & Pollak, 1964; Mikhailovsky et al., 1965). The analogous curves were also obtained for the polycrystalline hexaferrites (Mikhailovsky et al., 1966; Pollak et al., 1969). The magnetic field H_0 required to achieve the FMR in the case of a hexagonal ferrite magnetized in the easy direction, appears to be dozens times lower than when using ordinary, low-anisotropy ferrites. Thus, for the uniaxial ferrites, the applied bias field to achieve the resonance frequency ω_{res} is (Kittel, 1948)

$$H_0 = \omega_{res}/(\mu_0 \cdot \gamma) - H_A. \quad (4)$$

For high-coercivity magneto-uniaxial ferrites, the applied field H_0 may be zero or even negative (anti-parallel to the magnetization vector), and this broadens the frequency range of applications of ferrites. The anisotropy field is the main parameter for classifying hexagonal ferrites for applied engineering problems.

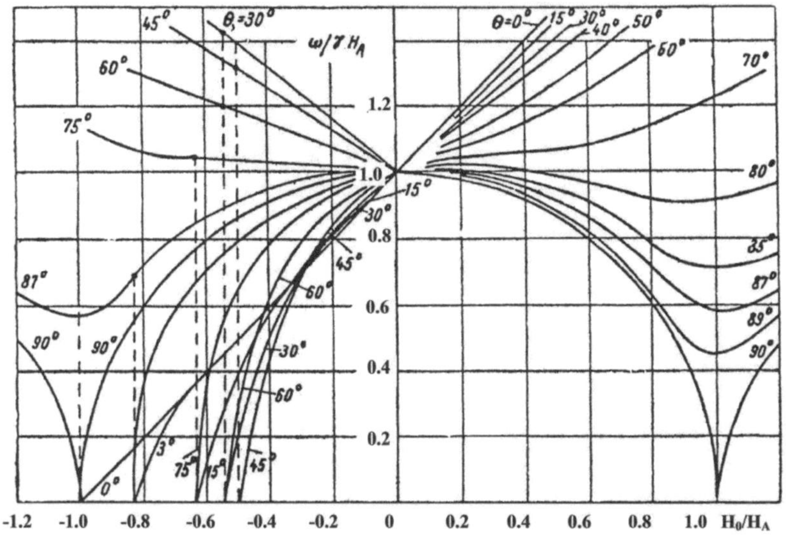

Fig. 4. Weiss-Pollak curves as conditions of gyromagnetic resonance in a single-domain particle of a magneto-uniaxial hexagonal ferrite (Mikhailovsky et al., 1965)

The research on microwave and mm-wave hexagonal ferrites started in the OPLF of the MPEI went through the three stages.

- The first stage included the attempts to synthesize, in the laboratory conditions, different types of ferrites with various fields of crystallographic anisotropy, test their charactristics, and build devices of EHF (30...300 GHz) frequency band on their basis.
- The second stage was focused on the improvement and optimization of the synthesized materials from the point of view of practical applications, as well as engineering of the advanced designs of microwave and mm-waved devices.
- The third stage was developing and producing industrial series of the engineered ferrites of different types and devices on their basis using the facilities of the electronics industry, including those at the leading enterprise NIIMD and the experimental plant of the MPEI.

The work on the synthesis of magneto-uniaxial ferrites was mainly done in two directions:
(1) synthesis of ferrites with different anisotropy fields to be able to design devices for different frequency bands, and (2) an optimization of technological processes, structure and stohiometry of ferrites to obtain ferrites with the best possible characteristics.

Practically all the uniaxial hexagonal ferrites synthesized and studied in the OPLF belong to one of two groups: either M-type, or W-type. M-type hexagonal ferrites are based on Barium ($BaFe_{12}O_{19}$) and/or Strontium ($SrFe_{12}O_{19}$) ferrites with partial isomorphic substitution of the Fe^{3+} ions by ions of dia- or paramagnetic metals.

Ferrite series	Type of ferrite	Concentration of dopant ions	$4\pi M_s$, G	H_A, kOe	ΔH, Oe
1 MΦ	*BaMnM*	0.4...0.0	368...380	16.7...17	55...13
2 MΦ	*BaNiZnM*	1.54...0.55	344...364	10.2...13.5	72...18
3 MΦ	*BaTiNiM*	0.5	360	13.0	61
4 MΦ	*BaTiCoM*	1.4...0.25	330...280	10.2...14.4	220...65
5 MΦ	*BaYbM*	0.8...0.5	330	13.5...14.5	32
6 MΦ	*BaLuM*	0.6	330	12.8	21
7 MΦ	*BaScM*	1.4...0.5	225...350	1.2...10.6	200...27
8 MΦ	*SrGaScM*	0.82...0.0	317...380	9.2...18.7	36...12

Table 1. Magnetic parameters for a number of series of synthesized monocrystalline M-type hexaferrites (Mikhailovsky et al., 2002)

Momocrystals of hexagonal ferrites were maily synthesized in OPLF by S.A. Medvedev, A.M. Balbashov, V.P. Cheparin, and A.P. Cherkasov (S.A. Medvedev et al., 1967, 1969; Mikhailovsky et al., 1965, 2002; Pollak et al., 1976). Most of the monocrystals are obtained by the method of spontaneous crystallization of high-temperature melt solution, and in a few cases by the method of non-crucible zone smelting. Results of magnetostaic and microwave measurements conducted on a number of series of synthesized monocrystalline hexaferrites are summarized in Table 1. The data is presented in the Gaussian Magnetic Unit System with 1 Oersted (Oe)=$1000/4\pi \approx 79.6$ A/m, and 1 Gauss (G) =10^{-4} T). The synthesized monocrystalline magneto-uniaxial hexagonal ferrites had the values of crystallograhic anisotropy field H_A = 0.075...7.9 MA/m (corresponding to 0.9...95 kOe). This allows for operating in the frequency range ~2.5...260 GHz both at the NFMR and the FMR. To achieve the latter, significantly reduced bias magnetization fields were applied (less than 3 kOe).
W-type hexagonal ferrites are mainly solid solutions of Me_2W ($Me_2BaFe_{16}O_{27}$), where Me is a bivalent metal, for example, Co_2W, Ni_2W, or Zn_2W. Mainly polycrystalline hexagonal ferrites with different values of anisotropy field have been synthesized with this structure; however, the ferrite Zn_2W was also synthesized as a monocrystal.
The monocrystals with the H_A fields ranging from 1.2 kOe (BaM ferrites doped by Sc, Lu, or Yb) to 120 kOe (BaM and SrM ferrites with Fe ions replaced by ions of Ga and Al) have been synthesized. The minimal FMR linewidth of about 10 Oe was achieved in experimental BaSr ferrites, when Mn ions were doped in the crystal lattice of the hexagonal ferrite, as this is typically done to reduce the linewidth in monocrystal ferrogarnets, e.g., YIG. In the pure BaM ferrite, Mn ions were introduced using the BaO-B_2O_3 solvent, while in Sc-doped ferrites, the solvent $NaFeO_2$ was chosen. As for Ti-containing ferrites, the comparatively narrow lines (~ 10 Oe) were achieved in only Ti-Zn ferrites, when the cooling speed of the crystallizing melt was below 2 ^{0}C/hour (Sveshnikov & Cheparin, 1969). The ion Fe^{2+} is known to be responsible for wider FMR line, so to reduce its contents, the monocrystals were grown by the method of non-crucible zone smelting at the oxygen pressure of 50 atmospheres.

Series	Type of ferrite	Dopant (x)	$4\pi M_s$, G	H_A, kOe	ΔH, kOe
1 ПФ	$BaO\cdot(6-x)Fe_2O_3\cdot xCr_2O_3$	2.5...0.0	900...4700	43.4...16.3	4.9...1.8
2 ПФ	$SrO\cdot(6-x)Fe_2O_3\cdot xCr_2O_3$	3.0...0.0	250...3400	52.2...16.2	5.0...0.6
3 ПФ	$BaO\cdot(6-x)Fe_2O_3\cdot 0.5x(CoO+TiO_2)$	0.65...0.45	3800...4400	9.1...12.1	5.0...1.9
4 ПФ	$BaO\cdot (6-x)Fe_2O_3\cdot 0.5x(ZnO+TiO_2)$	1.0...0.55	3600...3800	10.0...7.7	3.6...1.3
5 ПФ	$BaO\cdot (5.9-x)Fe_2O_3\cdot 0.5x(ZnO+TiO_2)$	1.9...0.45	3000...4400	7.2...13.4	4.8...1.4
6 ПФ	$BaO\cdot (5.9-x)Fe_2O_3\cdot 0.5x(NiO+TiO_2)$	1.0...0.45	3800...3900	11.5...14.1	5.4...4.0
7 ПФ	$BaO\cdot 6Fe_2O_3\cdot 2[xCoO\cdot(1-x)NiO \cdot 0.9Fe_2O_3]$	0.4...0.36	3540...2900	4.0...4.8	2.95...4.7
8 ПФ	$1.1BaO\cdot 6Fe_2O_3\cdot 2[xCoO\ (1-x)NiO\cdot 0.9Fe_2O_3]$	0.4...0.26	3520...4300	3.4...6.7	2.5...4.8
9 ПФ	$BaO\cdot 5.4Fe_2O_3\cdot 2[0.4CoO\cdot 0.6NiO \cdot 1.2Fe_2O_3]$		3900	6.0	3.4
10 ПФ	$SrO\cdot 6Fe_2O_3\cdot 2[0.4CoO\cdot 0.6NiO \cdot 0.9Fe_2O_3]$		4270	7.7	2.5
11 ПФ	$BaO\cdot 6Fe_2O_3\cdot 2[xCoO\cdot(1-x)ZnO \cdot 1.2Fe_2O_3]$	0.4…0.0	3000-5020	0.9...11.0	5.3...2.4
12 ПФ	$BaO\cdot(6-x)Fe_2O_3\cdot xCr_2O_3\cdot 2(ZnO\cdot 0.9Fe_2O_3)$	1.5…0.0	3900	16.1...10.0	3.3...1.95
13 ПФ	$BaO\cdot(6-x)\ Fe_2O_3\cdot xCr_2O_3\cdot 2(NiO\cdot 0.9Fe_2O_3)$	1.2...0.4	3900	18.5...14.6	3.5...2.2
14 ПФ	$SrO\cdot(6-x)Fe_2O_3\cdot xCr_2O_3\cdot 2(0.4CoO\cdot 0.6NiO\cdot 0.9Fe_2O_3)$	0.5...0.0	3900	6.8...7.7	3.0...2.5
15 ПФ	$BaO\cdot(6-x)Fe_2O_3\cdot xAl_2O_3\cdot 2(NiO\cdot 0.9Fe_2O_3)$	1.1...0.0	3900	18.0...13.3	4.6...2.2

Table 2. Parameters of some laboratory synthesized polyrcystalline hexaferrites (Mikhailovsky et al., 2002)

Polyrcrystalline hexaferrites were synthesized in both the MPEI, and in industry. The final goal was obtaining industrial series of magneto-uniaxial ferrites and devices on their basis. The experimental series of polycrystalline hexaferrites were engineered by S.A. Medvedev, A.M. Balbashov, and V.V. Kolchin (Polivanov et al., 1969).

It is known that partial substitution of Fe_2O_3 by Al_2O_3 in SrM or BaM ferrites due to the presence of Al^{3+} ions of varying concentration allows for comparatively sharp control of crystallographic anisotropy field of hexaferrites (De Bitetto, 1964; Qui et. Al., 2005). This effect is widely used in the world practice to synthesize hexaferrites with different K_1 (or H_A) values. The peculiarity of polycrystalline hexaferrites synthesized in Russia is using Cr_2O_3 , since it was found that Cr^{3+} allows for fine tuning of K_1 (or H_A) field to the desirable values. Besides, it has been noticed that the ferrites with Cr^{3+} have better microwave properties than those with Al^{3+} (Nedkov et al., 1988). However, it is more difficult to synthesize ferrites-chromites, since Chrome oxides are gaseous and require ferrite annealing at high pressure in different media. Besides, ferrites-chromites have the higher magnetic saturation and Curie temperature than their aluminate counterparts at the same concentration. The parameters of the polycrystalline hexagonal ferrites of different series synthesized and studied in MPEI are presented in Table 2. The highest achieved anisotropy field in the case of the Sr ferrite-chromite with substitution x=4.5 was H_A = 95 kOe was.

The optimization of the synthesis process was done to acheive the ferrites with the given and controllable anisotropy fields, with the highest-level texture (grain alignment), and the minimal possible NFMR line, determined by the statistical distribution of the anisotropy fields of the grains). As a result of optimization of grinding and burning, it was possible to get polyrcrystalline magneto-uniaxial ferrites with ΔH=0.6...1.0 kOe.

The polycrystalline hexaferrite „parametric series" (series of ferrites with the fixed values of the anisotropy field, differing by 1.0...1.5 Oe) with the increased thermal stability of H_A have been synthesized in industry (Petrova, 1980). These hexaferrites have been intended for the development of EHF devices, in particular, resonance isolators (Pollak et al., 1980). The parameters of such hexaferrites are shown in Table 3. These ferrites exhibit an enhanced thermal stability and low dielectric loss. It is important that all the ferrites of an individual parametric set belong to the same system, i.e., the classification group. An important requirement is using the same ferrite system for as wide anisotropy range as possible. Thus, the system $BaNi_2ScW$ was chosen for the range H_A = 5...12 kOe; the system $BaNi_2Cr_xW$ was used to provide the range H_A = 12...18 kOe; the system $SrNi_2Cr_xW$ allowed for getting H_A = 13...20 kOe. Ferrites-aluminates and ferrites-chromites with H_A = 18...30 kOe have been synthesized on the basis of both BaM and SrM. Aluminates with high density and high Curie temparature are preferable for H_A > 30 kOe. As is seen from Table 3, the present-day polycrystalline ferrites possess substantially better parameters, especially ferrite *04СЧА12*. For this ferrite, the anisotropy field is H_A =24 kOe, and the value of the resonance width has been achieved as small as ΔH< 0.5 kOe, the rectangularity of the hysteresis loop is M_r/M_s = 0.995, coercivity is H_c = 2 kOe, and the dielectric loss is as low as $tan\delta_\varepsilon$ = 6.0×10^{-4}.

Engineering and application of hexagonal ferrite films for the EHF (30-300 GHz) resonance and wideband devices operating without any bias magnetic field is an important advance in improvement and simplification of the manufacturing processes. These films are based on hexaferrite composites, which are the mixtures of hexaferrite powders of the particular contents with a glue-like base (host) material (Pollak, 1980). The powders are obtained by the grinding bulk hexaferrites that have already completely gone through the ferritization process (the metasomatic alteration of initial raw material ingredients into ferrite), and have a well-defined texture. The latter means that the hexaferrites have undergone the ferritization annealing twice, and before the second firing they have been pressed in a magnetic field. The average size of a particle in a powder is close to that of a single domain (~ 1-10 μm). The powder is then mixed with a bonding dielectric, which may be a polystyrene glue, glue *BF* (Russian-make), etc. Then the suspension is deposited on a substrate, and dried at room temperature and normal atmospheric pressure. To assure a high-rate texturing, samples must be dried in a magnetic field. Films have the minimum thickness on the order of 10 μm. They have a relative density of 50%, and their texturing is as good as of the bulk sintered polycrystalline hexaferrite plates.

Type of ferrite	$4\pi M_s$, G	H_A, kOe (f_o, GHz)	ΔH, kOe (f_o, GHz)	ε_r @ f_0=9.4 GHz
06СЧА3	3700	14 (55)	< 2 (55)	16
05СЧА4	4000	16(50)	< 2 (50)	16
05СЧА5	3000	18(65)	< 2 (65)	15
04СЧА11	2500	21(70)	< 2 (70)	15
04СЧА12	2100	24(75)	< 2(75)	15
04СЧА13	1600	27(80)	< 2.5(80)	15
03СЧА2	1500	31(100)	< 2.5(100)	15
03СЧА	1400	35(110)	< 2.5(110)	15

Table 3. Parameters of some industrially manufactured hexagonal ferrites

A mixture of a few types of hexaferrite powders differing by their anisotropy fields can be used to make multiphase composites. They typically have a greater width of the FMR, which is favorable for developing resonance isolators or other devices operating over a wider frequency range. Films based on hexaferrite composites exhibit higher coercivity, which allows for operating without any external bias magnets in the frequency range up to 100 GHz. Another important feature is their comparatively low permittivity, which provides better matching of films with the other dielectric elements in a microwave (mm-wave) transmission line. Besides, it is much easier and cheaper to manufacture such films than the bulk plates. The requirement of having an extremely small thickness is not difficult to satisfy, since the chip technology can be used for their manufacturing, and these films can be used in microwave chips, though there may be problems at the interfaces with other materials. Moreover, when dealing with polycrystalline hexaferrite powders, the control of the ferrite contents at different stages of their manufacturing, is substantially simplified. It is possible to do without making special test samples - plates of thickness less than 0.1 mm, or spheres of at least of 0.4 mm in diameter to apply the standard techniques for measuring intrinsic parameters of ferrites. Also, there is no necessity of texturizing samples for study, and no need in bias field for measurements.

3. Gyromagnetic applications of hexagonal ferrites

Hexagonal ferrites are traditionally applied in microwave and mm-wave engineering. These are different gyromagnetic devices for the EHF range (30...300 GHz). When using hexagonal ferrites, it is possible to reduce the external bias magnetic field by an order of magnitude, or remove it completely. Application of hexaferrites also solves a number of functional problems, which cannot be successfully solved using other types of ferrites. The primary attention in this work is paid to hexaferrite isolators, because isolators are of the greatest demand in general, and hexaferrite isolators, from our point of view, are the most promising as compared to other types of non-reciprocal isolating devices for telecommunication microwave and millimeter-wave systems. An important perspective on hexaferrite isolators is their application for transmission lines and broadcast telecommunication systems, when compact, low-weight, technologically simple, and inexpensive devices are of top priority. Some other examples of applications of hexaferrites in devices developed by the authors are presented below.

3.1 Resonance isolators

Resonance isolators for the 8-mm (*Ka*) waveband are known to be industrially developed and manufactured. They are based on the application of crystallographically isotropic (ordinary) ferrites, as well as on anisotropic (hexagonal) ferrites, inside standard metal rectangular waveguides (cross-section of 7.2 x 3.4 mm). Laboratory samples of *hexagonal ferrite resonance isolators* for operation in the frequency range up to 150 GHz have been designed and studied in the MPEI (Mikhailovsky et al., 1965, 2002; Polivanov et al., 1969; Pollak et al., 1976, 1980). The characteristics of resonance isolators are given in Table 4. They are made on both metal waveguides (MW) and dielectric waveguides (DW). Characteristics of devices are preserved at average power less than 200 mW and over the temperature range -40…+60 ^{0}C.

Frequency range, GHz	Bandwidth at - 3 dB level, %	Return loss, dB	Insertion loss, dB, less than	VSWR	Bias magnetic field, kOe	Transmission line
25 - 150	2…3	20…25	1…2	1.2	0…3	MW, DW
25 - 150	20…40	20…25	1…3	1.3	0…3	MW, DW
40…150	0.5	25…30	0.3…1	1.2	0…3	MW

Table 4. Typical parameters of resonance isolators based on magneto-uniaxial polyrcystalline hexaferrites (Mikhailovsky et al., 2002)

When developing these isolators for the EFH frequency range, many problems have arisen. One of the challenges is the tiny size of isolators. The cross-sectional dimensions b x a of standard metal rectangular waveguides are typically in the range of b = (0.55...3.4) mm and a = (1.1...7.2) mm, the thickness of ferrite slabs is t_f= (0.01... 0.1) mm, the thickness of dielectric plates is t_d = (0.15...1.0 mm), and the mounting dimensions in the waveguide are just d = (0.1...2.0 mm). These miniature isolators typically have high insertion loss which may reach up to 5 dB. Another problem is that to cover the wide frequency range of operation, many types of ferrites with different anisotropy field are needed, for example, the anisotropy field should be in a range of H_A = 0.4...8.0 MA/m (5...100 kOe) with a discrete of δH_A ~ 0.1 MA/m (1.0...1.2 kOe). At the same time, the values of M_S, T_C, and density decrease as H_A increases. The abovementioned difficulties of isolator design and manufacturing have been overcome in an elegant way. The insertion loss in comparatively narrowband isolators was reduced by the following solutions.

- The so-called effect of the "*bound modes*" (Korneyev, 1980) and the *magnetodynamic resonators* (MDR) with different values of anisotropy field (Korneyev & Pollak, 1982) were used. The effect of the "bound modes" is typically observed in bias magnetic fields that are insufficient for achieving resonances, and in comparatively big ferrite samples falls within the bandwidth much less than the FMR line. These isolators operate on a combination of FMR and volume resonance of the hexaferrite slab. These are typically short flange-like isolators with high isolation level in a narrow band, and the resonance frequency can be tunable (± 5...10%) by a low magnetic bias field while keeping good isolation.
- Another solution is an application of *a monocrystalline hexaferrite* resonator, e.g., a spherical resonator, with variable orientation of crystallographic axes with respect to

the external magnetic field (Pollak et al., 1976). Thus, in an isolator with a pure BaM ferrite (H_A=17.5 kOe), when changing its orientation in the limits of 0...60^0, the range of resonance frequency was 62...55 GHz, isolation=40...22 dB, and resonance linewidth was less than 150 MHz.

To broaden the frequency range of isolators (Pollak et al., 1976; Mikhailovsky et al., 2002) the multi-component composite hexaferrite materials with the size of hexaferrite particles not exceeding a few micrometers was proposed. This is a mixture of different polycrystalline ferrites as thin bulk plates or composite films with a spread of different values of anisotropy field H_A . The resultant wideband isolators have a bandwidth of up to an octave (see Fig. 5).

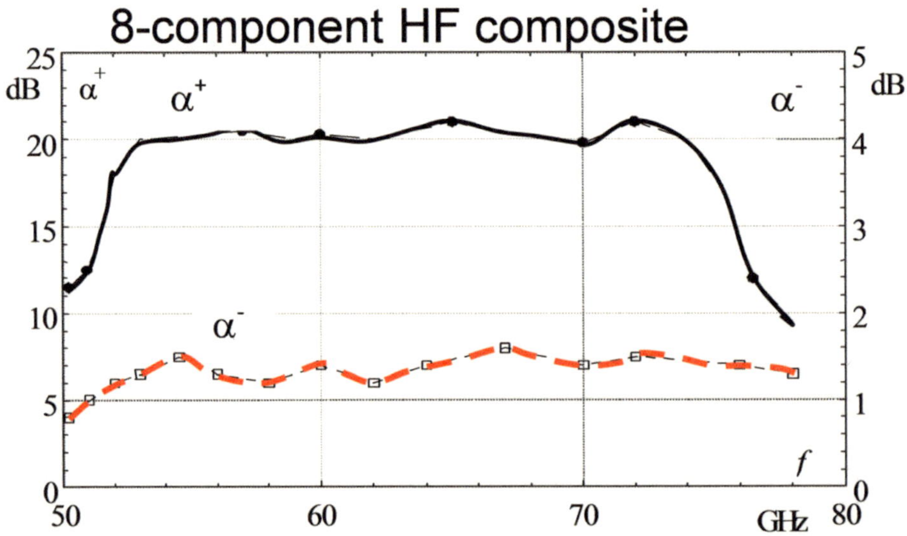

Fig. 5. Frequency characteristics of the eight-component hexagonal ferrite isolator

The return loss (RL) α^+ and the insertion loss (IL) α^- shown in Fig. 5 are calculated as

$$\alpha^{\pm} \approx \sigma V(\mu'' \pm \mu_a''), \tag{5}$$

where σ is the coefficient which depends on the parameters of the transmission line and frequency, V is the resonator volume, and μ'', μ_a'' are the imaginary parts of the tensor components

$$\bar{\bar{\mu}} = \begin{bmatrix} \mu & j\mu_a & 0 \\ -j\mu_a & \mu & 0 \\ 0 & 0 & \mu_{||} \end{bmatrix}, \tag{6}$$

and the hexaferrite slab is in the waveguide points with circular polarization of microwave magnetic field.

Another way to design wideband isolators is an application of a chain of monocrystalline magneto-uniaxial ferrite resonators with various values of H_A and/or different orientations of the internal magnetic field. For example, in an isolator for the fixed band 35.7...36.7 GHz with RL=15...40 dB, six spherical resonators made of Sr-Sc hexaferrite were used (H_A=9 kOe). Their orientation was 0...35^0 with respect to the bias field. Each sphere provided a

resonance absorption of 3...7 dB and an off-resonance loss less than 0.1 dB. Such isolators were implemented in the masers of traveling waves (Mikhailovsly et al., 2002).
Typically, hexaferrite resonators are placed upon a dielectric substrate, e.g. a ceramic slab with relative permittivity $\varepsilon \approx 9$, and this layered structure is fixed in the middle of the wide wall of the waveguide along the propagation direction. To make the isolators shorter, high-coercivity hexagonal ferrite plates or composite films of opposing (subtracting) magnetization are placed on two sides of a dielectric substrate.
Hexaferrite resonance isolators without bias magnets have been designed on various transmission lines (dielectric rod waveguides, dielectric reflecting waveguides, grooved waveguides, planar and cylindrical slot lines, spiral waveguides, and other specific types of the EHF transmission lines - see Fig. 6). The assurance of proper isolation is a crucial issue for practical realization of the required systems. Hexaferrite resonance isolators have been developed for a number of the abovementioned transmission lines.

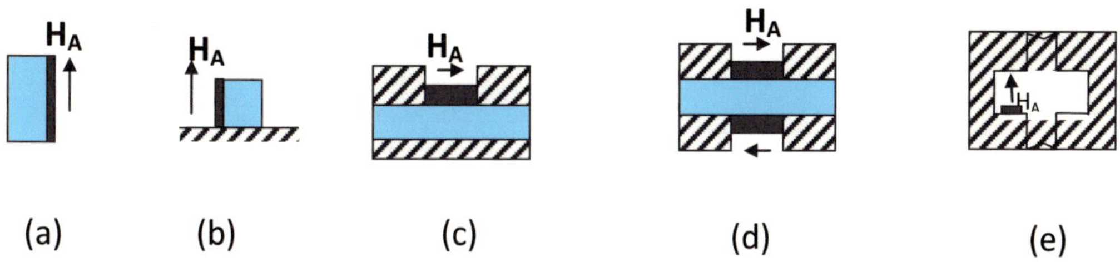

Fig. 6. Examples of hexagonal ferrite isolators on different transmission lines: (a) dielectric rod waveguides; (b) reflective dielectric rod line; (c) single-sided slot line; (d) double-sided slot line; and (e) waveguides with groove.

3.2 Off-resonance non-reciprocal devices

Herein, some examples of using high-coercivity *polycrystalline hexagonal ferrites* for the design of off-resonance non-reciprocal devices operating without any external bias magnetic field are considered. One of them is the Y-circulator. It is well-known that the circulator effect is possible only in gyrotropic media. Due to the presence of the off-diagonal component in the permeability tensor (5) of hexagonal ferrite, this is a gyrotropic, in particular, gyromagnetic medium. This component for magneto-uniaxial ferrites is

$$\mu_a \approx \frac{4\pi M_S H_\omega}{[(H_A + H_\omega)(H_A - H_\omega)]}, \tag{7}$$

where $H_\omega = \omega/\gamma$, even without any bias magnetic field H_0.
The design of such a circulator is analogous to that of traditional waveguide circulators (Tsankov et al., 1992). In the OPLF, the Y-circulator for 8-mm waveband was developed (Musial et al., 1972). It was using a cylinder made of a polycrystalline SrCrM ferrite with H_A=21 kOe, ΔH=1 kOe, H_C=1.5 kOe, $4\pi M_S$ = 3.4 kG, and $4\pi M_r$ =3.1 kG on the basis of the standard metal waveguide with the cross-section 7.2 x 3.4 mm^2 . The ferrite cylindrical was placed in the center completely coveing the cross-section. It was operating at frequencies above FMR.
In the short-wave part of the EHF band (above 100 GHz), the problem of extremely small cross-sections of standard transmission lines arises. Application of metal-dielectric waveguides, e.g., „hollow dielectric channel" lines (Kazantev & Kharlashkin, 1978) may

solve this problem. Thus, instead of the standard metal waveguides of a cross-section of 1.1 mm x 0.55 mm, the metal-dielectric waveguides of 10 mm x 10 mm cross-section have been used, and three- and four- port circulators based on polycrystalline hexaferrites operating in the regime below the resonance (H_0=0) have been designed (Avakyan et al., 1995). Fig. 7 schematically shows a three-port circulator. If the load at Port 2 is a receiver, this system works as a waveguide switch in a single-antenna radar. In the case of a matched-load termination, this is an off-resonance isolator. In the circulator under study, a hexaferrite sample was magnetized to saturation along the axis of the waveguide, and was completely closing the cross-section of the waveguide. The length of the sample (along the axis of propagation) assured the 45^0 rotation of the polarization plane. In the circuits designed for the frequency ranges of 80-130 GHz and 40-180 GHz, the high-coercivity industrial synthesized hexagonal ferrites 03СЧА (H_A=35.0 kOe; ΔH=3.5 kOe, H_C=4.0 kOe, $4\pi M_s$ = 1400 G) and 04СЧА2 (H_A=23.5 kOe; ΔH=3.5 kOe, H_C=5.0 kOe, $4\pi M_s$ = 1900 G) were used. VSWR was about 1.1 in the 25% frequency bandwidth, and isolation was more than 18 dB.

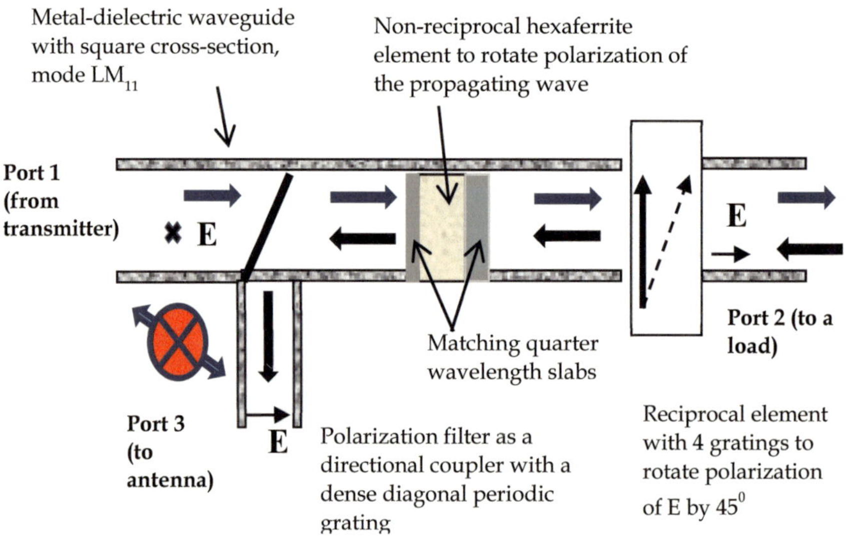

Fig. 7. Three-port circulator based on metal-dielectric waveguide with a hexaferrite slab

3.3 Bandpass and stopband filters

From the very beginning *monocrystalline hexaferrites* were specifically designed for applications in bandpass and stopband filters for the EHF range. Even nowadays, for many practical puropses EHF filters with required parameters can be designed only on the basis of monocrystalline hexaferrites.

An isolator with a monocrystalline hexaferrite is a *stopband filter* indeed. As is mentioned above, it provides signal suppression at the required frequency over a bandwidth of 10…40%. This can be achieved by the proper orientation of a spherical resonator made of a monocrystalline hexaferrite, when the bias magnetic field H_0 is fixed. Such filters were designed in MPEI using standard metal waveguides of cross-sections 7.2 mm × 3.4 mm, 5.2 mm × 2.6 mm, and 3.6 mm × 1.8 mm. They provide the rejection rate (defined as the difference between the power attenuation levels in the stopband and in the passband) of 20…40 dB in the frequency range 60…200 GHz (Mikhailovsky et al., 2002).

In magnetically tunable filters, a magneto-uniaxial hexagonal ferrite resonator is placed in a matched waveguide in such a way that the crystallographic axis would be parallel to the bias magnetic field ($H_A \| H_0$). Three methods have been tested for increasing the rejection rate of filters. First, this is an increase of the microwave field power density near the hexagonal ferrite resonator by placing it on a dielectric substrate in the waveguide. Second, an application of ferrite disk resonators with azimuth modes (Moiseyev & Pollak, 1982). Third method is the abovementioned effect of the "bound modes" due to combination of FMR and volume resonance of the hexaferrite resonator.
Crystals of magneto-uniaxial monocrystalline hexagonal ferrites with parallel orientation of H_A and H_0 have also been used in *bandpass filters* with magnetic tuning. The parameters of such filters are given in Table 5.

Frequency range, GHz	Band width at -3 dB level, MHz	Insertion loss, dB, less than	Isolation outside the pass band, dB, more than	Transmission line
25 - 38	400	8	30	MW
36 - 52	500	7	30	MW
52 - 78	300	12	30	MW
52 - 78	250	8	30	DW
78 - 119	250	10	23	MW
78 - 105	400	10	30	DW

Table 5. Parameters of designed bandpass filters

An original design based on the orthogonal reflective *dielectric waveguides* has also been used at frequencies up to 150 GHz (Khokhlov et al, 1984). Our studies have shown that filters built on dielectric waveguides are technologically simpler compared to the metal waveguides, and they provide low insertion loss, as well as excellent non-reciprocal and directional properties. Therefore they can serve as elements of the EHF frequency band.

3.4 Ferrite mixers and frequency-selective primary transducers for power meters

A number of novel measuring systems and devices for EHF band have been developed in the OPLF of MPEI. Their design has been based on the application of high-anisotropy hexagonal ferrite resonators that provide a substantially reduced bias magnetic field necessary for operation. First of all, these are the new functional frequency-selective devices for measuring power parameters of signals of medium and high intensity.

Frequency range, GHz	Sensitivity, μW	VSWR	Operating bias field, kOe	Weight with magnet, g
25… 37.5	<150*	< 1.4	2.5 … 6.5	2440
37.5 … 52.5	< 60	< 1.3	2 … 5.3	2440
52.5 … 75	< 30	< 1.9	1 … 10	2440
66 … 74.4	< 60	< 1.7	0	110

* Ferrite contents was not optimized

Table 6. Parameters of designed ferrite mixers for the EHF frequency range

The *ferrite mixers* can be used in the devices for measuring pulse and continuous power in the frequency range of 25 - 75 GHz (Mikhailovsky et al., 1965; 2002). Their parameters are summarized in Table 6. In these mixers, the resonance response proportional to the power level under measurement is induced in a coil surrounding a cylindrical ferrite resonator. These mixers have sensitivity up to 30 μW, and they are characterized by extremely high stability to power overload. They have been developed within a single magnetic system. For the 4-mm waveband, a mixer has been developed without any magnetic system at all.

Another perspective application of monocrystalline hexaferrite resonators is development of magnetically tuned *primary transducers*. These transducers are intended for converting microwave to low-frequency signals. The picture of a stripline device with a magnetic detector – a ferrite resonator surrounded by a spiral microcoil is shown in Fig.8 (a).

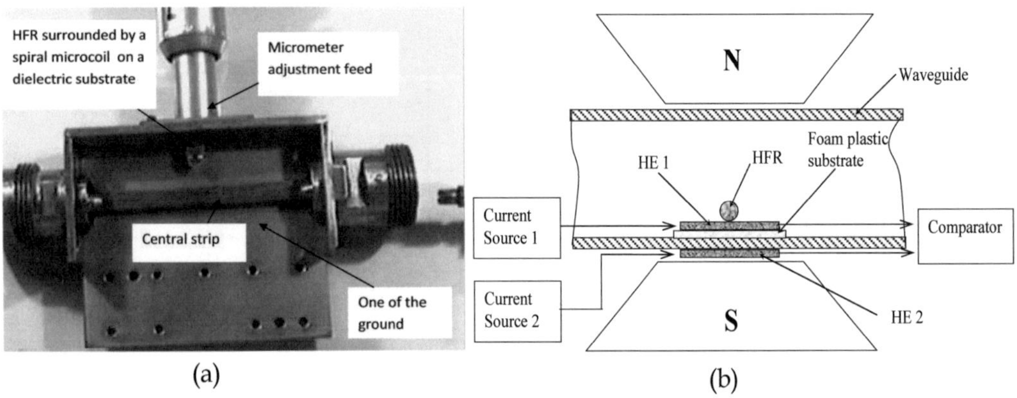

Fig. 8. (a) Stripline gyromagnetic frequency converter (top ground plane cover removed), and (b) primary transducer with two Hall elements

The magnetic detector was invented by L.K. Mikhailovsky (Mikhailovsky, 1964), and it has become the "heart" of a quantum cross-multiplier. This is a cross-non-linear element, which converts a microwave (mm-wave) carrier signal at the FMR (or NFMR) frequency down to harmonics of a pumping RF signal. Based on this element, a tunable single-frequency gyromagnetic converter was designed. It is now used for frequency-selective measuring of microwave power spectral density of short (nanosecond) pulse and noise signals.

However, a primary transducer with a spiral microcoil as shown in Fig. 8 (a) is effective only if it contains an extremely high-Q ferrogarnet monocrystalline resonators ($\Delta H<0.5$ Oe) and operates at comparatively low microwave frequencies (~300 MHz…18 GHz), requiring significant bias magnetic fields for keeping the ferrite resonator in magnetic saturation. For applications in the EHF band, the different design principles are needed. Thus, in early 1990s the authors of this Chapter proposed to use a monocrystalline magneto-uniaxial hexaferrite resonator (HFR) with comparatively narrow for hexagonal ferrites FMR line ($\Delta H\sim10$ Oe) in direct contact with a semiconductor element. The latter is able to detect variation of the temperature of ferrite resonators at the absorption of electromagnetic power passing through them inside a transmission line. Two types of semiconductor thermo-sensitive elements were used in experimental testing of the hexagonal ferrite primary transducers, designed for the 8-mm band: (1) two Hall-elements (HE) connected in a compensating schematic as Fig. 8 (b), and (2) a chip transistor (without housing), used in a diode regime, in a direct contact with an HFR. The designed transducers demonstrated a coefficient of power conversion of 10 μV/mW when using the Hall-element, and with chip transistors the coefficient of power conversion was about 1200 μV/mW. In all the cases, the linear dynamic range was in the range of 20-30 dB. Over the 4-mm band, the monocrystalline hexaferrites can have much smaller FMR line widths (since these will be resonators made of pure Ba-ferrites), so the expected parameters of such primary transducers or power converters are expected to be substantially better.

Also, an effect of the RF self-generation in a closed-loop system containing a ferrite resonator was detected and studied. This effect can also be used for frequency-selective microwave and mm-wave power measurements. The schematic is shown in Fig. 9, where a ferrite is affected by two signals - a microwave and a pumping RF (a few MHz), and the output signal of the crystal detector terminating the microwave transmission line is amplified by a narrowband RF amplifier and then is used as a feedback for the RF pumping of ferrite. Such system was built with a modulator which used a high-Q monocrystalline BaScM HFR for the 8-mm waveband.

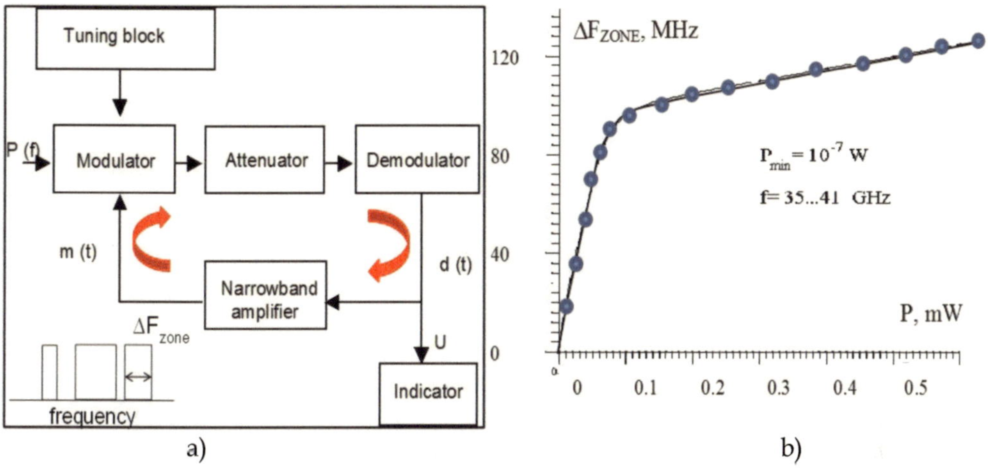

Fig. 9. (a) Autogeneration system with a modulator built with a HFR. (b)The width of a generation zone as a function of the spectral power density within an FMR line.

The system can be used as a threshold detector of power levels at the selected frequencies and also for frequency-selective power measurements. The pumping of the RF (1 MHz) signal is done by a piezoelectric element which modulates the resonance frequency of the hexaferrite resonator by varying its main axis orientation. The lower level of the measured power is determined by the sensitivity of the crystal detector, and the upper level depends on attenuation introduced by a calibrated attenuator in the feedback loop. The resonance frequency of the ferrite resonator is swept by a sawtooth current in the small bias magnetic system (H_0=0…3 kOe).

3.5 Absorbers of electromagnetic waves

Application of monocrystalline, polycrystalline, and dispersed conventional low-anisotropy garnet and spinel ferrites in shields, coatings, and various filtering devices of the EHF band is known to be limited, and in many cases, impossible, because of the necessity of applying intense bias magnetic fields (above ~10^6 A/m). For this reason, hexaferrites that have a high internal magnetic field are very desirable, since they exhibit natural ferromagnetic resonance (NFMR) even if no bias magnetic field is applied. In traditional devices of the SHF (3-30 GHz) and EHF (30-300 GHz) bands, mostly dense hexaferrite samples have been used. This limits the design possibilities for obtaining required frequency characteristics and other microwave or mm-wave parameters.

When monolithic hexaferrite samples are replaced by hexaferrite powders, an additional degree of freedom for engineering composites is created. Application of dispersed hexagonal ferrites allows for designing optimal devices and solving a number of technological problems, e.g., for absorbing the energy of electromagnetic fields and waves. The frequency characteristics of absorption loss in powders of doped hexagonal ferrites, taken from the functional series of the engineered materials, cover the frequency range from 4 to 40 GHz. The possibility of shifting the central frequency of absorption at the NFMR, varying the width of absorption, and modifying the shape of frequency characteristics is possible, for example, due to the variation of Scandium contents in the BaScM ferrites, as is shown in Fig. 10. An example of the absorption frequency dependence for a hexagonal ferrite thick film, which is made of a mixture of two different hexagonal ferrite powders in an epoxy resin base is given in Fig. 11. The frequency dependence is obtained based on the model proposed in (Pollak, 1977) for the effective electromagnetic parameters (permittivity and permeability) for composite materials containing highly anisotropic uniaxial hexagonal ferrite inclusions.

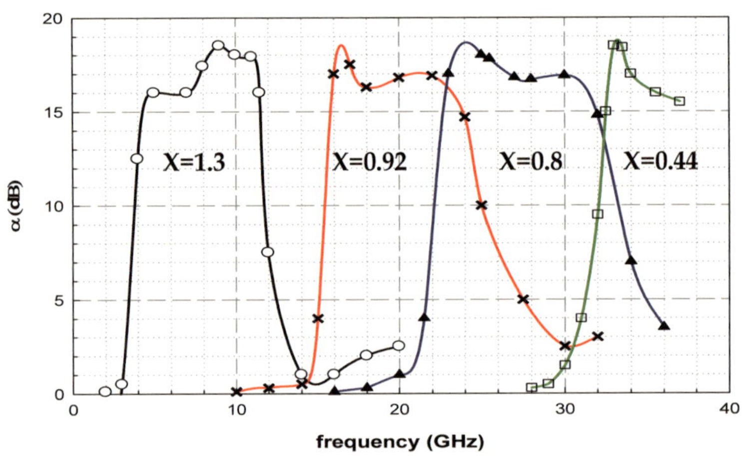

Fig. 10. Frequency characteristics of absorption loss in BaScM ($BaSc_xFe_{12-x}O_{19}$) ferrite

3.6 Phase shifters

Phase shifters are the only group of known gyromagnetic devices, where hexagonal ferrites have not received much attention yet. Application of hexagonal ferrites in phase shifters is substantially less popular compared to isolators and filters. Possibilities of using magneto-uniaxial and planar hexaferrites in laboratory samples of phase shifters in the frequency ranges of 30-35 GHz and 90-94 GHz have been reported in some publications, e.g., (Patton, 1988; Thompson, 1995). However, industrial designs and applications of such devices are still unknown. The reason for this is that to develop phase shifters, more narrowband materials with increased saturation magnetization and low dielectric loss are required.

However, it is clear now that a number of Russian-make hexaferrites (Catalogue, 2006) can be used for this purpose. The above mentioned examples of using hexagonal ferrites in non-resonance isolating devices with a fixed angle of polarization plane rotation (Musial, 1972) can apply to phase shifter design as well.

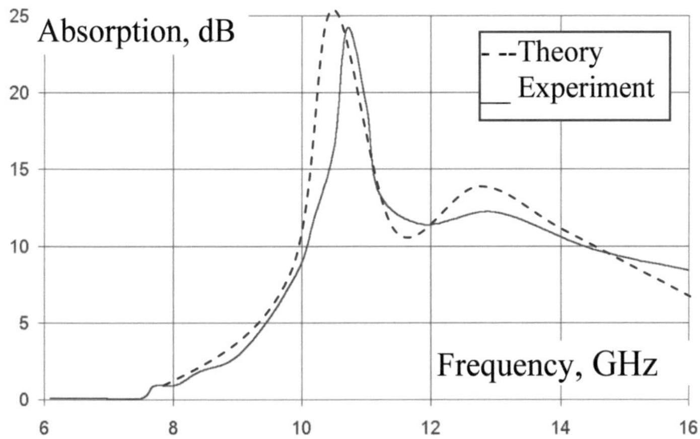

Fig. 11. Frequency characteristic of absorption of a thick film containing powders of hexagonal ferrites with two different compositions

3.7 Traveling-wave generators

The design of the EHF band *traveling-wave masers* (TWMs) requires creating built-in low-size non-reciprocal isolation at a given frequency and a certain magnetic field (the maser's operating magnetic field, including a zero magnetic field, is stringenly determined by the specifications). This problem was solved using the polycrystalline hexaferrite isolator. Another model of a TWM on rutile was developed using a chain of hexaferrite spherical resonators. For this purpose, multiple resonators assuring the necessary non-reciprocal per-unit-length isolation both at the fixed frequency and within a given frequency band have been designed. Tuning was achieved by variation of the orientation of the ferrite.

Traveling-wave tubes (TWTs) operating over the the EHF band belong to the class of the devices that definitely need, and we beleive that in future will widely use hexagonal ferrites. The traditional built-in absorbing filters protecting traveling-wave tubes from self-excitation cannot be used in the EHF band. Non-reciprocal isolation in the EHF band can be achieved only using hexaferrites. The experience of developing intra-tube non-reciprocal absorbers based on ferrite garnets in the centimeter waveband (Vambersky et al., 1973) adds optimism about the application of polycrystalline hexaferrites in the EHF band. The analysis of industrially manufactured TWTs shows that the specific delay systems needed for TWTs can be built into hexaferrite isolators. From our point of view, the most appropriate up-to-date design for realization of this idea is the delay system of the so-called *transparent TWT*. This is the output tube in the cascade of two TWT power amplifiers. For example, in the 8-mm waveband tube (which is an analogue of the tubes for 2-cm and 3-cm wavebands), providing an average output power up to 300 W in a frequency bandwidth of 1 GHz, and at the static bias field of 4 kOe, the application of polycrystalline ferrites can readily provide the required non-reciprocal isolation.

4. Societal applications of hexagonal ferrites

Some examples of applications of hexagonal ferrites in non-technical areas are given. These are applications in medical, agriculture, and transport monitoring, as well as in every-day electronic devices.

4.1 Microwave ovens

Microwave ovens are designed in such a way that there are protections against radiation leakage outside their enclosures. However, these measures are provided for the fundamental frequency and its second harmonic, while radiation at higher harmonics is not controlled or tested, although it may be substantial enough to cause EMC/EMI problems for the other electronic devices operating nearby.

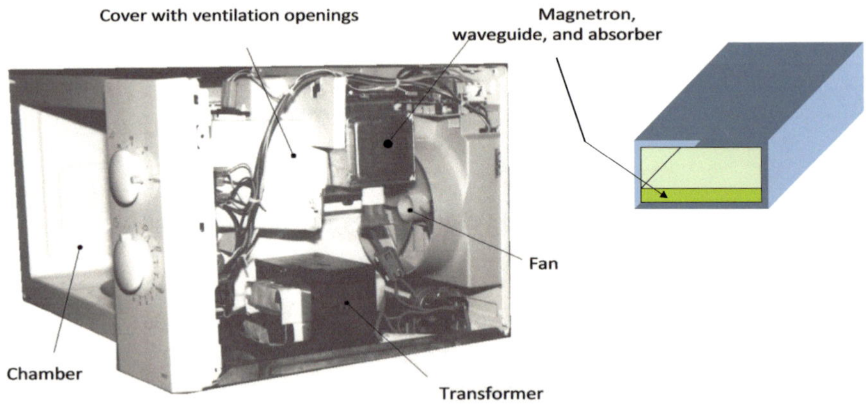

Fig. 12. Microwave oven with hexagonal ferrite absorber

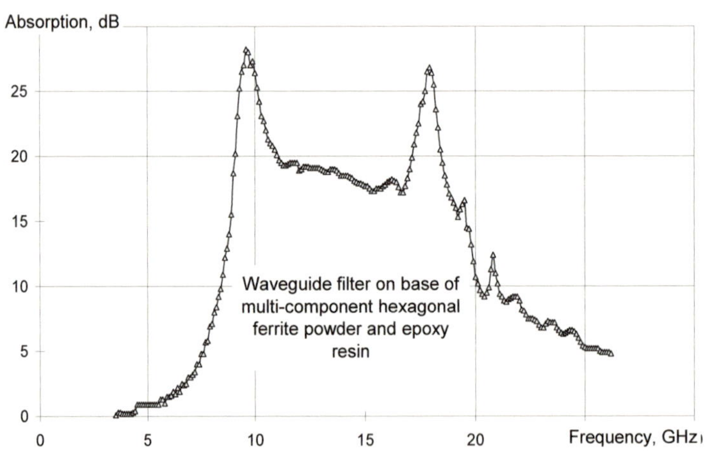

Fig. 13. Frequency characteristic of filter of harmonics for suppressing spurious radiation of microwave ovens

4.2 Protecting shields for high-speed electronic devices

The hexaferrite-based absorbers are recommended for protecting power cords, cables, individual intra-system blocks, enclosures, and antenna caps of modern high-speed electronic devices, including computers and cellular phones, whose operating frequencies fall into the microwave band (> 2 GHz). However, if these are active devices, leakage at the main (clock) frequency and its harmonics should be eliminated, as well as the susceptible circuits should be protected from external noise sources. Also, hexaferrites can be used in stealth-technology for creating non-reflecting surface coatings.

4.3 Medical applications – EHF therapy

Hexaferrite isolators, in addition to their known applications in traditional engineering systems, have been used in narrowband and wideband systems for EHF therapy (Avakian et al., 1995). For introducing the latest achievements and recommendations of EHF therapy, the development of non-reciprocal devices for the frequency bands of 42-95 GHz and 90-160 GHz was needed. Among the obvious requirements for the isolators to be included in medical devices for EHF therapy, are small size and weight, and low cost. From our point of view, only resonance isolators based on hexagonal ferrite composites, operating without any bias magnetic fields, can satisfy these criteria. Wideband isolators of this class have been developed for the frequency range of 37-118 GHz. The insertion (direct) loss in these isolators was 1.3-2.0 dB, while the return loss was 16-19 dB. Maximum return loss and minimum insertion loss have been noticed around 60 GHz. To satisfy the particular technical requirements of customers producing medical equipment, a number of isolators with a maximum of return loss at different frequencies within the above mentioned band have been designed.

4.4 Transport: radar systems for measuring motion parameters

The development of the EHF frequency range is very promising for small-size and highly accurate radars of local operation. Applications of isolated mirror and slot dielectric waveguides provide wide possibilities for integrated technology design of a microwave (mm-wave) system, which is much cheaper than using standard metal rectangular waveguides. The integrated microwave (or mm-wave) blocks unites an antenna, a pattern-forming circuit, and a signal processing device. All these advantages can be realized only when using hexagonal ferrite non-reciprocal isolators without external magnets. The latter can be manufactured using a film-sputtering technology. Thus, based on the mirror dielectric waveguide, the 8-mm wavelength block was tested within an automotive set for measuring parameters of motion and preventing traffic accidents. Based on the double-sided slot waveguide, two blocks were developed. The first was of the 5-mm wavelength Doppler measuring device to operate in the vibrometer (Bankov, 1999), and the second was developed for the 8-mm radar system with linear frequency modulator for the level gauge and other applications (Abdulkin et al., 1991). In addition to the given examples, it is important to also mention the potential advances of applying planar integrated EHF blocks on dielectric slot waveguides with hexaferrite isolation inside cellular network systems.

4.5 Agriculture: processing seeds before sowing

The application of special capsules for green-sprouting of seeds before their sowing is known. Typically, for this purpose, the biologically active porous materials, where seeds are placed together with a nutrient medium, are used. When a hexagonal ferrite particle, which is a miniature magnet (its size is less than 10 μm) is placed together with a seed, it substantially stimulates the process of green-sprouting. Hexagonal ferrite particles orient themselves along the Earth's magnetic lines, and this provides independence of the green-sprouting speed upon the seed's initial spatial orientation. The required magnitudes of the magnetic field intensity and coercive force are 0.5 T and 5 T, respectively.

5. Conclusion

A review of pioneering work conducted in the MPEI since 1950 on the theoretical and experimental study and development of hexaferrites and devices on their basis for different engineering and social applications is presented. As a result of the fundamental theoretical research led by Mikhailovsky, the founder of the OPLF, the magnetic bias field needed for the operation of the devices at higher microwave and mm-wave frequencies was „moved" to the crystal lattice of the gyromagnetic (ferrite) medium.For the first time in Russia, a new class of ferrite materials was synthesized: magneto-uniaxial hexagonal ferrites with high internal fields of crystallographic anisotropy. This allowed for the design of various gyromagnetic resonance devices operating without bias magnetic field or with low bias magnetization needed only for ferrite saturation and tuning of resonance frequency. These are the passive devices, such as resonance isolators, stopband and bandpass filters, circulators, matched loads, electromagnetic wave absorbers, and also cross-non-linear devices for mm-wave power measurements. Hexaferrites can be used for non-reciprocal isolators in masers and traveling-wave generators, and also for the design of frequency-selective microwave absorbing coats and filters that can solve numerous problems of electromagnetic compatibility and immunity. Over a hundred different types of polycrystalline and monocrystalline hexagonal ferrites having different composition have been synthesized in the OPLF, mainly for applications at 3-100 GHz. Based on those ferrites, over 20 different types of various composite electromagnetic wave „currentless" absorbers have been developed for the frequency range of 1.5 - 100 GHz.

Microwave and mm-wave devices of the future generation, whose development has been driven by modern wireless and radar technologies, should be mainly planar and low-loss, operating without huge external bias magnetic fields, and have more functional possibilities compared to conventional present-day devices. The authors are convinced that it would be impossible to solve these problems without using a natural physical advantageous feature of hexagonal ferrites - their high internal field of crystallographic anisotropy.

6. Acknowledgment

Koledintseva and Hanamirov dedicate this work to the memory of the colleague, Dr. Alexander A. Kitaitsev who passed away in November 2010, when the work on this Chapter has already begun. The authors of this review would like to express the deepest gratitude to the „fathers-founders" - Professors Leonard K. Mikhailovsky, Boris P. Pollak, and Vladirim P. Cheparin who for many decades lead the hexagonal ferrite research and synthesis in Russia. The authors are grateful to colleagues Dr. Tatyana S. Kasatkina, Dr. Irina E. Kabina, Dr. Sergey S. Egorov, Andrey A. Shinkov, and Andrey S. Fedotov, for useful discussions and participation in theoretical and experimental research.

7. References

Abdulkin, A.A.; Bankov, S.E.; Plescheyev, V.I.; Khanamirov, A.E. and Khryunov, A.V. (1991). Small-size radar for applications in transport and industry, *Proc. I Ukrainian Symposium "Physics and Technology of Millimeter and Submillimeter Waves"*, Vol. 2, Kharkov, pp. 77-78 (in Russian).

Avakian, R.S.; Aivazyan, M.T.; Khanamirov, A.E.; Kocharian, K.N.; Karneeva, S.S. & Sarkissian, S.A. (1995). Millimeter-wave non-reciprocal devices on hexaferrite and square MDW, *J. Electronic Measur. and Instrum.*, Vol. 9 (Sept. 1995), pp.161-162, ISSN 1000-7105.

Avakian, R.; Taube, A.A. & Teppone, M. (1996). The state-of-the-art of EMF-puncture devices, *Int. Journal of Oriental Medicine*, Vol. 7 (1L) (1996), pp. 34-44, ISSN 1044-0003.

Bankov, S.E. & Khanamirov, A.E. (1999). Microwave modules on dielectric slot waveguides, *Proc. 8th Int. Conf. on Spin Electronics*, Moscow (1999), pp. 458-463.

Catalogue *Microwave Materials*, Joint Stock Company "Enterprise Magneton", St. Petersburg, Russia, 2006.

De Bitetto, D.J. (1964). Anisotropy fields in hexagonal ferrimagnetic oxides by ferrimagnetic resonance, *J. Appl. Phys.*, Vol. 35, No. 12 (Dec. 1964), pp. 3482-3487, ISSN 0021-8979.

Dirac, P.A.M. (1975). *Spinors in Hilbert Space*, Kluwer Academic/ Plenum Publishers (Jan, 1975), ISBN-10: 0306307987.

Feynman, R.P.; Leighton, R.B. & Sands, M. (2005). *The Feynman Lectures on Physics*, 2nd ed., Vol. 2, Ch. 34, Addison-Wesley, 2005, *ISBN* 9780805390452.

Gurevich, A.G. & Melkov, G.A. (1996). *Magnetization Oscillations and Waves*, CRC Press, 1996, *ISBN*-10: 0849394600.

Harris, V.G.; Chen, Z.; Chen, Y.; Yoon, S.; Sakai, T. ; Gieler, A.; Yang, A.; He, Y. ; Ziemer, K.S.; Sun, N.X. & Vittoria, C. (2006). Ba-hexaferrite films for next generation microwave devices (invited), *J. Appl.Phys.*, Vol. 99, 08M911, 2006, *ISSN* 0021-8979.

Harris, V.G.; Geiler, A., Chen, Y.; Yoon, S.D. ; Wu, M.; Yang, A.; Chen, Z.; He, P.; Parimi, P.; Zuo, X.; Patton, C.E.; Abe, M.; Acher, O. & Vittoria, C. (2009). Recent advances in processing and applications of microwave ferrites, *J. Magn.Magn. Mater.*, Vol. 321, 2009, pp. 2035-2047, ISSN 0304-8853.

Kazantsev, Y.N. & Kharlashkin, O.A. (1978). Rectangular waveguides of the class „hollow dielectric channel", Radiotekhnika i Elektronika (*Journal of Communications Technology and Electronics)*, Vol. 23, No. 10, (Oct. 1978), pp. 2060-2068, ISSN 0033-8494.

Khokhlov, M.A.; Pollak, B.P. & Solomkin, A.A. (1984). Ferrite filters based on dielectric waveguides for the EHF band, *Inter-University Trans.* , No. 48 (1984), Moscow Power Engineering Institute, Moscow, pp. 125-131 (in Russian).

Kittel, C. (1948). On the theory of ferromagnetic resonance absorption, *Phys. Rev.*, Vol. 73 (1948), pp. 155-161, ISSN 1050-2947.

Kittel, C. (1949). On the gyromagnetic ratio and spectroscopic splitting factor of ferromagnetic substances, *Phys. Rev.*, Vol. 76 (1949), pp. 743-748, ISSN 1050-2947.

Korneyev, I.V. (1980). On the analysis of the "effect of bound waves" in a waveguide with a dielectric and anisotropic ferrite, *Trans. Moscow Power Engineering Institute*, No. 464 (1980), pp. 74-79 (in Russian).

Korneyev, I.V. & Pollak, B.P. (1982). Isolators-flanges of resonance type on the basis of hexagonal ferrites, Electronics Engineering, Ser. 1, Microwave Electronics, vol. 4 (340), 1982, pp. 59-61 (in Russian).

Landau, L.D. & Lifshitz, E.M. (1935). On the theory of the dispersion of magnetic permeability in ferromagnetic bodies, *Phys. Zeitsch. der Sowietunion, Vol.* 8, pp. 153-169, reprinted by *Ukrainian J. Phys.*, Vol. 53, Special Issue, pp. 14-22, Physico-Technical Institute, Academy of Sciences of the Ukrainan SSR, 2008, ISSN 2071-0194.

Landau, L.D. & Lifshitz, E.M. (1960). *Electrodynamics of Continuous Media: Landau and Lifshitz Course of Theoretical Physics,* Vol. 8, Pergamon Press, Addison-Wesley, Oxford, UK, ISBN 0080091059.

Maxwell, J.C. (1856). On Faraday's Lines of Force, *Trans. Cambridge Phil. Society*, Vol. 10, Part 1 (Feb 1856), pp. 155- 229.

Medvedev, S.A.; Cheparin, V.P. & Balbashov, A.M. (1967). Synthesis and properties of monocrystals of Scandium-doped Barium ferrite, *Proc. 5th USSR Meeting on Physical and Physico-Chemical Properties of Ferrites* (1967), Minsk, Belarus (in Russian).

Medvedev, S.A.; Pollak, B.P.; Cheparin, V.P.; Sveshnikov, Y.A. & Khanamirov, A.E. (1969). Development, research, and application of hexaferrite monocrystals as new microwave materials, *Reports of Scientific and Technological Conference on the Results of Scientific and Research Works in 1968-1969.* Radio Engineering, Ferrite Microwave Radio Physics, Moscow Power Engineering Institute, Moscow, USSR, pp. 80-89 (in Russian).

Mikhailovsky, L.K. ; Pollak, B.P.; Balakov, V.F. & Khanamirov, A.E. (1965). Properties and application of magneto-uniaxial ferrites at millimeter waves (Review), *Radiotekhnika i Elektronika (Radio Engineering and Electronics)*, Vol. 10, No. 10 (Oct. 1965), pp. 1739-1752 (in Russian).

Mikhailovsky, L.K. ; Pollak, B.P. & Sokolov, O.A. (1966). On the problem of ferromagnetic resonance in uniaxial single-domain ferromagnetic particle, *Fizika metallov i metallovedenie (Physics of metals and physical metallurgy)*, Vol. 21, No. 4, 1966, pp. 524-528 (in Russian).

Mikhailovsky, L.K. (1964). Method of absolute frequency-selective measurement of microwave magnetic field intensity and power in a pulse. Certificate of Authorship No. 163226 for the application No. 822530 of Mar. 02, 1963. USSR Bulletin of Inventions No. 12 (Dec. 1964), Moscow (in Russian).

Mikhailovsky, L.K. ; Pollak, B.P. & Khanamirov, A.E. (2002). Research and development of EHF hexaferrite devices in MPEI, *Proc. 9th Int. Conf. on Spin Electronics*, Moscow (Dec. 2002), pp. 559-573 (in Russian).

Mikhailovsky, L.K. (2002). Elements and objects of quantum gyrovector electrodynamics, *Proc. 11 Int. Conf. on Spin-Electronics and Gyrovector Electrodynamics, Section of Int. Conf. „Electromagnetic Fields and Materials"*, Dec. 20-22, 2002, Moscow (Firsanovka), Publ. UNC-1 MPEI(TU), pp. 20-61.

Moiseyev, A.N. & Pollak, B.P. (1982). Study of hexaferrite disk resonators of azimuth modes. *Trans. Moscow Power Engineering Institute*, Vol. 645 (1982), Moscow, USSR, pp. 83-93 (in Russian)

Musyal, Y.V. ; Benevolenskaya, N.B. &. Hanamirov, A.E (1972). Circulator with a hexagonal ferrite slab without magnets, *Voprosy Radioelektroniki (Problems of Radio Electronics)*,

No. 4 "Radio Engineering Measurements" (Apr. 1972), Moscow, USSR, pp. 18-20 (in Russian).

Nedkov, I.; Cheparin, W. & Khanamirov, A. (1988). Ferromagnetic resonance of polycrystalline Al-substituted M-type hexagonal ferrite, *Le Journal de Physique Colloques*, vol. 49, no. C8, Dec. 1988, pp. 945-946, ISSN 0449-1947.

Patton, C.E. (1988). Hexagonal ferrite materials for phase shifter applications at millimeter wave frequencies, *IEEE Trans. on Magn.*, Vol. 24, No. 3 (May 1988), pp. 2024-2028.

Petrova, I.I.; Ivanova, V.I.; Khanamirov, A.E. & Grigorieva, L.N. (1980). Polycrystalline hexagonal ferrites as the materials for solid-state electronics, *Trans. Moscow Power Engineering Institute*, No. 464 (1980), pp. 59-69 (in Russian).

Pollak, B.P.; Hanamirow, A.E.; and Korneew, I.W. (1976). Mono- and polycrystalline hexaferrites as materials for resonance microwave devices, *Nachrichtentechnik Electronik (Communication Electronics)*, Vol. 26, No. 7 (July 1976), pp. 245-250 (in German).

Pollak, B.P.; Kolchin, V.V. & Khanamirov, A.E. (1969). On the nature of ferromagnetic resonance linewidth in polycrystalline hexagonal ferrites, *Izvestiya vuzov (News of Universities), Physics*, No. 1 (Jan. 1969), pp. 24-27 (in Russian).

Pollak, B.P. & Kolchin, V.V. (1969). Peculiarities of ferromagnetic resonance in polycrystalline hexaferrites, *Reports of Scientific and Technological Conference on the Results of Scientific and Research Works in 1968-1969*. Radio Engineering, Ferrite Microwave Radio Physics, Moscow Power Engineering Institute, Moscow, USSR, pp. 131-138 (in Russian).

Pollak, B.P. (1977). Analysis of the peculiarities of the magnetic susceptibility tensor of polycrystalline hexaferrite. *Trans. Moscow Power Eng. Inst.*, Vol. 320 (1977), Moscow, USSR, pp. 45-53 (in Russian).

Pollak, B.P. ; Korneyev, I.V. ; Sobyanina, O.Y. & Petrova, I.I. (1980). Polycrystalline hexaferrite films as gyromagnetic resonators for non-reciprocal devices, *Proc. 5th Int. Conf. on Gyromagnetic Electronics and Electrodynamics*, Moscow, Vol. 3 (1980), pp. 143-151 (in Russian).

Polivanov, K.M.; Mikhailovsky, L.K.; Medvedev, S.A., Pollak, B.P. & Balakov, V.F. (1960). Magneto-uniaxial ferrites at microwave frequencies, *Ferrites, Physical and Physico-Chemical Properties, Reports of 3rd All-USSR Meeting on Ferrites,* Minsk, Belarus Academy of Sciences (1960), pp. 567-576 (in Russian).

Polivanov K.M. & Pollak, B.P. (1964). Resonance characteristics of magneto-uniaxial polycrystalline ferrite in microwave field, *Izvestiya AN SSSR (News of the USSR Academy of Sciences), ser. Physics*, Vol. 28, No. 3 (March 1964), pp. 470-480 (in Russian).

Polivanov, K.M.; Medvedev, S.A.; Khanamirov, A.E.; Kolchin, V.V. & Balbashov, A.M. (1969). Development, research, and application of polycrystalline hexaferrites as new microwave materials, *Reports of Scientific and Technological Conference on the Results of Scientific and Research Works in 1968-1969*. Radio Engineering, Ferrite Microwave Radio Physics, Moscow Power Engineering Institute, Moscow, USSR, pp. 120-130 (in Russian).

Qui, J.; Gu, M & Shen, H. (2005). Microwave sbsorption properties of Al- and Cr-substituted M-type barium hexaferrite, *Journal of Magnetism and Magnetic Materials*, Vol. 295, No. 3 (Sept. 2005), Elsevier, pp. 263-268, ISSN 0304-8853.

Rathenau, G.W.; Smit, J. & Stuyts, A.L. (1952). Ferromagnetic properties of hexagonal iron oxide splicing, *Zeitschrift fuerPhysik(Journal of Physics): A*, Vol. 133, No. 1-2 (Sept. 1952), pp. 250-260, ISSN 0044-3328.

Sixtus, K. J.; Kronenberg, K. J. & Tenzer R. K.(1956). Investigations on Barium ferrite magnets, *J. Appl. Phys.*, Vol. 27, No. 9 (Sept. 1956), pp. 1051-1057, ISSN 0021-8979.

Smit, J. & Wijn, H.P.J. (1959). *Ferrites: Physical Properties of Ferrimagnetic Oxides in Relation to Their Technical Applications,* Eindhoven, The Netherlands, Philips Technical Library (1959), 369 p.

Sveshnikov, Y.A. & Cheparin, V.P. (1969). Microwave properties of Titanium-Zinc Barium ferrites. *Reports of Scientific and Technological Conference on the Results of Scientific and Research Works in 1968-1969.* Radio Engineering, Ferrite Microwave Radio Physics, Moscow Power Engineering Institute, Moscow, USSR, pp. 101-106 (in Russian).

Taft, D.R. (1964). Hexagonal ferrite isolators, J. Appl. Phys., Vol. 35, No. 3 (March 1964), pp. 776-778, ISSN 0021-8979.

Thompson, S.B. & Rodrigue, G.P. (1995). The application of planar anisotropy to millimeter-wave ferrite phase shifters, *IEEE Trans. Microw. Theory Techn.*, Vol. 33, No. 11 (Nov. 1995), pp. 1204-1209, ISSN 0018-9480.

Tsankov, M.A. ; Ganchev, S.I. & Milenova, L.G. (1992), Higher-order mode waveguide circulators for millimeter wavelengths, *IEEE Trans. Magn.,* Vol. 28, No. 5, Part II (May 1992), pp. 3228-3230, ISSN 0018-9464.

Vambersky, M.V.; Kazantsev, V.I. & Pavlova, N.I. (1973). Problems of developing non-reciprocal intra-tube absorbers for microwave devices of the M-type, *Physics of Magnetic Phenomena,* Ashkhabad, Turkmenia, pp. 173-188 (in Russian).

Weiss, M.T. & Anderson, P.W. (1955). Ferromagnetic resonance in ferroxdure, Phys. Rev., Vol. 98, No. 4 (May 1955), pp. 925-926, ISSN 1050-2947.

Weiss, M.T. (1995). The behavior of ferroxdure at microwave frequencies, *IRE Conv. Rec.* Vol. 3, Part 8 (1955), pp. 95-108.

5

Local Structure of Relaxor Dielectric Ceramics

Yun Liu and Ray L. Withers
The Australian National University
Australia

1. Introduction

The electric field and frequency dependence of the dielectric permittivities of relaxor dielectrics makes them promising tuneable microwave component materials and gives rise to numerous potential applications, especially in the fields of electronics and telecommunications technologies. Relaxors of this type have been known for over half a century. The relaxation mechanisms underlying their unique dielectric behaviour, however, are still far from well understood. This lack of understanding is a significant impediment to the further development and optimization of their desirable dielectric properties (high dielectric constants, low dielectric losses and high field tuneabilities). Because such functional relaxors are almost invariably chemically and/or displacively disordered, an understanding of their local nanoscale crystal structure is an important starting point to gain a better understanding of their intrinsic dielectric properties.

Conventional Bragg scattering diffraction techniques, whether obtained via X-ray diffraction (XRD), neutron diffraction or electron diffraction, contain only 1-body information on the time and space averaged crystal structure as a result of the assumption of a regularly repeating real space unit cell and its corollary, sharp Bragg reflections only in reciprocal space (see *e.g.* Fig.1a). Correlated chemical and/or displacive deviations from that average structure *i.e.* information on local order, on the other hand, gives rise to additional, usually much weaker, structured diffuse scattering in addition to the strong Bragg reflections of the underlying average structure (see *e.g.* Fig.1b). The detection and reciprocal space mapping of such structured diffuse intensity distributions is an important step towards a structural understanding of the chemical and/or displacive disorder responsible and the relationship that this local order may, or may not, have with electric dipoles and intrinsic dielectric relaxation behaviour. Ultimately, real space modelling and fitting to the observed diffuse distribution is required in order to extract the local ordering responsible.

Electron diffraction is ideally suited to the detection and mapping task as a result of the strength of the interaction between fast electrons and matter, the ability to record planar cross-sections of reciprocal space from localized real space regions coupled with the ability to tilt over large angular ranges enabling the systematic exploration of reciprocal space. The strong interaction of electrons with matter, however, does mean that the effects of multiple scattering and dynamical diffraction need to be borne in mind, in particular at major zone axis orientations. It is thus often quite useful to deliberately take *off-axis* electron diffraction patterns in order to minimize the effects of multiple scattering. By such methods, useful and reliable structural information can be extracted from such data.

Figure 1a, for example, shows a typical <-110> zone axis electron diffraction pattern (EDP) of $CaCu_3Ti_4O_{12}$, a giant dielectric constant material. Only sharp Bragg reflections are observed, entirely compatible with the reported *Im*-3, non-polar, average structure space group symmetry of this material. On tilting away from this exact zone axis orientation, however, keeping an <002>* systematic row excited and choosing a longer exposure time, the weak but quite reproducible, highly structured diffuse intensity distribution shown in Fig.1b becomes apparent in this compound. Transverse polarized diffuse streaking of this type continues to exist no matter how far one tilts around the <002>* direction showing that the diffuse distribution takes the form of sheets of diffuse intensity perpendicular to the three <001> real space directions (Liu et al., 2005). It is shown in their work that this observed diffuse distribution results from correlated off-center displacements of Ti ions along one-dimensional <001> columns. While the giant dielectric constant of this material has now been shown to arise from an extrinsic rather than intrinsic origin, it nonetheless remains of interest that $CaCu_3Ti_4O_{12}$ is an incipient ferroelectric.

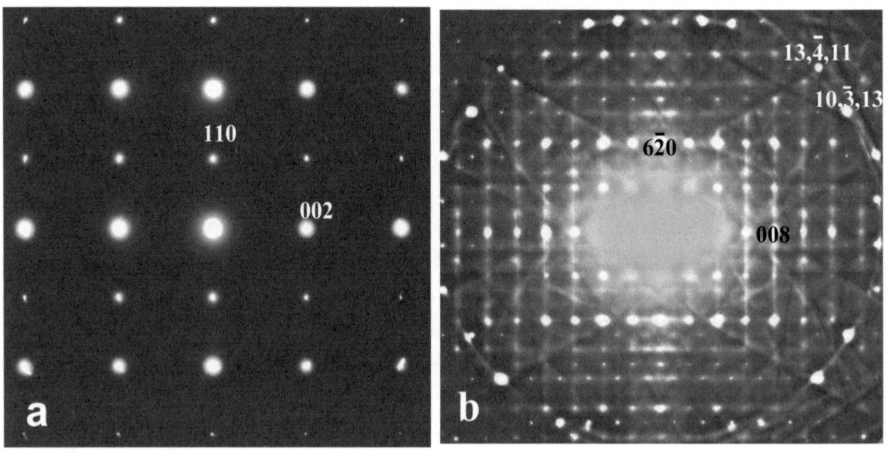

Fig. 1. Typical <110> (a) and close to <130> zone axis electron diffraction patterns of $CaCu_3Ti_4O_{12}$ ceramic (Liu et al, 2005)

For the remainder of this chapter, we will take the family of bismuth-based niobate pyrochlore electroceramics as an example to show how local nanoscale structure affects dielectric relaxation properties even though they have essentially the same average structures.

2. Average structure and dielectric properties of Bi-based niobate pyrochlores

The ideal $A_2B_2O_7$ cubic pyrochlore structure type (see Fig.2a) has space group symmetry *Fd-3m* (with *A* on *16d* at 1/2, 1/2, 1/2; *B* on *16c* at 000; O on *48f* at x, 1/8, 1/8 and O' on *8b* at 3/8, 3/8, 3/8, origin choice 2; note that x is typically ~ 0.30-0.34). It consists of two intergrown substructures: an $O'A_2$ tetrahedral corner-connected, anti-cristobalite type substructure (see Fig. 2b) and a B_2O_6 octahedral corner-connected substructure (see Fig. 2c). These two component substructures then intergrow to form the overall pyrochlore structure type (see Fig.2a).

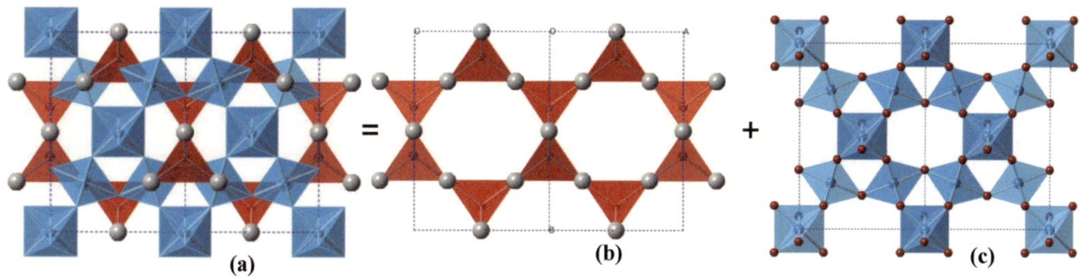

Fig. 2. Ideal cubic pyrochlore structure type (a) described by two constituent network substructures: corner-connected, oxygen-centered O′ A_2 tetrahedral anti-cristobalite structure (b) and corner connected, cation-centered BO_6 octahedral network (c).

While the bonding is strongest within each component sub-structure, the two sub-structures are not independent of one another. The *A* cations of the anti-cristobalite type substructure, in particular, are bonded reasonably strongly to the O ions of the B_2O_6 octahedral substructure. Breaking the ideal cubic pyrochlore structure up in this way is nonetheless a valid and quite useful means of investigating the crystal chemistry of the overall structure.

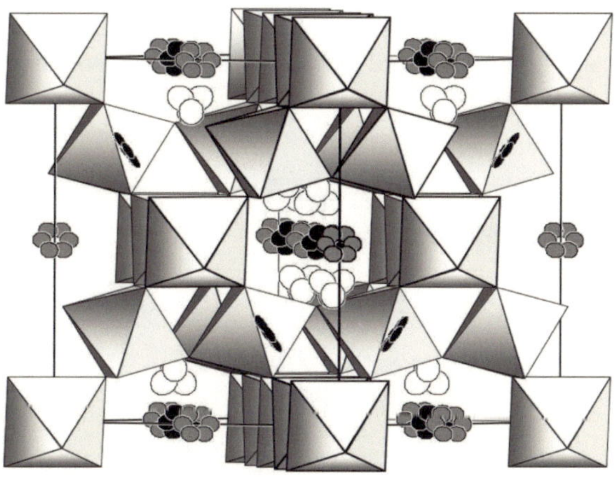

Fig. 3. The displacively disordered average structure of $Bi_{1.667}Mg_{0.70}Nb_{1.52}O_7$ projected along a close to <110> direction (see Nguyen et al., (2007a) for details). The corner-connected octahedral array represents the $(Mg_{0.24}Nb_{0.76})_2O_6$ octahedral sub-structure. The toroids of flat ellipsoids represents the six equivalent *96h A* site positions of the $O'(Bi_{0.833}Mg_{0.11})_2$ sub-structure. The tetrahedral clusters of ellipsoids represents the four equivalent 32*e* site positions for the displaced O′ oxygens.

The larger eight-coordinate *A* site of bismuth-based niobate pyrochlores is occupied mainly by Bi^{3+} ions and the smaller octahedral *B* site mainly by Nb^{5+} ions as would be expected. Surprisingly, however, a large and growing family of cubic pyrochlores have now been shown to exist with nominally too small M^{2+} metal ions, *M* = Zn, Ni, Mg *etc.*, occupying up to 25% of the large *A* cation sites. The nominal stoichiometry of such bismuth-based niobate pyrochlores is $(Bi_{1.5}M_{0.5})(M_{0.5}Nb_{1.5})O_7$. Examples of this type include $(Bi_{1.5}Zn_{0.5})(Zn_{0.5}Nb_{1.5})O_7$ (BZN), $(Bi_{1.5}Ni_{0.5})(Ni_{0.5}Nb_{1.5})O_7$ (BNN) and $(Bi_{1.5}Mg_{0.5})(Mg_{0.5}Nb_{1.5})O_7$ (BMN). Alternatively, three-valent metal ions M^{3+}, can half occupy the pyrochlore *B* site giving rise to bismuth-based niobate pyrochlores of nominal stoichiometry $(Bi_2)(M_1Nb_1)O_7$. Examples of this type include Bi_2InNbO_7 (BIN), Bi_2ScNbO_7 (BSN) and Bi_2FeNbO_7 (BFN) *etc.* Careful phase analysis studies of many such systems, however, shows that there are often vacancies on the pyrochlore *A* site and associated oxygen ion vacancies as well (Levin et al., 2002; Withers et al., 2004). Clearly there is considerable scope for local chemical and associated displacive disorder in these systems.

Average structure refinements of these bismuth-based niobate pyrochlores shows that they are all of the same cubic pyrochlore average structure type with similar cubic lattice parameters [a = 10.5465(9)-10.5633(9) Å for BZN(Levin et al., 2002; Withers et al., 2004), a = 10.5354(2) Å for BNN and a = 10.5607 (5) Å for BMN (Nguyen et al., 2007b), a = 10.5255(1) Å for BFN (Somphon et al., 2006), a = 10.792(2) Å for BIN and a = 10.660(9) Å for BSN (Liu et al., 2009)]. In addition, they all show considerable displacive disorder in the $O'A_2$ tetrahedral structure, usually modelled in terms of a split atom approach (see Fig.3 above). The nature of this disorder, however, is unclear from these average structure refinements. Do the disordered atoms hop independently of one another or do they move together in a correlated fashion and how might this affect the formation of electric dipoles and their dielectric behaviour?

From the dielectric properties point of view, frequency dispersive behaviour is clearly apparent and always observed in both the dielectric permittivity and dielectric loss spectra of these materials (see *e.g.* Fig.4) *i.e.* all of these bismuth-based niobate pryochlores are relaxor dielectrics.

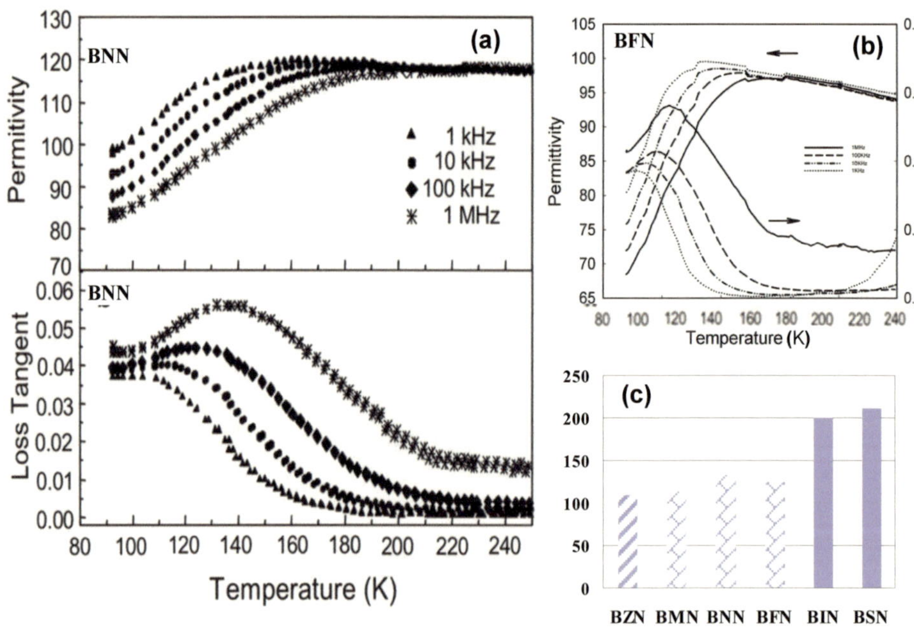

Fig. 4. Temperature dependece of the dielectric permittivity and loss tangent of BNN (a) and BFN (b) as well as the maximum temperature of loss tangent of Bi-based niobate pyrochlores (c) measured at 1 MHz.

Fig.4, for example, shows the temperature dependence of the dielectric permittivity and dielectric loss tangent of (a) BNN and (b) BFN at different frequencies (1 kHz, 10 kHz, 100 kHz and 1 MHz). It is noteworthy that the dielectric loss tangent of BNN first increases with decreasing temperature until a peak in the dielectric loss tangent (at T_m) and then decreases until it flattens off at a somewhat lower temperature into an apparent plateau region. Indeed, the dielectric loss tangent curve at 1 kHz is so flat at low temperature that a prior peak in the dielectric loss tangent is not distinguishable. It appears then that there may well be at least two peaks in the dielectric loss tangent in the case of BNN. Such asymmetric dielectric loss tangent curves such as that observed for BNN but not for BZN, suggest that BNN may well have different local structural order and hence a different dielectric relaxation mechanism to BZN. A similar trend is also observed in the cases of BMN (Nguyen et al., 2007a) and BFN (Fig. 4b).

Note that the dielectric permittivity is relatively flat at higher temperatures, but exhibits a typical step-like slowing down on cooling. Such behaviour is indicative of displacive motion requiring thermally activated hopping over a potential barrier, which then freezes out at a sufficiently low temperature. The frequency dependence of the peak position in the dielectric loss curve (see Fig.4), following a previous approach (Kamba et al., 2002; Nino et al., 2001), can in general be modelled using either the basic Arrhenius type equation $f_r = f_0 \exp(-Ea / kT)$ or the Vogel-Fulcher equation $f_r = f_0 \exp[-Ea / k(T - T_f)]$. Here f_r is the measuring frequency, f_0 is the relaxation frequency at infinite temperature, E_a represents an activation energy, k is the Boltzmann constant, T is the temperature of the peak position in the dielectric loss curve at the particular measuring frequency and T_f represents the freezing temperature of the relaxation. The Arrhenius type equation should be valid provided that the potential barrier is uniform for all hopping charges and that the charges hop independently of one another. The Vogel–Fulcher equation, however, also takes into consideration the possibility of correlated hopping.

It is noteworthy that the peak in the dielectric loss as a function of frequency is shifted to significantly higher temperatures in the cases of BIN and BSN relative to the other Bi-based pyrochlores *e.g.* T_m at 1 MHz = 211 K for BSN and 200 K for BIN but only 133 K for BNN, 125 K for BFN, 114 K for BMN and ~ 110 K for BZN (see Fig. 4c). The differing temperatures of the peak in the dielectric loss tangent as well its different behaviour below T_m from one Bi-based pyrochlore system to the next suggests the existence of different dipole activation energies, different local structural order and consequently different relaxation mechanisms in these compounds even through they have the same average structure.

2.1 $(Bi_{1.5}Zn_{0.5})(Zn_{0.5}Nb_{1.5})O_7$ (BZN) pyrochlore

2.1.1 Local structure of BZN

Figure 5 shows close to (a) <001>, (b) <118> and (c) <551> zone axis electron diffraction patterns typical of BZN collected at room temperature along with Monte Carlo simulated diffraction patterns at the same zone axis orientations in (d)-(f). No diffuse scattering was observed at either the exact <100> or <110> zone axis orientations. On tilting away from these exact zone axis orientations, however, a characteristic, highly structured and quite reproducible diffuse intensity distribution was always observed, as apparent from Fig.5. Fig.5b, for example, shows an electron diffraction pattern collected by tilting (~10°) away from the <001> orientation keeping a <2-20>* systematic row excited. Likewise, Fig.5c is obtained by tilting ~9° away from the exact <110> zone axis orientation keeping a <2-20>* systematic row excited.

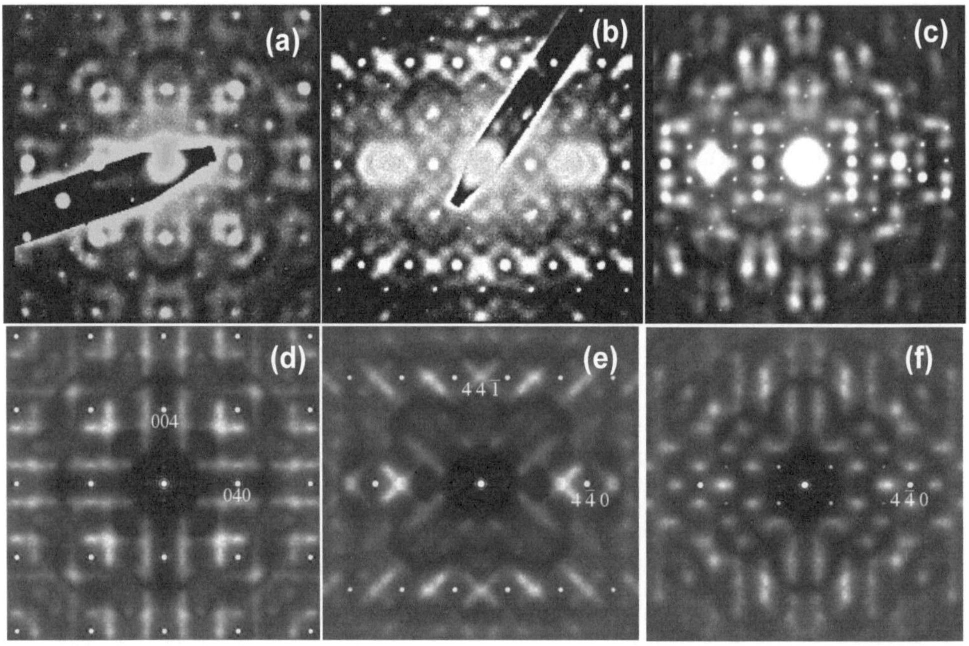

Fig. 5. Typical <001> (a), <118>(b) and <551> (c) zone axis electron diffraction patterns of BZN and corresponding Monte Carlo simulation result (d-f). See detail in ref (Withers et al., 2004).

Note that tilting progressively away from the <001> orientation shows the intensity of the diffuse streaking along the <w01>* and <0w1>* directions of reciprocal space become more limited, suggesting that the diffuse streaking occurs in one-dimensional lines running simultaneously along all three <001>* directions of reciprocal space. The intensity of the <801>* diffuse streaking is strongest along the <801>* direction of reciprocal space itself, however it is virtually absent along the orthogonal direction. This is due to a substantial 'size effect' contribution (Butler et al., 1992) that leads to the asymmetry of the diffuse distribution surrounding Bragg reflections e.g. <044>* in Fig. 5a [see ref (Withers et al., 2004) for details]. Furthermore, the diffuse streaks only run through the G±<001>* type positions and not through the Fd3m allowed Bragg reflections. The reciprocal space positioning of the observed diffuse distribution thus appears to be given by **G**±[10*l*]*, **G**±[*h*10]* and **G**±[0*k*1]* respectively, where **G** represents an allowed parent pyrochlore Bragg reflection. The strong diffuse blobs at the **G**±<001>* positions of reciprocal space (such as <441>* in Fig. 5b) originates from the intersection of two such diffuse streaks at this point. Now the question becomes what is the origin of this structured diffuse intensity distribution.

The bond valence approach is a powerful empirical tool which gives the relationship between the bond length and strength of chemical bonds between cations and anions and can be used for the assessment and prediction of structural stability, in particular disordered local structure. Using this approach, the apparent valence (AV) or bond valence sum (V(*A*)) of a cation A in terms of its bonds to all its coordinating anions X is given by $V(A)=\sum_X S_{A-X}=\sum_x \exp[(R_0-R_{A-X})/b]$ where R_0 and b (= 0.37 Å) are empirical parameters, available from the list in *e.g.* ref (Brese & O'Keeffe, 1991). For this bond valence sum calculation, a nominal stoichiometry of $(Bi_{1.5}Zn_{0.5})(Zn_{0.5}Nb_{1.5})O_7$ is assumed, ignoring the known slight non-stoichiometry (Levin et al., 2002; Withers et al., 2004). The average occupancy of the octahedral B site is thus $^3/_4$ Nb + $^1/_4$ Zn while that of the A site is $^3/_4$ Bi + $^1/_4$ Zn. Consider firstly the B_2O_6, octahedral corner-connected sub-structure (see Fig.2c). For a Nb-centred octahedra, the expected metal-oxygen bond length is then given by R_{Nb-O} = 1.911 - 0.37 ln(5/6) Å = 1.9785 Å using $R_0(Nb^{5+}-O^{2-})$ = 1.911 Å (Brese & O'Keeffe, 1991). Likewise, for a Zn-centred octahedra, the expected metal-oxygen bond length is given by R_{Zn-O} = 1.704 - 0.37 ln(2/6) = 2.1105 Å, using $R_0(Zn^{2+}-O^{2-})$ = 1.704 Å(Brese & O'Keeffe, 1991). The average metal ion-oxygen bond length should then be $^3/_4$ x 1.9785 + $^1/_4$ x 2.1105 = 2.0115 Å which is very close to the refined value of 2.0065 Å (Levin et al., 2002). The two predicted ideal bond lengths are each quite close to the average $R_{B\text{-}O}$ distance, with the ideal Nb-O bond length being slightly smaller than the average and the ideal Zn-O bond length being slightly larger than the average. Therefore, local displacements of the B and O sites away from their refined average positions can be expected to be relatively small whatever the local chemistry. Locally, one would expect Nb^{5+}-centred octahedra to shrink slightly by sucking in the 6 surrounding O ions from the average distance of 2.01 Å to the ideal distance of 1.91 Å while Zn^{2+}-centred octahedron could be expected to expand slightly slightly from the average distance of 2.01 Å to the ideal distance of 2.11 Å. These displacements are so small that we can largely neglect the contribution of the B_2O_6 sub-structure to the observed diffuse intensity.

A	B	AV(A)	AV(B)	AV(O)	AV(O')
Bi	Nb	2.461	4.635	1.969	2.377
Bi	Zn	2.461	2.649	1.307	2.3777
Zn	Nb	0.858	4.635	1.693	0.828
Zn	Zn	0.858	2.649	1.031	0.828
	$[Bi_4]$	$[Bi_3Zn]$	$[Bi_2Zn_2]$	$[BiZn_3]$	$[Zn_4]$
$AV(O')[A_4]$	2.377	1.990	1.603	1.215	0.828
	$[Bi_2Nb_2]$	$[Bi_2NbZn]$	$[BiZnNb_2]$	$[BiZnNbZn]$	
$AV(O)[A_2B_2]$	1.969	1.638	1.831	1.500	

Table 1. Bond Valence Sums (AV) for the BZN average structure, O' and O dependent on $[A_4]$ and $[A_2B_2]$ {Withers, 2004}

Now consider the $O'A_2$ anti-cristobalite substructure (see Fig.2b and Table 1 above). Each O' anion is tetrahedrally coordinated to 4 A cations at an average distance $R_{O'-A}=\sqrt{3}/8\ a$ = 2.287 Å. Using the same type of bond valence argument given above [see ref. (Withers et al., 2004) for details], one would expect an ideal $R_{O'-Bi}$ distance of 2.3505 Å if the O' ion is surrounded by 4 Bi's and an ideal $R_{O'-Zn}$ distance of 1.9605 Å if the O' ion is surrounded by 4 Zn's. Note that this time these two bond lengths differ quite significantly from each other and from the average O'-A distance. Thus considerable local strain relaxation of the $O'A_2$

sub-structure can be expected associated with the local distribution of Bi^{3+} and Zn^{2+} on the *A* sites. This strongly suggests that the disorder and subsequent structural relaxation in the $O'A_2$ substructure is largely responsible for the observed diffuse intensity distribution.

The best way to minimize this local strain relaxation is for the stoichiometry of the $O'A_4$ tetrahedra to have the average stoichiometry wherever possible *i.e.* $O'Bi_3Zn$ so that the 2 O' ions on either side of a central Zn^{2+} ion can simultaneously contract in towards it along a local <111> direction and away from the neighbouring Bi^{3+} ions. From Table 1, the O' ion can be relatively happily bonded only when it is surrounded by 3 Bi's and 1 Zn, *i.e.* from the local crystal chemical point of view it is of paramount importance wherever possible to achieve the average composition of $O'Bi_3Zn$ in each local tetrahedron (see Table 1 and Fig. 6). There are only two distinct types of inter-tetrahedral Zn-Zn separation vectors possible (shown in Figs.6a and b). The first is of $^1/_4$ <112> type (shown in Fig.6a) in which neighbouring O' ions centring the $O'Bi_3Zn$ tetrahedra must shift directly towards the significantly under-bonded Zn ions (solid line arrows). The Bi ions in the same tetrahedron cannot afford to lose the valence contribution from these O' ions, however, and hence will seek to follow the induced O' ion shifts. This induces shifts of the Bi ions perpendicular to the local O'-Bi-O' axis towards two of the six surrounding equatorial O ions (see Fig. 6c). The only other possible type is of $^1/_2$ <110> type Zn-Zn separation. Such a Zn-Zn separation (Fig. 6b), however, does not allow the Bi ions to follow the O' shifts and hence leads to a significantly under-bonded Bi ion and thus is energetically extremely unfavourable (see ref. (Withers et al., 2004) for details). This provides a clear crystal chemical rationale as to why $^1/_4$ <112> type Zn-Zn separation vectors should be strongly favoured and $^1/_2$ <110> type Zn-Zn separation vectors completely avoided in BZN-related pyrochlores. Finally, the difference in size between the Bi^{3+} and Zn^{2+} ions also leads to an additional "size effect" relaxation within each tetrahedron. From Table 1, it is apparent that Bi cations in the pyrochlore *A* site are always underbonded. Thus, when an O' ion moves towards a Zn ion (see Figs. 6a and c), the remaining three Bi cations in the tetrahedron will attempt to follow, consequently shortening the local Zn-Bi separation distance and lengthening the local Bi-Bi separation distance.

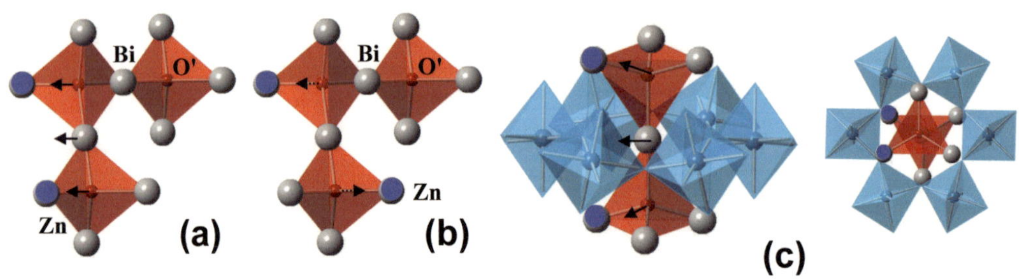

Fig. 6. (a) and (b) two possible arrangements of inter-tetrahedral Zn-Zn separation, and (c) the hexagonal prismatic enviroment of the *A* cations in the pyrochlore average structure type in projection along <1,-1,0> and <111> respectively. Zn ions: blue, Bi: grey. The displacements of the O′ ions (initially centring the two $O'Bi_3Zn$ tetrahedra shown) towards the Zn ions in each tetrahedra are shown by the solid lines while the induced shift of the central Bi ion represented by the dashed line.

Monte Carlo simulation using the local constraints described above was thus carried out [see ref. (Withers et al., 2004) for details] and the corresponding electron diffraction patterns calculated (Figs.5d-f) for comparison with the experimental ones shown in Figs.5a-c. From Fig. 5, it is clear that this local structure model fit very well with the observed electron diffraction data and gives a good idea of the local crystal chemistry. The constraints applied, as described above, were that the stoichiometry of each $O'A_4$ tetrahedron should be $O'Bi_3Zn$ as far as possible, that $^1/_2$ <110> type Zn-Zn separation vectors be avoided as far as possible and finally that the Bi-Zn separation distances within each tetrahedron should be smaller than the average while the Bi-Bi distances should be larger than the average.

In summary, structured diffuse scattering provides clear evidence for crystal chemically sensible, local short range ordering of Bi and Zn ions and associated strain induced structural relaxation in cubic BZN pyrochlores. The experimentally observed diffuse distribution is dominated by the displacive shifts of the Bi and Zn ions on the pyrochore *A* sites of the $O'A_2$ sub-structure induced by local Bi/Zn ordering. While such induced displacive shifts do create local dipole moments (see Fig.6 above), these dipoles arise from local metal ion ordering and would therefore be expected to be static at room temperature (and therefore not responsible for the observed dynamic dielectric relaxation behaviour) unless there are a certain percentage of vacancies on the pyrochlore *A* site enabling the remaining *A* site metal ions to dynamically re-distribute themselves if the energy barriers to such motion are not too high.

2.1.2 'Non-stoichiometry' and the nature of defects in BZN (Liu et al., 2011)

While the above model provides a good fit to the observed structured diffuse scattering of BZN, there remain other sources of potential disorder arising from the intrinsic non-stoichiometry of BZN that should also be considered from the dielectric properties point of view. Levin *et al* (Levin et al., 2002) for example, found the composition of their BZN sample to be $(Bi_{1.5}Zn_{0.42})(Zn_{0.50}Nb_{1.5})O_{\sim6.92}$. We have also synthesized 'non-stoichiometric' BZN samples as reported previously (Withers et al., 2004). It is thus clear that a small percentage of *A* site vacancies (Zn deficiency on the pyrochlore *A* site) and associated oxygen vacancies always co-exist (Levin et al., 2002; Withers et al., 2004; Levin et al., 2002a) in BZN pyrochlores and that the nominally ideal $(Bi_{1.5}Zn_{0.5})(Zn_{0.5}Nb_{1.5})O_7$ composition for BZN does not actually exist. How might this affect dielectric properties? As mentioned above, such non-stoichiometry might allow for local changes in the Bi/Zn/vacancy distribution and associated structural relaxation and hence changes in the local dipole distribution. On the other hand, one would expect any such re-distribution to occur on a rather slow time scale at room temperature.

An alternative possibility to explain dynamic dielectric relaxation behaviour is to consider the possibility of oxygen hopping and associated structural relaxation. This is best understood by considering the ideal pyrochlore structure type as a superstructure of the *Fm3m* fluorite structure type: six oxygens occupy the *48f* sites therein and one is located in the *8a* site per formula unit. The *8b* site of the fluorite structure type, however, is nominally empty in an ideal pyrochlore [see refs. (Heremans et al., 1995; Moon etal., 1988) for details]. The existence of these vacant oxygen sites suggests the possibility of an exchange of an oxygen from the *48f* site to this initially empty *8b* interstitial site (Heremans et al., 1995; Moon et al., 1988). This creates a Frenkel defect denoted by $O_0 \leftrightarrow O_i^{''} + V_o^{\cdot\cdot}$ with a high activation energy between 1.54-1.70 eV(Tan et al., 2009; Clyton et al., 2001) due to the high level of intrinsic oxygen Frenkel disorder induced by the Zn deficiency and associated oxygen vacancies. The activation energy of such Frenkel defects is almost identical to a strong peak observed in the photoluminescence (PL) spectra of BZN (Liu et al, 2011), suggesting a potential link between these two physical phenomena. Such an activation energy level (~1.67 eV) was also observed in (nominally) $Bi_{1.5}ZnSb_{1.5}O_7$, analogous to BZN, by Clayton *et al* (Clayton et al., 2001). The observed PL intensity is significantly reduced with increasing temperature due to the common PL quenching effect where non-radiative recombination is thermally activated and results in a decrease in the PL intensity. Once such a recombination occurs, the defects inducing the dipole moments vary and thus contribute to the observed dynamic dielectric relaxation. On the other hand, in the dilute limit, the concentration of defects then increases as a function of temperature (Kohan et al., 2000), which probably leads to an increase in the total number of defect-induced dipole moments and consequently to changes in the dielectric constant. This coincides with the trend observed in the dielectric-temperature spectra of BZN (Kamba et al., 2002). Volanti *et al* (Volanti et al., 2007) also suggest that oxygen vacancies and recombination of electrons and holes in the valence band lead to the formation of $[NbO_5 \cdot V_O^x]$, $[NbO_5 . V_O^{\cdot}]$ and $[NbO_5 \cdot V_O^{\cdot\cdot}]$ *etc* complex clusters in the octahedral framework as donor, donor-acceptor and acceptor defects. The oxygen complex clusters generate localized states in the band gap and inhomogeneous charge distribution in the cell, and thus contribute to dielectric relaxation.

Taking all of these observations into account, it is apparent that Frenkel defects and associated defect clusters have a significant impact on the dielectric relaxation behaviour of BZN pyrochlores. Such Frenkel defects not only occur at high temperature but also can exist at much lower temperature if the BZN has an intrinsically significant numbers of vacancies on the *A* and O' sites.

2.2 Bi_2MNbO_7 (*M*=Fe^{3+}, In^{3+} and Sc^{3+}) pyrochlores

Now consider bismuth-rich niobate pyrochlores of nominal stoichiometry $Bi_2M^{III}NbO_7$ e.g. Bi_2FeNbO_7, Bi_2InNbO_7 and Bi_2ScNbO_7. Electron probe microanalysis showed the chemical compositions of the resultant compounds to be $Bi_{1.89}Fe_{1.16}Nb_{0.95}O_{6.95}$ (BFN), $Bi_{1.98}In_{0.99}Nb_{1.01}O_7$ (BIN) and $Bi_{1.99}Sc_{0.96}Nb_{1.03}O_7$ (BSN) respectively. BIN and BSN are thus cubic pyrochlores with only Bi in the pyrochlore *A* sites while BFN appears slightly Bi-deficient and to require a small amount (~5%) of Fe^{3+} ions on the pyrochlore *A* site as well as the Bi ions.

Figure 7 shows typical (a) <11-2> and (b) <22-3> zone axis EDPs of the $Bi_{1.89}Fe_{1.16}Nb_{0.95}O_{6.95}$ (BFN) sample. Note the presence of a highly structured diffuse intensity distribution, this time in the form of transverse polarized diffuse streaking in reciprocal space running through particular parent Bragg reflections perpendicular to each of the <110> directions of real space *e.g.* along <3-11>*, <1-3-1>* and <111>* in (a) and [10,-4,4]*, [-4,10, 4]* and [668]* in (b) [see ref. (Somphon et al., 2006) for detail]. Diffuse streaking of this type forms part of **G**±{110}* sheets of diffuse intensity in reciprocal space running perpendicular to each of the <110> directions of real space. Note that the diffuse streaking along the [668]* direction in Fig. 7b runs through the [*hkl*]*, $h-k = 4J$, *J* an integer, parent Bragg reflections e.g. [2-20]* but not through reflections [424]*. Likewise the diffuse streaking running along the [10, -4, 4]* direction runs through the [*hkl*]*, $k+l = 4J$, e.g. [244]* but not through reflections [424]*. Such characteristic pseudo-extinction conditions are characteristic of anti-cristobalite-type, orientational disorder of the $O'Bi_4$ tetrahedra (Tabira et al., 2001; Withers et al., 1989) that make up the $O'A_2$ sub-structure of the ideal pyrochlore structure type because of the correlated displacement of heavily scattering Bi ions separated by 1/4 <110> [see Fig. 8a and ref. (Somphon et al., 2006) for more detail].

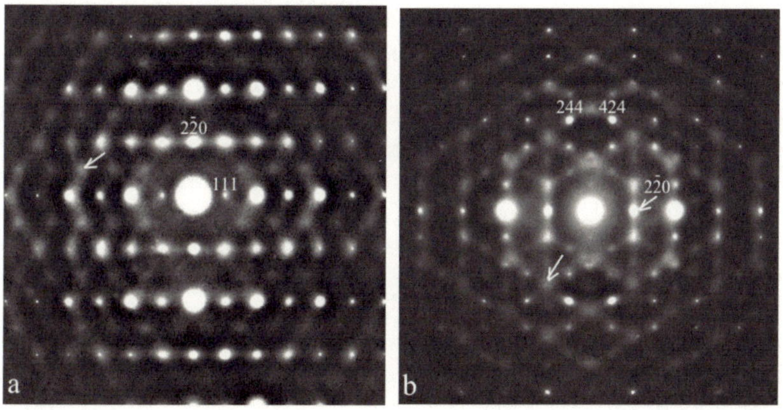

Fig. 7. Typical <11-2> and <22-3> zone axis electron diffraction patterns of BFN. See ref. (Somphon et al., 2006) for details.

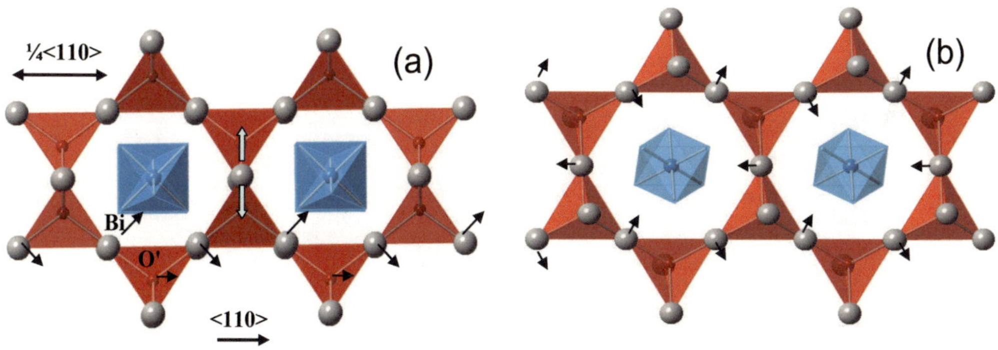

Fig. 8. (a) The characteristic anti-cristobalite-type displacive disorder of the O'Bi_2 sub-structure of the ideal pyrochlore structure type. The rotation of any one O'Bi_4 tetrahedron around the appropriate <1-10> tetrahedral edge automatically constrains all the tetrahedra in that particular <110> column to rotate but does not constrain the sense of tetrahedral rotation from one such <110> column to the next. Blue arrows indicates this Bi does not move. (b) the pattern of correlated O'Bi_4 tetrahedral rotation around the <111> axis responsible for the observed <111>* rods of diffuse intensity. Note that rotation of any O'Bi_4 tetrahedron around <111> automatically constrains all the tetrahedra in that particular <111> plane to rotate as shown in (a) but does not constrain the sense of tetrahedral rotation from one such <111> plane to the next. Note that each Bi ion moves essentially directly towards two of the O anions bonded to the B cations and directly away from two others.

The arrowed regions in Fig.7 show that virtual 'satellite reflections' in the form of quite distinct "peaks" in diffuse intensity arise whenever three diffuse streaks perpendicular to <110> intersect, e.g., at $\mathbf{G}\pm{}^1/_2[3\text{-}11]^* \equiv \mathbf{G}'\pm 1/2[\text{-}1\text{-}11]^*$ in Fig. 7a, and at $\mathbf{G}\pm(2J/7)[668]^* \equiv \mathbf{G}'\pm(2J/7[\text{-}1\text{-}11]^*$ in Fig. 7b (J an integer). Such peaks in the diffuse intensity are part of essentially continuous 1-D rods of diffuse intensity running along the $\mathbf{G} \pm$ <111>* directions of reciprocal space and imply the existence of correlated {111} planes of Bi displacements (shown in Fig. 8b) that are uncorrelated from one such {111} plane to the next.

The 'extinction conditions' characteristic of the {110}* sheets of diffuse intensity are also characteristic of the <111>* rods of diffuse intensity. This suggests that the displacive disorder responsible for these <111>* rods of diffuse intensity arises from an appropriate linear combination of tetrahedral edge rotations of the type shown in Fig.8a involving Bi displacements along the three <1-10> real space directions perpendicular to the particular <111>* rod of diffuse intensity and leading to resultant tetrahedral rotation around the <111> axis (as shown in Fig. 8b). According the analysis above, the mechanism adopted for local structural distortion away from the ideal pyrochlore structure type in BFN is clearly tetrahedral edge rotation of the $O'A_2$ sub-structure, either of the individual type shown in Fig.8a and/or of the correlated type shown in Fig. 8b.

Rather similar structured diffuse intensity distributions are also observed in the cases of BIN and BSN. Figure 9, for example, shows typical (a) <001> and (b) <11-2> zone axis electron diffraction patterns of BIN. Note the characteristic transverse polarized diffuse streaking perpendicular to the six <110> directions of real space *e.g.* along <220>* and <-220>* in (a) and along <3-11>* and <1-3-1>* in (b) *i.e.* the observed diffuse distribution again takes the form of sheets of diffuse intensity perpendicular to the six <110> directions of the average structure. As above, this again implies the existence of 1-D tetrahedral edge rotation of the $O'A_2$ tetrahedral sub-structure in BIN, in which the $O'Bi_4$ tetrahedra rotate as essentially rigid bodies with the O' remaining in the center of the tetrahedra (see Fig.8a above).

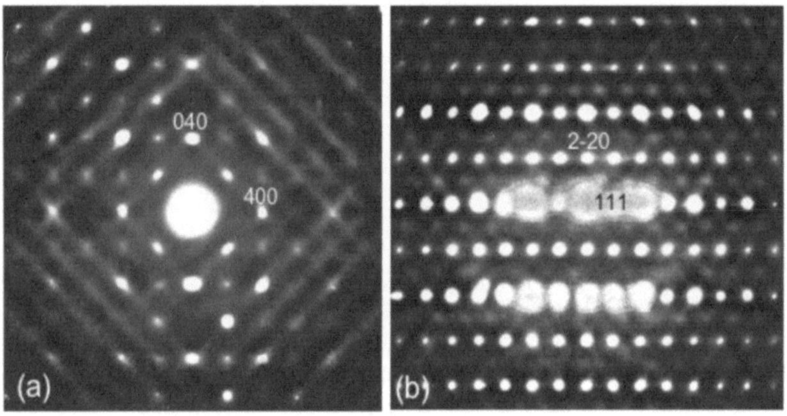

Fig. 9. Typical (a) <001> and (b) <11-2> zone axis electron diffraction patterns of BIN.

Figure 10 shows <-3,6,-1> (a) and (b) <5,1,-2> zone axis electron diffraction patterns of BSN. Again the characteristic, transverse polarized diffuse streaking running along a <*h*,-*h*, *l*>* direction of reciprocal space perpendicular to one or other of the six <110> directions of reciprocal space is clearly apparent *i.e.* the well-defined **G**±{110}* sheets of diffuse intensity also exist in the case of BSN as does the 1-D tetrahedral edge rotation of the $O'A_2$ tetrahedral sub-structure responsible.

According to the bond valence sum calculations based on the average structures of BFN, BIN and BSN (*e.g.* in Somphon et al., 2006), the O' anion co-ordinated to four Bi cations in the $O'A_2$ tetrahedral sub-structure is always over-bonded. The local O'-Bi bond lengths are thus happy to be slightly expanded as is achieved by large amplitude tetrahedral rotation of the type shown in Fig.8a. On the other hand, the significantly under-bonded Bi^{3+} ion cannot afford to lose much of the contribution to its bond valence arising from its two surrounding

O' ions. Therefore, the Bi ions must displace orthogonal to their local O'-Bi-O' axis and the $O'Bi_4$ tetrahedra must rotate and translate essentially as rigid units, as necessitated by the tetrahedral edge rotation patterns shown in Fig.8.

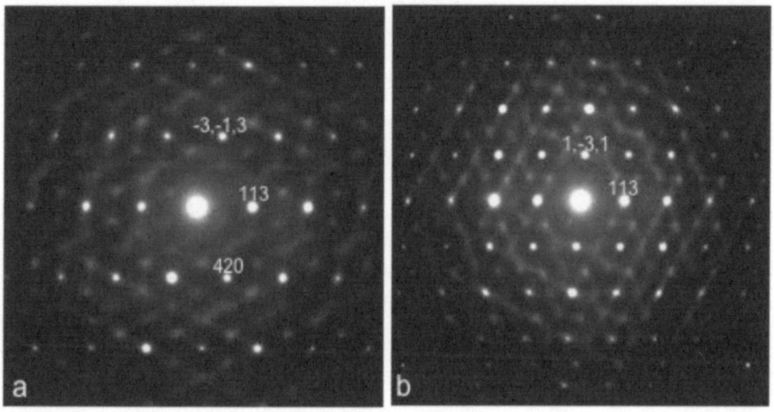

Fig. 10. Typical (a) <-36-1> and (b) <51-2> zone axis electron diffraction patterns of BSN.

The characteristic structured diffuse scattering observed in Figs. 7, 9 and 10 above (in the cases of BFN, BIN and BSN) arise from β-cristobalite-like, 1-d correlated rotations and associated translations along <110> of chains of corner-connected $O'Bi_4$ tetrahedra, as shown in Fig.8a above. It is important to note that the latter rigid body translation along <110> of a nominally +4 charged $O'Bi_2$ column of tetrahedra relative to the nominally -4 charged $M^{3+}Nb^{5+}O_6$ octahedral sub-structure, leads to a net dipole moment along the relevant <110> direction. The lack of transverse correlation from one such <110> chain to the next gives rise to the **G**±{110}* sheets of diffuse intensity. The observed structured diffuse distributions imply that such dipole moments exist along all six <110> directions, the local competition between the different possible polar directions presumably giving rise to the dielectric relaxation behaviour at low temperature and preventing the onset of long range hence ferroelectric ordering even at the lowest possible temperatures. By fitting experimental data to the measured temperature dependence of the dielectric spectra of BSN and BIN (Liu et al., 2009), reasonable activation energies, E_a, of 0.071 (0.072) eV and attempt jump frequencies of 6.48 (6.34) THz were obtained for BSN (BIN) respectively. We thus believe these intrinsically very low energy, correlated, chain rotation modes provide a key to understanding the structural origin of the dipoles as well as the slow dielectric relaxation behaviour of these Bi-based pyrochlores. On the other hand, it is noteworthy that the non-stoichiometry of BFN, like BZN, probably produces defect-induced dielectric relaxation behaviour like BZN.

2.3 BNN and BMN pyrochlores

Finally consider the cases of BNN and BMN of nominal stoichiometry $Bi_{1.5}MgNb_{1.5}O_7$ for BMN and $Bi_{1.5}NiNb_{1.5}O_7$ for BNN. Careful phase analysis shows that both of these systems are distinctly 'non-stoichiometric' with the as-synthesized materials having compositions of $(Bi_{0.833}Mg_{0.11}\square_{0.06})_2(Mg_{0.24}Nb_{0.76})_2O_7$ and $(Bi_{0.833}Ni_{0.125}\square_{0.04})_2(Ni_{0.25}Nb_{0.75})_2O_7$ (□ a vacancy) assuming no oxygen vacancies. In both cases, the Ni or Mg ions must occupy both the pyrochlore *A* and *B* sites while there are also significant concentration vacancies on the *A* sites, like BZN.

One might therefore expect a structured diffuse distribution like that observed above for BZN. Instead of this, in both cases structured diffuse distributions quite similar to those seen for BIN, BSN and BFN were observed *i.e.* transverse polarized diffuse streaking along the <*h*,-*h*,*l*>* directions of reciprocal space perpendicular to the six <110> directions of real space. Fig.11, for example, shows typical (a) <2,2,-3> and (b) close to <1,1,-2> zone axis electron diffraction patterns of BNN and BMN respectively. The observed highly structured diffuse distributions in both cases have distinct similarities to the structured diffuse distribution shown to be characteristic of BFN (Fig. 7), BIN (Fig.9) and BSN (Fig.10), and coincidentally, quite dissimilar to that characteristic of BZN.

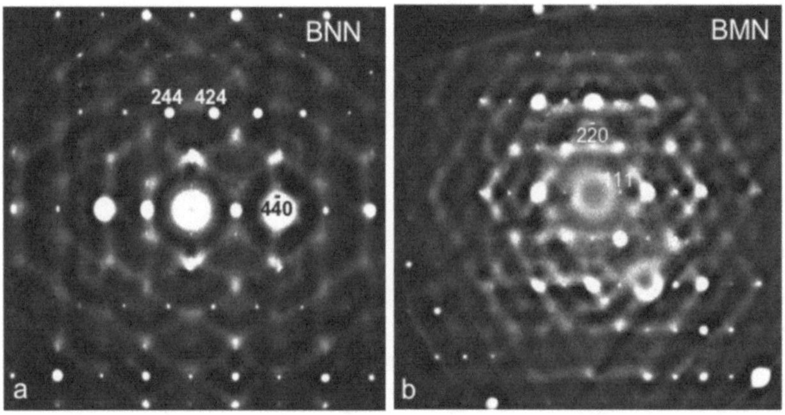

Fig. 11. Typical (a) <22-3> and (b) close to <11-2> zone axis electron diffraction patterns of BNN and BMN respectively.

Clearly the same β-cristobalite-like, 1-d correlated rotations and associated translations along <110> of chains of corner-connected $O'A_4$ tetrahedra must also be taking place in both BNN and BMN. However, when the concentration of the small, heavily under-bonded *M* = Mg or Ni on the pyrochlore A site is much more like 11–12% rather than the ~ 21% for BZN and where the concentration of □'s on the *A* sites is ~5%, the explanation for the observed highly anisotropic displacive disorder on the *A* site is not so obvious. The presence of ~5% □'s on the *A* sites, complicates matters considerably. When an O' ion is surrounded by either 4 Bi's of 3 Bi's and one Mg or Ni ion (the most likely local configurations), the centreing O' ion is either happily bonded or over-bonded implying that the significantly under-bonded Bi and heavily underbonded Mg or Ni ions cannot improve their under-bonding by moving closer to the O' ions but must instead move perpendicular to the local O'-A-O'-axis. When an O'ion is surrounded by 3 Bi's and one □ (a reasonably common likelihood), however, the centreing O' ion will be under-bonded and hence will move away from the □ and towards the remaining three Bi ions.

3. Chapter summary

All the above bismuth-based niobate pyroclores have the same displacively disordered *Fd-3m* average structure, as shown in Fig.2 above. The local structure of BZN as revealed by structured diffuse scattering, however, is quite different to the others. Two types of local order are observed in the studied pyrochlores:

1. Local (Bi/Zn) ordering on the O'A_2 sub-structure and associated displacive relaxation has been shown to give rise to the observed **G**±[10*l*]*, **G**±[*h*10]* and **G**±[0*k*1]* diffuse distribution. This diffuse distribution, however, is only observed in the case of BZN. One would not expect the metal ions to be mobile at room temperature, however, and hence this short range order is most probably not responsible for the dielectric relaxation that occurs at low temperature.
2. For the rest of the pyrochlores studied, **G**±{110}* sheets of diffuse intensity perpendicular to each of the six <110> directions of real space were observed implying that the same β-cristobalite-like, 1-d correlated rotations and associated translations along <110> of chains of corner-connected $O'A_4$ tetrahedra are taking place. It is important to note that the latter rigid body translation along <110> of a nominally +4 charged O'Bi_2 column of tetrahedra relative to the nominally -4 charged $M^{3+}Nb^{5+}O_6$ octahedral sub-structure, leads to a net dipole moment along the relevant <110> direction.
3. The observed {110}* diffuse sheets imply that these dipole moments exist along all six <110> directions simultaneously. The competition between these different local dipoles and their lifetimes may well be responsible for the observed dynamic dielectric relaxations. The low temperature peak in the dielectric loss tangent at T_m depends on the Bi content on the pyrochlore *A* site. The higher the Bi content, the higher T_m *i.e.* $T_{BZN}<T_{(BNN\ and\ BMN)}<T_{BFN}<T_{(BIN\ and\ BSN)}$.
4. The observed structured diffuse distribution is likely a widespread phenomenon in Bi-based pyrochlores and arises from their particular local crystal chemistry. The ratio of *M* to Bi on the pyrochlore A site plays an important role. The dielectric relaxation properties of such Bi-based niobate pyrochlores are dependent upon this local crystal chemistry.

4. Acknowledgment

The authors acknowledge financial support from the Australian Research Council (ARC) in the form of ARC Discovery Grants.

5. References

Brese N.E. and O'Keeffe M., (1991), Bond-valence paramters for solids, *Acta Cryst.B*, Vol 47, No 2, PP192. DOI: 10.1107/S0108768190011041

Butler B.D., Withers R.L. and Welberry T.R., (1992), Diffuse absences due to the atomic size effect, *Acta Cryst. A*, Vol 48, PP737. DOI: 10.1107/S0108767392003131

Clayton J., Takamura H., Metz R., Tuller H.L. and Wuensch B.J., (2001), The Electrical and Defect Properties of $Bi_3Zn_2Sb_3O_{14}$ Pyrochlore: A Grain-Boundary Phase in ZnO-Based Varistors, *J. Electroceram.*, Vol 7, PP113.

Heremans C., Wuensch B.J., Stalick J.K. and Prince E., (1995), Fast-Ion Conducting $Y_2(Zr_yTi_{1-y})_2O_7$ Pyrochlores: Neutron Rietveld Analysis of Disorder Induced by Zr Substitution, *J. solid state chem.*, Vol 117, No 1, PP108.

Kamba S., Porokhonskyy V., Pashkin A., Bovtun V., Petzel J., Nino J. Trolier-McKinstry C., S., Lanagan M.T., and Randall C.A., (2002), Anomalous broad dielectric relaxation in $Bi_{1.5}Zn_{1.0}Nb_{1.5}O_7$ pyrochlore, *Phys. Rev. B* , Vol 66, No 5, PP054106. DOI: 10.1103 /PhysRevB.66.054106

Kohan F., Ceder G., Morgan D. and Walle C.G.V.d., (2000), First-principles study of native point defects in ZnO, *Phys. Rev.B*, Vol 61, No 22, PP15019. DOI: 10.1103 /PhysRevB.61.15019

Levin I., Amos T.G., Nino J.C., Vanderah T.A., Randall C.A. and Lanagan M.T., (2002), Structural Study of an Unusual Cubic Pyrochlore $Bi_{1.5}Zn_{0.92}Nb_{1.5}O_{6.92}$, *J. Solid State Chem.*, Vol 168, No 1, PP69. doi:10.1006/jssc.2002.9681

Levin I., Amos T.G., Nino J.C., Vanderah T.A., Reaney I.M. and Randall C.A., (2002), Crystal Structure of the Compound $Bi_2Zn_{2/3}Nb_{4/3}O_7$, *J. Mater. Res.*, Vol 17, No 6, PP1406.

Liu Y., Withers R. L., Nguyen B. and Elliott K., (2009), Displacive disorder and dielectric relaxation in the stoichiometric bismuth-containing pyrochlores, $Bi_2M^{III}NbO_7$ (*M*=In and Sc), *J. Solid State Chem.*, Vol 182, No 10, PP2748. doi:10.1016 /j.jssc.2009.07.007

Liu Y., Withers R.L. and Wei X.Y., (2005), Structurally frustrated relaxor ferroelectric behavior in $CaCu_3Ti_4O_{12}$, *Phys. Rev. B*, Vol 72, No 13,PP134104. doi: 10.1103/ PhysRevB.72.134104

Liu Y., Withers R.L., Chen H., Li Q., and Tan H., (2011), Raman spectra, photoluminescence and dielectric relaxation in $Bi_{1.5}ZnNb_{1.5}O_7$ pyrochlore, *Curr. App. Phys.*, Vol xx, PP xxxx. doi:10.1016/j.cap.2011.03.014

Moon P.K. and Tuller H.L., (1988), Ionic conduction in the $Gd_2Ti_2O_7$–$Gd_2Zr_2O_7$ system , *Solid State Ionics*, Vol 28-30, No 1, PP470.

Nino J.C., Lanagan M.T., and Randall C.A., (2001), Dielectric relaxation in Bi_2O_3–ZnO–Nb_2O_5 cubic pyrochlore, *J. Appl. Phys.*, Vol 89, No 8, PP4512. doi:10.1063/1.1357468

Nguyena. H.B., Norén L., Liu Y., Withers R.L., Wei. X.Y. and Elcombe M.M., (2007a), The disordered structures and low temperature dielectric relaxation properties of two misplaced-displacive cubic pyrochlores found in the Bi_2O_3–$M^{II}O$–Nb_2O_5 (M=Mg, Ni) systems, *J. Solid State Chem.*, Vol 180, No 9, PP2558. doi:10.1016 /j.jssc.2007.07.003

Nguyen B., Liu Y., and Withers R.L., (2007b), The local crystal chemistry and dielectric properties of the cubic pyrochlore phase in the Bi_2O_3-M^{2+}O-Nb_2O_5 (M^{2+}=Ni^{2+} and Mg^{2+}) systems, *J. Solid State Chem.*, Vol 180, No 2, PP549. doi:10.1016 /j.jssc.2006.10.039

Somphon W., Ting V., Liu Y., Withers R.L., Zhou Q. and Kennedy B.J., (2006), Local crystal chemistry, structured diffuse scattering and the dielectric properties of $(Bi_{1-x}Y_x)_2(M^{III}Nb^{V})O_7$ (M=Fe^{3+}, In^{3+}) Bi-pyrochlores, *J. Solid State Chem.*, Vol 179, No 8 PP2495. doi:10.1016/j.jssc.2006.04.046

Tabira Y., Withers R.L. , Yamada T. and Ishizawa N. , (2001), Annular dynamical disorder of the rare earth ions in a $La_2Zr_2O_7$ pyrochlore via single crystal synchrotron X-ray diffraction, *Z. Krist.*, Vol 216, No 2, PP92.

Tan K.B., Khaw C.C., Lee C.K., Zainal Z., Tan Y.P. and Shaari H., (2009), High temperature impedance spectroscopy study of non-stoichiometric bismuth zinc niobate pyrochlore, *Materials Science-Poland*, Vol 27, No 3, PP825.

Volanti D.P., Cavalcante L.S., Paris E.C., Simões A.Z., Keyson D., Longo V.M., Figueiredo A.T.d., Longo E., and Varela J.A., (2007), Photoluminescent behavior of $SrBi_2Nb_2O_9$ powders explained by means of *β*-Bi_2O_3 phase, *Appl. Phys.Lett.*, Vol 90, No 26, PP261913. doi:10.1063/1.2753114

Withers R.L., Welberry T.R., Larsson A.-K., Liu Y., Norén L., Rundlöf H. and Brink F. J., (2004), Local crystal chemistry, induced strain and short range order in the cubic pyrochlore $(Bi_{1.5-\alpha}Zn_{0.5-\beta})(Zn_{0.5-\gamma}Nb_{1.5-\delta})O_{(7-1.5\alpha-\beta-\gamma-2.5\delta)}$ (BZN) , *J. Solid State Chem.*, Vol 177, No 1, PP231. doi:10.1016/j.jssc.2003.07.005

Withers, R.L., Thompson J.G. and Welberry T.R., (1989), The structure and microstructure of α-cristobalite and its relationship to β-cristobalite, Phys. Chem. Miner., Vol 16, No 6, PP 517.

6

Diffuse Dielectric Anomalies in $CoTiO_3$ at High Temperatures

C. C. Wang, M. N. Zhang, G. J. Wang, and K. B. Xu
*Laboratory of Dielectric Functional Materials, School of Physics and Material Science, Anhui University, Hefei
P. R. China*

1. Introduction

The ABO_3-type materials with ilmenite structure have been studied extensively as functional inorganic materials due to their weak magnetism and semiconductivity. They find wide applications such as pigment, gas sensor for ethanol, high performance catalysts, electrodes of solid oxide fuel cells, and microwave dielectric resonator, etc [1-7]. As a typical ilmenite structure material, Cobalt titanate, $CoTiO_3$ (CTO), has received considerable research interest in recent years due to its dielectric properties [8]. This material appears to be a promising high-k dielectric material in semiconductor devices like Metal-Oxide-Semiconductor Field-Effect Transistors and Dynamic Random Access Memories [9-11]. This promising prospect requires a full characterization of the dielectric properties of CTO in the case of both thin film and ceramics around room temperature. In our previous work [8], we reported the low-frequency (10^2~10^5 Hz) dielectric properties of CTO ceramics in the temperature range between 130 and 430 K. Two dielectric relaxations were found with the low-temperature one ascribed to the dipolar effect induced by charge-carrier hopping motions and the high-temperature one related to the defect dipolar polarization created by Co vacancies and Ti^{3+} ions. This work revealed some interesting dielectric features of CTO at low temperatures. Nevertheless, the high-temperature dielectric properties of this material have not been reported so far. Hence, the aim of this work is to investigate, in details, the dielectric properties of CTO ceramics at high temperatures ranging from room temperature to 800 ^oC. Two diffuse dielectric anomalies were found and their physical natures were discussed.

2. Experimental details

Single phased CTO ceramic samples used for dielectric measurements were prepared by solid-state reaction using high purity (99.99%) starting powders of Co_3O_4 and TiO_2. Details about the sintering processes were reported in our preceding paper [8]. The purity of the resultant compound was examined by powder X-ray diffraction (XRD) on a XD3 diffractometer. The structural properties of CTO at high temperatures were measured from room temperature to 900 ^oC using a MXP18AHF system with a high-temperature attachment and analyzed using Jade refinement. To modify the dielectric properties of CTO ceramics, excess Co_3O_4 powder was added into the resultant CTO powder. The mixture was thoroughly ground, palletized, and sintered at 1050 °C for 20 h. Annealing treatments were performed in flowing (200 ml/min) O_2 and N_2 (both with purity >99.999%) at 800 °C for 2 h. The temperature-dependent dielectric properties were obtained using a Wayne Kerr 6500B precise impedance analyzer with the sample mounted in a holder placed inside a PST-2000HL dielectric measuring system. The temperature variations were automatically controlled using a Stanford temperature controller with a heating rate of 3 °C/min. The system can provide a high temperature range from room temperature to 1000 °C. The ac measuring signal was 100 mV rms. Electrodes were made by printing Pt paste on both sides of the disk-type samples and then fired at 800 °C for 1 h in order to remove the polymeric component.

3. Results and discussion

Figure 1 shows a representative diagram of the variation in dielectric constant (ε', the real part of the complex permittivity) with temperature at various frequencies for an as-prepared CTO pellet (with diameter of ~ 12 mm and thickness of 1.82 mm). At first glance, $\varepsilon'(T)$ exhibits two diffuse dielectric anomalies located at 250 °C and 600 °C, respectively. Both anomalies are characterized by: (i) a broad peak in $\varepsilon'(T)$, with the peak getting smaller and shifting to higher temperature as the measuring frequency increases; and (ii) the low-temperature side of the peak shows strong frequency dispersion, whereas the high-temperature side almost exhibits frequency independence.

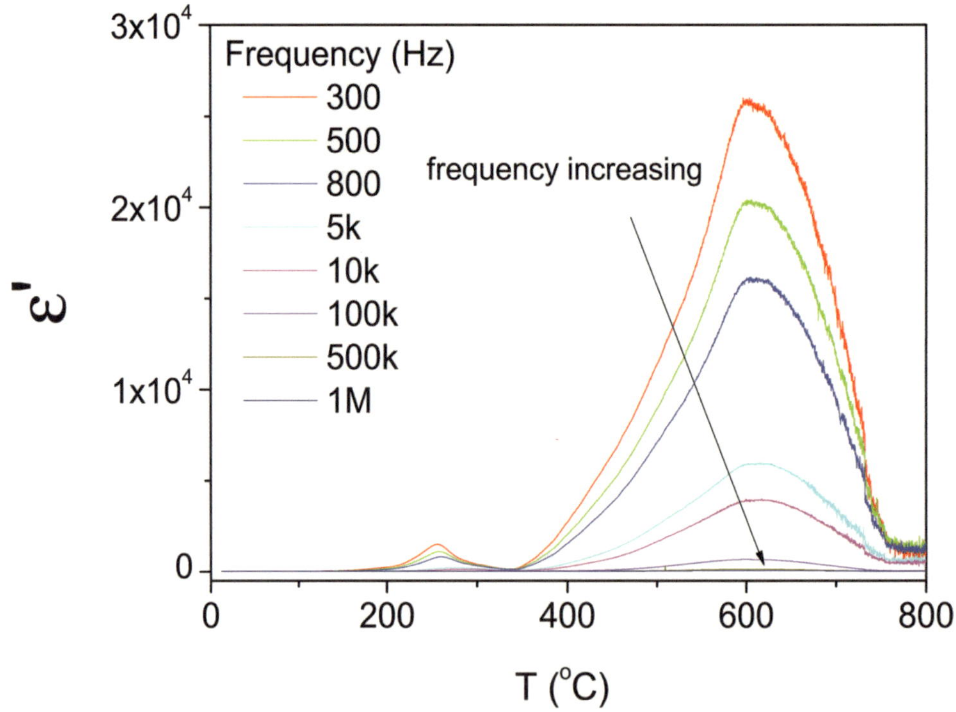

Fig. 1. Temperature dependence of dielectric constant at various frequencies for an as-prepared CTO ceramic sample. By CC Wang et al.

These behaviors are similar to those of relaxors [12]. It seems that the sample might be a relaxor, and therefore, the anomalies are expected to be related to ferroelectric phase transitions. To check out this possibility, we conducted high-temperature XRD measurements. Figure 2 presents the XRD patterns measured at several temperatures. Besides the fact that all the reflections move to lower two theta values with increasing temperature due to thermal expansion, no alien reflections were detected. This indicates that the ilmenite structure maintains to at least the highest measuring temperature of 900 °C. The Jade refinements reveal that the lattice parameter *a* shown in Fig. 3 exhibits two linear lines with an inflection point of 662 °C, suggesting that CTO undergoes a structural phase transition at the inflecting temperature. This transition temperature is much higher than that of the high-temperature anomaly indicating that the observed anomalies could not be associated with paraelectric-ferroelectric phase transitions, and also suggesting other mechanism underlies the anomalies.

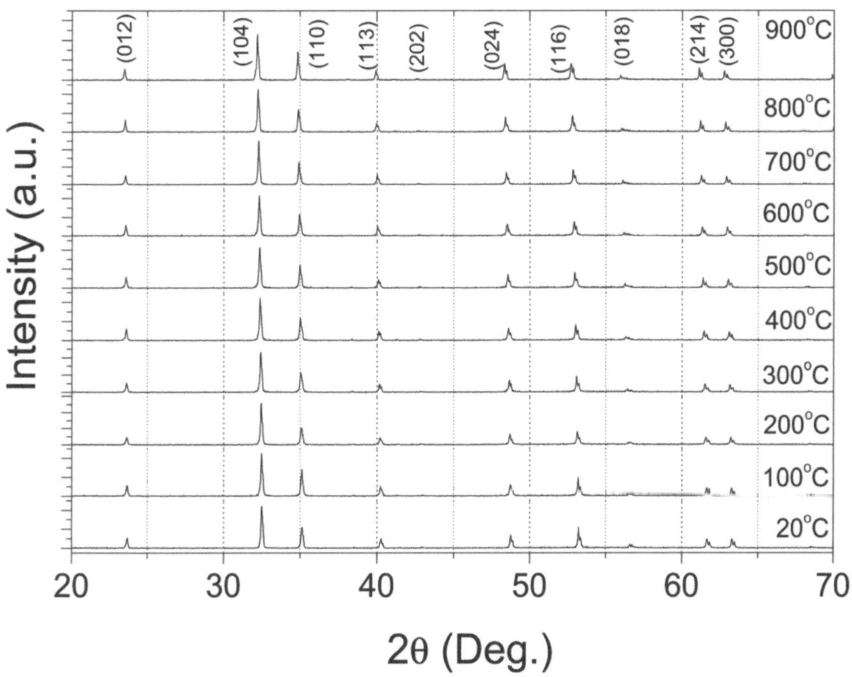

Fig. 2. XRD patterns of CTO sample at several temperatures from room temperature to 900 °C. By CC Wang et al.

To date, several mechanisms unrelated to true relaxor that can produce a relaxor-like anomaly have been proposed. These mechanisms can be classified into three types: (i) the dipole model associated with different mobile defects based on the universal feature that the anomaly is very sensitive to the oxygen vacancy especially for the titanate perovskites [13-15]; (ii) the Maxwell-Wagner model due to electrical inhomogeneity in the tested sample [16-19]; and (iii) the competitive phenomenon between the dielectric relaxation and the electric conduction of the relaxing species [20-22]. Before clarifying which mechanism underlies the observed anomalies, details about the nature of these anomalies are required. We thus take a careful examination of both ε' and loss tangent ($\tan\delta = \varepsilon''/\varepsilon'$, where ε'' is the imaginary part of the complex permittivity) at 300 Hz in a half-logarithmic representation as displayed in Fig. 4.

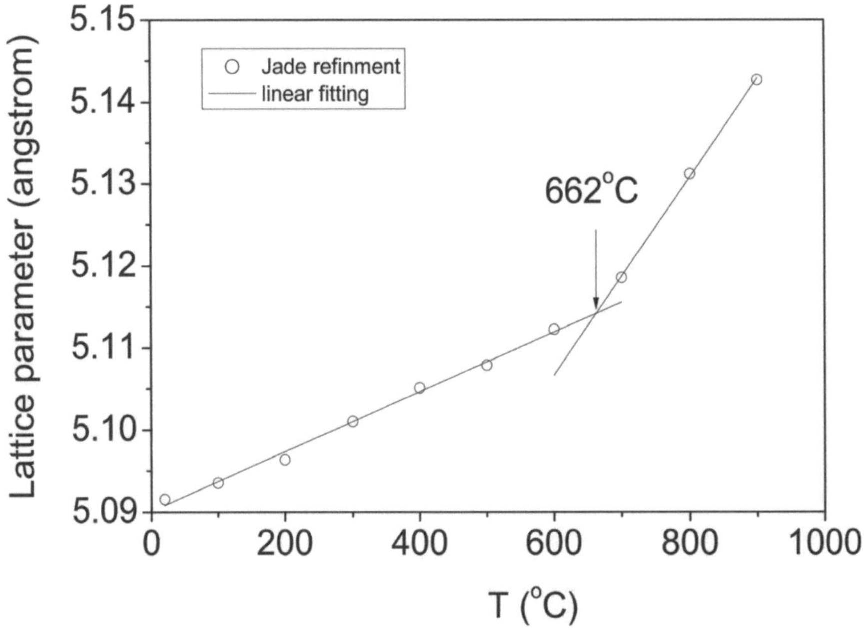

Fig. 3. Lattice parameter as a function of temperature for CTO obtained from the Jade refinements. The inflection point was observed at 662 °C. The straight lines are linear fitting results.By CC Wang et al.

From which some complicated features can be seen: (1) There are two steplike increases in the low-temperature side for each anomaly. The beginnings of the steplike increases were indicated by vertical lines and termed as A, B, C, and D from low- to high-temperature. (2) Corresponding to each beginning of the steplike increase, there exists a peak in loss tangent. Although the peak in $\tan\delta$ for D-steplike increase is invisible because of the increasing background as will be mentioned in the following feature, this peak will be well developed as later discussed by thinning the sample. (3) Corresponding to each anomaly, the $\tan\delta$ increases rapidly with increasing temperature. This feature is much more pronounced in the curve of $\varepsilon''(T)$ (not shown here). Since a steplike increase in $\varepsilon'(T)$ accompanied by a peak in $\tan\delta(T)$ at the beginning temperature of the increase (or by a peak in $\varepsilon''(T)$ at the middle temperature of the increase) is a hallmark for a Debye-type relaxation caused by dipolar or Maxwell-Wagner polarization. This implies that each anomaly contains two Debye-type relaxations. A diffuse dielectric anomaly consisted of two peaks in a thinning sample had been already reported by Stumpe et al in the single crystal of $BaTiO_3$ [16]. To identify whether the high-temperature anomaly in CTO is composed of two relaxations, we conducted dielectric measurements after two consecutive thinning processes by polishing the pellet evenly from both sides. After each process, dielectric properties were measured as a function of temperature. The results after each process as well as the result of the as-prepared sample are depicted in Fig.5 for comparison. Compared with the as-prepared sample, it can be clearly seen that the thinned sample exhibits distinct two steplike increases for both anomalies in the curve of $\varepsilon'(T)$ (Fig.5 (a)). Correspondingly, two peaks in loss tangent related to the increases can be well identified (Fig.5 (b)). This result indicates that each anomaly in CTO is truly composed of two relaxations.

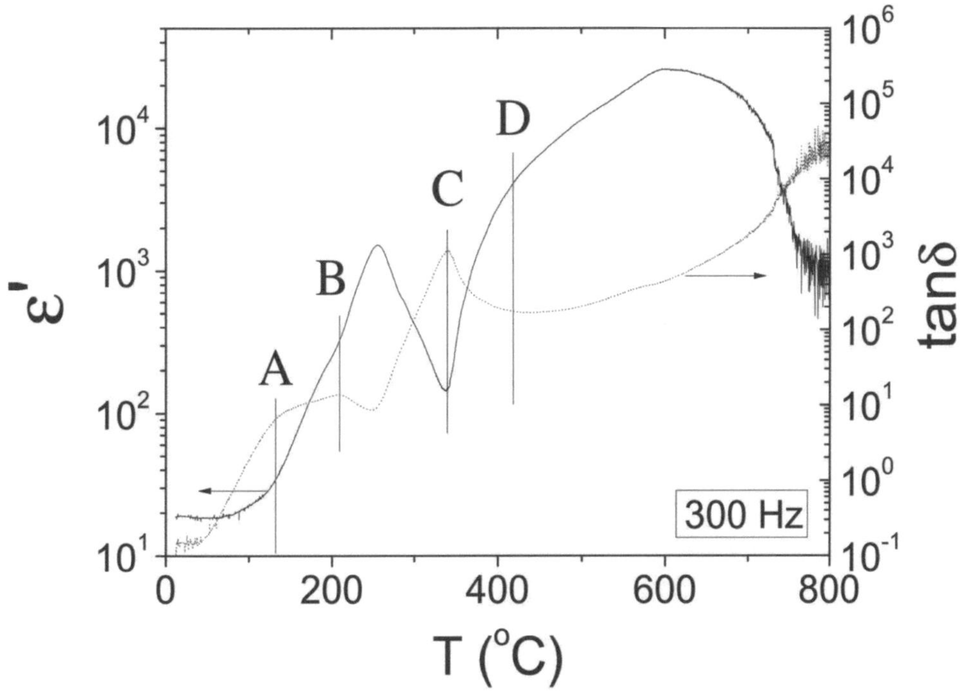

Fig. 4. Temperature dependences of ε' and $\tan\delta$ measured at 300 Hz shown in a half logarithmic scale, which clearly shows some complicated features. By CC Wang et al.

Relaxor-like anomaly was widely reported in various materials [23]. There is a wide consensus that the anomaly appeared in the temperature range of 400 - 900 °C in oxide materials, especially for those containing titanium, is related to oxygen vacancies [13]. This suggests that the high-temperature anomaly in CTO may be related to oxygen vacancy. In order to confirm this inference, the as-prepared sample used in Fig.1 was annealed firstly in O_2 and then in N_2 atmospheres. After each annealing treatment, dielectric properties were measured as a function of temperature. Figure 6 compares the results of $\varepsilon'(T)$ at 300 Hz before (as-prepared) and after O_2- and N_2-annealing treatments. One can clearly see that the O_2-annealing treatment greatly destroys the high-temperature anomaly, but the N_2-annealing treatment enhances this anomaly. But these annealing treatments give rise to opposite effects to the low-temperature anomaly, i.e., oxygen annealing strongly enhances the low-temperature anomaly, and after nitrogen annealing this anomaly even disappears. These results reveal that oxygen vacancies favor for the high-temperature anomaly and

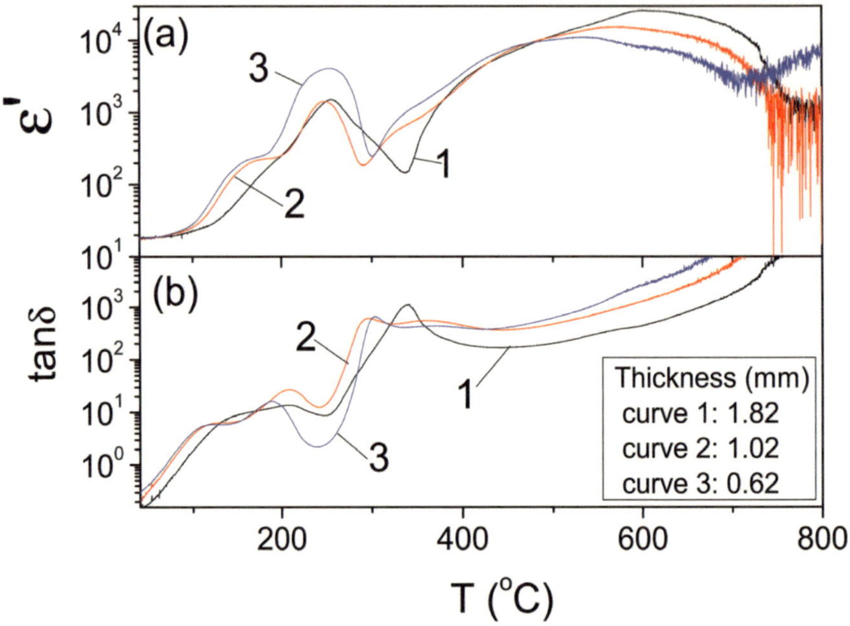

Fig. 5. Temperature dependences of ε' (a) and $\tan\delta$ (b) at 300 Hz for the as-prepared (curve 1) and polished (curves 2 and 3) CTO pellet. By CC Wang et al.

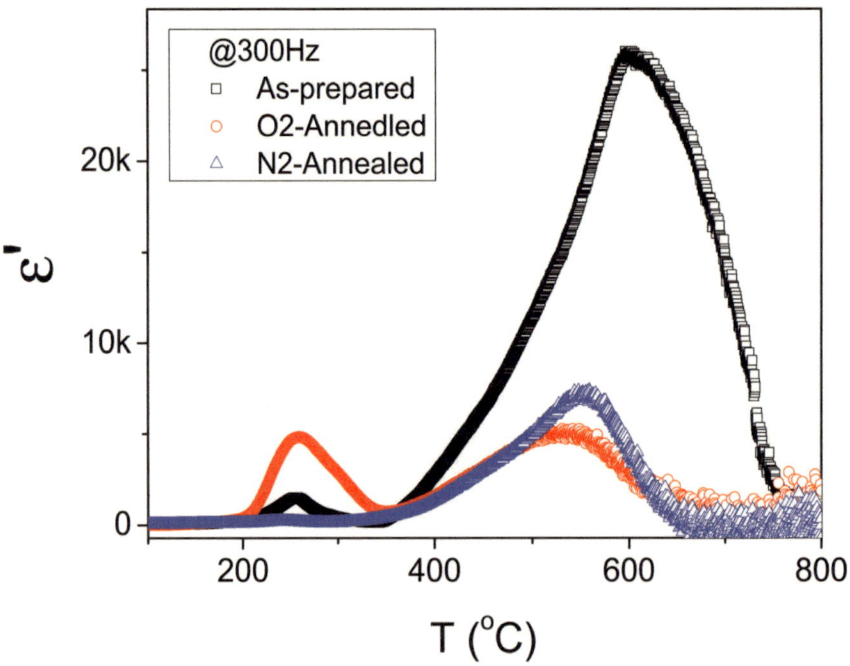

Fig. 6. Comparison of the temperature dependence of ε' obtained at 300 Hz for a CTO sample before (as-prepared) and after annealed in high-purity O_2 and N_2 at 800 °C for 2 h. By CC Wang et al.

seriously destroy the low-temperature anomaly. It therefore follows that the oxygen vacancy is truly at the origin of the high-temperature anomaly. It also strongly suggests that the low-temperature anomaly is closely related to positively charged relaxation species, considering the fact that the oxygen vacancies actually act as donors in the sample. Since the cobalt as a volatile element is easy to be lost during the sintering process as reported by many authors [8,24]. The ionization of cobalt vacancies creates holes [8], naturally, the Co vacancies could be suggested as the most probable origin of the low-temperature anomaly. To confirm this point, 5- and 10-wt% Co_3O_4 were added into the resulting CTO powder with the purpose of reducing the Co loss. Details about the preparation of the Co_3O_4-containing samples were given in the experimental procedure. The results of the real and imaginary parts of the dielectric permittivity for the pure and Co_3O_4-containing samples were shown in Fig. 7. From the real part (upper panel), it can be seen that the intensity of the low-temperature anomaly gradually decreases with increasing Co_3O_4 content. This feature can be clearly seen from the imaginary part (lower panel), which also reveals that the rapid increasing background becomes much more remarkable with increasing Co_3O_4 content (please note the logarithmic scale of ε''). These results confirm the suggestion that the low-temperature anomaly is associated with Co vacancies.

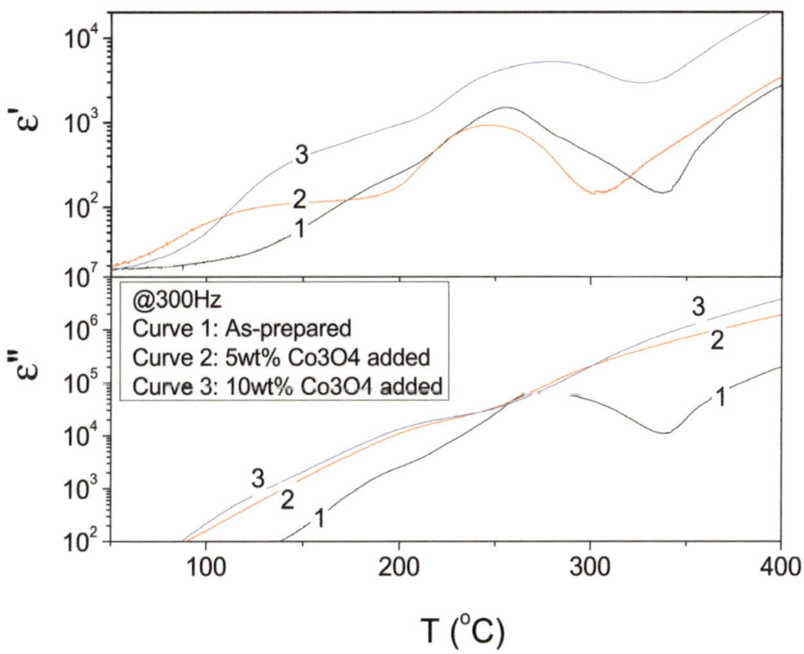

Fig. 7. Temperature dependences of ε' (upper panel) and ε'' (lower panel) of CTO samples containing various Co_3O_4 contents measured at 300Hz. By CC Wang et al.

It seems clear that the low- and high-temperature anomalies are associated to the cobalt and oxygen vacancies, respectively. However, a pertinent question is, why each of the anomaly contains two relaxation processes? Before answering this question, further information about the nature of the relaxing species is needed. A more sophisticated analysis of the frequency-dependent dielectric behavior can give some insight to this issue. This is done in

Fig.8 showing $\varepsilon''(f)$ for the as-prepared CTO at several selected temperatures. Perfect linear lines were observed in the double-logarithmic representation covering the entire measuring temperature range. This feature indicates that the dielectric behavior follows the universal power law[25], i.e.,

$$\varepsilon'' = B(T)\omega^{s-1} \tag{1}$$

where $B(T)$ and the frequency exponent s (with the value between 0 and 1) are temperature-dependent constants. The values of $s-1$ deduced from the straight lines were plotted in Fig. 9, from which one can see that $s-1$ decreases rapidly with increasing temperature, until at about 250 °C, it registers a peak, then decreases again and reaches a saturation value of -1 at temperatures above 400 °C. Since the value of s actually scales the extent of charger carriers been localized[26], e.g., in the case of s =0, Eq. (1) shows the usual reciprocal frequency behavior, and the system is nondispersive transport of free charge carriers process; for $s=1$, Eq. (1) reduces to $\varepsilon'' = \text{constant}$, the system has the feature of nearly constant loss relating to strictly localized carriers [27,28]; while for $0<s<1$, the system obeys the universal power law with confined hopping carriers. It, therefore, follows that the relaxing species for the low-temperature anomaly are confined carriers and for the high-temperature anomaly are free ones. The peak in $s-1$ might imply an alteration of the relaxing species changing from acceptor defects (cobalt vacancies) to donor defects (oxygen vacancies).

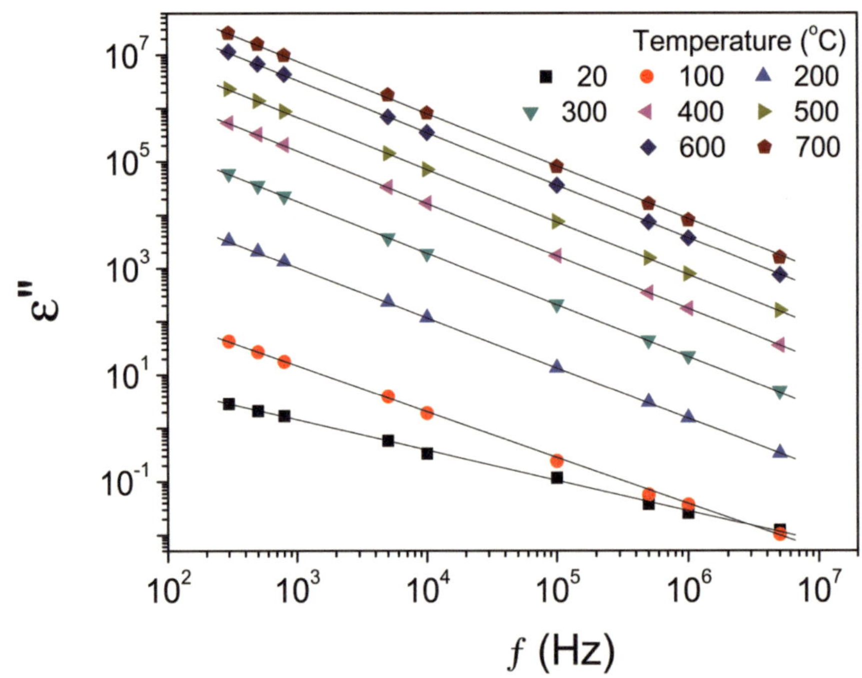

Fig. 8. Frequency dependence of ε'' for the as-prepared CTO sample at various temperatures. The straight lines are linear fitting results. By CC Wang et al.

Based on the nature of the relaxing species, the low- and high-temperature anomalies can be explained reasonably: First of all, the vacancy hopping motions between spatially fluctuating lattice positions not only produce long distance charge transport leading to notable conductivity but also give rise to dipolar effect. The former aspect results in a near-exponential increase in ε'' (or $\tan\delta$) as indicated in Fig. 4; the later makes a significant contribution to dielectric permittivity in the form of dipolar moment-reorientation relaxation under an alternative electric field. Both aspects can be described, as already confirmed, by the universal power law. Furthermore, the long distance transport of vacancies can be blocked by interfaces (e.g., grain boundaries) and sample surfaces, creating space charge there and Maxwell-Wagner relaxation. Therefore, the hopping vacancies can produce both the dipolar and the Maxwell-Wagner relaxations. This is the reason why each anomaly consists of two relaxation processes. Secondly, both relaxations can be described by Debye-like relaxation equations[19,29]:

$$\varepsilon' = \varepsilon_\infty + \frac{\varepsilon_s - \varepsilon_\infty}{1+(\omega\tau)^2} \tag{2}$$

$$\varepsilon'' = \frac{(\varepsilon_s - \varepsilon_\infty)\omega\tau}{1+(\omega\tau)^2} + \frac{\sigma}{\omega} \tag{3}$$

where ω is the angular frequency, τ is the mean relaxation time, σ is the electrical conductivity of the sample, ε_s and ε_∞ are the electric permittivity at low- and high-frequency limit, respectively. The relaxation strength, $\Delta\varepsilon$, defined as $\Delta\varepsilon = \varepsilon_s - \varepsilon_\infty$, is generally considered as a constant. However, in the case of relaxation processes associated with hopping carriers, the relaxation strength is expressed by[30] $\Delta\varepsilon = N\mu^2 / 3k_BT$, with μ the dipole moment and N the number of the hoping carriers, which varies with temperature following a thermally activated relation:[31]

$$N = N_0 \exp(E / k_BT) \tag{4}$$

where N_0 is the pre-exponential factor, E, the activation energy, and k_B, the Boltzmann constant. One thus has:

$$\Delta\varepsilon = N_0 \exp(E / k_BT)\mu^2 / 3k_BT \tag{5}$$

This equation predicates a nearly exponential decrease in $\varepsilon'(T)$. When the vacancy defect-induced Debye-like relaxation occurs, on one hand, $\varepsilon'(T)$ increases steplike and trends to a saturation value of ε_s; on the other hand, $\varepsilon'(T)$ decreases rapidly due to the relaxation strength decreases with temperature as predicated by Eq. (5). Therefore, a dielectric peak in $\varepsilon'(T)$, i.e. a dielectric anomaly, should be observed. For the Debye-like relaxation, the steplike increase in $\varepsilon'(T)$, and as a consequence, the anomaly shifts to higher temperatures for higher measuring frequencies, leading to the diffuse nature of the anomaly. So, we can come to the conclusion that the observed anomalies originate from vacancy defect-induced Debye-like relaxations with strongly temperature-dependent relaxation strength. Further evidence supporting this conclusion is the relation between the anomaly intensity and the temperature. Since the Debye-like relaxation occurs around the temperature where $\omega\tau = 1$ is achieved, the anomaly intensity (Ω) can then be described by the relaxation strength as seen from Eq. (2), viz.

$$\Omega \propto \Delta\varepsilon = N_0 \exp(E / k_B T)\mu^2 / 3k_B T \tag{6}$$

One has:

$$\Omega T \propto N_0 \exp(E / k_B T)\mu^2 / 3k_B \tag{7}$$

Therefore a straight line should be obtained if $\log(\Omega T)$ is plotted as a function of $1/T$. We truly found this linear behavior for the low-temperature anomaly as shown in the inset of Fig. 9. But for the high-temperature anomaly, Eq. (7) is not suitable. This is because that the relaxing species of the low-temperature anomaly are confined cobalt vacancies, whose number will exponentially decrease with increasing temperature as more and more vacancies become free ones. Hence, Eq.(7) works for the confined vacancy defect-induced anomaly. Whereas for the high-temperature anomaly, the number of the relaxing carriers is a constant, because the relaxing species were confirmed to be free ones. In this case, the anomaly intensity should be in proportional to the inverse temperature. This inference was confirmed as seen from the inset of Fig. 9. These results substantially support the point that the observed anomalies are associated with the relaxation processes induced by vacancy defects.

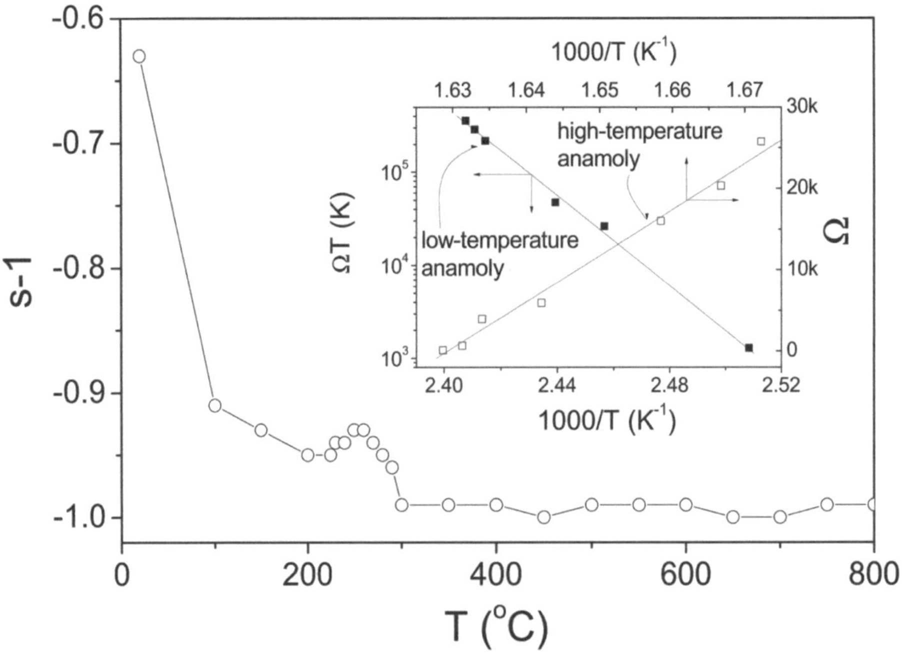

Fig. 9. Temperature dependence of the frequency exponent $s-1$ deduced from Fig. 8. The inset shows the anomaly intensity as a function of temperature for the low- and high-temperature anomalies. The straight lines in the inset are linear fitting results. By CC Wang et al.

4. Conclusions

In summary, two diffuse dielectric anomalies were observed in $CoTiO_3$ ceramics. The low-temperature anomaly situated at around 250 °C was found to be related to Co vacancies, while the high-temperature one appeared at about 600 °C was ascribed to O vacancies. The hopping motions of these vacancies firstly create a dipolar relaxation and then a Maxwell-Wagner relaxation as the hopping carriers blocked by the interfaces and surfaces of the samples. Both relaxations obey the Debye-like relaxation equations but with the relaxation strength depending strongly on the temperature. The appearance of the anomalies is a competition process between a $\varepsilon'(T)$ increasing process due to the Debye-like relaxation and a $\varepsilon'(T)$ decreasing process of the relaxation strength.

5. Acknowledgments

The authors thank financial support from National Natural Science Foundation of China (Grant No. 11074001). This work was supported in part by '211 Project' and Innovative Test Program for Undergraduates of Anhui University.

6. References

[1] K. H. Buchel and H. H. Moretto, P. Woditsch, *Industrial Inorganic Chemistry,* (second ed., Wiley-VCH, Germany, 2000).
[2] X. F. Chu, X. Q. Liu, G. Z. Wang, and G. Y. Meng, Mater. Res. Bull. 34 (1999) 1789.
[3] Y. Brik, M. Kacimi, M. ziyad, and F. Bozon-Verduraz, J. Catal. 202 (2001) 118.
[4] T. Kazuyuki, U. Yasuo, T. Shuji, I. Takashi, and U. Akifumi, J. Am. Chem. Soc. 106 (1984) 5172.
[5] N. Dharmaraj, H. C. Park, C. K. Kim, H. Y. Kim, and D. R. Lee, Mater Chem Phys.87 (2004) 5.
[6] H. T. Kim, S. H. Kim, S. Nahm, and J. D. Dyum, J. Am. Ceram. Soc. 82 (1999) 1901.
[7] P. S. Anjana and M. T. Scbastian, J. Am. Ceram. Soc. 89 (2006) 2114.
[8] C. C. Wang and L. W. Zhang, J. Phys. D: Appl. Phys. 40 (2007) 6834.
[9] T. S. Chao, W. M. Ku, H. C. Lin, D. Landheer, Y. Yang, and Y. Mori, IEEE Trans. Electr. Dev. 51 (2004) 2200.
[10] T. M. Pan, T. F. Lei, and T. S. Chao, J. Appl. Phys. 89 (2001) 3447.
[11] T. M. Pan, T. F. Lei, and T. S. Chao, Appl. Phys. Lett.78 (2001) 1439.
[12] L. E. Cross, Ferroelectrics 76 (1987) 241.
[13] C. Ang, Z. Yu, L. E. Cross, Phys. Rev. B 62 (2000) 228; and references cited therein.
[14] Z. Yu, C. Ang, P. M. Vilarinho, P. O. Mantas, J. L. Baptista, J. Appl. Phys. 83 (1998) 4874.
[15] A. I. Baranov, Ferroelectrics 285 (2003) 225.
[16] R. Stumpe, D. Wagner, D. Bauerle, Phys. Stat. Sol. (a) 75 (1983) 143.
[17] O. Bidault, P. Goux, M. Kchikech, M. Belkaoumi, M. Maglione, Phys. Rev. B 49 (1994) 7868.
[18] D. O'Neill, R. M. Bowman, J. M. Gregg, Appl. Phys. Lett. 77 (2000) 1520.
[19] G. Catalan, D. O'Neill, R.M. Bowman, J. M. Gregg, Appl. Phys. Lett. 77 (2000) 3078.
[20] B. S. Kang, S. K. Choi, C. H. Park, J. Appl. Phys. 94 (2003)1904.

[21] S. K. Choi, B. S. Kang, Y. W. Cho, and Y. M. Vysochanshii, J. Electroceramics 13 (2004) 493.
[22] B. S. Kang and S. K. Choi, Solid state Communi. 121 (2002) 441.
[23] C. C. Wang and S. X. Dou, Solid state Communi. 149 (2009) 2017.
[24] T. Ogawa, M. Ono, and M. Fujiwara, Jpn. J. Appl. Phys. 34 (1995) 5306.
[25] A. K. Jonscher, *Dielectric Relaxation in Solids* (Chelsea Dielectrics Press, London, 1983).
[26] C. C. Wang and L. W. Zhang, New J. Phys. 9 (2007) 210.
[27] M. Jain, Mater. Processes, 11 (1999) 317.
[28] A. S. Nowick and B. S. Lim, Phys. Rev. B 63 (2001) 184115.
[29] C. C.Wang, H. B. Lu, K. J. Jin, and G. Z. Yang, Mod. Phys. Lett. B 22 (2008) 1297.
[30] L. L. Hench and J. K. West, *Principles of Electronic Ceramics* (Jihn Wiley & Sons, New York, 1990).
[31] I. G. Austin and N. F. Mott, Adv. Phys. 18 (1969) 41.

7

Polymer Based Nanodielectric Composites

Daniel Tan and Patricia Irwin
GE Global Research Center
USA

1. Introduction

The rapid expansion of renewable energy applications demands higher efficiency and higher density energy storage and energy conversion systems (Arico et al 2005, Nourai et al 2005). Various DC-AC, AC-AC conversions are needed for solar and wind farms, while primary and secondary electrochemical devices are needed for transportation and telecommunication applications. In addition, military equipment and transport have been actively moving toward more electric systems. High energy density, power density and high temperature components are desired for applications such as active armor, electrochemical guns, directed energy weapons, more electric aircraft, electric launch platform, all electric warships, and so on (Sarjeant 1998, DARPA 2004). Passive components such as capacitors have been a limiting factor in full implementation of high energy density electrical systems. Figure 1 shows the development history of capacitor technology. Every generation of capacitors are primarily credited to the innovation and engineering of new dielectric materials. The advances of such a passive component and active components have provided great foundation for future power electronics and electric power.

Dielectric materials are categorized into organic polymers, inorganic ceramics, filled and unfilled polymeric thin films, ceramic films and nanodielectric composites. Polymer dielectric films show very high dielectric strength (>300kV/mm), lower dielectric loss (<0.01), and adequate mechanical flexibility. Single-phase polymers eliminate complexities arising from mixing and dispersion that are prevalent in the multi-phase systems, but, they are subject to low dielectric permittivity or constant (<4) and working temperature (<200°C). Ceramic dielectrics tend to have very high dielectric permittivity (>100) but relatively low dielectric strength (<50kV/mm). Increasing their dielectric strength is one of the challenges and goals in modern day dielectrics research. The presence of grain boundaries, porosity, impurities, surface defects, and chemical deterioration causes ceramic dielectrics to fail at relatively low field stresses (<10kV/mm). In addition, many ceramic dielectrics have high dissipation factor (>0.01), high capacitance loss (>30%) under high voltage stress (>3kV/mm) and/or piezoelectric effect associated with the ferroelectric type ceramics. Thin film dielectrics are an important area leveraging either polymer or ceramic materials. The thin films are usually in nanometer to submicron in thickness with very high breakdown strength. But, they are primarily useful for low voltage and small size microelectronic application. Scalability, reliability, cost and power level are great concerns.

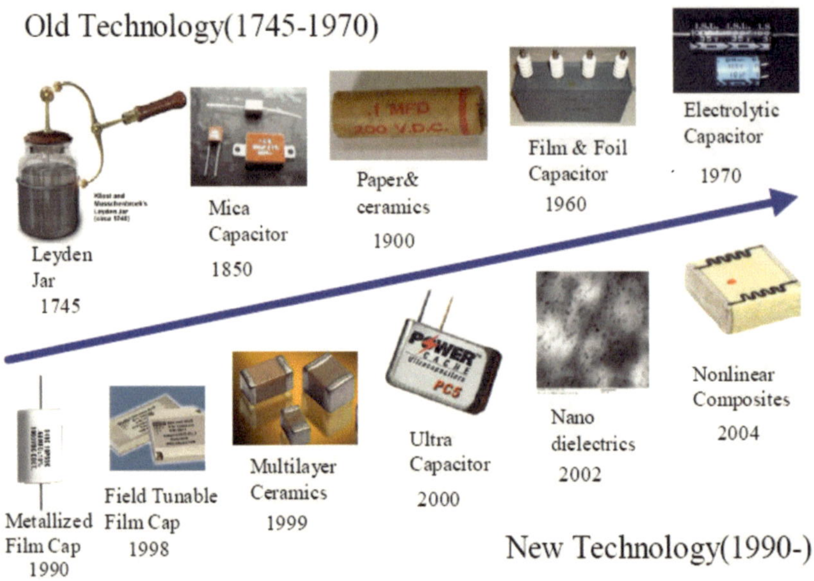

Fig. 1. Evolution of capacitor technology has a close relationship with materials development (Tan et al 2006).

Nanodielectric composites belong to a new type of materials engineered for improved functions, performance for dielectrics and electrical insulation (Lewis 19994). Certain ceramic materials can be selected to blend with polymers to provide synergy between the high breakdown strength polymer and high permittivity ceramic materials as shown in Figure 2. A number of research areas are being actively pursued to fully explore the advantages of the functional composites.

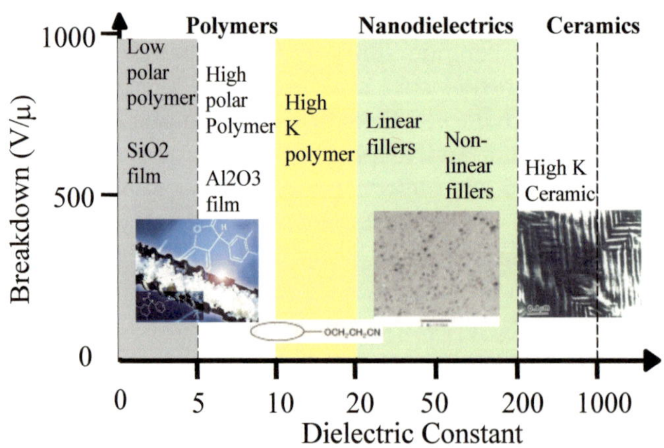

Fig. 2. Relationship between breakdown strength and dielectric permittivity of various dielectric materials (Tan et al 2010).

With properly selected nanoceramic fillers, dielectric properties of the polymers can be improved due to the absence of grain boundaries in ceramics. The chapter will review existing high performance polymer dielectrics and current nanodielectric engineering efforts. The effect of nanofiller type and size, distribution, particle-polymer interface and the host polymers will be discussed. A better understanding of the issues and fundamental mechanisms in nanodielectric engineering will lead to the successful development and implementation of high performance nanodielectrics.

2. High performance dielectric materials

An insulator is a material that resists the flow of electrons due to a full valence band and a large band gap between the valence band and the next higher state. A dielectric material is an electrical insulator that can be polarized by an applied electric field. The internal electric charges only slightly shift from their average equilibrium position resulting in dielectric polarization. Ceramics and polymer-based dielectrics are the most common materials of industrial interests, where the control of the storage and dissipation of electric and magnetic energy in these materials are of great interest. The following section discusses high performance dielectric materials with a focus on high temperature and high energy density capability.

2.1 High temperature dielectric materials

Generally speaking, voltage stress, complex dielectric permittivity, breakdown strength, mechanical strength, thermal stability and conductivity are the primary interests for dielectric materials. In many cases, these properties are mutually dependent. For example, the rise in temperature in the core of the capacitor causes an increase in dielectric losses and partial discharge, which can lead to premature failure. Often the self-generated internal heat can overshadow the ambient temperature, which then can lead to a more rapid degeneration and earlier failure. To meet the needs of high temperature devices and equipment, , a number of high temperature ceramics and polymer dielectrics have received intensive investigations. This includes Polycarbonate (PC), Polyphenylene sulfide (PPS), Fluorene Polyester (FPE), Diamond like carbon (DLC), polyetherimide (PEI), Polyetheretherketone (PEEK), Polyimide (Kapton® PI), Polytetrafluoroethylene (Teflon®), AlN, AlON, TiO_2, etc. For applications at or above 200°C, very limited choice of polymers and ceramic materials are available. Table 1. shows a number of high performance polymer and thin film dielectric materials being of interest for capacitor applications (Tan et al 2005). High temperature polymer films such as fluoropolyester (FPE), polyetherimide (PEI), and polytetrafluoroethylene (PTFE) not only meet the 200°C temperature requirement, but also have potential to be further engineered for higher performance. While stable to 260°C, PTFE has poor mechanical stability, difficulty in metallization and lower breakdown strength (<400kV/mm). FPE films are restricted to low energy density due to low permittivity. PEI has been successfully melt extruded into 5μm film. A great combination of dielectric, thermal and mechanical properties offers a good polymer matrix for inorganic fillers.
Ceramic dielectrics are very good high temperature dielectric materials by nature. They can be categorized into linear and nonlinear dielectrics by polarization mechanism as shown in Figure 3. Linear dielectric ceramics have lower dielectric permittivity and relatively higher breakdown strength than nonlinear material. The nanostructured TiO_2 is one example of the category, which has received intensive investigations due to the high energy density projection for capacitor applications [Chao et al 2010].

Properties	BOPP	PVDF	PC*	PPS*	PEEK*	Siloxane	PEI	PTFE	FPE	Kapton™ Polyimide	DLC	ALN
Temp. Range (°C)	90	125	130	150	150	>150	200	260	275	300	350	400
Dielectric Constant	2.2	10	30	3.1	3.2	8.6	3.2	2.1	3.3	3.3	3.3	9
Loss at 1kHz (10^{-3}) (25 °C)	0.2	15	1.3	0.5	4	60	2	0.5	2.6	2	4	5
Dielectric strength (V/μm)	700	400	350	400	430	380	650	296	520	420	800	>500
Tensile strength (ksi)	4.9	7.5	9.5	36	17.4	N/A	14	3	9.5	10.5	N/A	N/A
Film energy density (J/cc)	4.7	7.1	1.6	2.2	2.6	7.5	6	1.6	3.9	2.5	9.3	9.9
Film process	Extrusion or blow	Extrusion	Solvent cast	Extrusion	Extrusion	Solvent cast	Extrusion	Extrusion or blow	Solvent cast	Solvent cast	Ion beam	Sputtering

*Polycarbonate (PC), Polyphenylsulfide (PPS), Polyetheretherketone (PEEK)

Table 1. High temperature dielectric films for capacitor applications (Tan 2010)

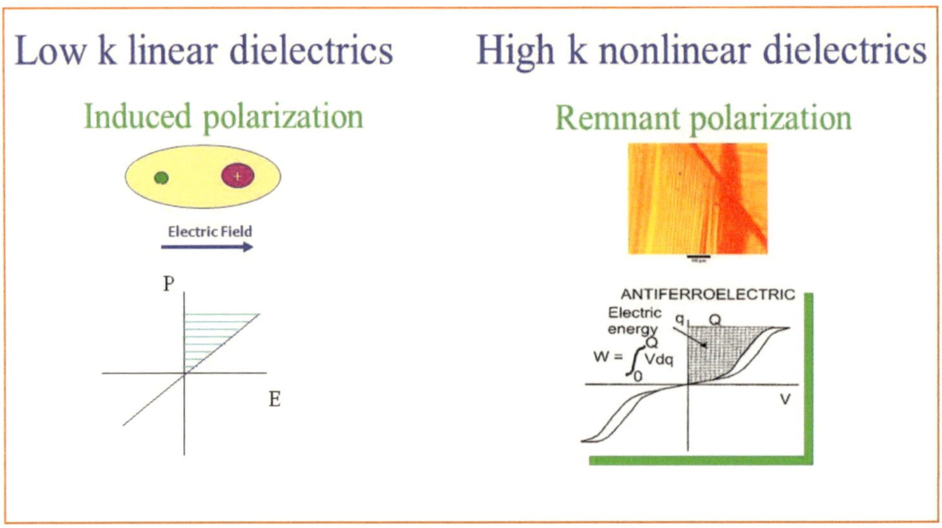

Fig. 3. Dielectric material classifications in terms of their polarization mechanism (Tan et al 2010).

Nonlinear dielectric ceramics exhibit a dipolar orientation mechanism resulting in a very high dielectric constant, but intend to fail easily at relatively low field stresses due to the grain boundary and inherent piezoelectric / electrostrictive effects. Ferroelectric and antiferroelectric materials are representative of this category. Their properties have been exploited by Sandia National Lab and other research labs to fabricate high energy density capacitors for neutron generators and power electronic inverter applications (Tuttle et al 2001, Campbell et al 2002). The hybridization of linear and nonlinear dielectric ceramics was also investigated by TRS Technologies, The Pennsylvania State University and Naval Research Laboratory (Hackenberger et al 2010, Gorzkowski et al 2007). Glass was added into ferroelectric ceramics to improve sintering behavior and to reach high density, however, the defects and microstructural control in the process created difficulties in reaching high breakdown strength and energy density.

In thin film dielectric approaches, ceramic films received great interest not only for their high breakdown strength appropriate for capacitor applications, but also their high dielectric permittivity needed for high frequency devices and memory applications. Among them are diamond-like carbon (DLC), aluminum oxynitride (AlON), stronium bismuth tantalate, and barium titanate, which are usually deposited on metal foil, Si wafers, or ceramics (Wu et al 2005, Bray et al 2006, Xu et al 1998, Kaufman et al 2010). The thin film technology is generally subjected to lower dielectric permittivity, low film quality in scale-up processes, processing difficulties and therefore low voltage applications.

2.2 High energy density dielectric materials

Energy density of a dielectric material is defined by $\varepsilon_0 \varepsilon E^2/2$, where ε_0 is the permitivity of free space, ε is the permitivity of the dielectric material, and E is the breakdown strength of the dielectric material. Either raising breakdown strength or dielectric permittivity will result in the increase of the energy density. A survey of dielectric materials has revealed that the combination of high breakdown strength and permittivity are not commonly found in dielectric materials (Figure 4). There have been two different research approaches to increase energy density so far. Increasing the breakdown strength appears to be emphasized the most because of the square relationship in the above definition. However, a great deal of difficulties has occurred particularly in the scale-up of materials. An increasing trend of effort is to seek high dielectric permittivity materials while maintaining breakdown strength of the dielectrics. This effort also finds similarity in other applications such as transistor gate, non-volatile ferroelectric memory, integral capacitors, and transmission lines.

Figure 5 shows the trend of increasing the dielectric permittivity that has been studied for microelectronics, power electronics and electronic actuation (Tan et al 2006). Some ceramics can have the dielectric permittivity of up to 10^3, but they need better engineering to augment breakdown strength. Some highly preferred polymers also show high dielectric permittivity, high breakdown strength and mechanical flexibility. However, a concern to be addressed is the increased dielectric loss. GE has developed some polymer dielectrics that have good combination of dielectric properties and self-clearing capability, which can extend the film life at voltages near the breakdown field of the films.

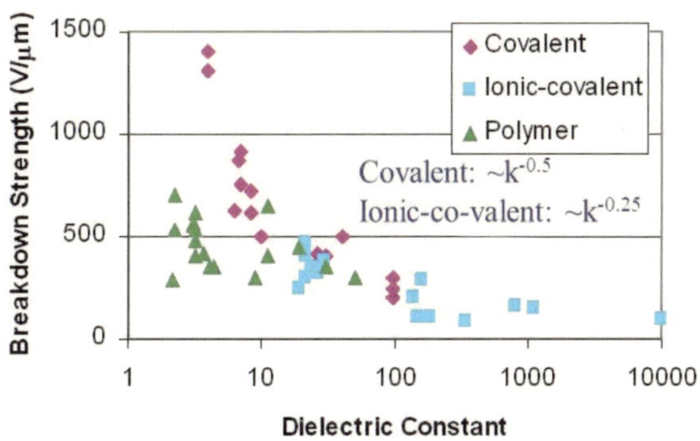

Fig. 4. A survey of dielectric materials with dielectric properties (Tan et al 2010).

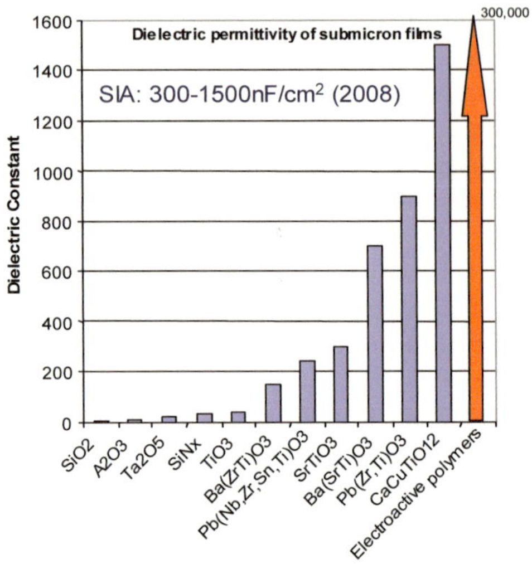

Fig. 5. Dielectrics of interests to microelectronics and electrical systems (Tan et al 2010).

Recently, various governmental agencies have been actively seeking dielectric materials that can offer a high energy density (>20J/cm^3) and high temperatures (200°C) for pulsed power and power conditioning applications. Figure 6 shows a theoretical design rule for dielectric films to meet a 25 J/cm^3 energy density requirement. If a dielectric material can have a dielectric permittivity of 10-20, a film with reasonable breakdown strength and thickness can be the great candidate for the purpose. If using a ceramic material with a permittivity of higher than 60, much lower breakdown strength is required. The film in need can be more than 10um in thickness. Currently, many institutions are actively investigating on dielectric films. The high performance dielectric materials have become one of the strategic thrust in material technology.

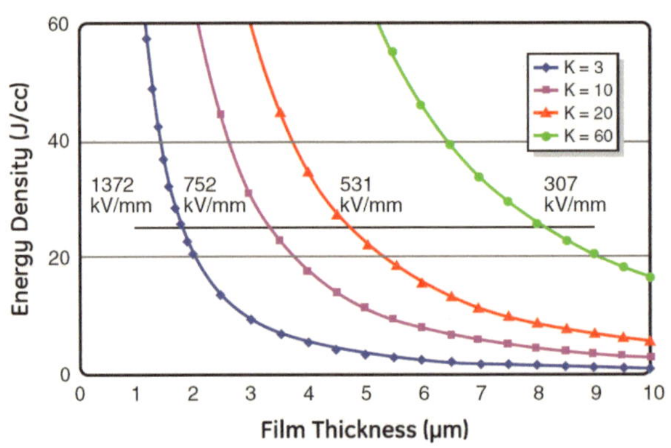

Fig. 6. Relationship of energy density, film thickness, dielectric constant and breakdown strength (Tan 2010).

2.3 Synergistic effect of nanodielectric composites

Polymer based nanodielectric materials have been investigated over the last 10 years forhigh energy density capacitors and higher corona resistant electrical insulation (Aulagner 1995).

Various investigators have pointed to the challenges of increasing breakdown strength. For example, a homogeneous dispersion of ferroelectric $BaTiO_3$ particles were prepared in polypropylene with an in-situ polymerization process (Guo et al 2007). Although the permittivity was increased by almost 3x over polypropylene, the breakdown strength was decreased. The breakdown mechanism was found very complicated particularly for the multi-phase systems. The mismatch of different phases and the various non-uniformity and defects introduced from processing were of critical consideration.

Depending on the filler-matrix interface, the size and shape, and agglomeration of fillers, the charge carriers are facing a more complicated environment under an electric field. Figure 7 shows a typical nanoparticle distribution in a Ultem™ polyetherimide (PEI). Agglomeration and aggregation are a great issue for the dry particles (left). Better particle dispersion could be achieved if solution prepared nanoparticles (right) are used, however, many desired materials are not available in solution or in large quantity. This would be an area of interest that need more in-depth investigations.

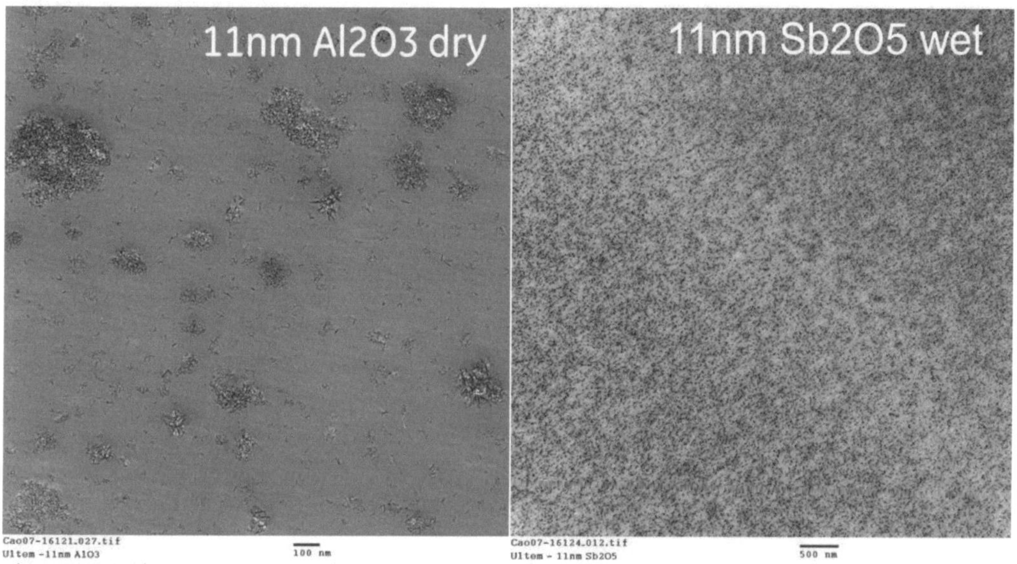

Fig. 7. Transmission electron microscope images of dry and wet nanoparticles in Ultem™ (Tan et al 2008).

2.3.1 Breakdown strength maintenance

In order to leverage the synergy of ceramic fillers and polymer matrices, the composite material physics should be better understood. Tan et al investigated the nano-filled polyetherimide films and emphasized the critical role of interface for the dielectric phenomenon (Tan et al 2007). In this study, Tan reported that ceramic fillers of higher breakdown strength do not impart the composite more endurance to high electrical field. Figure 8 depicts the dependence of intrinsic breakdown strength of various dielectric materials on their energy band gap. A linear increase with increasing band gap is well established. Yet data previously presented suggests that such a linear relationship cannot be correspondingly translated into nanofilled composites. A plausible explanation is that poor particle dispersion, agglomeration, voids or defects occurring at the particle-polymer interface might over shadow the positive effect of the nanoparticles.

Tan proposed a interfacial transition model as shown in Figure 9. The dielectric permittivity is increasing from polymer matrix to fillers through a complicated interfacial region. Even inside the interfacial region, the dielectric permittivity could include a free volume (void) layer and transition layer (surface state). In these low dielectric permittivity zones, field stress are higher and could trigger localized discharge causing free carrier movement. If a good dispersion and interface are established, this type of local failure might not happen. According to Qi et al (Qi et al 2004), the critical field (E_b) depends on the dielectric permittivity (ε), thermal conductivity (κ) of the dielectric with thickness (d): $E_b \sim \ln(4\varepsilon\kappa/\sigma_0 d^2)$. Higher dielectric permittivity and thermal conductivity associated with ceramic fillers help extend breakdown resistance. If localized heating occurs, it can be either dissipated in oxide fillers or more quickly transported through the samples due to shortened time permittivity ($d^2/4\kappa$). The thermal excitation of charges is thus restricted and breakdown strength is increased. It can be conjectured that better ceramic particle dispersion and elimination of interfacial defects could even improve the breakdown strength beyond that of the polymers. Improvments in particle processing, stabilization and dipersion control are required to realize this goal.

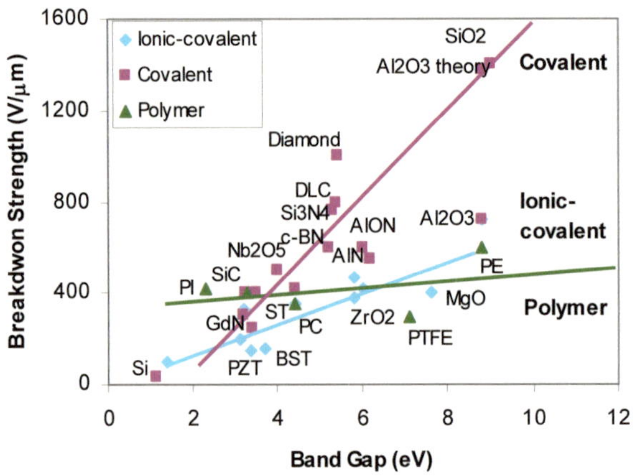

Fig. 8. Band gap dependence of breakdon strength of various dielectric materials (Tan et al 2007).

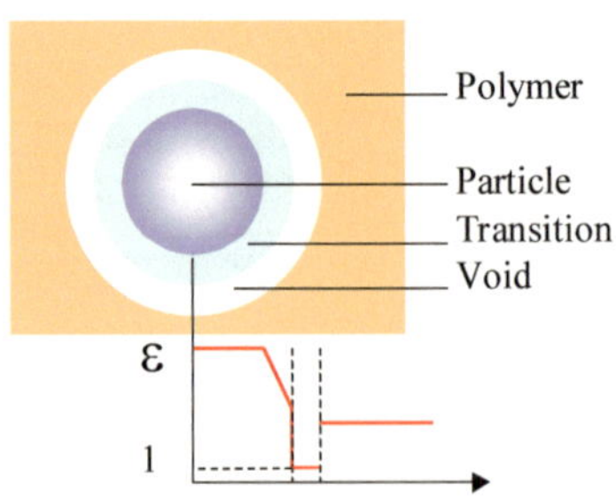

Fig. 9. A chematic model of the transition interface for a dielectric inclusion in a polymer (Tan et al 2007).

2.3.2 Dielectric permittivity augmentation

The increase in dielectric permittivity of polymer through addition of inorganic fillers is also challenging. For a 0-3 composite consisting of high permittivity ceramic and polymer, the logarithmic rule is most closely obeyed.

$$\log \varepsilon'_{composite} = \log \varepsilon'_{matrix} + \varphi_{filler} \log\left(\frac{\varepsilon'_{filler}}{\varepsilon'_{matrix}}\right) \tag{1}$$

The volume fraction (ϕ) of fillers randomly distributed in the composite cannot exceed the value for percolation point to avoid low mechanical strength and dielectric strength. Therefore, the permittivity of the composites ($\varepsilon'_{composite}$) is limited by the polymer permittivity (ε'_{matrix}). Filling a polymer matrix with high permittivity particles was known to result in a limited increase in permittivity of the resulting composite. For example, a ferroelectric type of ceramic materials having a dielectric permittivity of several hundred at room temperature is shown in Figure 10. A significantly high dielectric permittivity peak was dramatically suppressed when added to a polymer (polyetherimide). How to synergize filler and polymer, still remains a question.

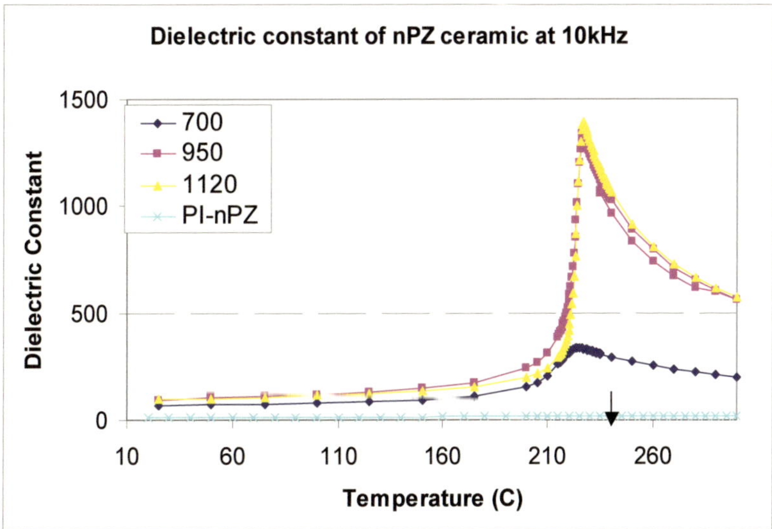

Fig. 10. The temperature dependence of the dielectric permittivity of lead zirconate ceramics and particle filled polyetherimide polymer (Tan 2010).

GE found that the higher permittivity base polymer is favored for achieving higher permittivity in nanodielectric composites. Figure 11 shows the pronounced effect in the engineering the dissimilar dielectrics. When dealing with fillers with various the shape, surface chemistry, dispersion, polarity, it still requires ingeneous thought and understanding of the physical properties of the materials. Figure 12 shows a number of materials with a wide range of dielectric permittivity and dielectric loss. Some of them are good candidates not only for their dielectric properties but also for their processing capability. With properly selected polymer matrices and inorganic fillers, one can expect to achieve good overall properties.

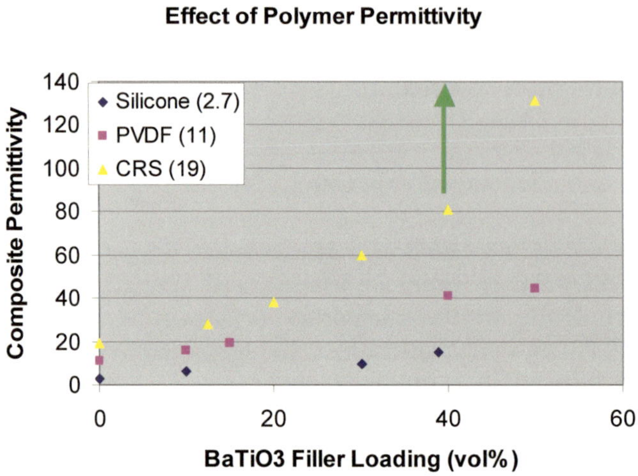

Fig. 11. The effect of the polymer matrix on the composite permittivity augmentation (Tan et al 2010).

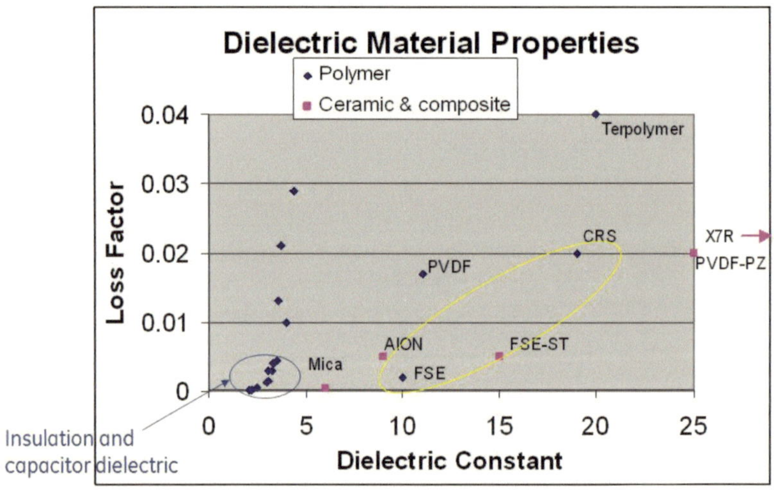

Fig. 12. Dielectric properties of various dielectric materials. High dielectric permittivity generally exhibits high dielectric loss (Tan et al 2010).

2.3.3 Nanofiller considerations

The simplest 2-phase system is a nanosphere filled polymer composite, which has received extensive investigation. A common phenomenon of such kind of composites is represented by Figure 13, where particles are of 20-100nm in diameter. It is not surprising that semiconductive and partially oxidized aluminum fillers lead to lower breakdown strength. However, for the insulative fillers with a breakdown strength of greater than 200kV/mm, the composite breakdown strength does not show much difference. It appears to slightly increase and saturate in the range of 500-600kV/mm for all insulating particles. It is interesting to see that ceramic particles of higher breakdown strength do not impart the composites more endurance to high electrical field.

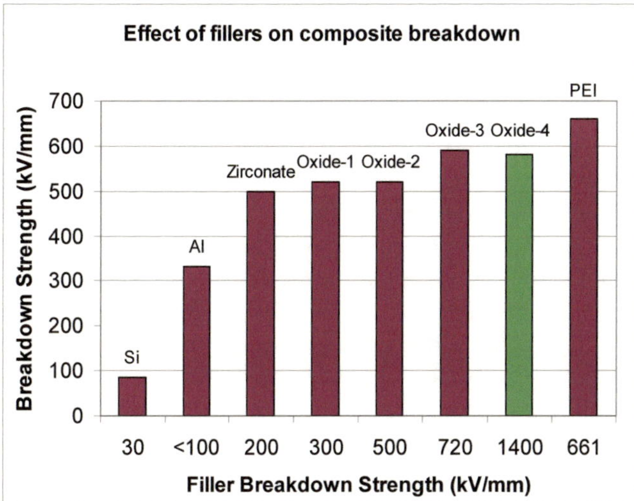

Fig. 13. The breakdown strength (BDS) of PEI composites increases with increasing BDS of 5 wt% various particles (Tan et al 2007).

When particles are smaller than 10nm, the interfacial fraction becomes dominant, which might result in interesting physical phenomena. Figure 18 depicts the increasing role of particle interface with decreasing the particle sizes (Raetzke et al 2006). With a 5 vol% particle added in a polymer, the 10nm particles will result in 40 vol% interfaces. The 5nm particles result in 95 vol% interface of the composites. When the fillers are spheroid, more complicated filler-matrix interaction and physical properties are to be expected, which requires more investigations.

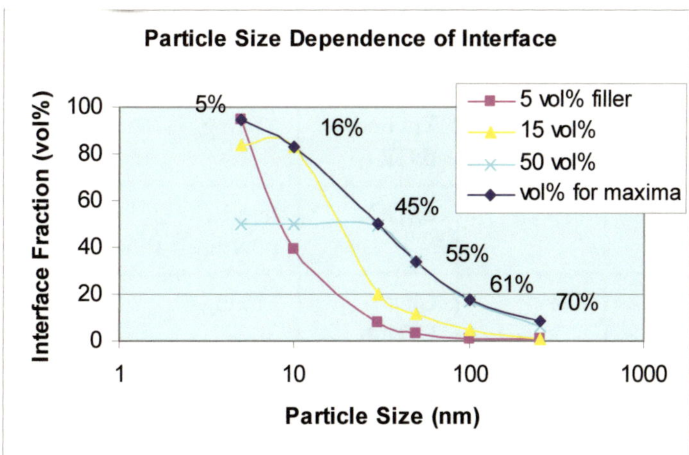

Fig. 14. The particle size dependence of the particle interfacial contribution at various loading concentration (Tan et al 2008).

2.3.4 Energy density accomplishment

In order to increase the dielectric constant of the composites, researchers mostly are interested in high permittivity nonlinear fillers like $BaTiO_3$ and TiO_2. Because of the difficulty in synthesizing very fine and clean particles, researchers have added high concentration of particles. Table 2 summarized the accomplishment of several representative researches. The maximum energy density from the nanodielectric engineering so far is 15 J/cc. Recently, very fine TiO_2 nanoparticles were synthesized using self-assembling method (Tuncer et al 2009). The breakdown strength of the composites was increased by >20%. Therefore, better results could be achieved after optimizing nanoparticle size, surface chemistry and processing method of composites.

For polar fillers that have high permittivity, one needs to consider their polar interaction with the matrix. The alignment of dipoles in the long-range order is known to give high permittivity due to collective response to the electric field such as those in ferroelectric materials (Figure 15). Correspondingly, it also results in lower breakdown strength. For example, dipolar ordering in PVDF polymer is relatively interrupted due to randomness in molecular chain configuration as opposed to those in ferroelectric ceramics. This is why PVDF ferroelectric polymer intrinsically has lower dielectric permittivity (11) but higher breakdown strength (>400kV/mm). Similarly, controlling dipolar ordering within short-range in nanodielectrics could result in both higher permittivity and breakdown strength. A balanced value in the two parameters would be eventually beneficial to the achievement in a high energy density. Such a complex dielectric system requires not only experimental verification, but also computer modeling and simulation for conceptual guidance.

Researcher	Polymer	Filler	Process	Energy Density (J/cm^3)
PennState	PVDF-CTFE PVDF-TrFE	-$BaTiO_3$ -PMNPT	Mixing	7 15
Georgia Tech	PVDF-HFP	Treated $BaTiO_3$	Mixing	6
Northwestern Univ.	PP	$BaTiO_3$	In-situ polymerization	9
GE	CR	Clean $BaTiO_3$	Mixing	15

Table 2. Status of energy density in nanodielectric composite investigation

2.4 Nanodielectric composite modeling

Due to the complexity of the composite microstructures, computational methods are usually required for study of the realistic multi-component microstructures (Ang et al 2003, Tuncer

2005, Todd et al 2005, Zhou et al 2008). Boundary integral technique (Azimi et al 1994, Cheng et al 1997) and finite element method (Ang et al 2003, Zhou et al 2008) are the commonly used computational methods, which numerically solve for interface charge density distribution and spatial potential distribution, respectively, in composite systems.

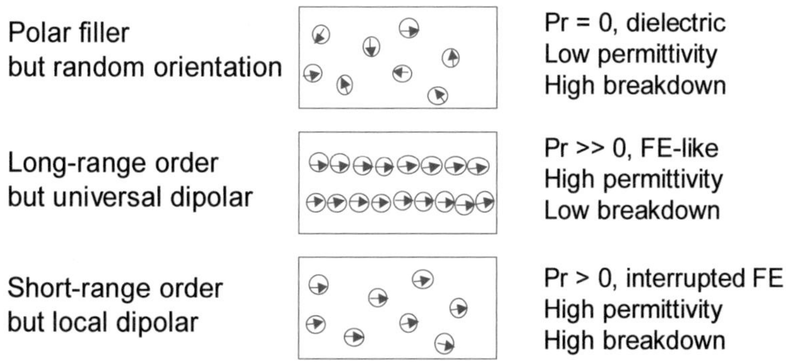

Fig. 15. Dipolar ordering model for nanofilled polymer composites (Tan et al 2007).

To avoid the complication associated with them, phase field method is recently employed to perform computational studies of composite materials (Zhang et al 2007, Wang 2010). It calculates heterogeneous distributions of polarization, charge density, local electric field, and effective dielectric permittivity of the composites, where inter-phase boundary conditions are automatically satisfied without explicitly tracking inter-phase interfaces. The phase field model of dielectric composites[9] is formulated in terms of polarization vector field **P**(**r**). The total system free energy F of a dielectric composite under externally applied electric field $\mathbf{E}^{ex}$ is:

$$F = \int \frac{P^2(\mathbf{r})}{2\varepsilon_0 \chi(\mathbf{r})} d^3r - \int \mathbf{P}(\mathbf{r}) \cdot \mathbf{E}^{ex} d^3r + \int \frac{d^3k}{(2\pi)^3} \frac{\left|\mathbf{n} \cdot \tilde{\mathbf{P}}(\mathbf{k})\right|^2}{2\varepsilon_0}, \tag{2}$$

where ε_0 is the permittivity of free space, $\chi(\mathbf{r})$ is the spatial position-dependent dielectric susceptibility that describes arbitrary multi-component composite microstructure, $\tilde{\mathbf{P}}(\mathbf{k})$ is the Fourier transform of the field **P**(**r**), and $\mathbf{n}=\mathbf{k}/k$ is a unit directional vector in **k**-space. The integrand of the first energy term in Eq. (2), i.e., $f(\mathbf{P})=P^2/(2\varepsilon_0\chi)$, is the non-equilibrium local bulk free energy density function that defines the thermodynamic properties of linear isotropic dielectrics.

The effective susceptibility tensor χ^{eff} of the composite is determined according to the anisotropic constitutive relation:

$$\langle P_i \rangle = \varepsilon_0 \chi_{ij}^{eff} \langle E_j \rangle, \tag{3}$$

where $\langle \mathbf{E} \rangle = \mathbf{E}^{ex}$. The effective anisotropic dielectric permittivity tensor is:

$$k_{ij}^{eff} = \delta_{ij} + \chi_{ij}^{eff}, \tag{4}$$

where δ_{ij} is Kronecker delta.

Wang et al (2011) did a series of computation for a composite containing well-dispersed particles of 20% volume fraction arranged into square or quasi-hexagonal lattices. The particle interactions are assumed not strong enough to significantly reduce the depolarization effect of individual particles, and the composites exhibit moderate effective susceptibility in all directions. However, with particles aligned into chains, particles strongly interact through electrostatic forces in the chain direction to effectively reduce the depolarization effect of the particles in each chain, resulting in significantly improved composite susceptibility along the chain direction as shown in Figure 16. In the transverse direction, on the other hand, the strong chain-chain interactions enhance the depolarization effect, leading to decreased composite susceptibility in direction perpendicular to the chains. Therefore, the composites exhibit significant susceptibility anisotropy, which results from the microstructure anisotropy of the filler particle arrangement. In fact, alignment of filler particles into chains establishes pseudo-1-3 connectivity as compared to 0-3 connectivity of dispersed particles (Newnham 1978), which forms continuous paths of high-susceptibility phase and enhances filler particle polarization under external electric field applied along the chain direction.

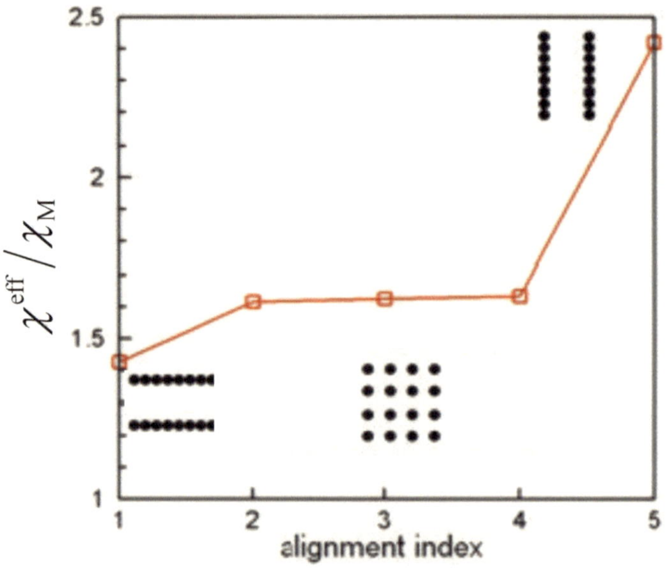

Fig. 16. Effect of filler particle microstructural arrangement on effective composite susceptibility in composites composed of 20% volume fraction of circular fillers with dielectric susceptibility $\chi_F/\chi_M = 10$. Insets illustrate the corresponding particle arrangements. Arrow indicates electric field (i.e., **E**) direction. χ^{eff} is the susceptibility component in **E** direction (Wang et al 2011).

Because of the issues associated with high volume fraction such as poor filler dispersion, porosity and defected filler-matrix interfaces, an appropriate range of filler volume fraction should be designed. To investigate the effect of filler particle volume fraction on the effective dielectric permittivity of composites, composites composed of different volume fractions of same-sized circular fillers with the same dielectric susceptibility $\chi_F/\chi_M = 10$ are considered.

In particular, the upper and lower bounds for χ^{eff} are obtained with parallel and series two-phase morphologies, respectively, and are given as

$$\chi_{upper}^{eff} = V\chi_F + (1-V)\chi_M, \qquad \frac{1}{\chi_{lower}^{eff}} = \frac{V}{\chi_F} + \frac{1-V}{\chi_M}, \tag{5}$$

where V is filler phase volume fraction. It is found that at low volume fraction (e.g., $V<30\%$) χ^{eff} is closer to the lower bound χ_{lower}^{eff}, while at high volume fraction (e.g., V>60%) χ^{eff} starts departing from χ_{lower}^{eff} to approach the upper bound χ_{upper}^{eff}. This transition behavior of χ^{eff} at low and high filler volume fraction with respect to χ_{lower}^{eff} and χ_{upper}^{eff} is associated with the increasingly strong particle interactions and gradual establishment of particle connectivity when filler volume fraction increases. Figure 17 presents the computational results for dispersed fillers, the aligned filler chains in parallel and in perpendicular to electric field direction, respectively. It can be found that the alignment of particles into chains is an effective way to establish filler connectivity in volume fraction range much lower than that required by randomly dispersed fillers. This may help alleviate the problems encountered during composite fabrication with high filler volume fractions.

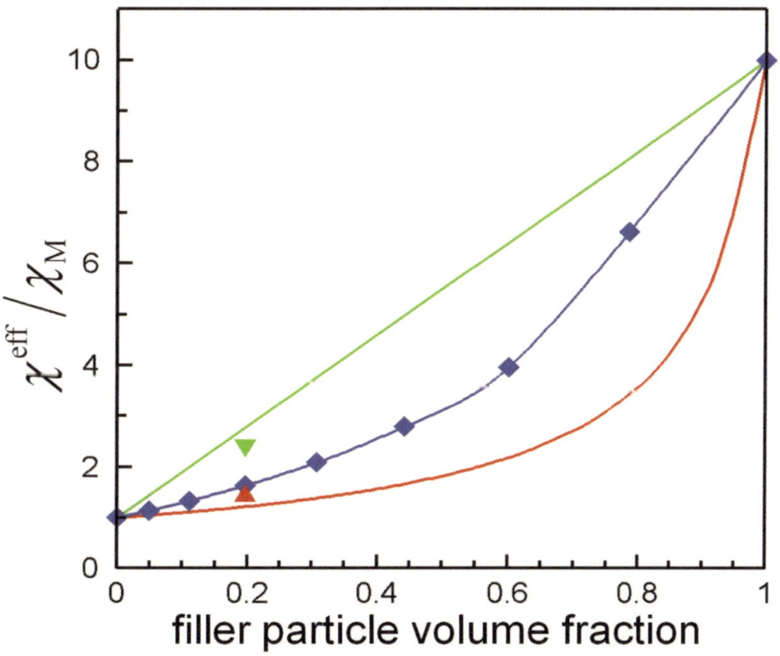

Fig. 17. Effect of filler particle volume fraction on effective composite susceptibility in composites composed of fillers with dielectric susceptibility $\chi_F/\chi_M = 10$. Green and red curves represent upper and lower bounds based on rule of mixtures for parallel and series two-phase morphologies, respectively. Blue curve represents computational results for composites composed of same-sized dispersed circular fillers. Green and red triangles (|, ʃ) correspond to composites with filler particles aligned into chains parallel and perpendicular to electric field direction, respectively, as shown in Fig. 16 (Wang et al 2011).

3. Conclusion and future interests

Synthesis of higher permittivity composites by mixing the various fillers (ceramic, semiconductor, metallic, organic) has been conducted over the last decade. Engineering the existing materials requires advances in designing, processing, forming and manufacturing technologies. Successes require (i) higher dielectric permittivity polymers, (ii) higher dielectric permittivity ceramics, (iii) proper dispersion, and (iv) good interfacial adhesion between filler and polymer. Fillers that have been studied include ceramics (SiO_2, Al_2O_3, TiO_2, $BaTiO_3$, $SrTiO_2$, PZT, PMN-PT, PLZT, MgO, Mica, layered silicate, Si_3N_4, BN), semiconductors (Si, SiC, ZnO, varistor particles, SnO_2, antimony modified SnO_2), metals (Al, Cu, Ag, Co-ACAC, Pt), carbon (black, CNTs). Great deal of knowledge has been acquired about nanodielectrics, however, there is still no breakthrough in increasing dielectric strength and permittivity simultaneously. In order to enhance dielectric properties, mechanical flexibility and thermal stability, further investigations are required, which can include the following subjects.

- What types of ceramic fillers and polymers are able to synergize the dielectric properties?
- What type of fillers, their size, shape and orientation are the best choice for composites?
- What kinds of filler surface chemistry promote filler dispersion and better properties?
- How does filler interface contribute to dielectric properties?
- How effective are the insulation coated nanofillers on dielectric property augmentation?
- What needs to be done to develop the polymers having high dielectric permittivity, high temperature and also low dielectric loss?
- How does the nanodielectric engineering improve both dielectric properties and thermal conduction properties?

4. Acknowledgment

Support from GE Energy Conversion AT program and DARPA DSO under Contract FA9451-08-C-0166 are greatfully acknowledged.

5. References

Ang, C.; Yu, Z.; Guo, R.; and Bhalla, A.S. (2003), J. Appl. Phys. 93, 3475.

Arico, A. S.; Bruce, Peter; Scrosati, B.; Tarascon, J.-M.; and Van Schalkwijk, W. (2005) Nature Materials, Vol.4, pp366-377.

Aulagner, E.; Guillet, J.; Seytre, G.; Hantouche, C.; Le Gonidec, P. (1995), IEEE 5th International Conf. On Conduction and Breakdown in Solid Dielectrics, p.423.

Azimi, M.E.; and Ghosh, P.K. (1994), Modelling Simul. Mater. Sci. Eng. 2, 1093.

Bray, K.R.; Wu, R.L.C.; Weimer, J.; Fries-Carr, S.; CARTS USA 2006 April 3-6, Orlando, FL.

Chao, S. and Dogan, F. (2010), Int. J. Appl. Ceram. Technol.,]] []]] 1–11.

Campbell, C.K. (2002) : *IEEE Trans. Components Packaging Tech.* Vol.25, No.2, p.211.

Cheng, H.; and Torquato, S. (1997), Phys. Rev. B 56, 8060.

DARPA report (2004), "Combat Hybrid Power System Component Technologies: Technical Challenges and Research Priorities". 2004 National Academy of Sciences. Available from http://books.nap.edu/catalog/.

Gorzkowski, E.P.; Pan, M.-J.; Bender, B. and Wu, C.C.M. (2007), Journal of Electroceramics, Volume 18, Numbers 3-4, 269-276.

Guo, N.; DiBenedetto, S.A.;; Kwon, D.K.; Wang, L.; Russell, M.T..; Lanagan, M.T.; Facchetti, A.; and Marks, T.J. (2007); J. AM. CHEM. SOC., 129, p.766.

Hackenberger, W. S. (2010), TRS Technologies, Available from http://www. trsceramics.com/

Kaufman, D. (2010), Argonne National Lab, Energy Technology Division, Available from http://www.transportation.anl.gov/publications/transforum/v2/ceramic.html

Lewis, T.J. (1994), *IEEE Trans on DEI*, Vol. 1, No.5, p.812.

Newnham, R.E.; Skinner, D.P.; and Cross, L.E. (1978), Mater. Res. Bull. 13, 525.

Nourai, A.; Martin, B.P.; and Fitchett, D.R. (2005), IEEE Power & Energy Magzine, March/April 2005, pp.41-46.

Qi, X.; Zheng, Z.; Boggs, S. (2004), IEEE Electrical Insulation Magazine, Vol. 20, No.6.

Raetzke, S.; Kindersberger, J. (2006), IEEJ Trans. FM, Vol 126, No.11, p.1004.

Sarjeant, W.J., J. Zirnheld, and F.W. Mcdougall, IEEE Transactions on Plasma Science, 1998. 26(5): p. 1368-1392.

Tan, Q.; Irwin, P.C.; and Cao, Y. (2006), *IEEJ Trans. Fundamentals and Materials*, Vol. 126, 1209.

Tan, D.Q.; Cao, Y.; and Irwin, P. (2007), International Conference on Solid Dielectrics, ICSD; Winchester; July 8-13, 2007, Article number 4290839, Pages 411-414

Tan, D. Q.; Cao, Y.; Irwin, P.; Shuman, K.; McTigue, C. (2008), "Interfacial study of Nanoparticle filled polyetherimide". Presentation at International conference on materials (IUMRS-ICM), July.

Tan, D.Q.; Chen, Q.; Cao, Y.; Irwin, P.; Heidger, S. (2010), "Polymer based nanodielectric composites for capacitors". Presentation at American Ceramic Soceity Electronic Materials and Applications 2010, Orlando, FL.

Tan, D.Q. (2010), "Intergrated High Energy Density Capacitors". GE internal report, March.

Tuncer, E. (2005), J. Phys. D: Appl. Phys. 38.

Tuncer, E.; Rondinone, J.; Woodward, J.; Sauers, I.; James, D.R.; Ellis, A.R. (2009), Appl Phys A 94: 843-852.

Tuttle, A.; Yang, P.; Gieske, J.H.; Voigt, J.A.; Scofield, T.W.; Zeuch, D.H.; and Olson. W.R. (2001): *J. Am. Ceram. Soc.*, Vol.84, p.1260.

Wang, Y. (2010), Appl. Phys. Lett. 96, 232901.

Wang, Y.; and Tan, D.Q. (2011), accepted by J.Applied Phys.

Wu, R.L.C.; Kosai, H.; Carter, M.; Rumler, M.; Fries-Carr, S.; Weimer, J.; Furmaniak, T.; Barshaw, E.; Scozzie, S.; Jow, R.; Garrison, R.; and Warnock, F. (2005), CARTS USA 2005 March 21-24, Palm Springs, CA.

Xu, B.; Moses, P.; Pai, N.G.; and Cross, L.E. (1998): *Appl. Phys. Lett.*, Vol.72, No.5, p.2.

Zhang, J.X.; Li, Y.L.; Schlom, D.G.; Chen, L.Q.; Zavaliche, F.; Ramesh, R.; and Jia, Q.X. (2007), Appl. Phys. Lett. 90, 052909.

Zhou, K.; Boggs, S.A.; Ramprasad, R.; Aindow, R.; Erkey, C.; and Alpay, S.P. (2008), Appl. Phys. Lett. 93, 102908.

8

Ceramic Based Intelligent Piezoelectric Energy Harvesting Device

Imran Patel[1,2]
[1]*British University in Egypt, Suez Desert Road, El Sherouk City*
[2]*Institute for Materials Research and Innovation, The University of Bolton, Bolton*
[1]*Egypt*
[2]*U.K.*

1. Introduction

In 2008, total worldwide energy consumption was 474 exajoules (474×10^{18} J) with 80 to 90 percent derived from fossil fuels(1). This is equivalent to an average power consumption rate of 15 terawatts (1.504×10^{13} W).

Most of the world's energy resources are from the sun's rays hitting Earth. Some of that energy has been preserved as fossil energy; some are directly or indirectly usable, for example via wind, hydro- or wave power. The term solar constant is the amount of incoming solar electromagnetic radiation per unit area, measured on the outer surface of Earth's atmosphere, in a plane perpendicular to the rays. The solar constant includes all types of solar radiation, not just visible light.

The estimates of remaining non-renewable worldwide energy resources vary, with the remaining fossil fuels totalling an estimated 0.4YJ (1YJ = 10^{24}J) and the available nuclear fuel such as uranium exceeding 2.5YJ. Fossil fuels range from 0.6-3YJ if estimates of reserves of methane are accurate and become technically extractable.

The twentieth century saw a rapid twenty-fold increase in the use of fossil fuels. Between 1980 and 2006, the worldwide annual growth rate was 2%(1). According to the US Energy Information Administration's 2006, the estimated 471.8EJ total consumption in 2004 was divided as follows, with fossil fuels supplying 86% of the world's energy, see **Figure** 1.

Coal fuelled the industrial revolution in the 18th and 19th century. With the advent of the automobile, airplanes and the spreading use of electricity, oil became the dominant fuel during the twentieth century. The growth of oil as the largest fossil fuel was further enabled by steadily dropping prices from 1920 until 1973.

2. Literature review

In today's world of sophisticated mechanical, electromechanical and electronic applications, polymers are playing an increasingly important part. One of the most significant advantages of polymeric materials over other materials is their complex structure, which can be physically or chemically tailored for specific applications. In year 2000, pioneering work carried out by HEEGER, MaCDIARMID and SHIRIKAWA on conducting polymers was rewarded with the Nobel Prize in Chemistry(2).

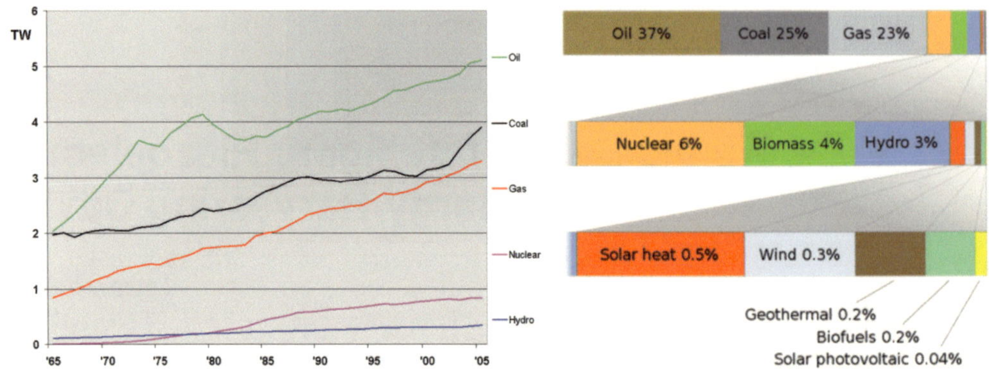

Fig. 1. Left, Rate of world energy usage in terawatts (TW), 1965-2005[1]. Right, Global energy usage in successively increasing detail[1]. Author of graph: Frank van Meirlo

Highly insulating polymers have been used extensively in electrical cable insulation since their invention in the first half of the 20th century. Some fluoropolymers were shown to store injected electrical charges for long periods of time[3], even at elevated temperatures. The discovery of piezoelectric and pyroelectricity in polyvinylidene fluoride (PVDF)[4] opened up wide market for electromechanical transducers applications. In recent years non-polar ferroelectrets[5] with high piezoelectric coefficient have received a lot of attention. Already a new class of non- polar ferroelectrets based on charge storing polymers have become available in the market[6].

Reducing size and power requirements of wearable microelectronics can make it possible to replace batteries with smart systems that capture energy from the user's environment[7]. The consumer reliance on wearable electronic devices is growing significantly, which is leading to the demand for decreased size and enhanced capabilities of micro-power generation devices. Until now batteries have been sufficient, but a nuisance in terms of excess weight, rechargeability, replacement and disposal.

An average person spends a significant part of the day on foot, dissipating abundant energy into the soles of the footwear. This wasted energy could be harnessed in an unobtrusive manner to power a variety of low power application systems, such as pagers, health monitors, I-pods and possibly mobile phones. Studies at Massachusetts Institute of Technology (MIT) explored the feasibility of harnessing wasted energy from a variety of body sources, however, heel strike during walking has shown to be the largest untapped source of wasted energy[8]. Scavenging most of the energy unobtrusively would be impossible, but a sizable fraction of it could provide sufficient energy to operate many of the personal micro-powered systems on the market today[9].

Another MIT study[10], as well as an independent work by ANTAKI *et al*[11] at another institution, supported and proposed a system of embedded piezoelectric materials and miniature controls. The research observed that a shoe having relatively large volume available in the sole would make an ideal test bed for exploring body energy harvesting.

A recent development demonstrated the feasibility of scavenged shoe power using a simple application circuit[12]. The design is a self-powered radio frequency (RF) tag that transmits a short range 12bit wireless identification (ID) code during walking, see **Figure 2.** PVDF and PZT staves installed in the sole of the shoes. Previously, this area of work relied on battery powered IR badges[13], however, PVDF or PZT staves could be configured into the sole of the shoes to provide the required power for the signal to be transmitted.

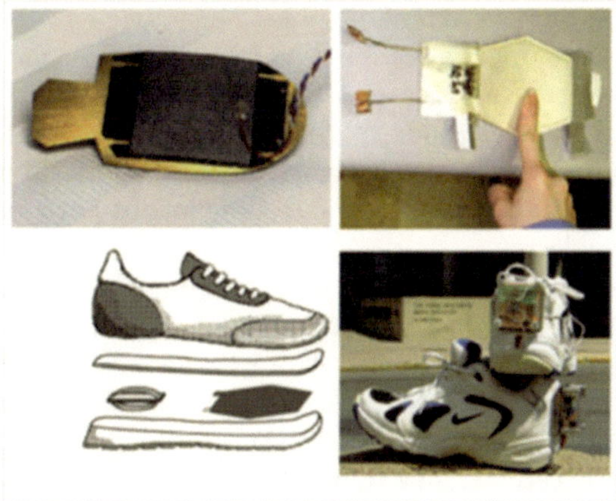

Fig. 2. PVDF and PZT staves installed in the sole of the shoes

It has been calculated that 5W of electrical power can be generated by a 52kg person at a brisk walking pace using a PVDF shoe insert(14). Similarly, another study used a free falling ball to impact a plate with a piezo-ceramic wafer attached to its underside, developed an electrical equivalent model of the PZT transforming mechanical impact energy to electrical power(15). This study also investigated the energy storage mechanism of the PZT with a bridge rectifier and a capacitor. Further studies examined use of a piezoelectric film in addition to ceramic-based piezoelectric material to provide power to light a bulb(16).

Since the discovery of piezoelectric properties, particularly of PVDF by KAWAI in 1969(17), piezoelectric polymer shaves been used in a wide range of sensor and actuator applications(18). PVDF has advantages over ceramic-based piezoelectric materials, in terms of mechanical flexibility, good acoustic coupling to aqueous media, and relatively low cost per unit area. A number of other polymers has been known to have potentially useful piezoelectric properties and these include some polymides(19), copolymers of vinylidene cyanide (VDCN)(20), polyureas(21), polyurethane(22) and polythioureas(23).

2.1 Introduction to Piezoelectricity

A piezoelectric material has the unique ability to interchange electrical energy and mechanical strain energy or force. Due to this characteristic of the material, it has been found to be very effective for use in dynamic applications involving vibration suppression, mechanical impact and sensing. However piezoelectric materials have been used in numerous other applications including sonar application, audio buzzers, air ultrasonic transducers and piezoelectric ceramic ignition systems. While these materials have found notable applications, piezoelectric devices with a view to energy harvesting have not found the market that will allow the production rates of the materials to grow substantially and to become cost effective in other applications. As a result, a number of developments have emerged in the fields of harvesting or scavenging wasted energy, even from the human body i.e. from walking and the expansion and contraction of human lungs.

2.1.1 Overview of the theory of Piezoelectric and Ferroelectrics

Since the piezoelectric effect exhibited by natural materials such as quartz, tourmaline, rochelle salt, etc. is limited in terms of usable power, polycrystalline ferroelectric ceramic materials such as barium titanate ($BaTiO_3$) and lead zirconate titanate (PZT) with improved properties have been developed over the last few decades. Likes of PZT ceramic are available in many variations and are still the most widely used materials for actuator applications today. Prior to polarization mechanism, PZT crystallites have symmetric cubic unit cells. At temperatures below the Curie temperature, the lattice structure becomes deformed and asymmetric. The unit cells exhibit spontaneous polarization i.e. the individual PZT crystallites are piezoelectric.

A traditional piezoelectric ceramic is a mass of perovskite structure (material with the same type of crystal structure as calcium titanium oxide - $CaTiO_3$), each consisting of a small, tetravalent metal ion, usually titanium or zirconium, in a lattice of larger divalent metal ions, usually lead or barium, and O_2- ions. Under conditions that confer tetragonal or rhombohedral symmetry on the crystals, each crystal has a dipole moment, see **Figure 3**.

To prepare a piezoelectric ceramic, fine powders of the component metal oxides are mixed in specific proportions then heated to form a uniform powder. The powder is mixed with an organic binder and is formed into structural elements having the desired shape (discs, rods, plates, etc.), for this research, a 120μm and 250μm diameter fibres were used. The elements are fired according to a specific time and temperature, during which the powder particles sinter and the material attains a dense crystalline structure. The elements are cooled, then shaped or trimmed to specifications, and electrodes are applied to the appropriate surfaces.

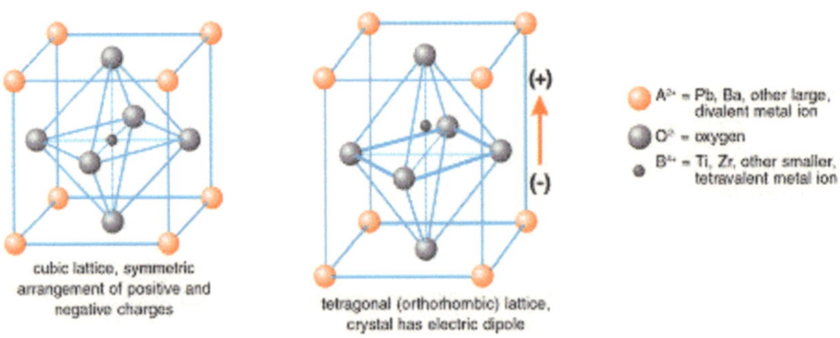

Fig. 3. Crystal structure of a traditional piezoelectric ceramic, left, temperature above curie point and right, temperature below curie point

Above a critical temperature, the Curie point, each perovskite crystal in the fired ceramic element exhibits a simple cubic symmetry with no dipole moment (**Figure 3** left). At temperatures below the Curie point, however, each crystal has tetragonal or rhombohedral symmetry and a dipole moment (**Figure 3** right). Adjoining dipoles form regions of local alignment called domains. The alignment gives a net dipole moment to the domain, and thus a net polarization. The direction of polarization among neighbouring domains is random, however, so the ceramic element has no overall polarization (**Figure 4** left). The domains in a ceramic element are aligned by exposing the element to a strong, direct current electric field, usually at a temperature slightly below the Curie point (**Figure 4** middle). Through this polarising treatment, domains mostly aligned with the electric field expand at the expense of domains that are not aligned with the field, and the element lengthens in the direction of the field. When the electric field is removed most of the dipoles are locked into a configuration of near alignment (**Figure 4** right). The element now has a permanent polarization, the remnant polarization, and is permanently elongated.

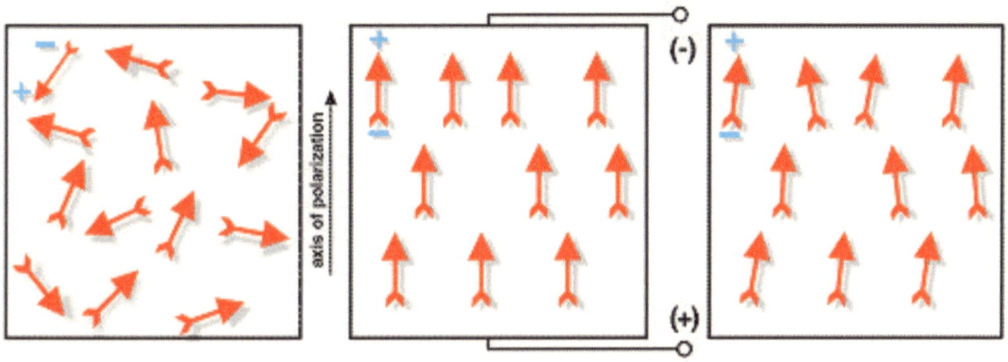

Fig. 4. Polarizing of a piezoelectric ceramic, left, random orientation of a polar domains prior to polarization, centre, polarization in DC electric field and right, remnant polarization after electric field is removed.

2.1.2 Working of piezoelectric materials

Mechanical compression or tension on a poled piezoelectric ceramic element changes the dipole moment, creating a voltage. Compression along the direction of polarization, or tension perpendicular to the direction of polarization, generates voltage of the same polarity as the poling voltage (**Figure 5b**). Tension along the direction of polarization, or compression perpendicular to the direction of polarization, generates a voltage with polarity opposite that of the poling voltage (**Figure 5c**). These actions are generator actions, where by the ceramic element converts the mechanical energy of compression or tension into electrical energy.

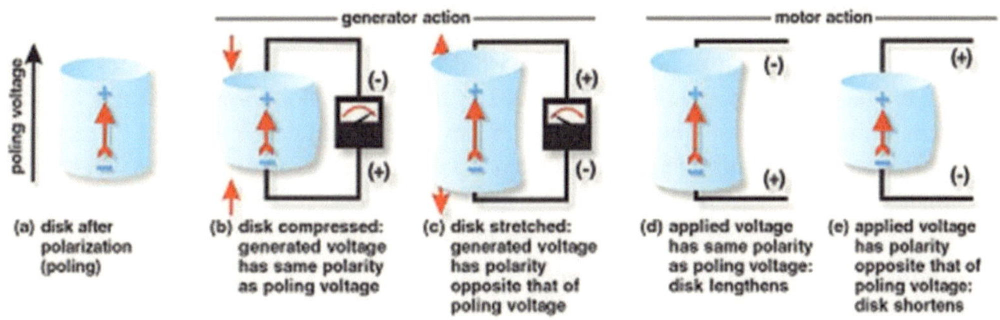

Fig. 5. Direct and indirect effect of piezoelectric material showing voltage generation when compressed and shape change when voltage is applied.

If a voltage of the same polarity as the poling voltage is applied to a ceramic element, in the direction of the poling voltage, the element will lengthen and its diameter will become smaller (**Figure 5d**). If a voltage of polarity opposite that of the poling voltage is applied, the element will become shorter and broader (**Figure 5e**). If an alternating voltage is applied, the element will lengthen and shorten cyclically, at the frequency of the applied voltage. This is motor action, where by electrical energy is converted into mechanical energy. The principle is adapted to piezoelectric motors, sound or ultrasound generating devices, and many other products. And it's this phenomenon which is exploited for energy, or voltage generation during various experiments that the piezoelectric materials are subjected to.

2.2 History of piezoelectricity

The very first experiments on the piezoelectric effect was performed on specially prepared crystals such as quartz, topaz, tourmaline, cane sugar and Rochelle salt by PIERRE and JACQUES CURIE in 1880[24]. The findings were termed *piezoelectricity, piezo* from a Greek word meaning to press or squeeze and *electricity* is the end product. Whilst the CURIE brothers predicted and proved that electricity generated from applied stress to give the direct piezoelectric effect, they did not prove the converse effect. The converse effect (namely that application of electrical field creates a mechanical stress) was later mathematically proven by LIPPMAN in 1881 using the fundamental thermodynamic principles.

Although the piezoelectric discovery was a topic of great scientific interest, it was not until 1916 during WWI, PAUL LANGEVIN[49] produced the first engineering use for a piezoelectric material in a form of ultrasonic submarine detection.

Since the second world war, development of piezoelectric devices and materials in the United States has been kept within the companies doing the development. New materials were developed such as the quartz crystals which were the first commercially exploited piezoelectric material, but scientists searched for higher-performance materials. In contrast, Japanese manufacturers shared their findings, quickly overcoming technical and manufacturing challenges and creating new markets. Since late 60s, Japanese efforts in materials research created piezoelectric ceramic materials competitive to the U.S. materials, but free of expensive patent restrictions[49]. Major Japanese piezoelectric developments include new designs of piezoelectric ceramic filters for radios and televisions, piezoelectric buzzers and audio transducers that can connect directly to electronic circuits, and the piezoelectric igniters, which generate sparks for small engine ignition systems (and gas-grill lighters) by compressing a ceramic disc. Ultrasonic transducers that transmit sound waves through air had existed for quite some time, but first saw major commercial use in early television remote controls. These transducers now are mounted on several car models, helping the driver determine the distance from the rear of the car to any objects that may be in its path.

2.2.1 Simple quantitative theory of piezoelectricity

Piezoelectricity is a material property that is manifested when voltage is produced by applying mechanical forces, and vice versa, the effect has been described as direct and converse[25]. Piezoelectricity has been described as coupling between a quasi-static electric field and dynamic mechanical motion. Equation (1) below represents constitutive[51] equation of linear piezoelectricity and is based on principle of energy conservation.

$$\begin{aligned} \varepsilon &= C\sigma + d^T E \\ D &= d\sigma + eE \end{aligned} \tag{1}$$

where σ and ε are stress and strain tensors that satisfy the condition of symmetry, respectively, and D and E represents electric flux density and the electric field, respectively. C is the elastic compliance matrix, d is the matrix of piezoelectricity coefficient, d^T is the transpose of d and e represents the dielectric permittivity matrix. There are also other forms of linear piezoelectric constitutive equations which are given in the literature(26).

2.3 Materials characteristics

The decisive important property for the piezoelectric effect is the change of polarization when applying a mechanical stress. This might either be caused by a re-configuration of the dipole-inducing surrounding or by re-orientation of molecular dipole moments under the influence of the external stress. Piezoelectricity may then manifest itself in a variation of the polarization strength and its direction:

1. Orientation of dipoles within the crystal,
2. Crystal symmetry and
3. Applied mechanical stress
4. Intrinsic and extrinsic properties

Ceramic fibres in the diameter range of 10-250µm were used in a study(27). When formed into composite fibre they possess all the qualities of conventional block ceramic such as electrical, mechanical and chemical properties with additional advantages of light weight and flexibility. The piezoelectric fibre composites (PFC)(28) consist of unidirectional aligned piezoelectric fibres in an epoxy matrix, sandwiched between two copper clad polyimide laminates. The PFC devices have higher piezoelectric efficiency than traditional bulk piezoelectric ceramic materials, due to their large length to area ratio(17).
Typically, when in fibrous form, crystalline materials have much higher strengths and the polymer shell of the PFC allow the fibres to withstand impacts and harsh environments far better than monolithic piezoelectric ceramic materials. **Table 1** shows a typical comparison values for the polymer and PZT based piezoelectric material. An energy harvesting system, using flexible fibre composite transducer, is capable of producing and storing typically 880mJ (40V) of energy from a 30Hz vibration over 13 seconds period.

Property	**Units**	**PVDF Film**	**PZT**
Density	10^3kg/m^3	1.78	7.5
Relative Permittivity	$\varepsilon / \varepsilon_0$	12	1200
d_{31} Piezoelectric Constant	(10^{-12})C/N	23	110
g_{31} Voltage Constant	(10^{-3})Vm/N	216	10
k_{31} Electromechanical Constant	% at 1 KHz	12	30
Acoustic Impedance	(10^6)kg/m^2-sec	2.7	30

Table 1. Comparison of piezoelectric materials

Piezoelectric properties of porous PTFE foams have already been reported some time ago(29,30). The high surface charge stability of these materials in particularly elevated temperatures range has been confirmed(31,32,33) and also in independent experiments(34,35). To some extent, the stability of the surface potential and of the resulting piezoelectric response were found to improve with increasing porosity to some extent(23). For a single PTFE foam film, a high piezoelectric coefficient was found(36), but the effect decreased by a factor of 2 when uncharged PTFE films were inserted between the charged films and the electrodes(37). Consequently, single layer foams were studied and it became clear that part of the piezoelectric effect in this case stems from change in the air gaps between porous electret and disk electrodes(27,38).

Further work has been conducted where PTFE foam layers have been incorporated into multilayered sandwiches(26,39), i.e. double layer sandwiches of one hard and one soft layer with an interface charge layer between them(40,41,42). If two or more separate foams are to be assembled into such a sandwich, then air pockets have to be avoided.

Since the discovery of piezoelectric PVDF(43), it has been widely used in a range of sensors and actuator applications(44). In addition, several other polymers have been shown to have potentially useful piezoelectric properties, such as polyamides(45), copolymers of vinylidene cyanide (VDCN)(46), polyureas(47) and polyurethane (PU)(48) mainly due to their lightweight and flexibility. The two layer approach has been investigated with the polypropylene (PP) and PU foams(49), whilst further research has been investigated on stacks of corona-charged porous and non-porous PTFE(50) and reported piezoelectric d_{33} coefficient values of up to 35pC/N and as high as 150pC/N has been obtained with single layer of porous PTFE(51).

Cellular PP foams are usually produced in a modified blow extrusion process(52,53). Prior to foam blowing, spherical voids of approximately 10μm are generated by gas injection into the polymer melt. The melt is extruded, cooled down and reheated for foam blowing. Foam formation is accompanied by biaxial orientation, which results in disk or lens shaped voids(54). In addition, most cellular PP foams have co-extruded outer layer of non voided polymer for higher surface smoothness, better thickness uniformity and improved electrode adhesion(55). Charging of PP is done by means of corona charging with high corona point voltages around 20kV(36,56).

The resulting surface charge leads to high electric fields across the thickness of the foam and thus internal breakdown in disk-shaped voids(57,58,59). After breakdown, the voids are charged to top and bottom polarities that are opposite to respective surface charge and electrode polarities of the foam(37). In addition to corona charging, electrode charging(60,61,62) and electron beam charging(63) have been reported. **Figure 6** below depicts the typical microstructure of PP foam.

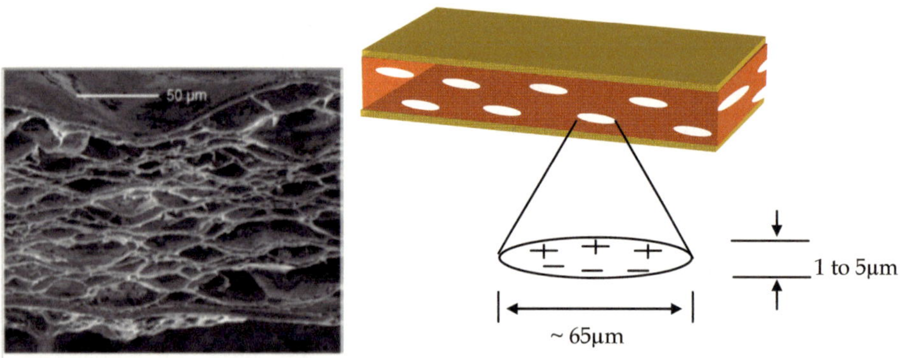

Fig. 6. SEM and schematic diagram of PP foam with elliptical voids

2.4 Poling methods

In order to understand the microscopic mechanism of charge storage, it is important to understand the characteristics of fundamental material parameters as well as the measurement of specific sample or device properties. Progress in the area of electrets research has been widely enhanced with the wide range of methods to characterise them with respect to their electrical, thermal, mechanical and chemical properties. To date, several characterisation techniques have been developed.

Over the past three decades, several techniques for obtaining space charge and polarisation depth profiles in insulating materials have been developed(64). These methods have been applied to a number of areas, such as accumulation of space charge in high voltage cable insulations(65), the development and optimisation pyroelectric and piezoelectric sensors(66,67) and basic research into mechanisms of charge storage(68) in electret polymers.

2.5 Production of PZT fibres

The nature of the ceramic monolithic piezoelectric material makes them brittle and vulnerable to accidental breakage during handling and bonding procedures as well as in the service intended. In addition, they have poor ability to conform to curved surfaces and are very dense and stiff causing mass loading and localized stiffness. These limitations have encouraged researchers to develop alternative methods of manufacturing the piezoelectric ceramics in order to make them more suitable for the next generation of piezoelectric applications.

In order to resolve the inadequacy of the monolithic piezoelectric ceramic material for many applications, the idea of using a composite material consisting of an active piezoelectric ceramic fibrous phase embedded in a polymeric matrix phase has been investigated by a number of researchers. In addition to this added strength of the base material, the flexibility of the polymer matrix allows the piezoelectric ceramic fibres to have greater conformability to curved surfaces and provides a protective shell around the piezoelectric material. This polymer shell allows the piezoelectric-fibre to withstand impacts and harsh environments far better than monolithic piezoelectric materials. The result of configuring the piezoelectric-fibre inside a polymer matrix is an actuator that can be incorporated into or bonded to more realistic structures.

This soft moulding process holds many advantages over the previously used die-and-fill, injection moulding or dicing techniques due to the moulds being reusable, allowing thousands of identical actuators to be made. In addition, the fabrication is relatively easy and cost effective because of the simplicity of the mould construction when compared to the other options for fibre construction. Once the brittle fibres are formed a polymer matrix is added to the remaining spaces to protect the fibres from breakage. Subsequent to this process is the addition of metal electrodes for poling of the device, later to be used as a means of applying an electric field or measuring the current produced during sensing.

Piezo ceramic fibres have been produced through a patented injection moulding process(69). The procedure combines PZT powder with a wax-based binder, and then the material is granulated as feedstock for the injection moulding process. Once this is completed the feedstock is heated to the specified viscosity and rapidly injected at high pressure into a cooled mould. Due to the incompressible nature of the material, when injected into the mould the fibres obtain a constant density throughout and remove of voids and internal defects that occur during dry pressing or low pressure forming methods. The homogeneous density of the material produces uniform microstructures, dimensions and electromechanical properties after firing.

After the production of the piezoelectric ceramic fibres an electrode layer is placed on the top and bottom of the fibres to facilitate the application of an electric field for the collection of current during sensing. The Active Fibre Composite (AFC) uses inter-digitised electrodes that allow the electrical potential to form along the length of the fibre; therefore capitalizing on the higher d_{33} piezoelectric coupling coefficient. A schematic diagram to illustrate the electric field developed along the fibres is shown in **Figure 7**. The metallic electrodes are normally formed using photolithography, which is a very precise process, however it is a time consuming and expensive process that etches copper strips onto a thin Kapton® film.

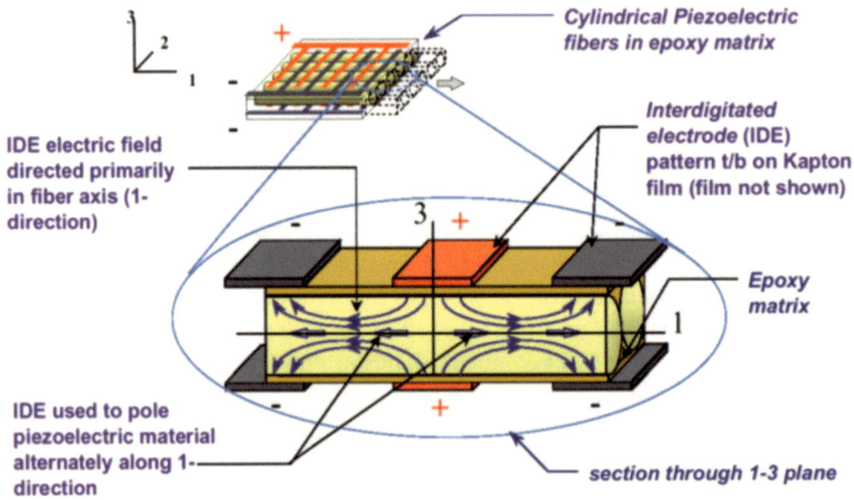

Fig. 7. Schematic of the cross section of an AFC actuator(164)

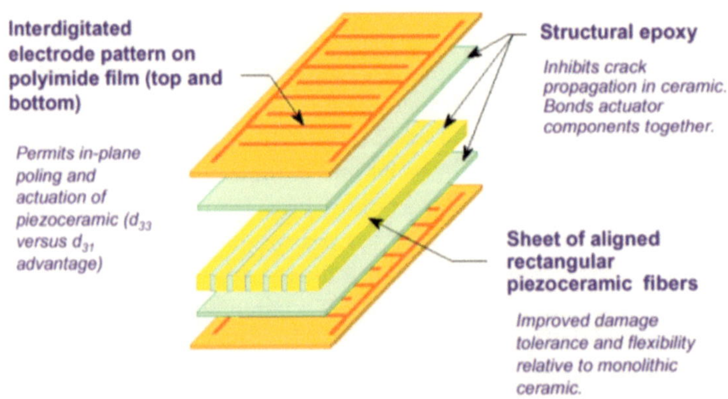

Fig. 8. Schematic showing the order of different layers in the macro-fibre composite actuator.

The final product is an actuator that has increased flexibility, durability and electromechanical coupling due to the inter-digitated electrodes that capitalize on the higher d_{33} coefficient. This type of technology has many possible structural applications; however, the assembly can be avoided and the fibre can be incorporated directly into any laminated composite material. One of the major considerations for the use of this property is to incorporate the fibres into the rotor blades of a helicopter. The helicopter blades experience large aerodynamic loads and in turn induce large vibration and noise issues(162). With the active fibre embedded inside the rotor blade the vibration and noise levels inside the helicopter could be significantly reduced providing increased comfort to the pilot and passengers of the aircraft. The layers of the MFC are shown in **Figure 8**. However, the MFC has one difference that greatly affects the manufacturing process and the performance of the actuator, it has rectangular fibres.

2.6 Production of PVDF Films

Rolls of piezoelectric PVDF film are produced in a clean room environment. The process begins with the melt extrusion of PVDF resin pellets into sheet form, followed by a stretching step that reduces the sheet to about one-fifth its extruded thickness. Stretching at temperatures well below the melting point of the polymer causes chain orientation of the molecules into parallel crystal planes. These are called "beta phase" materials. To obtain high levels of piezoelectric activity, the beta phase polymer is then exposed to very high electric fields to align the crystallites relative to the poling field(70).

Copolymers of PVDF are polarisable without stretching. Evaporative deposited metals are typically 500 to 1000Å in thickness, and almost any metal can be deposited. Popular metals are nickel, aluminium, copper, gold and alloys. Electrode patterns are made by sputtering through masks or by chemical etching continuous metallization using photo resists. Resolution to 25μm line widths has been achieved. Screen printed electrodes of conductive silver ink are much thicker, about 5-10μm, and can be applied in complex patterns to form multiple sensors on a single sheet. Foils are adhered with thin adhesive layers coupled to the piezoelectric film(165).

Generally, sputtered metals are for very high resolution arrays, pyroelectric applications requiring a low thermal mass, or for inertness, as with invasive medical applications. Fully metallised sheets can by carefully cut with a razor blade without shorting across the film thickness. Screened inks are very robust and compliant, withstand very high strains (>10%), can operate at high voltages without breakdown, and are easy to pattern on a continuous basis. However, un-metallised borders are required for cutting elements out of a sheet of screen printed electrodes, since there is a high likelihood of shorting across the films thickness with the thick inks. Foils may mechanically restrict the piezoelectric film from responding to externally applied stresses and strains in the plane of the film, but foils are useful in pure "thickness mode" operation(165).

After metallization, a wide variety of possible processing steps are followed to produce a packaged sensor. Generally, the piezoelectric film is laminated in a protective carrier then die cut to size, and packaged with lead wires or crimp connectors(165).

2.7 Production of PP foams

An electromechanical film is a thin porous PP film with biaxially oriented flat voids, with a lateral dimension of 10μm to 100μm and a vertical dimension of 1μm to 5μm(71), as shown in **Figure 6**. This internal structure is obtained through extrusion, biaxial stretching, and controlled inflation by patented pressure treatments(72). Under the effect of high electric field, during the manufacturing process, the heterogeneous foam film acquires a permanent space-charge, with the upper surface of the gas voids with the polarity opposite to the lower surface. The charged voids become perfectly oriented quasi-dipoles, as shown in **Figure 9**(73). These artificial dipoles are responsible for the macroscopic piezoelectric behaviour of these cellular polymers.

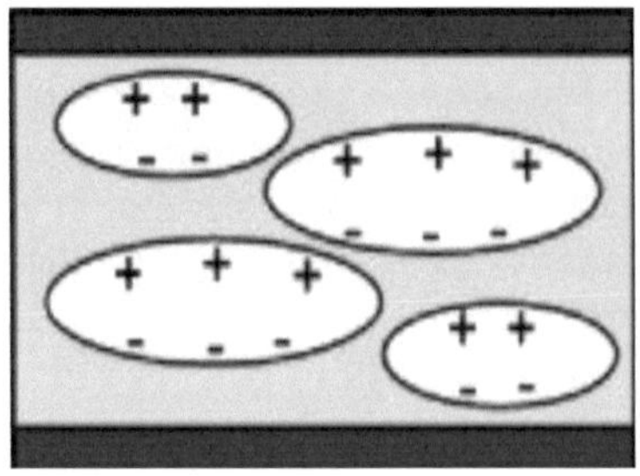

Fig. 9. Charged voids in the foam (*which produce the quasi-piezoelectricity*) represent the structure and the inhomogeneous charge distribution.

The transducer is obtained after evaporating or gluing electrodes to the external faces of the charged-foam film. From the application point of view, it is important to take into account that if the metal coating or the glue is too rigid, the piezoelectric response of the film (a charge) will be small, causing a shift of the piezoelectric resonance peak(74).

In summary, ferroelectrics polymers have some advantages compared to conventional piezoelectric materials based on ceramics or polar polymers. Apart from being thin and lightweight, the polymer piezoelectric material have high d_{33} coefficients, and may be obtained in almost any size and/or shape. They can be easily handled and they are low cost. Their relatively small specific acoustic impedance is an advantage for many transducer applications. Nevertheless, their main drawback is the reduced stability of the piezoelectric effect at temperatures over 50°C, and this can limit their application range.

2.8 Piezoelectric applications

Electret condenser microphones are still the most widely used applications for charge storing polymers, however, a number of other publications(75,76) have demonstrated the feasibility of using inorganic SiO_2 layers or SiO_2/SiO_3N_4 double layers as charge storing membranes. Generally, polymers are the materials of choice for large scale pyroelectric and piezoelectric applications. The main drawbacks of ferroelectric polymers as compared to ceramic materials are their relatively low piezoelectric and pyroelectric coefficients.

Significant advances have been realized in an effort to increase the piezoelectric coefficient of polymer films when results from electric dipoles at a macroscopic, rather than molecular level were considered. For example, KACPRZYK elaborated on a two-layer system of corona charged PP and PU films(77), and GERHARD-MULTHAUPT *et al.* investigated stacks of corona charged porous and non-porous PTFE(78) and reported piezoelectric d_{33} coefficient of up to 35pC/N. Even higher values of up to 150pC/N were obtained by KUNSTLER *et al.* with single layers of porous PTFE carrying a bipolar space charge(79). However, most significant new class of cellular space charge electrets is based on a concept reported by KIRJAVAINEN(80).

Voids are created by means of biaxially stretching a composite film consisting of a PP polymer host and inorganic fillers. After inflation, typical void dimensions are 30μm to 100μm in diameter and approximately 1μm to 5μm in height can be achieved. Similar films are also extensively used in the packaging industry, where the micro voids serve both as thermal insulation and give the polymer film an attractive glossy look(81). Subjecting these films to high electric fields is thought to initiate breakdown, with subsequent charge separation and charge trapping on the inner surface of the voids. This is supported by several recent experimental observations.

Micro discharges have been observed through their luminescence[82,83], and the bipolar charge distribution was visualized via the secondary emission yield in a SEM[84]. The induced microscopic polarization can be reversed by applying a sufficiently high electric field in the opposite direction[173] (a characteristic of ferroelectric material). Furthermore, electrical and electromechanical hysteresis was also observed[85], although this behaviour was recently shown not to be related to the polarization, but rather results from interfacial charge injection and charge generation by external gas discharges at the electrode edges[86]. As a result of their piezoelectric properties and switchable polarization, both of which are characteristic properties of ferroelectric materials, the term ferroelectrets has also been used[175].

Applying a mechanical stress to foam materials changes the vertical dimensions of the voids and therefore the dipole moment of the separated positive and negative charges trapped on the internal surfaces of the voids, resulting in macroscopic polarization changes, giving rise to a qausi-piezoelectric response. As these polymer foams are relatively soft perpendicular to their film plane, the resulting piezoelectric d_{33} coefficients are at least 1 order of magnitude larger than those found in conventional ferroelectric polymers, such as PVDF[87]. Through charging in a suitable dielectric gas atmosphere at elevated pressure[88], these coefficients were increased up to 790pC/N. The shape of the cavity can be optimized by means of a gas inflation treatment[89], resulting in even higher coefficient of up to 1500pC/N[90,91]. Biaxially stretched polymer foams are highly anisotropic, and as a result, exhibits large d_{33} coefficients but relatively small d_{31} values and little pyroelectricity[92].

Based on these piezoelectrets, several applications for large area transducers have been suggested and some are already commercially available. For example, using large mats of these films it is possible to monitor the motion of people and even identify individuals[93] based on single footsteps. Cellular PP foams have also been used for monitoring physiological processes in humans and animals[94]. Another promising area of applications is active vibration and noise control[95]. Numerous transducer applications have been suggested, demonstrated, and commercialized with different kinds of piezoelectric materials[96]. They cover a broad frequency range from quasi-static applications to the audio, super-audio, and ultrasound frequency range. Cellular polymers may find their niches in several of these applications, where the low cost and large-area processing of such polymers are advantageous. Therefore, cellular polymers have attracted a lot of interest in engineering studies, but they are not without their disadvantages[97].

2.8.1 Sensors for motion control and pressure measurement

Detection of motion control, for example, the traffic on the roads or the motion of humans in houses or factory buildings for medical or for safety observations is usually performed by video cameras, infrared detectors, or individual piezoelectric sensors. By placing cellular polymer films on the ground, surveillance of rooms and of the surrounding of machines can be easily achieved[98]. Due to the high transducer sensitivity it is possible to place the sensor film underneath different floor coatings such as polymers, woods, ceramics, or stones[194].

Pressure measurements with cellular polymers can also be used in sports facilities to optimize the training of athletes. One example is the implementation of cellular piezoelectric transducers in ski jumping, where cellular polymers allow for the force measurement during the jumpers takeoff[99]. Besides the monitoring of the pressure distribution on a floor, the pressure distribution in soles of shoes, for example, can be determined during running or walking[194,100].

For orthopaedic diagnostics, pressure monitoring on seats and backrests is very interesting, too. The results could lead to further optimization of office chairs or of seats in cars, trains, and aircraft. The sensitivity of cellular polymer transducers is also high enough for recording the respiration of human patients, even if the transducers are not directly fixed to the skin of the patient(101). Sensors made of cellular PP have been placed directly on the bed for long-term respiration monitoring. Usually, the sensor signals measured during the movement of the patient are superimposed on the respiration signal. Cellular polymer films also allow for measuring pressure distributions within the body itself.

2.8.2 Control panels and keyboards

Push buttons for keyboards, keypads, and control panels with small areas have been made with cellular piezoelectric polymer films. Keypads are commercially produced by the Screentec company in Finland, and are used for example, in interfaces for information systems in public transportation or as access readers for different applications(102). The high sensitivity allows the implementation of the sensors behind protective layers of different materials (with a thickness up to 5mm) to achieve vandal-proof control panels usable for cash dispensers and ticket machines or other systems in public transportation(201,103).

Based on soft and flexible polymer films, the construction of flexible keypads or keypads covering any shape, see **Figure 10**. In the long run, flexible keypads including electronic circuitry may become a reality, since organic electronic circuitry is rapidly developing towards commercialization. A mechanical pressure on one of the push buttons with a load of, for example, 2N leads to a sensor voltage signal of around 200mV with a rise time of only 30ms, which is easily detected with standard electronics.

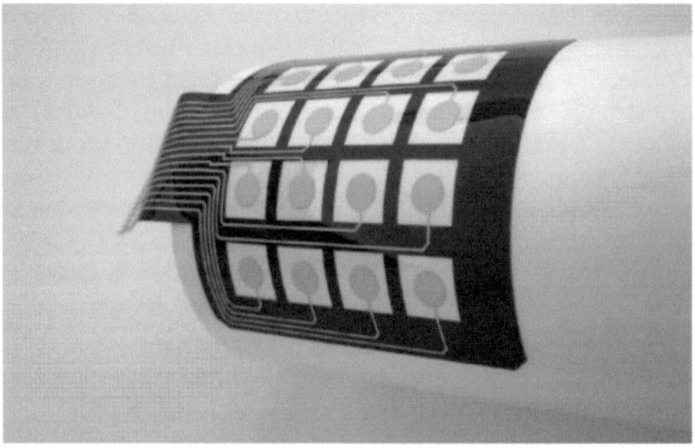

Fig. 10. Schematic sketch of a curved sensor arrangement(201)

2.8.3 Electro acoustics

There is a large potential for applications of internally charged cellular ferroelectrets in electro acoustic applications, where the frequency range from the audio to the ultrasound can be spanned. Microphones and especially pickups based on cellular PP sensors are already on the market, produced by the companies Emfit Ltd. and B-Band Ltd. in Finland(195,104). The commercial pick-ups show a bandwidth from 50Hz to 23kHz, a low distortion of 0.05%, and a good signal-to-noise ratio exceeding 90dB.

Based on optimized preparation methods including the formation of multilayer stacks, the sensitivity could be increased significantly(105). Piezoelectric coefficients between 100pC/N and 1000pC/N are now routinely achieved, depending on the preparation technique, thereby increasing the microphone sensitivity by a similar level(106). Cellular polypropylene films can also be used as microphones in the super audio frequency range(212) up to 40kHz. A bandwidth exceeding 315kHz was demonstrated, thereby opening routes for the detection (and generation) of airborne ultrasound(107). However, the strong sensitivity to pressure changes limits the use of cellular polymer hydrophones to shallow-water applications.

3. Experimental methodology and materials

In order to establish and compare the voltage output for all 3 piezoelectric materials, initially, vibration experiments were carried out using an electro mechanical shaker. The electro mechanical shaker vibrated the attached plastic ruler which had the piezoelectric material attached on the top surface of the ruler, frequency range from 0 to 120Hz were selected. The second part of the vibration experiment consisted of manually vibrating the piezoelectric materials using a metallic and plastic platform and comparing the energy outputs. Finally, this chapter will also compare and discuss the results output for all 3 piezoelectric materials when subjected to impact forces from a given height using 4.9g, 11.9 and 1.02kg mass using a free fall height experimental set up and a fixed structure.
Optical microscopy will be performed on the PZT-based 125μm piezoelectric material and Scanning Electron Microscope (SEM) will be carried out on the rest of the piezoelectric material to establish and verify manufacturers specifications.

3.1 Materials investigated

There are two common piezoelectric materials - polymer based PVDF and ceramic based PZT. The polymer materials are soft and flexible; however, they possess lower dielectric and piezoelectric properties than ceramics. Conventional piezoelectric ceramic materials are rigid, heavy and can only be produced in block form.
A variety of piezoelectric materials have been used in this study. Two ceramic materials, with active piezoelectric fibres of 250μm and 120μm diameters were embedded in a polymer matrix and encapsulated in copper-clad laminate, see **Figure 11** left and **Table 3**. A ceramic so called bimorph material consisting of two 250μm fibre diameter materials adhered to either side of a rigid metal centre shim material.
Laminated piezoelectric polymer material, PVDF, where two 125μm polyester laminates were bonded either side of a 28μm thick piezoelectric film element (28μm laminate), two un-laminated PVDF materials of 28μm and 52μm thicknesses (28μm and 52μm un-laminated, respectively), see **Figure 11** middle and **Table 2.**
Finally, a fully shielded, low mass, thin ribbon PP sample was used. The sample consisted of a sensing element constructed of elastic electret, 3 layers of polyester film. Aluminium electrodes with crimped connectors were used for connecting to electrodes and double-sided sticky tape for convenience, see **Figure 11** right and **Table 2**.
The dimensions and classification of the piezoelectric polymer and ceramic materials are given in **Tables 2** and **3**, respectively. The piezoelectric PZT fibre composites were obtained from Advanced Cerametrics Incorporated (ACI)(108). The PVDF polymer films were manufactured by Measurement Specialities Incorporated (MSI)(109) and piezoelectric PP polymer foam as manufactured and supplied by Emfit.

Material	Sample Width (a) (mm)	Sample Length (b) (mm)	Sample thickness (μm)
28μm laminated PVDF (LDT1-28μm)	16	41	205
28μm laminated PVDF (LDT2-28μm)	16	73	205
28μm laminated PVDF (LDT4-28μm)	22	171	205
28μm un-laminated PVDF (DT4-28μm)	22	171	40
52μm un-laminated PVDF (DT4-52μm)	22	171	70
65μm laminated PP	200	100	320

Table 2. Characteristics of Piezoelectric PVDF and PP Materials Used

Material	Sample width (c) (mm)	Sample Length (d) (mm)	Sample thickness (μm)
120μm fibre	15	140	190
250μm fibre	15	140	320
250μm fibre Bimorph	15	140	750

Table 3. Characteristics of Piezoelectric PZT Materials Used

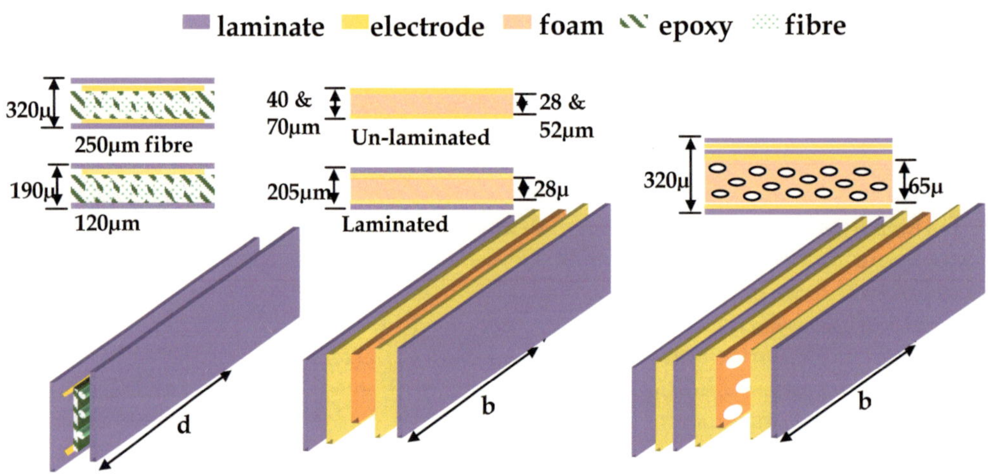

Fig. 11. Schematic representation of non symmetrical piezoelectric samples, left, Piezoelectric PZT fibre embedded in an epoxy with copper clad electrodes etched on to the inner surface of the laminate which acts as electrodes, middle, PVDF membrane film and right, porous PP with pores (app. 10μm diameter).

Ceramic materials add additional mass and stiffness to the host structure, especially when working with flexible/lightweight materials. This and their fragile nature limit possibilities for wearable devices, whereas the polymer membrane and the foam are very versatile, cost effective and able to conform to intricate shapes and designs.

This work has directly compared PZT, PVDF and PP piezoelectric materials for their ability to generate energy in terms of voltage output. Each category of material behaved differently under varying conditions. The initial vibration experiments using electro mechanical shaker and frequency between 0-120Hz have shown that the single layer of PZT 120μm and 250μm produced larger voltage output when compared with the double layer structures. However, this could have been due to inadequate bonding between the multilayer specimens. Similar conclusions can be drawn from the results for the single and multilayer PVDF specimens subjected to the same experimental procedures.

4. Conclusion and future work

Several experiments were conducted in order compare and contrast the voltage output between the PZT and PVDF piezoelectric material. Overall, the PZT bimorph produced largest voltage output under all experimental conditions, followed by the single layer 250μm sample and 120μm single layer specimen. PVDF based structures produced the lowest voltage output under these conditions.

Under impact testing using 4.9g and 11.9g mass from a fixed height, it was again shown that the PZT 250μm single layer specimen produced higher voltage output when both masses were dropped independently on to the specimen. However, PZT 120μm displayed the opposite effect, where the double layer generated marginally higher voltage in comparison to single layer. This could be due to the inconsistencies within the supplier batch of the material, as well as the lack of efficiency of bonding within the multilayer structure. In any case, the heavier mass i.e. 11.9g delivered larger voltage outputs for all the PZT and PVDF piezoelectric materials due to larger surface area contact between the heavier mass and the surface of the specimen, and the generating larger striking force.

In order to establish whether impact at various positions of the sample delivered any effect, the samples were subjected to impact analysis using an Instron machine with 1.02kg mass striking the sample on both ends and at the centre. No significant variation in the voltage output was observed due to the positional impact for most of the samples. However, for the PZT Bimorph structure the voltage output was three times greater when the impact was made at the edges than values obtained for the central impact. This is due to the presence of the metal shim between the layers, which produced additional vibrations within the structure. On the other hand the impact at the centre of the material may also have resulted in cancellation of the out-of-phase vibrations.

Notable variations were recorded for 3 of the 4 batches of the same PVDF material i.e. LDT1-28μm (smaller laminated) piezoelectric sample subjected to similar analysis. The higher voltage output was associated with the edge closest to the positive and negative terminals. This level of variation was evident for all PVDF configurations.

Controlled vibration of PZT and PVDF materials in contact with a flexible ruler indicated that generally, the metallic ruler provided largest amount of energy due to the better vibration properties compared with the plastic ruler. Again, PZT bimorph specimen displayed the largest voltage output out of the three materials investigated. It was also found that, generally, as the distance of the specimen from the fixed edge increased so did the voltage resonance and the collective total voltage output.

During temperature elevation experiments, it was found that as the temperature increased, the voltage output dropped for the PZT material, whereas the converse effect was observed with PVDF. This appears to be possibly due to reversion of the PZT material back to the post piezoelectric state, particularly as the polymer matrix in which the PZT piezoelectric fibres are embedded and consequently aids the temperature increase of the structure. As for the PVDF, the internal chains $-(C_2H_2F_2)_n$ are excited and become more mobile, hence, delivering voltage at an increased rate.

The subzero temperature experiment (from room to approximately -30°C) showed some interesting phenomena in terms of voltage output. For all of the three piezoelectric materials, energy output generally increased as the temperature began to drop. This may be the result of material shrinkage at sub-zero temperatures - causing compaction of the atoms with within the material and therefore increasing the efficiency of electron transfer processes.
All three materials showed positive and negative attributes in terms of energy output depending on the conditions. In terms of maximum voltage output, the PZT Bimorph specimen generated largest voltage output, however, at a cost of increased weight, lack of flexibility and due to additional manufacturing and production costs. The application PZT piezoelectric material will not be suitable at elevated temperature due to reverse polarisation process and the very nature of the fragile fibres.
The work reported in this thesis highlighted many aspects of various piezoelectric materials and their commercial and domestic applications. This research was aimed at comparing the 2 variants of piezoelectric materials, namely PZT-based and PVDF-based piezoelectric under normal room temperature, elevated and sub-zero temperatures. This work has opened many opportunities for future work, particularly in-house production of these materials which will provide huge flexibility. Future work will provide good opportunities to develop piezoelectric materials/structures with enhanced flexibility and voltage output of several magnitudes higher than those reported to date.
So far we have investigated piezoelectric materials which are produced by particular manufacturers with standard sizes of the specimen and limited material configurations. In terms of PZT, the piezoelectric materials that were available at the time were limited to 120μm and 250μm fibre diameters - as a single layer, bimorph and 4 layer structures. The future work should incorporate various fibre diameters and pitch optimisation when laying them onto the Kapton film. Multi-stacking of fibres within the host structure and variation in terms of several piezoelectric materials placed in the sandwich form - attached either with standard resin and / or other means, with metal shim or without metal shim.
The work on the PVDF piezoelectric materials could involve the study of the effect of material thickness to ascertain higher voltage output without compromising weight and flexibility. Also, the possibility of multi-layer structures should be considered, either with or without metal shim in between, as well as investigating variations of the dimensions and thickness of the sensing material.

5. References

[1] Energy - Consumption' A1 "Consumption by fuel, 1965 - 2008" (XLS). Statistical Review of World Energy 2009, BP. July 31, 2006. http://www.bp.com/liveassets/bp_internet/globalbp/globalbp_uk_english/reports_and_publications/statistical_energy_review_2008/
[2] GOSS LEVI, B (2000). *Nobel Prize in Chemistry Salutes the Discovery of Conducting Polymers,* Physics Today, 53, 19
[3] VAN TURNHOUT, J. (1975). *Thermally Stimulated Discharge of Polymer Electrets,* Elsevier, Amsterdam
[4] KAWAI, H. (1969). *The piezoelectricity of Polyvinylidene Fluoride,* Jpn. J. Appl. Phys., 8, 975-977 (1969).
[5] BAUER, S., NGERHARD-MULTHAUPT, R. & SESSLER, G.M. (2004). *Ferroelelectrts: Soft Electroactive Foams for Transducers,* Physics Today, 57, 37-34
[6] EMFIT LTD., Vaajakoski, Finland, http://www.emfit.com/

[7] Sheneck, N.S., Paradiso, J.A. (2001). *Energy Scavenging with Shoe-Mounted Piezoelectrics,* IEEE Micro

[8] Starner, T. (1996). *Human Powered Wearable Computing,* IBM Systems J., vol. 35, nos. 3 and 4, 618-629

[9] Marsden, J.P. & Montgomerey, S.R. (1971). *Plantar Power for Arm Prosthesis Using Body Weight Transfer,* Human Locomotor Engineering, Ins. Mechanical Engineers Press, London, 277-282

[10] Fletcher, R. (1996). *Force Transduction Materials for Human technology Interface,* IBM Systems J., vol. 35. 3 and 4, 630-638

[11] Antaki, J.F. et al. (1995). *A Gait Powered Autologous Battery Charging System for Artificial Organs,* Proc. 1995 American Society of Artificial Internal Organs Conf., Lippincott Williams & Wilkins, Philadelphia, M588-M595

[12] Kymissis, J. et al. (1998). *Parasitic Power Harvesting in Shoes,* Second IEEE Int'l Conf. Wearable Computing, IEEE CS Press, Los Alamitos, Calif., 132-139

[13] Want, R. et al. (1992). *The Active badge Location System,* ACM Trans. Information Systems, vol 10, no. 1, 91-102

[14] Starner, T. (1996). *Human-Powered Wearable Computing, IBM System Journal,* vol. 35, no. 3&4, pp. 618-629

[15] Umeda, M., Nakamura, K. & Ueha, S. (1996). *Analysis of Transformation of Mechanical Impact Energy to Electrical Energy Using a Piezoelectric Vibrator,* Japanese Journal of Applied Physics, Vol. 35, Part 1, No. 5B, May, pp. 3267-3273.

[16] Kymissis, J. et al. (1998). Parasitic Power Harvesting in Shoes, Second IEEE International Conference on Wearable Computing, pp. 132-139.

[17] Kawai, H. (1969). *The Piezoelectricity of Poly9vinylidene Fluoride).* Jpn. J. Appl. Phys. 8, 975-977

[18] Fukada, E. (2000). *History and Recent Progress in Piezoelectric Polymers,* IEEE Trans. Ultrason., Ferroelectr., Freq. Contr. 47, 1277-1290

[19] Newman, B. et al. (1992). *A New Class of Ferroelectric Polymers, The Odd-numbered Nylons.* Ferroelectrics 127, 229-234

[20] Miyata, S., et al. (1980). *Piezoelectricity Revealed in the Copolymer of Vinylidene Cyanide and Vinyl Acetate.* Polym. J. 12, 857-860

[21] Tasaka, S., et al. (1992). *Ferroelectric Polarization Reversal in Polyureas With Odd Number of CH_2 Groups.* JPN. J. Appl. Phys. 31, L1086-L1088

[22] Tasaka, S., et al. (1994). *Ferroelectric Behaviour in Aliphatic Polyurethanes.* Jpn. J. Appl. Phys. 33, 1376-1379

[23] Tasaka, S., et al. (1995). *Ferroelectric Behaviour in Aliphatic Polythioureas.* Ferroelectrics, 203-210

[24] Sodano, H.A. (2003). *Macro-Fibre Composites for Sensing, Actuating and Power Generation,* Blacksburg, Virginia, 1-12

[25] Schwartz, M. (2002). *Encylopedia of Smart Materials Vol. 1 and 2,* pp. 780-792

[26] Ikeda, T. (1990). *Fundemantals of Piezoelectricity,* Oxford Sci. Oxford

[27] Mohammadi, F., et al. (2003). "Power Generation from Piezoelectric Lead Zirconate Titanate Fiber Composites", *Symposium D," Electronics on Unconventonal Substrates--Electrotextiles and Giant-Area Flexible Circuits" as held at the 2002 MRS Fall Meeting,* pp. D5.5.1

[28] Williams, R.B. (2002). "An Overview of Composite Actuators with Piezoceramic Fibers", *Proceedings of 20th International Modal Analysis Conference,* pp. 0704

[29] Xia, Z. et al. (1997). *Ann. Rep., Conf. Electr. Insul. Diel. Phenom.,* IEEE, 471-474

[30] Cao, Y., et al. (1998). *IEEE Trans. Diel. Insul.* 5, 58-62
[31] Xia, Z., et al. (1999). *J. Physics. D:Appl. Phys.* 32, L83-85
[32] Xia, Z., et al. (1999). *Proc. 10th Int. Symp. Electrets,* IEEE, 23-26
[33] Xia, Z., et al. (2002). *Influence of Porosity on Stability of Charge Storage and Piezoelectricity for Porous PTFE Film Electrets,* these proceedings
[34] Schwodiauer, R., et al. (2002). *IEEE Trans. Diel. Electr. Insul.* 7, 578-586
[35] Schwodiauer, R., et al. (1999). *Proc. 10th Int. Symp. Electrets,* IEEE, 313-316
[36] Kunstler, W., et al. (2002). *Appl. Phys. A,* 70, 5-8
[37] Gerhard-Multhaupt, R., et al. (2000). *IEEE Trans. Diel. Electr. Insul.* 7, 480-488
[38] Wegener, M., et al. (2001). *Proc. 4th Int. Conf. Electr. Charges in Non-Conduct. Mater.,* Soc. Frans. Du Vide, Paris, 257-260
[39] Gerhard-Multhaupt, R., et al. (1999). *Proc. 10th Int. Symp. Electrets,* IEEE, 273-276
[40] Kacprzyk, R. & Motyl, E. (1994). *Proc. 8th Int. Symp. Electrets,* IEEE, 703-708
[41] Kacprzyk, R., et al. (1995). *J. Electrostatics,* 35, 161-166
[42] Kacprzyk, R., aet al. (1997). *J. Electrostatics,* 39, 33-40
[43] Kawai, H. (1969). *The Piezoelectricity of Poly(vinylidene Fluoride),* Jpn. J. Appl. Phys. 8, 975-977
[44] Fukada, E. (2000). *History and Recent Progress in Piezoelectric Polymers,* IEEE Trans. Ultrason. Ferroelectr. Freq. Contr. 47, 1277-1290
[45] Newman, B., et al. (1992). *A New Class of Ferroelectric Polymers, The Odd Numbered Nylons,* Ferroelectrics 127, 229-234
[46] Miyata, S., et al. (1980). *Piezoelectricity Revealed in the Copolymer of Vinylidene Cyanide ans Vinyl Acetate,* Polym. J. 12, 229-234
[47] Tasaka, S., et al. (1992). *Ferroelectric Polarization Reversal in Polyureas With Odd Number of CH_2 Groups.* Jpn. J. Appl. Phys. 31, L1086-L1088
[48] Tasaka, S., et al. (1994). *Ferroelectric Behaviour in Aliphatic Polyurethanes,* Jpn. J. Appl. Phys. 33, 1376-1379
[49] Kacprzyk, R. (1995). *Piezoelectric Properties of Nonuniform Electrets,* J. Electrostatics 35, 161-166
[50] Gerhard-Multhaupt, R., et al. (1999). *Priliminary Study of Multi Layer Space Charge Electrets With Piezoelectric Properties From porouse and Non Porouse Teflon Films,* In: *Proceedings, 10th Int. Symp. On Electrets,* IEEE Service Centre, Piscataway, NJ
[51] Kunstler, W., et al. (2000). *Piezoelectricity of Porouse Polytetrafluoroethylene Single and Multiple Film Electrets Containing High Charge Densities of Both polarities,* App. Phys. A 70, 5-8
[52] Kirjavainen, K. (1987). *U. S. pat.* 4,654,546
[53] Savolainen, A., et al. (1989). *Sci. – Chem.* A26, 583-59
[54] Gerhard-Multhaupt, R. (2002). *Voided Polymer Electrets – New Materials, New Challenges, New Chances,* 11th Int. Symp. On Electrets, 36-45
[55] Paajanen, M., et al. (2002). *Sens. Actuat.* 84, 95-102
[56] van Turnhout, J. (1999). in: *Proc. 10th Int. Symp. Electrets,* IEEE, 785-788
[57] Crichton, G.C., et al. (1989). *IEEE Trans. Electr. Insul.* 24, 335-342
[58] McAllister, I.W. (1992). *IEEE Trans. Electr. Insul.* 27, 1202-1207
[59] McAllister, I.W. (1997). *IEEE Trans. Electr. Insul.* 4, 456-461
[60] Lindner, M. et al. (2002). *J. Appl. Phys.* 91, 5283-5287
[61] Lindner, M. et al. (2002). *Dialectric Barrier Micro Discharges: Mechanism for the Charging of Piezoelectric Polymer Foams,* these proceedings
[62] Baur, S. et al. (2002). *Physics of Electromechanically Active Cellular Materials,* these proceedings

[63] Paajanen, M. et al. (2001). *IEEE Trans. Diel. Electr. Insul.* 8, 629-636
[64] Baurand, S. & Baur-gogonea, S. (2003). *Current practice in space charge and polarisation profile measurements using thermal techniques.* IEEE Trans. Diel. Electr. Insul. 10, 883-902
[65] Bambery, K.R. & Fleming, R.J. (1998). *Space charge accumulation in two power cable grades of XLPE.* IEEE Trans. Diel. Electr. Insul. 5, 103-109
[66] Baur, S. (1996). *Poled polymers for sensors and photonic applications.* J. Appl. Phys. 80, 5531-5558
[67] Wegener, M. et al. (2004). *Electric poling and electromechanical charecterisation of 0.1mm thick sensor films and 0.2mm thick cable layers from piezoelectric poly(vinylidene fluoride-trifluoroethylene).* IEEE Trans. Ultrason. Ferroelectric., Freq. Contr. 50, 921-931
[68] Mellinger, A. et al. (2004). *Photostimulated discharge in electret polymers: an alternative approach for investigating deep traps.* IEEE Trans. Diel. Electr. Insul. 11, 218-226
[69] Gentilman, R. Et al. (2003). *Enhanced Performance Active Fiber Composites,* Proceedings of SPIE's 10th Symposium on Smart Structures and Materials Conference, March 6
[70] Measurement Specialist Inc., 1000 Lucas Way, Hampton, VA 23666, Piezo Film Sensors Technical Manual, www.meas-spec.com
[71] Lindner, M. et al. (2004). Charged cellular polymers with "ferroelectretic" behavior, *IEEE Transactions on Dielectrics and Electrical Insulation* 2, 255–263
[72] Kirjavainen, K. (1987). lectromechanical film and procedure for manufacturing same, U.S. Patent No. 4,654,546
[73] Lindner, M. et al. (2002). Dielectric barrier microdischarges: mechanism for the charging of cellular piezoelectric polymers, *Journal of Applied Physics* 91, 5283–5287
[74] Neugschwandtner, G.S. et al. (2000). Large and broadband piezoelectricity in smart polymer-foam space-charge electrets, *Applied Physics Letters* 77, 3827–3829
[75] Amjadi, H. & Sessler, G.M. (1995). *Inorganic Electret Layers for Miniaturized Devices.* In: *Annual Report, Conference on Electrical Insulation and Dielectric Phenomena,* pp. 668-671, IEEE Service Centre, Piscataway
[76] Amjadi, H. (2000). *Electret Membranes and Backelectrodes for Application in Micromechanical Transducers.* J. Electrostatics 48, 179-191
[77] Kacprzyk, R. (1995). *Piezoelectric Properties of Nonuniform Elecetrets,* J. Electrostatics 35, 161-166
[78] Gerhard-Multhaupt, R. et al. (1999). *Preliminary Study of Multi Layer Space Charge Electrets with Piezoelectric Properties from Porous and non Porous Teflon Films.* In: *Proceedings, 10th Int. Symp. On Electrets,* IEEE Service Centre, Piscataway, NJ
[79] Kunstler, K. et al. (2000). *Piezoelectricity of Porous Polytetrafluoroethylene Single and Multiple Film Electrets Containing High Charge Densities of Both Polarities.* App. Phys. A 70, 5-8
[80] Kirjavainen, K. (1987). *Electromechanical Film and Procedure for Manufacturing Same.* US Patent No. 4,654,546
[81] Treofan, Reunheim, Germany, http://www.treofan.com
[82] Wegner, M. et al. (2002). *Corona-induced Partial Discharges, Internal Charge Seperation and Electromechanical Transducer Properties in Cellular Polymer Films.* In: *Proceedings, 11th International Symposium on Electrets,* 54-57, IEEE Service Centre, Piscataway
[83] Lindner, M. et al. (2002). *Dielectric Barrier Microdischarges: Mechanism for the Charging of Cellular Piezoelectric Polymers.* J. Appl. Phys. 91, 5283-5287
[84] Hillenbrand, J. & Sessler, G.M. (2000). *Piezoelectric Properties of a Polypropylene/air and Poly(vinylidene Fluoride)/air Composites.* In: *Annual Report, Conference on Electrical Insulation and Dielectric Phenomena,* 161-165, IEEE Service Centre, Piscataway

[85] Bauer, S. et al. (2004). *Ferroelectrets: Soft Electroactive Foams for Transducers.* Physics Today 57, 37-43
[86] Bauer, S. et al. (2004). *Do Ferroelectrets Always Behave Like Ferroelectrics?.* In: *Annual Report, Conference on Electrical Insulation and Dielectric Phenomina,* IEEE Service Centre, Piscataway
[87] Kressmann, R. (2001). *Linear and Nonlinear Piezoelectric Response of Charged Cellular Polypropylene.* J. Appl. Phys. 90, 3489-3496
[88] Paajanen, M. et al. (2001). *Charging of Celluar Space Charge Electret Films in Various Gas Atmosphere.* In: *Annual Report, Conference on Electrical Insulation and Dielectric Phenomena,* 24-27, IEEE Service Centre, Piscataway
[89] Wegener, M. et al. (2004). *Two Step Inlation of Cellular Polypropylene Films: Void Thickness Increase and Enhanced Electromechanical Properties.* J. Phys. D: Appl. Phys. 37, 623-627
[90] Zhang, X. Et al. (2004). *Piezoelectric d_{33} Coefficient of Cellular Polypropylene Subjected to Expansion by Pressure Treatment.* Appl. Phys. Lett. 85, 1226-1228
[91] _, *Improvement of Piezoelectric Activity of Cellular Polymers Using a Double Expansion Process,* J. Phys. D: Appl. Phys. 37, 2146-2150 (2004).
[92] Neugschwandtner, G.S. et al. (2001). *Piezo and Pyroelectricity of a Smart Polymer-foam Space-charge Electret.* J. Appl. Phys. 89, 4503-4511
[93] Suutala, J. & Roning, J. (2004). *Towards the Adaptive Identification of Walkers: Automated Feature Selection of Footsteps Using Distinction-sensitive lvq.* In: *Int'l. Workshop on Processing Sensory Information for Proactive Systems (PSIPS 2004),* Oulu, Finland, June 14-15
[94] Lekkala, J. et al. (1996). *EMF-force Sensor- a Flexibile and Sensitive Electret Film for Physiological Applications.* Medical & Biological Engineering & Computing 34, Supplement 1, Part 1
[95] Nykanen, H. et al. (1999). *Active Control of Sound Based on Utilizing EMFi-technology.* In:*Proceedings, ACTIVE 99,* 1159-1170, Ft. Lauderdale, Dec. 4-6
[96] _, *Piezoelectric materials in devices* (Ed.: N. Setter), EPFL Swiss Federal Institute of Technology, Lausanne (2002).
[97] Bauer, S. et al. (2004). *Phys. Today,* 57, 37–43
[98] Paajanen, M. et al. (2001). *IEEE Trans. Dielectr. Electr. Insul.,* 8, 629 –636
[99] Information of the company Emfit Ltd., Finland, http://www.emfit.com.
[100] Lekkala, J & Paajanen, M. (1999). In: *Proceedings of the 10th International symposium on Electrets* (Eds.: A. A. Konsta, A. Vassilikou-Dova, K. Vartzeli-Nikaki), IEEE, Piscataway, NJ, 743– 746
[101] Siivola, J. et al. (1993). *Med. Biol. Eng. Comput.,* 31, 634– 635
[102] Information of the company Screentec, Finland, http://www.screentec.com.
[103] Lekkala, J. (1997). *Ind. Horizons,* Dec., 12– 13
[104] Information of the company B-Band, Finland, http://www.b-band.com.
[105] Hillenbrand, J. & Sessler, G.M. (2004). *J. Acoust. Soc.* Am., 3267– 3270
[106] Paajanen, M. et al. (2002). In: *Proceedings of the 11th International Symposium on Electrets (Ed.: R. Fleming),* IEEE, Piscataway, NJ, 191 –194
[107] Kressmann, R. & Schallwandler, K. (2002). *Mit. Interner Polarisation auf Polymerbasis sowie Siliziummikromechanik, Shaker, Aachen,* (in German)
[108] Advanced Cerametrics Incorporated, P.O. Box 128, 245 North main Street, Lambertville, New Jersey 085300128.
[109] Measurement Specialists Incorporated, 100 Lucas way, Hampton, VA 23666.

9

Modeling and Identification of Parameters the Piezoelectric Transducers in Ultrasonic Systems

Pawel Fabijanski and Ryszard Lagoda
Warsaw University of Technology, Institute of Control and Industrial Electronics
Poland

1. Introduction

This chapter is dedicated to ultrasonic piezoelectric ceramic power transducers. These elements are now the most popular source of high power ultrasound and is used in many industrial applications. High power ultrasonic waves are generally used in such industrial processes as welding, acceleration of chemical reactions, scavenging in gas medium, echo sounding and underwater communication (sonar systems), picture transmission, and, above all, ultrasonic cleaning. In practice is now the most widely used the sandwich type power transducers.

Stage design power converters high power ultrasonic devices usually preceded by computer analysis of currents and voltages waveforms the elements of the system, particularly in semiconductor instruments of power. Competent representation requires the use of these waveforms of electrical models of piezoelectric ceramic transducers under the parameters of line with reality and allows to calculate the electrical operating parameters used in the layout of semiconductor switches, capacitors and reactors. Application to simulation circuit of the main generators of ultrasonic piezoelectric ceramic transducers correct model also allows analysis of different variants of control systems and regulation of voltage-frequency converters

For example the standard ultrasonic system for cleaning technology (Fig. 1) includes:

1. ultrasonic generator,
2. transducer or set of transducers,
3. cleaning tank.

Piezoelectric ceramic transducers placed in the tub generate ultrasonic waves that pass through the liquid and reach the element immersed in the tank. As a result, created in the liquid, with very high frequency, alternating areas of high and low pressure. In areas, where low pressure is forming millions of bubbles of vacuum. When the pressure in the alveoli increases and is high enough, bubbles implode, releasing enormous energy at the same time. This phenomenon is called cavitation. Emerging implosions work as a whole series of small cleaning brush. The phenomenon is spreading in all directions and causes intense but controlled detachment of particles of pollutants on the entire surface of cleaning detail. Washed away dirt particles collect on the surface of the cleaning solution from where they are blown into a nearby basin, and then be filtered and recycled.

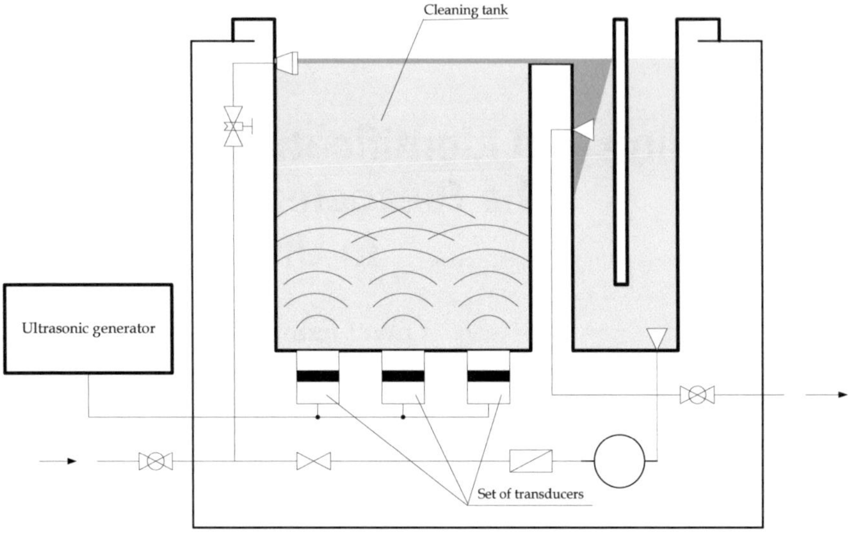

Fig. 1. Ultrasonic cleaning system

Ultrasonic cleaning is more effective in cleaning hard materials, than the cleaning of soft or porous materials. It was found that, the harder the surface, including the operation of ultrasound is more efficient. Hence, metals, glass, hard plastics well led by ultrasound and are ideally suited for ultrasonic cleaning.

2. Sandwich type transducer

In the technological equipment for cleaning, welding, etc. are generated ultrasound with high intensity and frequency from 20 kHz to 100 kHz. Currently, most teams or a single power are a source of ultrasonic piezoelectric ceramic transducers. The construction of such a transducer is shown in Fig. 2.

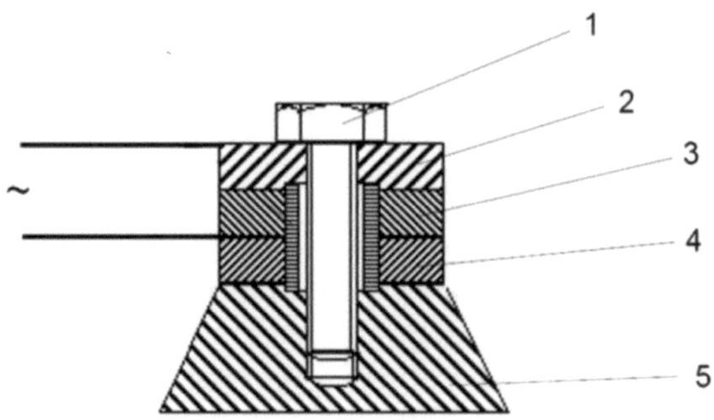

Fig. 2. Construction of sandwich type piezoelectric ceramic transducer;
1 - screw or pin settings (gripping), transmitter, 2, 5 - blocks of metal (eg. aluminum, iron, brass), 3, 4 - piezoelectric ceramic plates (cylindrical, annular)

Such transducers consists of two metal blocks (2, 5), between which are clamped to the material of the piezoelectric ceramic plate (3, 4). Metal blocks and plates are twisted with one or more screws (1). This construction has a much lower own resonant frequency compared to the same frequency of vibration plates, and what is more important allows you to generate high intensity ultrasound. Characteristic of ultrasonic power converters is that they work in a state of mechanical resonance. Thus, in this case the wave frequency of the supply voltage must be equal to the natural frequency of the transducer.

2.1 Mechanical model

Piezoelectric ceramic power converter in resonance state is a mechanically vibrating block, which can model the system with one degree of freedom shown in Fig. 3.
This model consists of mass M, which represents the mass of the whole converter, a damper with a coefficient of friction R and spring with a coefficient of mechanical sensitivity K.

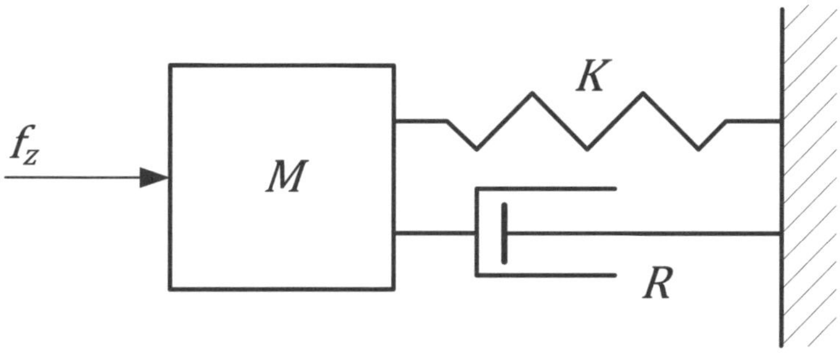

Fig. 3. The model of the mechanical oscillating system with one degree of freedom

In this system, there are four forces: the external force f_Z, the force of elasticity (Hook) f_K, the force of friction f_R and force of inertia f_M, which satisfy the equation:

$$f_Z = f_M + f_R + f_K \tag{1}$$

Assuming:
$K = \frac{1}{S}$, where S the coefficient of elasticity,
x - deviation from the equilibrium position,
v - linear velocity of particles (acoustic) dependence (1) can be written as:

$$f_Z = M\frac{d^2x}{dt^2} + R\frac{dx}{dt} + \frac{1}{K}x = M\frac{dv}{dt} + Rv + \frac{1}{K}\int_{-\infty}^{t} v d\tau \tag{2}$$

Assuming that the vibration exciting force F_Z is sinusoidal variable

$$f_Z = F_{ZM}\sin\omega t \tag{3}$$

where: F_{ZM} amplitude, and ω pulse of this force, the relationship (2) can be written as:

$$\underline{F_Z} = F_Z e^{j\omega t} = j\omega M\underline{V} + R\underline{V} + \frac{1}{j\omega K}\underline{V} \tag{4}$$

Complex mechanical impedance of the transducer is thus equal to:

$$\underline{Z_{mech}} = \frac{\underline{F_Z}}{\underline{V}} = R + j\omega M + \frac{1}{j\omega K} = R + j\left(\omega M - \frac{1}{\omega K}\right) = Z_{mech}e^{j\varphi} \tag{5}$$

The impedance module is equal to:

$$Z_{mech} = \sqrt{R^2 + \left(\omega M - \frac{1}{\omega K}\right)^2} = \sqrt{R^2 + X^2} \tag{6}$$

$$\varphi = arctg\left(\frac{X}{R}\right) \tag{7}$$

Finding a model similar to the mechanical model of electrical converter provides digital modeling of complete systems of generators supplying power electronic converters, the analysis found their current and voltage waveforms and to verify the different concepts of control algorithms of such systems.
The relationship (5) describing the complex mechanical impedance of the transducer is similar to the relationship describing the $\underline{Z_m}$ impedance of the electrical serial circuit $R_m L_m C_m$ shown on Fig. 4.

$$\underline{Z_m} = R_m + j\omega L_m + \frac{1}{j\omega C_m} = R_m + j\left(\omega L_m - \frac{1}{\omega C_m}\right) \tag{8}$$

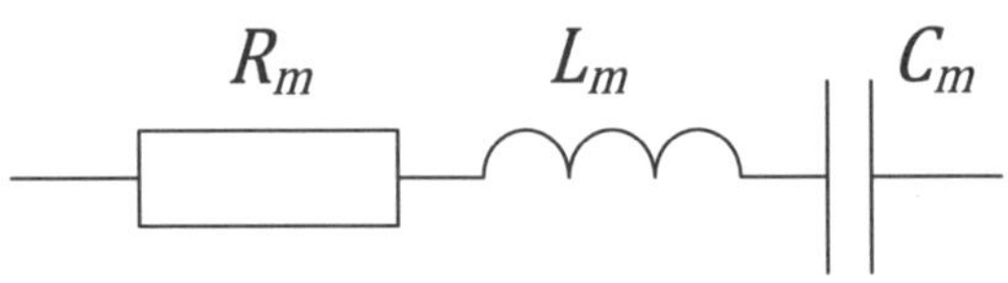

Fig. 4. Electrical serial circuit $R_m L_m C_m$

In this circuit the resistance R_m is equal to $R_m = R_s + R_a$, where R_s represents the mechanical losses of the converter, which are practically constant, and R_a the acoustic resistance, which is inversely proportional to the intensity I_a produced by the ultrasound transducer.

2.2 Electric model

Equation (5) shows that the linear velocity (molecular), vibrating transducer has the greatest amplitude when the mechanical impedance value Z_{mech} is the smallest. This condition occurs when the elastic force f_K balance the power of inertia f_M,

$$f_M + f_K = 0 \tag{9}$$

and pulse of external force f_Z, to compensate for the frictional force f_R will be equal to the pulsation resonant converter ω_m, where $\omega_m^2 = \frac{1}{M \cdot K}$.
In the piezoelectric phenomenon, the relationship between external force f_Z causing the deformation of the piezoelectric ceramic element and the intensity of the electric field E generated inside, is linear.
As for the specific thickness piezoelectric ceramic plate l we can write:

$$u = E \cdot l \tag{10}$$

then assuming a coefficient of proportionality k_p can be said that an external force k_p is proportional to the voltage that was applied to the plate.

$$f_Z = k_p u \tag{11}$$

Comparing the equation describing a serial electrical circuit $R_m L_m C_m$,

$$u = L_m \frac{d^2q}{dt^2} + R_m \frac{dq}{dt} + \frac{q}{C_m} = L_m \frac{di}{dt} + R_m i + \frac{1}{C_m} \int_{-\infty}^{t} i d\tau \tag{12}$$

of equation (2) can be seen that supplied to the circuit an electric charge is proportional to the deformation:

$$q = k_p x \tag{13}$$

Therefore meets the current relationship:

$$i = \frac{dq}{dt} = k_p \frac{dx}{dt} = k_p v \rightarrow v = \frac{1}{k_p} i \tag{14}$$

After inserting equation (11) and (14) to (2) is:

$$k_p u = \frac{M}{k_p} \frac{di}{dt} + \frac{R_m}{k_p} i + \frac{1}{k_p K} \int_{-\infty}^{t} i(\tau) d\tau \tag{15}$$

$$u = \frac{M}{k_p^2} \frac{di}{dt} + \frac{R}{k_p^2} i + \frac{1}{k_p^2 K} \int_{-\infty}^{t} i(\tau) d\tau \tag{16}$$

Since the components of the sum in equation (16) May dimension of tension is present in them can be replaced by permanent mechanical respective electrical components:

$$L_m = \frac{M}{k_p^2}, \quad R_m = \frac{R}{k_p^2}, \; C_m = k_p^2 K \tag{17}$$

2.2.1 Linear model

On the basis of equation (16) can be drawn, well-known and frequently used, an electric equivalent circuit oscillating piezoelectric ceramic transducer shown in Fig. 5.
Additional components C_e and R_e represent the electrical capacitance and resistance of the piezoelectric ceramic plate transducer.

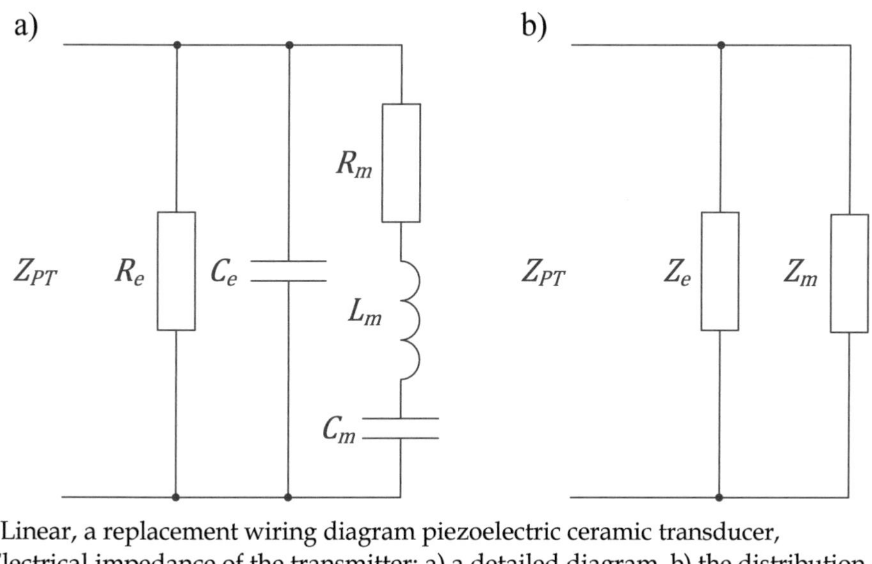

Fig. 5. Linear, a replacement wiring diagram piezoelectric ceramic transducer, Z_{PT} - Electrical impedance of the transmitter: a) a detailed diagram, b) the distribution of the branch of electrical and mechanical

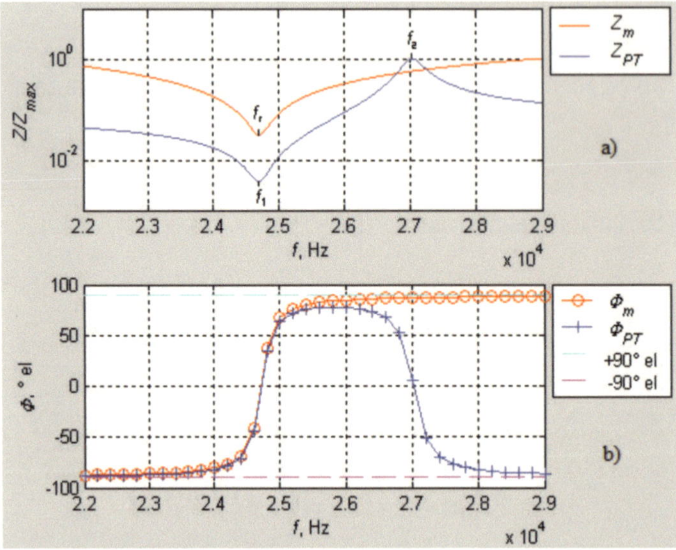

Fig. 6. Frequency characteristics of impedance modulus normalized and the phase of the mechanical industry Z_m and the entire transducerZ_{PT}

Examples of the frequency characteristics of a standard impedance modulus and phase of mechanical industry and the entire transducer is shown in Fig. 6.
When the frequency $f_1$reaches a minimum impedance Z_{PT} of the module. Near this frequency is the frequency of mechanical resonance f_r.
Resistance $R_m$can be divided into two parts:

$$R_m = R_s + R_a \tag{18}$$

where R_s represents the mechanical losses of the converter, which are practically constant, and R_a the acoustic resistance.
Presented in this chapter, a linear model of the piezoelectric ceramic transducer is not mapped correctly the actual frequency characteristics of the transmitter. Frequency response of linear model is much wider than in the actual transmitter. This is important in the process of creating digital models and design of ultrasonic generators.

2.2.2 Non-linear model

As mentioned above, the linear model presented in Fig. 5 does not correctly reproduces the frequency characteristics of the transmitter. This feature can be eliminated by supplementing the model with two non-linear electrical resistance R_{m1f} and R_{m2f} the accompanying in mechanical industries, as shown in Fig. 7.
The resultant mechanical resistance R_{mwf} in the industry is thus equal to:

$$R_{mwf} = \frac{(R_m + R_{m1f})R_{m2f}}{R_m + R_{m1f} + R_{m2f}} \tag{19}$$

Assuming that

$$R_{m1f} = R_{m1}\left(\frac{|f_r - f|}{f_r}\right)^n \tag{20}$$

$$R_{m2f} = \frac{R_{m2}}{\left(1+\frac{|f_r - f|}{f_r}\right)^m} \tag{21}$$

the current state of the branch of mechanical resonance, which is proportional to the speed of molecular v still is determined by the resistance R_m because, $R_{m1f}(f_r) = 0$ and $R_{m2f}(f_r) \to \infty$.

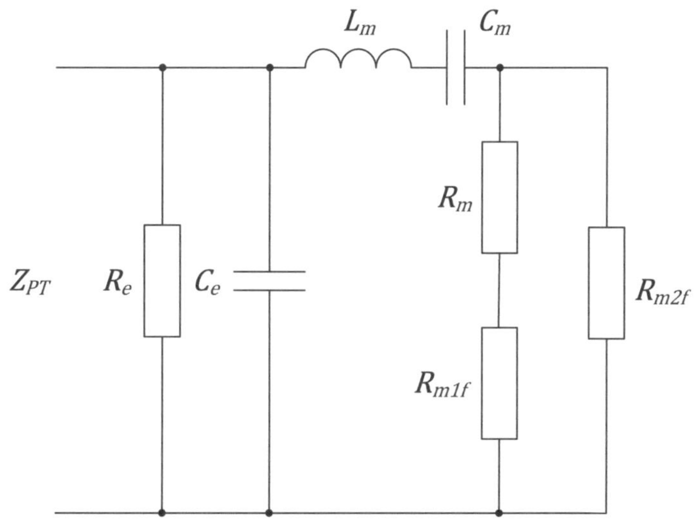

Fig. 7. Non-linear electrical model of piezoelectric ceramic transducer

3. Admittance characteristics of sandwich type transducers

The electrical characteristics of ceramic transducer resistance R_e and electrical capacitance C_e can be determined with high accuracy on the basis of electrical measurement. Changes in these parameters while the transducer can be considered to be negligible because of the transmitter power is assumed that the C_e and R_e are fixed. Parameters of dynamic branch $R_m C_m L_m$ of power piezoelectric ceramic transducer model electric are not physically measurable. We can only designate empirically. Identification algorithm is therefore based on numerical calculations. Input parameter for these calculations is the image of the actual characteristics of the transducer admittance.

Example characteristics of the actual and approximated admittance converter a resonance frequency around 25 kHz are shown in Fig. 8.

Assuming a digital model of the transmitter as shown in Fig. 5 the susceptance B, conductance G, admittance Y of the transducers in frequency ω function may be expressed by the following equations:

$$B(\omega) = ImY(\omega) = \omega C_e + \frac{\omega C_m(1-\omega^2 L_m C_m)}{\omega^2 C_m^2 R_m^2 + (1-\omega^2 L_m C_m)^2} \tag{22}$$

$$G(\omega) = ReY(\omega) = G_e + \frac{\omega^2 C_m^2 R_m}{\omega^2 C_m^2 R_m^2 + (1-\omega^2 L_m C_m)^2} \tag{23}$$

$$Y(\omega) = \sqrt{G(\omega)^2 + B(\omega)^2} \tag{22}$$

where $G_e = \frac{1}{R_e}$. For the resonant pulsation $\omega_m = \sqrt{\frac{1}{L_m C_m}}$ are:

$$B(\omega_m) = \frac{C_e}{\sqrt{L_m C_m}} \approx 0 \tag{25}$$

$$G(\omega_m) = \frac{1}{R_e} + \frac{1}{R_m} \approx \frac{1}{R_m} \tag{26}$$

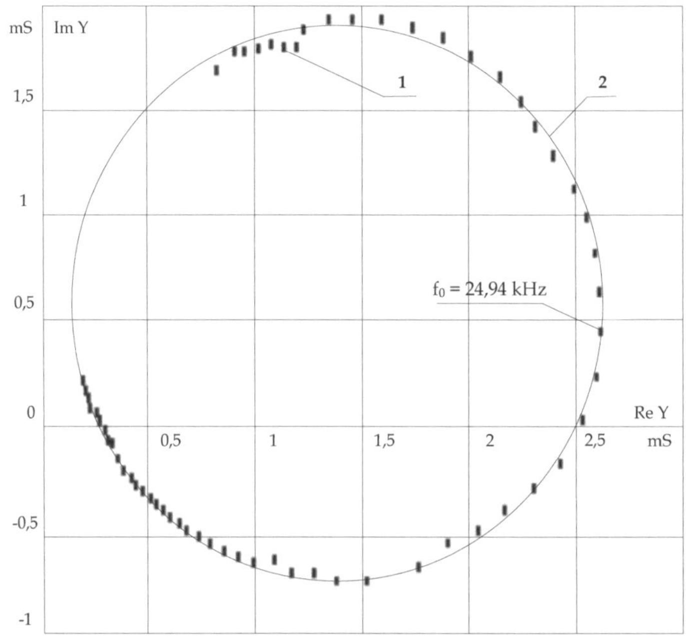

1. Real characteristic,
2. Equivalent characteristic

Fig. 8. Exemplary admittance characteristics of sandwich type transducers

4. The genetic algorithm in use to identify R_m, L_m, C_m parameters of the mechanical branch

As mentioned earlier parameters of dynamic branch $R_m L_m C_m$ of power piezoelectric ceramic transducer model electric are not physically measurable.

We can only designate empirically. Identification algorithm is therefore based on numerical calculations. Input parameter for these calculations is the image of the actual characteristics of the transducer admittance.

An example measurements of results for the transducer mechanical resonance frequency equal to about 43 kHz is locate in appendix in Table 1 and one-to-one correspondence real admittance characteristics of this transducer is shown in Fig. 9. The graph consists of points whose coordinates correspond to the conductance $G = Re(Y)$ and susceptance $B = Im(Y)$ of the converter, measured at a certain frequency.

Analysing the shape of this characteristic can be clearly observed that with increasing frequency (Figure accordance with the frequency increasing clockwise), outlines the main loop of the graph. In the general case, the image of the curve is an ellipse. This ellipse is interpolated electrical admittance characteristics of an ideal replacement transducer schedule shown in Fig. 9. It may be noted that the actual characteristics, in addition to the main loop also includes many smaller "loops" that testify to the presence of additional resonances in the transducer side. These resonances, however, will not occur in the adopted system replacement transmitter.

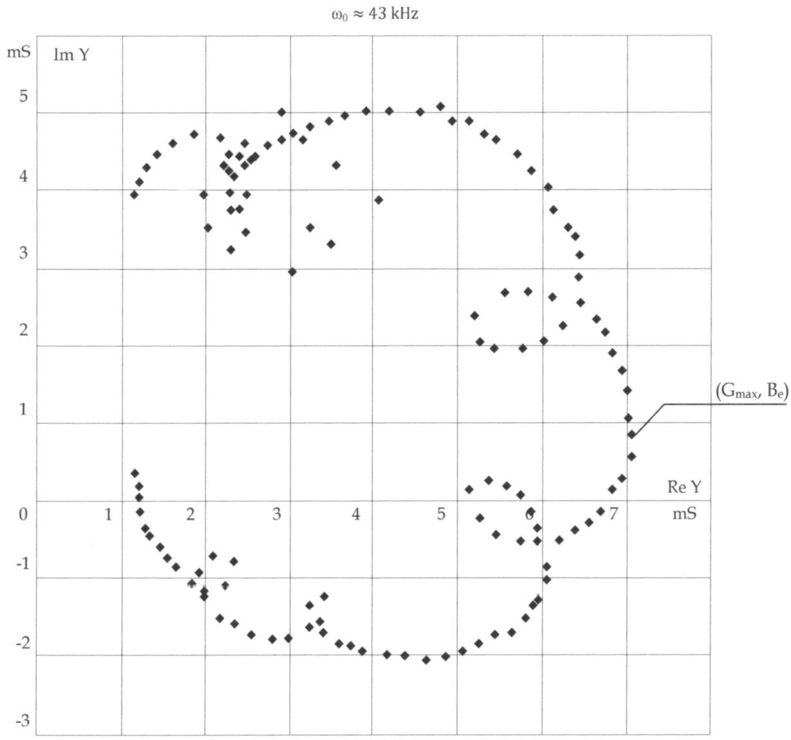

Fig. 9. Real admittance characteristics of the piezoelectric ceramic transducer

After eliminating these "loops" with the actual picture of the actual characteristics of the admittance characteristics of the transducer will be the figure presented in Fig. 10.
In Table 1 are marked in bold points, which missed the chart.

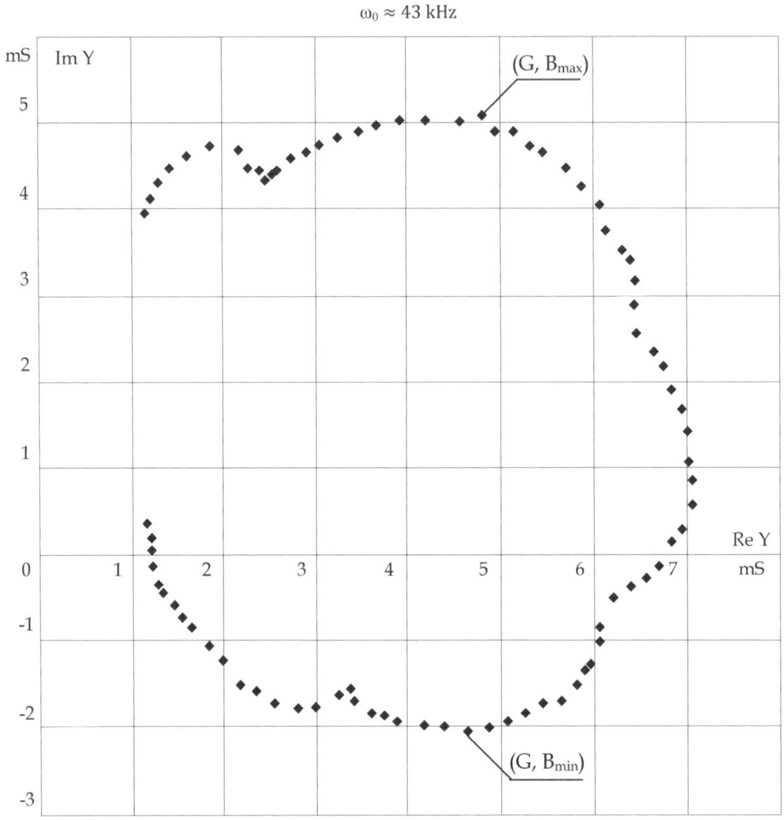

Fig. 10. The actual characteristics of the transducer after eliminating resonances fringe

The set of coordinates of the actual approximation characteristics transducer (Fig. 9) is a database for further calculations.

4.1 Calculation algorithm

Designation of alternative modes of dynamic parameters of the transmitter requires the implementation of numerical calculations and find such a set of $R_m L_m C_m$ parameters, which in a given error will allow mapping of the main loop, the actual characteristics of the transducer admittance presented Fig. 10.

One element of the identification of these parameters is a genetic algorithm. Generally one can say that the genetic algorithm is a regula - system learner who makes a certain number of iteration steps. It is a group of strict security procedures that are based on the fundamental mechanisms of biological evolution such as natural selection and inheritance. It works interactively with the environment in discrete time. The algorithm of this type of reproduction may take place subject to the diversity of the population:

- model with preload,
- measures niche

A special feature of this type is that the algorithm is not seeking a single optimal solution, but a group of cooperating the best solutions. At any time, the algorithm works evolutionary principle of survival, which is always available some of the best solutions at the moment.

As the proceedings algorithm solution to optimize and adapt to the conditions in which the algorithm works. For further calculations is always the best solutions are selected and rejected solutions are worse. Here there is a process of succession. In order to obtain optimal solutions group for further reproduction of the best solutions are selected at the time of the algorithm. It is a natural selection process occurs in nature. It is known that the probability of obtaining better result is greater if we use it to generate the best available solution than the use of inferior results. According to the law of nature and genetics "survive" the best and strongest.

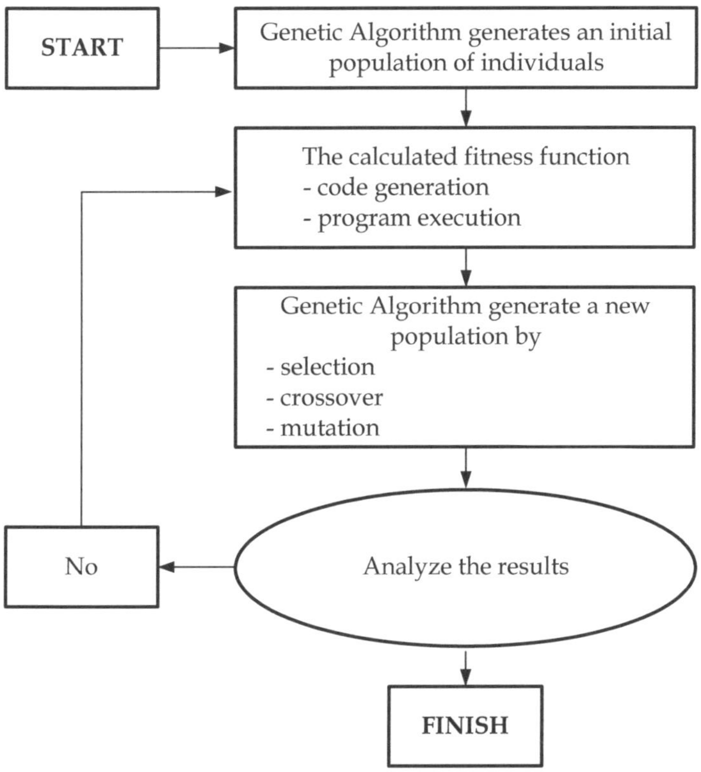

Fig. 11. Simply genetic algorithm

The genetic algorithms is a natural process of mating genes. It is no different genetic algorithm. There also are choosing the best "genes" of each solution and verify their combinations. Alongside the cross as a natural evolution is mutation, a random change in the gene. Both these processes are the values of the genetic operator.

Using the genetic algorithm, remember to keep the best balance between the transfer of genes to the next generation, and a draw solution space. Too broad conditions imposed solutions may give erroneous results in spite of every generation the best available solution at the moment. Genetic algorithm is an excellent tool to monitor and maintain the balance between these two dependencies.

The overall pattern of genetic algorithm is illustrated in Fig. 11.

Using a genetic algorithm to identify the parameters of dynamic model of the electrical branch of the transducer sandwich working near mechanical resonance is shown in Fig. 12.

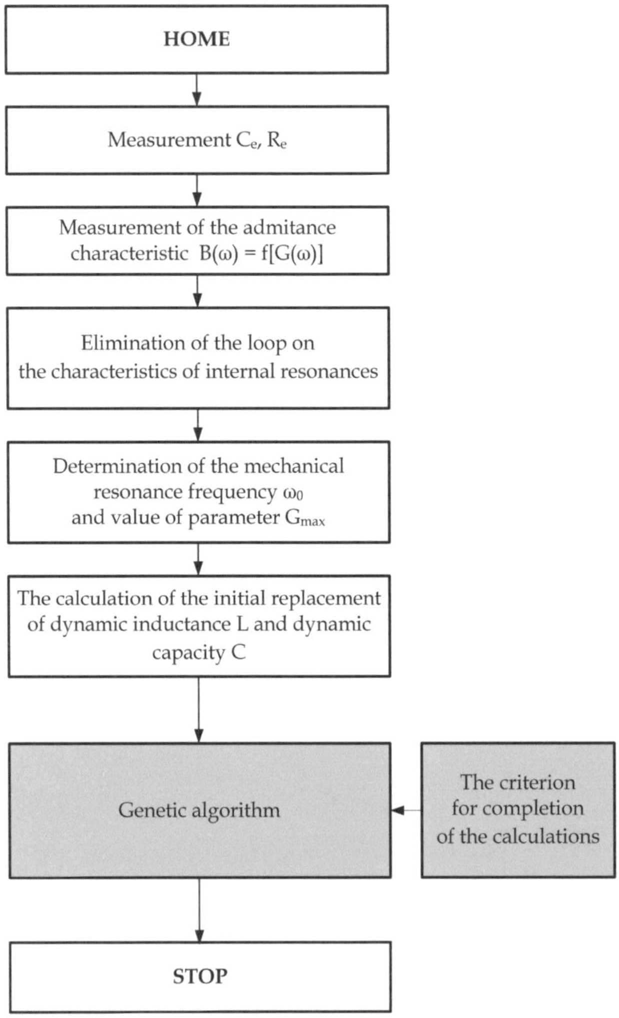

Fig. 12. Algorithm for parameter identification of dynamic model of the branch of electrical power ultrasonic transducer

The initial value of the dynamic resistance R_m is calculated by selecting the characteristics shown in Fig. 9 point with coordinates (G_{max}, B_e). In this point there is a mechanical resonance of the transducer and is a good approximation condition.

$$\omega_m^2 L_m C_m - 1 = 0 \tag{27}$$

The dependence (26) shows that the resonance:

$$G(\omega_m) = G_{max} = G_e + \frac{1}{R_m} \tag{28}$$

Therefore:

$$R_m = \frac{1}{G_{max} - G_e} \tag{29}$$

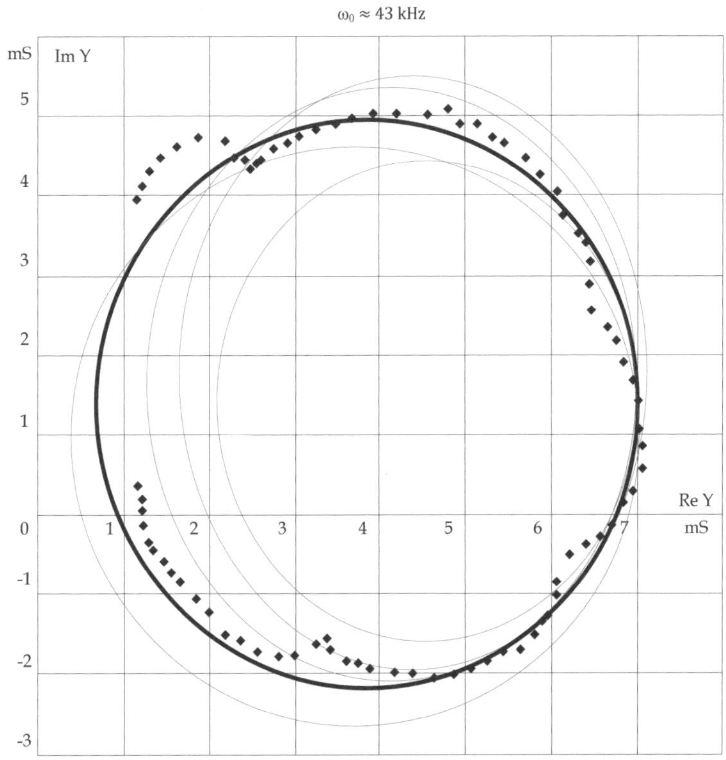

Fig. 13. Iterative improvement of the results obtained in numerical calculations and a computer simulation electric model of an sandwich type transducers

Choosing the actual characteristics of the admittance of any two points with coordinates (G_1, B_1), (G_2, B_2), which lie beyond the point of mechanical resonance $(\omega_1 \neq \omega_m, \omega_2 \neq \omega_m)$ can calculate the initial value of the replacement of mechanical capacity C_m.
Transforming the system of equations:

$$G_1 - G_e = \frac{\omega_1^2 C_m^2 R_m}{(1-\omega_1^2 L_m C_m)^2 + \omega_1^2 R_m^2 C_m^2} \tag{30}$$

$$G_2 - G_e = \frac{\omega_2^2 C_m^2 R_m}{(1-\omega_2^2 L_m C_m)^2 + \omega_2^2 R_m^2 C_m^2} \tag{31}$$

and taking into account that for G_{max} is equal to the pulsation vibrations ω_m approximate value of C_m is described by the formula:

$$C_m' = \frac{1-\frac{\omega_1^2}{\omega_m^2}}{\omega_1} \sqrt{\frac{(G_1-G_e)(G_{max}-G_e)}{1-\frac{G_1-G_e}{G_{max}-G_e}}} \tag{32}$$

$$C_m'' = \frac{1-\frac{\omega_2^2}{\omega_m^2}}{\omega_2} \sqrt{\frac{(G_2-G_e)(G_{max}-G_e)}{1-\frac{G_2-G_e}{G_{max}-G_e}}} \tag{33}$$

$$C_m = \frac{C'_m + C''_m}{2} \quad (34)$$

The initial value of the replacement of mechanical inductance L_m determined from the relationship:

$$L_m = \frac{1}{\omega_m^2 C_m} \quad (34)$$

After substituting the calculated value of the $R_m L_m C_m$ and the measurement C_e, R_e the relationship (22) and (23) is determined by numerical coordinates of the points of the electrical characteristics of the admittance model for the same pulse, for which measurements of actual performance. If the error resulting from a comparison of the approximation characteristic with real characteristic (Fig. 10) is greater than the accepted values, generate a new population of C_e, R_e, $R_m L_m C_m$ parameters, repeat the calculation of coordinates of the electrical characteristics of the model and then analyze the resulting error. Calculations should be continued until the error resulting from a comparison of the characteristics shown in Fig. 10 with the characteristics of the accepted model of the electrical transducer is smaller than the set value. Measurement error should be performed in the pulsation ω changing between points B_{max} and B_{min} (Fig. 10).
Procedure described above can be used and introduces the appropriate algorithm in the DSP program simulation.

```
//Program piezoelectric ceramic transducer
input {Circuit parameters [Re,Ce,R,L,C,u(t)];
        Simulation parameters (tp, tk,Δ,u(0))
        Evolution parameters (Num,Size, NumM, Initial, NumG, stop);)
Output (t, u; tp; uo =u(o);};

for (t,tk, t++){New function form;
start initial population;}
for (integer i=1,Num,i++){New population,
Select the best}
```

where:
Genotype dimension: Num,
Size of the generation created by mutation: Size,
Number of the best selected genotypes for mutation: NumM,
Number of the best selected genotypes for crossing: NumC,
initial parameter mutation range: Initial

4.1.2 Description of algorithm parameters

The algorithm builds an initial population from completely random values. Each algorithm result is therefore partly random. Values implicitly included in the program when it starts in most cases produce the desired result. In order to increase the effectiveness of the algorithm we can increase the population size or quantity. This implies, however, with prolonged time needed for calculations, and thus the waiting time for results. Increasing the likelihood may lead to an erroneous finding and losing the optimal solution. In the case of reduction in population size, increased the likelihood of crossing has a better chance of getting an individual with a desired adjustment even though his speech is less certain.

Similar algorithm behaves in a situation of increasing the probability of mutation, but in this case, the result is more random.

4.1.3 Description of program code

The main application file is interface.m. He is responsible for building the interface, retrieve data from the controls and calling the functions of the buttons.

The first step is to create an initial population for genetic algorithm.

```
// INITIAL POPULATION
startpop = zeros (popsize, 2);
for i = 1: popsize
   startpop (i,:) = individual (2, max_val); / of the initial random population
end
```

The creation of the initial population is to create a matrix of population size 'number of individuals' x 'dimension of the individual', where the number of individuals is the value declared in the program window, and the dimension is equal to the number of individual factors.

The population is filled with random individuals.

The next step is to assess the population. It is used here for the sum of squared differences of the values obtained and references. Since the genetic algorithm seeks to maximize, a function of adaptation used in this case is the inverse of the sum of the squares, in addition to the square have raised to accelerate the convergence of the algorithm.

Based on the resulting evaluation, subjects were assigned to be the appropriate probability of the next population. The draw of the population makes the next feature:

```
function [new] = sel (evaluation, current) / disproportionate Roulette
global popsize;
for j = 1: popsize
   number (j, 1) = round (100 * (score (j, 3)));
end
roulette = zeros (100.2);
for i = 1:100
   Roulette (i, 1) = i;
end
// 100-point roulette wheel
all = 100;
item = popsize;
while an element> 0; / complement vector numbers in accordance with the adaptation of individual
   for a = 1: all
      Roulette (a, 2) = element;
   end
   all = all-number (element, 1);
   element = element-1;
end
// Draw elements
items = Random (popsize, 100); / draws n = popsize position
for j = 1: popsize
   number (j, 1) = roulette ((positions (j, 1)), 2); / individual in this position
end
```

```
for i = 1: popsize
   new (i,:) = current ((number (i, 1 )),:); / choose a subject to reproduce
end
```

After drawing the new population are made on the operations of crossover and mutation.

```
// CROSSING
xn = fix (popsize / 2); / number of crossover operations
remainder = mod (popsize, 2); / population parity check
steam = select (popsize); / choice pairs of interbreeding
for i = 1: xn
      [Child1 Child2] = cross ((new (pair (2 * i-1, 1 ),:)),( new (pair (2 * i, a cross ),:)));
      pox (i,:) = Child1;% first child
      pox (i + xn,:) = Child2;% second child
end
   pox (2 * xn +1,:) = new (pair (xn, 1 2 * ),:); / last rewritten without change
end
```

Crossing begins by selecting pairs of individuals, in such a way that a given individual could occur in only one hand. If the population is odd that an individual will be rewritten without change. Then it is drawn for each pair whether the operation will cross (the probability of randomly selected cross <put the probability of crossover):

- if so, part of the code is converted into fish,
- if not, individuals are no changes.

According to the function below:

```
function [Child1, Child2] = cross (A, B); / crucifixion simple
global px;
or = rand (1);
if and> px / cross check whether
   Child1 = A;
   Child2 = B;
else / crossing straight
   Child1 (1) = A (1);
   Child1 (2) = B (2);
   Child2 (1) = B (1);
   Child2 (2) = A (2);
end
```

After the crossover operation for each individual of the population by analogy, we check if there is a mutation. If so, an individual code rate is changed to a random value according to the following function:

```
function [X] = mut (A) / mutation (replacing a randomly selected for another random value)
global max_val;
index = Random (1,2); / random mutation alleles
new = individual (1, max_val); / new value allele
A (index) = new;
X = A;
```

After mutation we get a new population, which also is subjected to evaluation of the adjustment operations, reproduction, crossover and mutation according to the number of iterations the algorithm set by the user.

Examples of results obtained in subsequent iterative steps for the transducer mechanical resonance frequency of 42.9 kHz is shown in Fig. 10.

When the next iterative step will be the condition for the completion of the calculations should be considered that the characteristics of the model shown in Fig. 5 parameters R_e, C_e and $R_m L_m C_m$ generated in the genetic algorithm in the last population of values, coincides with the actual characteristics of the interpolated transducer. The shape of this characteristic is shown in Fig. 14.

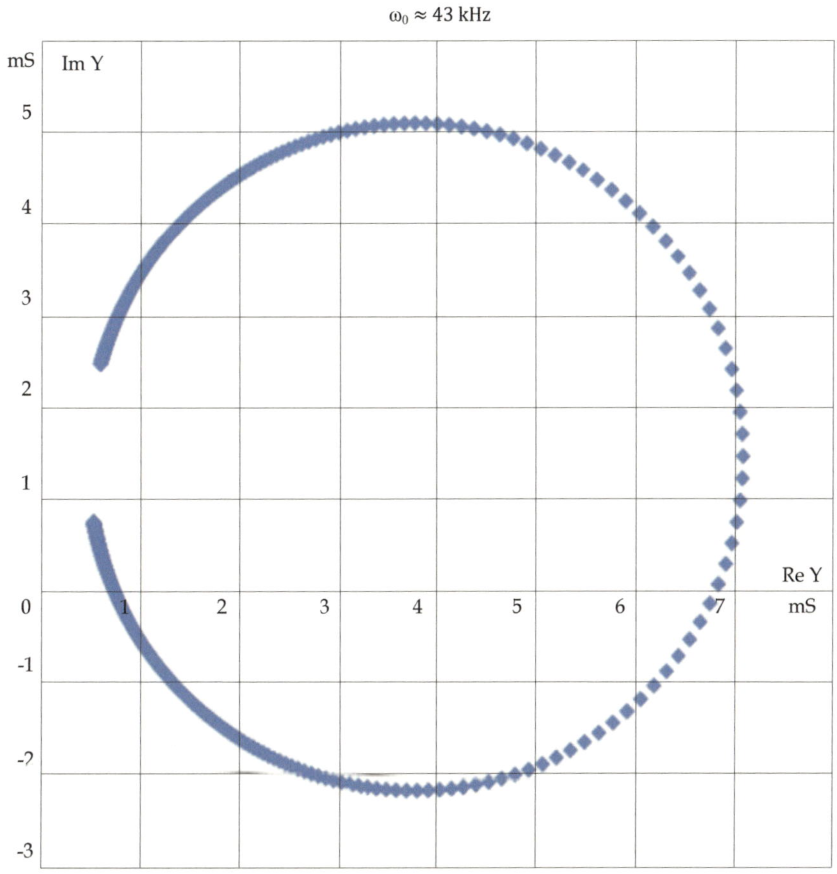

Fig. 14. Circle admittance obtained using genetic algorithm

4.2 Effect of the electrical equivalent circuit parameters on the shape characteristics of the transducer admittance

Location of figures which creates a characteristic admittance of the transducer coordinate system $B(\omega) = f[G(\omega)]$ can be determined by changing the value of capacity C_e and the loss of electrical resistance of ceramics R_e. The higher value of R_e the greater displacement figures to the right people along the conductance $G(\omega)$. If you forget in the pattern of replacement sensor resistance R_e ($R_e \rightarrow \infty$) will move to the left approximation of the characteristics. In this case, the graph will be almost tangential to the axis $B(\omega)$. Reduction in the capacity of C_e will move down the sheet, along the axis $B(\omega)$. Conductance G_{max}

value determines the width of the main loop of the graph. From relation (28) shows that the value of G_{max} decreases when the resistance R_m increases.
The impact of dynamic capacity C_m can analyze the shape characteristics of saving the relationship (22) in the form

$$B(\omega) \approx \frac{\omega C_m(1-\frac{\omega^2}{\omega_m^2})}{\omega^2 C_m^2 R_m^2+(1-\frac{\omega^2}{\omega_m^2})} \quad (35)$$

For small, ± 2% of the pulsation untuning mechanical resonance, it can be assumed that:

$$B(\omega) \approx \frac{1-\frac{\omega^2}{\omega_m^2}}{\omega C_m R_m^2} \quad (30)$$

From relation (30) shows that reducing the value of the dynamic capacity C_m will increase susceptance system. This also applies to the maximum B_{max} and minimum B_{min}. In this case, it means that the values of B_{max} and $|B_{min}|$ increased. The difference B_{min} - B_{min} is equal to the amount of main loop of the graph.
These rule changes in the electrical equivalent circuit parameters of the transducer should be used during the generation of successive populations of a set of values of R_e, C_e and $R_m L_m C_m$ genetic algorithm.

5. Conclusion

Linear model presented in Fig. 5 does not correctly reproduces the frequency characteristics of the transmitter. This feature can be eliminated by complementing the electrical model of two nonlinear resistances R_{m1f} and included R_{m2f} in the mechanical industry, as shown in Fig. 7 This is important in the process of creating digital models and design of ultrasonic generators.
The actual characteristics of the admittance is a source of transparent, condensed information about the properties and parameters of the piezoelectric ceramic transducer stimulated by the mechanical vibrations. Used to identify the dynamic parameters of $R_m L_m C_m$ and optimization of electrical parameters R_e, C_e occurring electric transducer model, genetic algorithm uses these data and represents one of many possible methods to use here. It has the advantage that it can be realized in automatic cycle.
Based on preliminary data obtained experimentally from measurements of electrical parameters (R_e, C_e) and measurements of voltage and current waveforms created image of the actual transducer admittance characteristics of $B(\omega) = f[G(\omega)]$. Genetic algorithm can find the optimal approximation of the characteristics, correction of the value of R_e, and C_e and calculation of parameters $R_m L_m C_m$ model of the electrical industry dynamic.
An important part of the described method is the proper preparation of input data. They should be given in the form of an ordered table of coordinates of points lying on the actual characteristics of the transducer according to those points that do not belong to the main loop of the graph. The process of eliminating points of the loop parasitic resonances can be implemented step by step analysis of the data included in the table of measurement results or by analyzing the image created by a set of characteristic points of $B(\omega) = f[G(\omega)]$.

Attempts to identify the parameters of a single transducer mechanical resonance frequency of 23.8 kHz and 43.2 kHz have confirmed the effectiveness of the genetic algorithm. Identification algorithm described above can be particularly useful in studies of larger groups of ultrasound transducers in high-power washing.
By introducing the necessary modifications can also be used to identify the parameters of transducers working in real conditions and to implement control systems that track the frequency of mechanical resonance.

6. Appendix

Table 1 shows selected results of detailed studies of the piezoelectric ceramic transducer mechanical resonance frequency of 42.9 kHz. Measurements were performed impedance meter HP 4192 type IMPEDANCE Analyzer. Indicated in bold letter omitted in the genetic algorithm performance admittance points beyond the main loop of the graph. The contents of Table 1 refers to Fig. 9, and Fig. 10.

f kHz	R kΩ	X kΩ	tg(φ)	φ rad	cos φ	sin φ	G mS	B mS
42,00	0,0721	-0,2338	3,2426	1,27166	0,2947	0,9556	1,204320	4,650189
42,04	0,0674	-0,2132	3,1659	1,26485	0,3012	0,9536	1,346922	4,223453
42,08	0,0703	-0,1925	2,7371	1,22052	0,3432	0,9393	1,674703	4,456680
42,12	0,0841	-0,1745	2,0750	1,12171	0,4341	0,9008	2,241050	4,963926
42,16	**0,1006**	**-0,1809**	**1,7988**	**1,06342**	**0,4859**	**0,8740**	**2,347906**	**4,296001**
42,20	**0,0921**	**-0,1763**	**1,9138**	**1,08930**	**0,4631**	**0,8863**	**2,328693**	**3,289479**
42,24	**0,0888**	**-0,1483**	**1,6705**	**1,03138**	**0,5136**	**0,8580**	**2,971594**	**3,192256**
42,28	**0,1148**	**-0,1358**	**1,1834**	**0,86919**	**0,6454**	**0,7638**	**3,630303**	**3,907909**
42,32	**0,1515**	**-0,1403**	**0,9255**	**0,74675**	**0,7339**	**0,6793**	**3,554097**	**3,487659**
42,36	**0,1501**	**-0,2015**	**1,3423**	**0,93051**	**0,5974**	**0,8019**	**2,378189**	**3,725892**
42,40	**0,1047**	**-0,2014**	**1,9231**	**1,09127**	**0,4614**	**0,8872**	**2,032128**	**3,739428**
42,44	**0,1432**	**-0,1502**	**1,0492**	**0,80942**	**0,6899**	**0,7239**	**3,323991**	**4,162891**
42,48	**0,1212**	**-0,1918**	**1,5831**	**1,00742**	**0,5340**	**0,8455**	**2,353507**	**4,431376**

f kHz	R kΩ	X kΩ	tg(φ)	φ rad	cos φ	sin φ	G mS	B mS
42,52	**0,1227**	**-0,1868**	**1,5228**	**0,98973**	**0,5489**	**0,8359**	**2,455637**	**4,700036**
42,56	**0,1041**	**-0,1800**	**1,7284**	**1,04629**	**0,5008**	**0,8656**	**2,408478**	**4,873619**
42,60	0,0992	-0,1666	1,6791	1,03365	0,5117	0,8592	2,639118	4,997220
42,64	0,0980	-0,1478	1,5079	0,98522	0,5527	0,8334	3,116930	4,959969
42,68	0,0973	-0,1351	1,3884	0,94660	0,5844	0,8114	3,510281	4,685777
42,72	0,0974	-0,1229	1,2610	0,90034	0,6213	0,7835	3,962806	4,035905
42,76	0,1005	-0,1081	1,0752	0,82162	0,6810	0,7322	4,613098	2,849480
42,80	0,1057	-0,0924	0,8738	0,71814	0,7530	0,6580	5,362597	1,944590
42,84	0,1139	-0,0752	0,6599	0,58331	0,8346	0,5508	6,115813	2,348010
42,88	0,1295	-0,0571	0,4408	0,41517	0,9150	0,4033	6,464511	2,602020
42,92	**0,1542**	**-0,0514**	**0,3331**	**0,32152**	**0,9488**	**0,3160**	**5,838341**	**1,397040**
42,96	**0,1582**	**-0,0704**	**0,4451**	**0,41880**	**0,9136**	**0,4067**	**5,274907**	**-0,305750**
43,00	0,1374	-0,0578	0,4207	0,39819	0,9218	0,3878	6,185462	-0,238180
43,04	0,1365	-0,0271	0,1982	0,19569	0,9809	0,1944	7,047725	-0,525000
43,08	0,1509	0,0070	-0,0463	-0,04622	0,9989	-0,0462	6,610636	-1,104520
43,12	0,1880	0,0084	-0,0449	-0,04483	0,9990	-0,0448	5,309238	-1,992950
43,15	0,1585	0,0133	-0,0838	-0,08362	0,9965	-0,0835	6,263456	-1,955190
43,17	0,1577	0,0283	-0,1798	-0,17789	0,9842	-0,1770	6,143423	-1,362000
43,20	0,1697	0,0661	-0,3894	-0,37135	0,9318	-0,3629	5,117705	-1,591100
43,24	0,2042	0,1018	-0,4985	-0,46245	0,8950	-0,4461	3,922151	-1,756280
43,28	**0,2579**	**0,1058**	**-0,4104**	**-0,38940**	**0,9251**	**-0,3796**	**3,319092**	**-1,271460**

f kHz	R kΩ	X kΩ	tg(φ)	φ rad	cos φ	sin φ	G mS	B mS
43,32	0,2395	0,1108	-0,4627	-0,43339	0,9075	-0,4200	3,438489	-0,718400
43,36	0,2638	0,1777	-0,6736	-0,59280	0,8294	-0,5587	2,607243	4,650189
43,40	**0,3530**	**0,2199**	**-0,6231**	**-0,55720**	**0,8487**	**-0,5288**	**2,040678**	**4,223453**

Table 1. Results of laboratory tests by the ultrasonic transducer sandwich-type of the resonance frequency around 43 kHz

7. References

Davis, L. (1991). *Handbook of Genetic Algorithms*, Van Nostrad Reinhold, New York

Gen, M. & Cheng, R. (1997). *Genetic Algorithms & Engineering Design,* John Wiley & Sons, New York

Rutkowski, L. (2009). Metody i techniki sztucznej inteligencji, Wydawnictwa Naukowo Techniczne PWN SA, ISBN 978-83-01-15731-9, Warsaw

Fabijański, P.& Łagoda, R. (2010). Genetic Identification of Parameters the Piezoelectric Transducers for Digital model of Power Converter in Ultrasonic Systems, In: *Piezoelectric Ceramics,* Ernesto Suaste-Gomez (Ed.) pp. 129-144, Scyio, ISBN 978-953-307-122-0, Rijeka, Croatia

Fabijański, P. & Łagoda, R. (2004). Theoretical and Experimental Analysis of Series Resonant Converter in Technical Cleaning System, Elektronika 5th International Conference MECHATRONICS 2004, No. 8-9, (September 2004) pp. 164-167, ISSN 0033-2089

Fabijański, P. & Łagoda, R. (2007). Digital Model of Series Resonant Converter with Piezoelectric Ceramic Transducers and Fuzzy Logic Control, In: *Adaptive and Natural Computing Algorithms - 8th International Conference, ICANNGA 2007, Warsaw, Poland 2007, Proceedings, Part I,* Beliczynski, B.; Dziclinski, A.; Iwanowski, M. & Ribeiro B., pp. 642-648, Springer, ISBN-13 978-3-540-71589-4, ISBN-10 540-71589-4, Springer Berlin Heidelberg New York

Fabijański, P. & Łagoda, R. (2007). Genetic identification of parameters the piezoelectric ceramic transducers for cleaning system, In: *Recent Advances in Mechatronics,* Jablonski, R.; Turkowski, M. & Szewczyk, R., pp. 16-21, Springer, ISBN-13 978-3-540-73955-5, Springer Berlin Heidelberg New York

Fabijański, P. & Łagoda, R. (2007). Series Resonant Converter with Piezoelectric Ceramic Transducers and Fuzzy Logic Control with Genetic Optimization, *Proceeding of Region 8 IEEE The International Conference on Computer as a Tool EUROCON 2007,* pp. 1884-1888, ISBN 1-4244-0813-X, Poland, September 2007, IEEE Catalog number: 07EX1617C, Warsaw

Fabijański, P. & Łagoda, R. (2008). Genetic Identification of Parameters the Sandwich Piezoelectric Ceramic Transducers for Ultrasonic Systems, *Proceeding of 13th International Power Electronics and Motion Control Conference EPE-PEMC2008,* CD-

ROM, ISBN: 978-1-4244-1742-1 (CD-ROM) , Poland, September 2008, IEEE Catalog Number CFPO834A-CDR, Poznań

Łagoda, R. & Fabijański, P. (2008). On Line PID Controller Using Genetic Algorithm and DSP PC Board, Proceding of 13th International Power Electronics and Motion Control Conference EPE-PEMC2008, CD-ROM, ISBN: 978-1-4244-1742-1 (CD-ROM) , Poland, September 2008, IEEE Catalog Number CFPO834A-CDR, Poznań

Wnuk, P. (2004). Genetic optimization of structure and parameters of TSK fuzzy models, *Elektronika 5th International Conference MECHATRONICS 2004*, No. 8-9, (September 2004), pp. 1-3, ISSN 0033-2089

Goldberg, D. (1999). Genetic Algorithms in Search, Optimization amd Machine Learning.Addison-Wesley, reading, MA

Man, K.F. & Tang, K.S. & Kwong, S. (1999). Genetic Algorithms: Concepts and Designs. Springer-Verlag London-UK

Hajela, P. & Lin, C.Y. (1992). Genetic search strategies in multicriterion optimal design.Structural Optimization,4: 99-107

Randy, L. & Sue, E. (2004). Practical Genetic algorithms. Published by John Wiley & Sons,Inc., Hoboken, Wiley-Interscience

Fathi, H. (2004). Design and simulation of a new controller for resonant operation of piezoelectric ultrasonic tools. Teheran: 19th International Power System Conference, 2004, p.1-6.

Fernandez, J. M. (2008). Adaptive control of ultrasonic motors using the maximum power point tracking method. Beijing: UFFC IEEE Conference, 2008, p.1823-1826.

Fröhleke, N. (2004). Inverter topologies for ultrasonic piezoelectric transducers with high mechanical q-factor. IEEE 35th Annual Power Electronics Specialists Conference, 2004, p.2736-2741.

Prokic, M.(2004). Piezoelectric transducers modeling and characterization. Le Locle: MP Interconsulting, 2004.

10

Influence of Dopants, Temperature and Atmosphere of Sintered on the Microstructure and Behavior of Lead Free Ceramics

Maria A. Zaghete[1], Francisco Moura[2], Alexandre Z. Simões[3], José A.Varela[1] and Elson Longo[1]

[1]*Instituto de Química, Universidade Estadual Paulista – UNESP – Campus de Araraquara (SP)*

[2]*Universidade Estadual Paulista,UNESP, Campus de Guaratinguetá, SP*

[3]*Laboratório Interdisciplinar de Materiais Avançados, Universidade Federal de Itajubá,Unifei-Itabira, MG*

Brasil

1. Introduction

Ceramics based on barium titanate ($BaTiO_3$) with modified electrical properties have been obtained by substitution of isovalent cations both in barium and titanium sites. Barium zirconium titanate solid solution has been widely researched owing to the diffusion phase transition characteristics caused by the substitution of Ti^{4+} for Zr^{4+}. It is well known that with increasing Zr^{4+} content the three transition points and the three corresponding ε_r maxima move closer together and finally coalesce into a single broad maximum (Nejman,1988; Hennings & Sehnell, 1982) . Undoped barium zirconium titanate ceramics, however, usually need a very high sintering temperature above 1400°C because of the slow diffusion velocity. Little amount of additives such as ZnO, CuO, and SiO_2 could greatly lower the sintering temperature by 100-300°C but cause the degradation of the dielectric properties at the same time owing to the deposition of vitreous phase along grain boundaries (Amador et al. 1998). It is well known that $BaTiO_3$ is cubic above 120-135 °C and belongs to the space group Pm3m (Oh1). At temperatures below 120 °C it is ferroelectric with P4mm (C4v1) structure, which further transforms to orthorhombic and rhombohedral structures at 5°C and -90 °C, respectively (Scott, 1974). The Curie temperature and the dielectric permittivity of barium titanate ($BaTiO_3$) can be adjusted in a large range in the $BaTiO_3$-$SrTiO_3$ solid solution (Miura et al., 1975; Bornstein, 1981).
Studies on the solid solution $BaTiO_3$-$BaZrO_3$ have shown some interesting characteristics in bulk materials (Ravez& Simon, 2000; Dobal et al., 2001). For example, in the $Ba[Ti_{(1-x)}Zr_x]O_3$ composition with x in the range $0.26 < x < 0.42$, the temperature dependence of the real part of the dielectric permittivity at the transition temperature is broad and frequency dependent (Ravez& Simon, 1999). The effect of doping on various physical and chemical properties of BZT is known, and this effect has been extensively exploited in piezoelectrics and ferroelectrics to improve their performance. Many aliovalent compositional alterations to BZT have been studied either with higher valence substitutions (donors), or with lower

valence ions (acceptors). Donor dopants, such as Va^{5+} and W^{6+}, induce cationic defects while occupying the B site of the perovskite lattice (Zhi et al., 2000; Moura et al., 2008). Such behavior may cause several effects on the dielectric behavior through interaction with domain walls (Shannigrahi et al. 1999; Bolten et al. 2000). Barium titanate compositions modified by Sn, Hf, Ce, Y and Zr (Payne& Tennery, 1965; Jing at al. 1998; . Yu et al. 2001) have been extensively studied for dielectric application and all exhibit a crossover from typical ferroelectric to ferroelectric-relaxor behavior at higher concentrations of substitution. Also was observed that remnant polarization and coercive field are affected by substitution due the electron-relaxation-mode in which carriers (polarons, protons, and so on) are coupled with existing dielectric modes. Dielectric relaxation in relaxor ferroelectrics has become a topic of considerable interest (Cross,1987; Cheng et al.1998; Tagantsev & Glazounov, 1998). To account for the relaxation phenomena two different approaches are commonly used. The traditional one, considers clusters of polar phase located in non-polar phase to be responsible for the unusual relaxation behavior. It was proposed that the relaxation polarization was attributed to the thermo activated reorientation of dipolar moments of clusters or the motion of the clusters boundaries (Smolensky et al. ,1984; Bovtoun & Leshchenko, 1997; Courtens, 1984). The second approach implies the existence of a structural (dipolar) glass state in relaxors. The literature reports the frequent use of impedance spectroscopy analyses to reveal and study the mechanisms responsible for the dielectric properties of polycrystalline materials (Sinclair et al.,2002; Fang & Shiau, 2004; West et al., 2004), particularly to obtain information on non intrinsic dielectric effects.

The purpose of using dielectric spectroscopy analysis instead of the traditional impedance spectroscopy approach is to demonstrate that dielectric complex diagrams of the frequency response of polycrystalline materials can sometimes reveal more about the relaxation processes involved with grain boundaries and dielectric dipolar relaxation than impedance diagrams can (Böttcher & Bordewijk, 1992; Jonscher,1990).

Although $BaTiO_3$ it is a well known lead-free ceramic, detailed ferroelectric and piezoelectric behavior was less studied although some piezoelectric properties has been reported (Kell & Hellicar,1956). On the other hand, the recent discovery about very high piezoelectric and electrostrictive properties of lead based relaxor ferroelectric single crystals (Park& Shrout,1997; Paik et al.,1999; Liu et al.1999) has stimulated a renewed interest in searching for high strain materials, especially environmentally friendly lead-free materials. It is known that doping is an effective way to improve the material performance in electroceramics. Zirconium doped $BaTiO_3$ single crystals were recently grown and show promising piezoelectric and electrostrictive properties (Rehrig et al., 1999; Zhi et al.2001, 2000) . In perovskites the relaxor behavior occurs mainly in lead-based compositions with more than one type of ion occupying the equivalent six coordinated crystallographic sites (Moura et al.,2008; Dmowski et al.,2002). In turn, the substitution of Ti^{4+} by Zr^{4+} promotes several interesting features in the dielectric behavior of Barium titanate ceramics. When the Zr content is less than 10 at.% the BZT ceramics show normal ferroelectric behavior and dielectric anomalies corresponding several phase transition: cubic to tetragonal (T1), tetragonal to orthorhombic (T2) and orthorhombic to rhombohedral (T3). The interest in high strain piezoelectric materials is increasing for electromechanical transducers and various related applications. $BaTiO_3$ (BTO) is known to have a large electromechanical coupling factor. The effect of doping on various physical and chemical properties of this material is known, and this effect has been extensively exploited in piezoelectric and ferroelectrics to improve their performance. Many aliovalent compositional alterations to

$Ba(Ti_{1-x}Zr_x)O_3$ have been studied either with higher valence substitutions (donors), or with lower valence ions (acceptors).

Recently, it was found that the solid solubility of Y doped $BaTiO_3$ could reach up to 12.2 mol% once yttrium ions have stable valence (Zhi et al., 2000). Donor dopant,such as W^{6+}, induces cationic defects while occupying the B site of the perovskite lattice (Carg & Goel, 1999; Shannigrahi et al.,1999). Such behaviour may cause several effects on the dielectric properties through interaction with domain walls (Bolten et al., 2000; Takahashi,1982). Most studies on $Ba(Ti_{1-x}Zr_x)O_3$ ceramics are focused on the temperature dependence of the dielectric permittivity and relaxor behaviour (Ravez & Simon,1997a; 1997b; Farhi et al., 1999). The DPT was found in solid solutions of $BaTiO_3$ and $BaZrO_3$ (BTZ) which is one of the most important compositions for dielectrics in multilayer ceramic capacitors (Hennings et al.,1982), because Zr ion has higher chemical stability than Ti (Hoffmann & Waser, 1999; Tsurumi et al., 2002) ion and the high permittivity of the $BaTiO_3$ ceramic is increased more by the addition of zirconium. The sintering temperature of BZT is 1400 °C, when Zr^{4+} is added the sintering temperature decrease and dielectric constant rising (Weber et al., 2001). When the ratio of Zr to Ti in BZT reaches 20%, the Curie temperature is at room temperature around 27C, which is the ideal material for preparing variable electric capacities (Hennings et al.,1982) . Chen suggested (Chen, 2000) that these features of the relaxor phenomena are not caused by elementary excitation or the typical dielectric sources, (space charge), but reminiscence of a broad first-order phase transition obscured by severe geometrical and compositional randomness. However, there are some theories considering the origin of high dielectric permittivity of relaxor as the space charges arise from defects or vacancies (Dixit et al.,2003; Wang et al.,2002). Ionic doping is a common method for optimizing the electric properties of ceramics. It is known that doping is an effective way to improve the material performance in electroceramics. Zr-doped $BaTiO_3$ single crystals were recently grown and show promising piezoelectric/electrostrictive properties (Rehrig et al.,1999; Zhi et al., 2001; 2000a; 2000b). Similarity to Zr modified BT, BZT:2W may be a promising material as lead free actuator. Because of that it is important to study and compreend how the properties of ceramics materials are significantly affected by temperature, defects, atmosphere flow. It is important to control the temperature due to its strong influence on the grain size, dielectric and ferroelectric properties (Tang et al., 2004; Simões et al., 2005; Marques et al.,2007). The medium and fine grained ceramics possess poor dielectric properties influencing the performance of the device. This is due the oxygen interdiffusion, chemical reaction, or structural defects in this particle size range . Properties of ceramic materials are significantly affected by temperature, defects, atmosphere flow. This is due to the oxygen interdiffusion, chemical reaction, or structural defects in this particle size range. In the present study, a deep investigation on temperature and concentration of vanadium and tungsten dependence of electrical properties of *BZT:10* ceramics prepared by mixed oxide method was performed. Due its better electrical properties BZT10:2V ceramics had their properties evaluated as a function of annealing atmosphere.

2. Influence of zirconium ions concentration on the crystal structure and electrical properties of the barium titanate ceramics

2.1 Preparation of solid solution $BaTiO_3$- $BaZrO_3$ by mixed oxide method

The addition of Zr strongly influenced the crystal structure and electrical properties of the barium titanate ceramics. The investigated systems were $Ba(Ti_{0.95}Zr_{0.05})O_3$, $Ba(Ti_{0.9}Zr_{0.10})O_3$

and $Ba(Ti_{0.85}Zr_{0.15})O_3$. In this study, barium zirconium titanate (BZT) ceramics were prepared by mixed oxide method. The powders were homogenized in a ball mill using isopropyl alcohol. All oxides were analytical grade: $BaCO_3$ (Vetec), ZrO_2 (Inlab), TiO_2 (Vetec). After drying, the powders were calcinated at 1200°C for 2 hours. After that powder were compacted in two steps, first by pressing one way then the pellets are pressed isostatically at 210 MPa, in the form of pellets with 6mm long and 8 mm of diameter to be

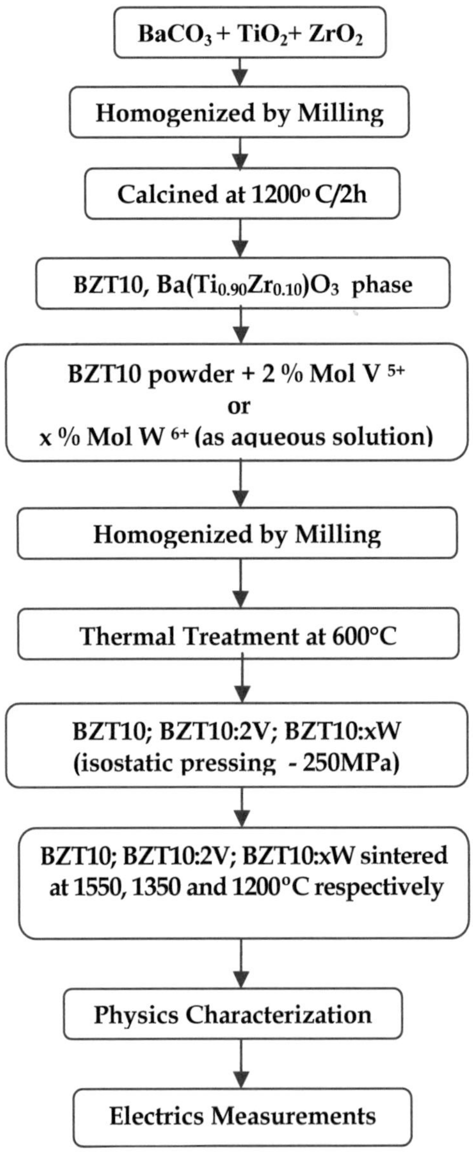

Fig. 1. Flowchart of the synthesis and characterization process of BZT10; BZT10:xV; BZT10:yW

submitted at dilatometric analyses. This mesurament were performed in a Nezstch dilatometer 402 E, up to 1550°C, at a heating rate of 5 °C/min in static atmosphere and the results of linear shrinkage rate ($d(\Delta l/l_0)/dT$) and linear shrinkage $\Delta l/l_0$ were determined. Considering the results of linear shrinkage rate, new samples were obtained from powders calcined at 1200 °C. The powders were compacted in the form of pellets with 1 mm of thickness and 12 mm of diameter, isostatically pressed at 210MPa. The pellets were sintered at 1550°C for 4 hours in static air and cooled to room temperature (5 °C/min).

2.2 Preparation of solid solution BZT10:xV and BZT10:yW by mixed oxide method

The dopants, as aqueous complex, were added to the calcined powders at 1200°C as described. Separately, vanadium or tungsten oxides were dissolved in nitric acid and complexed with citric acid. After that, ethylene glycol was added in the medium to obtain the stable aqueous complex. A volume, which correspond to the concentration of vanadium or tungsten necessary to get BZT:yV or BZT:xW in molar ratio was added to BZT powders. The BZT10 powder wet with doping solution was homogenized by grinding, dried and calcined ate 600°C to eliminated all the organic material (citric acid and ethylene glycol from dopant solution). After this process, the powders were compacted in the form of pellets with 1 mm of thickness and 12 mm of diameter, isostatically pressed at 210MPa and sintered at 1350°C for different concentrations of dopants (BZT10:1V to BZT10:6V) and BZT10:2V at different temperatures.

2.3 Techniques used in the characterization of solid solution BZT10; BZT10:xV and BZT10:yW

X-ray diffraction data were collected with a Rigaku Rint 2000 diffractometer under the following experimental condition: 50 kV, 150 mA, $20° \leq 2\theta \leq 80°$, $\Delta 2\theta = 0.02°$, $\lambda_{Cu}\ k_{\alpha}$ monocromatized by a graphite crystal, divergence slit = 2mm, reception slit = 0.6mm, step time = 10s. Raman measurements were performed using an ISAT 64000 triple monochromator. The samples also were characterized by dielectrics properties. Gold electrodes, for electrical measurements, were applied by evaporation through a sputtering system on polished surfaces of sintered discs. Ferroelectric properties were measured on a Radiant Technology RT6600A tester system equipped with a micrometer probe station in a virtual ground mode. The dielectric characterization was accomplished with HP 4192 impedance analyzer, and measurements of the capacitance as a function of temperature at a frequency of 10 kHz were performed. From the capacitance dependence temperature curves, the Curie temperature was determined. Piezoelectric measurements were carried out using a setup based on an atomic force microscope in a Multimode Scanning Probe Microscope with Nanoscope IV controller (Veeco FPP-100). In our experiments, piezoresponse images were acquired in ambient air by applying a small ac voltage with amplitude of 2.5 V (peak to peak) and a frequency of 10 kHz while scanning the sample surface. To apply the external voltage we used a standard gold coated Si_3N_4 cantilever with a spring constant of 0.09 N/m. The probing tip, with an apex radius of about 20 nm, was in mechanical contact with the uncoated samplesurface during the measurements. Cantilever vibration was detected using a conventional lock-in technique.

2.4 Results from characterization of the $Ba(Ti_{1-x}Zr_x)O_3$ systems.

The X-ray diffraction crystal structure of the $Ba(Ti_{0.95}Zr_{0.05})O_3$, $Ba(Ti_{0.9}Zr_{0.10})O_3$ and $Ba(Ti_{0.85}Zr_{0.15})O_3$ powders are represented in Figure 2 and show that all the ceramics composition were crystallized into a single-phase perovskite structure. This is a clear indication that the addition of Zr is forming a stable solid solution with the $BaTiO_3$ lattice. As reported in other studies (Yu et al.,2002; Dixit, et al., 2003; Dobal et al., 2002; Paik et al., 1999) depending upon the Zr content, BZT may have orthorhombic, rhombohedral and tetragonal structures at room temperature.

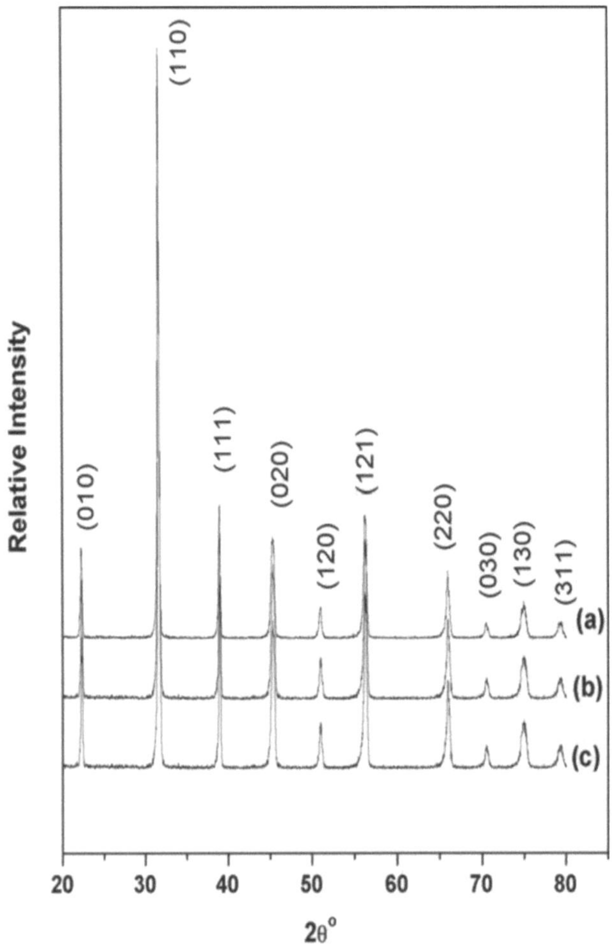

Fig. 2. X-ray diffraction pattern of the BZT powders with different concentrations calcinated at 1200 °C for 2 h: (a) $Ba(Ti_{0.95}Zr_{0.05})O_3$; (b) $Ba(Ti_{0.9}Zr_{0.10})O_3$ and (c) $Ba(Ti_{0.85}Zr_{0.15})O_3$.

The linear shrinkage rate ($d(l/l_0)/dT$) as a function of temperature for different Zr/Ti ratio are shown in Figure 3. It is noted that the sintering process is strongly influenced by Zr/Ti ratio leading to a reduction in the densification temperature. The maximum shrinkage rate occurred around 1410 °C for the composition $Ba(Ti_{0.95}Zr_{0.05})O_3$. Meanwhile, a reduction in the maximum shrinkage rate was observed for the $Ba(Ti_{0.85}Zr_{0.15})O_3$ composition, probably due to the segregation of ZrO_2 at the grain boundaries. The peak close to 980 °C indicates the sintering process intra and inter-agglomerates as a consequence of different surface area of the raw powders [ZrO_2 (5.67 m^2/g), $BaCO_3$ (1.35 m^2/g) and TiO_2 (53.71 m^2/g)].

Considering the results of linear shrinkage rate, new samples were obtained from powders calcined at 1200 °C and sintered in static air at 1550 °C for 4 hours in the tubular furnace.

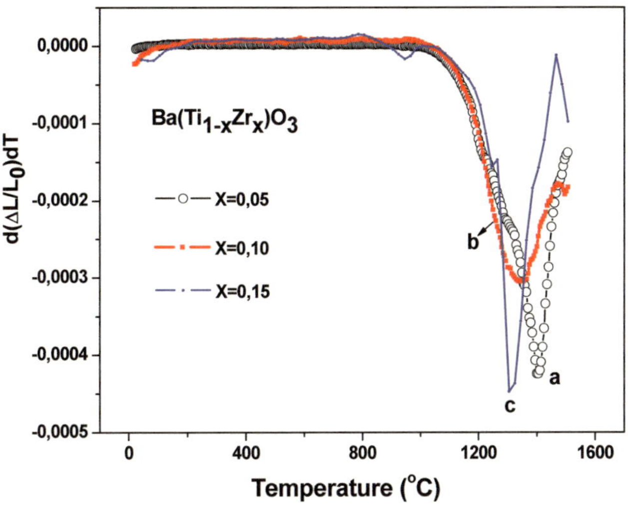

Fig. 3. Linear shrinkage rate as a function of temperature for BZT powders calcinated at 1200 °C for 2 h in different concentrations: (a) $Ba(Ti_{0.95}Zr_{0.05})O_3$; (b) $Ba(Ti_{0.9}Zr_{0.10})O_3$ and (c) $Ba(Ti_{0.85}Zr_{0.15})O_3$.

The Raman Spectra from these samples were presented in Figure 4. The directions of the phonon wave vectors are randomly distributed from one grain to another with respect to the crystallographic axes because the grains are random oriented in the precursor powders. The evolution of Raman spectra with Zr substituting on Ti sites shows some interesting changes with increasing Zr content the Raman line at 123 cm^{-1} has not been observed. Taking the mass ratio Zr/Ti = 1.9 into consideration this mode frequency is expected to be about 129 cm^{-1} for Zr replacing Ti sites, which will be reduced further by an increase in the ionic radius [$R(Ti^{4+}) = 0.0745$ nm, $R(Zr^{4+}) = 0.086$ nm]. The additional mode could therefore be associated with a normal mode involving Zr atoms. Since this mode disappears in the $Ba(Ti_{0.9}Zr_{0.10})O_3$ and $Ba(Ti_{0.85}Zr_{0.15})O_3$ compositions, it may be considered an indication of the orthorhombic to rhombohedral phase transition. Such observations could not be observed in the X-ray studies due to the different coherence length and time scale involved in the process.

Measurements on dielectric permittivity as a function of temperature reveal anisotropic behavior, Figure 5. The highest permittivity (ε_r= 14.500) is observed for the $Ba(Ti_{0.95}Zr_{0.05})O_3$ composition at 100 kHz. Three distinct phase transitions were observed and the Curie temperature reduces with the increase of zirconium content due the changes in crystal structure, as shown in Raman spectra, Figure 4. A strong reduction in the dielectric permittivity for the $Ba(Ti_{0.85}Zr_{0.15})O_3$ system is a consequence of changing in crystal structure. A sharp phase transition is indicative of a ferroelectric-relaxor behavior as observed in the literature (Yu et al.,2000; Weber et al.,2001).

Well saturated hysteresis loops with regular shape, typical of ferroelectric materials were evident for all investigated systems (Fig. 6a-c). There is no evidence of "imprint" phenomena indicating that ceramics present few defects as oxygen vacancies which pinning

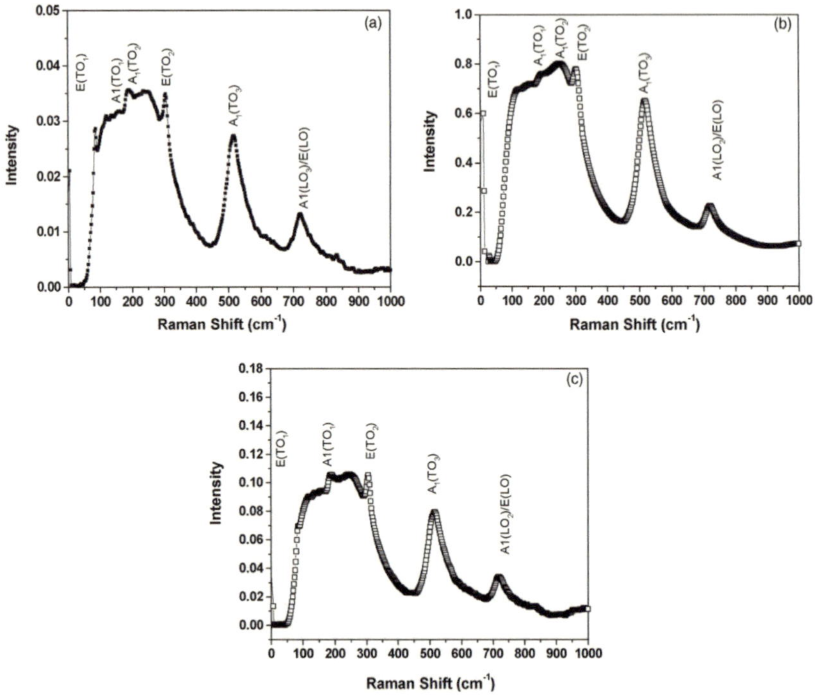

Fig. 4. Room temperature depolarized Raman spectra for BZT powders with different concentrations calcinated at 1200°C for 2h: (a) $Ba(Ti_{0.95}\,Zr_{0.05})O_3$; (b) $Ba(Ti_{0.9}Zr_{0.10})O_3$ and (c) $Ba(Ti_{0.85}Zr_{0.15})O_3$.

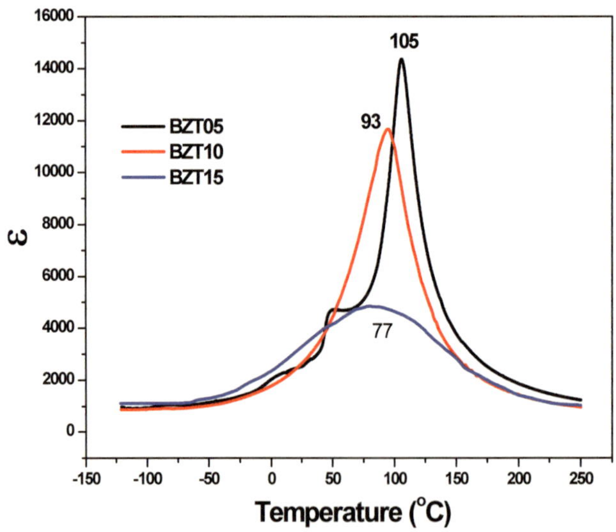

Fig. 5. Temperature dependence of dielectric permittivity (ε) for BZT pellets with different concentrations, sintered at 1550 °C/4h: (a) $Ba(Ti_{0.95}Zr\ _{0.05})O3$; (b) $Ba(Ti_{0.9}Zr_{0.10})O_3$ and (c) $Ba(Ti_{0.85}Zr_{0.15})O_3$.

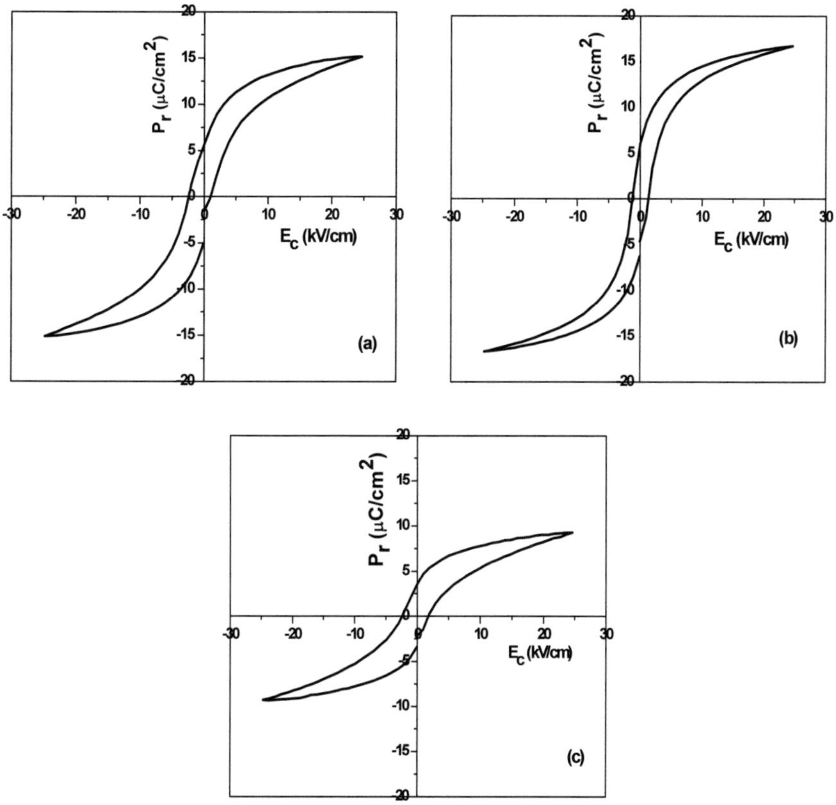

Fig. 6. P–E hysteresis loops for BZT pellets sintered at 1550 °C for 4 h in different concentrations: (a) $Ba(Ti_{0.95}Zr_{0.05})O_3$; (b) $Ba(Ti_{0.9}Zr_{0.10})O_3$ and (c) $Ba(Ti_{0.85}Zr_{0.15})O_3$.

the domain walls difficulting the saturation of hysteresis loops. The reduction in the remnant polarization for the $Ba(Ti_{0.85}Zr_{0.15})O_3$ system can be related to changes in crystal structure which can explain the relaxor behavior observed in Figure 5. The defects generated during sintering can be described by Equations 1 to 5. The substitution of Ti^{4+} for Zr^{4+} promotes defects type $\left[TiO_5.V_O^{\bullet}\right]$ that compete with the species TiO_6 raising the dielectric and ferroelectric properties of ceramics a maximum value when compared to species $\left[TiO_5.V_O^{\bullet}\right]$ that are much smaller than $\left[TiO_6\right]$.

$$ZrO_2 \xrightarrow{BaTiO_3} Zr_{Ti}^{x} + 2O_O^{x} \quad (1)$$

$$\left[ZrO_6\right]^{x} + \left[ZrO_5.V_o^{x}\right] \longrightarrow \left[ZrO_6\right]' + \left[ZrO_5.V_o^{\bullet}\right] \quad (2)$$

$$\left[BaO_{12}\right]^{x} + \left[BaO_{11}.V_O^{x}\right] \longrightarrow \left[BaO_{12}\right]' + \left[BaO_{11}.V_{Ba}^{\bullet}\right] \quad (3)$$

$$\left[BaO_{12}\right]^{x} + \left[BaO_{12}.V_{Ba}'\right] \longrightarrow \left[BaO_{12}\right]^{\bullet} + \left[BaO_{12}.V_{Ba}''\right] \quad (4)$$

$$\left[Ti\,O_6\right]^x + \left[TiO_5 . V_O^x\right] \longrightarrow \left[TiO_6\right]' + \left[TiO_5 . V_O^\bullet\right] \quad (5)$$

3.Temperature dependence and Dielectric response of tungsten or vanadium modified $Ba(Ti_{0.90}Zr_{0.10})O_3$ ceramics

3.1 Sample preparation of tungsten or vanadium modified $Ba(Ti_{0.90}Zr_{0.10})O_3$

As observed, BZT phase presents interesting behavior that need to be exploited. A much discussed variable is the influence of donor elements (W^{6+} or V^{5+}) whose effects depend on concentration and processing of the powder. The main motivation for using V^{5+} or W^{6+} as network modifier was the possibility of obtaining ceramics with high dielectric constant and shifted the phase transition to lower temperatures, in the range of 70 to 120°C. Move the sintering temperature to less than 1550°C used for ceramics $Ba(Ti_{1-x}Zr_x)O_3$, was also a goal to be achieved. The $Ba(Ti_{0.9}Zr_{0.10})O_3$ composition was chosen because among those studied showed a single phase transition Figure 5. In this study $Ba(Zr_{0.10}Ti_{0.90})O_3$ powders were prepared by mixed oxide method as described in the item 2.2. Temperature dependence and Dielectric response from vanadium modified $Ba(Ti_{0.90}Zr_{0.10})O_3$ ceramics

According to X-ray diffraction analysis the single phase crystallized in a perovskite structure were obtained. The X-ray diffraction patterns for the pure BZT and doped with different concentrations of vanadium sintered at 1350°C show a pure phase for concentrations up to 2mol % of dopant. On the other hand, when the vanadium content increases to (3-6 mol %) an intermediate phase of BaV_2O_6 has been obtained, Figure 7(a). Also, small traces of unreacted oxides such as TiO_2 and ZrO_2 are located at $2\theta = 27$ and 46 degree for the pure BZT:10 phase. Bragg reflection peaks are indicative of perovskite structure, mainly characterized by higher intense peak (110) at $2\theta = 31°$ and no apparent peak splitting is identified. This is a indication that the addition of vanadium up to 2wt% has formed a stable solid solution in the BZT matrix lattice at 1350°C. The X-ray diffraction pattern Figure 7(b) obtained from the BZT:2V ceramics sintered at various temperatures show that the single phase with a tetragonal perovskite structure with highest intensity peak at 31° was obtained. At low temperature (1200°C) although diffraction peaks corresponding to the tetragonal perovskite structure they are generally weak indicating a poor crystallinity of the ceramics in the (1 1 0) and (2 2 0) directions. At 1350°C the growth of (110) and (220) oriented grains was noted.

Raman spectrum located at 915 cm^{-1} indicates the presence of vanadium in the BZT structure as observed through X-ray diffraction. The substitution located in B site of perovskite BZT-10 reduces the distortion of the octahedral sites leading to a more organized symmetry when increasing the concentration of vanadium ions.

The degree of order-disorder of the atomic structure was observed by Raman spectroscopy in the longitudinal (LO) and transverse (TO) modes, Figure 8. These modes are associated with electrostatic and ionic forces from structure due to Ba^{2+} in $BaTiO_3$ (Domenico Jr et al.,1968; Chaves,et al.,1974) . For the perovskite $Ba(Zr_{0.10}Ti_{0.90})O_3$, Ba^{2+} ions are illustrated by bands on the spectra, A1 (TO1) and A1 (TO2) around 193 and 517 cm^{-1} (Dobal et al.,2001; Kreiselet al., 2004). The other modes OE1 (TO1) and E1 (TO2) located at 116 and 301 cm^{-1} are associated to tetragonal - cubic transition, whereas the A1 (LO3) mode was found at 720 cm^{-1} due the zirconium replacing titanium in B site. However, the coupling between the A1 (TO1) and A1 (TO2) modes reduces the intensity of A1 (TO2) mode (Dobal et al.,2001; Dixit et al.,2004).

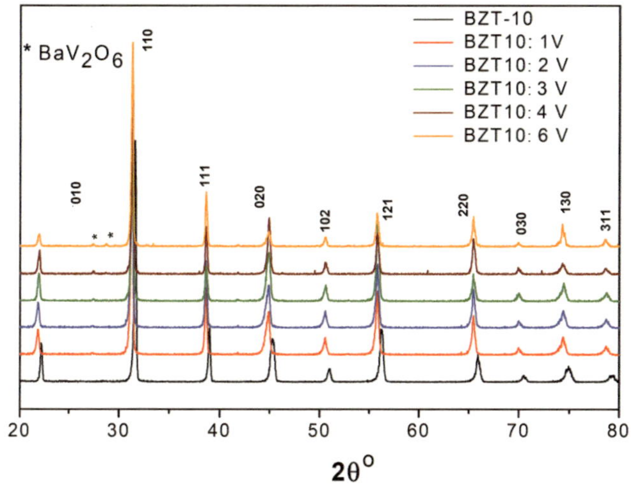

a)

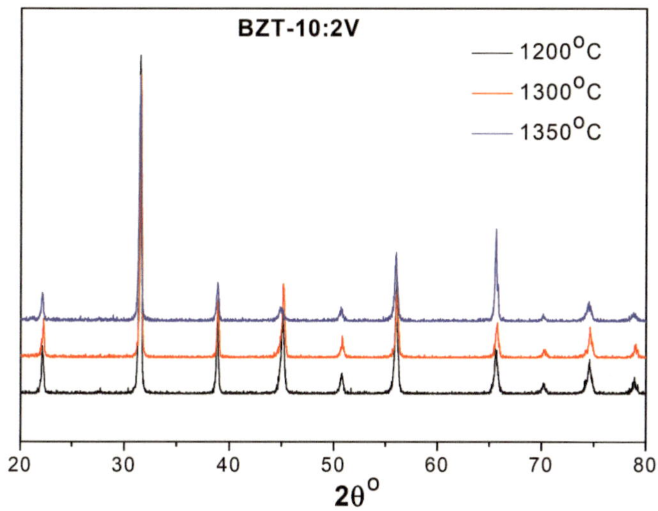

b)

Fig. 7. X-ray diffraction of BZT10 and BZT10:xV ceramics sintered : (a) at 1350 °C for 4 hours; (b) BZT10:2V ceramics sintered at diferentes temperatures.

The temperature dependence of dielectric permittivity for BZT:2V samples at 10 kHz is shown in Figure 9a. For coarse-grained sample (1350°C) the phase transition is observed at 92°C with maximum dielectric permittivity (ε_m) 15.000. In this sample, it was noted a structural phase transition which corresponds to the paraelectric (cubic) to ferroelectric (tetragonal) phase transition at Tc. For medium and small-grained samples only one phase transition is observed. However, the peak associated to phase transition temperature for the small-grained sample become broader than that of the big and medium grain size. These results suggest that the BZT ceramics with small grain size has the transition from a normal ferroelectric to 'relaxor-like' ferroelectric (Zhi, et al.,2002). The dielectric permittivity increases gradually with an increase at temperature up to the transition temperature (Tc) and then decreases. This increasing is caused by the relatively large ionic radius of the B ion which enhances the thermal stability of the BO_6 octahedra when compared to Ti^{4+} or Zr^{4+}.

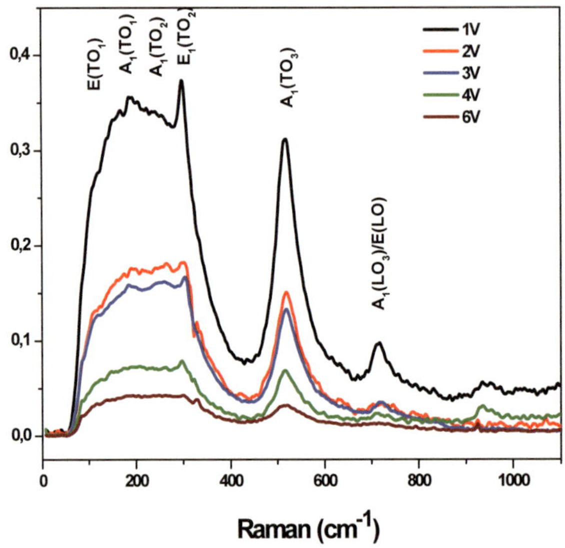

Fig. 8. Room temperature depolarized Raman spectra for BZT10:xV powders calcinated at 1200°C for 2hours

Also, the packing density of the BO_6 octahedra will be determined by the size of the B ion. Larger B ions give more closely packed octahedral which are therefore more stable. The V^{5+} enters in the B-site of the ABO_3 perovskite lattice leading to a charged [VO6]• defect which is associated with a barium vacancy in a local barium cluster $[V_{Ba}^{''}O_{12}]$. In fully or partly ionic compounds vacancies are charge balanced by other defects forming an overall neutral system. It can be assumed that particle charge compensation takes place at a nearest-neighbor barium cluster site in the $[BaO_{12}]$, because the resulting coulomb interaction is the most important driving force. This assignment is in accordance with first-principles calculation (Anicete-Santos et Al., 2005). Probably, an equilibrium can be reached between "free" [VO6]• centers and [VO6]• + $[V_{Ba}^{''}O_{12}]$ associated defects. We can consider the free [VO6]• clusters as responsible for ionic mobility and [VO6]• + $[V_{Ba}^{''}O_{12}]$ defect dipole complexes the main cause of electrical properties in the ceramic. Hence, charge transport will be considerably hindered. Below the Curie temperature, a high dependence of dielectric loss was observed while at elevated temperatures, vanadium doping stabilizes this dependence, Figure 9b. The possible formation of dipole complexes may result in a reduced dielectric loss at elevated temperatures. This reflects that good insulation resistance was maintained at high temperatures, which is important for high temperature piezoelectric applications. The dielectric loss shows strong dependence as temperature increases. At the same frequency region, the obtained values are 0.001 (1200°C) to 0.019 (1300 and 1350 °C). It is known that the increase of dissipation factor is due to extrinsic resonance behavior (Veith et al., 2000) This may be due to the defects (vacancy, movable ion, leaky grain boundary

(Hoffmann & Waser, 1997) that developed in the structure of the bulk material with the increase of sintering temperature. The sample sintered at low temperature possesses a significant difference in the dielectric loss peak suggesting a pinch-off of phase transition. In this sample, due the fine-grain microstructure the introduction of vanadium causes a more sensitive distortion of the perovskite lattice leading to a reduction in the oxygen octahedron interstices. The distortion of the perovskite lattice can strengthen the structure fluctuation of the BZT:2V ceramics which can be account for different behavior of diffusion phase transition characteristics in this sample. [ZrO6]x, [TiO6]x, and [VO6]• clusters are present in the BZT:2V lattice. As the temperature increases, there is a charge transference of [VO6]• to the Ti and Zr clusters, according to Equations (6) and (7):

$$[VO6]^{\bullet} + [ZrO6]x \rightarrow [VO6]x + [ZrO6]^{\bullet} \quad (6)$$

$$[VO6]^{\bullet} + [TiO6]x \rightarrow [VO6]x + [TiO6]^{\bullet} \quad (7)$$

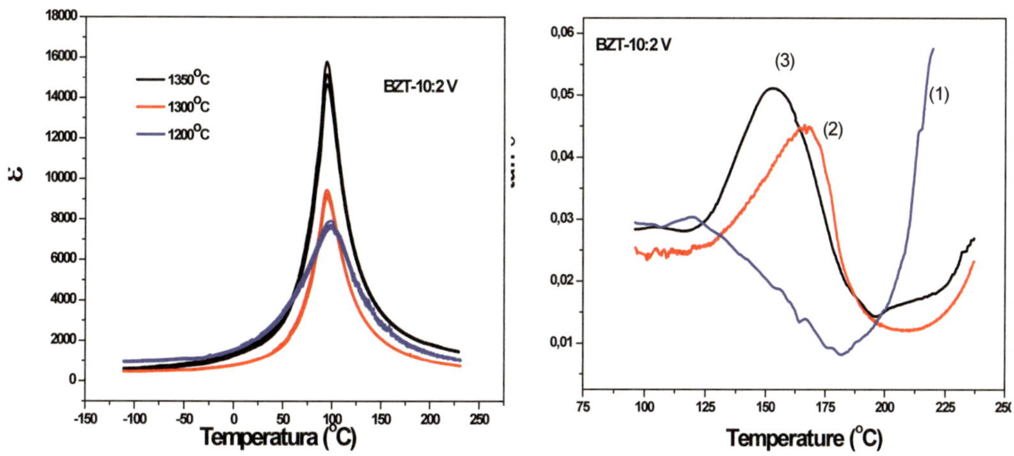

(A) (B)

Fig. 9. Temperature dependence of dielectric permittivity (A); and dielectric loss (B) at 10 kHz for BZT10:2V ceramics sintered at (1) 1200, (2) 1300 and (3) 1350 °C for 4 hours.

Scanning electron microscopy of vanadium doped BZT10 with different concentrations shows that there is formation of a liquid phase at the grain boundary with increasing concentration of vanadium, figure 10 (a) to (c). Scanning electron microscopy also allowed to observe the behavior of the sample BZT10:2V sintered at 1200 C, 1300°C and 1350°C where the increase in temperature promoted the diffusion of vanadium with the formation of defects that are represented in equations 6 and 7. These phenomena promote the grain growth and improves the dielectric permittivity.

4.Temperature dependence and dielectric response from tungsten modified $Ba(Ti_{0.90}Zr_{0.10})O_3$ ceramics

XRD patterns of BZT and BZT10:yW modified ceramics are shown in Figure11. Small traces of unreacted oxides such as TiO_2 and ZrO_2 are located at 2θ = 27 and 46 degree for the

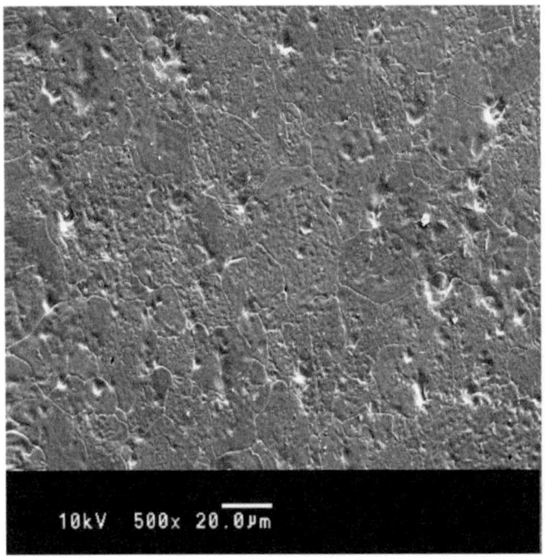

(a)

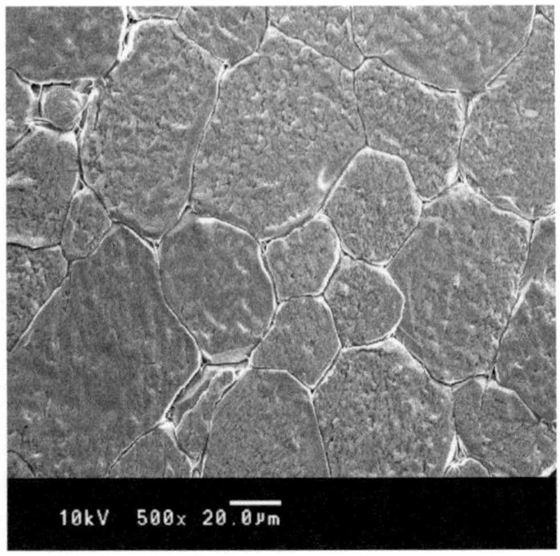

(b)

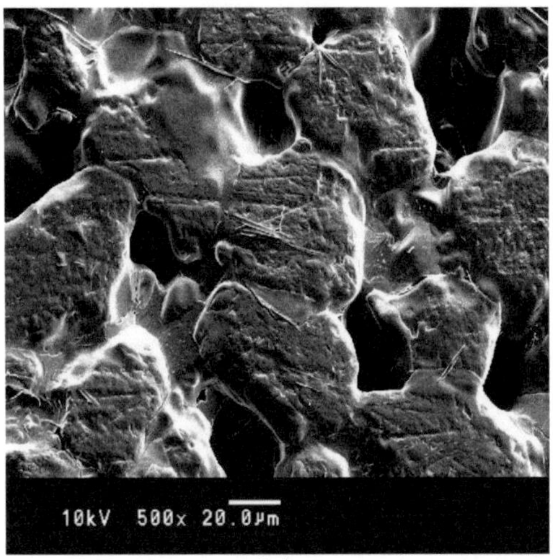

(c)

Fig. 10. SEM photomicrographs of the polished surface of BZT-10 ceramic modified with vanadium ions and sintered at 1350ºC for 4 h, (a) BZT10:1V; (b) BZT10:2V; (c) BZT10:6V

BZT10 phase. On the other hand, no secondary phases are evident in the BZT10:1W and BZT10:2W powders. This is a clear indication that the addition of tungsten is forming a stable solid solution in the *BZT* matrix lattice. As tungsten content increases, there was an intermediate phase of $BaWO_3$ was obtained. Bragg reflections peaks are indicative of perovskite structure, mainly characterized by higher intense peak (hkl-110) at $2\theta = 31^o$ and no apparent peak splitting is identified.

The evolution of Raman spectra in the tungsten substituted BZT ceramics displayed in Figure 12 shows some interesting order-disorder degree of the atomic structure at short range. The spectrum presents the stretching mode of A1(TO1) and A1(TO3) at around 193 and 517 cm^{-1}. The E1(TO1) and E1(TO2) modes that have been associated with the tetragonal-cubic phase transition were observed at 116 and 301 cm^{-1}. Whereas the A1(LO3)

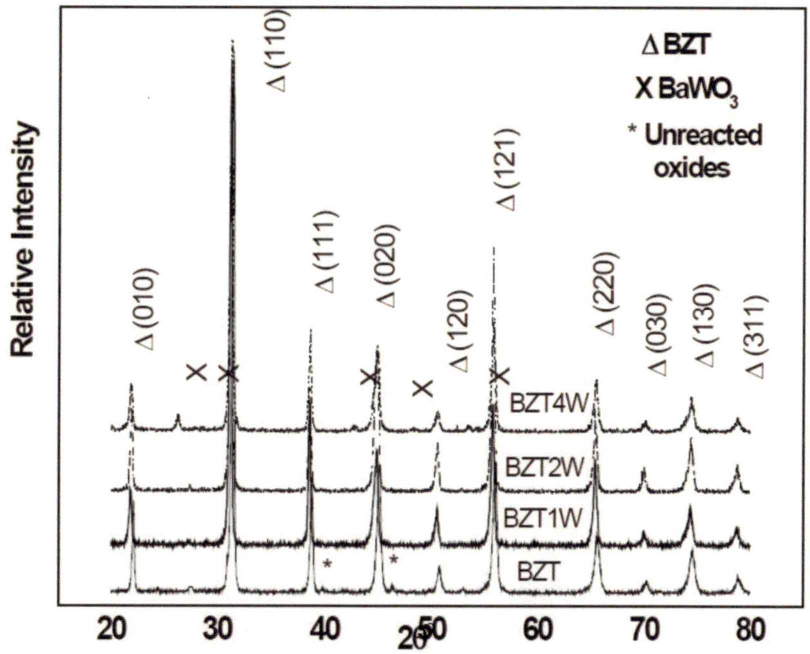

Fig. 11. X-ray diffraction for BZT10, BZT10:1W, BZT10:2W and BZT10:4W powders prepared from mixed oxide method, calcined at 900°C for 2 hours.

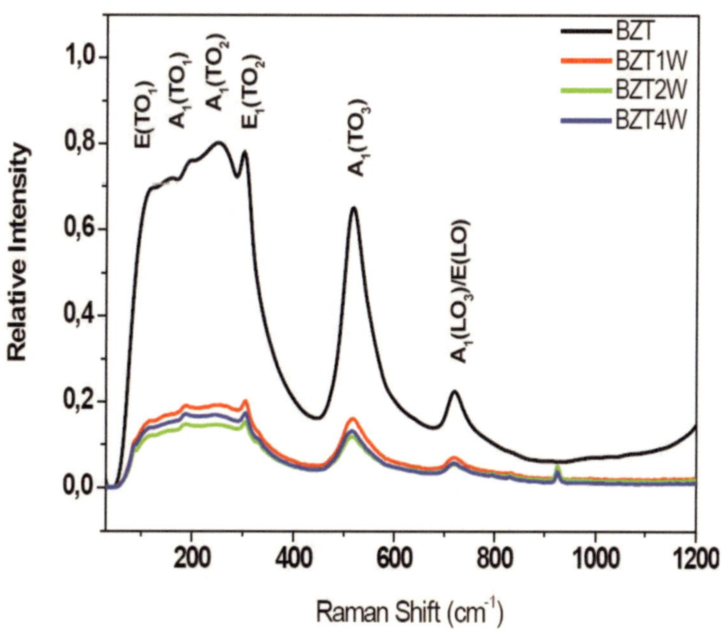

Fig. 12. Room temperature depolarized Raman spectra for BZT10, BZT10:1W, BZT10:2W and BZT10:4W powders prepared from mixed oxide method, calcined at 900°C for 2 hours.

mode was found at 720 cm^{-1} with zirconium (Zr) substituting on titanium (Ti) sites, as tungsten is incorporated in the BZT lattice, a Raman line at 925 cm^{-1} appears. Considering that tungsten substitutes the B-site of the lattice, as its content increases the relative intensity of bands reduces due to the distortion of octahedral sites. This results in diminished interstices in oxygen octahedron. Such observations could not be observed in the X-ray studies due to the different coherence length and time scale involved in the process.

The microstructure of those ceramics is shown in Figure 13. The grain growth is inhibited in the BZT10:4W sample due to the formation of $BaWO_3$ secondary phase in the grain boundary,Figure 13 d. Tungsten ion plays a role of donor in BZT because it possesses a higher valence than Ti or Zr ions suppressing the formation of oxygen vacancies. Considering that W^{+6} preferentially enter in B-sites, we expected an increase in oxygen vacancies concentration which affects the densification process. The small grain size can be interpreted by the suppression of oxygen vacancy concentration, which results in slower oxygen ion motion and lower grain growth rate as it was verified in the EPR spectra. No segregates in the grain boundaries of the BZT10:2W sample indicates the high solubility of WO_3 in the BZT matrix (Figure 13 c).

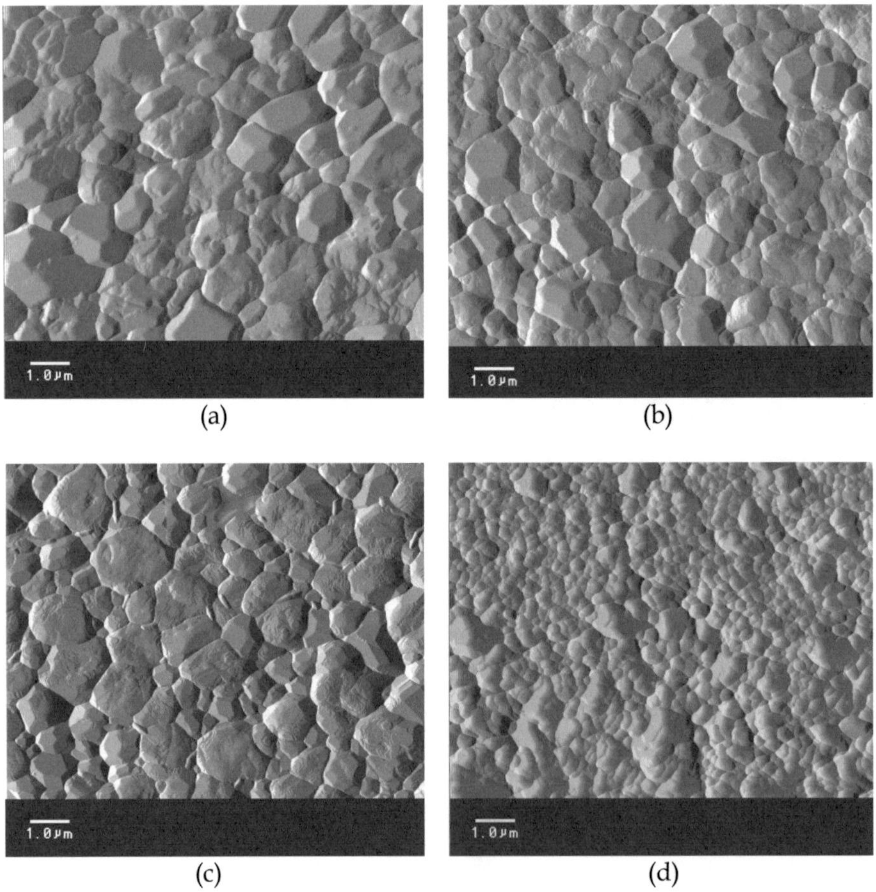

(a) (b)

(c) (d)

Fig. 13. Micrographies for (a) BZT10, (b) BZT10:1W, (c) BZT10:2W and (d) BZT10:4W ceramics prepared from mixed oxide method , sintered at 1200°C for 4 hours.

The temperature dependence of relative dielectric permittivity and dielectric losses measured at 10 KHz for selected BZT compositions are show in Figure 14. Substitution of W^{6+} in the B-site will lead to the distortion of the perovskite lattice leading to reduction of phase transition temperature up to tungsten content of 2 wt.% and broadening of dielectric peak. Since the ionic radius of W^{6+} is smaller (0.62Å) than that of Ti^{4+} (0.68Å) on B-site, increasing its amount would lead to a reduced contribution of overall atomic polarization. The dielectric permittivity increases gradually with an increase in temperature up to the transition temperature (T_c), Curie point, and then decreases. The region around the dielectric peak is broadened due to a disorder in the cations arrangement in one or more crystallographic sites of the structure (Lines & Glasss,1977; Wu et al.,2002). Large differences in the B valence results in a strong tendency for the material to disorder in one or more crystallographic sites leading to a microscopic heterogeneity in compounds with different Curie points. According to Wu et al.,2002 the dielectric permittivity consists of contributions of ionic and atomic polarization only. Therefore, the increase of dielectric permittivity may be attributable to increased ionic polarization. (Jin et al. 2001) proposed that the phase transition is caused by instability of the BO_6 octahedra. The relatively larger ionic radius of the B ion enhances the thermal stability of the BO_6 octahedra, when compared to Ti or Zr ions (Wu et al.,1983). This results in an increase of BO_6 volume octahedron and consequently decreasing its stability which results in a reduction of phase transition that involves structural inhomogeneity and existence of polar nanoregions. The interstices in oxygen octahedron will be diminished with the increase of metal vacancies in the B-site, and the c/a ratio will be lowered as well. Therefore, the dielectric polarization is depressed owing to the restraint of the B-site cations, and as a result the Curie point will lead to a diffuse phase transition. The dielectric permittivity is reduced with the increase in tungsten content leading to a typical relaxor behaviour, Figure 14a. At lower temperatures, a small dependence of dielectric loss was observed while at elevated ones, there is a significant dependence on it. Also, a small peak just below T_c was evident in the BZT, Figure 14b. It should be pointed out that this behaviour can be explained by the appearance of polar microregions in the samples. The dielectric relaxation peak can be attributed to the electron-relaxation-mode coupling mechanism in which carriers (or polarons, protons, and so on) are coupled with existing dielectric modes suggesting that the motion of carriers (or polaron hopping) is responsible for this behavior (Xu,1991). Poor insulation resistance was maintained at high temperatures indicating that these ceramics can be used for low temperature capacitor applications. In addition, the dielectric loss was much lower than that of the undoped BZT reported in our previous work (Wu et al.,1983; Moura et al.,2009; Moura et al.,2010; Marques et al., 2007) which is attributed to the decrease in space charge density as tungsten was incorporated in the BZT lattice. Two main mechanisms can be considered for substituting W^{+6} for Ti^{+4} in the BZT lattice. The compensation mechanism for substituting a ion^{6+} for a ion $^{4+}$ should reduce the oxygen vacancy concentration, leading to formation of Ba^{2+} vacancies. Another possibility is that the extra electrons become somewhat delocalized leading to some conductivity.

EPR Hamiltonian parameters were used to perfectly reproduce the observed sequence of symmetrical peak pattern of the fine structure, Figure 15. The substitution of W^{6+} in the Ti^{4+} site (Figure15b) causes a slight distortion in the spectra increasing disorder and symmetry changes in the BZT lattice (Figure 15a).

BZT10:2W spectra shows hyperfine bands typical for materials with spin 5/2 which can be attributed to the Zr presence as proposed in Figure 15b. (Abraham,et al.,1986). Moreover,

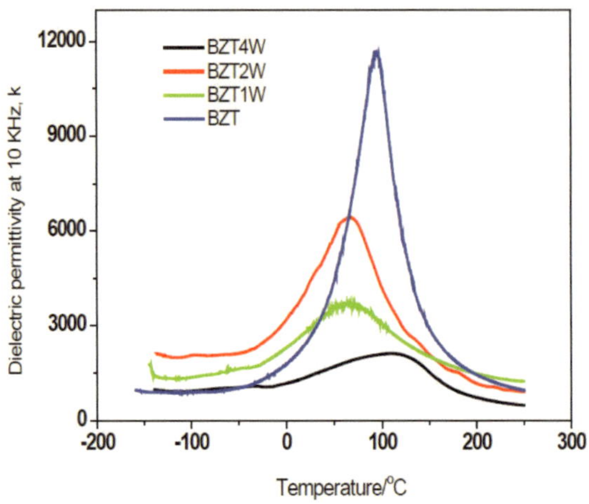

(A)

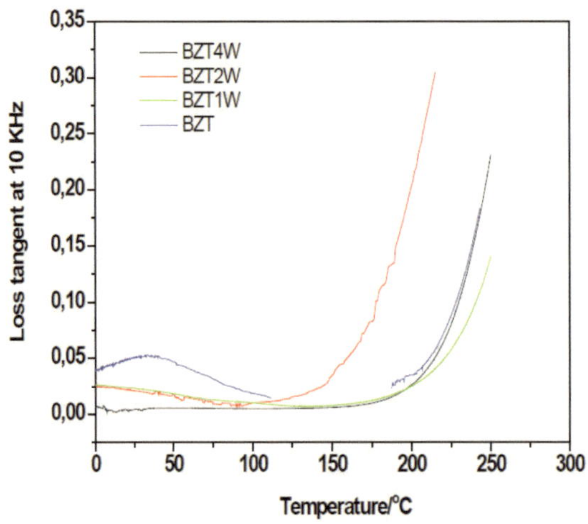

(B)

Fig. 14. Temperature dependence of dielectric permittivity (A); loss tangent (B) mesured at 10 KHz for BZT10, BZT10:1W, BZT10:2W, BZT10:4W ceramics prepared from mixed oxide method sintered at 1200°C for 4hours.

it suggests almost isotropic neighbourhood of the isolated Zr and Ti ions in the pattern. The obtained signal is typical for single ionized oxygen vacancies VO(, as observed by (Zhang et al.2004). The hyperfine bands in the spectra correspond to the expected line broadening attributed to dipolar interactions of tungsten in the host, according to reaction of defects as discussed in literature (Shannigrahi et al., 1999). Substitution of Ti^{4+} by W^{6+} causes distortion in the crystal structure changing lattice parameter. In this structure, [TiO5.Vox]c clusters are donor candidates and $[TiO_6]\ ^{x}c$ are acceptors candidates. $[TiO_5.Vo^{x}]c$ have shown two paired electrons ↑↓, $[TiO_5.Vo^{x}]c$ have shown one unpaired electron, while $[TiO_5.Vo^{x}]c$ have shown no unpaired electrons. The source of hyperfine bands can be a result of interaction between an ion with unpaired electrons with species which present complex vacancies. This is supported by the high symmetry as inferred from Raman studies. The main differences in the spectra correspond to the expected line broadening attributed to dipolar interactions of tungsten in the host according to equations 8 -15.

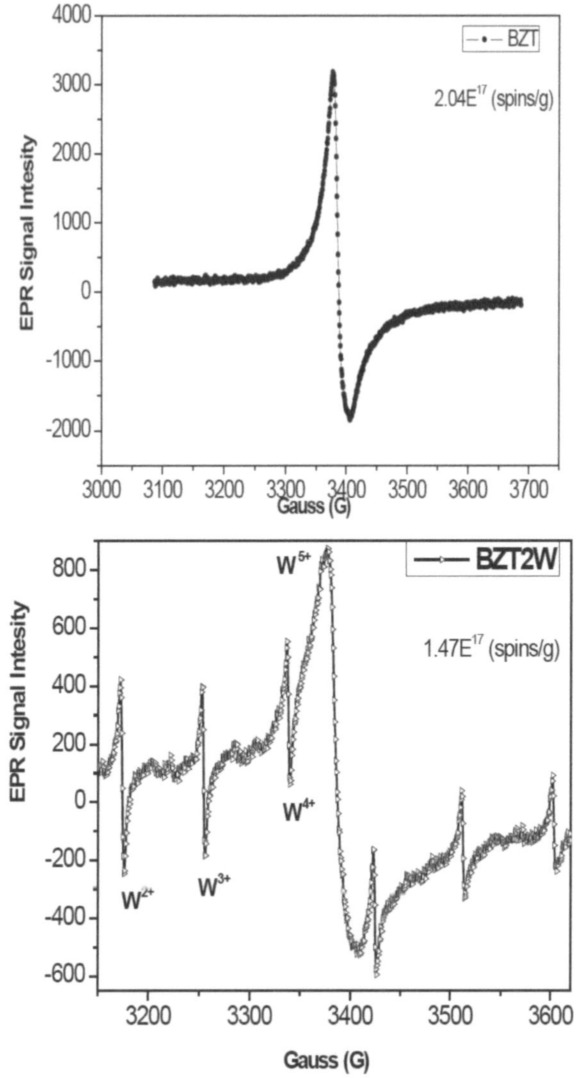

Fig. 15. Electron paramagnetic resonance measurements for BZT10 and BZT10:2W powders prepared from mixed oxide method, calcined at 900°C for 2 hours.

The corresponding defect reactions can be described as follows:

$$WO_3 \xrightarrow{BZT} W_{Ti}^{\bullet\bullet} + V_{Ba}'' + 3O_O^x \quad (8)$$

$$[MO_6]^x + \longrightarrow [MO_6]' + [MO_5 . V_o^{\bullet}] \quad (9)$$

For M = Zr or Ti

$$\left[V_{Ba}^{''}O_{12}\right] + \left[WO_6\right]^{\bullet\bullet} \longrightarrow \left[V_{Ba}^{x}O_{12}\right] + \left[WO_6\right]^{x} \quad (10)$$

$$\left[MO_5V_O^{x}\right] + \left[WO_6\right]^{\bullet\bullet} \longrightarrow \left[MO_5V_O^{\bullet}\right] + \left[WO_6\right]^{\bullet} \quad (11)$$

$$\left[MO_5V_O^{x}\right] + \left[WO_6\right]^{\bullet} \longrightarrow \left[MO_5V_O^{\bullet}\right] + \left[WO_6\right]^{x} \quad (12)$$

$$2\left[MO_6\right]^{'} + \left[WO_6\right]^{\bullet\bullet} \longrightarrow 2\left[WO_6\right]^{x} + \left[WO_6\right]^{x} \quad (13)$$

Despolarization of $\left[MO_6\right]^{x}$ clusters

$$\left[MO_5V_O^{*}\right].\,O_2 \longrightarrow \left[MO_5V_O^{\bullet}\right].O_2(ads) \quad (14)$$

$$\left[MO_5V_O^{*}\right].O_2 \longrightarrow \left[MO_5V_O^{\bullet\bullet}\right].O_2^{'} \quad (15)$$

High $\left[V_O^{\bullet}\right]$ concentration adsorbs O_2 leading to $\left[MO_5V_O^{\bullet\bullet}\right].O_2^{'}$ species.

Increasing tungsten concentration leads to $[MO_5V_O^{\bullet\bullet}]\cdot O_2'$ clusters. This species favour the creation of oxygen vacancies in $[TiO_6]$ or $[ZrO_6]$ sites, most of that being considered as complex vacancies in order-disorder structure. As a consequence, the oxygen vacancy-acceptor ion dipole may interact with polarization within a domain making its movement more difficult to switch. Ferroelectric hysteresis loop showed that the P_r and E_c were 3.0 μC/cm² and 1.25 kV/cm, respectively, Figure 16. The *P-E* loop was actually not closed

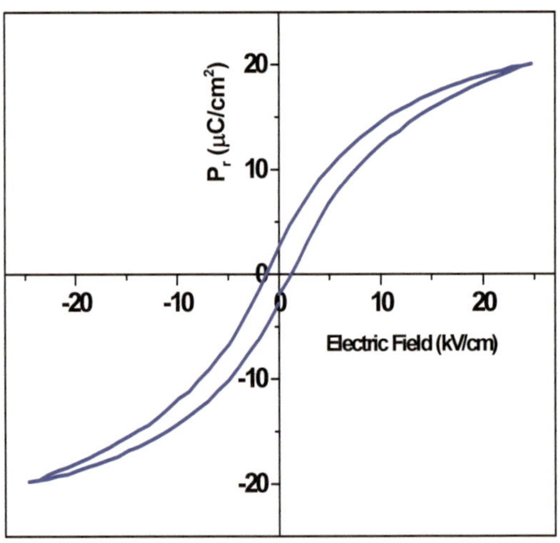

Fig. 16. Hysteresis loops for BZT10:2W ceramics sintered at 1200°C for 2 hours in a conventional furnace.

because the applied field was very high, up to 25.0 kV/cm. The obtained values pointed to the regular microstructure of the sintered specimen with small grains. The non saturation of the hysteresis loop can be resulted from the leakage current at high electric field. In this way, the charge compensation required by addition of W^{6+} ions could be achieved by reducing oxygen vacancies which induces changes in the leakage current. It also can be observed that the BZT10:2W ceramics is free of imprint phenomena which causes a shift in the coercitive field axis that leads to a failure in the capacitor. This failure can be caused by the defects as oxygen vacancies and space charges that leads to domain pinning and is almost absent in the BZT10:2W sample.

5. Influence of atmosphere on Dielectric response of vanadium modified $Ba(Ti_{0.90}Zr_{0.10})O_3$ ceramics

Having in mind that the substitution of vanadium on B-site broads the dielectric permittivity curves due the repulsion with their next nearest neighbors leading to a structure which is tetragonally distorted and improved ferroelectric response, we have examined the effects of annealing atmospheres (oxygen, air and nitrogen) on the electrical properties of BZT:2V ceramics. The dielectric permittivity for BZT10:2V the ceramics sintered at 1350°C in nitrogen showed higher values as compared to the ceramics sintered in air and oxygen atmospheres Figure 17. Regarding dielectric loss, the measurements suggest that low frequency loss values are significantly higher in nitrogen atmosphere. This can be explained by the higher space charge concentration, again arising due to higher oxygen vacancy concentration. Another observation that can be made is that dielectric properties (both dielectric permittivity and loss tangent) of oxygen and air do not improve appreciably whereas nitrogen sintered ceramics show a significant improvement in the dielectric properties. This again emphasizes the superior quality of nitrogen sintered ceramics. It is possible that this decrease in the permittivity for oxygen and air atmospheres is caused by space charge polarization which is inherently related to the nonuniform charge accumulation. The following equations (16-26) can be use to represent the dieletric permittivity dependence on sintering atmosphere.
For M = Zr ou Ti and c = complex

$$V_2O_5 \xrightarrow{Ba(Ti_{0,90}Zr_{0,10})O_3} 2V_{Ti}^{\bullet} + V_{Ba}^{''} + 5O_O^x \quad ou \quad 2V_{Zr}^{\bullet} + V_{Ba}^{''} + 5O_O^x \tag{16}$$

$$[MO_5.V_O^{\bullet}]_c + O_2 \rightarrow [MO_5.V_O^{\bullet}].O_{2(ads)} \tag{17}$$

$$[MO_5.V_O^{\bullet}].O_{2(ads)} \rightarrow [MO_5.V_O^{\bullet\bullet}].O_2^{'} \tag{18}$$

$$[V_{Ti}\,O_6]^{\bullet} + [BaO_{12}.V_{Ba}^{''}] \longrightarrow [V_{Ti}O_6]^x + [BaO_{12}.V_{Ba}^{'}] \tag{19}$$

$$[TiO_6]^x + [TiO_5.V_o^x] \longrightarrow [TiO_6]^{'} + [TiO_5.V_o^{\bullet}] \tag{20}$$

$$[TiO_6]^{'} + [TiO_5.V_o^{\bullet}] \longrightarrow [TiO_6]^x + [TiO_5.V_o^x] \tag{21}$$

$$[V_{Zr}\,O_6]^{\bullet} + [BaO_{12}.V_{Ba}^{''}] \longrightarrow [V_{Zr}O_6]^x + [BaO_{12}.V_{Ba}^{'}] \tag{22}$$

$$[ZrO_6]^x + [ZrO_5.V_o^x] \longrightarrow [ZrO_6]' + [ZrO_5.V_o^{\bullet}] \quad (23)$$

$$[ZrO_6]' + [ZrO_5.V_o^{\bullet}] \longrightarrow [ZrO_6]^x + [ZrO_5.V_o^x] \quad (24)$$

$$[MO_6]c^x + [MO_5.V_o^x]c \longrightarrow [MO_6]c' + [MO_5.V_o^{\bullet}] \quad (25)$$

$$[MO_6]c^x + [MO_5.V_o^x]c \longrightarrow [MO_6]c' + [MO_5.V_o^{\bullet}]c \quad (26)$$

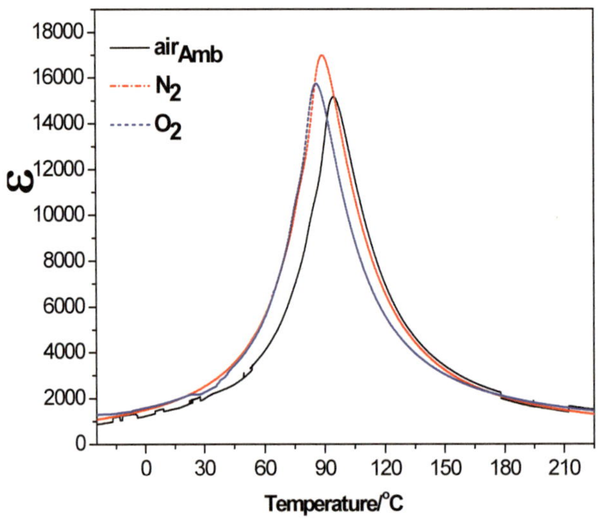

Fig. 17. Temperature dependence of dielectric permittivity at 10 kHz for BZT10:2V ceramics sintered at 1200 °C for 2 hours using air; O_2 and N_2 atmospheres.

The piezoeletric behavior at room temperature is shown in Figure 18a-c. Butterfly-shaped strain versus electric fields can be observed for different sintering atmospheres. The difference in the strain behavior might be attributed to different domain configurations. As usually observed in the relaxor-based "soft" piezoelectric materials, the hysteresis at low fields is attributed to domain motions. In the present work, the hysteresis could also be associated with the domain reorientation, which is prominent for a sample with a multidomain state. Above 30 kV/cm, the hysteresis-free strain is observed, implying a poling state free of domain wall motions induced by the high external electric fields. At 60 kV/cm, the highest electric field in the work, the piezoeletric coefficient is maximum. The oxidant and reducing atmosphere increase the piezoeletric behaviour in part due to domain reorientation. Beyond that point, it is possible that a modest bias field results in the transition from asymmetric to symmetric phase.

Inset shows the piezoforce microscopy of the BZT10:2V ceramics sintered under the air; O_2 and N_2 atmospheres. This field-induced phase transition may be ascribed to the pinching effect, that is, the consequent decrease in free energy difference among polymorphic phases.

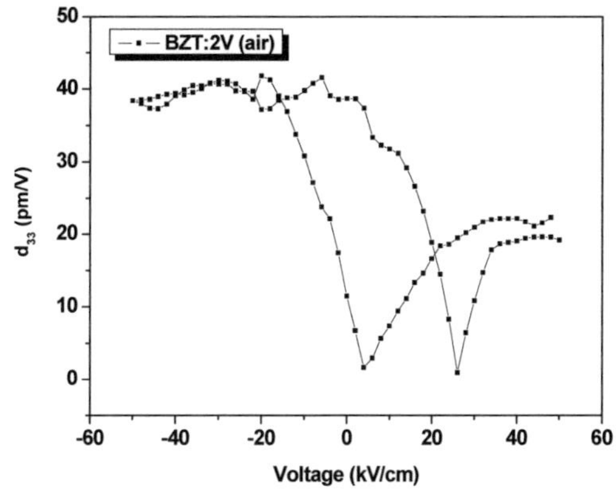

(a)

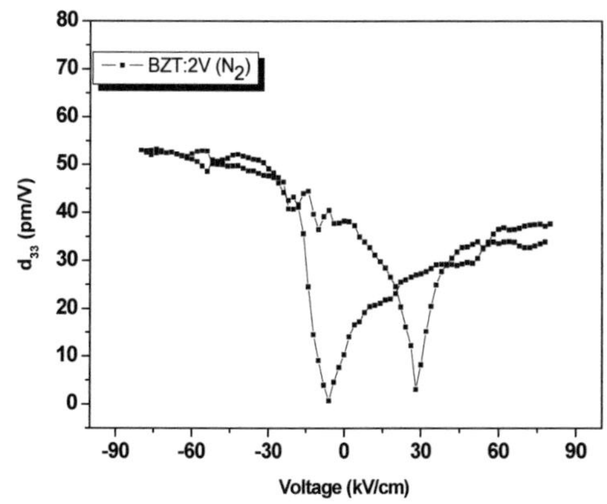

(b)

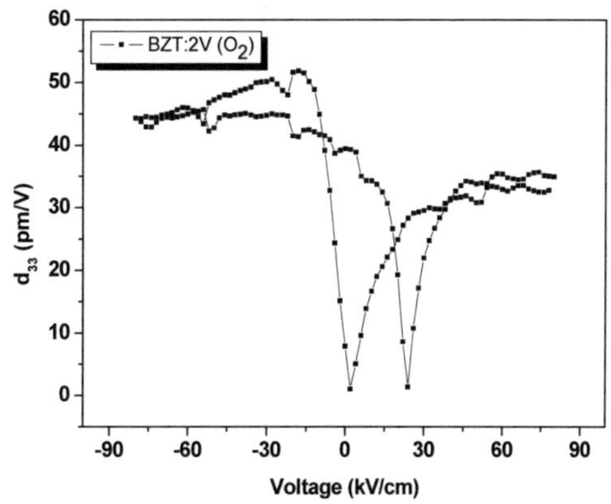

(c)

Fig. 18. Butterfly-shaped strain versus electric fields at room temperature for BZT10:2V ceramics sintered at 1200 °C for 2 hours using air; O_2 and N_2 atmospheres.

A careful inspection of the d_{33} -E plots reveals that there are two apparent linear regions at low fields (E<30 kV/cm) and high fields (E>70 kV/cm) and one transition region. That corresponds to domains reorientation induced by external electric fields. It is shown that BZT10:2V ceramic sintered in nitrogen atmosphere showed higher piezoeletric strain than the one sintered under air atmosphere. The piezoeletric coefficient was 43 pm/V, 40 pm/V and 30 pm/V for BZT10:2V ceramics sintered under nitrogen, oxygen and air atmospheres, respectively. This is a generally observed phenomenon in electronic ceramics, which might be attributed to the improved ceramic quality due to a small amount of impurity doping. It is shown that vanadium improves the piezoelectric strain. The improvement of piezoelectric response after doping can be associated with the better polarizability and the pinning effect.

6. Comparative study about behavior of BZT10, BZT10:2V and BZT10: 2W processed at different conditions

Figure 20 illustrates the XRD pattern for the BZT10, BZT10:2V and BZT10:2W ceramics sintered in a conventional furnace. The x-ray reflections show the single phase with a

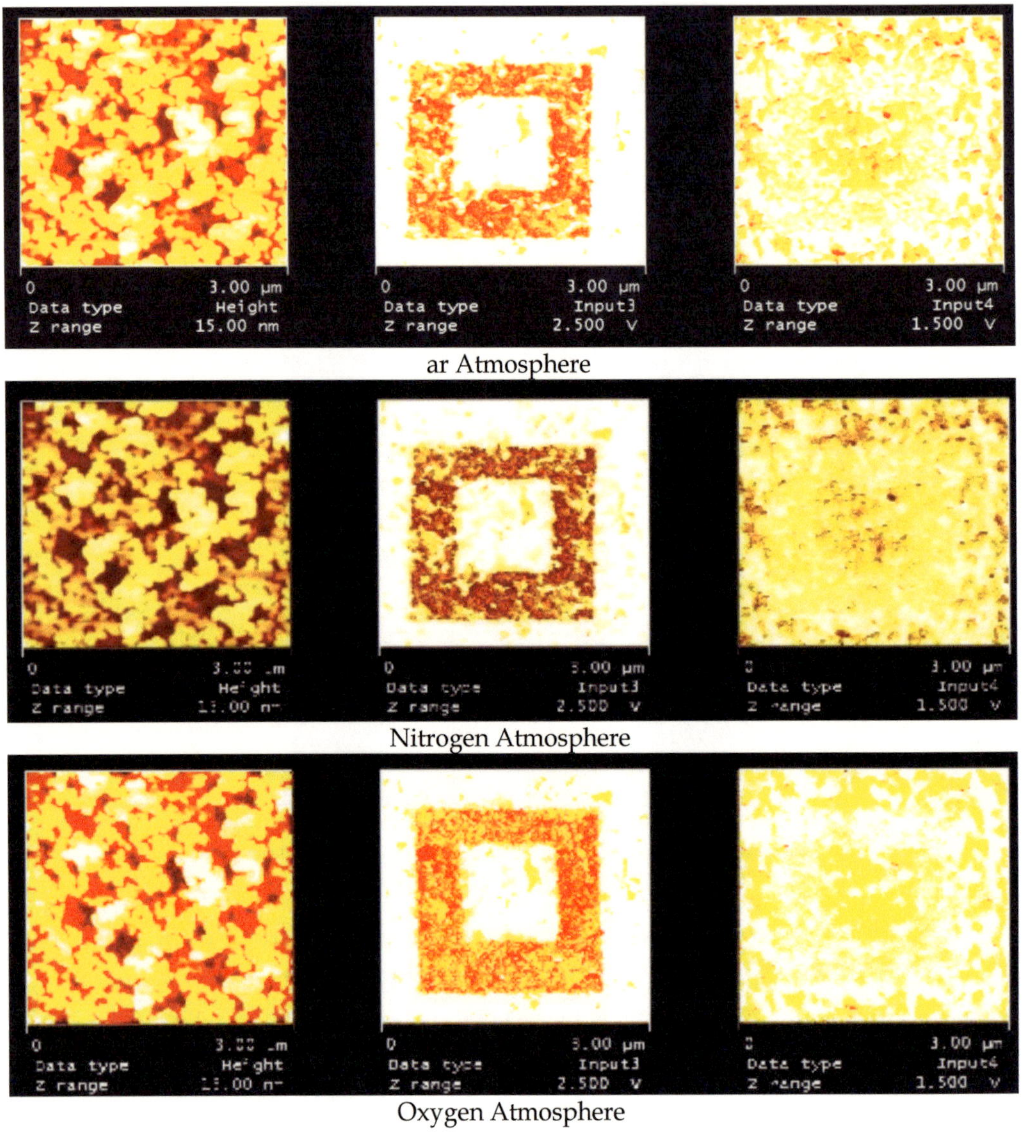

Fig. 19. Piezoelectric coefficient as function of voltage for BZT10:2V ceramics sintered at air; O_2 and N_2 atmospheres.

tetragonal perovskite structure was obtainedfor all the samples. This is a clear indication that V^{5+} and W^{6+} form a stable solid solution with the BZT10 lattice up to 2mol%. The peaks are indicative of a polycrystalline ceramic, mainly characterized by higher intense peak (hkl-110) at $2\theta = 31^{o}$. Vanadium and tungsten can either substitute titanium or zirconium in the lattice, depending on its concentration.

The substitution trend transition of dopant ions in the B-site of perovskite lattice can account for the influence of dopant on the sintering behavior of those ceramics, Figure 21. Due the fact that vanadium and tungsten play a role of donor in BZT10 lattice because it possesses a higher valence than Ti and Zr we noted a decrease of grain size, Figure 21b-c. This can be explained by the suppression of oxygen vacancy concentration, which results in slower oxygen ion motion and consequently lower grain growth rate. As a result, there is a substantial improvement in the diffusion during sintering which facilitates the densification process.

Temperature dependence of dielectric permittivity and dielectric loss determined is shown in Figure 22 (a) and (b), respectively. The BZT10:2V ceramics present the highest permittivity ($\varepsilon_r = 15111$) at Curie temperature. The dielectric permittivity increases gradually with an increase in temperature up to the transition temperature (T_c), Curie point, and then decreases. A normal ferroelectric-paraelectric phase transition was obtained for the BZT10 and BZT10:2V ceramics, Figure 22(a). Above the transition temperature, dielectric permittivity follows the Curie-Weiss law. The V^{5+} center occupies the B-site of the ABO_3 perovskite lattice leading to a charged $[VO_6]^{\bullet}$ defect which is associated with a barium vacancy in a local barium cluster $[V_{Ba}^{''}O_{12}]$. In fully or partly ionic compounds vacancies are charge balanced by other defects forming an overall neutral system. It can be assumed that particle charge compensation takes place at a nearest-neighbor barium cluster site in the $[BaO_{12}]$ because the resulting coulomb interaction is the most important driving force. This assignment is in accordance with first-principles calculation (Anicete-Santos et al.,2005). Alternatively, like in BZT10:2V an equilibrium between "free" $[VO_6]^{\bullet}$ centers and $[VO_6]^{\bullet}$ + $[V_{Ba}^{''}O_{12}]$ associated defects can be reached. Hence, charge transport will be considerably hindered. In this way, $[VO_6]^{\bullet}$ + $[V_{Ba}^{''}O_{12}]$ affects dielectric properties due the charge gradient in the structure arising from both species.

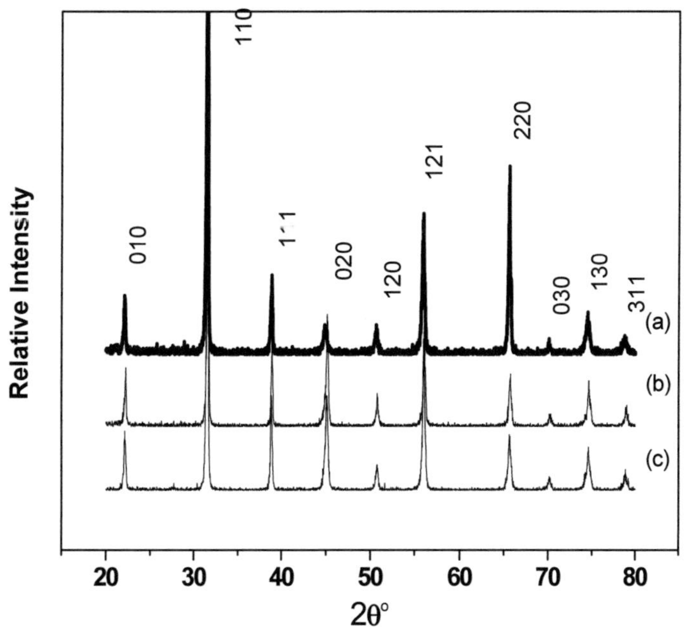

Fig. 20. X-ray diffraction data for ceramics sintered in a conventional furnace at different temperatures. (a) BZT10 -1550°C for 4 hours; (b) BZT10:2V- 1350°C for 4 hours; (c) BZT10:2W -1200°C for 2 hours.

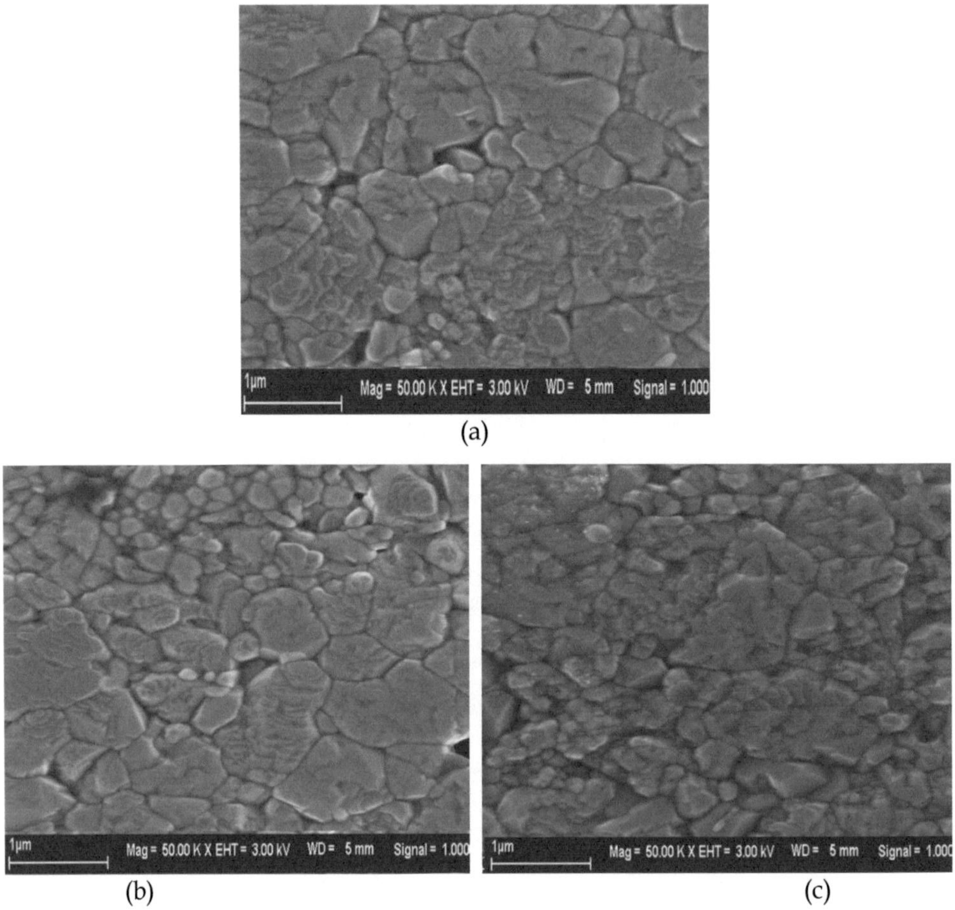

Fig. 21. Micrographies of the pellets ceramics sintered in a conventional furnace at different temperatures. (a) BZT10 -1550°C for 4 hours; (b) BZT10:2V- 1350°C for 4 hours; (c) BZT10:2W -1200°C for 4 hours;

That creates electron and hole polarons that can be designed as Jahn-Teller bipolarons, improving dielectric behavior due the charge trapping. For the BZT10:2W ceramic the dielectric anomaly is quite broad, covering the temperature range over 66°C. This compound exhibits diffuse dielectric permittivity anomalies around T_m. It can be also noticed that the maximum of dielectric permittivity ε_m and the corresponding maximum temperature, T_m, depend on the dopant. The magnitude of the dielectric permittivity decreases in this case, and the Curie temperature shifts toward lower temperature. This indicates that the dielectric polarization is of relaxation type in nature. The region around the dielectric peak is broadened due to a disorder in the cations arrangement in one or more crystallographic sites of the structure. Large differences in the B valence results in a strong tendency of disorder in one or more crystallographic sites of the structure leading to a microscopic heterogeneity in the compounds, with different Curie points. According to (Wu et al. 2002), the dielectric permittivity consists of contributions of ionic and atomic polarization only. Since the ionic radius of W^{6+} is smaller than that of Zr^{4+} and Ti^{4+}, increasing the amount of W^{6+} would lead to a reduced contribution of overall atomic polarization. The incorporation of tungsten ions into BZT10 structure would introduce some cationic vacancies to maintain the electroneutrality. These cationic vacancies also influence the dielectric permittivity, resulting in a reduced Curie point and decrease stability of perovskite structure. The relatively larger ionic radius of the B ion enhances the thermal stability of the BO_6 octahedra, when compared to Ti or Zr. Therefore, tungsten influences the BZT10 in the following way: Acting by repulsion with their next nearest neighbors. This results in an increase of a and reduction of c parameters of tetragonal phase. As a consequence, the BO_6 volume octahedron increases and consequently decreasing its stability which results in a reduction of T_c. The Figure 22b shows the temperature dependence of dielectric loss determined at 10 kHz. A small peak just below the Curie temperature was observed. At lower temperatures, a small temperature dependence of dielectric loss was observed while at elevated temperatures, vanadium stabilizes the dependence. The possible formation of dipole complexes may result in a reduced dielectric loss at elevated temperatures. This reflects that good insulation resistance was maintained at high temperatures, which is important for high temperature piezoelectric applications. The dielectric losses at low temperature appear to be stable but sensitive at high temperatures. The higher value of dielectric loss at elevated temperatures may be due to transport of ions with higher thermal energy. The sharp increase in dielectric loss may be due to the scattering of thermally activated charge carriers and the presence of defects. At higher temperature the conductivity begins to dominate, which in turn is responsible for the rise in dielectric loss that is associated with the loss by conduction. Also at high temperature (paraelectric phase) the contribution of ferroelectric domain walls to dielectric loss decreases. This behavior was also observed in some similar types of compounds.

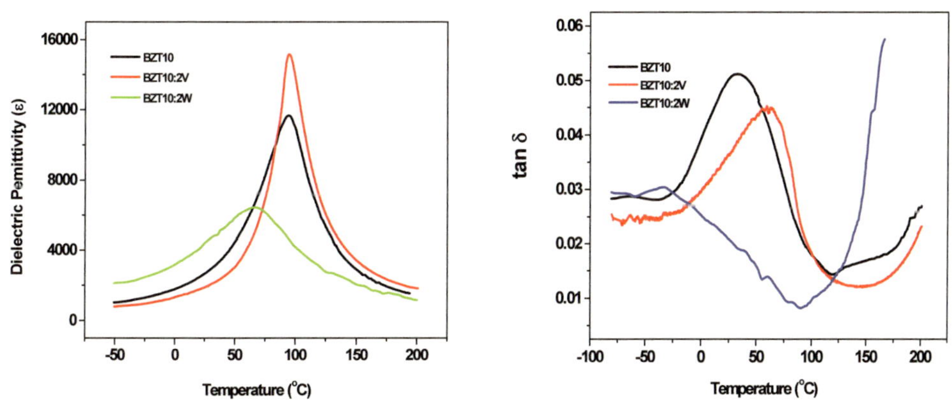

(a) (b)

Fig. 22. Temperature dependence of dielectric permittivity (a); dielectric loss (b) for BZT10, BZT10:2V and BZT10:2W ceramics. The measurements were performed at 10 KHz

For a normal ferroelectric the Curie-Weiss law is followed :

$$1/\varepsilon = (T - T_o)/C(T > T_C) \qquad (27)$$

where T_0 is the Curie temperature and C is the Curie-Weiss constant.

Figure 23 shows the inverse of dielectric permittivity as a function of temperature measured at 10 kHz. The fitting parameters for the BZT10, BZT10:2V and BZT10:2W were extracted from equation 27 and are $C = 6.8\text{x}10^{-6}$, $T_0 = 98°C$, $C = 5.2\text{x}10^{-6}$, $T_0 = 93°C$ and $C = 0.65\text{x}10^{-6}$, $T_0 = 75°C$, respectively (Table 1.). It is seen that the dielectric permittivity of BZT and BZT:2V ceramics follows the Curie-Weiss law at temperatures much higher than the $T_{\varepsilon m}$ (95 and 66 °C at 10 kHz).
Deviation from the Curie-Weiss law can be defined by

$$\Delta T_m = T_{cw} - T_{\varepsilon m} \tag{28}$$

where T_{cw} is a temperature at which ε starts to deviate from the Curie-Weiss law. Here, T_m = 95°C. In the literature, (Drougard & Huibregtse2010) a modified Curie-Weiss law was proposed to describe the diffuseness of the phase transition as

$$1/\varepsilon - 1/\varepsilon_m = (T - T_{\varepsilon m})^{\gamma} / C_1 \tag{29}$$

where γ and C_1 are assumed to be constant, and 1 < γ < 2. The parameter γ gives information on the character of the phase transition: For γ = 1, a normal Curie–Weiss law is obtained, γ = 2 describe a complete diffuse phase transition (Uchino et al.,1982) . The limiting values γ = 1 and γ = 2 reduce the expression to the Curie-Weiss law valid for the case of a normal ferroelectric and to the quadratic dependence valid for an ideal ferroelectric relaxor, respectively (Sinclair et al., 2002). From Figure 23, we obtained the parameter γ by fitting to the experimental data. The obtained values are for the BZT, BZT:2V and BZT:2W equal to 1.5, 1.4 and 2.0 respectively.The dipolar relaxation is clearly visible in the high frequency region. However, the low frequency region is dominated by the presence of grain boundary capacitances and likely the presence of deep trap states related to the Schottky-type barriers, which contribute to the total barrier-layer capacitive response of polycrystalline samples. With regard to the improvement in the dielectric properties, it can be inferred that both dopants exert a particularly marked influence on the grain's internal domain. Hence, they are able to increase the number of "active" domains, so that the cause of the strong increase in the dielectric properties is here mainly related to the number of active internal domains. The chemistry of dielectric internal domains most likely depends on the type of dopant to increase its effectiveness. In the particular context reported here, the grain boundary contribution is lower than the total dielectric response. For instance, the contribution of the grain boundary effect is more effective in the case of the sample modified with vanadium. In the case of the sample modified with tungsten, the contribution is not so significative and the total dielectric response decreased.
The piezoelectric behavior at room temperature is shown in Figure 24. The topography (TP), out-of-plane (OP) and in-plane (IP) piezoresponse images of the as-grown samples after applying a bias of -12V, on an area of 2 μm x 2 μm, and then an opposite bias of + 12V in the central 1 μm x 1 μm area. To obtain the domain images of the ceramics, a high voltage that exceeds the coercive field was applied during scanning. The contrast in these images is associated with the direction of the polarization. The PFM image indicates that the perpendicular component of polarization can be switched between two stable states: bright and dark contrast inside and outside of the square region. Higher PFM magnification images showed that the regions without piezoresponse exhibit a strong contrast in the PFM images. The white regions in the out-of-plane PFM images correspond to domains with the polarization vector oriented toward the bottom electrode hereafter referred to as down

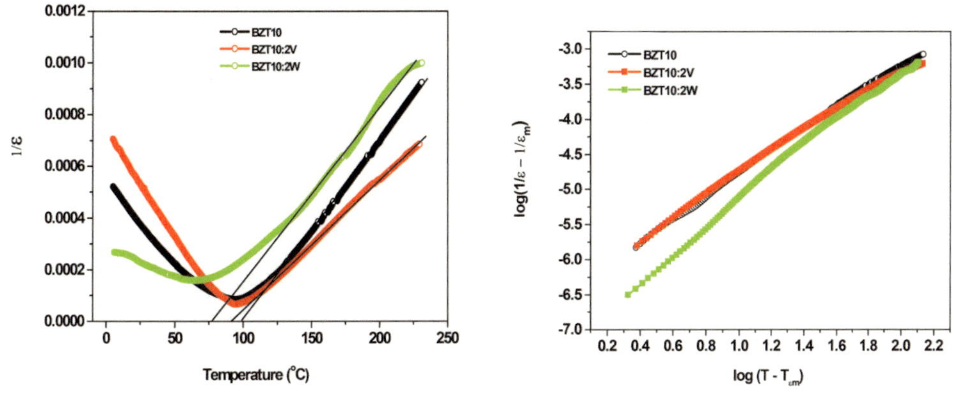

(a) (b)

Fig. 23. The inverse dielectric permittivity (10000/ε) as a function of temperature at zero electric field (a) ; $\ln(1/\varepsilon-1/\varepsilon_m)$ as a function of $\ln(T-T_m)$ (b); for BZT10, BZT10:2V and BZT10:2W ceramics.

Samples	BZT10	BZT10:2V	BZT10:2W
T_o(°C)	97.8	92.9	74.9
C (x10^{-6})	6.8	5.2	6.5
T_{cw}(°C)	125.5	113.4	131.5
$\Delta T_m = T_{cw} - T_{\varepsilon m}$(°C)	27.7	20.5	65.2
$T_{\varepsilon m}$(°C)	94.7	94.6	66.3
εm	11670	15160	6420
γ	1.5	1.4	2.0
$\Delta T_m = T_{\varepsilon m}$(100KHz) -$T_{\varepsilon m}$(1KHz)(°C)	571.6	1124.4	235

Table 1. The Curie-Weiss temperature (T_0); Curie-Weiss constant (C); temperature above which the dielectric permittivity follows the Curie-Weiss law (T_{cw}); temperature of maximum dielectric permittivity ($T_{\varepsilon m}$); maximum dielectric permittivity (ε_m); and diffuseness constant (γ).

polarization (Figures 24b, e and h) while the dark regions correspond to domains oriented upward referred to as up polarization. Grains which exhibit no contrast change is associated with zero out-of-plane polarization. A similar situation was observed when a positive bias was applied to the samples. We noticed that some of the grains exhibit a white contrast associated to a component of the polarization pointing toward the bottom electrode. On the other hand, in the in-plane PFM images (Figures 24 c, f and i) the contrast changes were associated with modifications in the in-plane polarization components. In this case, the white contrast indicates polarization pointing to the positive direction of the y-axis while dark contrast is given by in-plane polarization components pointing to the negative part of the y-axis.

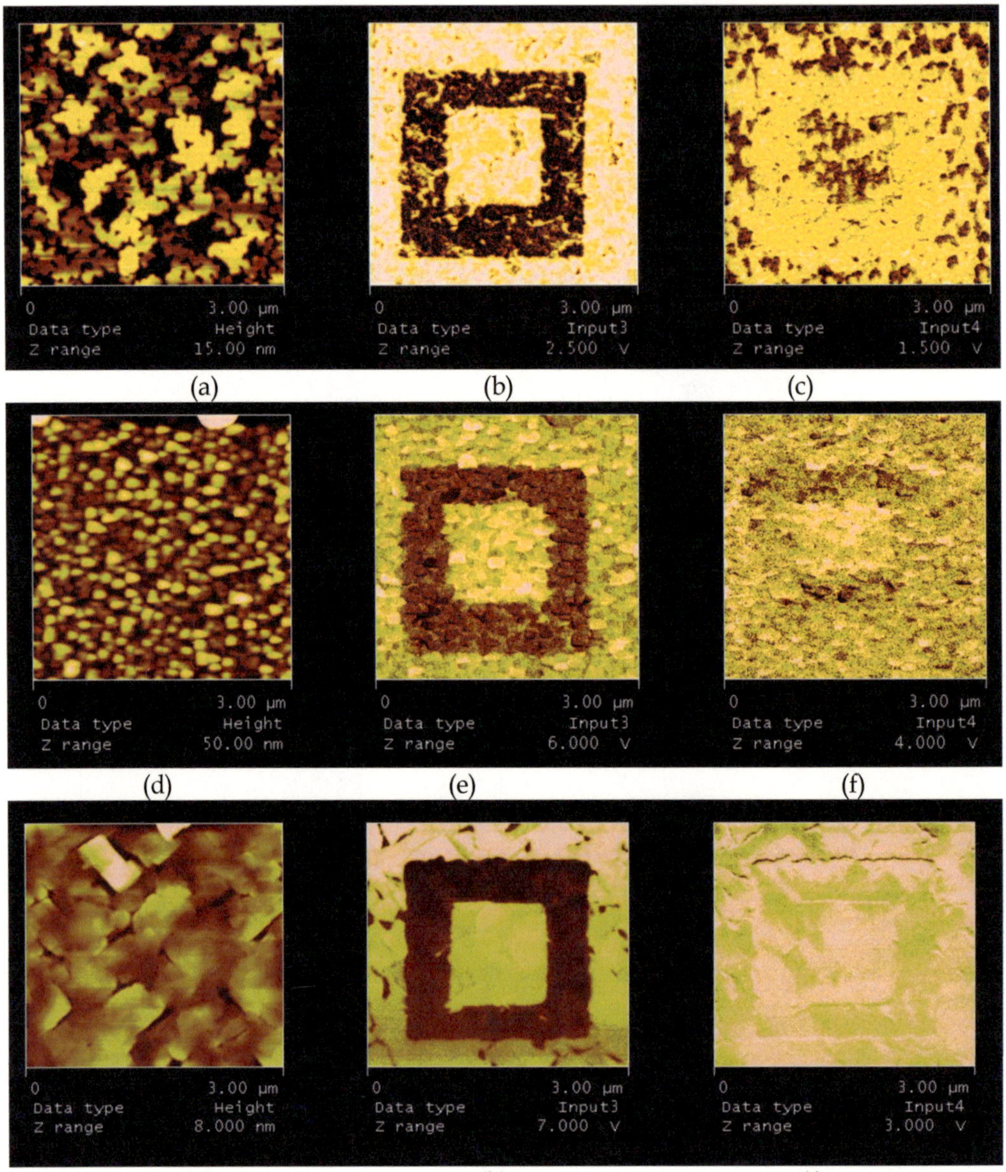

Fig. 24. Topography (TP), Out-of-plane (OP) and In-plane (IP) PFM images of (a) BZT10 (TP), (b) BZT10 (OP), (c) BZT10 (IP), (d) BZT10:2V (TP), (e) BZT10:2V (OP), (f) BZT10:2V (IP), (g) BZT10:2W (TP), (h) BZT10:2W (OP) and (i) BZT10:2W (IP)ceramics.

Considering the polycrystalline nature of our samples the effective piezoelectric coefficient depends on grain orientation. Therefore, as expected the piezoelectric response improves with addition of donor dopant due to the reduction of strain energy and the pinning effect of charged defects. The difference in the piezoresponse behavior might be attributed to different domain configurations. As usually observed in the relaxor-based "soft" piezoelectric materials, the piezoresponse at low fields is attributed to domain motions. In the present work, the high PFM contrast could also be associated with the domain reorientation, which is prominent for a sample with a multidomain state. Donor dopants such as vanadium and tungsten improve the piezoelectric response due the better polarizability and the pinning effect. PFM measurements also reveal a clear piezoelectric contrast corresponding to antiparallel domains on all locations tested. There is no reduction in the amplitude of the measured vibrations which is indicative that this phase is still polar and electric field-induced polarization switching still exists. Also, we noted that some of the crystallites apparently have not been switched and still exhibit a positive piezoresponse signal. This result, which can be explained by strong domain pinning in these crystallites, is direct experimental proof that repeated switching results information of unswitchable polarization, which, in turn, leads to the degradation of switching characteristics. Finally, it is shown in the PFM images that BZT10:2V and BZT10:2W ceramics showed higher piezoelectric strain than the BZT10. The piezoelectric coefficient was 43 pm/V, 40 pm/V and 30 pm/V for BZT10:2V, BZT10:2W and BZT10 ceramics, respectively (not shown in the text). This is a generally observed phenomenon in electronic ceramics, which might be attributed to the improved ceramic quality due to a small amount of impurity doping.

7. Conclusions

Single phase BZT powders were obtained from mixed oxide method at 1200°C for 2 hours. Raman spectroscopy indicated a change in the crystal structure with the increase of zirconium content. The transition temperature (ferroelectric to paraelectric phase) was found to be systematically reduced for Zr content equal to 15.0 mol.%. In this case, a typical relaxor behavior was observed. A decay in the dielectric permittivity and remnant polarization with the increase of zirconium content is indicative of changes in crystal structure and predominance of relaxor behaviour.

$Ba(Zr_{0.10}Ti_{0.90})O_3$:2V (BZT:2V) ferroelectric ceramics modified with vanadium were prepared from powders synthesized using the mixed oxide method. Single phase ceramics crystallized in a tetragonal structure was attained at 1350°C for 4 hours. Dielectric properties of BZT:2V ceramics have been investigated and at 10 KHz the maximum dielectric permittivity reached was 15000 at a Curie temperature of 94 °C. Dielectric constant of BZT:2V ceramics follows the Curie-Weiss law at temperatures much higher than $T_{\varepsilon m}$. The remanent polarization (*Pr*) and coercive electric field (*Ec*) of BZT:2V ceramics was found to be 8.5 μC/cm^2 and 2 kV/cm, respectively. The very high-dielectric constant and high at low temperatures of this material are promising for some practical applications.

The dielectric and piezoelectric properties of BZT10 based ceramics depends on the donor dopant. Single phase BZT10, BZT10:2V and BZT10:2W ceramics crystallized in a tetragonal structure was attained. Micrographies reveal that both dopants suppress the oxygen vacancy concentration which results in slower oxygen ion motion and consequently lower grain growth rate. A maximum dielectric constant was obtained for the BZT10:2V ceramics at a Curie temperature of 95°C.

The maximum sintering temperature is reduced after tungsten addition due to generation of oxygen vacancies. The tungsten doped BZT powders consist of soft agglomerates which were completely broken during milling process and densified at lower temperatures. Tungsten addition leads to the distortion of the perovskite lattice leading to typical relaxor behaviour. Dielectric properties have been investigated and a maximum dielectric permittivity for the undoped BZT10 reached 11500 at Curie temperature of 93°C. The

dielectric permittivity is reduced and shifted to lower temperatures up to a tungsten content of 2 mol%. Room-temperature EPR spectrum evidenced hyperfine bands resulting from interaction between species with unpaired electrons with species which present complex vacancies. The BZT10:2W ceramic showed a relaxor-like behaviour near phase transition which can be useful for low temperature capacitor applications.
Dielectric permittivity of BZT10 and BZT10:2V ceramics follows the Curie-Weiss law at temperatures much higher than $T_{\varepsilon m}$. BZT10:2W ceramic showed a relaxor-like behavior near phase transition based on empirical parameter γ. In BZT10, the movement of Ti^{4+} and Zr^{4+} ions reduces the space charge compensation leading to a low piezoresponse signal. Meanwhile, BZT:2V and BZT10:2W ceramics exhibited excellent piezoelectric We report the advances in the dielectric and piezoelectric properties of vanadium and tungsten doped BZT10 ceramics prepared by the mixed oxide method by using dielectric spectroscopy analyses and piezoresponse force microscopy. Ceramics with good electrical properties were obtained confirming the possibility of application of ferroelectric perovskite materials lead free such as barium zirconium titanate.
The effect of annealing atmosphere (air, N_2 and O_2) on the electrical properties of BZT10:2V was investigated. The results show that ceramic sintered in nitrogen atmosphere presents superior dielectric behavior at room temperature. The dielectric permittivity measured at a frequency of 10 kHz was equal to 16800 with dielectric loss of 0.030. The BZT10:2V ceramic sintered in air atmosphere restricting the movement of Ti^{4+} and Zr^{4+} ions and thus reducing the space charge compensation leading to a low piezoresponse. Meanwhile, BZT10:2V ceramic sintered in oxygen and nitrogen atmospheres exhibited excellent piezoelectric properties indicating that the environment of the oxygen ions is quite different from one sample to another. The piezoelectric coefficient is strongly improved by sintering the sample under nitrogen atmosphere. In this way, we reveal that BZT:2V ceramic sintered under nitrogen atmosphere can be useful for practical applications, including nonvolatile digital memories, spintronics and data-storage media.

8. Acknowledgments

The authors thank the financial support of the Brazilian research financing institutions: CAPES, FAPESP and CNPq.

9. References

Abraham, M.M.; Boatner, L.A. & Aronson, M.A. (1986), EPR observations of trivalent titanium in orthophosphate single crystals, *J.Chem. Phys,*vol. 85, n 1, pp. 1-6. ISSN 0021-9606.

Anicete-Santos, M. Cavalcante, L.S. Orhan, E. Paris, E.C. Joya, M.R. de Lucena P.R., Pizani, P.S. Leite, E.R. Varela, J.A. Longo, E. The role of structural order–disorder for visible intense photoluminescence in the $BaZr_{0.5}Ti_{0.5}O_3$ thin films, Chem. Phys, Vol.316, No 4 (2005). pp. 260-266. ISSN0301-0104.

Bolten, D. Bouger, U. Schneller, T. Grossnam, M. Lose, O., & Waser, R., Reversible and irreversible processes in donor-doped $Pb(Zr,Ti)O_3$. *Appl. Phys. Lett.*, Vol. 77, No 23, (2000). pp.3830–3832, ISSN 0003-6951.

Böttcher, C. J. F. & Bordewijk, P., (1992) Theory of Electric Polarization: Dielectrics in Time-Dependent Fields (Elsevier, Amsterdam) Vol. II.

Caballero, A. C.; Fernández, J. F.; Villegas, M. ; Moure, C.; Durán, P. ; Florian, P. & Coutures, J.P. (1998). Intermediate Phase Development in Phosphorus-Doped Barium Titanate. *J. Am. Ceram. Soc.*, 20 dec , vol. 8, n 6, pp.1499-1505, ISSN 0002-7820.

Carg, A. & Goel, T.C. Dielectric properties of NiZn ferrites by the precursor method, *Mater. Sci. Eng. B.*, Vol. 60, No 2, (1999). pp. 156–160, ISSN 0921-5107.

Chen, I-W. and Xue, L.A. Development of Superplastic Structural Ceramics. *J. Amer. Ceram. Soc.*, Vol. 73, No 9, (1990). pp. 2585-2609, ISSN 002-7820.

Chen, A.; Yu,Z.; Cross,L. E.; Guo, R. Y. & Bhalla, A. S., (2001). Dielectric relaxation and conduction in SrTiO3 thin films under dc bias. *Appl. Phys. Lett.*, vol. 79, n 8, pp. 818-820, ISSN 0003-6951.

Cheng, Z-Y. Katiyar, R. S. Yao, X. Bhalla, A. S. Temperature dependence of the dielectric constant of relaxor ferroelectrics, *Phys. Rev. B.*, Vol. 57, No 14, (1998). pp. 8166-8177, ISSN 0163-1829.

Courtens, E.. Vogel-Fulcher Scaling of the Susceptibility in a Mixed-Crystal Proton Glass. *Phys. Rev. Lett.*,Vol. 52, No 1, (1984). pp. 69-72, ISSN 0031-9007.

Cross, L. E. Relaxor ferroelectrics. *Ferroelectrics,* Vol. 76, n. 3-4, (1987). pp. 241-267, ISSN 1563-5228.

Dixit, A.; Majumder, S. B.; Dobal, P. S.; Katiyar, R. S. & Bhalla, A. S. (2004). Phase transítion studies of sol-gel deposited barium zirconaté titanaté thin films. Thin Solid Films, Vol. 447-448, pp. 284–288, ISSN 0040-6090

Dmowski, W. Akbas, M.K. Davies, P.K. Egami, T.Local structure of Pb(Sc1=2,Ta1=2)O_3 and related compounds, *J. Phys. Chem. Solids.*, Vol. 61, No 3, (2000) pp.229–237, ISSN 0022-3697.

Dobal, P. S.; Dixit, A.; Katiyara, R. S. (2010). Micro-Raman scattering and dielectric investigations of phase transition behavior in the BaTiO3–BaZrO3 system. *J. Appl. Phys,* Vol. 89, pp. 8085-8091, ISSN 0021-8979.

Domenico, M. D. Jr.; Wemple, S. H.; Porto, S. P. S. & Buman, P. R. Raman spectrum of single-domain BaTiO3. *Phys. Rev.*, v. 174, (1968) p. 522-530,

Fang, T. T. and Shiau, H. K. Mechanism for Developing the Boundary Barrier Layers of $CaCu_3Ti_4O_{12}$. *J. Am. Ceram. Soc., Vol.* 87, No 11, (2004). pp. 2072-2079, ISSN 1551-2916.

Farhi, R. Marssi, M. El. Simon, A. & Ravez, J. A Raman and dielectric study of ferroelectric $Ba(Ti_{1-x}Zr_x)O_3$ ceramics, *Eur. Phys. J. B.*, Vol. 9, No 4, (1999). pp. 599–604, ISSN 1434- 6028.

Fouskova, A. Cross, L. E. Dielectric properties of Bismuth Titanate. *J. Appl. Phys.*, Vol. 41, No 7, (1970). pp. 2834-2838, ISSN 1089-7550.

Hennings, D. & Sehnell, A. Diffuse Ferroelectric Phase Transitions in $Ba(Ti_{1-y}Zr_y)O_3$ Ceramics. *J.Am. Ceram. Soc,* Vol. 65, No 3, (1982) pp. 539-534, ISSN 0002-7820.

Hoffmann, S. Waser, R. Control of the morphology of CSD-prepared (Ba, Sr) TiO_3 thin films. *J. Eur. Ceram. Soc.*, Vol. 19, No 6-7, (1999). pp. 1339-1343, ISSN 0955-2219

Jin, D., Hing,P., & Sun, C.Q. Intense and stable blue-light emission of $Pb(Zr_xTi_{1-x})O_3$, *Appl. Phys. Lett., Vol.* 79 (2001). pp. 1082–1084. ISSN 0003-6951.

Jonscher, A. K. Dielectric characterisation of semiconductors. *Solid-State Electron.*, Vol. 33, No 6, (1990). pp.737-742, ISSN 0038-1101.

Kell, R.C. Hellicar,N.J. Structural transitions in barium titanate- zirconate transducer materials. Acusticu., Vol. 6, (1956). pp 235–238, ISSN 1610-1928.

Kreisel, J.; Bouvier, P.; Maglione, M.; Dkhil, B. & Simon, A. High-pressure Raman investigation of the Pb-free relaxor Ba(Ti0.65Zr0.35)O3. *Phys. Rev. B,* v1. 69 (2004)p. 092104.

Landolt Bornstein, (1981) Ferroelectric: oxides, New series, Group III, Vol. 16a Berlin, Heidelberg, New York : Springer,

Liu, S. F. Park, S. E. Shrout, T. R. and Cross, L. E. Electric field dependence of piezoelectric properties for rhombohedral $0.955Pb(Zn_{1/3}Nb_{2/3})O_3$-$0.045PbTiO_3$ single crystals. J. Appl. Phys., Vol. 85, No 1, (1999). pp. 2810-2814, ISSN 1089-7550.

Lines, M.E. & Glasss, A.M. (1977) *Principles and applications of ferroelectrics and related materials,* Oxford University Press, Oxford

Marques, L.G.A., Cavalcante,L.S., Simões,A.Z, Pontes, F.M., Santos-Junior,L.S., Santos,M.R.M.C., Rosa,I.L.V., Varela,J.A.,& Longo, E. ,Temperature dependence of dielectric properties for $Ba(Zr_{0.25}Ti_{0.75})O_3$ thin films obtained from the soft chemical method, *Mater. Chem. Phys,* Vol. 105, No 2, (2007). pp. 293–297, ISSN 0254-0584.

Merz, W.J. The effect of an electric field on the transitions of barium titanate. *Phys. Rev.,* Vol. 91, (1953). pp. 513-517, ISSN 0031-9007

Miura, S. Marutake, M. Unoki, H. Uwe, H. Sakudo, T.Composition Dependence of the Phase Transition Temperatures in the Mixed Crystal Systems near $SrTiO_3$. J. Phys Soc Jpn., Vol. 38, No 1, (1975).pp. 1056-1060, ISSN 0031-9015.

Moura, F. Simões, A. Z. Cavalcante, L. S. Zampieri, M. Zaghete, M. A. Varela, J.A. & Longo, E. Strain and vacancy cluster behavior of vanadium and tungsten-doped. *Appl. Phys. Lett.,* Vol. 92, No 3, (2008). pp. 032905-032907, ISSN 0003-6951.

Moura,F., Simões,A.Z., Aguiar,E.C., Nogueira,I.C., Zaghete,M.A., Varela, J.A. & Longo, E., Dielectric investigations of vanadium modified barium titanate zirconate ceramics obtained from mixed oxide method, *J. Alloys and Compds,* Vol. 479, No 4, (2009). pp. 280–283, ISSN 0254-0584.

Moura,F., Simões,A.Z., Paskocimas, C.A., Zaghete,M.A., Longo E. & Varela,J.A., Temperature Dependence of Electrical Properties of BZT:2V ceramics, *Mater. Chem. Phy,* Vol. 479, No 3 (2009). pp. 280–283, ISSN 0254-0584.

Neirman, S. M., The Curie Point Temperature of $Ba(Ti_{1-x}Zr_x)O_3$ Solid Solution., *J. Mater. Sci,* Vol. 23, No 11, (1988). pp. 3973-3980, ISSN 1573-4803.

Paik, D. S. Park, S. E. Wada, S. Liu, S. F. and Shrout, T. R. "E-field induced phase transition in ferroelectrics. *J. Appl. Phys.,* Vol. 5, No 2, (1999). pp. 1080-1083, ISSN 1089-7550.

Park, S. E. and Shrout, T. R. Ultrahigh strain and piezoelectric behavior in relaxor based ferroelectric single crystals. *J. Appl. Phys.,* Vol. 82, No 4, (1997). pp.1804-1815, ISSN 1089-7550.

Ravez J.& Simon A., "Raman spectroscopy and soft modes in the model ferroelastics", *Eur. Phys. J.: Appl. Phys.,* 11 (1997) 9–13.

Ravez, J. & Simon, A. (1997). The crystal *chemistry* of the higher tungsten oxides. *J. Solid State Chem,* Vol. 34, No 1, (1997). pp. 1199- 1203, ISSN 0022- 4596

Ravez, J. Simon, A. Spontaneous transition from relaxor to ferroelectric state in new lead-free perovskite ceramics. *Ferroelectrics., Vol.* 240, No 1, (1999). pp. 313-320, ISSN 1563-5228.

Ravez, J. Simon, A. Raman spectroscopy and soft modes in the model ferroelastics, *Eur. Phys. J: Appl. Phys.,* Vol. 11, No 3, (1997). pp. 9–13, ISSN 1089-7550.

Ravez, J. Simon, A. The crystal chemistry of the higher tungsten oxides, *J. Solid State Chem.*, Vol. 34, No 2, (1997). pp. 1199-1203, ISSN 0022-4596

Rehrig,P. W. Park, S. E. McKinstry, S. T. Messing, G. L. Jones, B. and Shrout, T.M. Piezoelectric properties of zirconium-doped barium titanate single crystals grown by templated grain growth. J. Appl. Phys., Vol . 86, No 3, (1999). pp. 1657-1651, ISSN 1089-7550.

Scott, J. F. Soft-mode spectroscopy: Experimental studies of structural phase transitions. *Rev. Mod. Phys*, Vol. 46, No 1, (1974) pp. 83-128, ISSN 0034-6861

Shannigrahi, R. Choudhary, R.N.P. Acharya, N. Electrohydrodynamic instability in 8CB (4'-n-octyl-4-cyanobiphenyl) liquid crystal, *Mater. Sci. Eng. B.*, Vol. 56, No 2, (1999). pp. 31–35, ISSN 0921-5107.

Shannigrahi,R. , Choudhary, R. N. P. & Acharya,N. Phase transition in $Ba_5RTi_3Nb_7O_{30}$ (R = Dy, Sm) ferroelectric ceramics, *Mater. Sci. Eng.*, B 56 (1999) 27-31.

Simões,A.Z., Ramirez,M.A., Riccardi,C.S., Ries, A., Longo,E., Varela,J.A.Influence of temperature on the dielectric and ferroelectric properties of bismuth titanate thin films obtained by the polymeric precursor method. *Materials Chemistry and Physics., Vol.* 92, No 2-3 (2005) pp. 373–378, ISSN 0254-0584.

Sinclair,D. C. Adams, T. B. Morrison,F. D. West, A. R. $CaCu_3Ti_4O_{12}$: one-step internal barrier layer capacitance. *Appl. Phys.Lett.*, Vol. 80, No 1, (2002). pp. 2153-2155 , ISSN 0003-6951.

Smolensky, G.A. ; Bokov,V.A. ; Isupov,V.A. ; Krainik, N. N. & Pasynkov, R. E., (1984) Ferroelectrics and Related Materials (New York: Gordon and Breach), ISBN 10 - 2881241077.

Tagantsev, A. K. Glazounov, A.E. Mechanism of polarization response in the ergodic phase of a relaxor ferroelectric. *Phys. Rev. B.*, Vol. 57, No 1, (1998). pp.18-21, ISSN 0163-1829.

Takahashi, S. Effects of impurity doping in lead zirconate-titanate ceramics. *Ferroelectrics.*, Vol. 41, No 2, (1982). pp. 277–290, ISSN 0015-0193

Tang, X.G. Wang, J. Wang, X.X. Chan, H.L.W. Tunabilities of sol-gel derived Ba (Zr0.2Ti0.8)O3 ceramics. *Solid State Communications.*, Vol. 131, No 3, (2004). pp. 163–168, ISSN 0038-1098

Tsurumi, T. Yamamoto, Y. Kakemoto, H. Wada, S. Chazono, H. Kishi, H. Dielectric properties of $BaTiO_3$-$BaZrO_3$ ceramics under a high electric field. *J. Mater. Res.*, Vol. 17, No 2, (2002). pp. 755-759, ISSN 0884-2914.

Uchino, K. Nomura, S.Critical Exponents of the Dielectric Constants in Diffused-Phase-Transition Crystals. *Ferroelectr. Lett. Sect.*, Vol. 44, No 1, (1982) pp. 55-61, ISSN 1563-5228.

Veith, G. M., Greenblatt, M., Croft, M. and Nowik, I. Fawcett, I. D., Properties of the perovskites, $SrMn_{1-x}Fe_xO_{3-\delta}$ (x=1/3; ½; 2/3), Solid State Sciences, Vol. 2, Issue 8, 2000, pp.821-831

Wang, Y.L. Li, L.T. Qi, J.Q. Gui, Z.L. Low-Temperature Facile Template Synthesis of Crystalline Inorganic. *Mater. Chem. Phys., Vol.* 76, No 3 (2002). pp. 250-254, ISSN 0254- 0584.

Weber, U. Greuel, G. Boettger, U. Weber, S. Hennings, D. Waser, R. Manufacturing Processes and Systems. *J. Am.Ceram. Soc.*, Vol. 84, No 4, (2001). 759-766.

WEBER, U.; GREUEL, G.; BOETTGER, U.; WEBER, S.; HENNINGS, D.; & WASER, R., Dielectric Properties of $Ba(Zr,Ti)O_3$-Based Ferroelectrics for Capacitor Applications, *J. Am. Ceram. Soc*, v. 84 (2001)p. 759–66

West, A. R. Adams, T. B. Morrison, F. D. Sinclair, D. C.'Novel High. Capacitance Materials: $BaTiO_3$:La and $CaCu_3Ti_4O_{12}$. J.Eur.Ceram. Soc., Vol. 24, No 6, (2004). pp. 1439-1448, ISSN 0

Wu,L.; Wei,C.; Wu,T. & Liu,H. Dielectric properties of modified PZT ceramics, *J. Phys. C: Solid State Phys.*, 16 (1983) 2803–2806.

Wu, L.; Wei,C. ; Wu, T. & Liu, H. "Effects of dye loading", *J. Phys. C: Solid State Phys.*, 16 (1983) 2803–2806.

Wu,Y.; Limmer, S.J. ; Chou,T.P. ; Nuguyen, C. & Guozhong, C. "Influence of tungsten doping on dielectric properties of strontium bismuth niobate ferroelectric ceramics", *J. Mater. Sci. Lett.*, 21 (2002) 947–949.

Xu, Y. H. 1991, *Ferroelectric Materials and their Applications*, Elsevier Science Publishers, Amsterdam.

Zhang, M.; Jin, Z.; Zhang,J.; Gou, X.; Yang, J.; Wang, X. & Zhang, Z. (2004). Effect of annealing temperature on morp hology, structure and photocatalytic behavior of nanotubed $H_2Ti_2O_4(OH)_2$, *J. Mol. Catal. A: Chem.*, vol. 217, pp 203–210

11

Glass Ceramics with Para, Anti or Ferroelectric Active Phases

Manuel Pedro Fernandes Graça and Manuel Almeida Valente
University of Aveiro / I3N - Physics Department
Portugal

1. Introduction

The formation of glass ceramics shows, at technology level the great advantage, when compared to single crystals and sintered ceramics, the possibility of their main properties (optical, electrical, mechanical, chemical) be controlled via the volume fraction of the active phase dispersed in the glass matrix. The optical transparency is one example. To maintain this physic characteristic, the process of nucleation and crystal growth requires a high control being achieved when the size of crystals dispersed in the glass matrix is not high enough to cause light scattering. However, for most electric applications it is necessary that the crystals have a size sufficient to present, for example, a ferroelectric response. This commitment is not easy to perform. Another condition that can maintain the optical transparency of the glass ceramic is the small difference between the refractive indices of crystals and glass matrix. If this difference is negligible it allows, regardless of the size of the crystals, to maintain the optical transparency of the glass ceramic.

In recent years there has been a growing interest in the preparation, characterization and technological implementation of glass and glass ceramics in new systems or substituting single crystals. However, it is important to note that, in general, the optical and electric properties of glass ceramics are not as good as their single crystals embedded in the matrix. This is because the glass ceramics present at least two distinct phases, the crystalline (considered the active phase) and the amorphous (support). The electric polarization of the crystals embedded in a glassy matrix is more difficult due to the low dielectric constant of the glass phase. Moreover, because of the growth processes of single crystals present extremely high economic costs, their substitution by glass ceramics is now an actual issue. Some glass ceramics have also the advantage of being a high density material and without porosity.

In this chapter it is discussed the preparation method paraelectric ($NaNbO_3$) and ferroelectric ($LiNbO_3$ and $KNbO_3$) crystal phases embedded in a glass matrix, through heat-treatments (HT) of a amorphous glass. The structural characterization was made using the techniques of differential thermal analysis, X-ray diffraction, scanning electronic microscopy and Raman spectroscopy. It was measured the dc and ac conductivity, thermally stimulated depolarization current (TSDC), complex impedance (Z^*) and dielectric constant in function of temperature and frequency. The analysis of the network former effect and the conditions of thermal treatments, namely the parameters temperature, time and the simultaneous presence of the external electric field is the main topic of this chapter. From all analysed systems the borate system shows that the crystallites precipitate in volume. For the silicate system with sodium, the $NaNbO_3$ particles crystallize at the sample surface. This behaviour does not occur when borate oxide is used as glass former or when the alkali ion is lithium. When using phosphate pentoxide the number of different crystalline phases that grows during the heat-treatment process is such that this glass former is, at this point of view, not suitable for prepare glass ceramics with one single crystal active phase. The electric and dielectric characteristic of all of the studied samples shows the important role of the thermal treatment conditions in their properties.

2. State of-the-art

Nowadays a lot of technological applications is based on the electrical characteristics of materials. From those characteristics, the ferroelectricity, antiferroelectricity and paraelectricity are the most important properties to be studied and controlled. Very briefly, the ferroelectric materials exhibit spontaneous polarization, i.e. a polarization in the absence of an external electric field. In the ferroelectric materials permanent electric dipoles exist, which origin is explained by the structural arrangement of the atoms in the unit cell. The most common ferroelectric material is barium titanate (BaTiO3), which spontaneous polarization is a consequence of the positioning of Ba^{2+}, Ti^{4+} and O^{2-} in the unit cell. In these materials, whose crystalline structure is a perovskite (ABO_3 - fig. 2.1), where the cation A has an atomic radius too large for a compact packaging, the oxygen atoms move from the expected positions. In the case of $BaTiO_3$, the Ba^{2+} ions are located in the corners (vertices) of the unit cell, which is of tetragonal symmetry (a cube that was slightly elongated in one direction). The apparent dipole moment results from the relative displacements of the Ti^{4+} and O^{2-} ions from their symmetrical positions. The O^{2-} ions are located near the centre (but slightly below) of each of the six faces, while the Ti^{4+} ion is displaced upward from the centre of the unit cell. Thus, a permanent ionic dipole moment is associated with each unit cell.

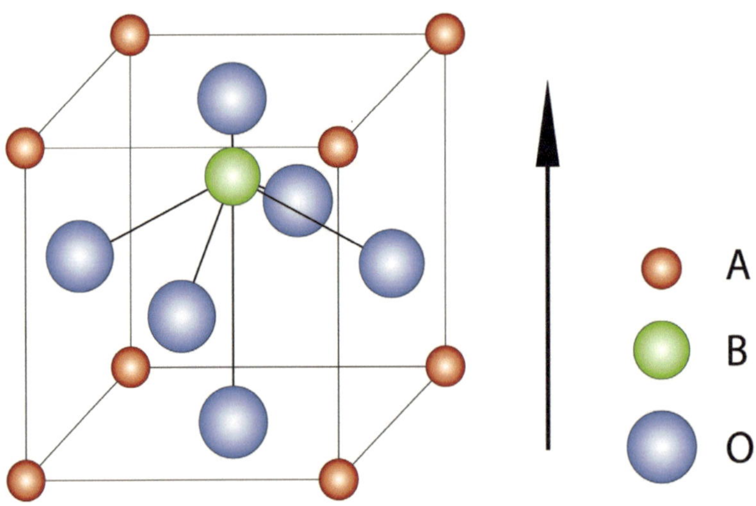

Fig. 2.1. Scheme of the elementary perovskite structure (ABO_3) with ferroelectric characteristic.

However, when $BaTiO_3$ is heated above its Curie ferroelectric temperature, the unit cell becomes cubic and all ions assume symmetrical positions within the cubic unit cell and the ferroelectric behavior ceases.

An antiferroelectric state is defined as one in which lines of ions in the crystal are spontaneously polarized, but with neighbouring lines polarized in antiparallel directions, so that the spontaneous macroscopic polarization is zero. In simple cubic lattices the antiferroelectric state is likely to be more stable than the ferroelectric state. The antiferroelectric state will not be piezoelectric.

Paraelectricity is the ability of some materials to become polarized under an applied electric field. Unlike ferroelectricity; this can happen even if there is no permanent electric dipole that exists in the material, and removal of the field results in the polarization in the material returning to zero. The mechanisms which give rise to paraelectric behavior are the distortion of individual ions (displacement of the electron cloud from the nucleus) and the polarization of molecules or combinations of ions or defects. Paraelectricity occurs in crystal phases in which electric dipoles are unaligned (i.e. unordered domains that are electrically charged) and thus have the potential to align in an external electric field and strengthen it. In comparison to the ferroelectric phase, the domains are unordered and the internal field is weak.

Sodium niobate ($NaNbO_3$) and Lithium niobate ($LiNbO_3$) are dielectric materials of the perovskite group. The $NaNbO_3$ has been the focus of special attention due to the interesting combination of the electrical and mechanical properties. At room temperature, $NaNbO_3$ is antiferroelectric, with pseudoperovskite orthorhombic crystalline structure with high dielectric constant (2000-3000) at Curie temperature (360°C). Ferroelectricity has been observed below 0°C. One of its potential uses is as piezo transducer operating in the microwave region (high frequency). However, this requires a ferroelectric behavior and therefore the addition of other niobates, such as $LiNbO_3$, $KNbO_3$ and $AgNbO_3$ in small quantities, as been studied because it induces in the $NaNbO_3$ a ferroelectric behavior at ambient conditions.

The main structural units of the $NaNbO_3$ crystal are the NbO_6 octahedrons, which are relatively rigid units and therefore there is, in the structure of $NaNbO_3$, a joint distortion of all these units. For this reason, the $NaNbO_3$ is characterized at room temperature, as antiferroelectric material, i.e. below its Curie temperature) has no electric dipole moment and no electrical hysteresis is observed. However the presence of an external electric field to induces a ferroelectric response on $NaNbO_3$ featuring piezoelectric characteristics of high interest for applications in high frequency devices.

Figure 2 shows the possible directions of twisting/distortion of the oxygen octahedron structure. The direction of distortion is privileged by the symmetry of the crystal structure. Thus, 3-axis of order 4 (A4) in the cubic phase, 6-axis of order 2 (A2) in the orthorhombic phase and 4-axis of order 3 (A3) in the rhombohedral phase.

The increase in temperature causes a distortion in the NbO_6 octahedron structure, which define the crystal structure of $NaNbO_3$ in different axis of symmetry (Fig. 2.2b) giving rise to six successive phase transitions. These changes are low symmetry structural variations of the perovskite structure.

The first studies about the structure and electrical properties of crystal $NaNbO_3$ were conducted by Matthias and Remeika in Bell Laboratories. Nowadays, it is considered, at the structural level, one of the more complicated perovskites by having with increasing temperature, a sequence of phase transitions. These phase transitions are shown in table 2.1.

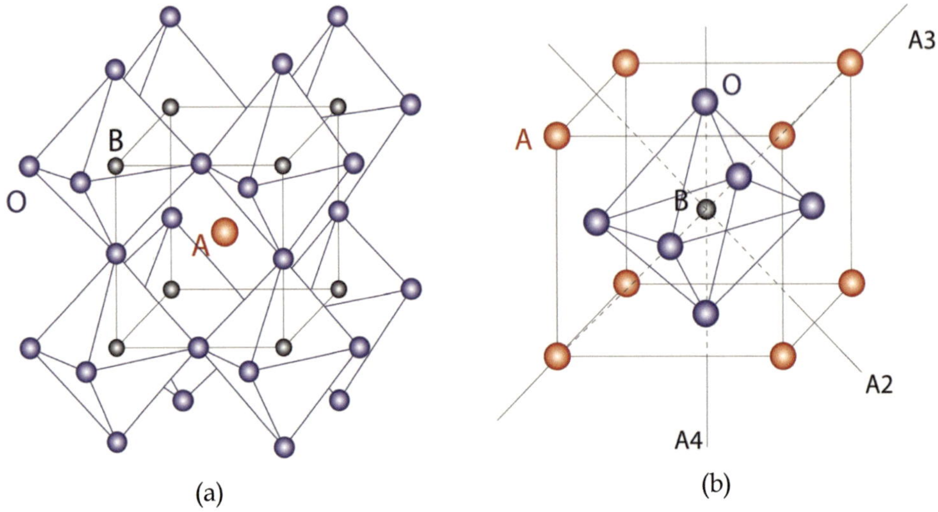

Fig. 2.2. a) Scheme of the elementary perovskite structure (ABO_3) b) Directions of distortion of the oxygen octahedron structure (A2, A3, A4) .

Temperature (°C)	Simetry	Unit cell	Electric characteristic
	Rhombohedric	$a = b = c, \alpha = \beta = \gamma < 90°$	ferroelectric
~ 80	↓		
	Monoclinic	$a \neq b > c, \alpha = \beta = 90°, \gamma < 90°$	Antiferroelectric
370	↓		
	Orthorhombic	$a \neq b \neq c, \alpha = \beta = \gamma = 90°$	Antiferroelectric
480	↓		
	Orthorhombic	$a \sim b \sim c, \alpha = \beta = \gamma = 90°$	Antiferroelectric
520	↓		
	Orthorhombic	$a < b < c, \alpha = \beta = \gamma = 90°$	Antiferroelectric
575	↓		
	Tetragonal	$a = b < c, \alpha = \beta = \gamma = 90°$	Antiferroelectric
640	↓		
	Cubic	$a = b = c , \alpha = \beta = \gamma = 90°$	Paraelectric

Table 2.1. $NaNbO_3$ characteristics in function of the temperature

As in $NaNbO_3$, in $LiNbO_3$ the unit that repeats is NbO_6 octahedra linked by sharing a vertice, along the c axis and in the form of a helix. For temperatures below the Curie temperature (T_c = 1210 ° C) the structure of $LiNbO_3$ consists of layers of oxygen atoms, parallel to each other in a hexagonal compact (HC) distorted configuration, where the Li^+ and Nb^{5+} ions are surrounded by the oxygen octahedra. In this structure, one third of the octahedral interstices are occupied by Li ions, 1/3 by the Nb ions and the remaining interstices (1/3) by structural voids. The morphological appearance of these crystals is pyramidal. This crystal structure is represented in the following figure (Fig. 2.3).

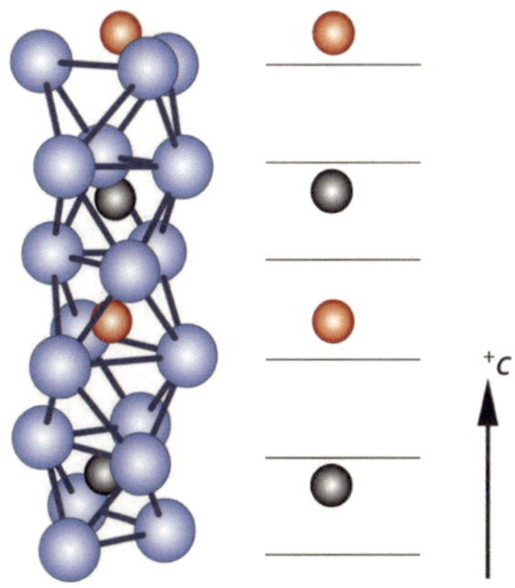

Fig. 2.3. $LiNbO_3$ atomic model ((• - oxigen; • - niobium; • - lithium, the horizontal lines represent layers of oxygen atoms).

The stability of the crystalline structure depends on the ionic radii of their constituents. When the cations are too small, such as lithium, when compared with the oxygen ion the maximum packaging gives rise to a distorted crystal structure. Thus, the $LiNbO_3$ structure is similar to that of a perovskite, but distorted due to the low ionic radius of lithium, and therefore the unit cell contains no center of symmetry. The fact that the unit cell does not have a center of symmetry gives rise to a dipole moment which causes a spontaneous polarization that characterizes this class of materials (ferroelectric). The crystal structure of $LiNbO_3$ is no longer distorted when placed at temperatures higher than T_c. In the next table are summarized the most relevant properties of $LiNbO_3$ monocrystal, near-stoichiometric ([Li] / [Nb] = 0.95), obtained by the Czochralski method.

The properties that $LiNbO_3$ present, make it suitable for various applications in technological systems, namely at the optical level, presenting itself as an excellent candidate for optoelectronic devices due to its high birefringence (short for modification of refractive index due to variations in the direction of the beam from a strong incident field), high efficiency in nonlinear optics, especially in the second harmonic generation (SHG). It is currently used as a waveguide amplifier, modulator and optical switch systems, in nonvolatile memories, Surface-Acoustic-Wave (SAW) devices and frequency doublers.

Both $NaNbO_3$ and $LiNbO_3$ crystals are usually grown by the Czochralski method. However, due to the technical and economical disadvantages that the Czochralski method presents, various processes for the preparation of these crystals, namely the $LiNbO_3$, has been tested.

The fact that the preparation of single crystals of $NaNbO_3$, $LiNbO_3$ and other niobates, the usual growth techniques (for example: Czochralski), be a difficult and costly process, scientific interest to study structural, electrical and optical glass and glass ceramics containing niobate crystals. In 1979, Nassau et al. showed the preparation of a $LiNbO_3$ glass, using the twin-roller quenching method. The melting temperature used was 1550 °C and a cooling rate of the molten was estimated at ~ 10^7 °C/s. The formation of the glass, with the $LiNbO_3$ molar composition was possible because niobium ion is an intermediate specie. In any glass, the units that describe its structure can be divided into three categories, defined according to its structural function, network former; network modifier and/or intermediate specie. The network formers are units that without the addition of other components can form glass. Examples are SiO_2, B_2O_3, P_2O_5 and GeO_2. The network modifiers, do not form glass by itself, but are often combined with a former in order to facilitate the glass formation. Examples of modifying elements are the alkaline ions (Li, Na, K, etc.) and alkaline earth ions (Mg, Ca, etc.). Intermediate species are elements that can either have the role of network former or network modifier (ex.: Al, Nb, etc.).

Melting temperature (ºC)	1260
Curie temperature (ºC)	1210
Density (at 25ºC) (g/cm^3)	4.64
Ordinary refraction index, n_0	2,296 (a)
Extraordinary refraction índex, n_e	2,208 (a)
Electro-optic coefficient, r_{33} (m/V)	30x10^{-12}
Transparency window [μm]	0,4 – 5
Resistivity, ρ (along the c axis and at T(K)) [Ω.cm]	log ρ=(7150/T)–2,823 ρ (300 K)= 10^{+21}
Dielectric constant (ε´) – along the c axis (at 25 ºC)	> 1000 (at 1kHz) ~ 80 (at 100 kHz)
Dielectric losses (tan δ) – along the c axis (at 25 ºC and 100 kHz)	≈0
Coercive field (at 1210ºC) (V/m)	20
Spontaneous polarization (P_s [x10^{-2}Cm^{-2}] at 300 K)	70
Piezoelectric coefficient (d_{33} [pC/N])	6
The ordinary refractive index, n_0 is defined as $n_0=c/v_\perp$ and the extraordinary, n_e, as $n_e=c/v_{//}$, where c is the speed of light in vacuum and $v_\perp$ e $v_{//}$ the perpendicular and parallel components of the velocity of the wave that propagates in the material, respectively.	

Table 2.2. Main properties of $LiNbO_3$ monocrystal

According to A. Paul a glass is a material that shows the elastic behavior characteristic of the crystalline state and the viscous behavior of a liquid. The most common properties of glass are transparency to visible radiation, mechanical stability, biological inert and electrical insulator. However, due to the possibility of controlling the microstructure, from changing the initial composition or by applying treatments and therefore controlling the process of nucleation and crystallization [james95; rüssel97], the properties of glass can be modified. The initial chemical composition is a factor, controllable, allowing to mould some of the properties of the glass. The formation of glass ceramics, for example by thermal treatment of base glass, shows at the technological level the great advantage, relatively to single crystals and sintered ceramics, the possibility of their properties (optical, electrical, mechanical, chemical, etc.) be controlled via the volume fraction of the active phase dispersed in the network (matrix). For example, to maintain optical transparency, the process of nucleation and crystal growth requires a high control being achieved when the size of crystals dispersed in the glass matrix is not high enough to cause light scattering. However, for most electrical applications it is necessary that the crystals present a size sufficient to show, for example, a ferroelectric response. This commitment is not easy to perform. Another condition that can maintain the optical transparency of the glass ceramic is the difference between the refractive indices of crystals and glass matrix. If this difference is negligible it allows, regardless of the size of the crystals, to maintain the optical transparency of the glass ceramic. In recent years there has been a growing interest in the preparation, characterization and implementation technology of glass and glass ceramics. However, it is important to note that, in general, the optical and electrical properties of glass ceramics are not as good as their single crystals. This is because the glass ceramics present, at least, two distinct phases, the crystalline (considered the active phase) and the amorphous phase. The electric polarization of the crystals embedded in a glassy matrix is not so easier due to the low dielectric constant of the glass phase.

In glasses, the crystal growth and crystal orientation can be generally achieved through different processes, such as: mechanical deformation; thermodynamic control, kinetic control (electrochemical induced nucleation). The use of thermodynamic control is the most common. However, control of crystallization, with the desirable crystalline phase is usually difficult because crystallization is a complex process affected by various factors such as composition, surface condition, heat treatment parameters, etc.

2.1 Glasses with $NaNbO_3$ and $LiNbO_3$ crystallites

A significant amount of research in the field of preparation and crystallization of glass and glass ceramics containing $LiNbO_3$ prepared by melt quenching currently exist. In 1980, Prasad and colleagues published a study about the preparation of glass ceramics of the system SiO_2-Li_2O-Nb_2O_5. The choice of SiO_2 as a network former was because it is desirable that the glass former does not contain cations that can change the crystal structure of $LiNbO_3$. The SiO_2 satisfy this condition due to the reduced value of atomic radius and coordination number. They showed that transparent glasses without crystalline phases are achieved for a molar amount of SiO_2 between 32 and 39% and using a equal molar amount of Li_2O and Nb_2O_5.

The glass-forming region of the glass system (SiO_2-Al_2O_3)-Li_2O-Nb_2O_5, using the fusion method is presented by Todorovic and colleagues. They found that when the molar ratio $[Nb_2O_5]/[SiO_2] > 1$ and the molar amount of $[Li_2O] > 25\%$, the glasses obtained are transparent and homogeneous. The application of thermal treatments to this glass, allowed the crystallization of $LiNbO_3$ particles with a size between 150 and 350 nm, while maintaining optical transparency. These glass ceramics have a refractive index of 1.84-1.93 and a dielectric constant of 155-185, measured at 1 kHz and room temperature. They observed that increasing the temperature and/or thermal treatment time leads to an increase in particle size and consequent opacity of the samples.

Zeng and colleagues showed that the composition $50SiO_2$-$25Li_2O$-$25Nb_2O_5$ (mole%) gives, through the melt quenching method, a transparent glass, and observed the devitrification/crystallization ($LiNbO_3$ phase crystallization) with thermal treatment at 800 ° C for 30 minutes. Applying an electric field during the thermal process, and cooling the melt to room temperature, gives rise to the appearance of dark brown zones in the glass, which indicates the occurrence of chemical and structural transformations. It is suggested the occurrence of an oxidation-reduction reaction activated by the external electric field. A similar study was conducted by Gerth in a glass composition $35SiO_2$-$45Li_2O$-$20Nb_2O_5$ (mole %), applying to the melt a dc electric current (50 mA). From the obtained results it is suggested that the presence of a external electric field favors nucleation and subsequent crystallization in areas close to the cathode ($Nb^{5+} + e^- \leftrightarrow Nb^{4+}$). They found that the $LiNbO_3$ crystallites are isotropically distributed and with the c-axis perpendicular to the electrode. One of the purposes of applying an electric field during the devitrification of the glass, is to control the crystal orientation, which is a very important parameter for glass-ceramics with ferroelectric properties. Ding and colleagues found that the introduction of K_2O, replacing the Li_2O, in the system SiO_2-Li_2O-Nb_2O_5 favors, during the heat treatment process, the growth of $LiNbO_3$ particles along the c axis. These particles were only detected in the glassy surface.

Using P_2O_5 as network former Chowdari and colleagues fabricated a series of glasses of the system P_2O_5-Li_2O-Nb_2O_5, by melt quenching, identifying the glass formation region. The analysis of the electrical conductivity of these glasses showed ionic conductivity at the temperature of 25 ° C, of about $2x10^{-6}$ S/cm attributed to the high cation mobility. The silicate glasses with $LiNbO_3$, prepared by Prasad shows that the mechanism of electrical conductivity is ionic due to the high mobility of Li^+. The doping of niobiophosphate glasses with iron is presented by Allen and colleagues. The nucleation and crystal growth of ferroelectric $LiNbO_3$, in the glass matrix composition [(50-x)P_20_5-50Li_20-xNb_20_5]: yFe_20_3, with 0 <x <50 (mole%) and y = 2 (mole%) occurs preferentially in samples with high concentration of niobium. In samples with lower amount of niobium is observed the formation of lithium phosphate phases, of different complexity. The analysis of the thermally stimulated depolarization currents (TSDC), of these glasses, shows the presence of at least two depolarization mechanisms dependent on the niobium concentration. The lower temperature is attributed to dipolar relaxation and the second to interfacial relaxation mechanisms.

Singh and colleagues, show the preparation of lithium borate glasses with niobium. This study shows that the introduction of more than 15% of Nb_2O_5 to the 42.5Li_2O -57.5B_2O_3 composition (mole%) gives origin to crystalline phases. The Nb^{5+} ions inserted will occupy sites where lithium ions were and thus the number of empty sites will increase to maintain the charge neutrality. According to Tuller, this type of structural rearrangement, where there are more empty sites than ions available to fill favors the increase of ionic conductivity. In glasses where the amount of $LiNbO_3$ is above 5%, there is an increase of the conductivity attributed to the increased mobility of Li^+ ion, which is related to the increased number of empty sites. However, if the random network model, proposed in 1932 by Warren and Zachariasen for silicate glasses, is suitable for lithium borate glasses is expected that the lithium ion will occupy the empty spaces in a random manner. If this occurs then it is expected a distribution of activation energies for the conduction process of lithium ion. Note that the concept proposed by Zachariasen and Warren was one of several, which does not entirely explain the experimental observations. In 1959, Porai-Koshits proposes the existence of a polymer-crystalline structure, i.e., the glassy network does not present an absolute disorder. This theory is the one that sets, more precisely, the experimental observations.

Huang and colleagues showed that using the melt quenching method, the molar composition $45Li_2O+50B_2O_3+5M_nO_m$ (%mole), where M = Al, Ti, Zr, V, Nb, Ta, Cr, Mo and W, and using Li_2CO_3 and H_3BO_3 as a starting material, transparent and homogeneous glasses without crystalline phases are achieved.

The application of thermal treatments to lithium borate glasses with niobium, with the aim of crystallize the $LiNbO_3$ phase is presented by Prasad and colleagues for the system (100-x)-$LiBO_2$-xNb_2O_5 (mole %) with $5 < x < 25$. The as-prepared glass was submitted to heat treatments at temperatures between 300 and 800 °C and the $LiNbO_3$ phase was detected for treatments at temperatures above 500 °C. In glasses with higher niobium content the heat treatment process also favors the formation of the Li_3NbO_8 phase. In these glasses it appears that the value of the dielectric constant increases with increasing volume fraction of $LiNbO_3$. In glasses treated at temperatures above 500 °C, there is the presence of spherical particles, with an average size of 25 nm, which increases with the increase of the thermal treatment temperature. The pyroelectric and ferroelectric effects were observed on these glass-ceramics at room temperature. The glass with x = 10 presents second harmonic generation (SHG) when the glass is irradiated with a $\lambda = 1064$ nm laser and it was observed that the SHG intensity depends on the angle of incidence.

Komatsu and colleagues showed the possibility of prepare glass-ceramics with $LiNbO_3$ crystallites using tellurium oxide has network former. The tellurium glass matrix has the advantage over the SiO_2, by possessing a refractive index of ~ 2.2 very close to the refractive index of many ferroelectric crystals and also a lower melting temperature. The glasses prepared by Komatsu and coworkers, contains the following molar composition $50TeO_2$-(50-x)Li_2O-xNb_2O_5 (mole %). The precipitation of $LiNbO_3$ crystals is obtained using heat treatments with two temperature steps. All the glasses after the heat treatments are no longer transparent. They found that the higher is the ratio $[Li^+]/[Nb^{5+}]$ less difficult is the precipitation of the $LiNbO_3$ crystals. Shankar and colleagues showed that the composition containing 50% molar of TeO_2 and an equal amount of Li_2O and Nb_2O_5, after heat treatment at 200 °C for 12h, presents $LiNbO_3$ crystals, which precipitates mainly on the glass surface.

Ding and colleagues [ding94] present the study of the crystallization of the glass composition $55TeO_2$-$25Li_2O$-$20Nb_2O_5$ (mole %) submitted to an ultrasonic treatment (28 kHz, 150 W), which promotes the nucleation and growth of $LiNbO_3$ crystals on the glass surface.

Using the method of embedding crystals in a glass matrix, obtaining transparent glass ceramics has been attempted by Kim and colleagues in the glass composition $80TeO_2$-(20-x)Li_2O-xNb_2O_5 (mole%), where $LiNbO_3$ crystals were incorporated. They found that adding to the glass 15%, in weight, of $LiNbO_3$ after melting at 950 °C, a transparent glass-ceramic containing lithium niobate crystalline phase was obtained.

Unlike glasses containing $LiNbO_3$ phase, few studies are published to our knowledge, on glasses containing $NaNbO_3$. Herczog and colleagues, showed the preparation of glasses of the system $xSiO_2$-(1-x)(Na_2O-Nb_2O_5), verifying that the decrease of x from 24% to 14% translates into an increase in the optical transparency. The precipitation of the $NaNbO_3$ crystalline phase is obtained by heat treatment at temperatures above 750 °C. Karapetyan and colleagues present a study of the preparation of silicate glass ceramics with $NaNbO_3$ based on the molar composition $32SiO_2$-$26Li_2O$-$20Nb_2O_5$-22(GeO_2 + Ti_2O + Ta_2O_5 + ZrO_2 + BaO + Sb_2O_5), which gives rise, using the melt quenching method, to transparent glasses.

They found that the total substitution of Li_2O for Na_2O does not favor the formation of a homogeneous glass. The glassy state is maintained until a replacement up to 5% of Li_2O by Na_2O. This glass was submitted to heat treatments with ion exchange, by placing it with $NaNO_3$ powder into a steel crucible. After heat treatments at 650 °C for 4h, is was observed the formation of the $NaNbO_3$ phase, verifying that this crystallization is mainly at the surface but, with increasing the treatment time (> 16h) the inner zone of the sample becomes translucent indicating the presence of inside particles.

3. Experimental methods

3.1 Glass preparation

The preparation of the transparent glasses containing lithium or sodium and niobium ions was performed using the melt quenching method and after, LiNbO3 or NaNbO3 crystal phases were precipitated in the glass matrix by heat-treatment process (HT).
The melt quenching method was used to prepare transparent glasses of the following ternary systems:

A. SiO_2-Nb_2O_5-X_2O (X = Li, Na);

B. B_2O_3-Nb_2O_5-X_2O (X = Li, Na);

The raw materials used were: silicon oxide (SiO_2), boron oxide (B_2O_3), lithium carbonate (Li_2CO_3), niobium oxide (Nb_2O_5) and sodium carbonate (Na_2CO_3). The molar composition used was chosen in order to validate the following criteria: i) transparent glasses must be obtained; ii) the glasses should not present initial crystalline phases, iii) glasses with similar molar compositions. Thus the composition prepared was: $60AO_2$-$30X_2O$-$10Nb_2O_5$ (% mole) (A = SiO_2, B_2O_3, X = Li, Na).
The preparation of glasses followed the basic process outlined in Figure 3.1.
The calcination step was performed in order to remove the CO_2 from the Na_2CO_3 and Li_2CO_3 raw materials, getting Na_2O and Li_2O, respectively. The last step, which is the glass annealing, held immediately after the melt was poured into the molds or pressed between stainless steel plates is intended to minimize the stresses of the glass network, preventing its disruptions. The parameters: temperature, calcination time, temperature and melting time and temperature and time of heat treatment (annealing) are characteristic of each composition. Therefore, these parameters are described in the description section of each composition.
The preparation of the glass with the molar composition $60SiO_2$-$30Li_2O$-$10Nb_2O_5$ start from the weight of the appropriate amounts of the oxides (SiO_2 and Nb_2O_5) and lithium carbonate (Li_2CO_3). These oxides were mixed in a planetary ball-mill system for 1 hour, obtaining a homogeneous mixture. The next step was the calcination of the mixture, at 700 °C for 3 hours. After this treatment, the mixture was placed in a platinum crucible, heated to 1450 °C and held at that temperature for 30 minutes, after which the melt was poured into a clean stainless steel plate. A second plate was used, immediately to press the melt. After this process the glass was heat treated (annealed), with the aim of reducing the mechanical stresses in the glass, in a preheated oven at 350 °C for about 4 hours, after which the system is turned off leaving the glass cool down naturally within the oven.
The preparation of glasses with a molar composition $60B_2O_3$-$30Li_2O$-$10Nb_2O_5$ followed the same process (Fig. 3.1). In summary form, the reagents (B_2O_3, Li_2CO_3 and Nb_2O_5) were

mixed and after calcinated with the same conditions used for the silicate composition. The melting temperature used was 1100 °C, and the melted was kept at this temperature for 30 minutes. After, the melt was poured into a clean stainless steel plate and immediately compressed with a second plate. The obtained glasses, transparent and colorless, and with a thickness between 1-2 mm were annealed at 300 °C, maintaining this temperature for 4 hours, then letting them cool in the furnace, slowly.

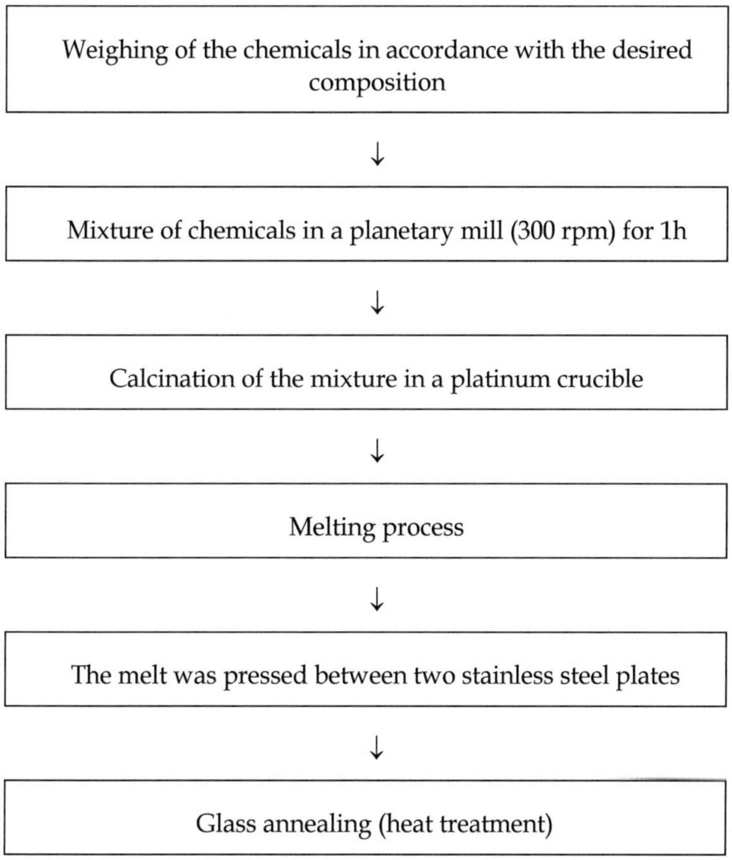

Fig. 3.1. Diagram of the glass preparation method used.

In order to obtain glass ceramics with $LiNbO_3$ crystallites controlled heat-treatments (HT) to the base glass were performed, previously polished in order to obtain samples with a thickness of about 1 mm and with faces the more parallel possible.

The preparation of the glass composition $60SiO_2$-$30Na_2O$-$10Nb_2O_5$ (mole %) followed the same procedure (Fig. 3.1). In a brief, the reagents (SiO_2, Na_2CO_3 and Nb_2O_5) were mixed in a planetary mill for 1 hour in order to obtain a homogeneous mixture. This mixture, placed in a platinum crucible was maintained at a temperature of 650 ° C for 3 hours in order to decompose the Na_2CO_3 into Na_2O, releasing the CO_2. After this step the mixture was heated to 1400 °C, remaining at this temperature for 30 minutes. The melt was then poured into a clean stainless steel plate being used a second plate to press the melt, obtaining transparent and colorless glasses with a thickness of 1-2 mm. This samples were then annealed at 250 °C (3 hours - annealing), and cooled inside the oven slowly to room temperature.

The preparation of the glass composition $60B_2O_3$-$30Na_2O$-$10Nb_2O_5$ followed the process shown previously. It must be noted that the melting temperature of the mixture of starting oxides (B_2O_3, Na_2O and Nb_2O_5) used was 1200 °C for 30 minutes.

3.1a Glass-ceramics preparation

The transparent glasses prepared by melt quenching were submitted to controlled thermal treatments in order to obtain glass ceramics with the $LiNbO_3$ or $NaNbO_3$ crystalline phases. Figure 3.2 shows the heat treatment program used, held in a horizontal tube furnace. The threshold temperature parameter (Tp - Fig. 3.2) was determined based on the thermal behavior information obtained from each glass composition.

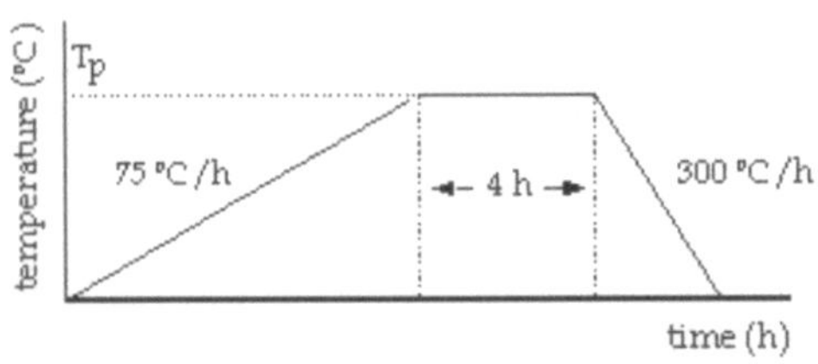

Fig. 3.2. Scheme of the heat treatment program applied to the glasses.

3.2 Structural and morphological characterization techniques

The techniques used for structural characterization of the glasses and glass-ceramics were the differential thermal analysis (DTA), the X-ray powder diffraction (XRD), Raman spectroscopy and scanning electron microscopy (SEM).

3.2a Differential Thermal Analysis (DTA)

The differential thermal analysis (DTA) has aimed to determine the value of the glasses transition temperature (T_g) and to identify the temperature zone undergoing structural changes, including crystallization (phenomenon associated with the detection of exothermic peaks or bands). In the system used, a Linseis Aparatus, the powder samples (± 40 mg) were carefully packed in a alumina crucible. The thermal analysis was performed using a heating rate of 2.5 °C/min and high purity Al_2O_3 was used as reference.

3.2b X-ray diffraction (XRD)

The X-ray diffraction was the technique that allowed us to detect and identify the presence, or not, of crystalline phases embedded in the glass network. The X-ray diffraction was performed at room temperature on a Phillips X'Pert system, where the X-ray production is performed on a Cu ampoule, operating at 40 kV and 30 mA, emitting the monochromatic $K\alpha$ radiation (λ = 1, 54056 Å - graphite monochromator). In this system the sweep is continuous, from 10.025 up to 89.975 ° (2θ) with a speed of 1.5 degrees per minute and with a step of 0.02 °. The identification of the crystalline phases was based on the database provided by the JCPDS (Joint Committee on Powder Diffraction Standards).

3.2c Raman spectroscopy

Raman spectroscopy is a technique sensitive to local structure and is therefore particularly useful for the study of the materials structure. This technique is nondestructive, allows the study of optical properties, vibrational, structural composition, crystallinity and/or amorphous state of samples. This analysis was performed on a spectrometer T64000, Jobin Yvon SPEX using an argon laser operating at 514.5 nm. The Raman spectrum was obtained with a back-scattering geometry (back-scattering) between 100 and 2000 cm^{-1}. The amplitude of the lens used was of 50x which allows a laser spot diameter on the sample of about 5 mm.

3.2d Scanning Electron Microscopy (SEM)

The main purpose of using this technique was to verify the existence of particles in the heat-treated samples and also analyzing their shapes, morphology, size, size distribution and orientations. It must be noticeD that using this technique the samples must possess three main characteristics: support high vacuum; support the energy of the incident electron beam and their surface must be conductive. Once the studied samples are electrically insulators, it was needed to perform a carbon deposition on the sample surface. In all samples it was observed the free surface and the fractured surfaces characteristics. Electron microscopy systems used were: Hitachi S4100-1 and Philips XL30.

3.3 Electric and dielectric characterization techniques

The study of electrical and dielectric properties of the glasses and glass ceramics was performed by measuring the dc conductivity (σ_{dc}) and ac conductivity (σ_{ac}) as function of the temperature (200 K - 370 K), the complex impedance (Z *= Z '+ jZ'') as a function of temperature (260 K - 315 K) and frequency (10^{-3} Hz - 10^{6} Hz). All measurements were made using a bath cryostat system. For electrical measurements the sample surfaces, which present a thickness of about 1 mm were painted with silver paint, thus forming the electrodes on which are attached with the same paint the electrical terminals. In the following items, it will be described, in a summary form, the description of the systems of measurement used and the mathematical basis used for the analysis of the results. It also presented the mathematical formalism necessary to calculate the different quantities that can be obtained from the measurements.

3.3a Dc conductivity measurement

To measure the dc conductivity (σ_{dc}) it was used a system that, at a given and stable temperature, applies to the sample a known dc voltage (V) through a variable voltage source and measuring the current (I) through an electrometer (Keithley 617). Due to the high resistivity values of this samples a two terminal model was always used. The current measurements were made from the lowest temperature and the rise of the temperature was done using a heater controlled by a IT54-Oxford Research. The calculation of the σ_{dc}, based on this measurement method, can be achieved through the following deduction: based on Ohm's law for electrical circuits (where V represents the voltage, R the resistance and I the intensity of current that flows through the resistance) and knowing that the electrical resistance of a material is given by equation 1:

$$R = \rho_{dc} \frac{d}{A} = \frac{1}{\sigma_{dc}} \frac{d}{A} \qquad (1)$$

where ρ_{dc} is the dc resistivity of the material, σ_{dc} the dc conductivity of the sample, d the thickness and A the area of the electrodes, it is concluded that the conductivity can be calculated using equation 2:

$$\sigma_{dc} = \frac{I}{V}\frac{d}{A} \tag{2}$$

According to Hakim and colleagues, the variation of the dc conductivity with temperature, in glasses, can be adjusted through an Arrhenius equation (eq.3):

$$\sigma_{dc} = \sigma_0 e^{\left(-\frac{E_A}{kT}\right)} \tag{3}$$

where σ_o is the pre-exponential factor, E_A is the activation energy, k is the Boltzmann constant (1.380658 x 10^{-23} J/K) and T the temperature. Using the logarithmic function on the equation 3, the activation energy (E_A) can be calculated through the slope of ln(σ_{dc}) versus 1/T (eq. 4):

$$\ln(\sigma_{dc}) = -\frac{E_A}{kT} + C \tag{4}$$

Sometimes, the ln (σ_{dc}) versus 1/T is not linear In these cases a better fit can be obtained using Mott's law (Eq. 4b):

$$\ln(\sigma_{dc}T) = -\frac{E_A}{kT} + C \tag{5}$$

Being the activation energy (E_A) calculated from the slope of ln(σ_{dc}T) versus 1/T.

3.3b Dielectric spectroscopy measurements

The complex impedance measurement (Z *= Z ' +j Z'') has temperature and frequency function allows us to conduct studies about the dependence of the ac conductivity, dielectric constant, dielectric loss and dielectric relaxation phenomena with those two variables. The measurement of Z*, was performed in the frequency range from 10 mHz to 32 MHz, by using a Solartron SI 1260 Impedance/gain-phase analyzer, which reads the real part (Z') and the imaginary part (Z'') of the sample impedance. The complex permittivity ($\varepsilon^*=\varepsilon'+j\varepsilon''$) was calculated based on the values of Z * and knowing that Z *= 1 / (μ ε *) (where μ = jωC_0, $j=\sqrt{-1}$, ω is the angular frequency and C_0 the empty space capacity ($C_0 = \varepsilon_0(A/d)$, ε_0 is the vacuum permittivity 8.854 x10^{-12} Fm, A the area of the electrode and d the sample thickness).

The measured impedance was normalized by calculating Z^*_{rel} (eq. 6):

$$Z^*_{rel} = Z^*(A/d) \tag{6}$$

The ac conductivity (σ_{ac}) for a given frequency and temperature was calculated using eq. 7:

$$\sigma_{ac} = \omega\varepsilon_0\varepsilon'' \tag{7}$$

The dependence of σ_{ac} with temperature for a fixed frequency, allowed us, by using the Arrhenius expression (Eq. 3), to calculate the ac conductivity activation energy ($E_{A(ac)}$).

3.3c Fit the experimental data to physical models

The need to fit the experimental complex impedance (Z*) data, as function of frequency and temperature, to physical models is of great interest because the resulting information can enable a quantitative interpretation of the measurements. Usually, physical models used in such settings are associated with equivalent electrical circuits. Thus, this section aims to present and discuss the physical models and numerical method used to fit the experimental data obtained by impedance spectroscopy.

The adjustment of this type of experimental data to theoretical functions is usually done through computational processes associated with numerical methods. Currently there are some commercial software that allows the realization of these settings (e.g. LEVM written by RS Macdonald, Equivalent Circuit written by BA Boukamp, Curvefit written by Graphpad Software, Inc, WinFit written by Novocontrol Inc). However, factors such as the theoretical model associated with the algorithm, the description of the method for calculating the adjustable parameters, the existence of limitations in terms of amount of experimental data that can be used in the adjustment process and the processing time led us to develop software that could make these adjustments.

In the developed software the computational analysis of the experimental data was performed using the complex non-linear least squared deviations method (CNLLS). One advantage of this method is to allow the use of complex models containing many parameters to adjust, enabling calculation of the uncertainties associated with each parameter used. The ability to adjust both the real and imaginary part, which does not occur in other methods, such as the Kramers-Kronig (KK) method, or the Havriliak-Negami (HN) method, which assumes that knowing one part (real or imaginary) the other is calculable. However, this mathematical approach introduces, in the calculated part errors that result in a poor final adjustment.

The CNLLS method was used based on the algorithm idealize by Levenberg-Marquardt and based in the Gauss-Newton numerical method. This method requires, necessarily, an estimation of the initial parameters to adjust. From these values the algorithm modifies all of them, or just some (this is one advantage), and simultaneously evaluates the result of the modification. If the result fit the experimental values better the modifications are accepted, otherwise the first remains. This process is repeated iteratively until the sum of the squared deviations is minimal. However, this algorithm has some limitations, and in some cases does not converge to an optimal fit. This can be assigned to factors such as:

- A bad estimation of the initial parameters;
- A high noise level associated with the experimental values;
- Optimizing the fit for the whole spectrum, does not take in care if the fit is not good in a certain region of the spectrum;

3.3d Physical model

The analysis of the results of the impedance spectroscopy is normal the use of electrical equivalent circuits that can represent the dielectric behavior. The choice of a particular electrical circuit must be made according to the experimental data. The electrical circuit represents a physical model. However, it is important to mention the possibility of different equivalent circuits, fit the same experimental spectrum. Therefore, one should not assume that a given equivalent circuit, which produces a good experimental data adjustment, represents the exact physical model of the sample [EISnet]. The model used in the developed algorithm, shown schematically in fig. 3.3, was chosen based on the characteristics of the experimental spectra of the glasses and glass ceramics studied.

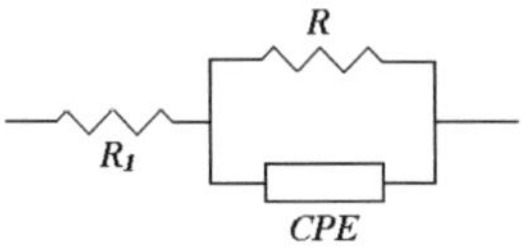

Fig. 3.3. Equivalent electric circuit.

R and R_1 represent electric resistances and CPE a constant phase element. This constant phase element is characterized by keeping constant the angle of the impedance as a function of frequency, i.e. the ratio between the real and imaginary part of the impedance is constant across the all frequency range. The impedance of this intuitive element (Z_{CPE}) can be represented by $Z_{CPE} = \frac{1}{Q_0 (j\omega)^n}$, where Q_0 and n are frequency independent parameters, but usually are temperature dependent. The parameter n varies between 0 and 1, when n = 1 the CPE is reduced to a capacitance element and when n =0 to a resistive element.
However, the main problem in using, the CPE element in a equivalent circuit, which may be associated with the existence of a heterogeneous charge distribution, is that the parameter Q_0, obtained through the iterative process, does not present capacity (Farad) units. This issue was discussed by Hsu and Mansfeld for the case of an electrical circuit formed by the parallel between a resistance and a CPE and they concluded that eq. 8 present the correction of Q_0 for the real capacity closer value.

$$C_{CPE} = C_0 (\omega_{\max})^{n-1} \tag{8}$$

where $\omega_{\max}$ is the angular frequency where max is related with the Z'' peak, which is independent of the exponent n while $C_0 = Q_0 \sin\left(\frac{n\pi}{2}\right)$.

The impedance expression of the CPE element (eq. 9) can be rewritten, taking into account the Euler expressions (eq. 10)

$$Z_{CPE} = \frac{1}{Q_0}\left(\omega e^{j\frac{\pi}{2}}\right)^{-1} = \frac{1}{Q_0}\omega^{-1}\left(\cos\left(n\frac{\pi}{2}\right) - jsen\left(n\frac{\pi}{2}\right)\right) \tag{9}$$

where $n\frac{\pi}{2}$ represents the impedance angle.

$$\begin{cases} j = e^{j\frac{\pi}{2}} \\ e^{j\theta} = \cos(\theta) + jsen(\theta) \end{cases} \tag{10}$$

Through an analysis of the equivalent circuit, the equivalent impedance can be described by equation 11:

$$Z_{eq} = Z_{R_1} + \left[\frac{Z_R . Z_{CPE}}{Z_R + Z_{CPE}} \right] \quad (11)$$

where $Z_{R1}=R_1$, $Z_R=R$ e $Z_{CPE} = \frac{1}{Q_0 (j\omega)^n}$. Therefore, rewritten equation 11:

$$Z_{eq} = R_1 + \frac{R}{1 + RQ_0\omega^n \left(\cos\left(n\frac{\pi}{2} \right) + jsen\left(n\frac{\pi}{2} \right) \right)} \quad (12)$$

The equation 12 represents the function that will be used to fit the experimental data (Z_{eq} represents the Z* function, mentioned previously). It must be noted that this function contains four parameters (R_1, R, Q_0 and n), to adjust.

3.3e Measuring the Thermally Stimulated Depolarization Currents (TSDC)

The measurement of the thermally stimulated depolarization currents (TSDC) enables the study of electrical and dielectric properties of solids as a function of temperature, particularly the phenomena of dipolar polarization and depolarization. It was Bucci and colleagues [chen76], who introduced this technique as a tool for the study of ionic conductivity in materials, been much used in the study of electrical properties of glasses, glass ceramics and polymers.

Heating a dielectric material, polarized by applying an external electric field, the response time of the dipoles and free charges increases allowing to achieve the polarization equilibrium in a very short time. If the applied electric field is maintained during the decreasing of the temperature until a temperature value so low that the relaxation times are very high (order of hours), the electronic configuration can be considered "frozen" and should not change when the electric field is removed. After, when the sample temperature is raised it will occurs a gradual release of charges, giving rise to a electric current, which initially increases with temperature and decreases thereafter with the same due to the reduction of these charges.

This whole process, shown schematically in fig. 3.4, is based on the following steps:

1. apply to the sample, at a polarization temperature (T_p), a static electric field (E), during the sufficient time (Δt) to orient the dipoles in the direction of the field;
2. cool the sample, with the applied electric field applied, until the temperature T_0, where dipolar movements can be considered negligible;
3. Remove the electric field and heat the sample with a constant rate (β). The current from the sample must be then read as a function of the temperature. This current comes from the dipole moments, when they return to their equilibrium sate (often called dipolar relaxation), and from the gradual release of electric charges with the sample temperature increasement.

The most important advantages of this technique are:

- materials that have multiple polarization processes, each process should produce a isolated TSDC peak current TSDC.
- requires no external electrical field applied during the measurement process decreasing, the effects of electrical conduction.

- allows quantitative measures of: impurity concentration, molecular motions, characterization of macromolecular states; characterization of local environments; chemical degradation; microstructure anisotropy.

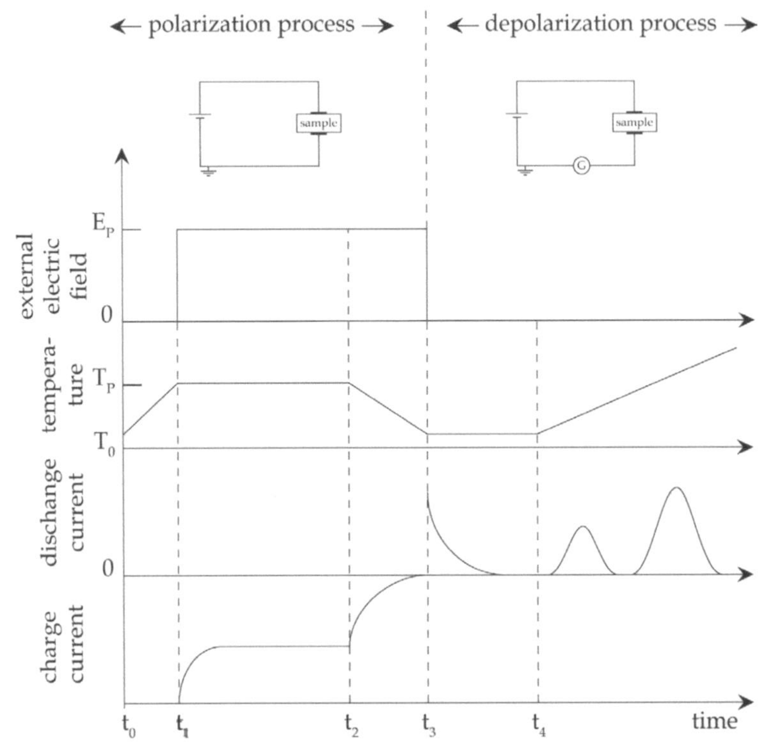

Fig. 3.4. Schematic diagram of the TSDC method.

The main disadvantages of this technique are the high sensitivity and high resolution that is necessary to have in the data acquisition system. The system used in these glass samples consists on a bath cryostat allowing to vary the temperature of the sample between 70 K and 400 K, approximately, in a temperature controller capable of controlling a linear variation of temperature versus time (IT54 - Oxford Research), in a stable voltage source for polarizing the samples (High Voltage supply, model PS325 - SRS), a high sensitive electrometer (Keitley-617, 10^{-14}A to 10^{-4}A) and a register device.

3.3f TSDC analysis (relaxation parameters – calculation process)

Most of the parameters involved in the TSDC process are microscopic, so it becomes impossible to measure them directly. This creates difficulties in linking the results and their analysis. To a first order process, the dependence of the depolarization current, i(T), with temperature is given by equation 13:

$$i(T) = A\exp\left(-\frac{E_a}{k_B T}\right)\exp\left[-B\int_{T_0}^{T}\exp\left(-\frac{E_a}{k_B T'}\right)dT'\right] \tag{13}$$

where Ea is the activation energy, k_B the Boltzmann constant, and A and B constants. However, the mathematical treatment of this expression is a difficult task, being the main problem the integration, which leads to series that converge to infinity, so it is necessary to use approximations.
In the part of the i(T) curve referred to the temperatures below the TSDC peak, and because the heating rate (β) is constant, the integral (eq. 13) is small and can be approximated to zero. Therefore eq. 13 can be simplified (see eq. 14), making it possible to determine the value of the activation energy (Ea) related to the process that gives rise to the peak current.

$$i(T) = Ae^{\left(-\frac{E_a}{k_B T}\right)} \tag{14}$$

This approach is legitimate and proved by several researchers through an asymptotic expansion of the exponential integral, in which for the simple case of the Arrhenius expression is (eq.15):

$$\int_0^T \exp(-y)dy = \frac{\exp(-y)}{y^2}\left(1 - \frac{2!}{y} + \frac{3!}{y^2}\cdots\right) \tag{15}$$

where $y = \frac{E_a}{k_B T}$. Mathematically manipulating equation 15 we can obtain equation 16:

$$\ln(i(T)) = \ln(A) - \frac{E_a}{k_B T} \tag{16}$$

ln(A) represents a constant and E_a the activation energy calculated from the slope of the linear zone in the graphic representation of ln i(T) versus 1/T. This is the method which, although it has the drawback of using just the values of the initial part of the curve, is the most used due to the advantages that presents.
To make the best possible theoretical fitting of the experimental data with the theoretical model represented by eq. 13, it is necessary to solve the integral. According to Chen and colleagues, for the case where $T = T_0 + \beta$, where β represents the linear heating rate, the integral can be written as follows (eq. 17):

$$\int_{T_0}^{T} e^{\left(-\frac{E_a}{k_B T}\right)} dT = F(T,E_a) - F(T_0,E_a) \tag{17}$$

where $F(T,E_a) = \int_0^T e^{\left(-\frac{E_a}{k_B T}\right)} dT$. Since F(T, E_a) is a function that grows considerably with increasing temperature, we can neglect F(T_0,E_a). One way to calculate F(T, E_a) is via an asymptotic expansion of the integral, based on eq. 13. Thus the function F(T, E_a) can be rewritten (eq. 18):

$$F(T,E_a) = T\exp\left(-\frac{E_a}{k_B T}\right)\sum_{n=1}^{N}\left(\frac{k_B T}{E_a}\right)^n (-1)^{n-1}(n!) \tag{18}$$

For N=1, we can write an approximation (eq. 19):

$$i(T) = A\exp\left(-\frac{E_a}{k_B T}\right)\exp\left(-B\frac{k_B T^2}{E_a}\exp\left(-\frac{E_a}{k_B T}\right)\right) \tag{19}$$

Where A, B and E_a (A and B are constants and E_a is the activation energy) are fitting parameters.

We used eq. 19 to fit the experimental spectra of the glasses presented in this chapter. The value of the parameters A, B and E_a were obtained as follows: parameters A and E_a are obtained by linear regression associated with eq.16 model. The parameter B can be calculated, according to J. Vanderschueren and J. Gasiot, based on the peak temperature (T_m) of the depolarization current and the activation energy (E_a) (eq. 20).

$$T_m = \left[\frac{E_a}{k_B B}\exp\left(\frac{E_a}{k_B T_m}\right)\right]^{1/2} \tag{20}$$

The calculus of the relaxation time associated with the TSDC current peak was obtained according to Neagu and co-workers work. They present a method for calculating the relaxation time, assuming that these depolarization processes can be explained based on eq. 22 (another way of writing eq. 13).

$$\tau(T) = \frac{P_0}{j(T)}\exp\left(-\frac{1}{\beta}\int_{T_0}^{T}\frac{dT'}{\tau(T)}\right) \tag{22}$$

where P_0 represents the polarization in the sample during the application of the external electric field, j(T) represents the current density, β the constant heating rate ($\beta = \frac{dT}{dt}$) and T_0 the initial temperature of measurement.

At temperatures below the glass transition, which is our case, we can assume that the dependence of τ with temperature can be described by the Arrhenius equation (eq. 23):

$$\tau(T) = \tau_0 \exp\left(\frac{E_a}{k_B T}\right) \tag{23}$$

τ_0 is the pre-exponential factor and represents the relaxation time at infinite temperature. Thus, knowing that the maximum current occurs when $\frac{d\tau(T)}{dT} = -\frac{1}{\beta}$, expression 24 can be achieved.

$$\tau_0 = \frac{1}{\beta}\frac{k_B T_m^2}{E_a \exp\left(\frac{E_a}{k_B T_m}\right)} \tag{24}$$

Thus, known the value of τ_0, the relaxation time τ can be obtained (eq. 23).

4. Results analysis

4.1 Silicate system

4.1.1 SiO_2-Li_2O-Nb_2O_5 composition

The composition $60SiO_2$-Li_2O-$10Nb_2O_5$ (named in the text as 60SiLi) gives origin to a transparent and yellow sample. This as-prepared sample was heat-treated using the program described in fig. 3.2. The dwell temperatures were (T_P) chosen in accordance to the DTA results. This thermal analysis showed the presence of one exothermic phenomenon center at 710 °C and one endothermic phenomenon at 960 °C. The glass transition temperature is approximately 665 °C. Thus, the HT process was performed with T_P temperatures of 600, 650 and 700°C.

The XRD results (Fig. 4.1) showed the presence in the sample HT at 650 °C of the $LiNbO_3$ and $Li_2Si_2O_5$ crystal phases. The exothermic phenomena detected by DTA should be assigned to this crystallization process. The spectrum of the sample HT at 700 °C shows also the presence of Nb_2O_5 phase.

At a macroscopic level, it was observed that the samples HT at 650 °C became translucid and opaque with the HT at 700 °C. The translucent appearance (and/or opaque) is a optical characteristic that may indicate the presence of particles dispersed in the glass matrix. However, it is important to note that an optically transparent glass may already contain particles dispersed in the glass matrix. According to Todorovic and colleagues, a glass ceramic is transparent when: i) the particles dispersed in the matrix have a minimum size such that the scattering of the visible light, in the particle-matrix boundary, is minimum; ii) the difference between the value of the refractive index of the particles and the glass matrix is minimal. The latter case is visible in transparent silicate glasses containing $Li_2Si_2O_5$ particles, whose refractive index ($n \sim 1.5$) is very close to the index of SiO_2 glass ($n \sim 1.4$). So, knowing that the XRD samples of the as-prepared and HT at 600 °C (fig. 4.1), revealed the presence of $LiNbO_3$ and $Li_2Si_2O_5$ crystalline, it is therefore reasonable to assume that the translucent appearance observed in these samples is related to the presence of $LiNbO_3$ crystallites, which are characterized by having a refractive index of ~ 2.2, quite distant from the SiO_2. The XRD diffractogram of the sample HT at 650 °C shows that the number and intensity of the XRD peaks indexed to the $LiNbO_3$ phase are superior to those related to the $Li_2Si_2O_5$ phase. This suggests the possible presence of a greater amount of $LiNbO_3$ particles, than $Li_2Si_2O_5$, dispersed in the glass matrix. However, the sample treated at 600 °C, which is optically transparent, particles were observed by SEM with an average size of 1 μm, approximately (fig. 4.3), but not detected by XRD (fig. 4.1). This phenomenon may indicate that those particles have a incipient crystallinity or present an amorphous nature.

The Raman spectra of all 60SiLi samples studied is shown in fig. 4.2. Analyzing the spectra it can be observed the existence, in the samples as-prepared and HT at 600 °C, two large bands centered at 870 and 265 cm^{-1}. In the samples HT at 650 and 700 °C it were detected vibration bands at 690, 630, 465, 439, 414, 370, 334, 280, 239 e 165-188 cm^{-1}. According to the literature the bands at 630, 439-437, 370, 335-334, 280, 265, 239 e 180 cm^{-1}, are assigned to vibrations of the NbO_6 octahedrons associated with the $LiNbO_3$ crystal phase.

Most research about the insertion of niobium ions in a glass matrix, is accomplished by comparison with the properties of the crystal whose structure is known, such as $LiNbO_3$. Most of the crystals containing niobium are formed by NbO_6 octahedral with different degrees of distortion. However, the detection of niobium crystals vibrations associated with NbO_4 tetrahedral is rare, justified by the fact that Nb^{5+} presents a size too large to be inserted in a tetrahedron of oxygen ions. Cardinal and colleagues reported that the progressive introduction of Nb_2O_5 in boron-phosphate glasses gives rise to a Raman band close to 900 cm^{-1}, attributed to vibrations of free NbO_6 octahedral and with increasing the concentration of Nb_2O_5 shifts to lower wave numbers, suggesting the formation of a network of NbO_6 octahedrons linked by the vertices. Thus, most authors assign the Raman band between 800 and 940 cm^{-1}, detected in glasses containing niobium, to vibrations of isolated NbO_6 octahedrons. However, this analysis does not consider the possibility of the niobium ion be structurally inserted in the glass matrix as a network former. Studies of Alekseeva and coworkers in glasses of the system K_2O-Nb_2O_5-SiO_2, suggest that for molar quantities of Nb_2O_5 below 20%, the Raman band between 800 and 950 cm^{-1} should be related to vibrations of NbO_4 tetrahedrons. For higher concentrations of Nb_2O_5, the band is assigned to vibrations of NbO_6 octahedrons. Increasing the degree of distortion of these octahedral, the band should shift to higher wave numbers. This band can be also linked to the NbO6 octahedral containing at least one Nb-O terminal bond. Lipovski relate this band to vibrations of non-bridging Nb-O bonds. The same trend was observed in SiO_2:Nb_2O_5 binary glasses, where a certain amount of niobium is introduced as network former. Therefore, the detection in Raman spectra (fig. 4.2) of the band at 870 cm^{-1} indicates that some niobium ions are probably introduced into the glass matrix as network formers. Considering that some niobium ions are in the glass matrix as network modifiers, the increase of the HT temperature promotes a decrease in their number due to the inclusion in the $LiNbO_3$ crystal structure. Consequently the volume ratio between the particles and the glass matrix increases.

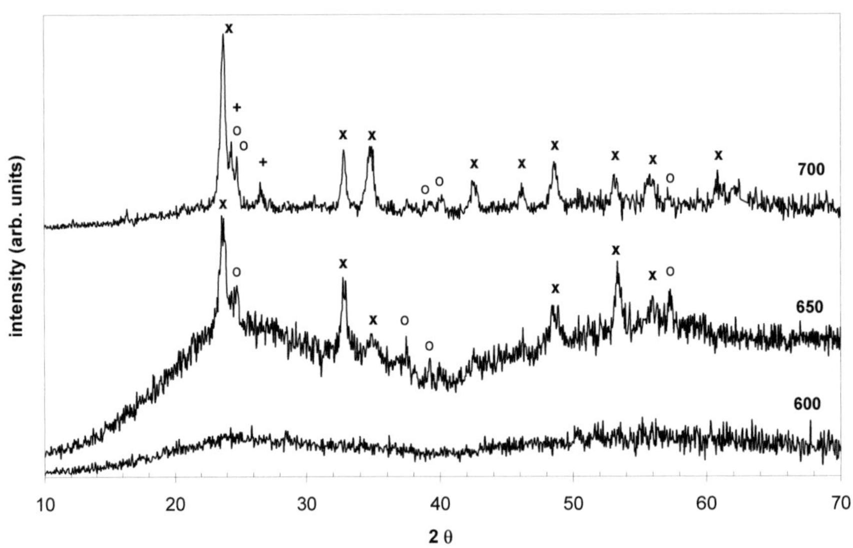

Fig. 4.1 DRX spectra of the 60SiLi heat-treated samples (x $LiNbO_3$; o $Li_2Si_2O_5$; + Nb_2O_5).

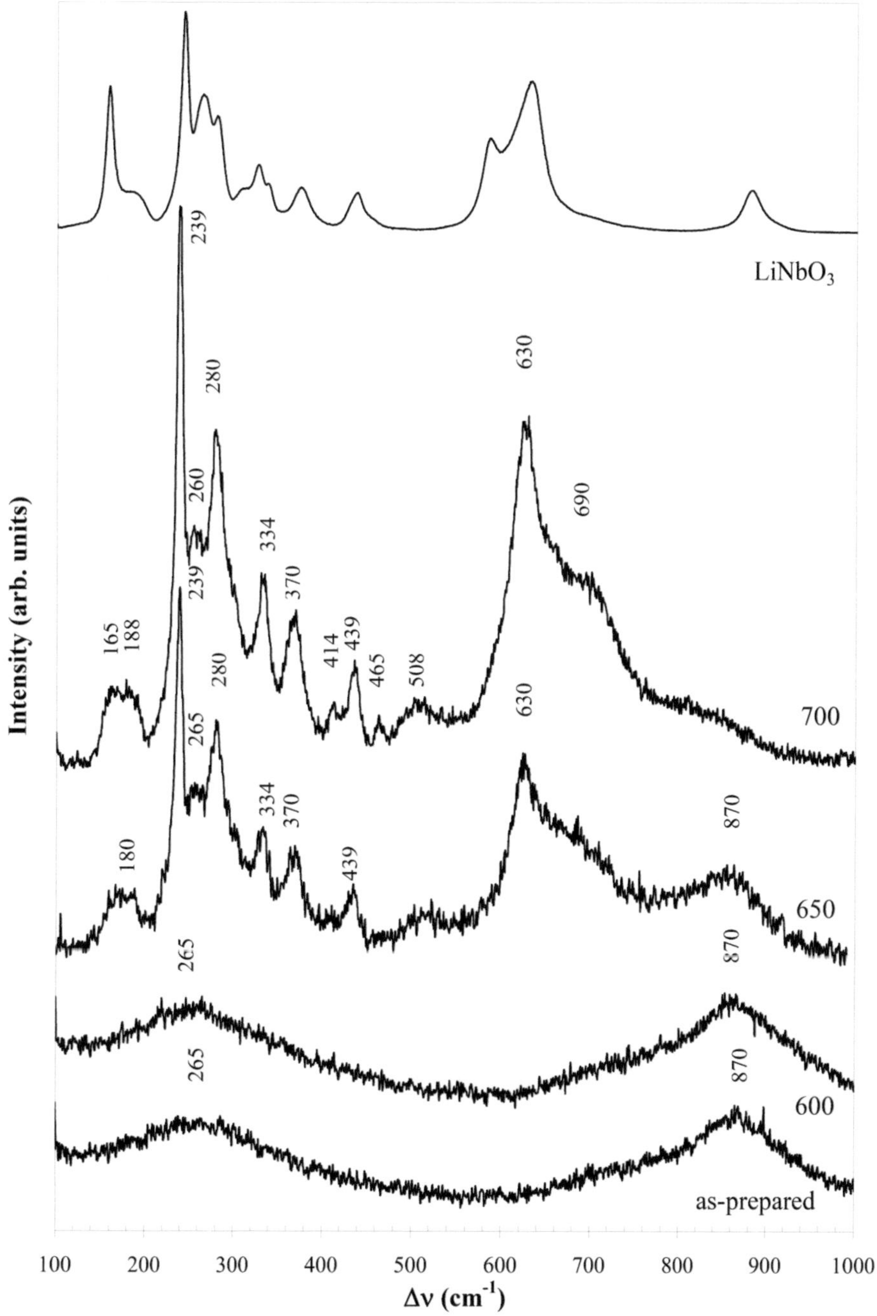

Fig. 4.2. Raman spectra of all 60SiLi samples and of high purity $LiNbO_3$ powder.

The morphological study was complemented using the SEM microscopy. Fig. 4.3 shows the surface micrographs of the samples as-prepared and HT at 650 and 700 °C. The micrographs of the as-prepared samples did not reveal the presence of particles in the free and in the fracture surfaces, as expected. However, with the increase of the HT temperature it was observed an increase in the number of particles in the surface and in the fracture. The size of these particles is between 900 nm and 1μm.

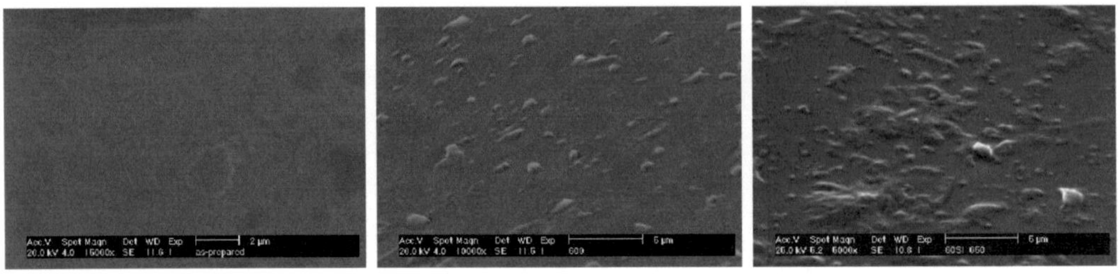

(a) (b) (c)

Fig. 4.3. SEM micrographs of the 60SiLi samples: (a) as-prepared; (b) HT at 600 °C; (c) HT at 650 °C.

The dependence of the dc conductivity (σ_{dc}) with the measurement temperature, for all the samples of this composition is presented in fig. 4.4. It can be observed that σ_{dc} increases, in all samples, with the increase of the temperature. At room temperature, 300 K, it was observed that σ_{dc} decreases with the increase of the treatment temperature (table 4.1). The decreased σ_{dc}, with the increase of the HT temperature can be associated to the formation of the crystalline phases (fig. 4.1). The formation of these phases promotes a decrease in the number of the ions inserted in the glass matrix has network modifier ions (Li^+ and Nb^{5+}) leading to a decrease in the conductivity. On the other hand, knowing that the conductivity of lithium silicate glasses at room temperature, is approximately 10^{-9} S/m and knowing that $LiNbO_3$ presents a high resistivity (table 2.2), and that the conductivity of $Li_2Si_2O_5$ at room temperature is ~ 10^{-12} S/m, it is clear that increasing the amount of crystalline phases, particularly of LiNbO3, it contributes to the increase of the electrical resistivity. The value of the dc activation energy ($E_{a(dc)}$ - table 4.1), calculated using the Arrhenius expression, are similar in the as-prepared and 600 °C HT samples, decreasing when the HT increases to 650 °C. This decrease indicates a diminish in the height of the potential barriers of the glass network, which makes the conduction process less difficult resulting in a higher mobility of the charge carriers. As it can be seen the conductivity decreases, with the increase of the HT temperature, indicating that the factor number of charge carriers will be the dominant factor for this process.

The ac conductivity (σ_{ac}) measured at 1 kHz and 300 K, shows a maximum for the sample HT at 600 °C (table 4.1). This behavior can be discussed using the potential barriers model, which assumes that the ions move by jumps, in a non-random manner. The ac activation energy ($E_{a(ac)}$) calculated through the dependence of the σ_{ac} with the temperature (fig. 4.5) is approximately constant for all samples (table 4.1). Knowing also that the increase of the treatment temperature promotes the formation of crystals dispersed in glass matrix, suggests that the $E_{a(ac)}$ is not very dependent on the presence and/or quantity of particles in the glass matrix. In all samples, increasing the measurement temperature, the σ_{ac} increases. It was also found, in all samples, that increasing the frequency the value of σ_{ac} increases and the value of $E_{a(ac)}$ decreases (fig. 4.6).

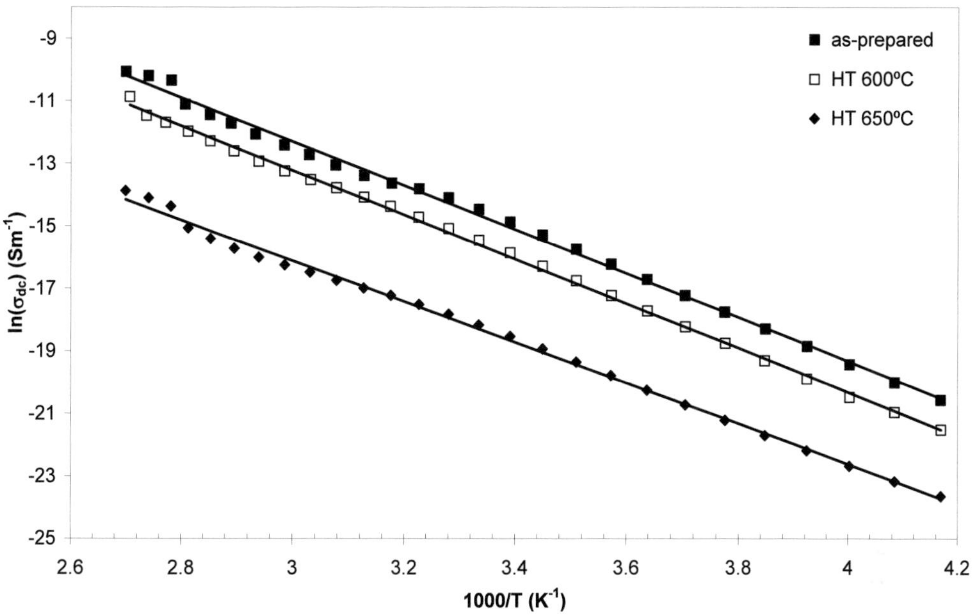

Fig. 4.4. σ_{dc} temperature dependence for all 60SiLi samples.

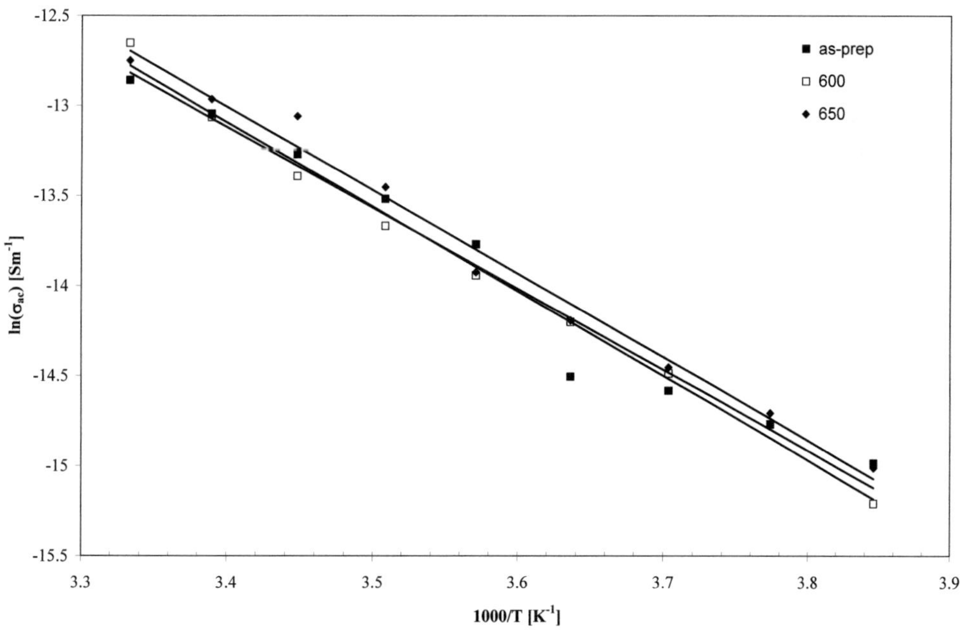

Fig. 4.5. σ_{ac} temperature dependence for all 60SiLi samples, at 1 kHz.

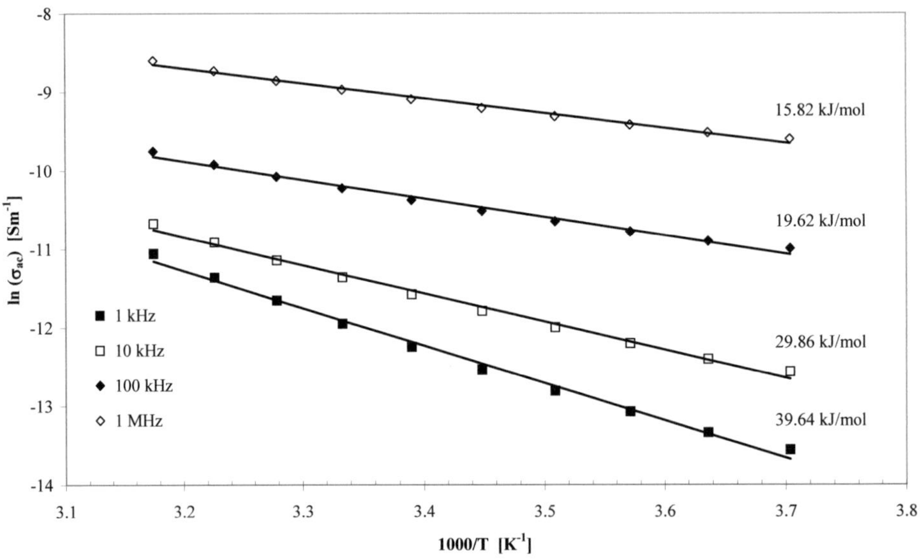

Fig. 4.6. σ_{ac} temperature dependence for all 60SiLi samples, at 1 k, 10 k, 100 k and 1 MHz.

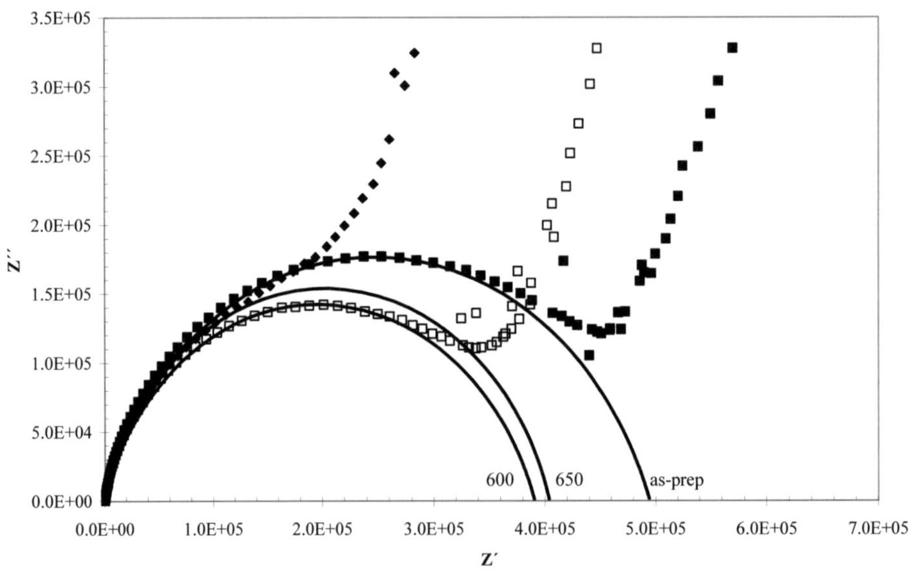

Fig. 4.7. Z´´ versu Z´ spectra for all 60SiLi samples, at 300 K.

Sample	σ_{dc} (x10^{-8}) [Sm^{-1}]	$E_{a(dc)}$ [kJ/mol]	σ_{ac} (x10^{-7}) [Sm^{-1}]	$E_{a(ac)}$ [kJ/mol]
As-prepared	29,4 ± 0,4	58,47 ± 0,66	26,02 ± 0,58	37,37 ± 2,56
600	19,4 ± 0,2	59,03 ± 0,41	31,98 ± 0,71	39,04 ± 1,20
650	1,30 ± 0,01	54,10 ± 0,63	29,01 ± 0,69	38,52 ± 1,62

Table 4.1. dc conductivity (σ_{dc}), at 300 K, dc activation energy ($E_{a(dc)}$), ac conductivity (σ_{ac}), at 300 K and 1 kHz, ac activation energy ($E_{a(ac)}$) of the as-prepared, HT at 600 and 650 °C samples.

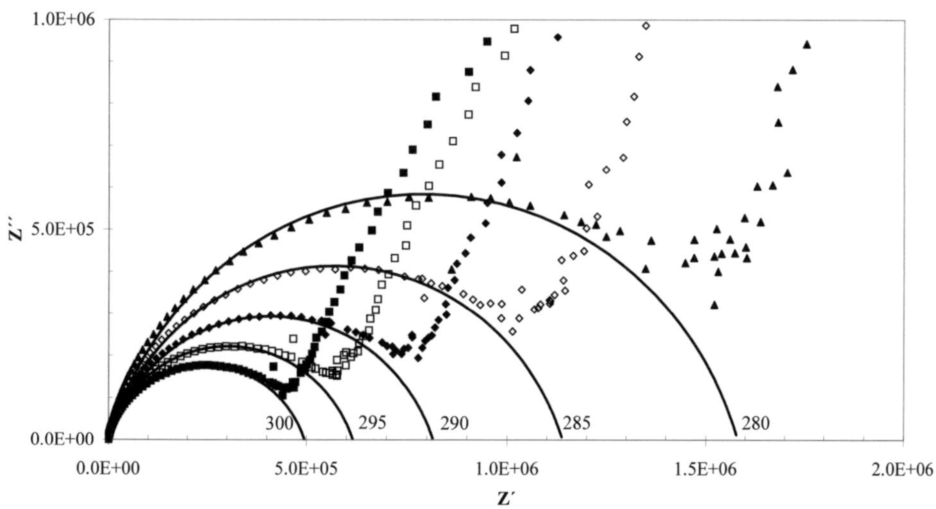

Fig. 4.8. Z´´ versus Z´, of the as-prepared samples at several measuring temperatures.

Figure 4.7 shows the Cole-Cole spectrum, using the impedance formalism (Z*) for all samples, in the frequency range from 10^{-2} Hz and 10^{6} Hz and at 300 K.

This spectrum shows semi-arches whose centers are below the Z' axis, indicating the existence of a distribution of relaxation times, which in these glasses should be related to the presence of various components such as the glass matrix, the crystalline phases and the dipoles from other electrical units, such as the network modifier ions, which contribute to dielectric response. In the low frequencies region (< 10 Hz), and for all samples, a Maxwell / Wagner polarization type is observed and attributed to interfacial polarization between the sample surfaces and the electrodes.

The theoretical fitting of impedance data for all these samples showed that the electrical circuit formed by the parallel between the resistance (R) and the CPE element (fig. 3.3) can describe the experimental data (these results are represented in fig. 4.7 by the lines). However, the correlation between the calculated values and the structural characteristics of the samples is not entirely clear due to the presence of the CPE empirical element, which can be associated with a relaxation time distribution. Typically, this element is related to the presence of electrical inhomogeneities in the sample surface. The main problem in using the

CPE element is that the parameter Q_0 does not have capacity dimensions. In fact, the parameter Q_0, has units of $\Omega^{-1}m^{-2}S^n$, which implies that its physical meaning depends on the value of the exponent n.

For these 60SiLi samples the value of the parameter n is around 0.80, which indicates that the CPE element is approaching a capacitor. It was verified that the behavior of the C_{CPE} capacitor (table 1-annex), with the increase of the HT temperature is similar to the behavior observed for ε '.

The values of the n parameter, which is associated with the Kohlrausch exponent (β) by the expression β =1-n are in all the samples and measuring temperatures near 0.80 (table 1-annex), which shows a minimal dependence with the presence of crystalline phases. According Guiyan and coworkers the value of n is associated with the surface characteristics of the samples. Thus, we suggest that the surface characteristics have a significant influence on the dielectric behavior. Ngai states that the parameter n should increase with increasing ion-ion interactions. Consequently, in these samples, this type of interaction does not undergo major changes for the different HT conditions. In all samples the value of the parameter R decreases with increasing temperature measurement (fig. 4.8), which agrees with the dependence of the dc and ac conductivity with temperature, indicating that the increase in temperature promotes an increase in the mobility of the charge carriers. This mobility increasement is related to the observed decrease in the τ_Z, revealing an increase in the ability of the electric units to accompany the dielectric field ac.

To study the evolution of the relaxation time (τ) with the temperature, it was used the representation of the dependence of the complex part of the impedance (Z'') with the frequency. This representation facilitates the visualization because the value of τ, is related with the frequency value at which occurs the maximum of Z''. Figures 4.9 and 4.10 show this trend.

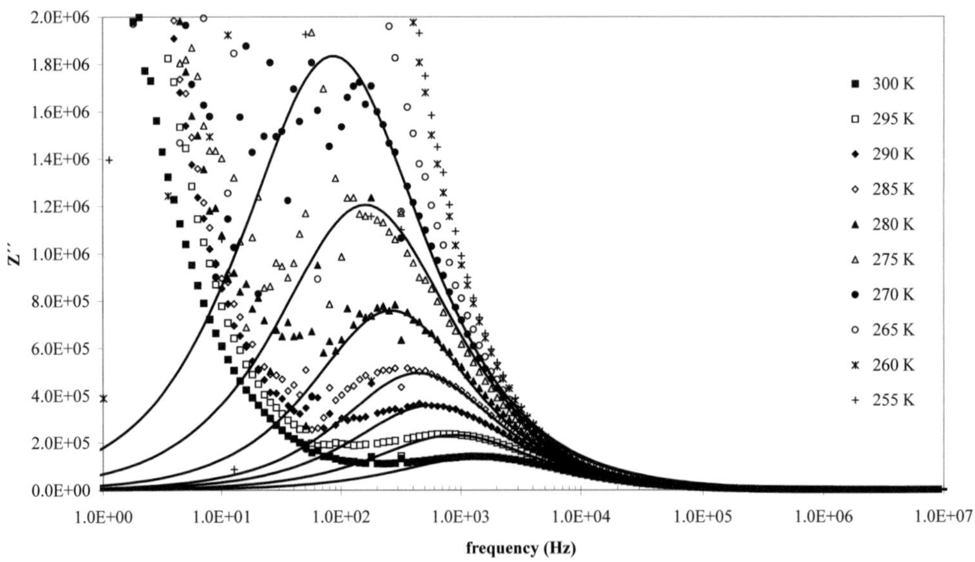

Fig. 4.9. Temperature dependence of Z´´ with frequency for several measuring temperatures (sample HT at 600 °C).

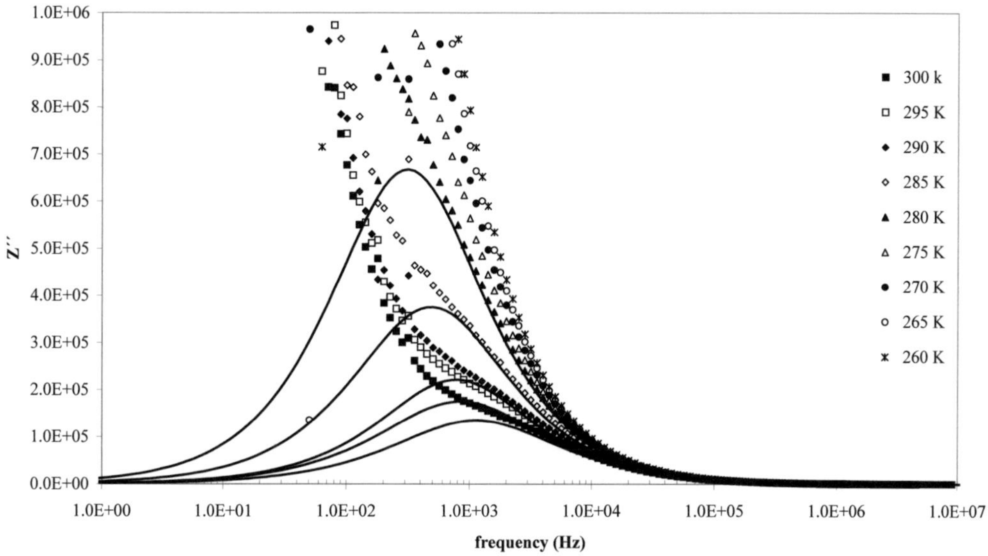

Fig. 4.10. Temperature dependence of Z´´ with frequency for several measuring temperatures (sample HT at 650 °C).

Based on the impedance data the dielectric permittivity was calculated. Table 1-annex shows the values of ε* and of the dielectric loss (tan δ) for all temperatures of measurement. In table 4.2 we highlight the values of the dielectric permittivity and dielectric loss measured at 300 K and 1 kHz. Based on those values it is observed that the value of dielectric constant (ε ') increases with the increase of the HT temperature. The sample HT at 650 °C presents the lowest value of dielectric loss.

Sample	ε´	ε´´	**tanδ**
As-prepared	30,23 ± 0,67	46,24 ± 1,03	1,53 ± 0,05
600	35,07 ± 0,78	56,84 ± 1,26	1,62 ± 0,05
650	48,20 ± 1,15	51,55 ± 1,23	1,07 ± 0,04

Table 4.2 Real and imaginary part of the dielectric permittivity and the dielectric loss value measured at 1 kHz and 300 K.

Figures 4.11 to 4.13 show the TSDC spectra of the as-prepared sample and of the samples HT at 600 and 650 °C, respectively. For a better view of the TSDC spectrum, is presented in each figure a zoom of the low temperature zone. All samples were polarized at the temperature of 350 K for 10 minutes, applying an electric field of 50 kV/m. In all spectra, the lines represent the theoretical fitting based on the process described in section 3.3.3.a.
The as-prepared glass presents only one depolarization peak thermally activated while the samples heat treated show two peaks, the first center at low temperatures (< 230 K) and the second at higher temperatures (> 290 K). The characteristics of these peaks, visible in the figures, are registered in table 4.3. It is observed that the temperature of the peaks shift to

higher temperatures with the increase of the thermal treatment temperature. The current value of both peaks, also decreases from the as-prepared sample to the sample HT at 650 °C. The low temperature TSDC peak (P1) is attributed to localized movements of the network modifier ions. This movement gives rise to a orientated dipole formed by the network ion (Li + and / or Nb5 +) and the non-bridging oxygen (NBO-) to which it is connected. Note that in this situation the network modifier ion, structurally inserted in the glass matrix, is not in the more favorable energetically position. According to Agarwal and colleagues, applying an electric field it results in a ion transfer in the field direction, thus creating electrical heterogeneity, i.e., electric dipoles. With increasing temperature, these ions tend to return to its original position thereby creating a depolarization current that give origin to the low-temperature peak (P1). The activation energy associated with the orientation of this dipoles consists mostly on the energy required for the movement of these ions to different positions within the hollow structure in which it is.

The values obtained for the activation energy associated with P1 (table 4.3), are lower than the values obtained for the dc process ($E_{a(dc)}$ - table 4.1) but close to those obtained for ac conductivity ($E_{a(ac)}$ - table 4.1) which suggests a possible relationship. Thus, the decrease in the P1 peak current (table 4.3), with the increase of the HT temperature, is due to the decrease of the number of dipoles related with the network modifier ions structurally inserted in the glass network. The temperature of the peak current corresponds to the transition temperature of the "freezing" state (non-equilibrium state), to the equilibrium state, analogous to what occurs in the glasses with the glass transition temperature. The shift of the temperature of the peak P1, with the increase of the HT temperature (table 4.3) should be, according to Agarwal results, associated with the reduction of the network modifier ions number.

In this type of glasses, containing alkali ions, a second depolarization peak, centered at higher temperatures are attributed to dipolar depolarization from the movements of the charge carriers in limited pathways, which result from the presence in the glass of micro-heterogeneities. According to Hong these movements are mainly associated with the jumps of the network modifier ions within, between different positions, in the structural vacancies where it is. Thus, the dipoles formed between these ions and the NBO closer to each ion, it must produce a higher activation energy when compared with the one related with P1. This movement is generally associated with the dc conductivity, resulting in a spatial polarization (space-charge polarization). However, Agarwal states that this process must have activation energy lower than the dc activation energy, which in this case it was not verified. Thus, based on the values of the activation energy, we do not ascribe the peak P2 to space-charge depolarization but to the depolarization of the interfacial dipoles formed between the sample surfaces and the electrodes. The decrease of the temperature at which the peak P2 is detected, with the increase of the HT temperature, can be related to the observed decrease of the dielectric loss factor, which should reveal an increase in the number of dipoles. It is suggested that, in the as-prepared sample, this P2 peak should be detected at temperatures higher than the upper limit of our measurement system. Besides the P2 peak and with increasing temperature, is observed in all samples, an increasing of the current. This current, commonly referred as background current, was fitted to an Arrhenius expression (table 4.4) allowing to obtain the activation energy of this process, which decreases with the increase of the HT temperature (similar to that observed in the dc conductivity), indicating that in this temperature range the charge carriers present a higher mobility.

Analysing the TSDC relaxation times, it was assumed that the value of τ, calculated for each current peak, which corresponds to the average value of the distribution of relaxation times, it is verified that the value of τ_{P1} (table 4.3) presents a maximum for the as-prepared sample. We believe that this maximum can be associated with the fact that this sample has the largest number of network modifier ions structurally inserted in the glass matrix and therefore a larger distribution of relaxation times is expected. The increase of τ_{P1}, from sample HT at 600 to the sample HT at 650 °C, is related to the presence of the LiNbO3 crystal whose dipoles are difficult depolarization, and therefore a higher relaxation time should be measured. The decrease of τ_{P2} (table 4.3), with the increase of the HT temperature, suggests that the dipoles that give rise to this peak are not related with the $LiNbO_3$ crystal, because, knowing that the volume ratio between these crystals and the glass matrix increases with the increase of the HT

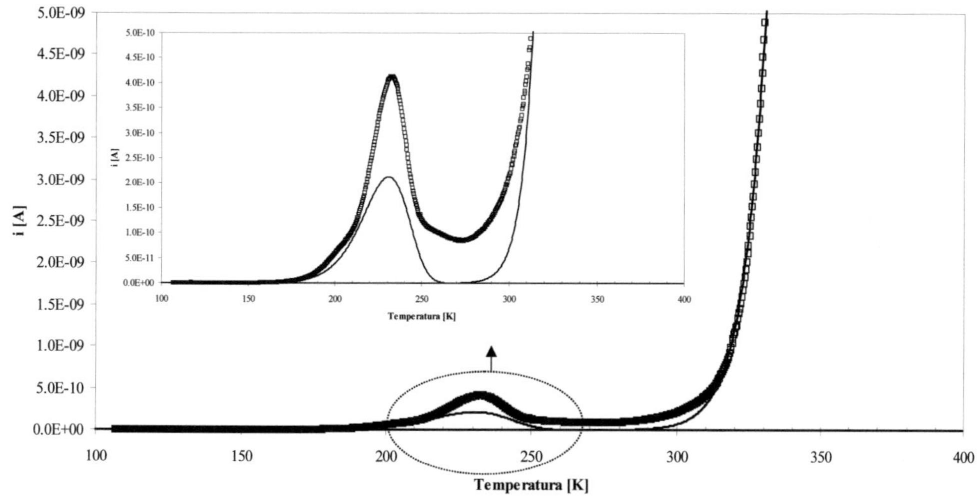

Fig. 4.11. TSDC current in function of the temperature for the as-prepared glass.

sample	T_{P1} (K)	$i(T_{P1})$ ($x10^{-10}$) (A)	E_a [kJ/mol]	$\tau_{0\,(P1)}$ ($x10^2$) [s]	$\tau_{(P1)}$ ($x10^3$) [s]
As-prepared	232,3	4,134	31,574 ± 0,198	4,159 ± 0,065	2,132
HT at 600°C	221,4	2,243	44,123 ± 0,265	1,262 ± 0,025	1,385
HT at 650°C	203,7	0,138	31,259 ± 0,151	2,626 ± 0,049	1,659
	T_{P2} (K)	$i(T_{P2})$ ($x10^{-10}$) (A)	E_a [kJ/mol]	τ_0 (P_2) ($x10^2$) [s]	$\tau_{(P2)}$ ($x10^3$) [s]
As-prepared					
HT at 600°C	352,3	17,883	84,081 ± 0,959	1,044 ± 0,014	1,841
HT at 650°C	261,4	0,948	50,569 ± 1,843	1,646 ± 0,027	1,685

Table 4.3 Peak temperature (T_p), peak current ($i(T_p)$), activation energy (E_a), pre-exponential parameter (τ_0) and the relaxation time (τ), calculated at the low (P_1) and high (P_2) peak temperatures.

temperature and that these crystals are difficult to depolarization, the relaxation time would be larger than the τ_{P1}. Thus, this relaxation time (τ_{P2}) should, as indicated, be associated with the depolarization produced by interfacial dipoles sample surface-electrode.

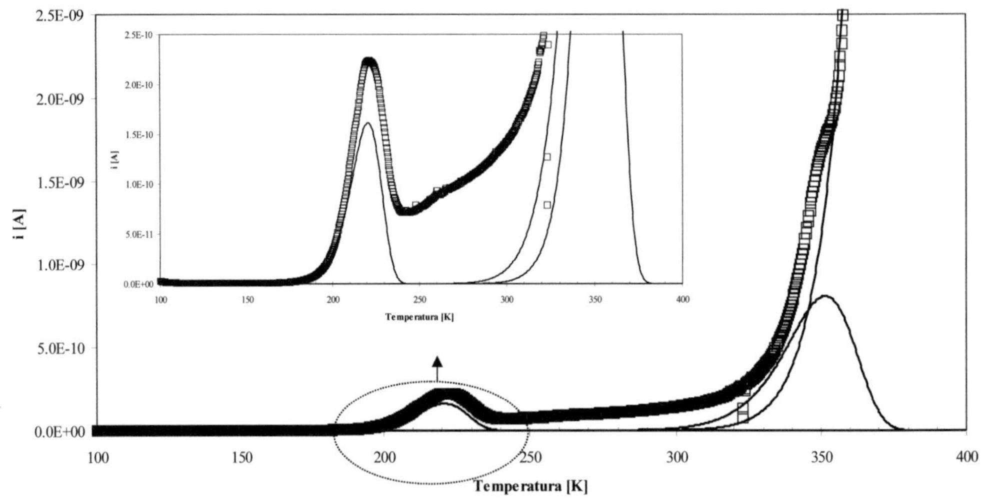

Fig. 4.12. TSDC current in function of the temperature for the sample HT at 600 °C.

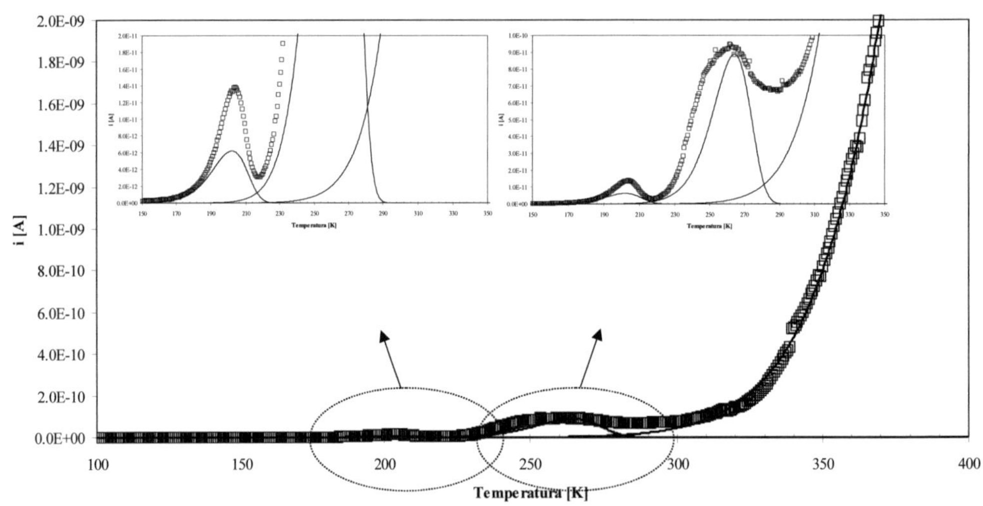

Fig. 4.13. TSDC current in function of the temperature for the sample HT at 650 °C.

sample	ln (A) [A]	Ea [kJ/mol]
As-prepared	22,928	115,775 ± 0,637
600	16,609	108,232 ± 0,936
650	-3,621	80,431 ± 0,605

Table 4.4. Parameters of the Arrhenius equation used in the adjust of the TSDC high temperature range. ln (A) represents the logarithm of the pre-exponential factor and Ea the activation energy for the as-prepared glass and samples heat treated at 600 and 650 ° C.

4.1.2 SiO_2-Na_2O-Nb_2O_5 composition

The DTA result of the 60SiO2-30Na2O-10Nb2O5 (60SiNa) glass did not reveal the presence of exothermic effects, indicating the occurrence of crystallization processes. Thus, the first temperatures chosen for the HT were similar to those used in the 60SiLi composition. Therefore in this composition the HT were performed at the temperatures of 650, 700, 750 and 800 ° C. At the macroscopic level it was observed that with the increase of the HT temperature, the samples initially transparent became translucent (sample HT at 700 °C) and after they became opaque with the treatments at 750 and 800 °C. This opacity is due to the formation of a white film on the sample surface. Through mechanical polishing this film can be removed, resulting in a glass, similar to the as-prepared. The XRD spectra (fig. 4.14) highlight the presence in the sample HT at 800 °C, the characteristic peaks of the naNbO3 crystalline phase. The main peak of this phase (2θ ~ 32.8) is already detectable in the sample treated at 750 °C. In the remaining samples, treated at 650 and 700 ° C, this peak is not well detected. The XRD of the samples treated at 750 and 800 °C, after being polished in order to remove the white film (samples 750B and 800B), are similar to the spectrum of the sample

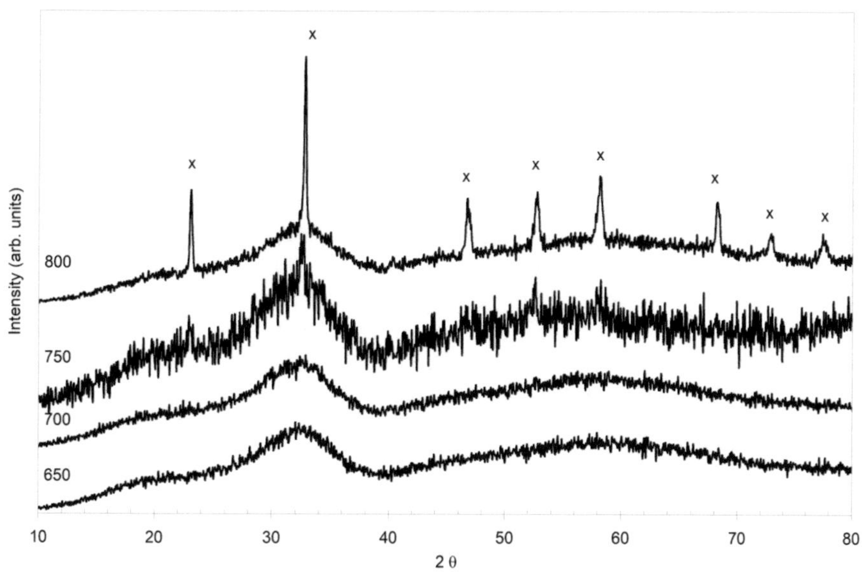

Fig. 4.14. XRD spectra of the 60SiNa samples treated at 650, 700, 750 and 800 °C (x $NaNbO_3$).

HT at 650 °C. Thus it was found that the white film is due to the crystallization of NaNbO3 crystals. In these samples the XRD peak width at half height decreases with the increase of the HT temperature, indicating that the amount of crystalline phase precipitated in the glass matrix increases with the increase of the treatment temperature.
Figure 4.15 shows the Raman spectra of all samples. In the samples spectra it was observe the existence of a band at 900-905 cm^{-1}. A broad band centred at 288 cm^{-1}, which disappears with the increase of the HT temperature, was observed in the as-prepared and 650 °C HT samples. The spectrum of the sample HT at 700 °C shows a band at 635-637 cm-1, which increases in intensity with the increase of the HT temperature. Similar behavior was observed for the band at 257-260 cm-1. In the sample HT at 750 °C, Raman bands were detected at 291, 148 and 124 cm-1. In the Raman spectrum of the sample HT at 800 °C were observed all bands detected in the sample HT at 750 °C and also bands at 600, 455, 240, 215, and 192 cm-1.
The SEM micrographs (fig. 4.16) showed the presence of particles with a size of about 200 nm dispersed on the surface of the sample HT at 650 °C. These particles, with a cubic morphology increase in size and number with the increase of the HT temperature up to 750 ° C. The sample treated at 800 °C shows, comparatively to the sample treated at 750 °C, a decrease in the number of particles but an increase in their size. These particles were only detected in the samples surface
Figure 4.17 presents the optical appearance of 750 and 800 °C HT samples cross section. In these images it is visible that the optical thickness of the white film increases with the increase of the HT temperature.
The dependence of the dc conductivity (σ_{dc}) with temperature for all samples is shown in figures 4.18 and 4.19. In all samples, there is an increase in the σ_{dc} with the increase of the measurement temperature, characteristic of a thermal stimulated process. The σ_{dc} measured at 300 K (table 4.5), decreases with the increase of the heat-treatment temperature up to 750 °C. The sample treated at 800 °C showed a σ_{dc} increasement. The σ_{dc} registered for the samples 750B and 800B (polished) is greater than that recorded for the samples HT at 750 and 800 °C.
The increase of the measurement temperature promotes an increase in energy of the charge carriers, which makes their jumping movement (hopping) through the potential barriers associated with the glass matrix easier, justifying the increase of the conductivity. At 300 K the decrease of σ_{dc} with the increase of the HT temperature up to 750 °C (table 4.5), can be attributed to the decrease of the number of network modifier ions structurally inserted in the glass matrix (Na^+ and Nb^{5+}), main responsible for the charge carrier. It must be noticed that the non-observation in the Raman spectrum, of the band between 800-850 cm-1 (fig. 4.15), associated with the vibrations of Nb-O-Si bonds indicates that the niobium ions are inserted in glass matrix as network modifiers. The Raman spectrum of the crystalline $NaNbO_3$ at room temperature, is characterized by the presence of bands associated with the vibration of octahedral NbO_6 units, linked by the vertices, which is the case of the band at 630 cm^{-1} observed in the spectrum of the sample HT at 700 °C. This band, attributed to vibrations of NbO_6 octahedral connected by Nb-O-Nb bonds, indicates the early formation of the $NaNbO_3$crystalline phase. The band in the region 900-905 cm-1 observed in all samples of the composition 60SiNa, but not detected in the spectrum of the $NaNbO_3$ crystal, is typical of the vibrations of NbO_6 isolated units. The decrease in the intensity of these bands, with the increase of the HT temperature, shows a progressive disappearance of these isolated units. The bands at 192, 148 and 124 cm^{-1}, observed in the samples HT at 750 and

800 °C are assigned to NbO_6 vibrations in the orthorhombic phase. The bands at 288 and 257-260 cm^{-1} are also assigned to vibrations of NbO_6 octahedral. The increase in the intensity of these bands, with the increase of the HT temperature, is associated with the formation of $NaNbO_3$ crystallites.

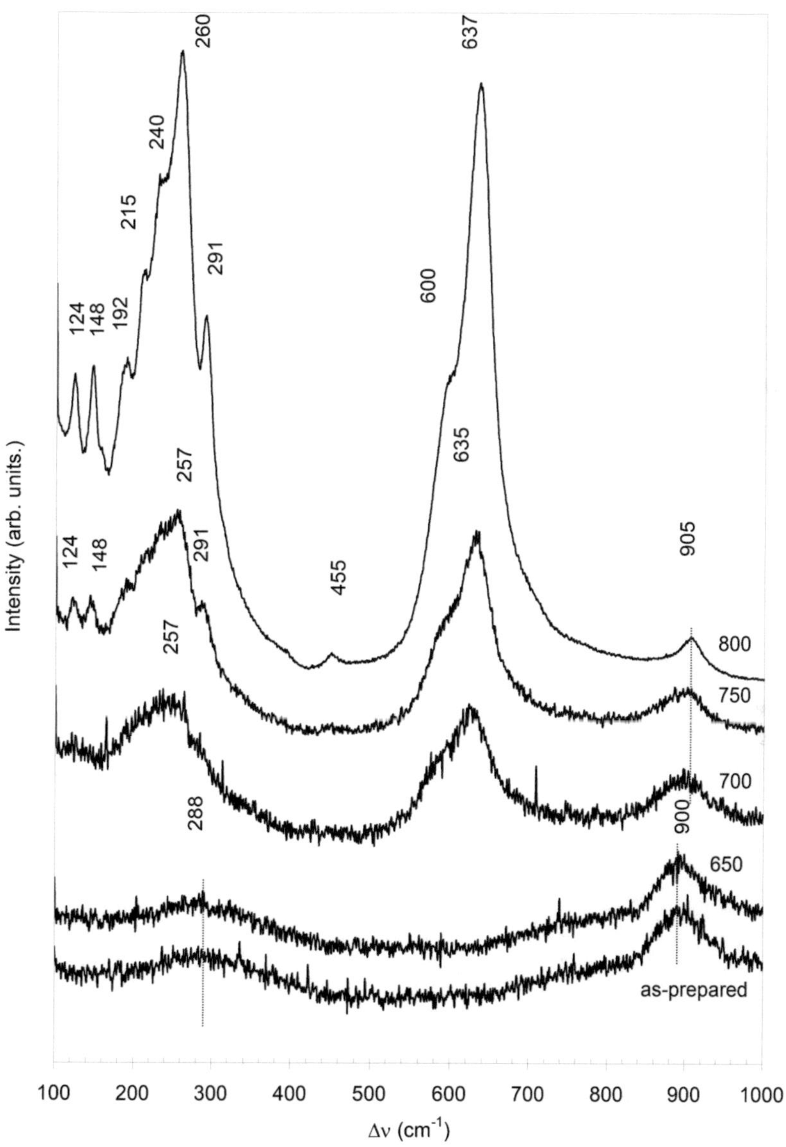

Fig. 4.15. Raman spectra of the 60SiNa samples.

The calculated $E_{a(dc)}$ showed that it decreases with the increase of the HT temperature up to the sample HT at 750 °C. The observed decrease of the $E_{a(dc)}$, with the increase of the HT temperature, indicates a decrease in the height of the potential barriers, which strengthens the proposed justification for the main responsible factor for the σ_{dc} behavior, which is the number of network modifier ions structurally inserted in the glass matrix. The σ_{dc} is higher in the sample treated at 800 °C than in the sample treated at 750 °C indicating the presence of a higher number, in the sample HT at 800 °C, of charge carriers. The SEM micrographs (fig. 4.16) shows that the particles observed in the sample surface of the sample HT at 800 °C, when compared with the surface of the sample HT at 750 °C, presents a higher size but in a lower number. A comparative analysis between these two samples suggests that the

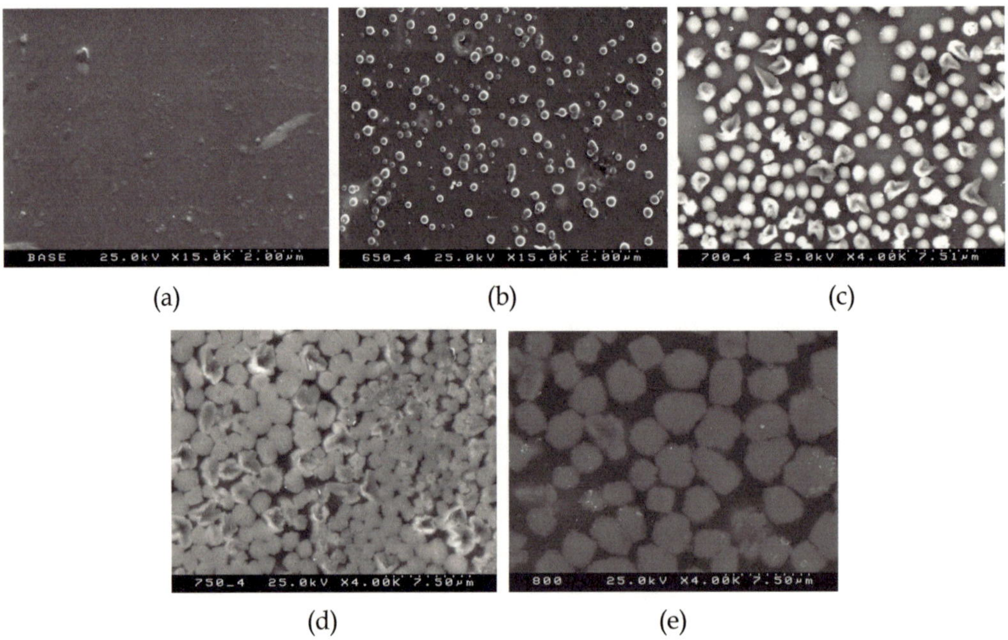

Fig. 4.16. SEM micrographs of: (a) as-prepared; (b) HT 650; (c) HT 700; (d) HT 750; (e) HT 800 °C.

Fig. 4.17. Optical photographs of the cross section of the 60SiNa samples surface treated at 750 and 800 °C (a - crystalline zone (film), b - glass matrix). 20x magnification.

volume ratio between particles and glass matrix decreases from the sample HT at 750 to the sample HT at 800 °C. Thus, the sample HT at 800 °C, should have a greater number of network modifier ions, structurally inserted in the glass matrix, justifying the increase in the conductivity and as well in the dielectric constant value (table 4.6). The samples that, after the heat treatment, were polished in order to remove the white surface film (sample 750B and 800B), show an increase in the dc conductivity from the sample 750B to the sample 800B, suggesting the existence of a higher number of charge carriers in the second sample. When compared with the correspondent samples, not polished, the 750B and 800B samples show a higher conductivity which can be related with the absence on the surface of $NaNbO_3$ crystals which will facilitate the conduction process. The samples 750B and 800B have a $E_{a(dc)}$ similar to the sample HT at 750 °C but below the value obtained for the sample HT at 800 °C. The ac conductivity (σ_{ac}), at 300 K and 1 kHz, decreases with the increase of the HT temperature up to 750 °C (table 4.5). Samples 750B and 800B have similar σ_{ac} but different from the value measured in the correspondent samples without being polished. With the increase of the measuring temperature there is, in all samples, an increase in the σ_{ac}. The ac activation energy ($E_{a(ac)}$), calculated based on the dependence of $\ln(\sigma_{ac})$ with $1/T$ (fig. 4.20), is maximum for the as-prepared sample. The samples HT at 650 and 700 °c have similar values of $E_{a(ac)}$. The same behavior was observed for the samples HT at 750 and 800 °C (table 4.5). The samples 750B and 800B present values of $E_{a(ac)}$ similar but higher than the correspondent samples without the surface polishing.

Figure 4.20 shows the variation of ac conductivity with the temperature of measurement for the as-prepared sample, at several frequencies. It is observed that with the increase of the frequency the value of σ_{ac} increases and the value of $E_{a(ac)}$ decreases. This behavior is similar to all the other samples.

In materials where the conduction process is essentially the ionic, the ac conduction mechanism can also be reproduced by the "jump" model of the charge carriers over potential barriers. From the dependence of the σ_{ac} with the temperature of measurement (fig. 4.20), we obtained the values of Ea(ac) and it was observed that it decreases from the as-prepared sample to the sample HT at 650 °C (table 4.5). In the other samples the value of $E_{a(ac)}$ is very similar, i.e., it varies between 33 and 35 kJ/mol. This indicates that with the increase of the thermal treatment temperature, there is an initial decrease in the height of the potential barriers, which then remain almost independent of the treatment temperature. From this we can deduce that the samples heat treated, the network modifier ions structurally inserted in the glass matrix present a similar mobility suggesting that the observed decrease of σ_{ac}, with the increase of the HT temperature, is due to the decrease of those ions number. It is also verified that in each sample the value of $E_{a(ac)}$ is less than the value of $E_{a(dc)}$ for the same sample. This shows that the electrons have to overcome different potential barriers when they relax (ac process) or when they participate in a long range process (dc process).

The optical photographs (fig. 4.17) show that increasing the thermal treatment temperature promotes an increase in the thickness of the surface film. These structural characteristic suggests that the electrical equivalent circuit model, shown in figure 4.22, can describe the electrical and dielectric properties of these samples.

In the dielectric analysis, the electric circuit comes down to a combination of three capacitors in series: two related with the sample surfaces and the third with the bulk characteristics. Knowing that the thickness of the samples are about 1.0 mm, the thickness of the film is about 150μ m (observed in sample HT at 800 °C), the dielectric constant (ε ') of $NaNbO_3$ is

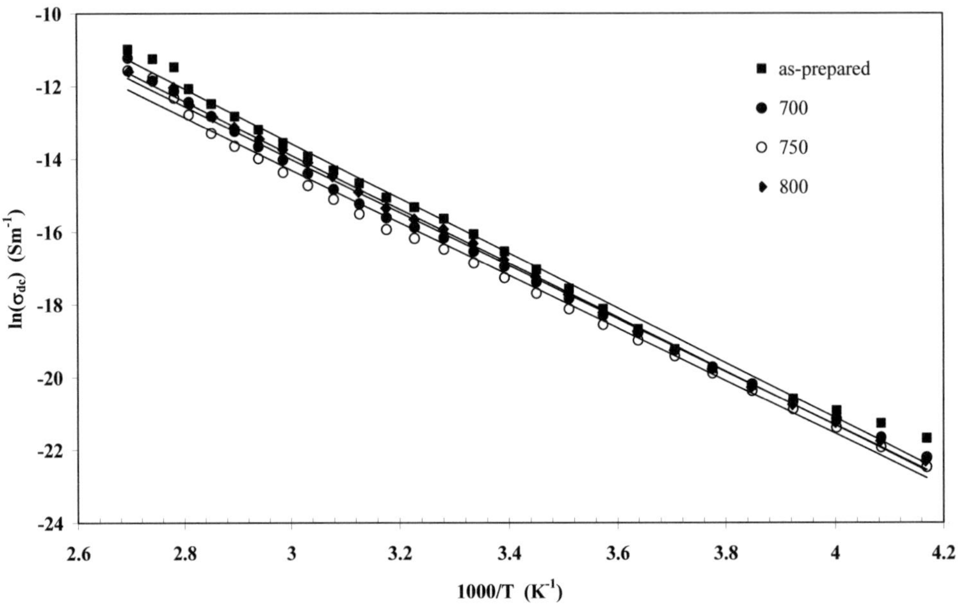

Fig. 4.18. ln(σ_{dc}) versus 1000/T for the 60SiNa samples: as-prepared and HT at temperatures between 650 and 800 °C.

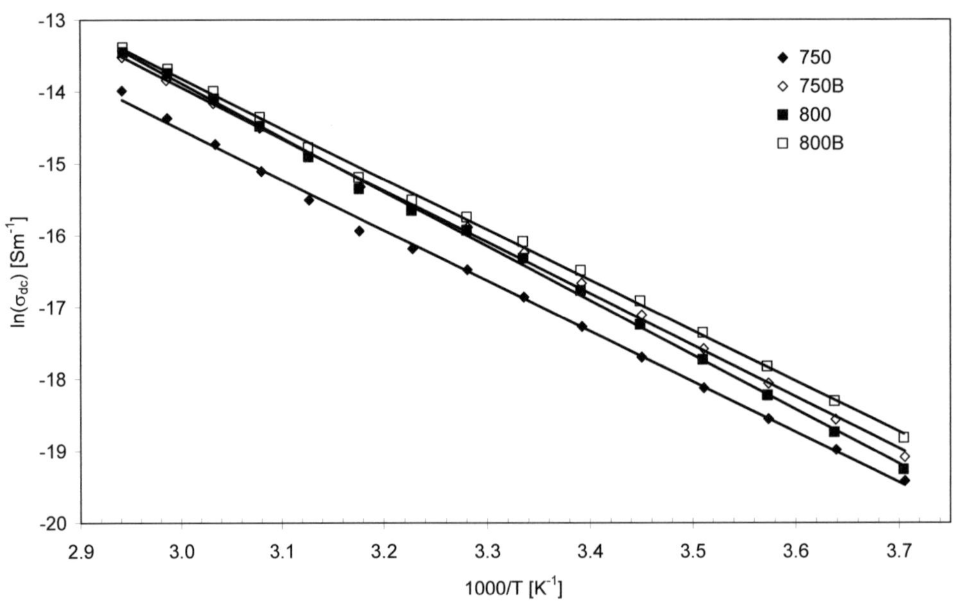

Fig. 4.19. ln(σ_{dc}) versus 1000/T for the 60SiNa samples HT at 750 and 800, with and without the white film (this samples are named 750B and 800B, respectively).

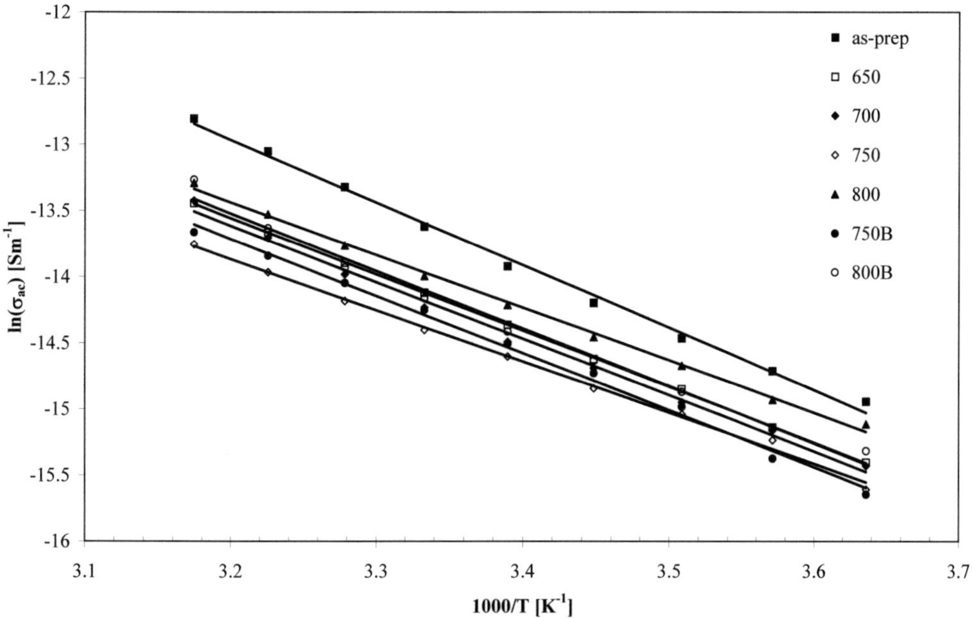

Fig. 4.20. ln(σ_{ac}) versus 1000/T for all the 60SiNa composition samples.

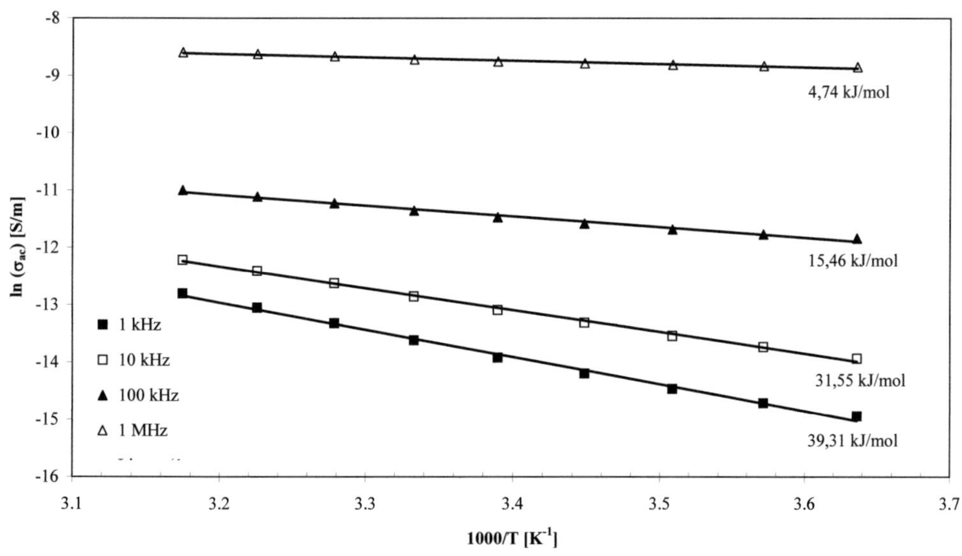

Fig. 4.21. ln(σ_{ac}) versus 1000/T for the as-prepared sample, at several frequencies.

Sample	σ_{dc} (x10^{-8}) [Sm^{-1}]	$E_{a(dc)}$ [kJ/mol]	σ_{ac} (x10^{-7}) [Sm^{-1}]	$E_{a(ac)}$ [kJ/mol]
as-prepared	10,55 ± 0,24	64,63 ± 0,91	12,2 ± 0,27	39,31 ± 0,98
650	7,25 ± 0,27	62,48 ± 0,74	7,37 ± 0,28	35,27 ± 0,17
700	6,56 ± 0,16	59,65 ± 0,71	6,59 ± 0,16	35,43 ± 0,97
750	4,76 ± 0,15	58,34 ± 0,67	5,57 ± 0,17	32,13 ± 0,66
800	8,17 ± 0,24	62,90 ± 0,67	8,36 ± 0,28	33,05 ± 0,64
750B	8,82 ± 0,23	59,63 ± 0,59	6,47 ± 0,17	35,83 ± 0,96
800B	10,30 ± 0,33	58,38 ± 0,63	7,07 ± 0,22	36,12 ± 1,50

Table 4.5. dc conductivity (σ_{dc}), at 300 K, dc activation energy ($E_{a(dc)}$), ac conductivity, at 300 K and 1 kHz, ac activation energy ($E_{a(ac)}$) for all 60SiNa samples.

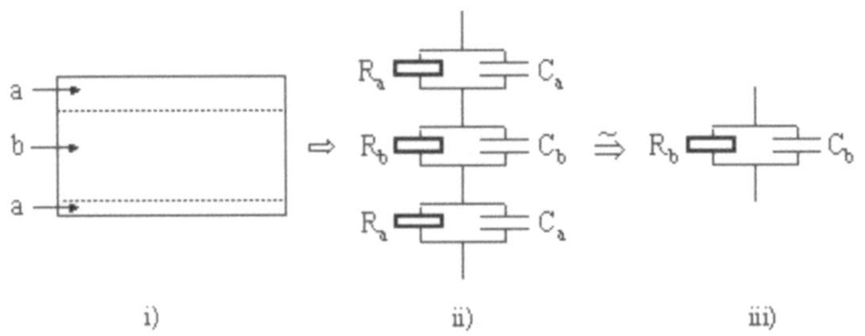

Fig. 4.22. Equivalent electric circuit model: i) sample (a – surface; b – bulk); ii) electric model where R_a and C_a represent the resistance and the capacity related with the sample surface characteristics, R_b and C_b the resistance and capacity related to the bulk characteristics; iii) approximate model.

approximately 600 at 300 K and 1 kHz, being much higher than the ε ' of the bulk sample zone, assumed to consist mainly on glass matrix ($\varepsilon'_{SiO_2} \approx 4.0$), it is reasonable consider that the dielectric behavior can be controlled by the characteristics of the bulk. Briefly, the analysis of the electrical circuit consisting of three capacitors in series shows that the equivalent capacity is $C_{eq} = \dfrac{C_a C_b}{C_a + 2C_b}$. If we consider that $C_a >> C_b$ than $C_{eq} \approx C_b$.

The decrease of the dielectric constant (ε'), with the increase of the HT temperature up to 750 ° C (table 4.5), can be related to the decrease of the number of dipoles existing in the bulk. The similarity between the values of ε' of the samples treated at 750 and 800 °C with and without the surface film indicates that the surface film, associated with $NaNbO_3$ crystal precipitation does not exert great influence on the dielectric properties, which corroborates the equivalent electrical circuit model proposed (fig. 4.22). The dielectric constant of the sample HT 800 °C is, at room temperature, the highest of all samples of this composition (table 4.5). This result should be expected due to the $NaNbO_3$ layer thickness increase, decreasing the thickness of the bulk zone and therefore increasing the capacity. Furthermore, according to the analysis of the dc and ac conductivities, it is assumed the existence in the bulk zone of the sample HT at 800 °C of a larger number of network modifier ions, structurally inserted in the glass network, resulting in an increase of the ε' value. The remotion of the white surface film of the sample treated at 800 °C leads to a decrease of ε ', justified by the sample thickness decreasement.

Sample	ε´	ε´´	tanδ
As-prepared	25,21 ± 0,57	21,86 ± 0,49	0,87 ± 0,03
650	23,50 ± 0,88	13,26 ± 0,50	0,56 ± 0,03
700	22,01 ± 0,55	11,85 ± 0,29	0,54 ± 0,02
750	20,19 ± 0,63	11,65 ± 0,30	0,50 ± 0,02
800	28,47 ± 0,96	15,04 ± 0,51	0,53 ± 0,03
750B	20,37 ± 0,52	11,65 ± 0,30	0,57 ± 0,02
800B	21,37 ± 0,68	12,71 ± 0,40	0,60 ± 0,03

Table 4.5. Real and imaginary part of the permittivity and the dielectric loss, for all samples, measured at 300 K and 1 kHz.

Figure 4.23 presents the Z´´ versus Z′ spectrum, measured at 300 K, for all samples. This spectrum shows semi-arches whose centers are under the Z´ axis, which indicates the existence of a distribution of relaxation time. The existence of this distribution must be connected to the fact that the dielectric response of these samples presents several contributions. Among these contributions we have the glass matrix, the crystalline phase and the dipoles associated with other electrical units (ions) present in the matrix. For frequencies < 10 Hz the samples show dispersion in the impedance spectra. This phenomenon, attributed to the presence of interfacial dipoles sample/electrode (Maxwell-Wagner effect). The theoretical fitting of the frequency dependence of the complex impedance (Z*), using the CNLLS algorithm, reveals that the electrical circuit formed by the parallel between a resistance (R) and a CPE element is an equivalent circuit which describes reasonably well the measured data. However, the correlation between the characteristic parameters of this circuit (R, Q_0 and n - table 2-annex) and the structural characteristics of the samples is not evident due to the presence of the CPE element. This element is a parameter dependent on the frequency and associated with a distribution of relaxation times is usually related to the presence of electrical inhomogeneities in the sample surface. In these samples, these surface heterogeneities can be related to the simultaneous presence on the surface of particles and glass matrix. Nevertheless, the behavior of the parameter R, with increasing annealing temperature (fig. 4.24), is similar to that observed in electrical conductivity (table 4.5). Thus, the increase of the R parameter with the increase of the HT temperature up to 750 °C can be associated with the decrease of the number of network modifier ions and consequent increase in the number of surface particles. Therefore, and according to the electric model assumed (fig. 4.22), an increase in the resistivity of the bulk is mainly justified by a decrease in the number of modifier ions structurally inserted in the glass matrix. Please note that the logarithmic representation of the parameter R as a function of measurement temperature (fig. 4.24), for all samples is almost linear. However, it increases with the increase of the measuring temperature, which indicates that this temperature increasement promotes an increase of the charge carriers energy, thereby increasing their mobility. This hypothesis is also supported by the decreased of τ_Z (fig. 4.28), in the same experimental conditions (table 2-annex), indicating an increased capacity of the dielectric units to follow the ac electric field. The calculated values of all parameters are registered in table 2-annex. From those results, it was also observed that the Q_0 parameter is between 3.78×10^{-9} and 6.16×10^{9} and the n parameter is between 0.80 and 0.85. Moreover, the value of the C_{CPE} capacitor is between 1.0 nF and 1.8 nF, approximately.

One approach to analyze the dielectric relaxation phenomena can be by using the Z´´ versus frequency representation (figs. 4.25 to 4.27). In these figures the lines represent the result of the CNLLS adjustment of the experimental data to the R//CPE equivalent electrical circuit. The frequency at which the Z´´ peak appears is associated with the parameter τ_Z (table 2-annex). This parameter corresponds to the most probable relaxation time or, in other words, to the relaxation time distribution average. It was observed that the value of τ_Z for the samples 750B and 800B is higher than that the one for the correspondent samples with the white film. This indicates that the dipoles associated with the $NaNbO_3$ crystallites, have a relaxation time higher than the dipoles present in the sample bulk. The τ_Z of the sample HT at 750 °C is smaller than the one of the sample HT at 800 °C, suggesting the existence of a larger number of dipoles associated with the $NaNbO_3$ crystalline phase in the sample HT 750. This assumption is consistent with the SEM micrographs, which shows a larger number of particles in the sample HT 750. Nevertheless, the values of τ_Z are very high (~10^{-4} s - table 2-annex) when compared with the relaxation times of dipoles associated with network modifier ions structurally inserted in the glass matrix (~ 10^{-11}s), showing the large influence of the glass matrix on the electrical properties.

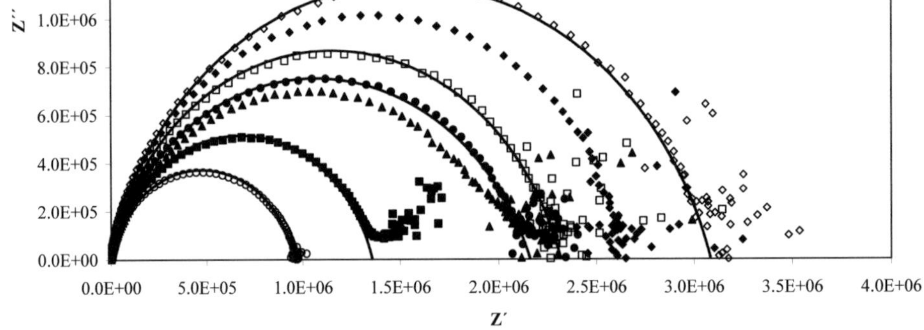

Fig. 4.23. Z´versus Z´´, at 300K, for all 60SiNa samples.

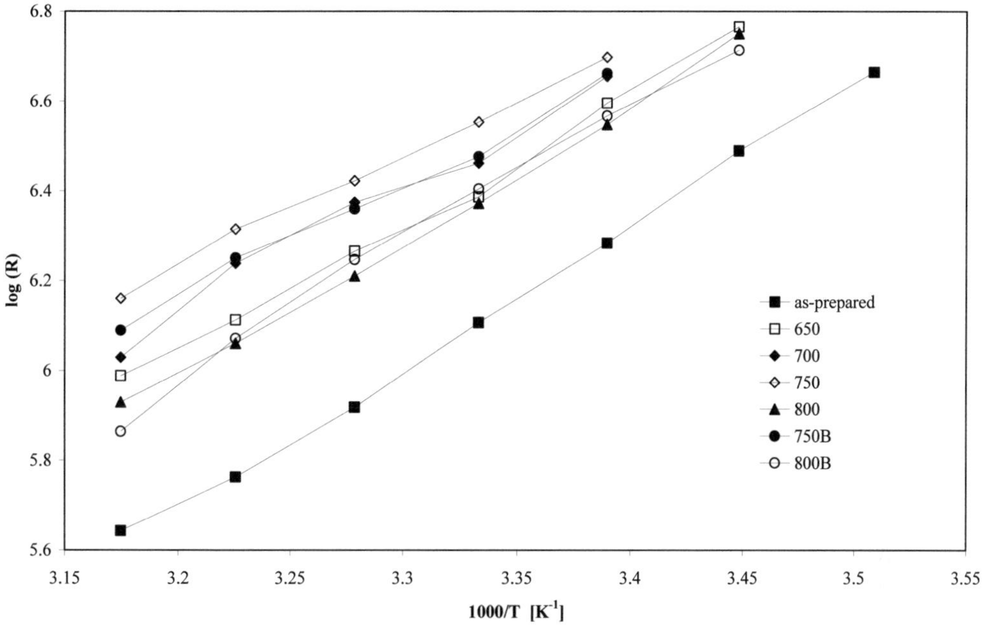

Fig. 4.24. log(R) versus 1000/T for all 60SiNa samples.

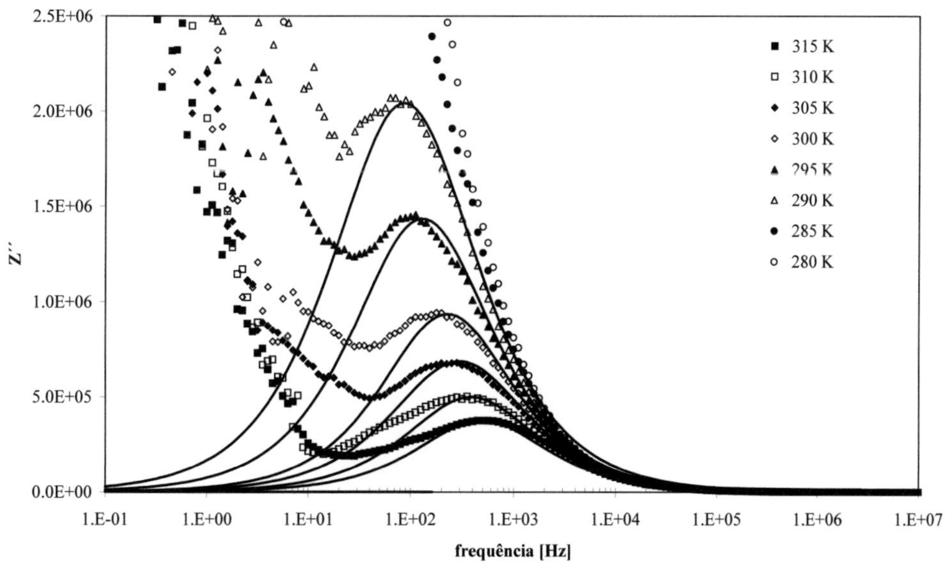

Fig. 4.25. Z´´ versus frequency of the sample HT at 650 °C, for several measuring temperatures.

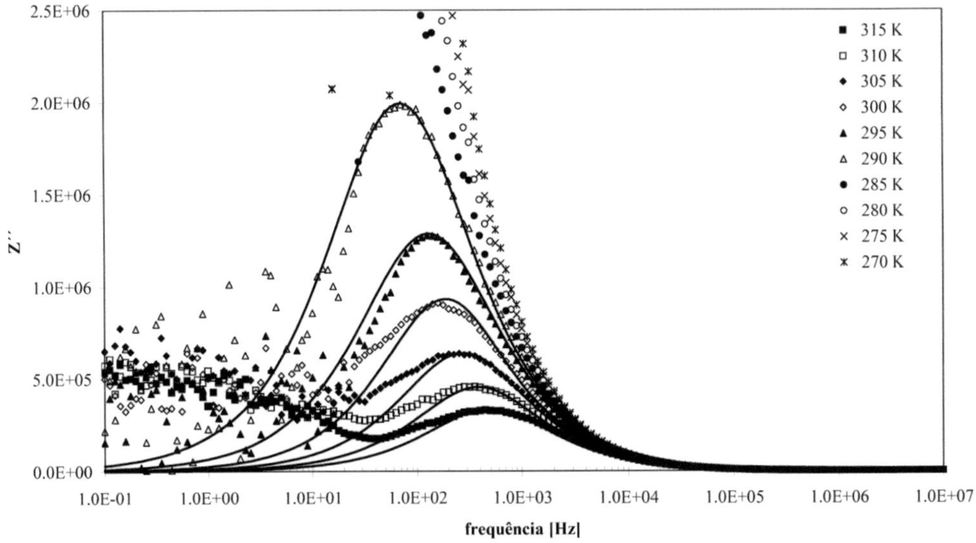

Fig. 4.26. Z´´ versus frequency of the sample HT at 800 °C, for several measuring temperatures.

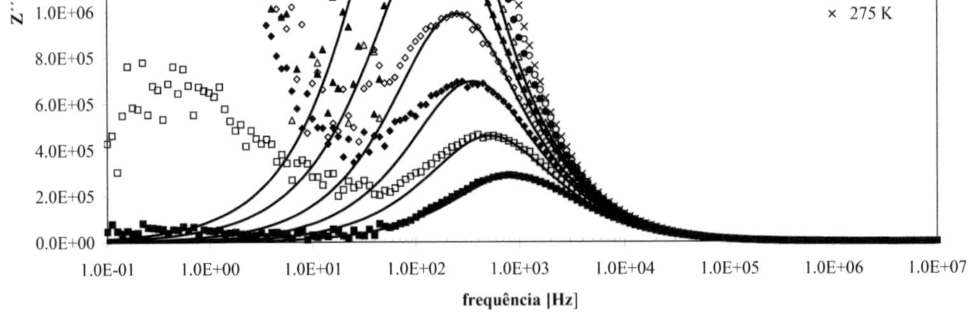

Fig. 4.27. Z´´ versus frequency of the sample 800B, for several measuring temperatures.

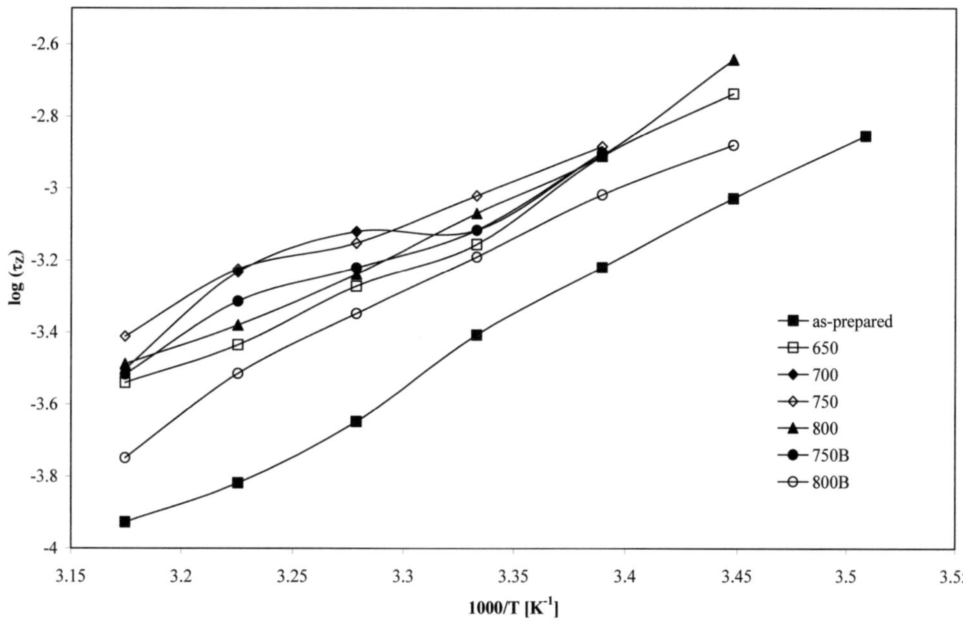

Fig. 4.28. ln(τ_Z) versus 1000/T for all 60SiNa composition samples.

Figures 4.29 to 4.33 show the TSDC spectra of samples of the 60SiNa composition. All samples were polarized at the temperature of 350 K, for 10 minutes, applying a dc electric field of 100 kV/m. The heating rate used was 4 °C/min. In all spectra, the lines represent the theoretical fit result. The TSDC results showed that the samples HT at 650, 750 °C and 800B present two thermally stimulated depolarization peaks. In the other samples, it was registered only one current peak. However, it is important to note that in these samples after reaching the peak, the current decreases very slowly, i.e., the slope is smaller than one would expect theoretically. This suggests the possible existence of at least one second peak current masked by the high value of the first. The hypothesis of the presence of depolarization mechanisms not observed is also justified by the bad adjustment of the theoretical model used to the experimental data. This bad adjustment indicates the presence of other depolarization mechanism(s), which corroborates the results obtained as a function of the frequency in which it was detected the existence of a relaxation times distribution. The characteristics of these peaks can be found in table 4.6.

It was observed that the temperature at which the peak current occurs increases with the increase of the HT temperature, except for the sample HT at 750 °C which shows the minimum value of current. The temperature of this peak decreases from sample 750B to sample 800B. The obtained values (table 4.6) do not allow a correspondence between the relaxation peak to a particular type of dipole. It is known that in these glasses and glass ceramics, the "units" that can give rise to peaks of thermally stimulated depolarization are the network modifier ions (Na^+ and Nb^{5+}), the particles and the glass network itself. We therefore believe that the most likely hypothesis that can explain the origin of the peaks observed is the dipoles formed by the network modifier ions because the glass matrix should present a thermally stimulated relaxation time very high, not giving rise to relaxation peak but to a broad band. The particles, which are associated with $NaNbO_3$ crystals should also not contribute significantly to the depolarization current peaks because in the sample 800B, which does not contain the surface film (i.e., does not contain $NaNbO_3$ particles), were also detected two depolarization peaks.

The activation energy (E_a) for each current peak, can be found in table 4.6. For the low temperature current peak, the value of E_a is larger in the as-prepared sample than in the sample HT at 650 °C. With the increase of the HT temperature up to 750 °C, there is a decrease in E_a. In the samples where two depolarization peaks were observed (HT 650, HT 750 and 800B) the activation energy associated with the first peak is always higher than that associated with the second.

The relaxation time (τ), associated with the low current TSDC peak decreases from the as-prepared glass to the sample HT at 650 °C and the increases with the increase of the HT temperature up to 750 °C (table 4.6). The relaxation time associated with the first peak of depolarization is always smaller than the relaxation time associated with the peak depolarization of the higher temperature.

With increasing measurement temperature the depolarization current increases in a way that can be adjusted to an Arrhenius expression. The characteristics of this adjustment are recorded in table 4.7. The E_a associated with this high temperature process is superior to any other TSDC process detected.

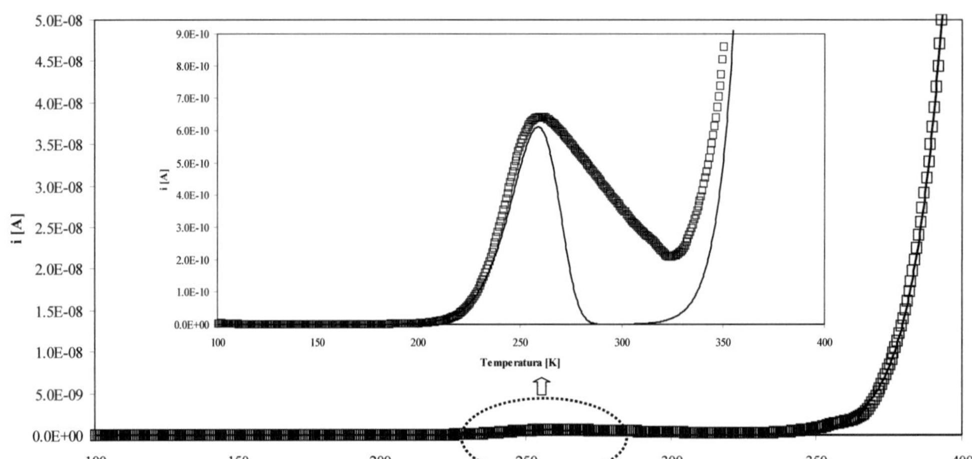

Fig. 4.29. TSDC spectra of the as-prepared sample (60SiNa composition).

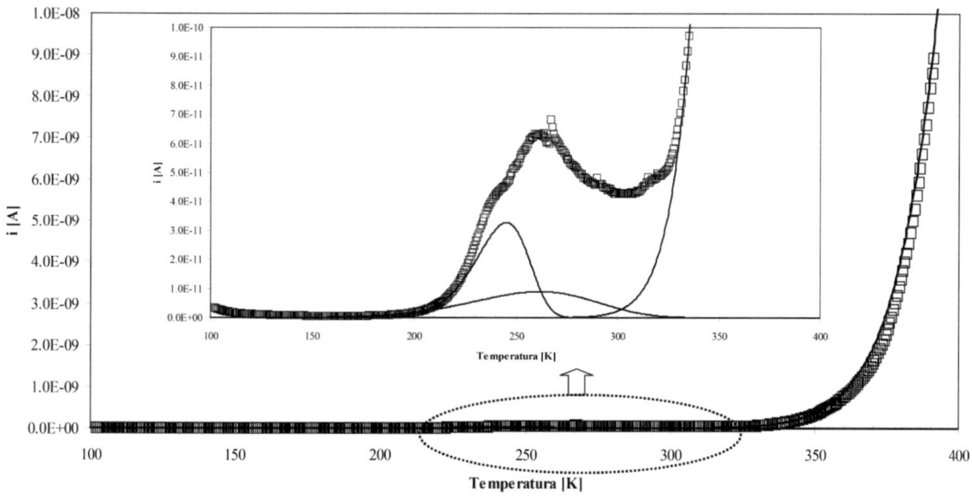

Fig. 4.30. TSDC spectra of the HT 750 °C sample (60SiNa composition).

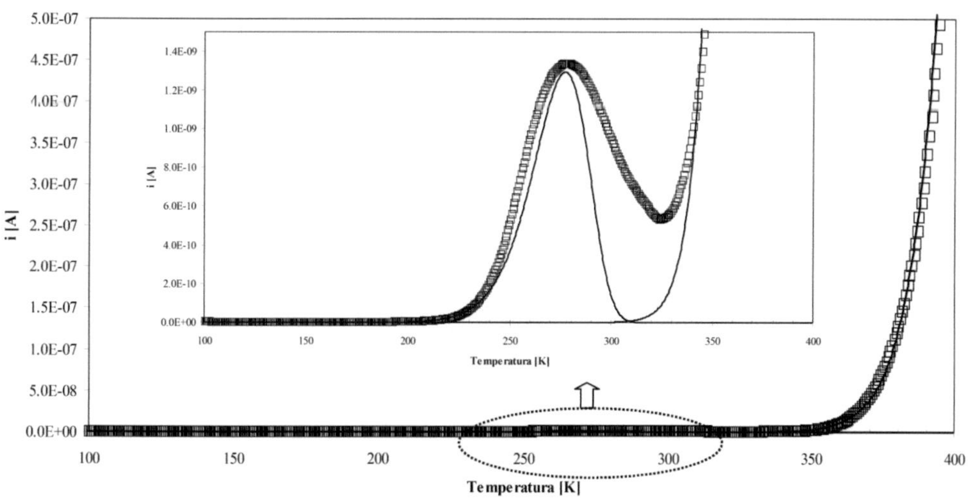

Fig. 4.31. TSDC spectra of the HT at 800 °C sample (60SiNa composition).

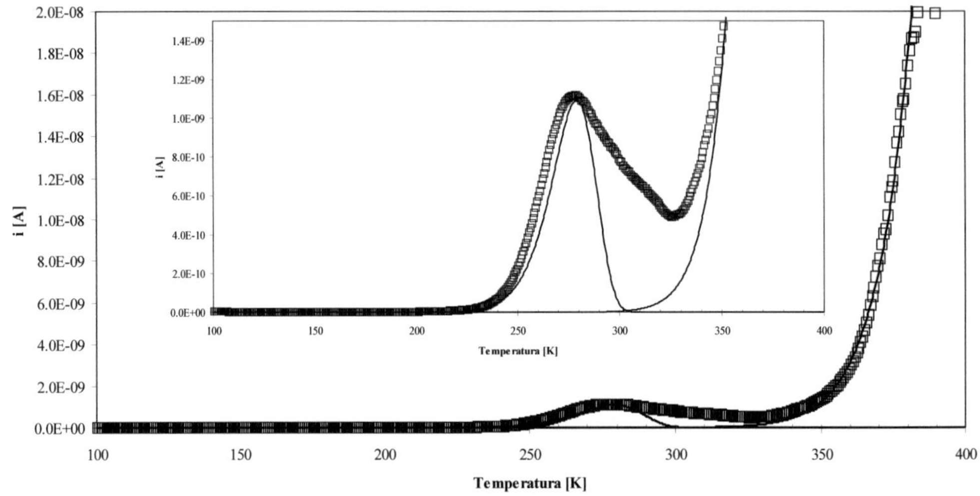

Fig. 4.32. TSDC spectra of the 750B sample (60SiNa composition).

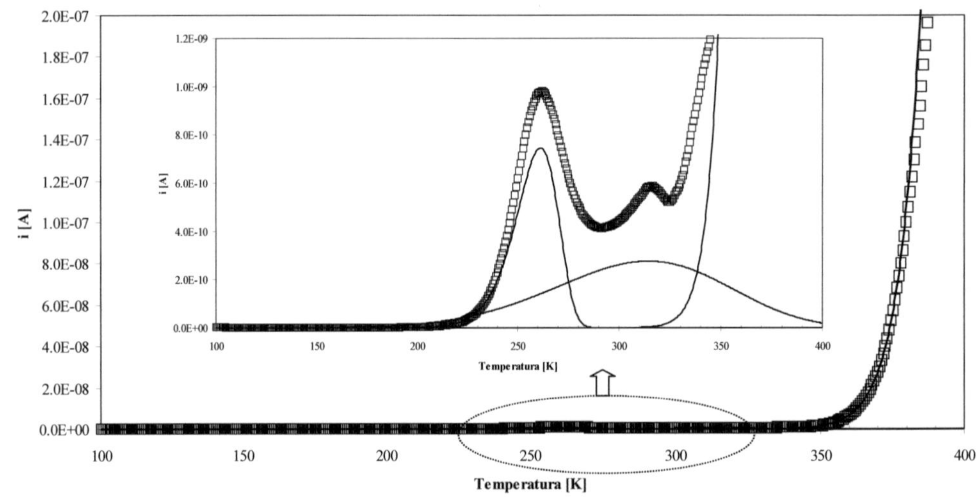

Fig. 4.33. TSDC spectra of the 800B sample (60SiNa composition).

Sample	T_{P1} (K)	$i(T_{P1})$ x10^{-10} (A)	Ea [kJ/mol]	$\tau_{0\,(P1)}$ (x10^{2}) [s]	$\tau_{(P1)}$ (x10^{3}) [s]
as-prepared	259,8	6,43	45,17 ± 0,59	2,30 ± 0,04	1,86
650	263,9	5,11	51,27 ± 0,15	1,64 ± 0,03	1,69
700	271,7	5,40	47,72 ± 0,21	2,34 ± 0,04	1,93
750	246,3	0,47	36,72 ± 0,38	3,43 ± 0,05	2,06
800	278,3	13,37	50,80 ± 0,36	2,12 ± 0,03	1,90
750B	279,0	11,16	57,44 ± 0,37	1,42 ± 0,02	1,69
800B	262,0	9,80	52,69 ± 0,34	1,45 ± 0,02	1,63
	T_{P2} (K)	$i(T_{P2})$ x10^{-10} (A)	E_a [kJ/mol]	$\tau_{0\,(P2)}$ (x10^{3}) [s]	$\tau_{(P2)}$ (x10^{4}) [s]
as-prepared	--	--	--	--	--
650	317,2	1,82	11,35 ± 3,56	7,20 ± 0,05	1,11
700	--	--	--	--	--
750	263,6	0,63	12,71 ± 1,01	4,16 ± 0,04	0,72
800	--	--	--	--	--
750B	--	--	--	--	--
800B	315,9	5,89	16,61 ± 0,19	3,98 ± 0,03	0,75

Table 4.6. Peak temperature (T_P), peak current ($i_{(T_P)}$), activation energy (E_a), pré-exponential parameter (τ_0) and relaxation time (τ), calculated on the low temperature peak (T_{P1}) and in the high temperature peak (T_{P2}).

Sample	ln (A) [A]	E_a [kJ/mol]
as-prepared	20,79 ± 0,28	122,88 ± 0,92
650	20,67 ± 0,45	119,80 ± 1,35
700	26,60 ± 1,04	143,66 ± 3,26
750	8,92 ± 0,23	89,63 ± 0,69
800	26,62 ± 0,39	134,44 ± 1,23
750B	13,07 ± 0,44	97,73 ± 1,30
800B	33,92 ± 0,11	157,87 ± 0,33

Table 4.7. Parameters of the Arrhenius equation used in the adjust of the TSDC high temperature range. ln (A) represents the logarithm of the pre-exponential factor and E_a the activation energy

4.2 Borate system

4.2.1 B_2O_3-Li_2O-Nb_2O_5 composition

The thermal analysis of the transparent and incolor glass with the molar composition $60B_2O_3$-$30Li_2O$-$10Nb_2O_5$ (60BLi) presents two exothermic phenomena (529 °C and 565 °C) and one endothermic (872 °C). The glass transition temperature is approximately 524 °C. From these results heat-treatments were performed at 450, 500, 550 and 600 °C. The obtained samples became translucent with the HT at 500 °C and opaque for the treatments at temperatures above 500 °C.

The HT at 550 °C favors the formation of $LiNbO_3$ and $Li_2B_4O_7$ crystals as indicated by the XRD result (fig. 4.34). The identification, in the XRD spectrum of the sample treated at 500 °C, of one peak that coincides with the main peak of the $LiNbO_3$ crystal, suggests that the particles with dimensions of, approximately, 1 μm observed by SEM (fig.4.36) correspond to $LiNbO_3$ crystallites. In the sample HT at 600 °C the XRD spectra reveal the formation of the $LiNb_3O_8$ phase. This result indicates that the exothermic phenomenon at 565 °C is due to the formation of the $LiNb_3O_8$ phase. It must be noticed that this sample (HT at 600 °C) does not presents the XRD diffraction peaks characteristics of the $LiNbO_3$ crystal.

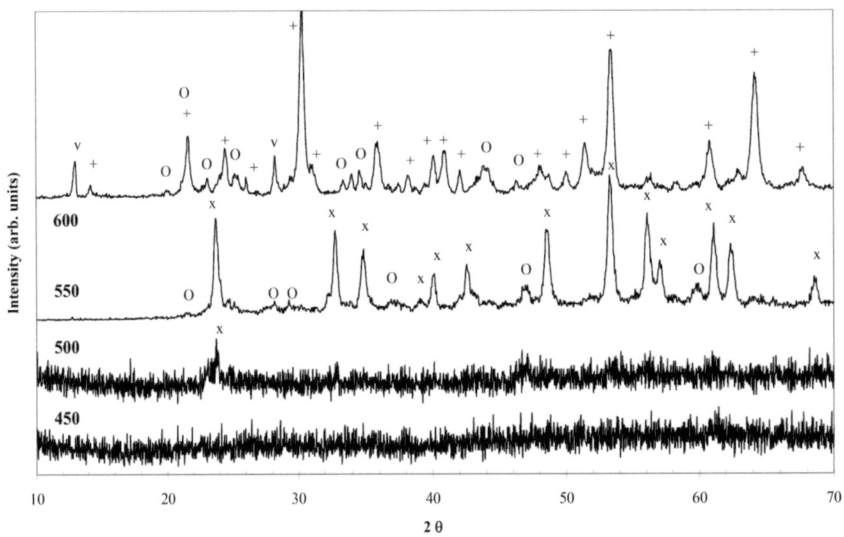

Fig. 4.34. XRD spectra of all 60BLi samples (x $LiNbO_3$; O $Li_2B_4O_7$; + $LiNb_3O_8$; v phase not identified).

The results of Raman spectroscopy (fig. 4.35) of the as-prepared glass and HT at 450 °C and 500 °C show the presence of a broad band centered at 870 cm^{-1} which may, in accordance with Aronne and co-workers, be associated with NbO_4 tetrahedral vibrations. However, some authors attribute the band between 800-940 cm^{-1} to vibrations of isolated NbO_6 octahedral and the shift to higher wave numbers is related with an increase in the degree of distortion. In samples treated at higher temperatures (550 and 600 °C) the band centered at 870 cm^{-1} disappears, indicating the displacement of the niobium ions from the matrix to the $LiNbO_3$ and $LiNb_3O_8$ structures. The vibration bands characteristics of the $LiNbO_3$ crystal were detected in the sample HT at 550 °C (640, 438, 370-373, 333-336, 280, 240 and 170 cm^{-1}).

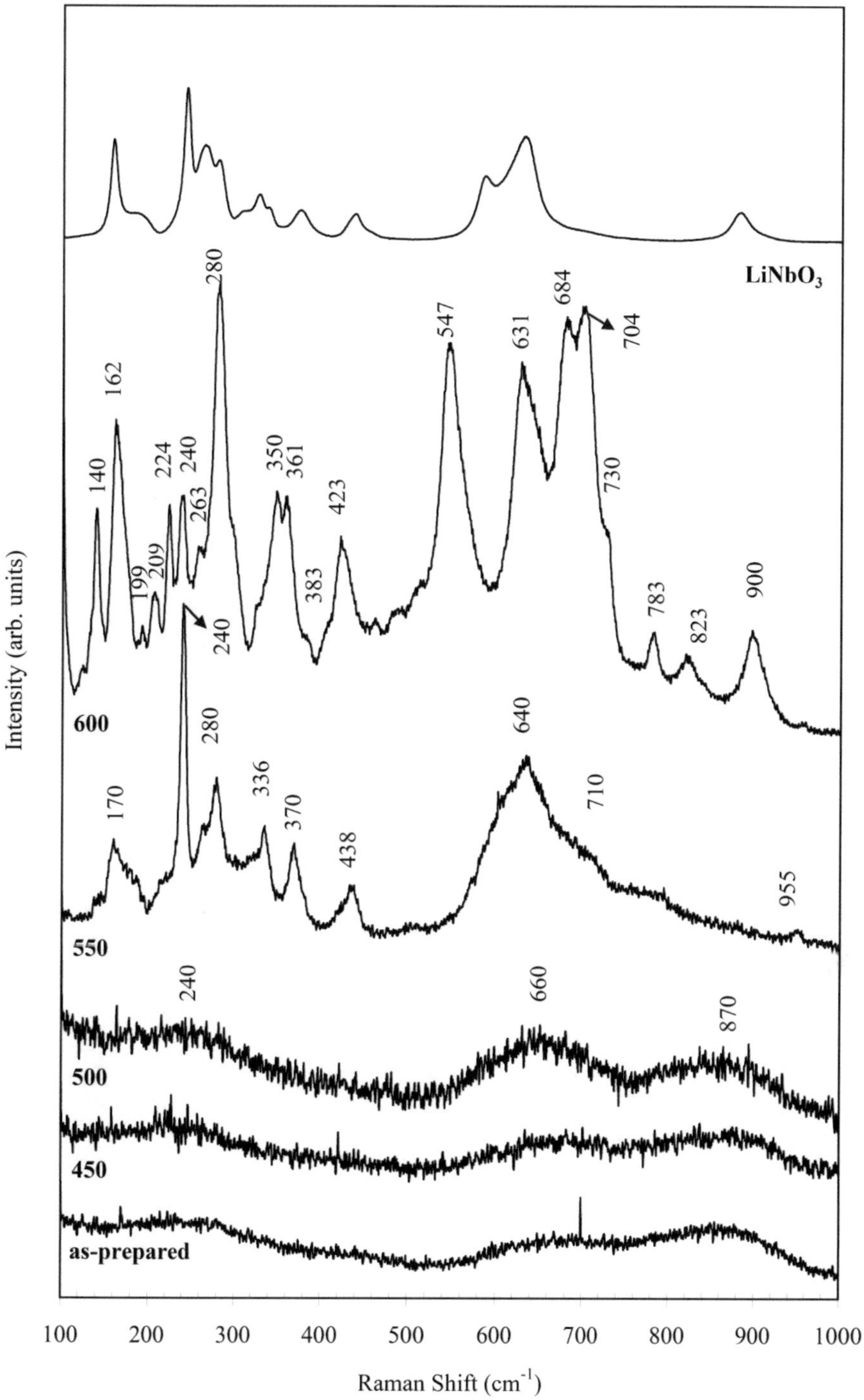

Fig. 4.35. Raman spectra of the 60BLi samples and of high-purity $LiNbO_3$ powder.

In the samples treated at temperatures below 550 ° C, the presence of a Raman band near the 660 cm^{-1} indicates the presence of pentaborate groups. In samples where the XRD detected the presence of lithium borate crystalline phase, the Raman spectroscopy detected the presence of bands centered at 690-710 cm^{-1}, attributed to the metaborate groups. It is known that in B_2O_3 glasses, the addition of network modifier ions result in the formation of tetrahedrons. The non-detection by Raman of the bands close to 930, 808 and 770 cm^{-1} show that the structure of the glass does not contains, respectively, tetraborate units, triborate units or boroxol rings.

It was verified by the SEM micrographs (fig. 4.36) that the as-prepared glass does not present particles. Particles with a size of ~ 500 nm were observed on the free and fracture surfaces of the samples HT at 450 ° C. With the increase of the thermal treatment temperature the size of these particles increases but their number decreases

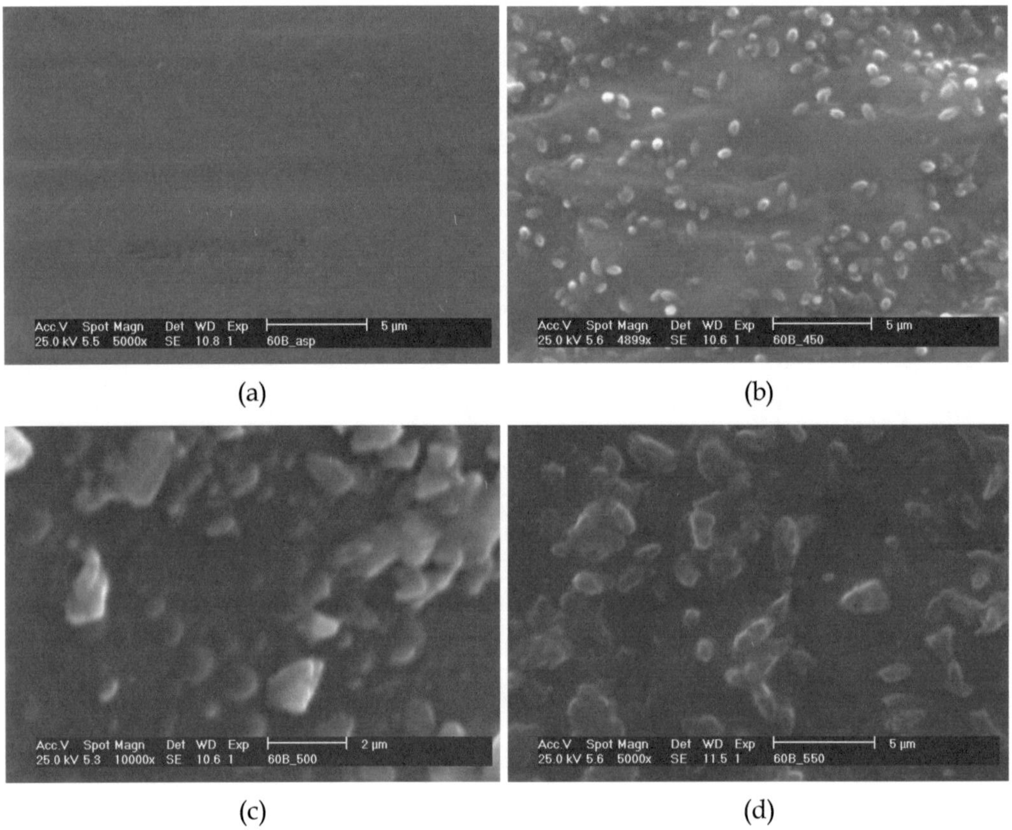

Fig. 4.36. SEM micrographs of the 60BLi samples: (a- as-prepared; b – HT at 450 °C; c.- HT at 500 °C; d – HT at 550 °C).

The figures 4.38 and 4.39 showed that with the increase of the HT temperature the σ_{dc} (fig. 4.37) and σ_{ac} (fig. 4.38) decreases (table 4.8). This behavior is related with the decrease of the number of network modifying ions, responsible for the ionic conductivity, which is characteristic of this type of glasses. However, the electrical conductivity also depends on the mobility of the charge carriers and therefore on the height of the network potential barriers. This characteristic is related with the activation energy, which for the dc conduction process revealed to be similar for the as-prepared, HT at 450 and HT at 500 °C samples (table 4.8). Thus, the decrease of the σ_{dc}, with the increase of the HT temperature, should be attributed to the decrease of the charge carriers number. These carriers should be related mainly to the number of Li^+, because from the results of Raman spectroscopy some Nb^{5+} ions, are introduced as network formers in the as-prepared, HT at 450 and HT at 500 °C samples, due to the presence of band center at 870 cm^{-1} (fig. 4.36), which is related with the presence of NbO_4 tetrahedral. The formation of the $Li_2B_4O_7$ crystal, which is characterized by a low electrical conductivity ($\sigma \sim 10^{-10}$ Sm^{-1}), and the presence of high resistivity $LiNbO_3$ crystals, indicates that the presence of this crystallite contributes to the observed decrease of the conductivity.

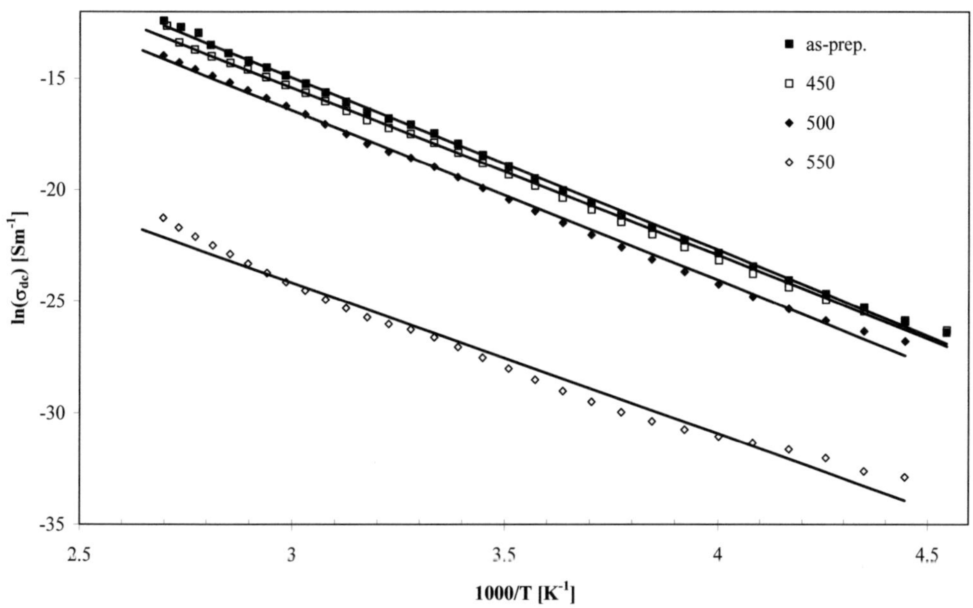

Fig. 4.37. σ_{dc} versus 1000/T for all 60BLiNa samples.

The decrease of the σ_{ac} (fig. 4.38), with the increase of the HT temperature, should be related with the decrease of the number of charge carriers because the activation energy ($E_{a(ac)}$ - table 4.8) is almost constant for the HT samples.

Figure 4.39 shows the σ_{ac} dependence with the temperature at several frequencies for the as-prepared sample. This behaviour is similar for all samples. It is observed that the increase of the frequency promotes an increase of σ_{ac} and a decrease in the $E_{a(ac)}$.

The "jump" of the potential barriers by the ions, or their oscillation between structurally close sites in response to an ac electric field, contribute to the behavior of the dielectric constant (ε '). Knowing that the ε' of the as-prepared sample, of the HT at 450 and HT at 500 °C (table 4.9) present similar values, the increase of ε' for the sample HT at 550 °C should be assigned to the presence of $LiNbO_3$ crystallites, being the main contribution to the ε' value, because with the increase of the HT temperature there is a decrease in the number of

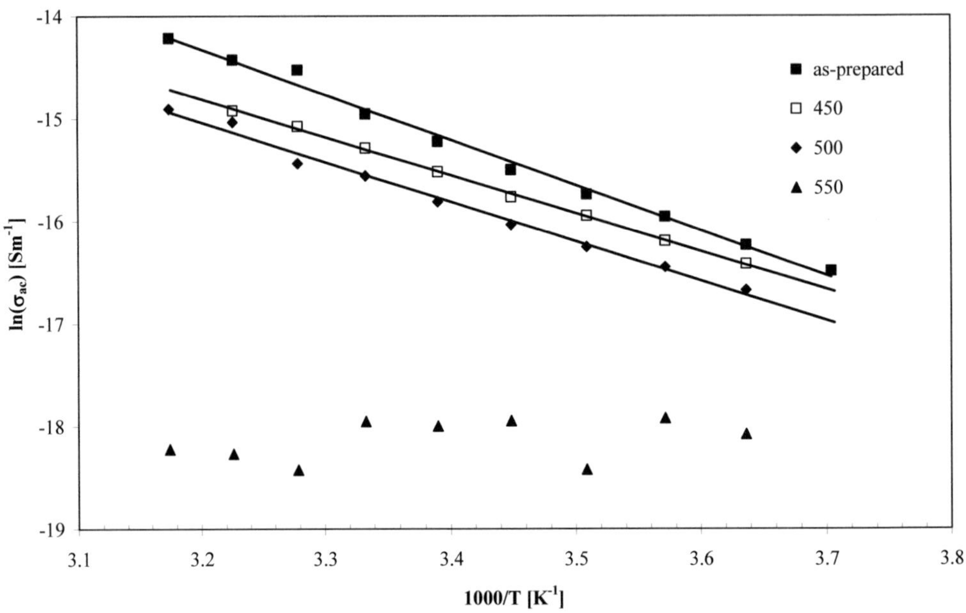

Fig. 4.38. σ_{ac} versus 1000/T for all 60BLiNa samples.

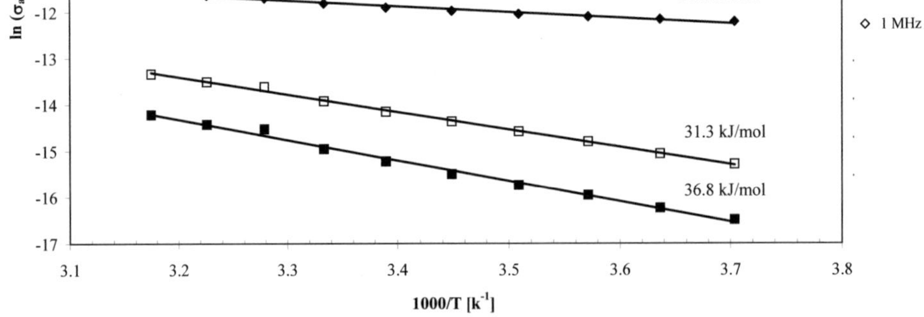

Fig. 4.39. σ_{ac} versus T, for several frequencies, of the as-prepared sample.

Sample	σ_{dc} (x10^{-8}) [Sm^{-1}]	$E_{a(dc)}$ [kJ/mol]	σ_{ac} (x10^{-7}) [Sm^{-1}]	$E_{a(ac)}$ [kJ/mol]
As-prepared	259,05±3,48	62,54±0,67	32,05±1,12	36,78±1,11
450	171,86±2,37	62,40±0,60	22,95±1,07	30,87±0,44
500	58,17±0,74	63,34±0,58	17,53±5,71	32,19±1,16
550	1,17±0,02	66,11±1,10	0,997±0,047	--

Table 4.8 dc conductivity (σ_{dc}) at 300 K, dc activation energy ($E_{a(dc)}$), ac conductivity (σ_{ac}), at 1 kHz and 300 K and ac activation energy ($E_{a(ac)}$).

network modifier ions structurally inserted in the glass matrix and thus their contribution to the dipole moment decreases. The decrease of the ε' from the sample HT at 500 to the sample HT at 550 °C (table 4.9), is associated with an increase of the volume ratio between the crystals and the glass matrix and therefore it can be suggested that in this glass network the $LiNbO_3$ crystallites present a preferential grow orientation. The presence, in this sample, of the $Li_2B_4O_7$ ferroelectric phase, which has a ε' ~ 80, smaller than that the one of $LiNbO_3$ (> 1000), should not influence significantly the dielectric behavior.

The increase of ε' in all samples, with the increase of the measurement temperature indicates that the increase of the temperature facilitates the dipole orientation. The same type of dependence was observed with the relaxation time τ_Z (table 3-annex), which decreases with the increase of the measurement temperature, indicating a decrease in the mobility difficulty of the electric dipoles. The decrease of σ_{ac}, related with the decrease of the charge carrier number, explains the decrease of the tan δ value (table 4.9).

The impedance spectra (fig. 4.41), show single semi-circles whose center are under the Z' axis, which indicates that the ac response can be represented by an equivalent electrical circuit comprising the parallel between a resistance (R) and an CPE element. The correlation between the circuit parameters (table 3-annex), and the sample characteristics are not entirely clear due to the presence of the CPE element, usually associated with the presence of a distribution of relaxation times and related to the presence of electrical heterogeneities on the samples surface. However, the increase of the parameter R, calculated at 300 K, with the increase of the HT temperature, is consistent with the decrease of the conductivity (table 4.8).

In the sample HT at 550 °C, the Z* formalism (fig. 4.42), the ε* (dielectric permittivity) formalism or the M * (dielectric module) formalism did not reveal the formation of semi-arches, associated with the presence of dielectric relaxation mechanism(s). This behavior may be due to the increase of the volume ratio between the crystallites (mainly $LiNbO_3$) and the glass matrix, as confirmed by XRD (fig. 4.34), by the SEM micrographs (fig. 4.36) and by the results of electrical conductivity (table 4.8). In the samples where it was detected the presence of a dielectric relaxation mechanism, it was found that the value of τ_Z increases, with the increase of the HT temperature (table 3-annex), which is another indication of the increased amount of $LiNbO_3$ crystals in the glass matrix, because the electric dipoles associated with $LiNbO_3$ are difficult to depolarize at room temperature.

The spectra represented in fig. 4.42 can be adjusted by the Curie-Von Schweidler model. This model assumes that for periods of time longer than two decades, it can follows a functional form of the type $i(t) \propto t^{-s}$. The inverse Fourier transform of the above expression, leads to impedance expressions of the type (eq. 25)

$$Z'' = k.f^{-n} \tag{25}$$

where Z" is the imaginary part of complex impedance, k a constant of proportionality, and f the frequency and n a adjustable parameter that, according to Jonsher, is associated with the type of polarization system (dipoles, charges, etc.). For the 550 °C HT sample spectra (fig. 4.42) it was observed that the parameters k and n (table 4.10), decrease with the increase of the temperature of measurement. However, this decrease, for the parameter n, is very smooth.

Figures 4.42 and 4.43 show the result of the TSDC measurements on the as-prepared and in the sample HT at 550 °C, respectively. All samples were polarized at the temperature of 350 K, for 10 minutes, by applying an electric field of 100 kV/m. In all spectra, the lines represent the theoretical fit.

The TSDC results show that with the increase of the HT temperature, the depolarization current peak temperature (TP1) shifts to higher temperatures (table 4.11), indicating that the electrical units, responsible for this depolarization, requires a higher energy to depolarize when the HT temperature is increased. The value of relaxation time (τ_{P1}) associated with this current peak, increases from sample HT at 450 to the sample HT at 550 °C. This behavior is similar to that observed in dielectric relaxation, which shows a progressive difficulty of the electrical units to return to their normal position. The value of the peak current i_{P1} (table 4.11) decreases, with the exception of the sample HT at 450 °C, with the increase of the treatment temperature. The sample HT at 450 °C show a value of i_{P1} lower than the one of the sample HT at 500 °C, which could be related to the value of E_a (table 4.11) which has a maximum for that sample. However, with increasing the HT temperature it is verified that the value of E_a decreases, which indicates, based on the model of potential barriers, a larger easiness of the ionic motion.

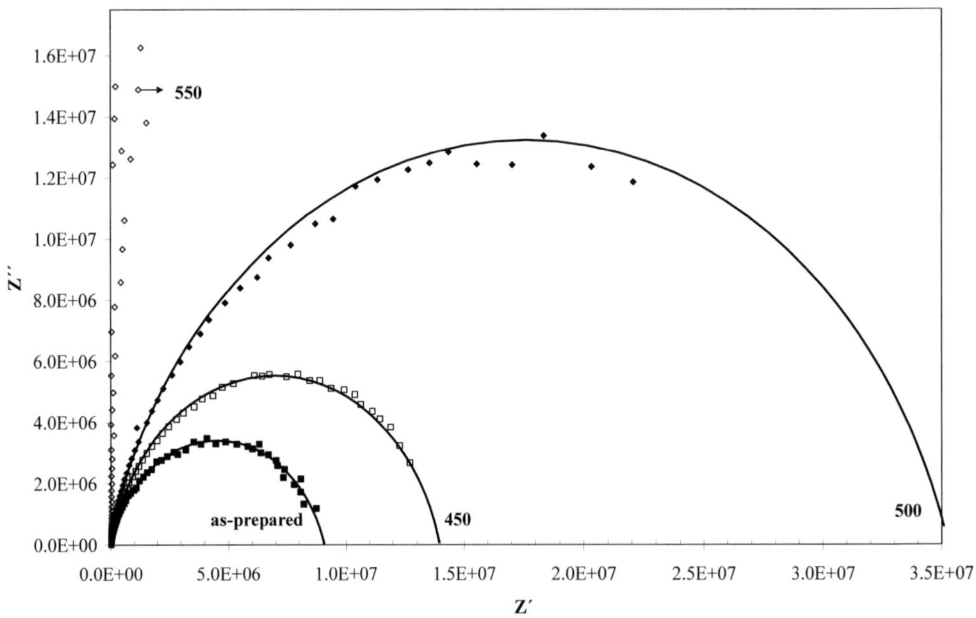

Fig. 4.40. Z'' versus Z' for all 60BLi samples, at 300 K.

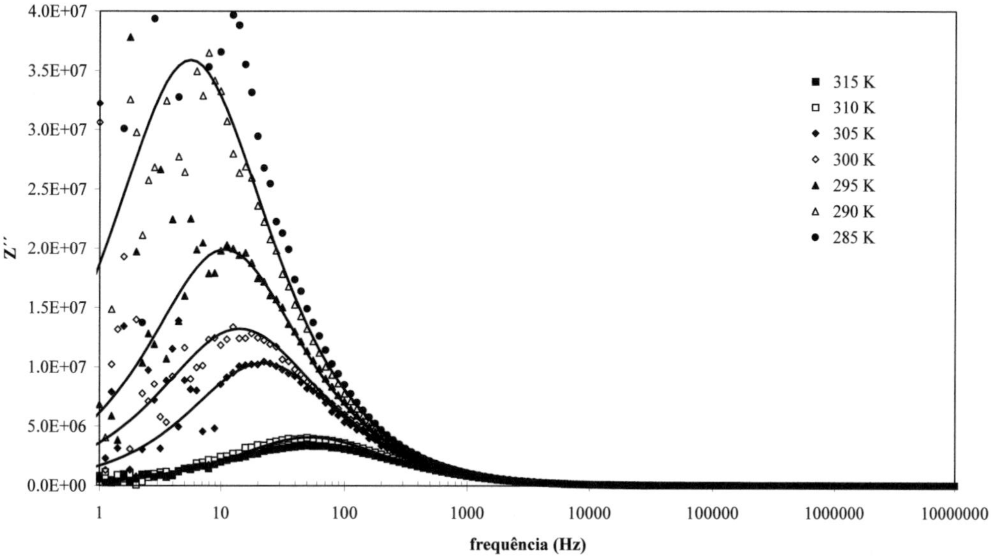

Fig. 4.41. Z´´ versus frequency, for the sample HT at 500 °C.

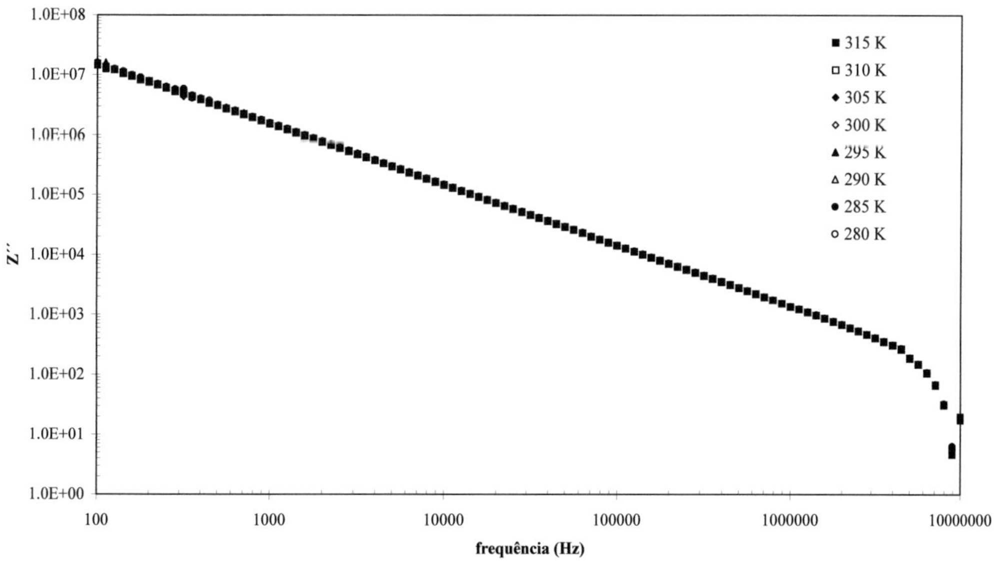

Fig. 4.42. Z´´ versus frequency, for the sample HT at 550 °C.

Sample	ε´	ε´´	tanδ
as-prepared	18,83±0,66	5,76 ± 0,20	0,31±0,02
450	17,16±0,76	4,13 ± 0,18	0,24±0,02
500	17,93±0,59	3,15 ± 0,10	0,18±0,01
550	11,39±0,53	0,18 ± 0,01	0,02±0,001

Table 4.9. Real and imaginary parts of the permittivity and the dielectric loss, at 300 K and 1 kHz.

Temperature [K]	k (x10^{8})	n
315	1,677	1,015
310	1,715	1,016
305	1,730	1,017
300	1,743	1,017
295	1,750	1,017
290	1,762	1,018
285	1,764	1,017
280	1,769	1,018
275	1,774	1,018

Table 4.10. k and n parameters for the sample HT at 550 °C, at several measurement temperatures

According to Rysiakiewicz and co-workers studies in lithium borate glasses, it is possible to detect two TSDC peaks, one related with different types of motion of the alkali ion. The TSDC peak at low temperatures can be attributed to the movement of the Li^+ ions around the non-bridging oxygen ions. In borate glasses where the amount of alkaline ion is high, a second TSDC peak can be observed and attributed to translational movements of the Li^+ ions, for example, in conduction paths. Thus, we assign the P_1 peak, observed in all samples to the depolarization of the dipoles formed by the Li^+ structurally inserted in the glass matrix. The increase in the T_{P1} value, decreased of i_{P1} (with the exception of the sample HT at 450 °C) and the increase of the relaxation time, corroborates this hypothesis, because the results of σ_{dc} and σ_{ac} show that with the increase of the HT temperature, the number of free ions decreases justifying, therefore, the decreasing of i_{P1}.

The second depolarization peak, only observed in the sample HT at 550 °C (fig. 4.43) should be associated with the presence of the $LiNbO_3$ and $Li_2B_4O_7$ crystalline phases. Assuming, based on the results of XRD, Raman and dielectric characterization of the sample HT at 550 °C, that the volume amount of $LiNbO_3$ is higher than the one of $Li_2B_4O_7$, the P_2 peak can be attributed to the dipoles associated with the $LiNbO_3$ crystallites. These dipoles are by nature difficult to depolarization, which justifies the fact that $\tau_{P2} > \tau_{P1}$. The fact that $T_{P2} > T_{P1}$ for the sample (table 4.11), shows that the energy required to depolarize the dipoles associated with the second peak is greater than that required by the first.

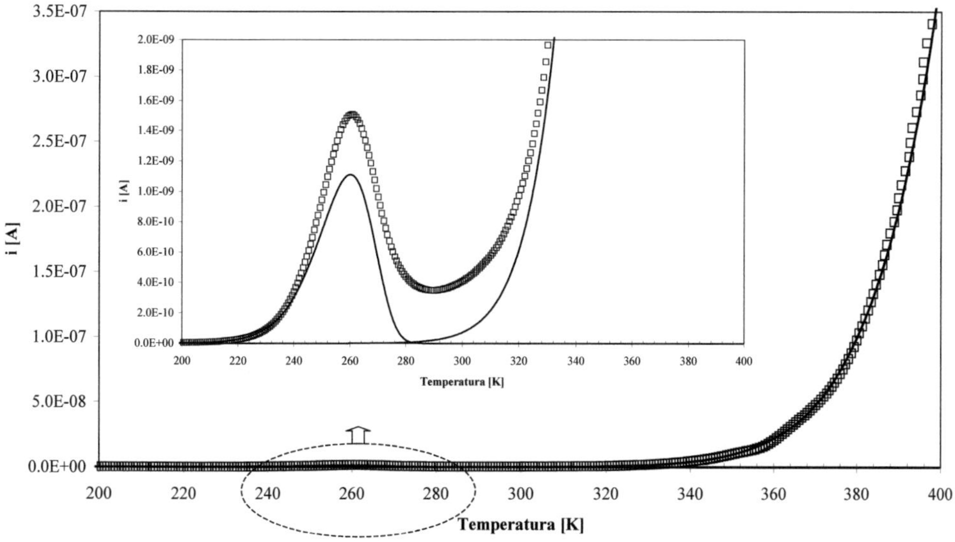

Fig. 4.43. TSDC spectra of the as-prepared sample (the lines represent the theoretical adjust).

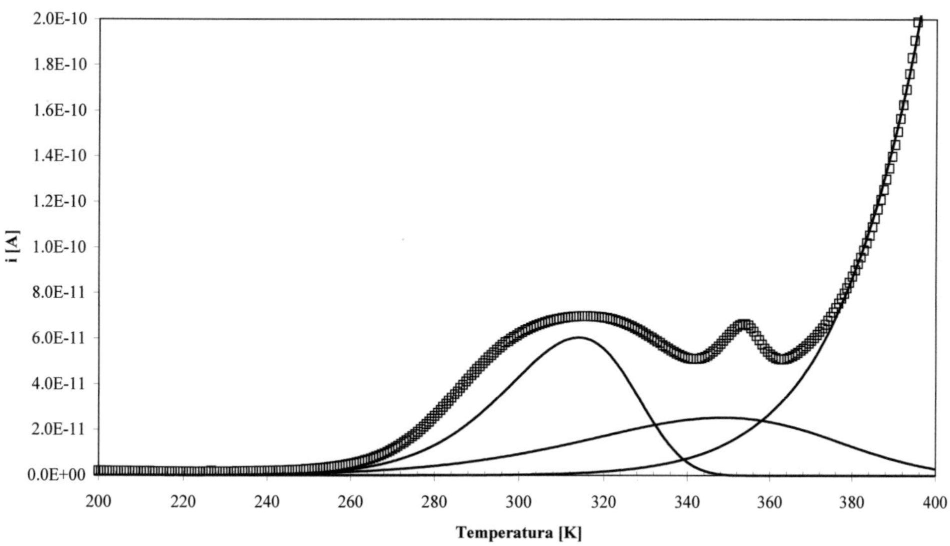

Fig. 4.43. TSDC spectra of the sample HT at 550 °C (the lines represent the theoretical adjust).

sample	T_{P1} (K)	$i(T_{P1})$ x10^{-9} (A)	Ea [kJ/mol]	$\tau_{0\ (P1)}$ (x10^2) [s]	$\tau_{(P1)}$ (x10^3) [s]
as-prepared	260,8	1,509	55,35 ± 0,45	1,194 ± 0,021	1,533
450	263,5	1,063	58,85 ± 0,30	1,004 ± 0,017	1,472
500	271,8	1,222	53,47 ± 0,62	1,619 ± 0,026	1,724
550	315,4	0,069	50,91 ± 0,21	3,499 ± 0,041	2,437
	T_{P2} (K)	$i(T_{P2})$ x10^{-11} (A)	Ea [kJ/mol]	$\tau_{0\ (P2)}$ (x10^2) [s]	$\tau_{(P2)}$ (x10^3) [s]
as-prepared	--	--	--	--	--
450	--	--	--	--	--
500	--	--	--	--	--
550	353,4	6,647	30,97 ± 0,62	17,54 ± 0,15	5,03

Table 4.11. TSDC peak temperature (T_p), peak current ($i_{(Tp)}$), activation energy (E_a), pre-exponential parameter (τ_0) and relaxation time (τ), calculated on the low temperature peak (T_{P1}) and in the high temperature peak (T_{P2}).

sample	ln (A) [A]	Ea [kJ/mol]
as-prepared	11,01 ± 0,16	85,77 ± 0,48
450	16,50 ± 0,24	102,31 ± 0,75
500	8,12 ± 0,29	77,81 ± 0,88
550	2,32 ± 0,25	65,89 ± 0,81

Table 4.12. Parameters of the Arrhenius equation used in the adjust of the TSDC high temperature range. ln (A) represents the logarithm of the pre-exponential factor and Ea the activation energy.

4.2.2 B_2O_3-Na_2O-Nb_2O_5 composition

The DTA analysis to this glass composition ($60B_2O_3$-$30Na_2O$-$10Nb_2O_5$ – named in this text as 60BNa) reveled the presence of two exothermic effects at 570 °C and 683 °C, the glass transition temperature (T_g) is about 550 °C and two endothermic effects at 710 and 786 °C. Thus, the HT temperatures chosen were: 475, 500, 550, 600 and 700 ° C.
From the XRD results (fig. 4.44) the HT at 500 °C gives rise to a glass-ceramic, with the $NaNbO_3$ crystalline phase. For treatments above 600 °C there is also the formation of Nb_2O_5 and $Na_2B_4O_7$ phases. In opposition to what occurs in the composition 60SiNa, the use of borate favors the bulk crystallization. We think that the reason for this phenomenon is the different structure of the two arrays. In pure borate glasses the boron ions present triangular coordination and, contrary to what occurs in SiO_2 glasses, the addition of network modifier ions (R_2O) is reflected, in an early stage, to a strengthening of the glass network, i.e., an increase in the number of cross links, to a maximum value occurring at the concentration of

~ 30% (mole) of R_2O. At this concentration of R_2O the borate glass shows the maximum number of boron ions tetrahedrally coordinated. With further increase of R_2O, these units and decreases and for molar concentrations above 70 % the boron ions are only with triangular coordination.

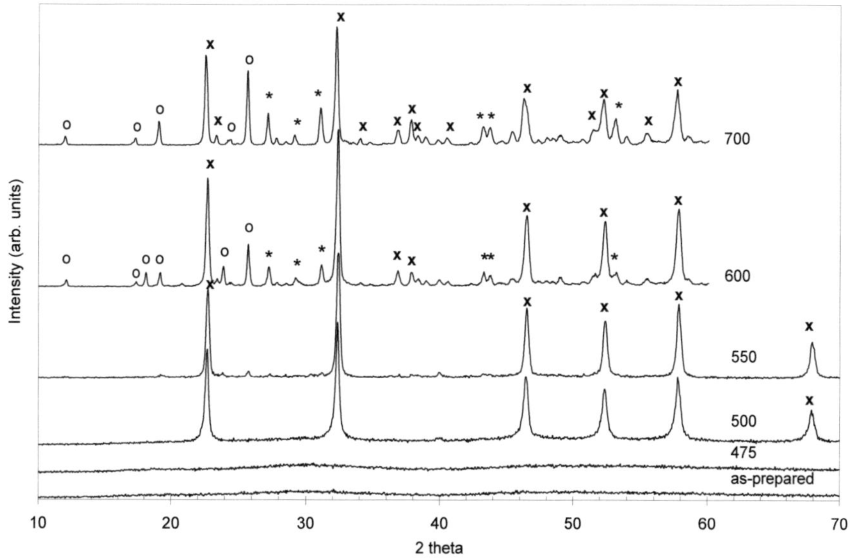

Fig. 4.44. XRD patterns of the 60BNa samples (x $NaNbO_3$; o Nb_2O_5; * $Na_2B_4O_7$).

Analyzing the Raman spectroscopy results (fig. 4.45) of the as-prepared and of the sample HT at 475 °C, it was observed the presence of a broad band at 900 cm^{-1}, which shifts to lower wave numbers with the increase of the HT temperature to 500 °C (885 cm^{-1}), and decreasing in intensity. This band, which according to Cardinal can be assigned to vibrations of isolated NbO_6 octahedrons in the glass network, is clearly detected in the sample treated at 550 °C. The decrease in the intensity of this band with the increase of the thermal treatment temperature can be related to a decrease in the number of isolated NbO_6 octahedrons, mainly due to the formation of $NaNbO_3$ particles. The detection in the Raman spectra of the sample treated at 550 °C, the band at 620 cm^{-1}, attributed to vibrations of NbO_6 octahedrons connected by oxygen bridges (Nb-O-Nb) reflects the presence of the $NaNbO_3$ phase. The bands at 450, 280, 250 and 210 cm^{-1} are due to NbO_6 vibrations associated with the $NaNbO_3$ crystalline phase. The vibrations at 194 and 123cm^{-1} are assigned to nbO6 vibrations in the orthorhombic phase. Note that in the samples where SiO_2 is the glass former, these bands were also detected. In the samples HT at 500 and 550 °C, the presence of the band centred at 780 cm^{-1}, reveals the existence in the vitreous network, of tetraborate units. So, the presence in the glass network of boron ion with triangular and tetrahedral coordination should favour the bulk crystallization.

The following figure (fig. 4.46) show the micrographs of the surface and fracture of the HT samples. From those micrographs it was verified that the sample HT at 475 °C presents, in the free and fracture surface, particles with a maximum size of 600 nm. With the increase of the HT temperature to 500 °C, the number of particles increases. In the sample HT at 500 °C the particles have a size of ~ 300 nm. The surface of the sample HT at 550 °c shows two types of particles that differ in size and number. The larger particles, with a size ~ 3 μm have a lower number when compared with the particles with a size of ~ 100 nm.

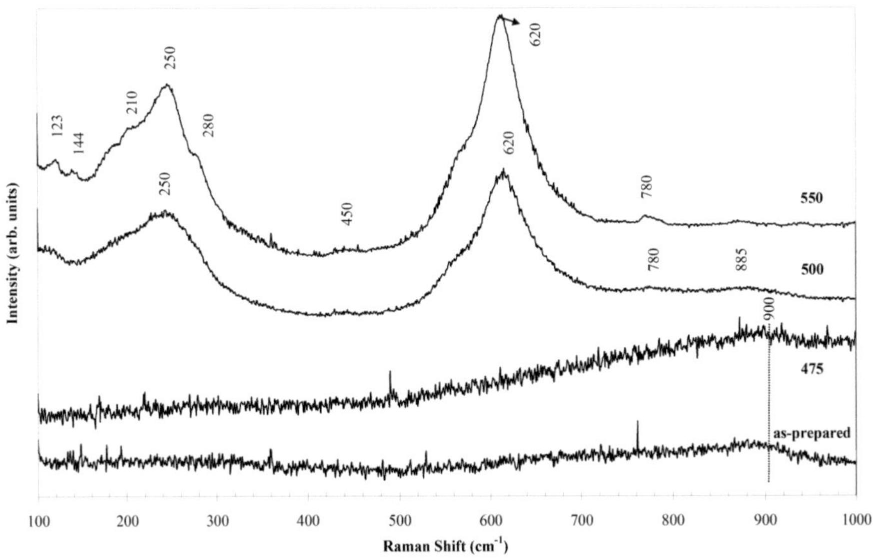

Fig. 4.45. Raman spectra of the 60BNa composition samples.

Figure 4.47 shows the dc conductivity in function of the temperature. It shows that σ_{dc} increases with the increase of the measurement temperature measurement, suggesting that the increase in temperature induces an increase in the mobility of the charge carriers. The increase of the HT temperature promotes a decrease of σ_{dc} (table 4.13), which should be related to the decrease of the charge carriers number, since the Ea(dc) (table 4.13), decreases from the as-prepared to the 550 °C HT sample. In these glasses, the charge carriers are the network modifier ions (Na^+ and Nb^{5+}), structurally inserted in the glass matrix, that, by participating in the formation of the $NaNbO_3$ crystalline structure, become no longer available to contribute to the conductivity. On the other hand, the increase of the volume ratio between the crystalline phase $NaNbO_3$ ($NaNbO_3$ presents, according Molak and co-workers, a electrical conductivity of ~ 10^{-7} S/m at 300 K) and the glass matrix, comproved by the XRD and SEM results, indicates that the number of "free" network modifier ions is the main factor in the electrical conductivity of these glass ceramics. For the same sample, it was verified that the $E_{a(ac)}$ is lower than the $E_{a(dc)}$ (table 4.13), which shows that the electrical units have to overcome different potential barriers when they relax (ac process) or when they participate in a long range process (dc conductivity). The σ_{ac} values, recorded at room temperature (table 4.13), are similar to those obtained by Singh and colleagues, in $NaNbO_3$ sintered ceramics (<10^{-8} S/m), indicating that the increase of the ac conductivity is due to the

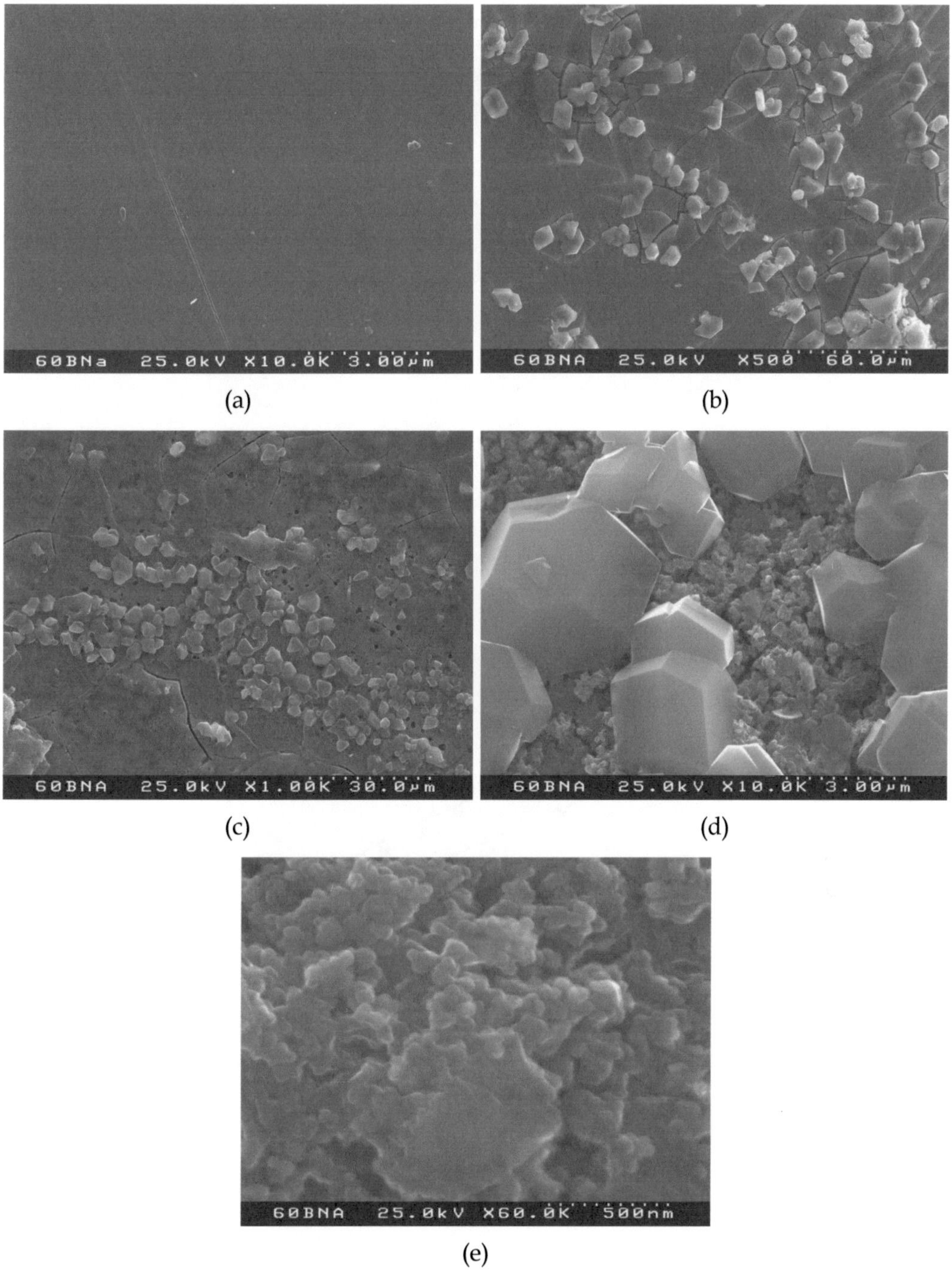

Fig. 4.46. SEM micrographs of the 60BNa samples:: a- as-prepared; b – HT at 475 °C; c.- HT at 500 °C; d – HT at 550 °C; e – HT at 550 °C - fracture.

increase of the quantity of $NaNbO_3$ crystals embedded in the glass matrix. Moreover, the decrease of the $E_{a(ac)}$ (table 4.13), with the increase of the HT temperature, shows a decrease in the height of the potential barriers and hence an increase of the electrical units mobility. The σ_{ac} behavior, related to the number of $NaNbO_3$ crystals embedded in the glass matrix, is consistent with the results obtained for the 60SiNa composition.
The ac conductivity (σ_{ac}), measured at 1 kHz and 300 K is similar in the as-prepared glass and in the sample HT at 475 °C, increasing with the increase of the HT temperature (table 4.13). The ac activation energy ($E_{a(ac)}$), calculated based on the dependence of σ_{ac} with temperature (fig. 4.48), revealed similar values for the as-prepared and the sample HT at 475 °C, decreasing for the sample HT at 500 °C, which has a value close to that obtained for the sample treated at 550 °C.

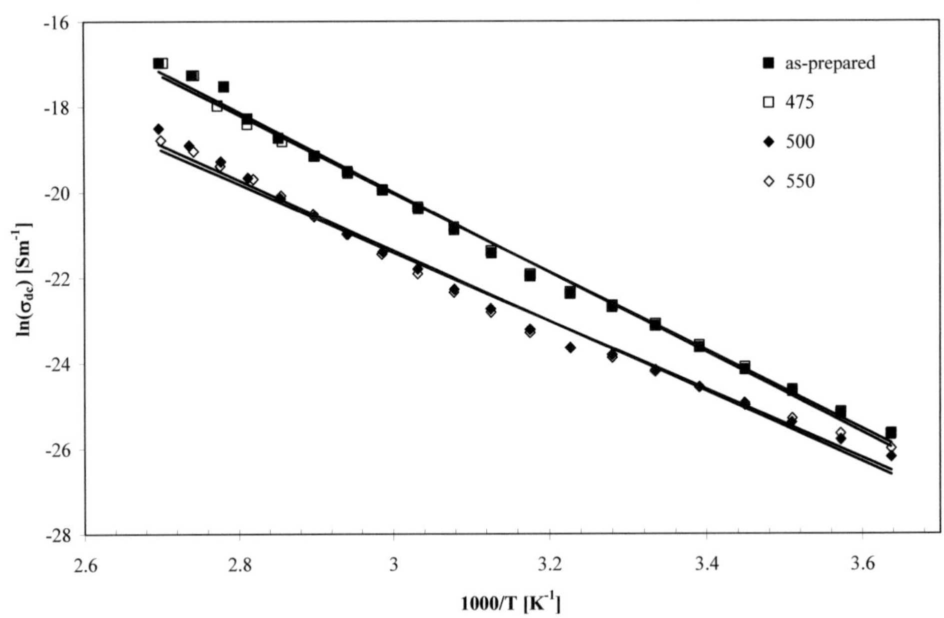

Fig. 4.47. ln(σ_{dc}) versus 1000/T for the 60BNa samples.

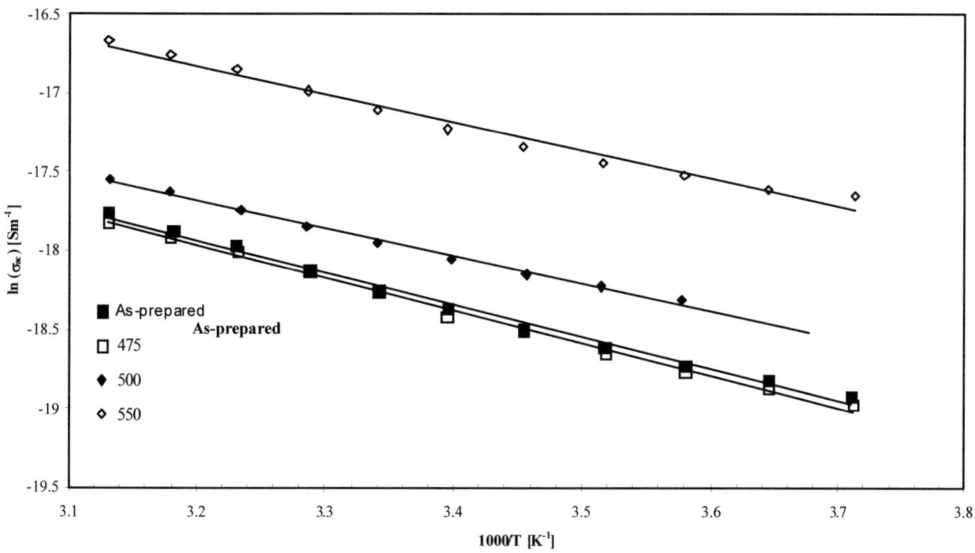

Fig. 4.48. ln(σ_{ac}) versus 1000/T for the 60BNa samples.

sample	σ_{dc} (x10^{-11}) [Sm^{-1}]	$E_{a(dc)}$ [kJ/mol]	σ_{ac} (x10^{-8}) [Sm^{-1}]	$E_{a(ac)}$ [kJ/mol]
As-prepared	9,03 ± 0,15	77,90 ± 1,33	1,19 ± 0,05	17,06 ± 0,44
475	9,44 ± 0,17	76,27 ± 1,17	1,17 ± 0,07	17,12 ± 0,37
500	3,20 ± 0,06	68,42 ± 1,69	1,60 ± 0,07	14,56 ± 0,37
550	3,08 ± 0,05	66,53 ± 1,91	3,71 ± 0,12	14,91 ± 0,68

Table 4.13 dc conductivity (σ_{dc}), at 300 K, dc activation energy ($E_{a(dc)}$), ac conductivity (σ_{ac}), at 300 K and 1 kHz and the ac activation energy ($E_{a(ac)}$).

Singh and coworkers present, for $NaNbO_3$ ceramic pellets sintered, values of dielectric constant (ε') of ~ 20, at 1 kHz and 305 K, much lower than the value of 600 measured in $NaNbO_3$ single crystals. The reason for this high difference can be the existence, either in glasses or in ceramics, of crystals with different dipolar orientations, which lowers the dipolar moment. In these glasses (60BNa), the increase of the HT temperature promotes an increase of ε' (table 4.14) which should be associated with the number of $NaNbO_3$ crystals precipitated in the glass matrix. The values obtained (between 10 and 36, table 4.14) are not comparable with those of the single crystals, indicating a random growth of crystals in the glass matrix. The dependence of ε', with the temperature of measurement is, in all samples, practically constant (fig. 4.49). This shows that, in the temperature range used, the $NaNbO_3$ does not suffers structural modifications, remaining antiferroelectric. It must be noticed that the dielectric losses (tan δ) are < 0.05, close to the value presented by Singh (~ 0.01), for the $NaNbO_3$ ceramic. The dielectric results of all samples, exemplified in figures 4.50 and 4.51, did not reveal the presence of dielectric relaxation mechanisms, in the frequency and

temperatures used, even using other dielectric formalisms (ε*, Y* or M*) beyond the Z*. This behavior is similar to that observed in the sample HT at 550 °C of the 60BLi system (fig. 4.42).
These data can be fitted to the Curie-Von Schweidler model (eq. 25), which corroborates the absence of dielectric relaxation in this frequency range. The results of these adjustments (table 4.15) shows that the n parameter is ~1.0, for all samples, and remains almost constant with the measuring temperature, which, according Jonsher, indicates that the electrical units responsible for this behavior are the same in all samples

Sample	ε´	ε´´	**tanδ (x 10-3)**
As-prepared	10.34 ± 0.43	0.15 ± 0.01	14.21 ± 0.84
475	11.77 ± 0.66	0.052 ± 0.003	4.46 ± 0.35
500	18.49 ± 0.77	0.13 ± 0.01	7.11 ± 0.41
550	36.94 ± 1.53	0.78 ± 0.03	21.06 ± 1.23

Table 4.14. Real and imaginary parts of the permittivity and the dielectric loss, at 300 K and 1 kHz.

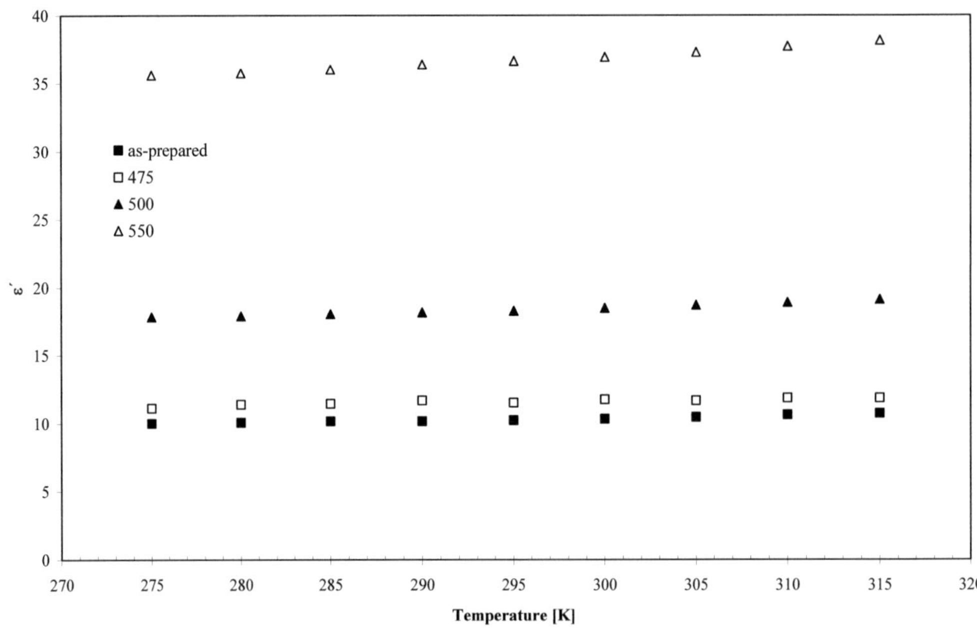

Fig. 4.49. ε´ versus temperature, measured at 1 kHz, for all 60BNa samples.

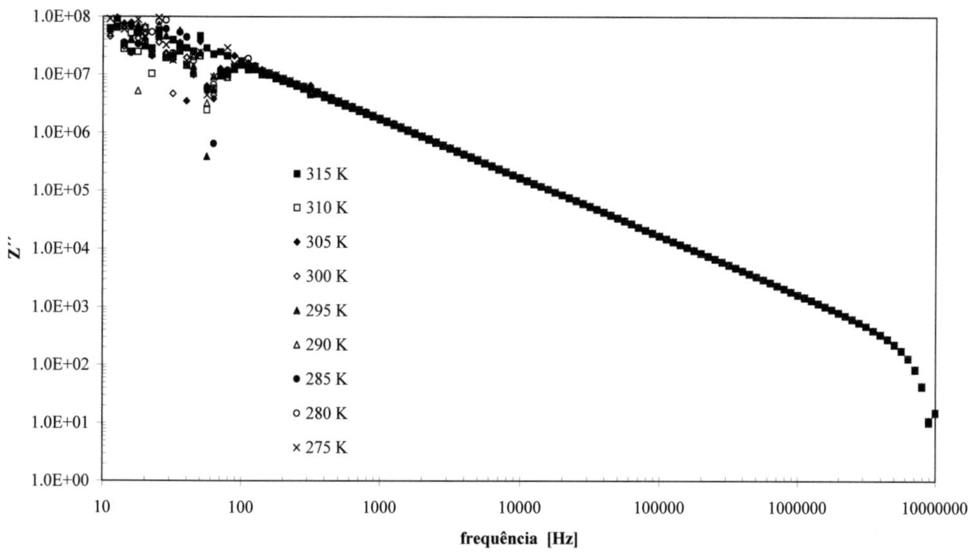

Fig. 4.50. Z´´ versus frequency for the as-prepared sample, at several temperatures.

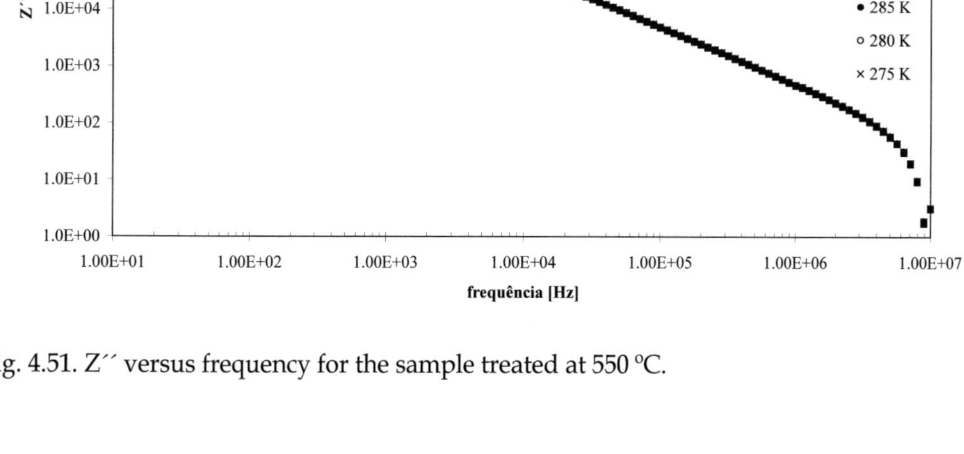

Fig. 4.51. Z´´ versus frequency for the sample treated at 550 °C.

Temper. [K]	As-prepared k (x10⁹)	As-prepared n	HT 475 k (x10⁹)	HT 475 n	HT 500 k (x10⁹)	HT 500 n	HT 550 k (x10⁸)	HT 550 n
315	1,735	1,005	1,653	1,015	0,952	1,001	4,579	0,998
310	1,761	1,006	1,653	1,015	0,970	1,002	4,653	0,999
305	1,786	1,007	1,678	1,016	0,983	1,003	4,730	1,000
300	1,802	1,008	1,674	1,016	0,998	1,004	4,775	1,000
295	1,837	1,008	1,718	1,018	1,012	1,005	4,834	1,001
290	1,853	1,009	1,714	1,018	1,021	1,005	4,880	1,001
285	1,874	1,010	1,765	1,020	1,032	1,006	4,928	1,002
280	1,896	1,010	1,737	1,019	1,042	1,006	4,959	1,002
275	1,906	1,000	1,777	1,020	1,049	1,006	4,993	1,002

Table 4.15. k and n parameters for the sample HT at 550 °C, at several measurement temperatures

The TSDC measurements reveal the presence of depolarization mechanisms indicating the possibility of dielectric relaxation mechanisms at low frequencies. In this particular case the 100 Hz represent, for this glass, the maximum frequency from which, due to sample-electrode interfacial polarization and the decreased sensitivity of the measuring apparatus, one begins to observe dispersion in measured values. Figures 5.3.7 to 5.3.10 present in the region of high frequency (> 1 MHz), a decrease of Z'' which suggests the possible presence of relaxation phenomena in the region of high frequencies. Note that the 30 MHz is the upper limit of the measuring apparatus used.

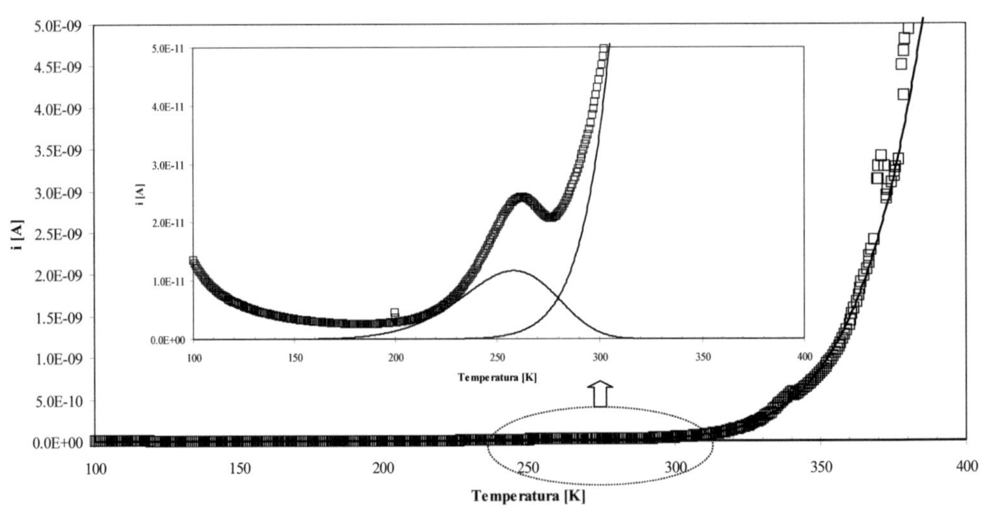

Fig. 4.52. TSDC spectra of the as-prepared glass.

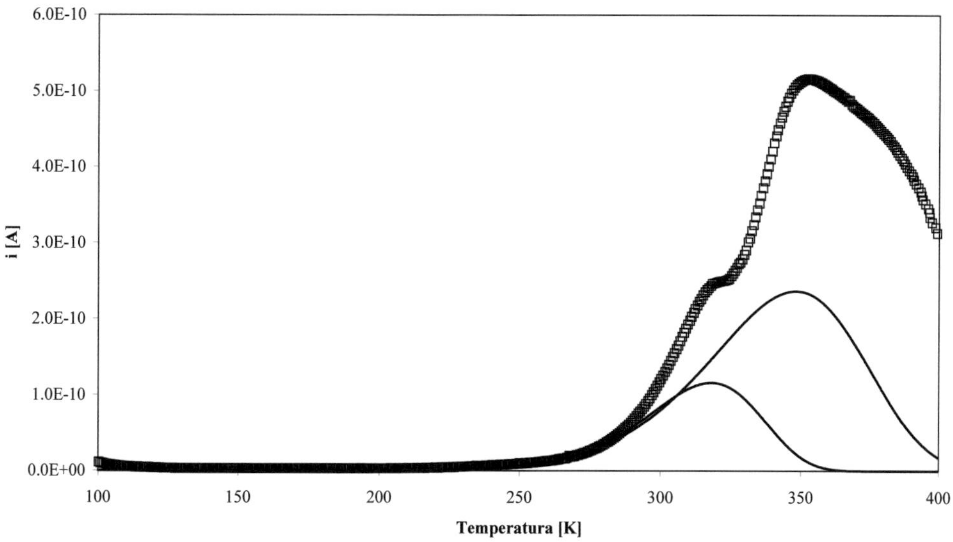

Fig. 4.53. TSDC spectra of the sample treated at 550ºC.

Sample	T_{P1} [K]	$i(T_{P1})$ x10^{-11} [A]	Ea [kJ/mol]	$\tau_{0\,(P1)}$ (x10^{2}) [s]	$\tau_{(P1)}$ (x10^{3}) [s]
as-prepared	262,3	2,43	21,68 ± 0,14	14,65 ± 0,16	3,96
475	281,0	1,14	22,46 ± 0,09	16,77 ± 0,18	4,39
500	305,8	24,11	34,17 ± 0,19	8,91 ± 0,09	3,41
550	320,3	24,87	38,86 ± 0,47	7,65 ± 0,08	3,29
	T_{P2} [K]	**$i(T_{P2})$ x10^{-10} [A]**	**Ea [kJ/mol]**	**$\tau_{0\,(P2)}$ (x10^{3}) [s]**	**$\tau_{(P2)}$ (x10^{3}) [s]**
as-prepared	--	--	--	--	--
475	--	--	--	--	--
500	--	--	--	--	--
550	352,3	5,16	33,12 ± 0,59	1,51 ± 0,01	4,67

Table 4.16. TSDC peak temperature (T_p), peak current ($i_{(Tp)}$), activation energy (E_a), pré-exponential parameter (τ_0) and relaxation time (τ), calculated on the low temperature peak (T_{P1}) and in the high temperature peak (T_{P2}).

The TSDC results revealed the presence, in the sample HT at 550 °C (fig. 4.53), of two depolarization peaks (P_1 and P_2) and in the other samples only one peak. The first TSDC peak (P_1) is attributed to localized movements of the network modifier ions structurally inserted in the glass matrix. With the increase of the HT temperature it was observed a shift to higher temperatures of the depolarization peak current (T_{P1}, table 4. 16), which indicates an increase in the difficulty of the dipolar units to relax. The presence of a second thermally stimulated depolarization current peak (P_2), in the sample treated at 550 ° C, can be attributed to dipolar depolarization associated with the particles embedded in the glass matrix. The non detection of this peak in the samples treated at lower temperatures should be related with the temperature at which this depolarization mechanism will occur, probably above 400 K.

Sample	ln (A) [A]	Ea [kJ/mol]
as-prepared	-1,26 ± 0,20	56,02 ± 0,57
475	2,04 ± 0,12	71,28 ± 0,36
500	-6,44 ± 0,05	44,24 ± 0,16
550	--	--

Table 4.17 Parameters of the Arrhenius equation used in the adjust of the TSDC high temperature range. ln (A) represents the logarithm of the pre-exponential factor and Ea the activation energy.

5. Conclusion

The main conclusions obtained from this study are:

- The molar composition 60M-30Li_2O-10Nb_2O_5 (mole%), with M = SiO_2, B_2O_3, prepared by melt-quenching, give origin to transparent glasses without crystalline phases;
- the silicate system is the one that allows, using the heat treatment process, to obtain glass ceramics where the $LiNbO_3$ crystalline phase can be precipitated has single or main crystal phase;
- The increase of the HT temperature in the 60SiLi composition 60SiLi promotes to an increase in the volume ratio between the $LiNbO_3$ particles and the glass matrix;
- The Raman spectroscopy of the 60SiLi samples, detected in the region of ~ 850 cm^{-1}, a band assigned to vibrations of the Nb-O-Si bond, indicating that probably some niobium ions are inserted in the glass matrix as network formers;
- The dc conduction process in the 60SiLi samples is dominated by the number of charge carriers, associated with the network modifier ions structurally inserted in the glass network;
- In all samples the ε' increases with the increase of the measurement temperature, indicating an decrease in the difficulty of the dipoles in follow the ac field;

The Z′ versus Z′′ spectra has the form, for most samples of the 60SiLi and 60BLi composition, of a semi-arc whose center lies below the axis Z', indicating the existence of a relaxation times distribution. The fit of these spectra was based on a CNLLS

sample	Temp. (K)	R [MΩ]	Q_0 ($x10^{-9}$) [$\Omega^{-1}m^{-2}s^n$]	n	τ_z ($x10^{-4}$) [s]	C_{CPE} ($x10^{-9}$) [F]	ε´	ε´´	tan δ
As-prepared	300	0,49	7,58	0,78	1,29	1,04	30,23	46,24	1,53
	295	0,63	7,28	0,78	1,65	1,05	28,62	38,40	1,34
	290	0,82	6,44	0,79	2,13	1,06	26,66	30,70	1,15
	285	1,14	5,90	0,80	3,06	1,11	24,80	23,93	0,96
	280	1,58	5,16	0,81	4,21	1,13	22,96	18,60	0,81
	275	4,27	5,06	0,77	15,60	1,06	18,69	8,92	0,48
	270	5,59	4,89	0,77	17,51	1,06	18,32	8,25	0,45
	265	7,28	4,26	0,79	19,60	1,10	17,57	6,85	0,39
	260	10,60	3,15	0,82	25,71	1,06	16,57	5,52	0,33
600	300	0,38	7,72	0,80	1,09	1,18	35,07	56,84	1,62
	295	0,62	6,61	0,81	1,80	1,20	30,84	37,69	1,22
	290	0,96	5,78	0,82	2,74	1,23	28,01	27,15	0,97
	285	1,30	4,99	0,83	3,61	1,22	25,90	20,59	0,80
	280	2,03	5,19	0,82	6,01	1,27	23,87	15,65	0,66
	275	3,54	5,92	0,76	9,95	1,06	22,26	12,08	0,54
	270	5,54	6,52	0,75	18,50	1,21	20,80	9,09	0,44
	265						19,52	6,86	0,35
	260						17,59	4,41	0,25
650	300	0,37	9,81	0,80	1,40	1,53	47,05	59,27	1,26
	295	0,47	8,99	0,81	1,88	1,67	41,85	41,61	0,99
	290	0,56	6,38	0,84	2,04	1,65	39,73	43,59	1,10
	285	0,97	5,84	0,83	3,25	1,49	33,19	25,55	0,77
	280	1,69	4,42	0,85	5,07	1,39	27,75	15,93	0,57
	275	2,51	3,90	0,86	7,51	1,43	25,72	12,19	0,47
	270						23,90	9,39	0,39
	265						22,37	7,29	0,33
	260						21,02	5,37	0,26

Table 1-annex. Electrical equivalent circuit parameters (R, Q_0, n), relaxation time (τ_z), capacitor C_{CPE}, real part (ε ') and imaginary (ε'') of the permittivity and the dielectric loss (tan δ) off all 60SiLi samples, at various temperatures of measurement.

sample	Temp. (K)	R [MΩ]	Q_0 ($x10^{-9}$) [$\Omega^{-1}m^{-2}s^n$]	n	τ_z ($x10^{-4}$) [s]	C_{CPE} ($x10^{-9}$) [F]	ε´	ε´´	tan δ
As-prepared	315	0,44	6,17	0,82	1,19	1,17	32,14	49,32	1,53
	310	0,58	4,92	0,84	1,52	1,20	29,72	38,57	1,30
	305	0,83	5,10	0,83	2,25	1,21	27,43	29,44	1,07
	300	1,28	6,15	0,81	3,91	1,28	25,21	21,86	0,87
	295	1,92	6,04	0,80	6,03	1,29	23,23	16,17	0,70
	290	3,09	6,16	0,77	9,36	1,18	21,54	12,28	0,57
	285	4,62	4,88	0,80	14,01	1,25	20,14	9,40	0,47
	280						18,90	7,34	0,39
	275						17,93	5,82	0,32
650	315	0,97	5,08	0,84	2,89	1,35	27,81	26,09	0,94
	310	1,30	4,85	0,83	3,68	1,27	26,39	20,92	0,79
	305	1,84	5,21	0,82	5,34	1,24	24,81	16,60	0,67
	300	2,44	4,45	0,83	6,98	1,28	23,50	13,26	0,56
	295	3,94	5,12	0,80	12,20	1,29	22,12	10,41	0,47
	290	5,83	5,33	0,78	18,31	1,24	21,09	8,04	0,38
	285						19,69	6,40	0,32
	280						18,84	4,79	0,25
	275						17,95	3,68	0,20
700	315	1,07	7,99	0,76	3,15	1,12	27,51	26,58	0,97
	310	1,73	9,50	0,73	5,85	1,18	25,85	20,08	0,78
	305	2,37	7,13	0,76	7,58	1,21	23,68	15,23	0,64
	300	2,90	4,12	0,83	7,65	1,17	22,01	11,85	0,54
	295	4,52	4,39	0,81	12,50	1,18	21,04	9,15	0,44
	290						19,82	7,50	0,38
	285						18,91	5,79	0,31
	280						17,78	4,68	0,26
	275						16,97	3,59	0,21
750	315	1,45	4,83	0,82	3,88	1,17	23,81	20,93	0,88
	310	2,07	5,35	0,81	5,96	1,21	22,67	17,53	0,77
	305	2,65	4,69	0,81	7,04	1,12	21,54	14,27	0,66
	300	3,58	4,59	0,80	9,53	1,11	20,37	11,65	0,57
	295	4,99	3,76	0,83	13,01	1,16	19,51	9,00	0,46
	290						18,27	7,22	0,40

sample	Temp. (K)	R [MΩ]	Q_0 ($x10^{-9}$) [$\Omega^{-1}m^{-2}s^{n}$]	n	τ_z ($x10^{-4}$) [s]	C_{CPE} ($x10^{-9}$) [F]	ε´	ε´´	tan δ
	285						17,29	5,59	0,32
	280						16,34	3,79	0,23
	275						15,48	2,89	0,19
	315	0,85	5,98	0,85	3,25	1,79	33,67	30,39	0,90
	310	1,15	5,49	0,85	4,17	1,69	31,90	23,99	0,75
	305	1,62	5,16	0,85	5,77	1,65	30,12	18,94	0,63
	300	2,36	4,90	0,85	8,51	1,68	28,47	15,04	0,53
800	295	3,53	5,72	0,80	12,30	1,46	26,96	12,09	0,45
	290	5,63	6,37	0,78	22,71	1,62	25,57	9,49	0,37
	285						24,50	7,64	0,31
	280						23,24	5,90	0,25
	275						22,17	4,91	0,22
	315	1,23	4,04	0,85	3,04	1,14	23,81	20,93	0,88
	310	1,78	6,49	0,77	4,85	1,05	22,67	17,53	0,77
	305	2,29	4,90	0,80	6,01	1,09	21,54	14,27	0,66
	300	2,99	4,27	0,82	7,65	1,10	20,37	11,65	0,57
750B	295	4,59	4,49	0,80	12,50	1,13	19,51	9,00	0,46
	290						18,27	7,22	0,40
	285						17,29	5,59	0,32
	280						16,34	3,79	0,23
	275						15,48	2,89	0,19
	315	0,73	4,24	0,85	1,78	1,13	26,45	31,17	1,18
	310	1,18	4,32	0,84	3,06	1,19	24,43	21,59	0,88
	305	1,77	3,84	0,85	4,48	1,18	22,68	16,02	0,71
	300	2,54	3,78	0,84	6,44	1,16	21,37	12,71	0,60
800B	295	3,69	4,13	0,82	9,59	1,12	20,18	9,89	0,49
	290	5,17	3,52	0,83	13,20	1,14	19,17	7,94	0,41
	285						18,14	6,27	0,35
	280						17,22	4,76	0,28
	275						16,55	4,01	0,24

Table 2-annex. Electrical equivalent circuit parameters (R, Q_0, n), relaxation time (τ_z), capacitor C_{CPE}, real part (ε ') and imaginary (ε'') of the permittivity and the dielectric loss (tan δ) off all 60SiNa samples, at various temperatures of measurement.

sample	Temp. (K)	R [MΩ]	Q_0 ($x10^{-9}$) [$\Omega^{-1}m^{-2}s^n$]	n	τ_z ($x10^{-3}$) [s]	C_{CPE} ($x10^{-9}$) [F]	ε´	ε´´	tan δ
As-prepared	315	2,83	4,60	0,81	0,77	1,16	22,01	12,07	0,55
	310	3,92	4,16	0,82	1,05	1,17	20,94	9,78	0,47
	305	4,57	4,39	0,81	1,27	1,17	20,28	8,86	0,44
	300	9,08	3,37	0,82	2,21	1,06	18,83	5,76	0,31
	295	15,06	3,59	0,81	4,30	1,22	17,92	4,40	0,25
	290	21,30	2,85	0,84	5,68	1,21	17,15	3,34	0,19
	285	31,00	2,30	0,86	7,58	1,13	16,56	2,63	0,16
	280						16,06	2,12	0,13
	275						15,51	1,60	0,10
450	315	8,82	2,67	0,87	2,07	1,13	18,69	5,49	0,29
	310	7,39	3,20	0,84	1,85	1,16	18,52	5,95	0,32
	305	10,62	3,35	0,83	2,84	1,19	17,97	5,11	0,28
	300	13,99	2,82	0,85	3,54	1,19	17,16	4,13	0,24
	295	23,18	3,48	0,79	6,37	1,12	16,61	3,28	0,20
	290	32,78	2,85	0,82	8,84	1,15	16,09	2,57	0,16
	285						15,68	2,14	0,14
	280						15,13	1,67	0,11
	275						14,79	1,33	0,09
500	315	9,86	4,66	0,76	2,79	1,08	19,97	6,04	0,30
	310	10,42	3,33	0,85	3,06	1,32	19,46	5,32	0,27
	305	25,77	2,69	0,86	7,23	1,34	18,38	3,56	0,19
	300	35,46	3,23	0,82	11,42	1,36	17,93	3,15	0,18
	295	50,91	2,63	0,84	14,50	1,33	17,20	2,45	0,14
	290	95,46	2,55	0,82	31,82	1,32	16,69	1,96	0,12
	285						16,26	1,57	0,10
	280						15,88	1,29	0,08
	275						15,50	1,03	0,07
550	315						11,82	0,22	0,02
	310						11,57	0,21	0,02
	305						11,48	0,29	0,03
	300						11,39	0,18	0,02
	295						11,42	0,27	0,02
	290						11,42	0,29	0,03
	285						11,42	0,18	0,02
	280						11,35	0,30	0,03
	275						11,28	0,25	0,02

Table 3-annex. Electrical equivalent circuit parameters (R, Q_0, n), relaxation time (τ_z), capacitor C_{CPE}, real part (ε ') and imaginary (ε'') of the permittivity and the dielectric loss (tan δ) off all 60BLi samples, at various temperatures of measurement.

- algorithm associated with an equivalent circuit model composed by the parallel between an R and a CPE;
- In the samples of the 60SiLi composition the parameter R has a similar behavior to that of the dc conductivity and the parameter n, being close to 0.80, allowed to approach the empirical element CPE to a capacitor;
- In the samples where it was not observed using various representations (Z*, Y*, M*, etc.), the formation of semi-circles, the spectra of Z* was adjustable to the Curie-Von Schweidler model.
- The TSDC results of the 60SiLi composition reveal the presence of at least two depolarization mechanisms. The current peak detected at lower temperatures (<250 K) is due to the depolarization of the dipoles associated with the "free" network modifier ions. The second depolarization peak, centered at temperatures >250 K, is due to dipolar depolarization from the movements of the electrical units responsible for the charge carrier in limited paths. The presence of depolarization current peaks at temperatures higher than the second peak must be associated with the dipoles formed between the sample surface and the electrodes;
- In the 60BLi composition, the precipitation of $LiNbO_3$ crystallites occurs with the HT at temperatures above 500 °C;
- The increase of the HT temperature favors the increase in size and the decrease in the number of particles.
- The results of the Raman spectroscopy in the 60BLi samples, show that the as-prepared glass, composed by pentaborate groups, undergoes a structural transformation, with the increase of the HT temperature, to a metaborate structure type;
- In the 60BLi samples the dc and ac conduction processes depend mainly on the number of network modifiers ions structurally inserted in the glass network. The dielectric analysis showed that the main contribution to the dipole moment is the amount of $LiNbO_3$ crystals embedded in the boron matrix;
- The equivalent electrical circuit formed by the parallel between R and CPE, adjusts the dielectric response of the 60BLi samples.
- It was found in the 60BLi samples that the increase of the relaxation time (τ_Z), with the increase of the HT temperature, indicates the increase of the volume ratio between particles of $LiNbO_3$ and the glass matrix;
- The preparation of glass ceramics containing $NaNbO_3$ crystals is possible through heat treatment of the as-prepared glass of the compositions 60M-30Na_2O-10Nb_2O5 (mole%), with M = SiO_2, B_2O_3;
- The $NaNbO_3$ crystalline phase was detected in the 60SiNa composition samples treated at temperatures above 750 °C and in the 60BNa composition for temperatures above 500 °C;
- In the 60SiNa composition it was not detected the presence of other crystalline phases with increasing the HT temperature up to 800 ° C;
- In the samples of the 60BNa composition, the HT for temperatures above 600 °C also favors the formation of the secondary phases (Nb_2O_5 and $Na_2B_4O_7$);
- The crystallization in the 60SiNa composition is superficial and can be removed by a mechanical polishing process. In the borate matrix, the crystallization occurs in volume and the increase of the HT temperature promotes an increase in the volume amount of particles embedded in the glass matrix. This increase is a consequence of the decrease of the number of "free" network modifier ions which also causes the observed decrease in the conductivity.

- In the glasses and glass ceramics of the 60SiNa composition, the dielectric behavior is explained by the bulk characteristics of the samples. The values of ε' depend on the number of dipoles existing in the bulk, i.e., the number of network modifier ions structurally inserted in the glass matrix. In the 60BNa composition, the behavior of ε' depends on the amount of particles $NaNbO_3$;
- The dielectric behavior of the 60SiNa samples shows the presence of semi-arcs, with the Z′ versus Z′′ representation, centered below the Z′ axis, indicating the existence of a relaxation times distribution.
- The results of the dielectric measurements of the 60BNa samples do not reveal the presence of dielectric relaxation mechanisms, in the frequency and temperature range used;
- The fact that the TSDC results reveal the presence of depolarization mechanisms indicates the possibility of the existence of dielectric relaxation mechanisms at lower frequencies.
- The behavior of ε' with the temperature of measurement is, in all 60BNa samples, practically constant showing that, in the temperature range used, the $NaNbO_3$ does not presents structural changes, remaining antiferroelectric.

6. Acknowledgment

The authors would like to thanks the Portuguese Foundation for Science and Technology (FCT) for the financial support and to Aveiro University, namely the laboratory of Non-Crystalline Solids and Disorder Systems (Physics Department) for all the laboratory support.

7. References

M.M. Aboulleil, F.J. Leonberger, J. Am. Ceram. Society, 72 (1989) 1311-1320.

A.K. Agarwal, D.E. Day, J. Am. Ceram. Soc. 65 (2), 111-117 (1981).

A.K. Agarwal, D.E. Day, J. Am. Ceram. Soc. 65 (5), 231-237 (1981).

R. Akagi, N. Ohtori, N. Umesaki, J. Non-Cryst. Solids, 293-295 (2001) 471-476

J.S. Andrade, A.G. Pinheiro, I.F. Vasconcelos, J.M. Sasaki, J.A.C. Paiva, M.A. Valente, A.S.B. Sombra, J. Phys. Cond. Matter, 11 (1999) 4451-4460.

E.B. de Araujo, J.A.C. de Paiva, M.A.B. de Araujo e A. Sergio Bezerra Sombra, Physica Scripta. Vol. 53, 104-107 (1996).

E.B. de Araujo, J.A.M. de Abreu, R.S. de Oliveira, J.A.C. de Paiva, A.S.B. Sombra, Canadian Journal of Physics, 75 (1997) 747-758.

A.C:V de Araujo, I.T. Weber, W.D: Fragoso, C.M. Donegá, J. of Alloys and Compounds 275-277 (1998) 738-741

A. Aronne, V.N. Sigaev, B. Champagnon, E. Fanelli, V. Califano, L.Z. Usmanova, P. Pernice, J. Non-Cryst. Solids 351 (2005) 3610–3618

F.Bahri,H.Khemakhem,M.Gargouri,A.Simon,R.V.Mühll,J.Ravez, Solid State Sci. 5 (2003) 1229-1234.

B.A Boukamp, Solid State Ionics, 18-19, 136-140 (1986).

B.A Boukamp, Solid State Ionics, 62, 131-141 (1993).

P. Braunlich, " Thermally stimulated relaxation in solids", Springer, Berlin, 1979.

R.K. Brow, "Introduction to Glass Science" Cer103 (2004), University of Missouri-Rolla, EUA (www.umr.edu/~brow/index.html).

T. Cardinal, E. Fargin, G. Le Flem, S. Leboiteux, J. Non-Cryst. Solids 222 (1997) 228-234.

S.H. Carr, "Electrical Properties of Polimers", Academic Press, 1982, N.Y.

F. Cavalier, Le contrôle global de VIRGO, Mémoire d'habilitation soutenu le 14/09/2001, Le Laboratoire de l'Accélérateur Linéaire - Université Paris-Sud, http://www.lal.in2p3.fr/presentation/bibliotheque/publications/2001/web/habilitation.html

R. Chen, J. Mater. Sci., 11, 1521-1541 (1976).

K. Chou, J. Non-Cryst. Solids, 110 (1989) 122-124.

L.M. Cadillon M. Costa, "Propriedades eléctricas de vidros com alguns iões de terras raras", Tese de Doutoramento, Universidade de Aveiro, Aveiro (1995).

M. Cutroni, A. Mandanici, Solid Sate Ionics, 105 (1998) 149-157.

A Dan, D. Chakravorty, J. Mater. Res., 15(6), 1324-1330 (2000)

C.N.W. Darlington, K.S. Knight, Physica B 266 (1999) 368-372.

M.M. El-Desoky, S.M.Salem, I.Kashif, J. Mat. Sci, 10 (1999) 279-283.

Y. Ding, Y. Miura, S. Nakaoka, T. Nanba, J. Non-Crist. Solids, 259, 132-138 (1999).

A Doi, Solid State Ionics, 107, 81-88 (1998)

H.L. Downing, N.L. Peterson, H. Jain, J. Non-Cryst. Solids, 50(2), 203-13 (1982)

AM. Efimov, J. Non-Cryst. Solids, 253 (1999) 95-118.

V.C. Farmer, "The infrared spectra of minerals", Mineralogical Society, London, 1974.

K. Fukumi, S. Sakka, J. Mater. Sci., 23 (1988) 2819-2823.

T. Fuss, C.S. Ray, N. Kitamura, M. Makihara, D.E. Day, J. Non-Cryst. Solids 318 (2003) 157-167.

K.Gerth, C.Rüsell, R.Kending, P.Schleevoigt, H.Dunken, Phys.Chem. Glasses, 40(3) (1999) 135-9.

M.P.F.Graça, M.A.Valente, M.G.F.Silva, J. Non-Cryst. Solids, 325 (2003) 267-274.

M.P.F. Graça, M.G.F. da Silva, A.S.B. Sombra, M.A. Valente, Journal of Non-Crystalline Solids 353 (2007) 1390-1394

MPF Graça, MGF Silva, ASB Sombra and MA Valente, Journal of Non-Crystalline Solids, 352 (42-49) 2006, Pages 5199-5204.

M.P.F. Graça, M.G.F. Silva, M.A. Valente - Journal of Non-Crystalline Solids, 351 (33-36) (2005) 2951-2957

M.P.F. Graça, M.A. Valente, M.G. Ferreira da Silva - Journal of Materials Science, 41 (2006) 1137-1141

M.P.F. Graça, M.G. Ferreira da Silva, M.A. Valente, Advanced Materials Forum III, 2005, pp274

M.P.F. Graça, M.G. Ferreira da Silva, A.S.B. Sombra and M.A. Valente Physica B 396 (2007) 62-69

G.H. Haertling, J. Am. Ceram. Soc., 82 (4), 797-818 (1999).

R.M. Hakim, D.R. Uhlmanh, Phys. Chem. of Glasses, 12 (5), 132-138 (1971).

A. Halliyal, A.S. Bhalla, R.E. Newnham, Materials Research Bulletin, 18, 1007-1019 (1983).

E. Hecht, "Óptica", Fundação Calouste Gulbenkian, 1991.

A. Herczog, J. Am. Ceram. Soc. 73(9) (1990) 2743-2746.

S. Hirano, T. Yogo, K. Kikuta, Y. Isobe, J. Mater. Sci., 28 (1993) 4188-4192.
C. Hong, D.E. Day, J. Am. Ceram. Soc., 64:2, 61-67 (1981).
C. Hong, D.E. Day, J. Mater. Sci., 14, 2493-2499 (1979).
P. Huang, X. Huang, Solid State Ionics 36 (1989) 59-63.
T. Hungría, L. Pardo, A. Moure, A. Castro, J. of Alloys and Compounds 395 (2005) 166–173.
R. Jenkins, R.L. Snyder, "Introduction to X-ray powder diffractometry" , John Wiley, N.Y. 1996.
A.K. Jonscher, "Dielectric relaxation in solids", Chelsea Dielectrics Press, London, 1983.
V.D. Jovic, B.M. Jovic, J. Electroanalytical Chemistry 541 (2003) 13-/21; V. D. Jovic, "Determination of the correct value of Cdl from the impedance results fitted by the commercially available software ", Research Solutions and Resources (http://www.consultrsr.com).
Y.D. Juang, S.B. Dai, Y.C. Wang, W.Y. Chou, J.S. Hwang, M.L. Hu, w.S. Tse, Solid State Communications 111 (1999) 723-728
G.O. Karapetyan, V.V. Loboda, D.K. Tagantsev, J. Non-Cryst. Solids, 283, 114-118 (2001).
C. Kim, Y. Hwang, H.K. Kim, J.N. Kim, Phys.Chem. Glasses, 44(2) (2003), 166-9.
H.G. Kim, T. Komatsu, R. Sato and K. Matusita , J. Non-Cryst. Solids, 162, 201-204 (1993).
Y.S. Kim, R.E. Tressler, J. Mat. Sci., 29, 2531-2535 (1994).
T. Komatsu, R. Ike, R. Sato, K. Matusita, Phys. Chem. Glasses, 36(5) (1995) 216-221
H.G. Kim, T. Komatsu, R. Sato, K. Matusita, J. Mat. Sci. 31 (1996) 2159-2164.
T. Komatsu, H. Tawarayama, H. Mohri and K. Matusita , J. Non-Cryst. Solids, 135, 105-113 (1991).
A.Koné, B.Barrau, J.L.Souquet, M.Ribes, Mat. Res. Bull, 14 (1979) 393-399.
F.Kremer, A.Schönhals, "Broadband Dielectric Spectroscopy", Springer, Germany, 2002.
C.D.E. Lakeman, Y. Xia, J. Kim, X. Wu, H.G. Eckert, F.F. Lange, J. Mater. Res. 13(6) (1998) 1596-1606.
S. Lanfredi, L. Dessemond, ACM Rodrigues, Journal of the European Ceramic Society, 20 (7) (2000) 983-990
G. Li, H. Ma, Y. Jiao, S. Chen, J. Serb. Chem. Soc. 69 (10) (2004) 791-805.
R.J.C. Lima, P.T.C. Freire, J.M. Sasaki, A.P. Ayala, F.E.A. Melo, J.M. Filho, K. C. Serra, S. Lanfredi, M.H. Lente, J.A. Eiras, J. Raman Spect. 33 (2002) 669-674.
S. Lin, Y. Tanaka, M. Aono, T. Suzuki, J. Appl. Phys., 36, 3510-3514 (1997).
A A Lipovskii, V.D. Petrikov, V.G. Melehin, D.K. Tagantsev, B.V. Tatarintsev, Solid State Comunications, 117, 733-737 (2001).
A.A. Lipovskii, D.K. Tagantsev, A.A. Vetrov, O.V. Yanush, Optical Materials 21 (2003) 749-757
P.B. Macedo, C.T. Moynihan and R. Bose, Phys. Chem. Glasses, 13:6 (1972) 171-179.
J.R. Macdonald, "Impedance spectroscopy", John Wiley & Sons, New York, 1987.
D. Maniu, T. Iliescu, I. Ardelean, S. Cinta-Pinzaru, N. Tarcea, W. Kiefer, J. of Molecular Structure 651–653 (2003) 485–488
M.L.F. Nascimento "Condutividade elétrica de vidros boratos, silicatos e sílico-sulfatos de íons alcalinos", Tese de Mestrado (2000), Univ. de São-Paulo, Brasil (http://www.teses.usp.br/teses/disponiveis/43/43133/tde-09122002-

161054/publico/dissertacao.pdf)
B.T. Matthias, J.P. Remaika, Physical Review, 82 (5), 1951, 727-731
F. Mehran, B.A. Scott, Solid State Communications, 11, 15-19 (1972).
A.M.Milankovic and D.E.Day, J. Non-Cryst. Solids, 162 (1993) 275-286.
A. Molak, J. Kubacki, Cryst.Res. Technol, 36 (8-10) (2001) 893-902.
A Molak, M Paluch, S Pawlus, J Klimontko, Z Ujma, I Gruszka, J. Phys. D: Appl. Phys. 38 (2005) 1450–1460.
H.I. Motulsky and A. Christopoulos, Fitting models to biological data using linear and nonlinear regression. A practical guide to curve fitting, GraphPad Software Inc., San Diego CA, http://www.graphpad.com, (2003).
K. Nassau, C.A. Wang, M. Grasso, J. Am. Ceram. Society, 62 (1978) 503-510.
K. Nassau, C.A. Wang, M. Grasso, J. Am. Ceram. Society, 62(1-2) (1978) 74-79
J.M.F. Navarro, "El Vidrio" (CSIC-Fundación Centro Nacional del Vidrio, Madrid 1991).
E.R. Neagu, R.M. Neagu, C. Botez, P. Pissis, Fizica Starii Condensate, 1999-2000, p. 249-259.
T.J. Negran, A.M. Glass, Phys. Chem. of Glasses, 20, 140-141 (1979).
Kia L. Ngai and Ronald W. Rendell in "Handbook of conducting polymers", vol II, Marcel Dekker, N.Y., 1986.
K.L. Ngai, S.W. Martin, Physical Review B, 40(15) (1989) 10550-56.
T. Nozawa, S. Miyazawa, J. Appl. Phys.s, 35, 107-113 (1996).
J. Palma, J.R Jurado, P. Duran, C. Pascual, Bol. Soc. Esp. Ceram. Vidrio, 30 (6), 472-475 (1991).
A. Paul, "Chemistry of Glasses", Chapman & Hall, London, 1982
N. Syam Prasad, K.B.R. Varma, J. Non-Cryst. Solids 351 (2005) 1455–1465.
E. Prasad, M. Sayer, H.M. Vyas, J. Non-Cryst. Solids, 40, 119-134 (1980).
K.J.Rao, K.C. Sobha, S. Kumar, Proc. Indian. Acad. Sci. (Chem. Sci.), 113 (5-6) (2001) 497-514
L. Rebouta, J.C. Soares, M.F. Da Silva, J.A. Sanz-Garcia, E. Dieguez, F. Agullo-Lopez, Mat. Res. Soc. Symp. Proc., 244 (1992) 311-316; Luis M.F. Rebouta, "Localização de dopantes e caracterização microscópica de dopantes em Niobato de Lítio", Tese de Doutoramento, Lisboa (1992)
R. Keding, C. Rüssel, J. Non-Cryst. Solids, 219 (1997) 136-141
E.Rysiakiewicz-Pasek, 10th International Symposium on Electrets, 1999 IEEE
M.V. Shankar, K.B.R. Varma, J. Non-Cryst. Solids, 243 (1999) 192-203.
Z.X. Shen, X.B. Wang, M.H. Kuok, S.H.Tang, J. Raman spectroscopy, 29 (1998) 379-384.
G. Shirane, R. Newnham, R. Pepinsky, Phys. Ver. 96(1) (1954) 581-588
N. Shibata, M. Horigudhi, T. Edahino, J. Non-Cryst. Solids, 45 (1981) 115-126.
C.J. Simmons, O.H. El-Bayoumi, "Experimental Techniques of Glass Science", The American Ceramic Society, Ohio, 1993.
K. Singh, P.R. Gandhi, B.M. Chaudhari, Solid State Ionics, 28-30 (1988) 752-755.
W.J. Smothers, Y. Chiang, "Handbook of Differential Thermal Analysis", chemical Publishing Company, Inc., New York (1966).
J.Tanaka, M. Tsukioka, Y. Miyazawa, Y. Mori, S. Shimazu, Solid State Comm. 34 (1980) 221-224

J.P. Fillard, J. Van Turnhout, "Thermally stimulated processes in solids:new prospects", Elsevier Scientific, 1977; P. Braunlich, " Thermally stimulated relaxation in solids", Springer, Berlin, 1979.

M. Todorovic, L. Radonjic, Ceramics International, 23 (1997) 55-60.

N. Umesaki, N. Iwamoto, M. Tatsumisago, T. Minami, J. Non-Cryst. Solids, 106 (1988) 77-80.

B.H. Venkatareman, N.S. Prasad, K.B.R. Varma, V. Rodriguez, M. Maglione, R. Vondermuhll, J. Etourneau, App. Phys. Lett. 87 (2005) 091113

E.M. Vogel, J. Am. Ceram. Society, 72 (1989) 719-724

G.H. Weiss, J.T. Bendler, M. Dishon, J. Chem. Phys. 83-3, 1424-1427 (1985); R.S. Weis, T.K. Gaylord, Applied Physics A, 37 (1985) 191-203.

A. J. Moulson, J. M. Herbert, "Electroceramics, Materials _ Properties _ Applications", 2nd Edition, Wiley (2003), England.

D. Xue, K. Kitamura, Ferroelectrics, 296 (2003) 1-9.

T. Yagi, M. Susa, K. Nagata, J. Non-Cryst. Solids 315 (2003) 54–62.

Y.I Yuzyuk, P Simon, E Gagarina, L Hennet, D Thiaudi`ere,V I Torgashev, S I Raevskaya, I P Raevskii, L A Reznitchenko, J L Sauvajol, J. Phys.: Condens. Matter 17 (2005) 4977–4990.

HC Zeng, K Tanaka, K Hiaro, N Soga, J. of Non-Cryst. Solids, 209 (1997) 112-121.

N D Zhigadlo, M Zhang, E K H Salje, J. Phys.: Condens. Matter 13 (2001) 6551–6561.

12

Evaluation on Structure Modification and Properties of $(Ba_{1-x}Sr_x)(Ti_{1-y}Zr_y)O_3$ Ceramics by using Rietveld Method

Hong-Hsin Huang[1], Hsin-Hsiung Chiu[2],
Nan-Chung Wu[2] and Moo-Ching Wang[3]
[1]*Cheng Shiu University,*
[2]*National Cheng Kung University,*
[3]*Kaohsiung Medical University*
Taiwan

1. Introduction

The $BaTiO_3$, BT, is representative ferroelectric prototype because of its excellent electrical property [Maison, 2003; Yu et al., 2004]. After the discovery of ferroelectricity in BT, research from various laboratories all over the word start studies of solid solutions of BT with other perovskites such as $ATiO_3$ and $BaBO_3$ where A=Ca, Sr, Pb, Mn and B=Zr, Sn, Hf, Mn. [Lemanov, 2007] This research field is of great basic and applied interest since the experimental results should facilitate developing of theory and fining of ferroelectric solid solutions with properties optimal for application. The phase development and dielectric behavior of various perovskite ferroelectric ceramics was reported, which may be applied to several micro- or nano-positioning devices such as deformable mirrors, microactuators, miniaturized transducers, multilayer capacitors (MLCs), PTC thermistors, piezoelectric transducers, and a variety of electro-optic devices [Haertling, (1999); Uchino, (1998); Polli, (2000)].

BT has been studied extensively in solid solution with $SrTiO_3$, ST, to form a nonlinear ferroelectric with high dielectric constant ceramic $Ba_{1-x}Sr_xTiO_3$, BST, for application as integrated storage capacitors in giga bit dynamic random access memory (DRAM) due to their low dielectric loss, loss leakage current, low temperature coefficient of dielectric constant and the composition dependent Curie temperature. [Rout & Panigrahi (2006)] It is well known that these properties critically depend on their composition and structural characteristics. [Bhalla et al., (2000); Dutta et al. (1994); Wei et al. (2008), Ezhilvalavan & Tseng (2000), Carlson et al. (2000); Zimmermann et al. (2001); Loachim et al. (2007)] Formation mechanism of BST from a mixture of BT and ST had been investigated by Kolar et al. [Kolar et al. (1982)] who reported that the formation of BST was governed by preferential diffusion of Ba^{2+} ion into ST. The preferential diffusion disturbs sintering in BT-ST powder mixture, whereas formation of a liquid phase results in exaggerated grain growth. In addition, two-stage sintering process was reported by Kolar et al, the similar process was also reported by Rout and Panigrahi. [Rout & Panigrahi, (2006)] The

intermediate phase such as Sr_2TiO_4, $SrTi_3O_7$, [Rout & Panigrahi, (2006); Bera & Rout (2005)] $BaTi_xO_{2x+1}$, and $(Ba,Sr)_2TiO_3$ [Kolar et al. (1982)] were observed or not for BST formation.
The $BaTi_{1-x}Zr_xO_3$, BTZ, solid solution formed from $BaTiO_3$ and $BaZrO_3$ (BZ) has received much attention and most studies are focused on the temperature dependence of the dielectric constant, the nature of phase transition, and the relaxor behavior of this material. [Yu et al., 2002; Bera & Rout, 2005] The formation mechanism of BTZ solid solution was explored by Bera and Rout [Bera & Rout, 2005] who deduced the diffusion of BT into BZ, where the BT acts as diminishing core. Currently, BTZ has been chosen as an alternative material to replace BST in the fraction of ceramic capacitors due to the $[ZrO_6]$ cluster is chemically stable than those of $[TiO_6]$. [Qin et al., 2008; Tsurumi et al., 2011; Tsurumi et al., 2011, Yu et al., 2002] The dielectric properties and phase transformation temperature, Tm (from ferroelectric to paraelectric), of BTZ are strongly dependent on Zr content into the lattice. [Badapanda et al., 2009] Yu et al. [Yu et al., 2002] had reported the dependence of Zr content on structure and dielectric properties, who also demonstrated the excellent dielectric properties were found owing to various additions of Zr into the BT lattice.
The phase formation kinetics and mechanism of $SrTiO_3$-$SrZrO_3$ (STZ) solid solution through solid-oxide reaction had reported by Bera and Rout. [Bera & Rout, 2005] They showed that the ST was formed at lower temperature of 800°C with lower activation energy of 42.274 kcal/mole than SZ. SZ formation started at 1000°C with higher activation energy of 65.78 kcal/mole. However, STZ formation started at 1500°C onward with very high activation energy of 297.52 kcal/mole. It is also concluded that the solid solution formed coherently with SZ lattice by diffusion of Ti into the SZ due to the ST plays a diminishing role and the intensity of SZ increases with the sintering temperature increasing as shown in XRD patterns.
Few lectures was reported to investigate the structure and dielectric properties of $(Ba_{1-x}Sr_x)(Ti_{1-y}Zr_y)O_3$, BSTZ, ceramics. The dielectric properties and tunability of $(Ba_{0.80}Sr_{0.20})(Ti_{1-x}Zr_x)O_3$ multilayered thin films as a function of Zr content was reported by Wang et al. [Wang et al., 2004] The dielectric properties of $(Ba_{1-x}Sr_x)(Ti_{0.5}Zr_{0.5})O_3$ ceramics as a function of Sr content was explored by Bera, and Rout [Bera & Rout, 2007] who showed the solid solution of $(Ba_{1-x}Sr_x)(Ti_{0.5}Zr_{0.5})O_3$ ceramics remains cubic phase up to $x<0.6$ and becomes tetragonal in the range of x=0.6-1.0. However, the effect of both Sr and Zr contents on the structure and dielectric properties was not reported. In this study, the structure and dielectric properties of BSTZ ceramics were investigated as a function of both Sr and Zr contents.

2. Experimental procedure

The intermediate source ceramics of BT and BZ were separately prepared via the method of solid-state reaction. The starting materials, $BaCO_3$ (99.9 pct, Katayama, Japan), TiO_2 (99.9 pct, Katayama), $SrCO_3$(99.9 pct, Katayama, Japan), and ZrO_2 (99.9 pct, Showa, Japan) powders with a stoichiometric composition of $BaTiO_3$, $BaZrO_3$, and $SrTiO_3$, respectively, were ball milled with deionized water for 10 h and then calcined at 1200°C for 2 h to form the single-phase BT and BZ powders. The BT and BZ powders were blended into several compositions according to the stoichiometric composition of $BaZr_xTi_{1-x}O_3$, with x = 0, 0.05, 0.10, and 0.15. After drying, the binder of 1 wt% PVA was added to form a disk in size of 12 mm in diameter and 0.8-mm thickness under a pressure of 3.9 MPa. These samples were sintered at 1350°C for 1.5 h in a furnace and then cooled to room temperature. The structure of the samples was analyzed by X-ray diffraction (D/Max V, Rigaku, Tokyo, Japan), Cu Kα (λ = 1.54056 Å) with a scanning rate ranging from 0.25 to 0.5 deg/min at a step of 0.02°. The structure model was calculated using the Rietveld method according to the XRD patterns. The microstructure of ceramics were analyzed by TEM (model 3010, JEOL), and the Ag-pasted samples were used as electrodes after being sintered at 780°C for 2 h and stabilized for 30 days. The dielectric constant of BZT was measured by using an impedance analyzer (HP4263A, Yakogawa Hewlett Packard, Tokyo, Japan) with an applied AC voltage of 1 V and a frequency of 1 kHz at temperatures ranging from -40°C to 140°C. The dielectric constant (ε_r) was calculated using Eq.(1):

$$\varepsilon_r=(C \cdot d)/(\varepsilon_0 \cdot A) \qquad (1)$$

where C is the capacitance, d the sample's thickness (m), A the area of Ag electrode (m^2), and ε_0 the permittivity of the free space (8.854×10^{-12} F/m). The hysteresis loop was measured at 10 kHz and 5 V by using a modified Sawyer–Tower circuit.

3. Results and discussion

3.1 Structure analysis

For $(Ba_{1-x}Sr_x)(Ti_{1-y}Zr_y)O_3$ ceramics preparation, the intermediate materials, $BaTiO_3$, $BaZrO_3$ and $SrTiO_3$ ceramics, were respectively prepared and sintered at various temperatures, which were analyzed by XRD to choice the co-sintering temperature at which the complete phase formation was finished. Fig. 1(a) shows the BT ceramic was formed through the reaction $BaCO_{3(s)}+TiO_{2(s)}=BaTiO_{3(s)}+CO_{2(g)}$ at temperature of 1000°C onward. The BZ ceramic could form through the reaction $BaCO_{3(s)}+ZrO_{2(s)}=BaZrO_{3(s)}+CO_{2(g)}$ at temperature of 1100°C onward as shown in Fig. 1(b). And the ST ceramic was formed through the reaction $SrCO_{3(s)}+TiO_{2(s)}=SrTiO_{3(s)}+CO_{2(g)}$ at temperature of 1000°C onward as shown in Fig.1(c). It suggested that the complete solid solution in BT, BZ, and ST systems was obtained at 1200°C. The activation energy for BT formation is lower than that of BZ. The BT formation started at 700°C, however, the BZ formation started at 800°C. [Bera & Rout, 2005] It is also found that the temperature of complete BZ formation is higher than BT. The ST formation is also started at temperature of 800°C with an activation energy similar to that of BZ. [Rout & Panigrahi, 2006] However, it was found that the sintering situation was similar to that of BT as shown in Figs. 1(a) and 1(c). These results demonstrate the sintering temperature of 1200°C is chosen for BSTZ ceramics sintering, which is high enough for complete solid-state reaction to form a solid solution. The reports of Rase et al. [Rase & Roy, 1955], Dry's et al. [Dry's & Trzebiatowski, 1957], and Kwestroo et al. [Kwestroo & Paping, 1959], the solid solutions of $BaTiO_3$, $BaZrO_3$ and $SrTiO_3$ ceramic could form completely to be the source materials of BSTZ ceramics. The formation temperature of BTZ was also reported by Bera and Rout [Bera & Rout, 2005] who pointed out the temperature is 1300°C due to loose powders and from $BaCO_3$, TiO_2 and ZrO_2. In this study, a compressed pressure of 3.0 MPa was used to form a disc for ceramics sintering.

The bulk density of sintered ceramics was affected by many factors, such as sintering temperature, atomic mobility, particle size, etc. The atomic effect was investigated for ST and BZ addition in BT. Fig. 2 shows the temperature dependence of bulk densities of BT, $(Ba_{0.90}Sr_{0.10})(Ti_{0.85}Zr_{0.15})O_3$, and $(Ba_{0.85}Sr_{0.15})(Ti_{0.85}Zr_{0.15})O_3$ ceramics. It suggests two facts, one is the higher sintering temperature is needed for higher both ST and BZ additions, the other is higher temperature dependence owing to ST additions. For BST solid solution, the bulk density decrease with the Sr content increasing although the lattice constant of BST ceramics decrease. [Muta et al., 2004] The similar trend was found by Bera and Rout [Bera & Rout, 2007] who also found the bulk density decreased and the apparent porosity increased with Sr content increasing even although the ceramics were sintered at 1600°C for 1 h. They also found that the $(Ba_{1-x}Sr_x)(Ti_{0.5}Zr_{0.5})O_3$ ceramics remained cubic up to 50 at% substitution, only peaks were shifted towards high angle yielding the decrease in lattice parameter due to the substitution of bigger Ba^{2+} by smaller Sr^{2+} in the structure.

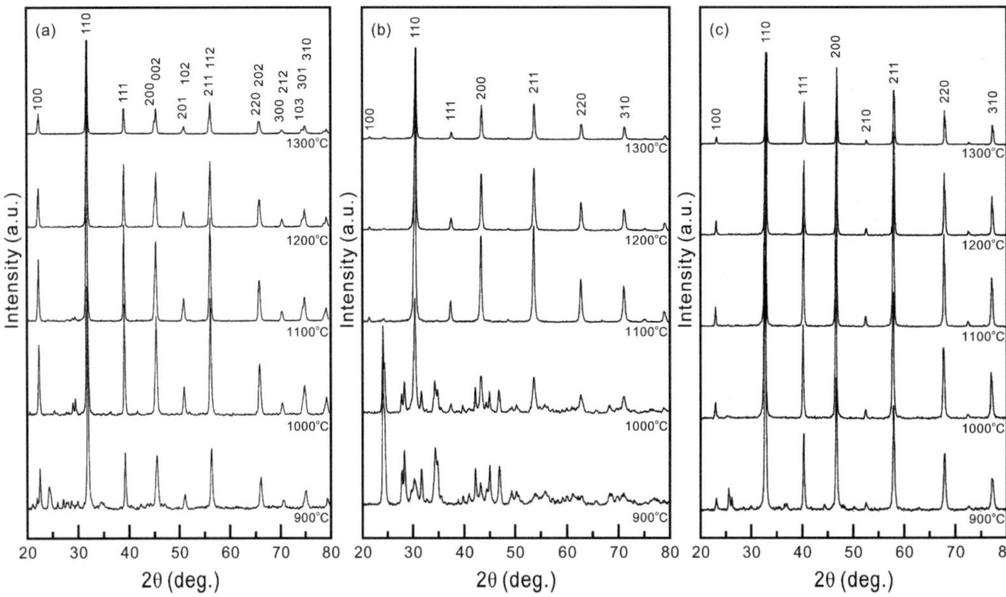

Fig. 1. XRD patterns of (a) BT, (b) BZ and (c) ST as a function of sintering temperature.

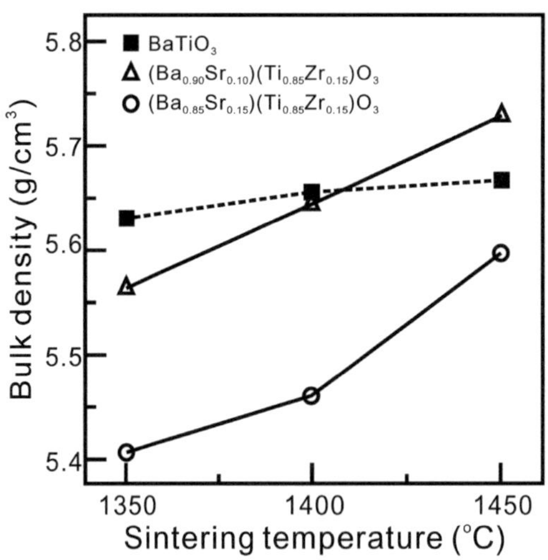

Fig. 2. Dependence of bulk density on sintering temperature for various $(Ba_{1-x}Sr_x)(Ti_{1-y}Zr_y)O_3$ ceramics.

The structures of BT and BTZ ceramics with various Zr substitutions were analyzed by XRD and shown in Fig. 3(a). It is suggested that no intermediate phase formed even 15% BZ was added into BT. However, the structure of BT was changed owing to the BZ addition. Hennings et al. [Hennings et al., 1982] had reported the structure changed from tetragonal to cubic phase with a lattice constant, *a*, of 4.052Å as the BZ content reached 25%. The influence of BZ addition on BT lattice constant was studied and showed in *c* axis reflections such as (103), (301), and (310), which was analyzed between 74° -76° as shown in Fig. 3(b). It shows that the lattice constant of BT increases with the BZ content increasing and the peaks of (310) and (301) were both merged and shifted to (103). The Zr substitution yields that the lattice constant of BTZ increase and the tetragonality decreased. Bera and Rout [Bera & Rout, 2005] had reported the larger lattice constant was obtained owing to the Zr^{4+} content increasing.

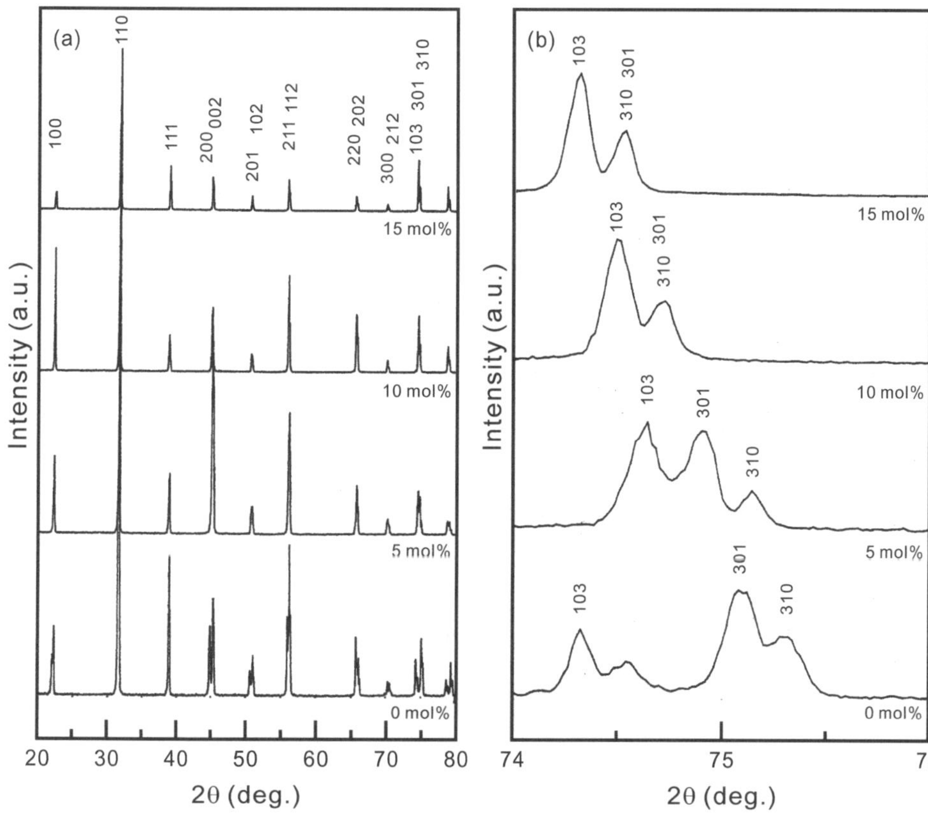

Fig. 3. (a) The XRD pattern of BTZ as a function of BZ content and (b) Enlarge view in 2θ of 74-76°.

The microstructure of BZT was investigated and reported in the previous study [Huang et al., 2008]. The mixture of tetragonal and cubic phases of BZT was found in the TEM analysis, which is difficult to be detected by using XRD as shown in Fig. 3(a). The bright-field (BF) and dark-field (DF) of the BZT ceramic are shown in Figs. 4(a) and 4(b), denoting grains precipitated in the matrix. The selected area electron diffraction (SAED) analyses of the $Ba(Zr_{0.15}Ti_{0.85})O_3$ crystal marked as A and B are also shown in Figs. 4(c) and 4(d), respectively. The indexing of the SAED patterns reveals that A is a cubic phase and matrix B is a tetragonal phase, which demonstrates that the BZ content induces cubic phase formation in the matrix of the tetragonal one; this demonstrates the coexistence of both tetragonal and cubic phases in the BTZ ceramics. The result of TEM analysis demonstrates that the XRD analysis is not sensitive to second phase formation resulting that numerous false conclusions may be arrived at if the ceramic is considered as a single phase.

Hennings et al. (1982) reported that a diffuse phase transition observed near the Curie temperature of BZT ceramics is shown to be caused by coexisting ferroelectric and paraelectric phases, which can be described by a normal distribution of Curie temperatures. In addition, a small difference in the Curie temperature was observed, therefore, two reasons were deduced: a mechanical stress distribution in the material or variation in chemical composition caused by sintering process. For a BTZ solid solution formation, the substitution of Ti^{4+} by Zr^{4+} results in mechanical stress formed in the BZT ceramics; consequently, a lattice constant shift was detected. The mixture of BT and BZ results in the increase in entropy of BTZ ceramics. [Kamishima et al., 2008] This increase of entropy gives the decrement of Gibbs free energy. [Kamishima et al., 2008] Therefore, the coexistence of cubic and tetragonal phases of BZT is possible. Owing to the limitations of XRD, the simulation of the Rietveld method should prove to be a powerful tool both for determining the existence of the cubic phase and for reducing inaccuracies in the lattice determination.

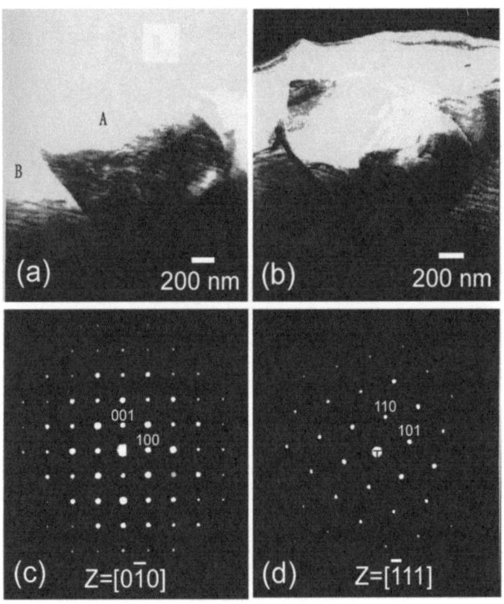

Fig. 4. TEM analysis of $Ba(Ti_{0.85}Zr_{0.15})O_3$: (a) BF, (b) DF, (c) SAED of A grain, and (d) SAED of B grain.

The ratio of the cubic/tetragonal phase and their lattice parameters are of utmost concern. The Rietveld method was usually used to simulate the precise structure of ceramics to determine the content of tetragonal and cubic phases. In order to obtain the exact lattice constants of BZT, the XRD patterns were measured at a very slow scanning rate (0.25 to 0.5 deg/min by a step of 0.02 deg) and calibrated with Si powders. The lattice constant and the phase ratio were calculated by Rietveld method and the results were listed in Table 1. For this calculation, the reliability index of the weighted profile R factor, R_{wp}, values of 17.45%, 15.95%, and 17.92% were obtained, which means the simulation is reliable; in addition, the goodness-of-fit indicator, s, values of 1.4, 1.1, and 1.3 were obtained, which means a good fitting is accomplished. The phase ratio changes which examined as a function of BZ content listed in Table 1, where the cubic phase in the BTZ solid solution is 36.73% for 5 mol% BZ addition. Furthermore, when the BZ content is increased to 15 mol%, the cubic phase ratio is 64.9%. The content of the cubic phase increases along with the increasing BZ content, although the increment decreases as the BZ level rising. The density of the bulk crystal estimated by the Rietvel method is very close to the measured value has been demonstrated by Souma and Ohtaki. [Souma, & Ohtaki, 2006] The higher the density, the lower the vacancy or void in the ceramics, resulting in both a higher dielectric constant and a lower Curie temperature. The dielectric behavior should be predictable, based on a combination of the existence of the cubic phase, the c/a ration of the tetragonal phase, and the theoretical density for ceramics with various BZ contents.

Composition	Tetragonal phase		cubic phase	Ratio of cubic to tetragonal phases
	a	c	a	
$BaTiO_3$	3.99423 ±0.00009	4.03457 ±0.00009		0.00:100.0
$Ba(Ti_{0.95}Zr_{0.05})O_3$	4.00715 ±0.00017	4.02142 ±0.00020	4.02230 ±0.00018	36.73: 63.27
$Ba(Ti_{0.90}Zr_{0.10})O_3$	4.02353 ±0.00018	4.02906 ±0.00049	4.02473 ±0.00027	56.23:43.77
$Ba(Ti_{0.85}Zr_{0.15})O_3$	4.03943 ±0.00024	4.04569 ±0.00027	4.03765 ±0.00009	64.87:35.13

Table 1. The lattice constants and phase ratios of cubic and tetragonal phases of BT and BTZ.

For BTZ ceramics, the relation between the tetragonality and the BZ content is shown in Fig.5. It reveals that the tetragonality of BZT dramatically decreases (from 1.0101 to 1.0035) as the content of BZ is increased from 0 to 5 mol%, but then only slightly decreases (from 1.0035 to 1.0016) when the BZ content is increased from 5 to 15 mol%. Arlt [Arlt, 1990] has reported that the formation of the ferroelectric domain fundamentally reduces the homogeneous stress within a given grain. As this happens, the inhomogeneous stress mainly forms in the grain boundaries, where a large internal stress occurs. When the tetragonality decreases, it leads to reduced internal stress due to the formation of 90 deg ferroelectric domains resulting that the dielectric constant increases and the c/a ratio decreases to constrain spontaneous polarization. By increasing the BZ content, the tetragonality of the tetragonal phase approaches 1, and the lattice constants of a and c of the tetragonal phase come close to the lattice constant of a value of the cubic phase.
When the ST was added into the BT, the structure of BST was analyzed by XRD and shown in Fig. 6. Fig. 6(a) shows that the formation of BST ceramics without any intermediate product. The similar reflections were found for BST ceramics with the ST content in the range of 0-15 mol%. The Ba^{2+} ion was replaced by smaller radius ion of Sr^{2+} resulting in the reflection peaks shift toward higher angle. In addition, the crotched reflections of reflection peaks of BST were still maintained as shown in Fig. 4(b). In comparison with Fig. 3, the

addition of ST in A site only affects the lattice constant but little effect on the tetragonality of BT. Keller & McCarthy [Keller & McCarthy, 1982] had reported the tetragonal phase transfer completely to cubic phase with a lattice constant of 3.965 Å when the $SrTiO_3$ content is higher than 40%. Because the ion radius of Sr^{2+} is smaller than that of Ba^{2+}, the lattice constant decreases with the ST content increasing. The Ba-rich $(Ba_{1-x}Sr_x)TiO_3$ ceramics have perovskite tetragonal structure. [Remmel et al., 1999] Sr-rich $(Ba_{1-x}Sr_x)TiO_3$ ceramics have perovskite cubic structure. [Izumskaya et al., 2010] The lattice constant of BST versus composition had reported by Baumert et al. [Baumert et al., 1997] who showed the cubic phase was found when the ST content is higher than 24% which is lower than that of Keller. Anyway, the perovskite tetragonal structure remains with the ST content in the range of 1-15 mol% addition.

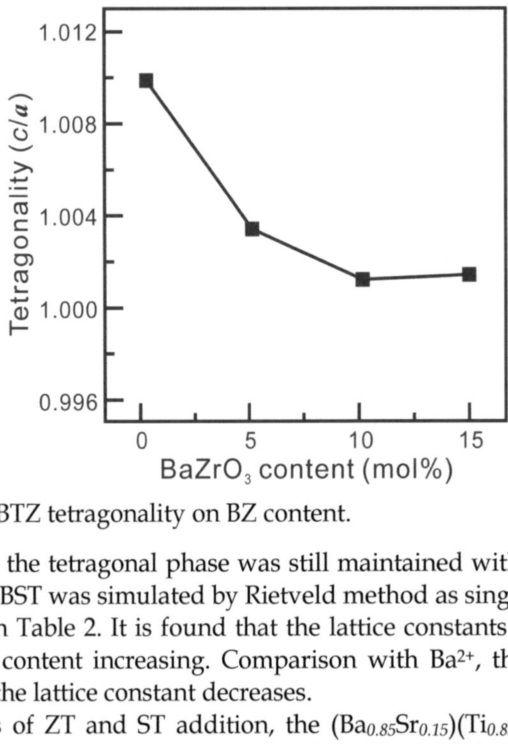

Fig. 5. Dependence of BTZ tetragonality on BZ content.

As shown in Fig. 6(b), the tetragonal phase was still maintained with various ST additions. The lattice constant of BST was simulated by Rietveld method as single tetragonal phase and the result was listed in Table 2. It is found that the lattice constants of both *a* and *c* of BST decrease with the ST content increasing. Comparison with Ba^{2+}, the ion radius of Sr^{2+} is smaller resulting that the lattice constant decreases.

Combining the effects of ZT and ST addition, the $(Ba_{0.85}Sr_{0.15})(Ti_{0.85}Zr_{0.15})O_3$ ceramics was prepared and its lattice constants was simulated as listed in Table 3. For Rietveld simulation, the best fitting result was obtained when coexistence of cubic and tetragonal phases was considered. The tetragonality of perovskite tetragonal $(Ba_{0.85}Sr_{0.15})(Ti_{0.85}Zr_{0.15})O_3$ ceramic is 1.00229 combining the enlargement and shrinkage effect of BZ and ST addition, respectively. In addition, the volume percentage of cubic phase with 43.56% was found.

Comparison of Tables 1-3, the lattice constants of BSTZ ceramics are larger than BT indication the effect of Zr substitution possessing dominate role to enlarged the lattice. The tetragonality of BSTZ is also close to that of BTZ.

3.2 Dielectric properties

Fig. 7 shows the dependence of the dielectric constant of BZT and BST ceramics on the BZ and ST content, respectively. It is found that the Curie temperature decrease from 125°C to

56°C as the BZ content increase from 0 to 15%. It is found that the reduction in Curie temperature is 4.66°C per mol% BZ addition. Masuno et al. [Masuno et al., 1972] have pointed out that the dependence of the Curie temperature on the BZ content consistently shows a slope of -5°C/mol%. The results are consistent with the values obtained by Masuno et al. [Masuno et al., 1972], however, the effect of the amount of cubic phase on the dielectric properties needs further investigation. The largest Curie temperature decrease with BZ and ST contents is considered to be due to a decrease in the pseudo-Jahn-Teller effect, [Kristoffel & Konsin, 1967] where the interaction between phonons and electrons results in a B-site atom shift in a BO_6 octahedral structure. The interaction is reduced because the increased BZ and ST contents and the overlap of the dπ and pπ orbits cause the Curie temperature to decrease. [Bersuker, 966] It is also found that the Curie temperature decrease with ST is larger than that with BZ. Currently, BZT has chosen as an alternative material to replace BST in the fabrication of ceramic capacitors, since $[ZrO_6]$ clusters are chemically more stable than those of $[TiO_6]$.[Badapanda et al., 2009]

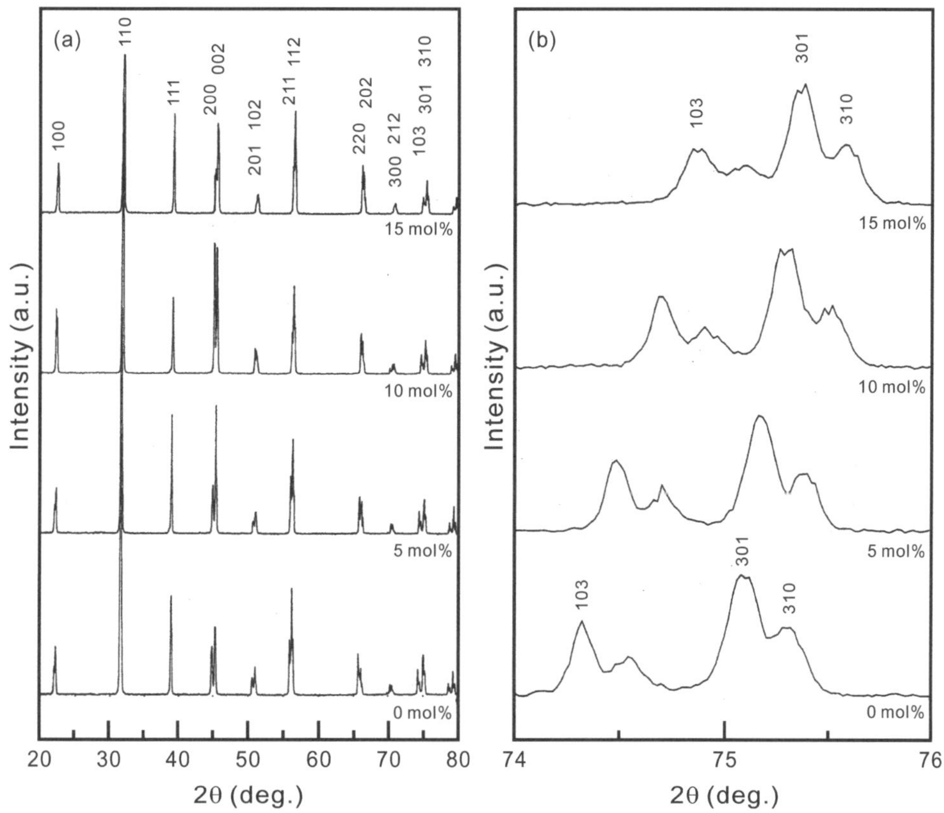

Fig. 6. (a) The XRD pattern of BTZ as a function of BZ content and (b) Enlarge view in 2θ of 74-76°.

The Curie temperature of BSZT with various ZT and ST contents is higher than 50°C indicating that the ceramics possess ferroelectric properties at room temperature. The hysteresis loop (P vs E) of BT and $(Ba_{1-x}Sr_x)(Ti_{1-y}Zr_y)O_3$ ceramic are shown in Fig. 7 which shows that the remnant polarization, P_r, may be enhanced by the addition of ZT and ST. Yu et al. [Yu et al., 2002] report that the P_r of $Ba(Zr_{0.15}Ti_{0.85})O_3$ is 0.08 μC/cm², which is quit smaller, although its Curie temperature is close to room temperature. The profiles of P vs E are related to the crystalline structure of ceramics at room temperature. [Dixit et al., 2004] The higher the BZ content, the higher the cubic phase ratio. It could be deduced from this that the dielectric properties can be further enhanced by increasing the tetragonal phase ratio in the ceramics. The relation of P_r and relative dielectric constant, ε_s, can be express as Eq.2:

$$P_r = (1/4\pi)\cdot(\varepsilon_s - 1)\cdot E \quad (2)$$

According to the measured results as shown in Fig.7, the values of ε_s were 1586, 2934, 2048, and 14340 for BT, BZT, BST and BSZT, respectively.

Composition	*a*	*c*
$BaTiO_3$	3.99423±0.00009	4.03457±0.00009
$(Ba_{0.95}Sr_{0.05})TiO_3$	3.99081±0.00008	4.02680±0.00009
$(Ba_{0.90}Sr_{0.10})TiO_3$	3.98698±0.00019	4.01723±0.00020
$(Ba_{0.85}Sr_{0.15})TiO_3$	3.97919±0.00018	4.00489±0.00019

Table 2. The lattice constants of tetragonal phase.

Composition	Tetragonal phase		cubic phase	Ratio of cubic to tetragonal phases
	a	*c*	*a*	
$(Ba_{0.85}Sr_{0.15})(Ti_{0.85}Zr_{0.15})O_3$	4.01699 ±0.00030	4.02619 ±0.00036	4.01768 ±0.00015	43.56: 56.44

Table 3. The lattice constants of $(Ba_{0.85}Sr_{0.15})(Ti_{0.85}Zr_{0.15})O_3$ ceramic.

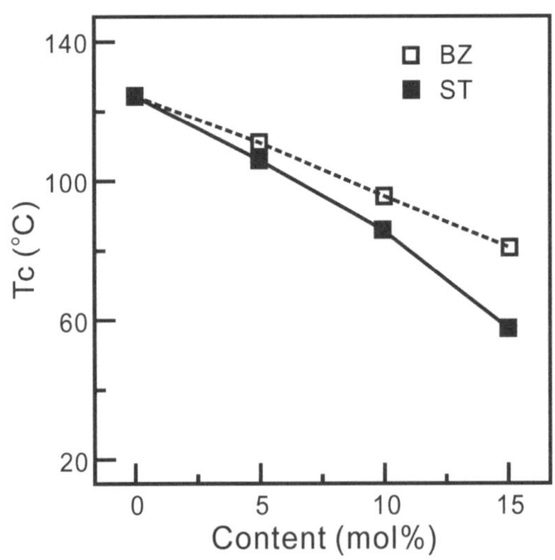

Fig. 7. Dependence of Curie temperature on BZ/ST contents.

As shown in Fig. 8, the Sr and Zr substitutions induced the Pr increase in 1.5-3 times. However, the Pr value of BSTZ ceramics increases dramatically for 24 times of that of BT. It is found that the restrictionism in lattice by using both Zr and Sr substitutions should yield the huge increase in dielectric properties.

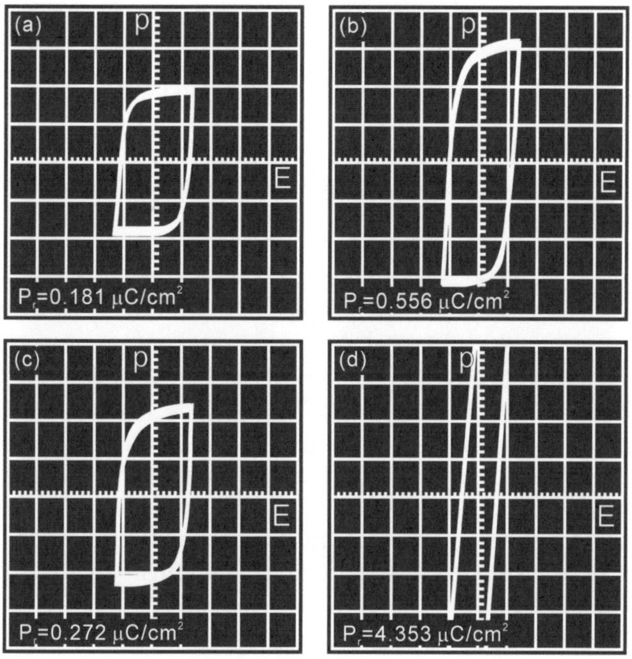

Fig. 8. Hysteresis loops of (a) $BaTiO_3$, (b) $Ba(Ti_{0.85}Zr_{0.15})O_3$, (c) $(Ba_{0.85}Sr_{0.15})TiO_3$ and (d) $(Ba_{0.85}Sr_{0.15})(Ti_{0.85}Zr_{0.15})$.

4. Conclusions

The structure of $BaTiO_3$ ceramics was modified by both $SrTiO_3$ and $BaZrO_3$ additions. Substitution with Sr in A site, the lattice constants of BST ceramics decrease but the tetrogonality increases from 1.01010 to 1.05617 for 0-15 mol% addition. The remnant polarization, Pr, value increases for 1.5 times. Substitution with Zr in B site, the lattice constant of BTZ ceramics increase but the tetragonality decreases to 1.00155 for 15 mol% addition. The Pr also increases for 3 times. Substitutions of both Sr and Zr in both A and B sites, the lattice constants of BSTZ ceramics are still larger than that of BT indicating Zr substitution possessing dominate role, and the tetragonality of decreases to 1.00229 close to that of BTZ ceramics. Under these effects of both enlargement and shrinkage, the Pr dramatically increases for 24 times. It is concluded that the substitutions Sr and Zr in both A and B sites should increase effectively the dielectric properties of BSTZ ceramics.

5. References

Arlt, G. (1990). The influence of microstructure on the properties of ferroelectric ceramics, *Ferroelectrics,* Vol.104, pp.217-27.

Badapanda, T., Rout, S.K., Cavalcante, L.S., Sczancoski, J.C., Panigrahi, S., Longo, E. & Li, M.S. (2009). Optical and dielectric relaxor behavior of $Ba(Zr_{0.25}Ti_{0.75})O_3$ ceramic explained by means of distorted clusters. *J. Phys. D: Appl. Phys.*, Vol.42, 175414 (9pp).

Baumert, B.A., Chang, L.H., Matsuda, A.T., Tsai, T.L., Tracy, C.J., Gregory, R.B., Fejes, P.L., Cave, N.G., Chen, W., Taylor, D.J., Otsuki, T., Fujii, E., Hayashi, S. & Sun, K. J. (1997). *Appl. Phys.*, Vol.82 p.2558.

Bera, J. & Rout, S.K. (2005). On the formation mechanism of $BaTiO_3$-$BaZrO_3$ solid solution through solid-oxide reaction. *Materials Letter,* Vol.59, No. 1, pp.135-138.

Bera, J. & Rout, S.K. (2005). $SrTiO_3$-$SrZrO_3$ solid solution: Phase formation kinetics and mechanism through solid-oxide reaction. *Materials Research Bulletin,* Vol. 40, No.7, pp.1187–1193.

Bera, J. & Rout, S.K., (2007). Synthesis of $(Ba_{1-x}Sr_x)(Ti_{0.5}Zr_{0.5})O_3$ ceramics and effect of Sr content on room temperature dielectric properties. *J. Electroceram.*, Vol. 18, pp.33-37.

Bersuker, I.B. (1966). The Jahn-Teller Effect. *Phys. Lett. A,* Vol.20, pp.589-590.

Bhalla, A.S., Guo, R. & Roy, R. (2000). The perovskite structure-a review of its role in ceramic science and technology. *Mater. Res. Innovat.* Vol.4, pp.3-26.

Carlson, C. M.; Rivkin, T. V.; Parilla, P. A.; Perkins, J. D.; Ginley, D. S.; Kozyrev, A. B.; Oshadchy, V. N. & Pavlov, A. S. (2000). Large dielectric constant ($\varepsilon/\varepsilon_0$>6000) $Ba_{0.4}Sr_{0.6}TiO_3$ thin films for high-performance microwave phase shifters, *Appl. Phys. Lett.*, Vol.76, No.14, pp.1920-1922.

Dixit, A., Majumder, S.B., Dobal, P.S., Katiyar, R.S. & Bhalla, A.S. (2004). Phase transition studies of sol–gel deposited barium zirconate titanate thin films, *Thin Solid Films,* Vol.447-448, pp.284–88.

Dry's, M. & Trzebiatowski, W. (1957). The system strontium oxide-titanium dioxide, *Roczniki Chem.*, Vol.31, No.2, pp.489-496.

Dutta, P. K.; Asiaie, R.; Akbar, S. K.; Zhu, W. (1994). Hydrthermal Synthesis and Dielectric Properties of Tetragonal $BaTiO_3$. *Chem. Mater.*, Vol.6, p.1542-48.

Ezhilvalavan, S. & Tseng, T. Y. (2000). Progress in the developments of $(Ba,Sr)TiO_3$ (BST) thin films for Gigabit era DRAMs, *Mater. Chem. Phys.*, Vol.65, pp.227-248.

Haertling, G.H., (1999). Ferroelectric Ceramics: History and Technology. *J. Am. Ceram. Soc.*, Vol.82, No.4, pp. 797-818.

Hennings, D., Schnell, A. & Simon, G. (1982). Diffuse ferroelectric phase transitions in $Ba(Ti_{1-y}Zr_y)O_3$ ceramics, *J. Am. Ceram. Soc.*, Vol.65, No.11, pp.539-544.

Huang, H.-H., Chiu, H.-H., Wu, N.-C. & Wang, M.-C. (2008). Tetragonality and Properties of $Ba(Zr_xTi_{1-x})O_3$ Ceramics Determined Using the Rietveld Method. *Metallurgical and Materials Transactions A,* Vol. 39A pp.3275-3282.

Izumskaya, N., Alivov, Y. & Morkoc, H. (2010). Oxides, oxides, and more oxides: high-k oxides, ferroelectrics, ferromagnetics, and multiferroics. Critical Reviews in Solid State and Materials Sciences, Virginia Commonwealth University, Richmond.

Kamishima, K., Nagashima, Y., Kakizaki, K., Hiratsuka, N., Watanabe, K., Mise, T., Naganuma, H. & Okamura S. (2008). Simple process Synthesis of $BaTiO_3$-(Ni, Zn, Cu)Fe_2O_4 Ceramic Composite, *J. Phys. Soc. Jpn.*, vol. 77, No. 6, 064801 pp.1-4.

Keller, L. & McCarthy, (1982). ICDD Grant-in-Aid, JCPDS File no. 33-150, North Dakota State University, Fargo, North Dakota, USA.

Kolar, D., Trontelj M. & Stadler Z. (1982). Influence of Interdiffusion on Solid Solution Formation and Sintering in the System $BaTiO_3$-$SrTiO_3$. *Journal of the American Ceramic Society,* Vol.65, No.10, pp.470-474.

Kristoffel, N.N. & Konsin, I.P. (1967). Pseudo-Jahn--Teller effect and the second order phase transitions in crystals. *Phys. Status Solidi,* Vol.21, No. 1, K39-K43.

Kwestroo, W. & Paping, H.A.M. (1959). The systems BaO-SrO-TiO_2, BaO-CaO-TiO_2, and SrO-CaO-TiO_2, *J. Am. Ceram. Soc.,* Vol. 42, No.6, pp.292-299.

Lemanov, V.V. (2007). Barium Titanate-Base Solid Solutions. *Ferroelectrics,* Vol.354, pp. 69-76.

Loachim, A., Toacsan, M. I., Banciu, M. G. Nedelcu, L., Vasiliu, F., Alexandru, H.V., Berbecaru, C., & Stoica, G. (2007). Barium strontium titanate-based perovskite materials for microwave applications. *Prog. Solid State Chem.* Vol.35, pp.513-520.

Maison, W.; Kleeberg, R.; Heimann, R.B. & Phanichphant, S. (2003). Phase Content, Tetragonality, and Crystallite Size of Nanoscaled Barium Titanate Synthesized by The Catecholate Process: effect of calcinations temperature. *Journal of the European Ceramic Society,* Vol.23, pp.127-132.

Masuno, K., Murakami, T. & Waku, S. (1972). Dielectric ceramics with boundary-layer structure for high frequency application. *Ferroelectrics,* Vol.3 pp.315-19.

Muta, H., Kurosaki, K. & Yamanaka, S. (2004). Thermoelectric properties of doped $BaTiO_3$-$SrTiO_3$ solid solution. *Journal of Alloys and Compounds,* Vol.368 22-24.

Polli, A.D., Lange, F.F., Levi, C.G. (2000). Metastability of the Fluorite, Pyrochlore, and Perovskite Structures in the PbO–ZrO_2–TiO_2 System", *J. Am. Ceram. Soc.,* Vol.83, No.4, pp.873-881.

Qin, W.F., Xiong, J., Zhu, J., Tang, J.L., Jie, W.J., Zhang, Y. & Li, Y.R. (2008). Dielectric Charateristics of BST/BZT/BST Multilayer. Surface Review and Letters, Vol.15, No.1-2, pp.195-200.

Rout, S.K. & Panigrahi S. (2006). Mechanism of phase formation of $BaTiO_3$-$SrTiO_3$ solid solution through solid-oxide reaction. *Indian Journal of Pure and Applied Physics,* Vol.44, No.8, pp.606-611.

Rout, S.K. & Panigrahi, S. (2006). Mechanism of phase formation of $BaTiO_3$-$SrTiO_3$ solid solution through solid-oxide reaction. *Indian Journal of Pure & Applied Physics,* Vol. 44, pp. 606-611.

Rase, D.E. & Roy, R. (1955). Phase Equilibria in the System BaO-TiO_2. *J. Am. Ceram. Soc.,* Vol.38, No.3, pp.102-113.

Remmel, T., Gregory, R., Baumert, B., (1999). JCPDS Int. Cent. Diff. Data 38-45.

Souma, T. & Ohtaki, M. (2006). Synthesis and Rietveld analysis of $Zn_{4-x}Cd_xSb_3$ bulk crystals in the Zn-rich region, *J. Alloys Compd.,* Vol.413, No.1-2, 289-297.

Tsurumi, T., Yamamoto, Y., Kakemoto, H. & Wada, S. (2011). Dielectric properties of $BaTiO_3$–$BaZrO_3$ ceramics under a high electric field. *J. Mater. Res.,* Vol.17, No. 4 pp.755-759.

Uchino, K. (1998). Materials issues in design and performance of piezoelectric actuators: an overview. *Acta Materialia,* Vol.46, pp. 3745–3753.

Wang, C., Cheng, B.L., Wang, S.Y., Lu, H.B., Zhou, Y.L., Chen, Z.H. & Yang, G.Z. (2004). Improved dielectric properties and tunability of multilayered thin films of $(Ba_{0.80}Sr_{0.20})(Ti_{1-x}Zr_x)O_3$ with compositionally graded layer. *Applied Physics Letters,* Vol. 84, No. 5, pp.765-767.

Wei, X., Xu, G., Ren, Z.H., Wang, Y.G., Shen, G. & Han, G.R. (2008). Composition and shape control of single-crystalline $Ba_{1-x}Sr_xTiO_3$ (x=0-1) nanocrystals via a solvothermal route. *J. Cryst. Growth,* Vol.310, pp.4132-4137

Yu, J., Paradis, P.-F., Ishikawa, T., Yoda, S., Saita, Y., Itoh, M. & Kano, F. (2004). Giant Dielectric Constant of Hexagonal $BaTiO_3$ Crystal Grown by Containerless Processing. *Chemistry of Materials,* Vol.16, No. 21, pp.3973-75.

Yu, Z., Ang, C., Guo, R., & Bhalla, A.S. (2002). Dielectric properties and high tunability of $Ba(Ti_{0.7}Zr_{0.3})O_3$ ceramics under dc electric field. *Appl. Phys. Lett.,* Vol. 81, No. 7, pp.1285-87.

Yu, Z., Ang, C., Guo, R. & Bhalla, A.S. (2002). Piezoelectric and strain properties of $Ba(Ti_{1-x}Zr_x)O_3$ ceramics, *Journal of Applied Physics,* Vol.92, No.3, pp.1489-1493.

Zimmermann, F., Voigts, M., Weil, C., Jakoby, R., Wang, P., Menesklou, W. & Tiffée, E. I. (2001). Investigation of barium strontium titanate thick films for tunable phase shifters. *J. Euro. Ceram. Soc.,* Vol. 21 pp.2019-2023.

13

Structural, Morphological, Magneto-Transport and Thermal Properties of Antimony Substituted $(La,Pr)_{2/3}Ba_{1/3}Mn_{1-x}Sb_xO_3$ Perovskite Manganites

Neeraj Panwar[1], Indrani Coondoo[1], Vikram Sen[2,3] and S. K. Agarwal[2]
[1]*Departamento de Ceramica e Engenharia do Vidro & CICECO, Universidade de Aveiro*
[2]*National Physical Laboratory, (CSIR), Dr. K. S. Krishnan Marg, New Delhi*
[3]*Bhagwati Institute of Technology & Science, Masuri - Ghaziabad, UP*
[1]*Portugal*
[2,3]*India*

1. Introduction

Magneto-resistance (MR) is generally defined as the relative change in the electrical resistivity of a material upon the application of magnetic field. Mathematically, MR can be defined by the following equation,

$$MR(\%) = \left[\frac{\rho_0 - \rho_H}{\rho_H} \right] \times 100 \qquad (1)$$

where ρ_H and ρ_0 are the electrical resistivities with and without magnetic field respectively. Electrical resistivity can increase or decrease upon the application of magnetic field and accordingly MR is defined as positive or negative respectively. By using the above definition observed MR could be ~10^6% or even higher, but if we define MR by the following relation,

$$MR(\%) = \left[\frac{\rho_0 - \rho_H}{\rho_0} \right] \times 100 \qquad (2)$$

then maximum MR will be ~100%.

W. Thomson (Lord Kelvin) in 1857, for the first time, discovered magneto-resistance[1]. All metals exhibit MR due to Lorentz force but the value of MR is only few percent even after the application of higher magnetic field. For example, Cu metal exhibits 1% MR at room temperature under 7 Tesla applied magnetic field. In the case of ferromagnetic metals like Fe and Co, MR is more (~ 15%) as compared to that of nonmagnetic metals such as gold (Au). Semimetal bismuth (Bi) shows about 18% MR in a transverse magnetic field of 0.6T and 250% MR at room temperature under the application of 5 Tesla magnetic field[2].

In search of higher MR values near room temperatures, the last two decades have witnessed the discovery of an enormous suppression in electrical resistivity upon the application of magnetic field in certain class of materials called manganites. The observed change was so large that a special name "colossal magneto-resistance (CMR)" had to be coined for this phenomenon to distinguish it from the already existing giant magneto-resistance (GMR) in the magnetic multilayers. For example, Chahara et al.[3], R. von Helmolt et al.[4], Jin et al.[5], Xiong et al.[6] have reported ~ 10^6 % MR under the application of 6T magnetic field. Searle and Wang et al.[7, 8] were the first to report MR studies in $La_{1-x}Pb_xMnO_3$ single crystal manganites. The renewed surge of interest in manganites in the 1990s started with the experimental observation of large magneto-resistance (MR) studies in $Nd_{0.5}Pb_{0.5}MnO_3$ by Kusters et al.[9] and in $La_{2/3}Ba_{1/3}MnO_3$ by R. von Helmolt et al.[4]. MR effect was found to be as high as 60% (using eq. 2) in thin films at room temperature, and it was exciting to observe that this value was higher than found in artificial magnetic/nonmagnetic multilayers i.e. Giant magneto-resistive materials (GMR)[10] , allowing for potential applications in magnetic recording. Thereafter Chahara et al.[3] on $La_{3/4}Ca_{1/4}MnO_3$ and Ju et al.[11] on $La_{1-x}Sr_xMnO_3$ films produced similar results. Jin et al.[5] observed 127,000% around 77K using Eq. (1) for defining MR in thin films of $La_{0.67}Ca_{0.33}MnO_3$ and Xiong et al.[6] 100,000% in $Nd_{0.7}Sr_{0.3}MnO_3$ film at 60K.

1.1 Crystal structure

Rare-earth manganites possess a perovskite crystal structure (Fig. 1a) with the general formula $R_{1-x}A_xMnO_3$, where R is any trivalent rare-earth ion like La^{+3}, Pr^{+3}, Nd^{+3}, Eu^{+3} etc. and A is divalent alkaline-earth ion (Ca^{+2}, Sr^{+2}, Ba^{+2} etc.) whereas x represents the amount of doping of such divalent ions at the rare-earth site and equal number of Mn^{+3}-ions are converted into Mn^{+4}- ions. The mixed valency of manganese ions can also be controlled by varying the oxygen content[12,13]. In the perovskite structure each manganese (Mn) ion is surrounded by six oxygen ions forming an octahedron (Fig. 1a).

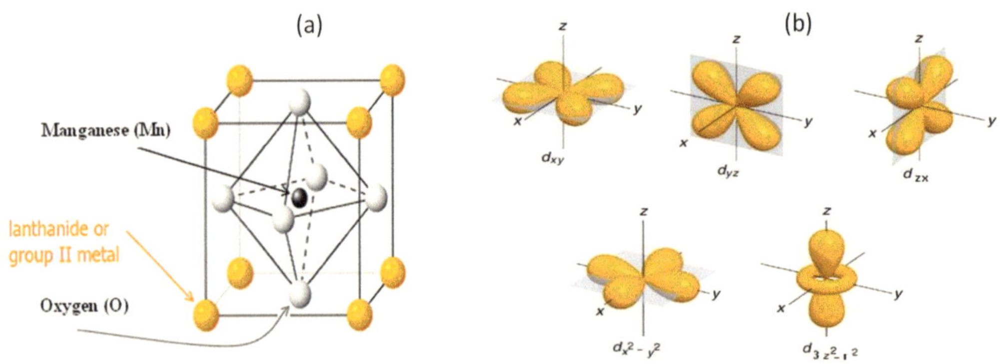

Fig. 1. (a) arrangements of ions in the perovskite structure of manganites. (b) shape of different d-orbitals.

In MnO_6 octahedron overlapping of manganese 3d orbitals with the 2p orbital of oxygen splits the five-fold degeneracy of the manganese 3d orbitals due to their different shapes (Fig. 1b). This is called crystal field splitting (CFS). Thus in an MnO_6 octahedron three degenerate orbitals (d_{xy}, d_{yz}, d_{zx}) with lower energy (because their lobes are oriented between the O^{-2} ions) are called t_{2g} levels and two degenerate orbitals ($d_{x^2-y^2}$, $d_{3z^2-r^2}$) with higher energy (their lobes are towards O^{-2} ions and hence large Coulomb repulsion) called e_g

levels[14].Due to crystal field splitting the energy difference between t_{2g} and e_g levels is ~ 1.5 eV[15]. All electrons in these energy levels get aligned parallel to each other due to strong on-site Hund's coupling, leading to a total spin (S) of 2 for Mn^{+3} and S = 3/2 for Mn^{+4} ions respectively. The three electrons of Mn^{+4} - ion occupy t_{2g} states whereas fourth electron of Mn^{+3} goes to the e_g state. The e_g electrons are more itinerant than the t_{2g} electrons and can hop from one Mn-site to the other. However, according to Jahn-Teller (JT) theorem this configuration is not stable and the degeneracy of the e_g levels are further removed[16]. The oxygen ions surrounding the Mn^{+3} -ion slightly readjust their positions, creating an asymmetry in different directions which ultimately breaks the degeneracy. The breaking of degeneracy due to orbital- lattice interaction is called Jahn-Teller splitting (Fig. 2). Only those ions like Mn^{+3} which have odd number of electrons in the e_g levels can undergo Jahn-Teller distortion. Therefore Mn^{+3} is a Jahn-Teller ion whereas Mn^{+4} is not. In the case of manganites there are 21 degrees of freedom (modes of vibration) for the movement of oxygen and Mn ion[17]. Out of these only two types of distortion are relevant for the splitting of e_g levels i.e. Q_2 and Q_3 (Fig. 3) [18]. Q_2 is a basal plane distortion (defined as Q_2 mode) in which one diagonally O pair is displaced inwards whereas the other pair is displaced outwards. On the other hand, Q_3 is a tetragonal distortion which results in elongation or distortion of MnO_6 octahedron. JT distortion can be static or dynamic.

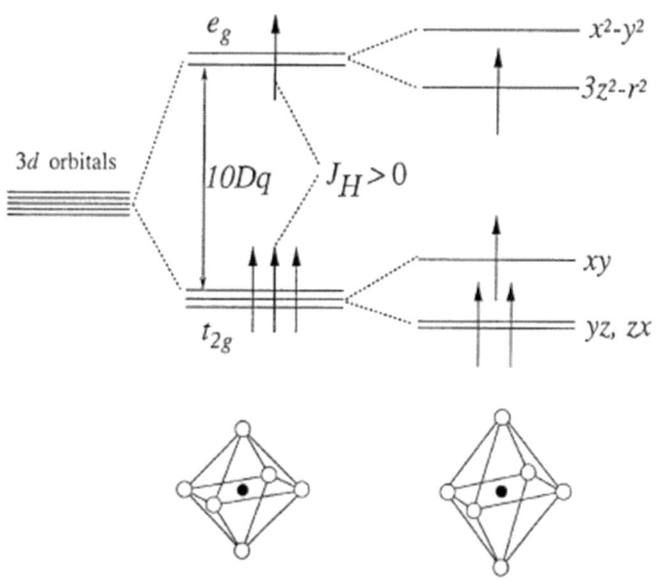

Fig. 2. Crystal field splitting of the five-fold degenerate atomic 3d levels into lower t_{2g} and higher e_g levels. Jahn-Teller distortion further lifts the degeneracy (from Tokura, 2000 [19]).

Perovskite manganites not only exhibit the colossal magnetoresistance (CMR) effect but also other important phenomena like charge ordering, phase separation at nanoscale owing to the subtle balance between charge, spin, lattice and orbital degrees of freedom[19]. As mentioned earlier, electrical resistivity of these manganites can change dramatically in response to an applied magnetic field. The occurrence of magneto-resistance can be explained by Zener's theory of double exchange along-with the formation of Jahn-Teller polarons[20]. In order to understand magnetoresistance behavior and other peculiar properties

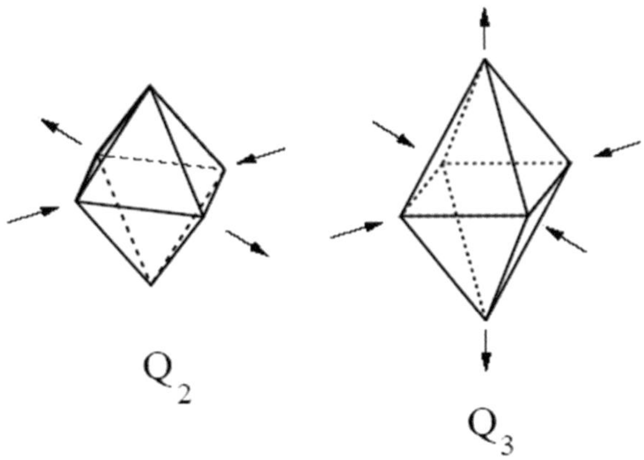

Fig. 3. Mode of vibrations Q_2 and Q_3 for the splitting of e_g doublet.

exhibited by these materials, substitution engineering is carried out either at the rare-earth site or at the manganese site. However, out of the two, the last is more appropriate as it directly affects the mechanism happening among different MnO_6 octahedra. In this regard, several authors have substituted different transition metal ions like Fe^{+3}, Al^{+3}, Ga^{+3}, Zn^{+2}, Cu^{+3}, Ni^{+2} at the Mn-site and have shown that Mn-site substitution severely affects the conduction mechanism as well as the magnetic properties[21-28]. However, fewer results exist for higher valent ion substitution at Mn-site to see its effect on the transport properties[29-33]. Substitutions at Mn-sites in manganites, irrespective of their electronic and magnetic nature, lower the ferromagnetic transition temperature but to different extents except in the case of Ru-doped $Pr_{0.5}Sr_{0.5}MnO_3$ where it is found to increase with Ru[32]. In this work, we have attempted to substitute Sb^{+5} ion on the Mn-site in two different manganite materials *viz.* $La_{2/3}Ba_{1/3}MnO_3$ (LBMO) and $Pr_{2/3}Ba_{1/3}MnO_3$ (PBMO)[34-38]. Sb^{+5} ion has been chosen because its ionic size is also comparable to that of manganese ion in the six-fold co-ordination. Besides, by virtue of its closed shell configuration, Sb^{+5} ion is not expected to introduce any additional magnetic coupling. It is also not expected to participate explicitly in the mechanism between Mn^{+3}/Mn^{+4} ions, however, due to its filled shell structure it can hinder the electron transfer. We have carried out structural, morphological, electrical, magnetoresistive and thermal measurements on the aforementioned compositions and the results will be discussed here.

2. Experimental procedure

Bulk polycrystalline manganite samples of $La_{2/3}Ba_{1/3}(Mn_{1-x}Sb_x)O_3$ and $Pr_{2/3}Ba_{1/3}(Mn_{1-x}Sb_x)O_3$ ($0 \leq x \leq 0.03$) [referred hereafter as $(La, Pr)_{2/3}Ba_{1/3}(Mn_{1-x}Sb_x)O_3$] were synthesized following the conventional solid-state reaction method. Powders of La_2O_3, Pr_6O_{11}, $BaCO_3$, Sb_2O_5 and MnO_2 were mixed in nominal stoichiometric ratios, grounded properly and calcined several times between 900-1100 °C for 15 h with intermediate grindings. Finally, the pellets were made from the calcined powders and sintered at1260 °C for 20 h. X-ray diffraction (XRD) was carried out with Cu K_a radiation (λ = 1.54 $\AA$) between 20° and 80° with a step size of 0.05° in 2θ mode to ascertain the phase purity of the synthesized materials. Electrical resistivity measurement in the absence and presence of a magnetic field was carried out using a four-probe technique with the field (0.6 T) applied parallel to the direction of the current flow. Thermoelectric power $S(T)$ and thermal conductivity $\kappa(T)$ measurements were carried out simultaneously from 350 to 10K in a closed cycle refrigerator by using a direct-heat pulse technique. Specific heat $C_P(T)$ was measured from 350 to 77K with a high resolution ac calorimeter, using chopped light as heat source. Further details of the thermal measurements techniques can be found in ref.[39].

3. Results and discussion

3.1 Structural studies

Fig. 4 shows the x-ray diffractograms of $La_{2/3}Ba_{1/3}(Mn_{1-x}Sb_x)O_3$ and $Pr_{2/3}Ba_{1/3}(Mn_{1-x}Sb_x)O_3$ ($0 \le x \le 0.03$) series. All samples are single phase in nature. It is also clear that Sb^{+5} ion occupies the Mn-site in LBMO and PBMO. Substitutional criteria of valence, ionic size and the coordination assured the preferential occupation of Mn^{+4} ion site by Sb^{+5} ion. The parent compound LBMO possesses the cubic structure [lattice parameter = 3.9094(± 0.09%) Å] at room temperature. In refs.[40-42] the structure of LBMO has been reported cubic. In some reports neutron diffraction study has been carried out as a function of temperature on LBMO perovskite manganite and it has been deduced that it crystallizes in the rhombohedral structure (closer to cubic, space group R3c) at room temperature[43,44]. However, the final sintering temperature in our case is 1260 °C while it is much higher as reported in ref. [45](e.g.1470 °C used by Beznosov et al. and 1400 °C by Moutis et al.) and such a high temperature can change the lattice structure from cubic to rhombohedral. With the subsequent doping of Sb^{+5} ion the lattice parameters increase successively [e.g. 3.9180 (± 0.011%) Å for 3% Sb-doped sample] which indicates the occupation of Mn^{+4} ion by Sb^{+5} in LBMO. The increase in the lattice parameter is also evident as the main peak shifts towards lower 2θ values with increasing Sb-content. The lower value of 2θ implies higher d values and consequently higher lattice parameter. The increase in the lattice parameters can also be understood in terms of the increase of the mean ionic radius at Mn site [Mn^{+3} (0.645 Å)/Sb^{+5} (0.60 Å)/Mn^{+4} (0.53 Å), ionic radii are in six- fold coordination] with increase in Sb^{+5} content.

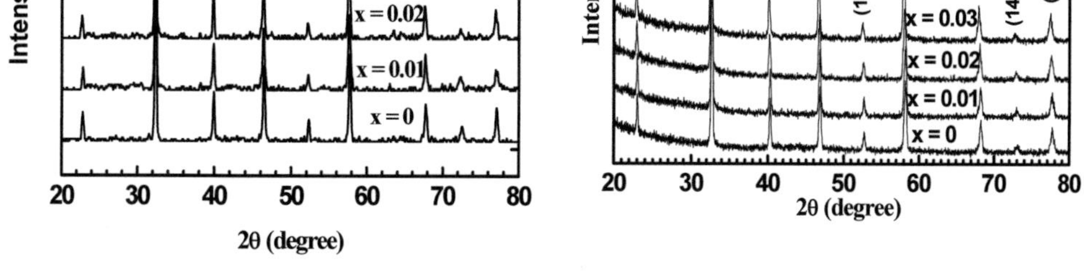

Fig. 4. X-ray diffraction patterns of (a)$La_{2/3}Ba_{1/3}(Mn_{1-x}Sb_x)O_3$ and (b) $Pr_{2/3}Ba_{1/3}(Mn_{1-x}Sb_x)O_3$ ($0 \le x \le 0.03$) samples.

The lattice parameters were calculated using the powder X software. Similar results were also observed in case of Sb-substituted PBMO samples except that the crystal structure here is orthorhombic.

3.2 Morphological studies

The SEM micrographs of bulk LBMO, 3Sb-LBMO, PBMO, 3Sb-PBMO are shown in Fig. 5 (a-d). The SEM micrographs clearly show the grain and grain boundaries. The grains show good connectivity among each other and the grain size is found to increase with doping in both cases. It is assumed that Sb-doping helps in promoting the grain growth.

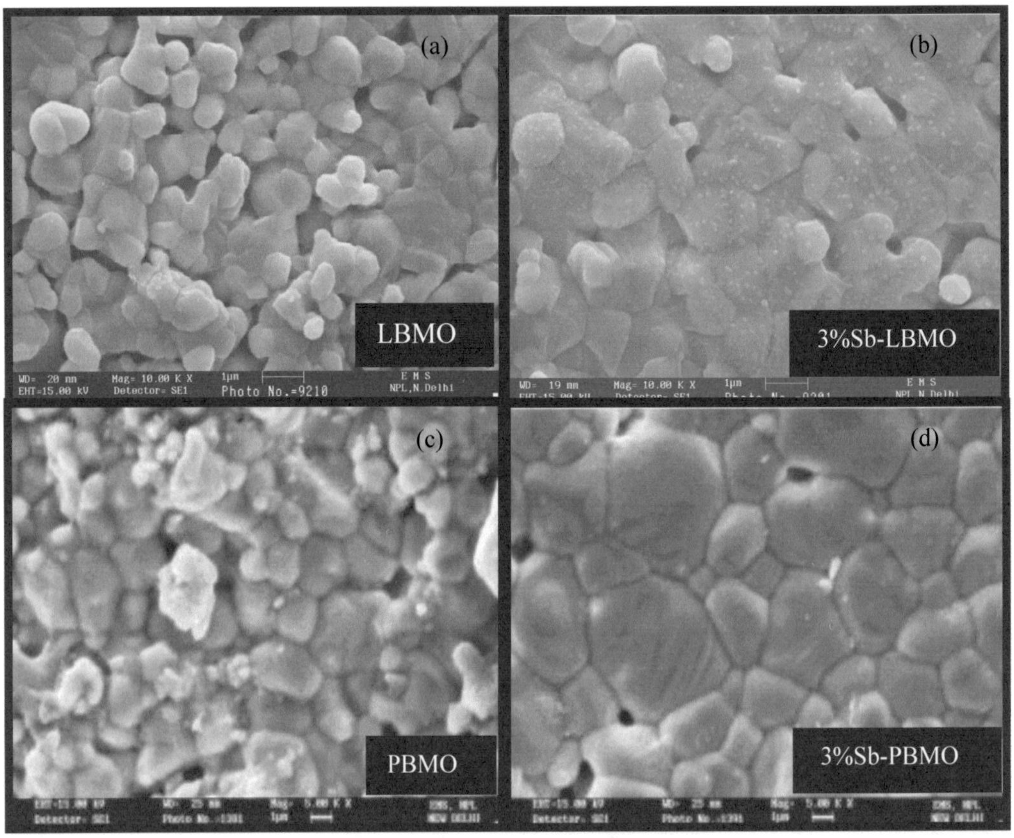

Fig. 5. SEM micrographs of (a) $La_{2/3}Ba_{1/3}MnO_3$, (b) $La_{2/3}Ba_{1/3}Mn_{0.97}Sb_{0.03}O_3$ (3Sb-LBMO), (c) $Pr_{2/3}Ba_{1/3}MnO_3$, (d) $Pr_{2/3}Ba_{1/3}Mn_{0.97}Sb_{0.03}O_3$ (3Sb-PBMO).

3.3 Electrical resistivity measurement

Figure 6 shows the electrical resistivity variation with temperature (ρ-T) of $La_{2/3}Ba_{1/3}(Mn_{1-x}Sb_x)O_3$ and $Pr_{2/3}Ba_{1/3}(Mn_{1-x}Sb_x)O_3$ ($0 \leq x \leq 0.03$) samples. Undoped samples LBMO and PBMO exhibit two insulator-metal (I-M) transitions. The higher temperature transition (T_{P1}) is sharp and at ~ 340 K and 194K whereas the lower temperature transition (T_{P2}) is broader and at ~ 250K and 160K respectively. In case of LBMO, Ju et al.[44] reported two transitions at the same temperatures T_{P1} and T_{P2} but the value of the resistivity at T_{P2} in their sample was lesser than that at T_{P1} but in our case the situation is reverse. The reason for this is the grain size difference in two cases. While the average grain size is ~ 5 μm reported by Ju et al.[44], it is much smaller (0.5 μm) in our case (Fig. 5a). Smaller grain size leads to extra strain at the grain boundary, which eventually leads to the enhancement of the resistivity. However, the same transition temperatures confirm that oxygen content is close to 3.0 in our sample as in the other report Ju et al.[40] have studied the effect of oxygen variation on the electrical and magnetic properties of LBMO and observed that only the samples having oxygen content close to 3.0 show two transitions in the resistivity-temperature behavior.

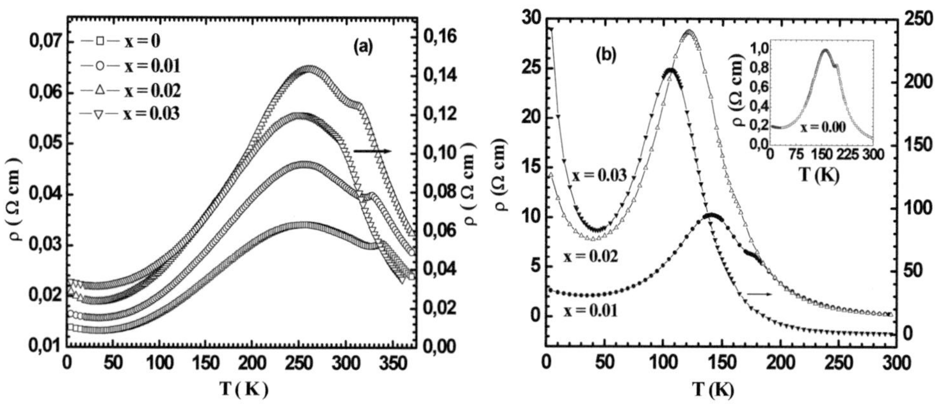

Fig. 6. Electrical resistivity variation with temperature of (a) $La_{2/3}Ba_{1/3}(Mn_{1-x}Sb_x)O_3$ and (b) $Pr_{2/3}Ba_{1/3}(Mn_{1-x}Sb_x)O_3$ ($0\leq x \leq 0.03$).

With Sb-doping at the Mn-site the two transitions, in both the cases, shift to lower temperatures successively with the overall increase in the resistivity. However, T_{P1} shifts to lower temperatures at a faster rate than T_{P2} and almost disappear with higher Sb content. T_{P1} is a result of the competition between double-exchange and the super-exchange (SE) mechanisms generally observed in all perovskite manganites. However, T_{P2} results out of the grain boundary effects that in turn arise due to the larger ionic size mismatch between the ions present at the rare-earth site in LBMO and PBMO (La^{+3}/Ba^{+2} and Pr^{+3}/Ba^{+2} respectively). Due to such ionic size difference the lattice comes under strain but since this is not a situation where the mechanical energy is minimum (which should be the case in equilibrium) and because of the presence of the grain boundaries in the polycrystalline sample, the lattice transfers the extra strain to the grain boundaries to attain a minimum energy configuration. This transferred strain at the grain boundary tilts the MnO_6 octahedra of the LBMO/PBMO lattices present there and thus weakens the electron transfer probability t ($t = t_0\cos\theta$, which depends upon the angle θ between two neighboring Mn-ions). Therefore the two transitions one from the grain (T_{P1}, where the Mn -ions are more parallel and the electron transfer is easier and hence stronger double exchange mechanism) and the other, T_{P2}, from the grain boundary (weaker double exchange) separate from each other. When Sb^{+5} ion is doped at Mn^{+4}- site, it does not participate in the exchange mechanism between Mn^{+3}/Mn^{+4} ions because of its closed shell electronic structure ($4d^{10}$). However, it does decrease the available number of sites where the electron can hop from the

e_g orbital of Mn^{+3}, resulting in some sort of localization of the carriers and the two transitions shift to lower temperatures with increasing of Sb content. It is worthwhile to mention here that ionic size mismatch between the ions present at rare-earth site [Pr^{+3} (1.179Å) and Ba^{+2} (1.47 Å)] in case of PBMO is larger than in LBMO [La^{+3} (1.210Å) and Ba^{+2} (1.47Å)]. Therefore, T_{P2}, is lower in case of PBMO. However, T_{P1}, depends upon the average ionic radius at the rare earth site which is smaller for PBMO that's why T_{P1} is also smaller in this case.

3.4 Electrical resistivity at low temperature (< 50*K*)

At low temperatures, transport mechanisms such as weak localization[46,47], electron-electron interaction[47,48], Kondo effect[49,50,51] *etc.* can cause an upturn in the resistivity with decreasing temperature, whereas scattering mechanisms such as electron-electron scattering[52,53], electron-phonon scattering[53,54], electron-magnon scattering[55,56], magnon-magnon scattering[57] increase resistivity with increasing temperature in metals and alloys. Figure 7 shows the temperature dependence of resistivity of Sb-doped LBMO and PBMO systems in the temperature range $4K \leq T \leq 50K$. It is clear that electrical resistivity exhibits an increasing trend with decreasing temperature reminiscent of semiconducting behavior for all compositions. There is a resistivity minimum which is lowest for the undoped samples and shift to high temperatures with antimony substitution. In order to analyze the electrical resistivity behavior at low temperatures we tried to fit the data using the following equation:

$$\rho(T) = \rho_0 - \rho_1 T^{1/2} + \rho_2 T^2 + \rho_5 T^5 \quad (3)$$

where $\rho_0=1/a$ and $\rho_1= b/a^2$ are constants. '*a*' is temperature independent residual conductivity; '*b*' is the diffusion constant and is due to weak localization effect[46]. The other two terms viz. $\rho_2 T^2$ and $\rho_5 T^5$ arise due to the electron-electron and electron-phonon scattering[53]. The fitting parameters ρ_0, ρ_1, ρ_2, ρ_5 for the two systems are shown in Table 1.

$La_{2/3}Ba_{1/3}(Mn_{1-x}Sb_x)O_3$ $(0\leq x\leq 0.03)$				
Sample	ρ_0 (Ω cm)	ρ_1 (Ω cm $K^{-1/2}$)	ρ_2 (Ω cm K^{-2})	ρ_5 (Ω cm K^{-5})
x = 0	0.01436	0.00023	1.8885×10^{-7}	6.89×10^{-13}
x = 0.01	0.01723	0.00033	3.3297×10^{-7}	6.9613×10^{-13}
x = 0.02	0.02282	0.00093	1.269×10^{-6}	3.184×10^{-13}
x = 0.03	0.03529	0.00081	9.8659×10^{-7}	1.8152×10^{-12}
$Pr_{2/3}Ba_{1/3}(Mn_{1-x}Sb_x)O_3$ $(0\leq x\leq 0.03)$				
x = 0	0.2175	0.0082	$8.9x10^{-6}$	$4.83x10^{-11}$
x = 0.01	3.0015	0.1922	$1.7x10^{-4}$	$5.26x10^{-10}$
x = 0.02	19.044	2.4068	$2.53x10^{-3}$	$7.18x10^{-10}$
x = 0.03	362.885	61.8983	$7.69x10^{-2}$	$1.09x10^{-7}$

Table 1. Various fitting parameters for $(La/Pr)_{2/3}Ba_{1/3}(Mn_{1-x}Sb_x)O_3$ $(0\leq x\leq 0.03)$ systems using equation (3).

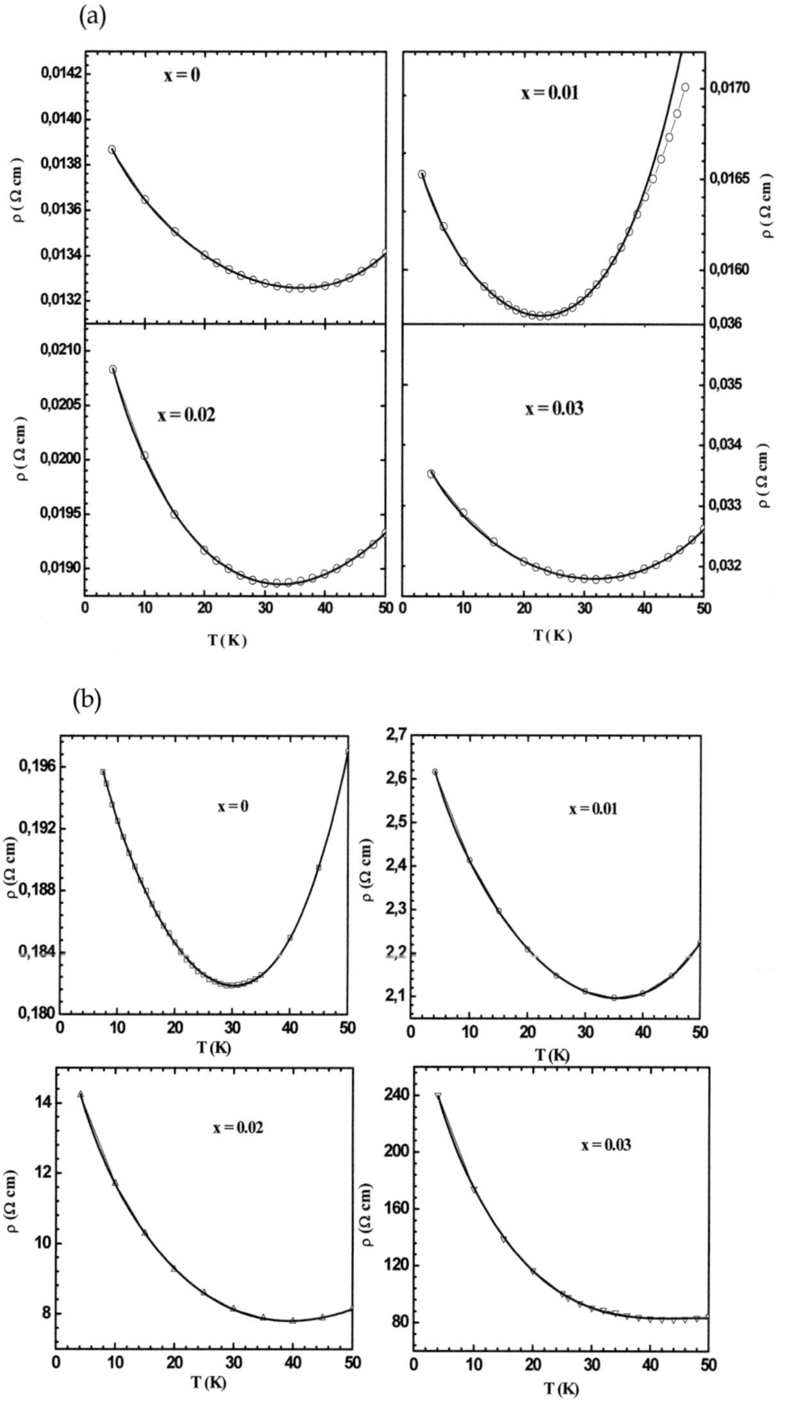

Fig. 7. Electrical resistivity variation with temperature below 50*K* (a) $La_{2/3}Ba_{1/3}(Mn_{1-x}Sb_x)O_3$ and, (b) $Pr_{2/3}Ba_{1/3}(Mn_{1-x}Sb_x)O_3$ ($0 \leq x \leq 0.03$). Solid line shows the fitting with the Eq. (3).

It is observed that the value of fitting parameters ρ_0, ρ_1, ρ_2, ρ_5 increases with doping. This indicates that the weak localization, electron-electron and electron-phonon scattering increases with doping and making pristine material more disordered. This is also obvious from the fact that random distribution of ions $La^{+3}/Pr^{+3}/Ba^{+2}/Mn^{+3}/Mn^{+4}/Sb^{+5}$ leads to random variation in the potential experienced by the electrons so the electrons get trapped due to less kinetic energy at low temperatures. The other two terms i.e. electron-electron and electron-phonon scattering will increase because of the difference in ionic sizes. It is worth mentioning here that in addition to these effects the Coulomb blockade (CB) effect[58] could also contribute to the observed electrical resistivity upturn as the large ionic size mismatch between La^{+3}/Pr^{+3} and Ba^{+2} ions leads to the strong disorder at the grain boundaries and eventually increases the Coulomb charging energy E_C leading to increase in resistivity. This would get further enhanced for grains with smaller sizes ($E_C \propto 1/d$). In this case grain size d, however, has increased with Sb-doping (figure 5), which implies that E_C should decrease and hence the resistivity upturn should also decrease but here a reverse situation is observed. The CB effect, therefore, cannot account for the resistivity upturn for the Sb-doped samples, even though it might be one of the possible sources of localization in the pristine sample case.

3.5 Transport mechanism in the metallic region

We used the following Eq. to fit the metallic region of the temperature dependent electrical resistivity data

$$\rho = \rho_0 + \rho_{2.5}T^{2.5} \tag{4}$$

here ρ_0 is the temperature independent residual resistivity. $\rho_{2.5}T^{2.5}$ represents the electrical resistivity due to electron-magnon scattering processes in the ferromagnetic phase. We found that in the metallic regime electrical resistivity data for Sb doped samples best fit the Eq. (4). The quality of these fittings, in general, is evaluated by comparing the square of the linear correlation coefficient (R^2) obtained for each equation. The values of R^2 were found as high as 99.9% for Eqn. (4) which confirms the applicability of electron-magnon scattering process. The obtained fitted parameters are given in Table 2. It is observed that value of both ρ_0 and $\rho_{2.5}$ increases with doping. This reflects the dominating nature of the electron-magnon scattering mechanism with increasing Sb content. Similar results were observed for the Sb-doped PBMO manganites.

Sample	ρ_0 (Ω cm)	$\rho_{2.5}$ (Ω cm$K^{-2.5}$)	R^2
x = 0	0.11296	8.1315x10^{-6}	0.99994
x = 0.01	0.31814	0.00005	0.99905
x = 0.02	3.18173	0.00015	0.99962
x = 0.03	33.90211	0.00103	0.99952

Table 2. Fitting parameters obtained using Eq. (4) after fitting the metallic region of $La_{2/3}Ba_{1/3}(Mn_{1-x}Sb_x)O_3$ ($0 \le x \le 0.03$) system.

3.6 Transport mechanism above T_{P1}

Conduction in manganites at higher temperature ($T \geq T_P$) is governed either by variable range hopping (VRH) or by small polaron hopping (SPH) depending on the Debye temperature (θ_D)[59,60]. If $T > \theta_D/2$, the conduction mechanism is governed by small polarons and these polaronic models can be either adiabatic or non-adiabatic. Since the Debye temperature of the LBMO system is ~400 K [61] so the polaronic model can be applied to the insulating region above T_{P1}. The adiabatic small polaron hopping model is given by[59, 60]:

$$\rho = \rho_0 T \exp(E_\rho/k_B T) \tag{4}$$

where ρ_0 is the residual electrical resistivity, E_ρ is the activation energy and k_B is the Boltzmann's constant. We estimated the activation energy E_ρ and it is found to increase with Sb content as shown in Table 3. Similar results were observed for Sb-doped PBMO system.

Sample	E_ρ (meV)	E_S (meV)	α	W_H (meV)	E_P (meV)
x = 0	91.2	21.693	-0.8491	69.507	139.014
x = 0.01	98.02	14.016	-0.6444	84.004	168.008
x = 0.02	112.92	7.903	-0.457	105.001	210.002
x = 0.03	126.15	5.869	-0.3941	120.281	240.562

Table 3. Activation energies E_ρ (from electrical resistivity) and E_S (from Thermoelectric Power), W_H (= E_ρ - E_S) corresponds to polaron binding energies for Sb-doped LBMO series.

3.7 Magnetoresistance studies

Generally, application of the external magnetic field enhances spin ordering, promotes the charge transfer and thus suppresses the electrical resistivity. Alignment of the spins of the neighboring Mn-sites favors the electronic motion. Figure 8 (a) & (b) shows the temperature dependent magnetoresistance behavior of all the samples from 77K to 300K, for an applied magnetic field of 0.6T. For Sb-doped LBMO samples a sharp peak is observed in all the samples near T_{P1} and below it MR increases continuously reflecting the important role of grain boundaries in producing LFMR (in case of single crystals MR decreases very fast below T_{P1}). With Sb doping the MR is found to increase. Magnetoresistance for 3% Sb-doped sample $La_{2/3}Ba_{1/3}Mn_{0.97}Sb_{0.03}O_3$ is 6% under the application of 0.6 T magnetic field which shows that this material may be useful from the application point of view. In case of Sb-doped PBMO (Fig. 8b), $Pr_{2/3}Ba_{1/3}(Mn_{1-x}Sb_x)O_3$ series, besides a usual peak near T_{P1} there is the development of an additional peak with increasing Sb-content. For 3% Sb sample low temperature peak MR is larger than that the high temperature peak. Also, in case of Sb-doped PBMO, MR value at the respective T_{P1} decreases with Sb which is reverse to that of Sb-LBMO system. The difference of MR below T_{P1} may be attributed to the varying grain boundary effects in both the cases. To identify this we carried out the MR measurement with magnetic field variation at 77K and the results are shown in figure 9 (a) & (b). MR behavior of Sb-doped LBMO series exhibits the trend of spin polarized tunneling proposed by Hwang et al.[62] with lower applied field MR increases sharply and at higher values of the field it gets saturated. Whereas in case of Sb-doped PBMO, it seems that spin dependent scattering and not the spin polarized tunneling dominates the process.

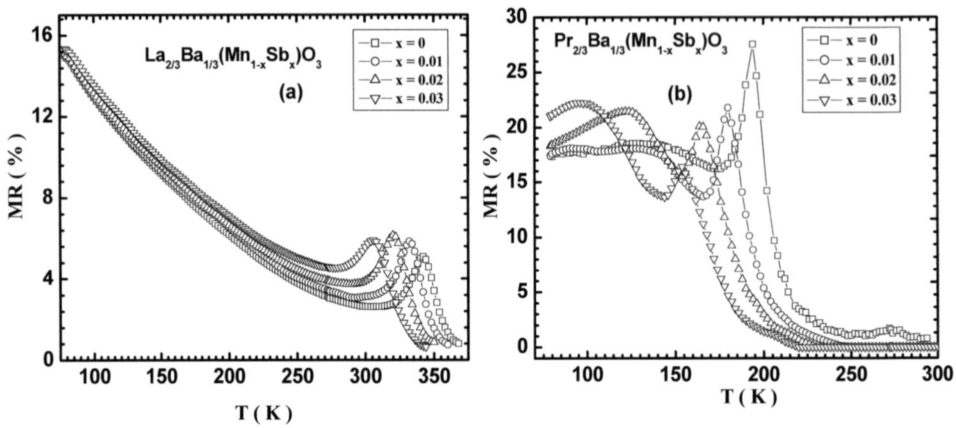

Fig. 8. Magnetoresistance variation with temperature of Sb-LBMO and Sb-PBMO samples under 0.6 T magnetic field.

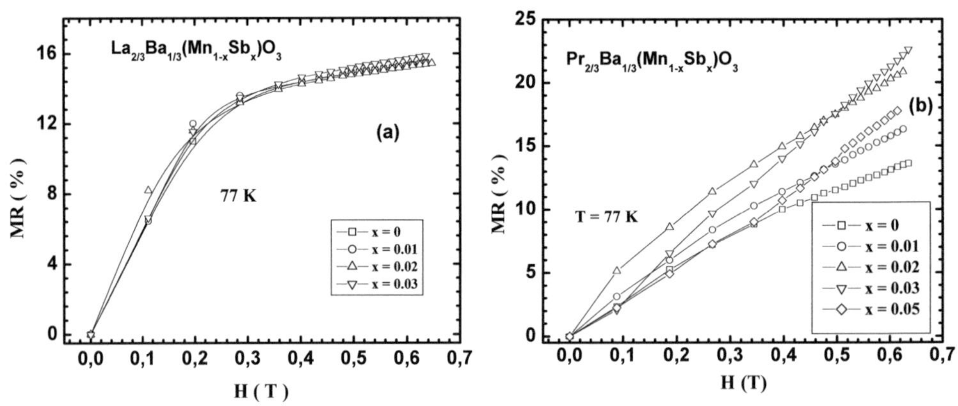

Fig. 9. Magnetoresistance variation with magnetic field of Sb-LBMO and Sb-PBMO samples at 77K.

3.8 Thermoelectric power

Thermoelectric Power (TEP) is especially suited to explore the carrier dynamics because it is expected to be very sensitive to the local moments[63] and thus it monitors the ferromagnetic transition and provide new insight into the dynamics of manganite systems. $S(T)$ is less affected by the presence of grain boundaries and hence the grain boundary effects that are observed in the resistivity temperature measurements, can be masked. The thermoelectric power data for the two series is shown in Figure 10. In the case of Sb-doped LBMO series, a crossover of thermopower is observed from positive to negative at temperature $T^* = 264$, 258, 250, 241 and 210 K respectively with increasing Sb-content whereas in Sb-PBMO series only two samples *viz.* pristine PBMO and 1% Sb-doped exhibit such behavior. Further, in

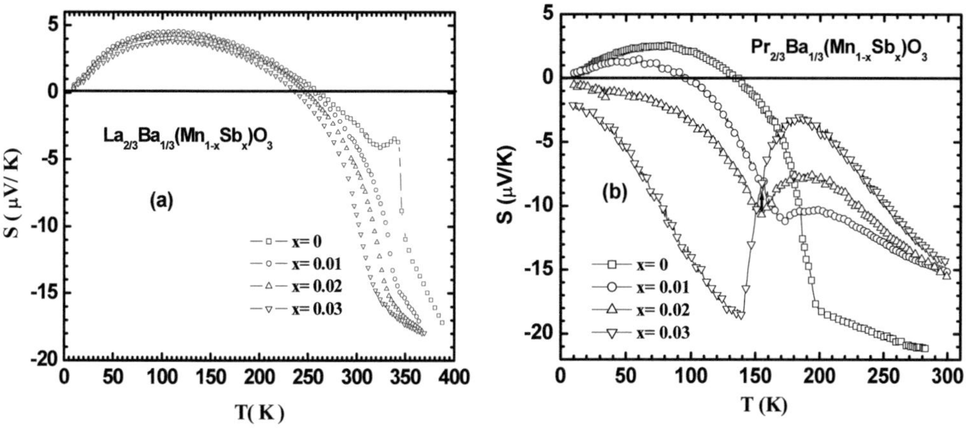

Fig. 10. Thermopower variation with temperature of Sb-LBMO and Sb-PBMO samples.

case of Sb-LBMO, insulator-metal transition T_S is very close to T_{P1} for all the samples whereas for Sb-PBMO, the two transitions are closer for PBMO sample only and as the doping of Sb increases at Mn-site a peak starts developing near the phase transition and separation between T_{P1} and T_S increases. The development of the thermopower peak and separation of T_{P1} and T_S can be related to the increase in magnetic inhomogeneity due to increase in spin entropy by the weakening of double exchange interaction between Mn^{+3}/Mn^{+4} ions in the sample with Sb-doping that could lead to a phase separation. In the ferromagnetic state, the magnitude of the observed thermoelectric power in the range of a few μV/K seems to be reminiscent of the metallic system. In order to explain the change of sign of thermopower from hole-like behavior (positive) to electron-like behavior (negative) in the metallic ferromagnetic (FMM) state we adopt the model given by Asamitsu et al.[64] that is based on the vanishing of the exchange interaction J. This interaction arises due to the excitation of the electrons from the valence band (VB) t_{2g} to the conduction band (CB) e_g. The mobility of electrons in the conduction band is high, leading to a negative thermopower. The degeneracy of the e_g band seems to be gradually lifted as the temperature is lowered below T^* and the thermopower changes its sign from negative to positive with the increase of spin polarization. Based on thermoelectric data, we have attempted to explore various mechanisms governing the thermoelectric power in different temperature regions. It would be worthwhile mentioning here that the effect of Jahn-Teller distortion in manganites results in strong electron-phonon coupling and hence the formation of polarons[65]. Therefore, charge carriers in the insulating region above T_P are not itinerant and transport properties are governed by thermally activated carriers (polarons)[66]. We, therefore, fitted the $S(T)$ data above T_S (T_{P1}) fig. 10, with the Mott's polaron hopping relation[66]:

$$S = k_B/e\,[(E_S/k_BT) + a] \tag{5}$$

where e is the electronic charge; k_B is the Boltzmann constant, E_S is the activation energy required to activate the hopping of the carriers and a is a constant related to the kinetic energy of the polarons[67]. The value of $a < 1$ suggests the conduction due to small polarons hopping (SPH) whereas $a > 2$ corresponds to the case of large polaron hopping (LPH)[66]. The fitting for two systems are shown in figure 11. The electrical resistivity activation energy E_ρ and the thermopower activation energy E_S, are shown in Table 3 for Sb-doped LBMO manganites and from there it is clear that E_ρ is much larger than E_S. Such a large difference in the values of two activation energies is the indication of the applicability of the small polaron-hopping (SPH) model in the insulating region above T_{P1} which is further corroborated by the value of constant a (less than 1). In the framework of SPH model, E_ρ is the sum of the activation energy required for the creation of the carriers and activating the hopping of the carriers and E_S is the energy required to activate the hopping of carriers only. Hence E_S is much smaller than E_ρ. Based on E_ρ & E_S, both the polaronic energy $W_H = E_\rho - E_S$ and the polaronic formation energy $E_P = 2W_H$ are also given in Table 3 for Sb substituted LBMO.

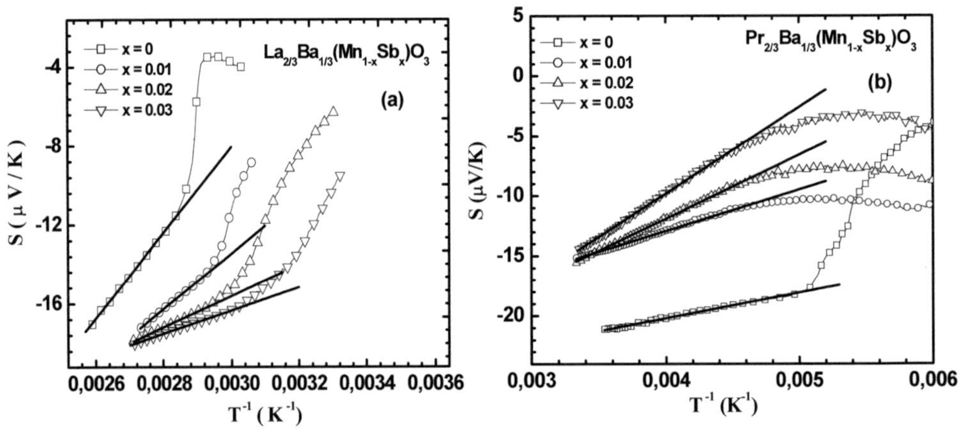

Fig. 11. Thermopower fitting above T_S of Sb-LBMO and Sb-PBMO samples using Eq. (5).

Similar to $\rho(T)$ data, several factors, such as impurity, complicated band structure, electron-electron, electron-magnon scattering etc also affect $S(T)$ data in the FM metallic regime. $S(T)$ data in this intermediate temperature FM metallic region has been analyzed by the following relation[68,69]:

$$S = S_0 + S_{3/2}T^{3/2} + S_4T^4 \tag{6}$$

here S_0, $S_{3/2}$, and S_4 are fitting parameters. S_0 has no physical origin but inserted to account for the low temperature data, $S^{3/2}T^{3/2}$ is attributed to single magnon scattering process[68,70]. The term S_4T^4 dominant in high temperature region near T_C, is thought to arise from spin wave fluctuation in the FM phase[68,70]. Figure 12 shows the fitting of Eq. (6) in the metallic region of the Sb-doped LBMO samples and the fitting parameters are given in Table 4. From Table 4, it is clear that $S_{3/2}$ is nearly five orders of magnitude larger than that of S_4, implying that the second term i.e. electron-magnon scattering in Eq.(6) dominates the transport mechanism in the FM metallic regime below T_{P1}. The hole-like to electron-like cross over in thermoelectric power arises due to the fact that $S_4 < 0$. Then due to higher power of T, the term S_4T^4 will dominate for large T making thermopower negative. Hence, based on the electrical resistivity and thermopower data fittings we can conclude that insulating region above T_{P1} is governed by the small polaron hopping (SPH) model whereas electron-magnon scattering is dominant in ferromagnetic metallic region.

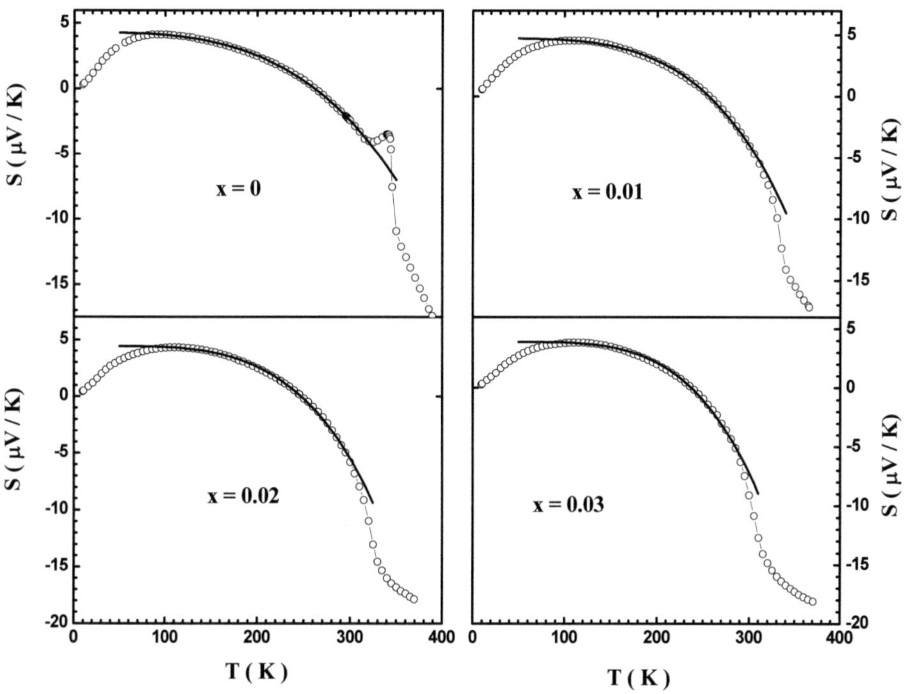

Fig. 12. Thermopower fitting of $La_{2/3}Ba_{1/3}(Mn_{1-x}Sb_x)O_3$ ($0 \leq x \leq 0.03$). Solid line represents the fitting with Eq. (6).

Sample	S_0 ($\mu V/K$)	$S_{3/2}$ ($\mu V/K^{2.5}$)	S_4 ($\mu V/K^5$)
x = 0	4.5365± 0.03756	-0.00036± 0.00002	- 6.3773x10^{-10} ± 9.951x10^{-12}
x = 0.01	4.7905± 0.06958	-0.00009± 0.00004	- 1.0287x10^{-9}± 2.2475x10^{-11}
x = 0.02	4.43458± 0.08659	0.00003± 0.00006	- 1.212x10^{-9}± 3.5867x10^{-11}
x = 0.03	3.69632± 0.0923	0.00028± 0.00006	- 1.4882x10^{-9}± 3.8002x10^{-11}

Table 4. Electron-magnon fitting parameters of thermoelectric power of Sb-LBMO series in the FM region with Eq. (6).

3.9 Thermal conductivity

Figure 13 shows the temperature dependent thermal conductivity (κ) measurements of Sb-doped manganites. The magnitude of $\kappa(T)$, typically of non-crystalline materials (bad metals), lies in the range of 5-70 mW/cm K. For a crystalline solid, such a low value of thermal conductivity in manganites can be attributed to the disorder due to strong Jahn-Teller effect. Magnitude of κ decreases with decreasing temperature down to T_κ (close to T_S and T_{P1}), which is unusual since the high temperature thermal conductivity of the crystalline insulators is mostly a decreasing function of temperature and cannot be attributed to high temperature electron or phonon processes. Also the thermal conductivity decreases with Sb-doping in the whole temperature range. As discussed earlier, number of Mn^{+3} -ions increases with Sb-doping and since Mn^{+3} is a Jahn-Teller ion it causes the distortion in $Mn^{+3}O_6$ octahedra which eventually scatters the phonons and causes κ to decrease with Sb-doping.

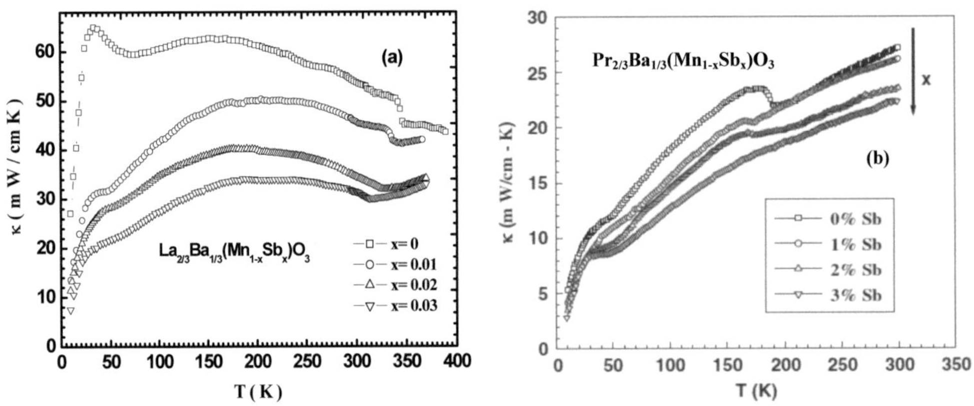

Fig. 13. κ (T) variation of Sb-LBMO and Sb-PBMO samples.

At high temperatures thermal conductivity of the crystalline solids is expected to follow the temperature dependency according to the formula $\kappa \propto aMC_P\theta_D/\ \gamma^2 T$, where a is the lattice constant, M is the mass per atom, C_P is the specific heat, θ_D is the Debye temperature, and γ is the Gruneisen constant with a value ranging from 2-3 for solids. These parameters are considered to be a weak function of temperature at high temperatures. As a result, κ is expected to increase with decreasing temperature. However, in the present case κ decreases monotonically with temperature for all the samples, which is peculiar. Neutron scattering experiments in these manganites, however, have shown that the Gruneisen constant γ is much higher (~180) than usual and decreases remarkably with temperature[71]. Thus, high temperature behaviour of κ in the paramagnetic state could be understood in terms of the local anharmonic lattice distortions associated with small polarons[72]. An anomaly in thermal conductivity below T_{P1} has been observed for all the studied samples. It may be noticed that this anomaly is closely related to the phase transition at T_{P1}. Below T_{P1}, κ first increases for pure and 1% doped samples and change in slope for the other doped samples which may be attributed to the reduction in phonon-phonon scattering (Umklapp process) as a result of JT distortions which become delocalized alongwith the charge carriers. The small peak observed in thermal conductivity below $35K$ in all the samples is an indication of the crossover from Umklapp to defect limited scattering[73,74]. In general, the total thermal conductivity for ordinary metals and semimetals is sum of the electronic and lattice terms. The electronic thermal conductivity κ_e can be evaluated using Wiedemann-Franz (W-F) law $\kappa_e \rho / T = L$. Here ρ is the electrical resistivity and the Lorentz number L is $2.45\text{x}10^{-8}$ $W\Omega/K^2$. Therefore, using the above W-F formula electronic thermal conductivity at any particular temperature (say 250*K*, where the value of electrical resistivity is shown in Table 5), is very small to the total thermal conductivity κ at that temperature. Hence the measured κ is suggested to arise from phonons, the contribution of charge carriers being negligible.

Sample	ρ_{250K} (mΩ- cm)	κ_e (m*W*/cm-*K*)	ΔS (in *R*)	ΔC_{mag} (J/mol-*K*)
x = 0	30.48	0.201	0.0917	12.065
x = 0.01	45.95	0.1333	0.194	17.45
x = 0.02	64.2	0.0954	0.1265	12.71
x = 0.03	119.97	0.0511	0.1231	10.116

Table 5. Electronic thermal conductivity, entropy and magnon contribution to the specific heat for Sb-doped LBMO series.

3.10 Specific heat

The observed specific heat measurements [$C_P(T)$] carried out from 85 K to 380 K for the Sb-doped manganites is shown in Figure14. It increases with temperature for all pure as well as doped samples. There is not much effect of Sb doping up to ~ 250K. However all the samples show a peak in its $C_P(T)$ plot just below the corresponding paramagnetic-ferromagnetic transition temperature (T_C). For clarity, each curve has been offset by 10 J/mole K. The transition temperature, defined as the peak position, being lower than the

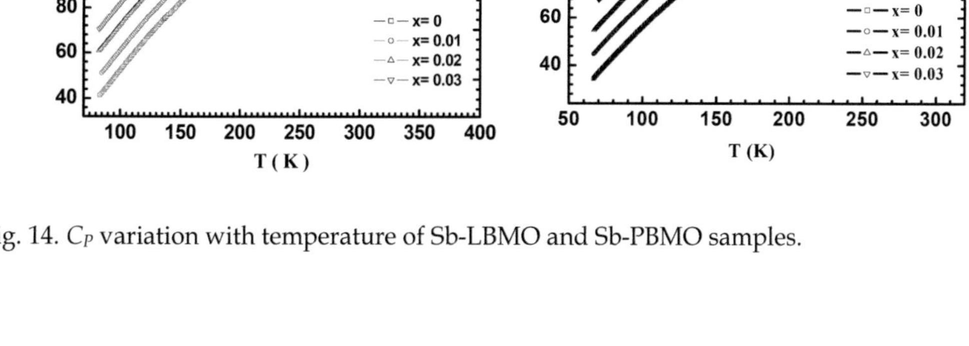

Fig. 14. C_P variation with temperature of Sb-LBMO and Sb-PBMO samples.

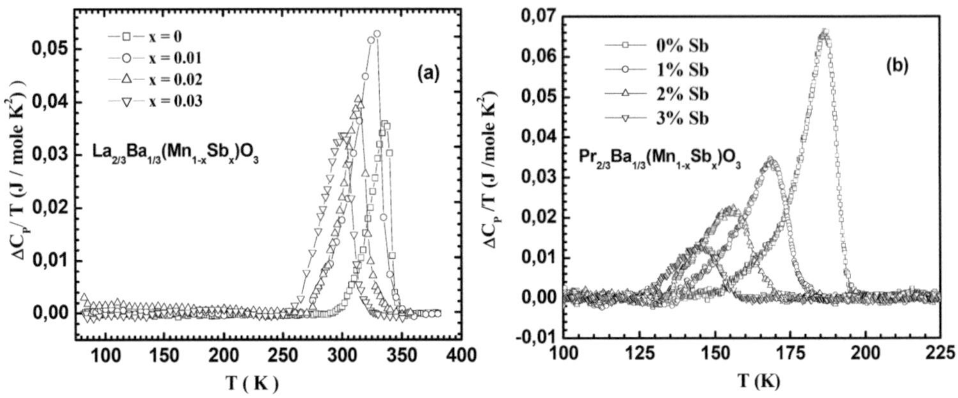

Fig. 15. Specific heat jumps with background subtracted for Sb-LBMO and Sb-PBMO samples.

corresponding I-M transition temperature T_{P1} reflects that the specific heat anomaly is related to the magnetic ordering in the sample due to paramagnetic-ferromagnetic transition. This specific heat anomaly shifts to lower temperatures and becomes progressively broader with Sb-doping. However, no such anomaly corresponding to T_{P2} could be observed in our measurements, which again confirm that thermal measurements are insensitive to the grain boundaries. Considerable broadening in the $C_P(T)$ curves with Sb substituted samples implies the increase in local structural disordering or/and magnetic inhomogeneity.

In order to separate the lattice contribution and to estimate the excess specific heat (ΔC_P) due to magnetic ordering a smooth background line (tangential to both end points and shown in Fig.14(b) for pristine PBMO manganite) was subtracted from the data of every sample and the obtained curves are plotted in Figure 15. Integration of $\Delta C_P/T$ gives the excess entropy (ΔS) under the C_P anomaly of the specific heat curve. Considerable reduction in the value of ΔS is observed with the increase in Sb content (Table 5). This confirms that magnetic inhomogeneity increases with Sb. The value of ΔS is also less than the theoretical value $R\ln 2$ for a paramagnetic-ferromagnetic transition (where R is the ideal gas constant). This discrepancy can be attributed to the magnetic inhomogeniety or partially canted spins in the ferromagnetic state.

4. Conclusions

Substitution effect of Sb^{5+} ion on the structural, morphological, magneto-transport and thermal properties (including thermoelectric power, specific heat and thermal conductivity) of $La_{2/3}Ba_{1/3}MnO_3$ (LBMO) and $Pr_{2/3}Ba_{1/3}MnO_3$ (PBMO) manganites have been carried out. Structural measurements on polycrystalline samples synthesized showed that lattice parameters increase with Sb-doping. The grain size is also found to increase with Sb. Electrical resistivity variation with temperature of these managnites showed two insulator-metal like transitions (T_{P1} and T_{P2}) which suppressed to low temperatures with an overall increase in electrical resistivity with Sb substitution. Thermal measurements and their analyses show that high temperature insulating region was governed by the hopping of small polarons and the metallic region below T_{P1} was dominated by electron-magnon scattering process. Intrinsic magneto-resistance (MR due to grain) at T_{P1} (high temperature insulator-metal transition) was suppressed whereas extrinsic MR (due to grain boundary) got enhanced with Sb. While in the case of Sb-LBMO, MR kept on increasing below a peak near T_{P1} in all samples whereas in Sb-PBMO system second peak appeared in MR with its value larger than that at the high temperature peak. MR ~7% was observed in 3% Sb-LBMO sample at room temperature under 0.6T magnetic field which indicates the potential application of these materials as low magnetic field sensors.

5. Acknowledgement

The authors are thankful to Prof. Y. K. Kuo (National Dong Hwa University Taiwan), Prof. G. L. Bhalla (University of Delhi, India), Prof. Ashok Rao (Manipal University, India) and Prof. D. K. Pandya (I.I.T. Delhi, India) for scientific discussions. One of the authors (SKA) would like to thank the Council of Scientific and Industrial Research (CSIR, India) for providing the financial grants under the Emeritus Scientist Scheme.

6. References

[1] W. Thomson, Proc. Roy. Soc. Lond. 8 (1857) 546.
[2] F. Y. Yang, K. Liu, D. H. Reich, P. C. Searson and C. L. Chien, Science 284 (1999) 1335.
[3] K. Chahara, T. Ohno, M. Kasai and Y. Kozono, Appl. Phys. Lett. 63 (1993) 1990.
[4] R. von Helmolt, J. Wocker, B. Holzapfel, M. Schultz and K. Samver, Phys. Rev. Lett. 71 (1993) 2331.
[5] S. Jin, T. H. Tiefel, M. McCormack, R. A. Fastnacht, R. Ramesh and L. H. Chen, Science 264 (1994) 413.
[6] G. C. Xiong, Q. Li, H. L. Ju, S. M. Bhagat, S. E. Lofland, R. L. Greene, and T. Venkatesan, Appl. Phys. Lett. 67 (1995) 3031.
[7] C. W. Searle and S. T. Wang, Can. J. Phys. 47 (1969) 2703.
[8] C. W. Searle and S. T. Wang, Can. J. Phys. 48 (1970) 2023.
[9] R. M. Kusters, D. A. Singleton, D. A. Keen, R. AcGreevy and W. Hayes, Physica B 115 (1989) 362.
[10] S. S. P. Parkin, Annu. Rev. Mater. Sci. 25 (1995) 357.
[11] H. L. Ju, C. Kwon, Q. Li, R. L. Green and T. Venkatesan, Appl. Phys. Lett. 65 (1994) 2108.
[12] Y. G. Zhao, M. Rajeswari, R. C. Srivastva, A. Biswas, S. B. Ogale, D. J. Kang, W. Prellier, Z. Chen, R. L. Green and T. Venkatesan, J. Appl. Phys. 86 (1999) 6327.
[13] W. Prellier, A. M. Haghiri-Gosnet, B. Mercey, P. Lecoeur, M. Hervieu, C. Simon and B. Raveau, Appl. Phys. Lett. 77 (2000) 1023.
[14] P. A. Cox, "Transition Metal Oxides: An Introduction to their Electronic Structure and Properties", Clarendon Press, Oxford (1995).
[15] W. E. Pickett and D. J. Singh, Phys. Rev. B. 53 (1995) 1146.
[16] H. A. Jahn and E. Teller, Proc. Roy. Soc. London. Series A, Mathematical and Physical Sciences (1943-1990), 161 (1937) 220.

[17] R. Orbach and H. J. Stapleton, Electron Spin-Lattice Relaxation, Chapter 2 of "Electron Paramagnetic Resonance" Ed. S. Geschwind (Plenum Press, New York, 1972).
[18] J. Kanamori, J. Appl. Phys. (Suppl.) 31 (1960) 14S.
[19] Y. Tokura (Ed.), Colossal Magnetoresistive Oxides, (Gordon and Breach, London, U. K., 2000).
[20] A. J. Millis, B. I. Shraiman and R. Mueller, Phys. Rev. Lett. 77 (1996) 175.
[21] K. H. Ahn, X. W. Wu, K. Liu, C. L. Chien, J. Appl. Phys. 81 (1997) 5505.
[22] J. W. Cai, C. Wang, B.-G. Shen, J.-G. Zhao, W.-S. Zhan, Appl. Phys. Lett. 71 (1997) 1727.
[23] J. Blasco, J. Garcia, J.M. de Teresa, M.R. Ibarra, J. Perez, P.A. Algarabel, C. Ritter, Phys. Rev. B 55 (1997) 8905.
[24] Y. Sun, X. Xu, L. Zheng, Y. Zhang, Phys. Rev. B 60 (1999) 12317.
[25] E. Sotirova, X. L. Wang, J. Horvat, T. Silver, K. Konstantinov, H. K. Liu, Supercond. Sci. Tech. 15 (2002) 346.
[26] J. Yang, W. H. Song, Y. Q. Ma, R. L. Zhang, B. C. Zhao, Z. C. Sheng, G. H. Zheng, J. M. Dai, Y. P. Sun, Phys. Rev. B 70 (2004) 092504.
[27] S. Pal, E. Bose, B. K. Chaudhuri, H. D. Yang, S. Neeleshwar and Y. Y. Chen, J. Magn. Mag. Mater. 293 (2005) 872.
[28] R. Ang, Y. P. Sun, J. Yang, X. B. Zhu and W. H. Song, J. Appl. Phys. 100 (2006)073706.
[29] X. Liu, X. Xu and Y. Zhang, Phys. Rev. B 62 (2000)15112.
[30] L. S. Lakshmi, K. Dorr, K. Nenkov, V. Sridharan, V. S. Sastry, K. -H. Muller, J. Magn. Mag. Mater. 290-291(2005) 924; V. Sridharan, L. S. Lakshmi, R. Govindraj, R. Nithya, D. V. Natarajan and T. S. Radhakrishnan, J. Alloys Comp. 326 (2001) 65.
[31] Y. H. Liu, B. -X. Huang, R. -Z. Zhang, X. -B. Yuan, C. -J. Wang and L. M. Mei, J. Magn. Mag. Mater. 293 (2005) 872.
[32] J. S. Kim, D. C. Kim, G. C. McIntosh, S. W. Chu, Y. W. Park, B. J. Kim, Y. C. Kim, A. Maignan and B. Raveau, Phys. Rev. B 66 (2002) 224427.
[33] I. O. Troyanchuk, M. V. Bushinsky, H. Szymazak, K. Barner, A. Maignan, Eur.Phys. J. B, 28 (2002) 75.
[34] S. K. Agarwal, Neeraj Panwar, Vikram Sen and D. K. Pandya, J. Phys. D: Appl. Phys. 41 (2008)105004.
[35] V. Sen, Neeraj Panwar, G.L. Bhalla and S.K. Agarwal, J. Alloys Compnd. 439 (2007) 205.
[36] V. Sen, Neeraj Panwar, G.L. Bhalla and S.K. Agarwal, J. Phys. Chem. Solids 68 (2007) 1685.
[37] V. Sen, Neeraj Panwar, Ashok Rao, C.K. Hsu, Y.K. Kuo and S.K. Agarwal, Solid State Communications 145 (2008) 86.
[38] V. Sen, G. L. Bhalla, Neeraj Panwar, W. K. Syu, N. Kaurav, Y. K. Kuo, Ashok Rao and S. K. Agarwal, Physica B 405 (2010) 1.
[39] Y.-K. Kuo, C.S. Lue, F.H. Hsu, H.H. Li and H.D. Yang, Phys. Rev. B 64 (2001) 125124.
[40] H. L. Ju, J. Gopalakrishnan, J. L. Peng, Q. Li, G. C. Xiong, T. Venkatesan, R. L. Greene, Phys. Rev. B 51 (1995) 6143.
[41] Y. S. Nam, H. L. Ju and C.W. Park, Solid State Commun. 119 (2001) 613.
[42] W. Zhong, W. Chen, C.T. Au, and Y.W. Du, J. Magn. Magn. Mater. 261. (2003) 238–243.
[43] L. S. Lakshmi, K. Dorr, K. Nenkov, V. Sridharan, V.S. Sastry and K. H. Muller, J. Mag. Magn. Mater. 290–291 (2005) 924.

[44] H. L. Ju, H. Sohn, Solid State Commun. 102 (1997) 463.
[45] A. B. Beznosov, V. A. Desnenko, E. L. Fertman, C. Ritter and D. D. Khalyavin, Phys. Rev. B 68 (2003) 054109;N. Moutis, I. Panagiotopoulos, M. Pissas and D. Niarchos, Phys. Rev. B 59 (1999) 1129.
[46] P. A. Lee and T. V. Ramakrishnan, Rev. Mod. Phys. 57 (1985) 287.
[47] B. L. Altshuler and P.A. Lee, Phys. Tod. 41(1988) 36.
[48]B. L. Altshuler and A. G. Aronov, in electron-electron interactions in Disordered system, (Eds.) A. L. Efros and M. Pollak (North-Holland, New York,1985).
[49] J. Kondo, Prog. Theor. Phys. 32 (1964) 37.
[50] J. Kondo, Solid State Physics, 23 (1969) 183.
[51] L. Kouwenhoven and L. Glazman, Phys. World 14 (2001) 33.
[52] C. Kittel, Introduction to Solid State Physics (John Wiley & Sons, Singapore, 1995).
[53] J. M. Ziman, Electrons and Phonons (Oxford University Press, London, 1967).
[54] N. W. Aschroft and N. D. Mermin, Solid State Physics (Saunders, Philadelphia,1967).
[55] I. Mannari, Prog. Theo. Phys. 22 (1959) 325.
[56] T. Kasuya, Prog. Theo. Phys. 16 (1956) 58.
[57] M. Viret, F.Ott, J. P. Renard, H. Glattli, L. P. -Gaudart and A. Revcolevschi, Phys.Rev. Lett. 93(2004) 27402.
[58] M. G.-Hernandez, F. Guinea, A. de Andres, J. L. Martinez, C. Prieto and L. Vazquez, Phys. Rev. B 61 (2000) 9549; A. Maignan, C. Simon, V. Caignaert and B. Raveau, J. Appl. Phys. 79 (1996) 7891.
[59] N. F. Mott, E. A. Davis, Electronics Process in Noncrystalline Materials, Clarendon, Oxford,1971.
[60] D. Emin, T. Holstein, Studies of small-polaron motion IV. Adiabatic theory of Hall Effect, Ann. Phys. 53 (1969) 439.
[61] J. J. Hamilton, E. L. Keatley, H. L. Ju, A. K. Raychaudhuri, V. N. Smolyaninova, R. L. Greene, Phys. Rev. B 54 (1996)14926.
[62] H. Y. Hwang, S. -W. Cheong, N. P. Ong and B. Batlogg, Phys. Rev. Lett. 77 (1996) 2041.
[63] F. J. Blatt, P. A. Schroeder, C. L. Foiles and D. Greig, in Thermoelectric Power of Metals (Plennum, New York), 1976.
[64] A. Asamitsu, Y. Moritomo, and Y. Tokura, Phys. Rev. B 53 (1996) 2952.
[65] A. J.Millis, P. B.Littlewood, B. I.Shraiman, Phys.Rev. Lett. 74 (1995) 5144.
[66] N. F. Mott and E. A. Davis (Eds.), Electronic Processes in Noncrystalline Materials, Clarendon, Oxford, 1979.
[67] K. Sega, Y. Kuroda and H. Sakata, J. Mater. Sci. 33 (1998) 1303.
[68] A. Banerjee, S. Pal, S. Bhattacharya, B. K. Chaudhuri and H. D. Yang, Phys. Rev. B 64 (2001) 104428.
[69] P. Mandal, Phys. Rev. B 61 (2000) 14675.
[70] A. Urushibara, Y. Moritomo and Y. Tokura, Phys. Rev. B 51 (1995) 14103.
[71] P. Dai, J. Zhang, H. A. Mook, S.-J. Lion, P. A. Dowben, and E. W. Plummer, Phys. Rev. B 54 (1996) 3694.
[72] Y. Sun, X. Xu, and Y. Zhang, Phys. Rev. B 63 (2000) 054404.
[73] D. W. Visser, A. P. Ramirez, M. A. Subramanium, Phys. Rev. Lett. 78 (1997) 3947.

[74] P. G. Klemens, in: F. Seitz, D. Turnbull (Eds.), Solid State Physics, Academic Press, New York, 1958, p 7.

14

Metal Oxide ZnO-Based Varistor Ceramics

Mohammad Reza Meshkatoddini
Faculty Member of Shahid Abbaspour PWUT University of Technology (SAUT), Tehran, Iran

1. Introduction

The metal oxide ZnO-based varistors are non-linear ceramic resistors which are largely used to protect the electric and electronic circuits and components against overvoltages. These varistors, which are among the most non-linear discovered materials, are used in lightning arresters owing to their strongly non-linear characteristics I(V). (Figure 1).

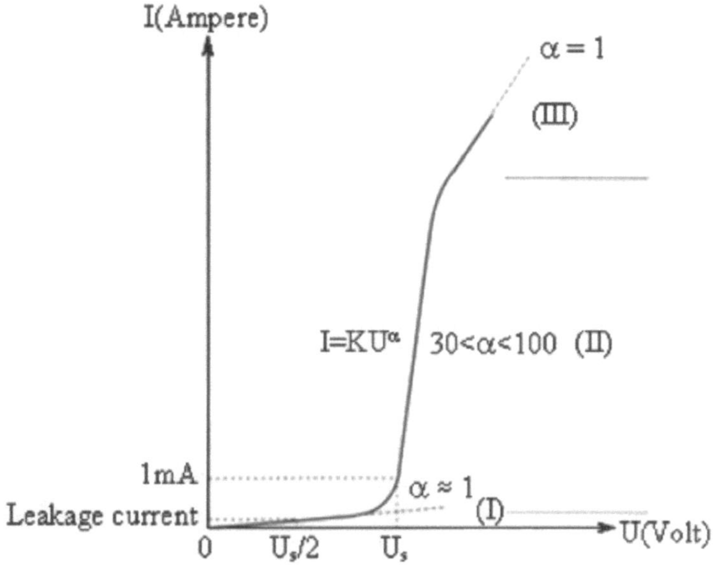

Fig. 1. Current versus Voltage characteristic in a ZnO-Based Varistor.

A varistor is a type of resistor with a significantly non-ohmic current-voltage characteristic. The name is a portmanteau of variable resistor, which is misleading since it is not continuously user-variable like a potentiometer or rheostat and is capacitor rather than resistor at low field. The most famous type of varistor is metal oxide varistor (MOV), which is also called as ZnO varistor. These varistors are used to protect circuits against excessive voltages. They have become more and more important during the past four decades due to their highly non-linear electrical characteristics and their large energy absorption capacity. They are normally connected in parallel with an electric device to protect it against the overvoltages. They contain a mass of zinc oxide grains in a matrix of other metal oxides sandwiched between two plasma sprayed metal electrodes. The ZnO grains have dimensions in the range of 10μm to 100μm. The boundaries between the grains form double potential barriers with Schottky junctions having conduction voltages in the range of 3.5V. The boundary between each grain and its neighbor forms a Zener-like diode junction. ZnO grains are separated by these "active" grain boundaries of nanometers thickness. Then the mass of randomly oriented grains is electrically equivalent to a network of back-to-back diode pairs, each pair in parallel with many other pairs. When a small or moderate voltage is applied across the electrodes, a small thermally activated reverse leakage current flows through the diode junctions. When a large voltage is applied, the diode junctions break down from the avalanche effect, and large current flows. The result of this behavior is a highly nonlinear current-voltage characteristic, in which the MOV has a high resistance at low voltages and a low resistance at high voltages.

Three regions can be distinguished in the current voltage characteristics of the ZnO varistor. At low voltages, the insulating barriers between the grains result in a very high and almost Ohmic resistivity, which is called the pre-breakdown or Ohmic region. At a certain voltage, called the threshold or breakdown voltage, the system enters the breakdown region in which the current increases abruptly, and the dependence of current on voltage is described by the empirical relation:

$$I = k\,V^{\alpha} \qquad (1)$$

From which the parameter α is equal to:

$$\alpha = d\,[\log(I)]\;/\;d\,[\log(V)] \qquad (2)$$

This parameter is a measure of the element nonlinearity, which varies with voltage. At higher current densities, the voltage starts to increase again resulting in an upturn region of the I-V characteristic. This voltage increase gradually becomes linear with current, i.e. Ohmic, and is associated with the resistivity of the ZnO grains, i.e. the voltage drop in the ZnO grains.

Among their electric properties the most important ones are:

- The threshold voltage: It can be defined as the value of the voltage across the varistor, corresponding to a current of 1mA passing through it. From this voltage value, the varistor starts to change from the insulating state into the conducting state.
- Energy capacity: It is the maximum capacity of the energy absorption of a varistor without any damage, while the discharge current due to an overvoltage passes through it.

The other properties (chemical, mechanical...) are closely related to the two properties quoted above.

We have tried in our works to accomplish several statistical studies on these varistors, to find suitable ways to control their main characteristics such as the nonlinearity coefficient and conduction threshold voltage.

These varistors are composed of zinc oxide and some other metal oxides, which provide the desired characteristics for these varistors. The microstructure of the varistor ceramics develops while sintering ZnO powder doped with small amounts of additives such as Bi_2O_3, Sb_2O_3, Mn_3O_4, Co_3O_4, Cr_2O_3 and others, at a temperature in the range of 1100 to 1300°C.

The typical microstructure of a ZnO-based varistor is shown in figures 2 and 3. It is composed of ZnO matrix grains doped with Co, Mn, and Ni. These grains are n-type semiconductors. Both Bi_2O_3-rich and $Zn_7Sb_2O_{12}$ spinel phases are also usually present at the grain boundaries of the ZnO, but the presence of a $Bi_3Zn_2Sb_3O_{14}$ phase is possible as well. ZnO-ZnO grain boundary, rich of Bismuth, which is a highly resistive phase, is the main cause of the varistor effect, while spinel-ZnO junctions do not contribute to the nonlinear effect.

Fig. 2. Typical microstructure of a ZnO varistor taken by electronic microscope. (ZnO=Zinc oxide grain, Bi=Bismuth, Sp=Spinel phase).

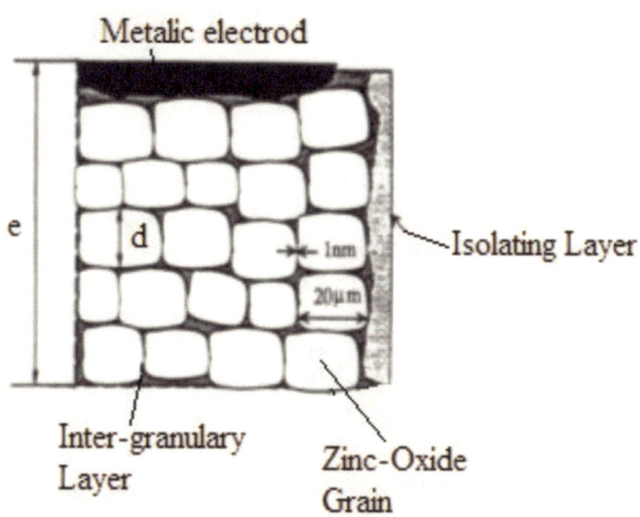

Fig. 3. A Model of ZnO-ZnO grain boundaries in a zinc oxide varistor.

The size of ZnO grains (d in Figure 2) determines the number of ZnO grain boundaries between the electrodes of the varistor. As mentioned before, the typical breakdown voltage for a non-ohmic ZnO-ZnO grain boundary is around 3 Volts, and hence, the size of the grains determines the overall breakdown voltage of varistor and then the length of the varistor column in a lightning arrester.

Sb_2O_3 is the standard additive for inhibiting the ZnO grain growth. The inhibition of ZnO grain growth is normally attributed to the existence of $Zn_7Sb_2O_{12}$ spinel-type particles, formed during heat treatment in ZnO grain boundaries.

To decrease dimensions of the varistors, and at the same time, to save the raw materials, many tests have been carried out on various formulations in order to apprehend an increase in the threshold electric field. There are many researches on this subject by adding additive oxides such as the lithium oxide, the magnesium oxide, the antimony oxide, etc. But one often runs up against the same difficulty; plus the threshold voltage is raised, plus the energy capacity decreases. For example when the threshold voltage is around 100V/mm, the capacity for energy absorption is in the order of 100 to 120 J/cm^3, but when the threshold voltage is about 350 V/mm, the capacity for absorption of energy falls down to 30 J/cm^3.

Another type of varistor has been proposed, which is made by adding a certain percentage of some rare earth oxides such as praseodymium, Pr_6O_{11}, to the traditional composition. The analysis of the results of the electric characteristics of the various studied samples has made it possible to highlight a threshold voltage of about 300 to 400 V/mm, and a capacity for absorption of energy about 90 to 120 J/cm^3. The increase in the height of potential barrier and the inhibition of the growth of the grains during the sintering cycle explain this physical phenomenon.

For a high energy-absorption capacity, a micro-structural homogeneity (uniform ZnO grain-size distribution; uniform distribution of phases along the grain boundaries of ZnO; no or at least very little fine porosity) is required, that allows a uniform current and hence an energy distribution throughout the whole varistor.

2. Manufacturing of ZnO-based varistors

Normally the varistors are prepared by traditional method used for electro-ceramics (Figure 4).

The principal chemical formulation is made up of about 95% ZnO, plus Bi_2O_3, Sb_2O_3, Co_2O_3, MnO_2, Cr_2O_3 and NiO as additives. All these oxides are mixed in a plastic earthenware jar with balls with zirconium, and pure ion-free water distilled during 24 hours. Rare earth oxide (Pr_6O_{11} or Nd_2O_3) can be added too in the principal composition. The powder is obtained after drying and 150μm sifting. An appropriate dimension of the blocks to be made and tested in an experimental procedure can be 26mm in diameter and 2mm of thickness, but industrial varistors have much bigger dimensions, up to tens of centimeters as diameter and height. They are sintered during a period in the range of 2 hours. Electrodes are deposited on two surfaces of the samples to provide electric connections and to measure I(V) characteristics. These characteristics are measured while continuous currents up to 10mA pass through the samples, and impulses of great amplitude by using 4/10 and 8/20μs impulse generators up to tens of kA are applied.

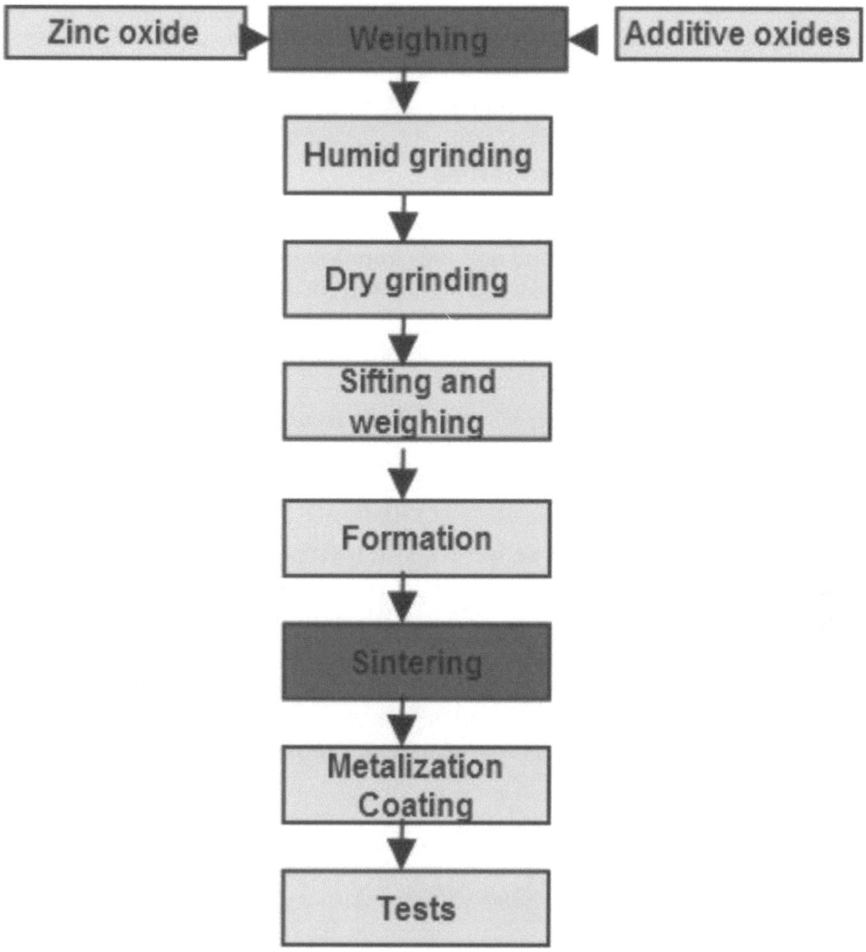

Fig. 4. Traditional procedure to manufacture a ZnO-based varistor.

Some examples for experimental composition of the samples are given in Table I. The composition S_1 can be modified by the addition of small amounts of Pr_6O_{11} or Nd_2O_3 to obtain the compositions S_2 and S_3, respectively.

Sample	Composition
S_1	95.0 % ZnO + 2.5 % Sb_2O_3 + 0.5 % Co_3O_4 + 0.5% Cr_2O_3 + 0.5 % MnO_2 + 0.5 % NiO + 0.5 % Bi_2O_3
S_2	99.9 % S_1 + 0.1 % Pr_6O_{11}
S_3	99.9 % S_1 + 0.1 % Nd_2O_3

Table I. Some experimental compositions of ZnO varistor samples.

Reagent-grade oxides are mixed in appropriate ratios for each composition, and disks are cold pressed at a pressure up to hundreds of MPa. Electrodes are coated on parallel surfaces of the sintered samples. Polished cross sections of the samples are prepared, and the microstructures of the samples are examined using Scanning Electron Microscope (SEM).

3. Measurement of electrical properties

The current versus voltage (I-V) characteristics of the varistor samples are measured using a dc power supply up to tens of mA, and a 4/10μs impulse generator up to 100 kA, to identify the upturn voltage and the current energy absorption capacity. Energy-absorption capacity (A) is the maximum amount of lightning energy absorbed and/or passed through a varistor when it explodes. To measure this capacity, impulse currents are applied to the samples with increasingly larger amplitudes.
The current I(t) and the voltage V(t) waveforms are recorded with a storage oscilloscope. The energy absorption is calculated as follows.

$$A = I(t)\ V(t)\ dt \tag{3}$$

This parameter is calculated for all samples, and the average energy absorption is used to estimate the result. As an example, the electrical properties of some experimental varistor samples are given in Table II. As can be observed, the introduction of small amounts of rare earth oxides (REO) into composition S_1, with a threshold voltage (V_{1mA}) of 280V/mm, increased the threshold voltage of sample S_2, doped with Pr_6O_{11}, and sample S_3, doped with Nd_2O_3, to slightly above 300 V/mm. But what is more significant, is that doping with REO strongly increased the energy absorption capacity of samples S_2 and S_3 in comparison with sample S_1, from 52 to 112 J/ cm^3.

Sample	α	V_{1mA} (V/mm)	V_{GB} (V)	A J/cm^3
S_1	42	280	1.9	52
S_2	40	312	1.5	107
S_3	52	302	2.5	112

Nonlinear coefficient (α), Threshold voltage (V_{1mA}), Breakdown voltage per grain boundary (V_{GB}), and Energy absorption capacity (A) of varistor samples.

Table II. Average Current-Voltage characteristics

Further investigation shows that when the Pr_6O_{11} content increases, the threshold voltage increases as well but the coefficient of non-linearity α decreases.

4. Study of the varistors' microstructure

The microstructures of some investigated samples are presented in Figure 5. Phase composition and the distribution of phases in samples S_1, S_2, and S_3 are evident from micrographs 5(a) to 5(f). The analyses confirms presence of the $Zn_7Sb_2O_{12}$ spinel-type phase containing Cr, Mn, Co, and Ni and also Bi_2O_3-rich phases with Zn, Sb, Cr, Mn, Co, and Ni detected at the ZnO grain boundaries of all the samples.

We can also see that the Bi_2O_3-rich phase, although present in all samples and more noticeable in sample S_3. The analysis also confirms the presence of the $Bi_3Zn_2Sb_3O_{14}$ (called as pyrochlore-type phase), at the grain boundaries of ZnO in sample S_1. In samples S_2 and S_3, a new phase is determined containing oxides of Pr and Nd, respectively.

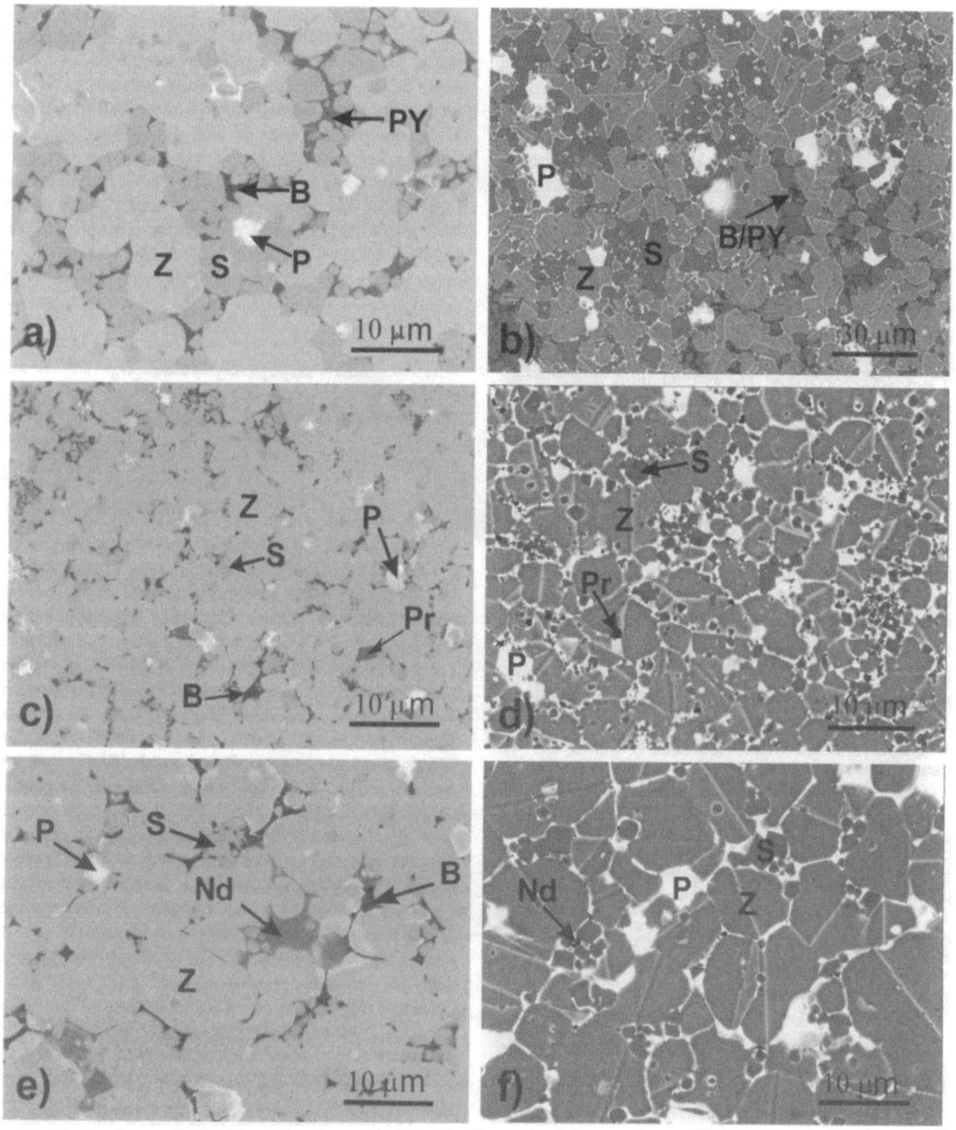

Key: Z=ZnO phase; B=Bi_2O_3-rich phase; S=$Zn_7Sb_2O_{12}$ spinel-type phase; PY=$Bi_3Zn_2Sb_3O_{14}$ (pyrochlore-type) phase; Pr=Pr-containing phase; Nd=Nd-containing phase; P=pore.

Fig. 5. Images from SEM of microstructures of varistor samples sintered at 1200^{0}C: (a) S_1; (b)S_1(etched), (c) S_2; (d) S_2(etched); (e) S_3; (f) S_3(etched).

The Pr-containing phase in particular is relatively fine-grained and distributed everywhere along the grain boundaries of the ZnO, while the grains of Nd-containing phase are larger and localized. While the grains of spinel phase are large in sample S_1, they are significantly smaller in samples S_2 and S_3. The formation of significant spinels in compositions with large amounts of added Sb_2O_3 has been observed by different researchers.
Thus micro-structural observations show a strong influence of REO doping on the ZnO and spinel grains, which is clearly evident from the micrographs in Figure 5. Doping of the composition with Pr_6O_{11} results in a significant decrease in the ZnO grain size and doping with both Pr_6O_{11} and Nd_2O_3 has a similar effect on the spinel phase; the size of spinel grains. Average size of the different phases of the varistor specimens are given in Table III.

Sample	ZnO		Spinel		Pores	
	D(μm)	δ (μm)	D(μm)	δ (μm)	D(μm)	δ (μm)
S_1	6.7	2.9	2.1	1.5	6.3	5.9
S_2	4.7	1.9	0.7	0.4	2.1	0.9
S_3	8.3	3.6	1.3	0.7	3.6	2.0

Table III. Average size D (μm) of ZnO grains, spinel grains, and pores of varistor samples S_1, S_2, and S_3 with corresponding deviations δ (μm).

These observations indicate that doping with REO has a strong influence on the mechanism of formation of the $Zn_7Sb_2O_{12}$ spinel phase. There are many reports in the literature about the formation of the spinel phase in the ZnO-Bi_2O_3-Sb_2O_3-based varistor compositions. Depending on the Sb_2O_3 / Bi_2O_3 ratio, the spinel phase forms either by the direct reaction of Sb_2O_3 with ZnO or by the decomposition of the $Bi_3Zn_2Sb_3O_{14}$ (named as pyrochlore phase) according to the following reactions:

$$ZnO + Sb_2O_3 + O2 \text{ ----> } ZnSb_2O_6 \quad (4)$$

$$ZnSb_2O_6 + 6\ ZnO \text{ ---> } Zn_7Sb_2O_{12} \quad (5)$$

$$2\ Bi_3Zn_2Sb_3O_{14} + 17\ ZnO \text{ ---> } 3\ Zn_7Sb_2O_{12} + 3\ Bi_2O_3 \quad (6)$$

The increase in threshold voltage (V_{1mA}) can be ascribed to the smaller ZnO grain size. However, the increase in V_{1mA} is much smaller than could be expected from the large decrease in the ZnO grain size and suggests that the increase in the number of non-ohmic grain boundaries is not proportional to the increase of all ZnO-ZnO grain boundaries due to the smaller ZnO grain size in this sample. It is evident from Table II that the average breakdown voltage of the grain boundary (V_{GB}) in sample S_1 is 1.9 V, while in sample S_2 it is only 1.5 V. Sample S_3 also has a significantly higher V_{1mA} than sample S_1, despite the fact that it has a larger ZnO grain size than sample S_1. The V_{GB} in sample S_3 is 2.5 V which indicates that a larger fraction of grain boundaries in this sample has a non-ohmic character.

Sample S_3 also has a higher nonlinear coefficient α of 52 than samples S_1 and S_2 with α equal to 40.

Doping with REO significantly improves the energy characteristics of samples. The low energy-absorption capacity of sample S_1 can be attributed to the large amount of spinel phase in this sample.

The spinel phase forms large grains along the grain boundaries of ZnO, and so insulating chains of spinel phase significantly interrupt the current flow by narrowing the effective conduction section of the varistor. This leads to current localization and local overload, and hence a low energy-absorption capacity due to a non-uniform energy distribution.

The analysis of the whole results obtained during the tests of the samples, manufactured with various percentages of praseodymium and neodymium oxides makes it possible to suggest that:

- The presence of rare earth oxides improves the homogenization of the size of the grains in material.
- The increase of potential barrier height in the grain boundary supports a rise in the threshold electric field of the varistor.
- A good capacity of energy absorption is resulted compared to the traditional varistors.

Doping with Pr_6O_{11} and Nd_2O_3 appears to be promising for the preparation of ZnO-based varistors with a high breakdown voltage and also high energy-absorption capacity. This can be a successful step because our objective is to have smaller and lighter surge arresters in power network. This aim involves such varistors, which have high conduction threshold voltage, while their energy absorption capacity remains enough high. In such conditions we will be able to use a smaller number of varistors to make a high-voltage arrester. Consequently this will provide smaller and lighter arresters.

Of course we have to respect the necessary outer creepage distance of the arrester housing, according to the pollution level of the location where the arrester is to be used.

5. Computation of nonlinear properties in ZnO ceramics

Different computational methods are used for investigation of the non-linear behaviour of zinc-oxide varistors. In a ZnO varistor, when a voltage is applied between the electrodes, the majority of the grain boundaries show a strong non-linear behavior, but a certain number of grains do not present, under the applied voltage, a high non-linear characteristic or are nonconducting. Under a known voltage level, several current paths occur from one electrode to the other, which are called as the current percolation paths. The number of grains on each path crossing by the current is a statistical parameter. It is shown that the distribution of this statistical number depend on the block thickness and percentage of nonconducting grains in the varistor.

Using a Monte Carlo method in our research works, we have realized that the number of ZnO grains providing the percolation path fits a lognormal distribution especially in thin varistors. We have also proposed a binomial direct approach for this problem. It is found that the direct approach could be satisfying too.

Both approaches show that the threshold voltage and the nonlinearity coefficient of the varistors can be controlled, to some degree, by the fraction of nonconducting grains. These results help us to have a better understanding of the varistors' behavior and enable us to make more realistic electric models for these elements.

Few works can be found, which have experimentally studied the individual grain boundaries in the varistor. Most of the Schottky junctions give a nonlinearity coefficient which is normally in the range of 30-70 for a normal varistor, whereas the actual α of a good ZnO material junction can be in the range of 150. Even it can attain values greater than 200 in certain grain to grain microvaristors. Figure 6 shows the typical variation of the current density as a function of the barrier voltage, for a single barrier in a varistor.

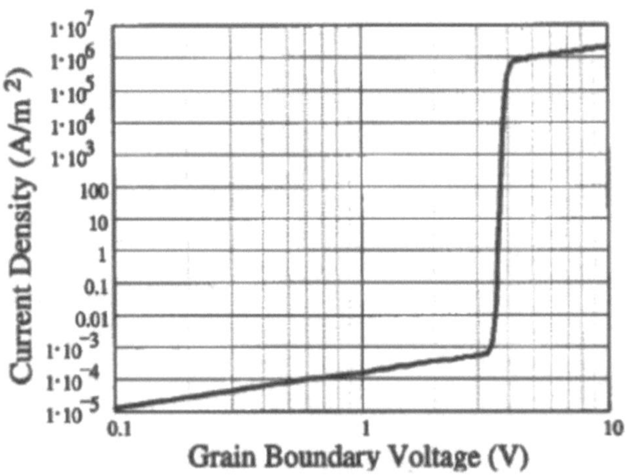

Fig. 6. The grain boundary current density vs. grain boundary voltage.

In Figure 7 the variation of the non-linearity coefficient α as a function of the varistor barrier voltage, for a single potential barrier is observed. This curve is deduced computationally, using Maple software, from the slope of the current-voltage characteristic of a single grain boundary as in Figure 6.

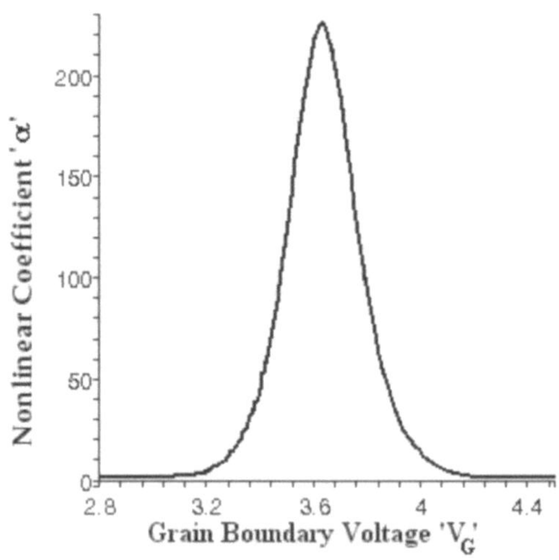

Fig. 7. Non-linearity coefficient α as a function of the varistor barrier voltage, for a single potential barrier.

5.1 Model of the varistor's microstructure

In Figure 8 a simplified model of the varistor's microstructure is observed. We use this model for computer simulation.

If the ZnO element thickness is D and the average grain thickness is d, then the minimum number of grain boundaries between the electrodes is L=D/d.

5.2 Monte Carlo method

As we read in the literature, the Monte Carlo is a technique that provides approximate solutions to problems expressed mathematically. Using random numbers and trial and error, it repeatedly calculates the equations to arrive at a solution. Then using random numbers or more often pseudo-random numbers, as opposed to deterministic algorithms, uses this algorithm for solving various kinds of computational problems.

Monte Carlo methods are extremely important in computational physics and related applied fields. Interestingly, the Monte Carlo method does not require truly random numbers to be useful. Much of the most useful techniques use deterministic, pseudo-random sequences, making it easy to test and re-run simulations. The only quality usually necessary to make good simulations is for the pseudo-random sequence to appear "random enough" in a certain sense. They must either be uniformly distributed or follow another desired distribution when a large enough number of elements of the sequence are considered. Because of the repetition of algorithms and the large number of calculations involved, Monte Carlo is a method suited to calculation using a computer, utilizing many techniques of computer simulation.

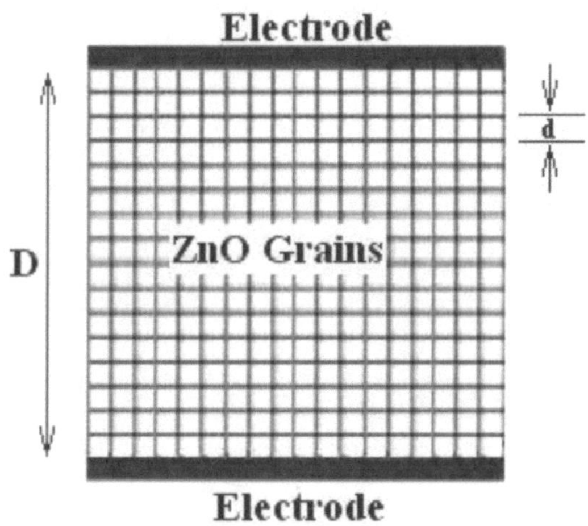

Fig. 8. Simplified micro-structural model of varistor for computer simulation.

Using a Monte Carlo algorithm, we follow a stochastic procedure to compute the number of the conducting grains on the current path in the varistor model as a statistical parameter. The flowchart of the used program is observed in Figure 9. In this diagram the letters K and N denote, respectively, the iteration number and the variable for the number of each layer in micro-structural model of the varistor. B is the number of active grain boundaries through which the current passes in going from one electrode to the other. As well, we define the probability of a non-conducting grain boundary as P. For P=0, all grain boundaries are always active. It is obvious that the existence of non-conducting grains results in a longer path for current across the ZnO element, which depends on the fraction of non-conducting grains. We undertake a statistical analysis of the effect of L (the number of ZnO grain layers across the varistor) and P (the probability of a non-conducting grain boundary) on the non-linear characteristics of the varistor as characterized by α.

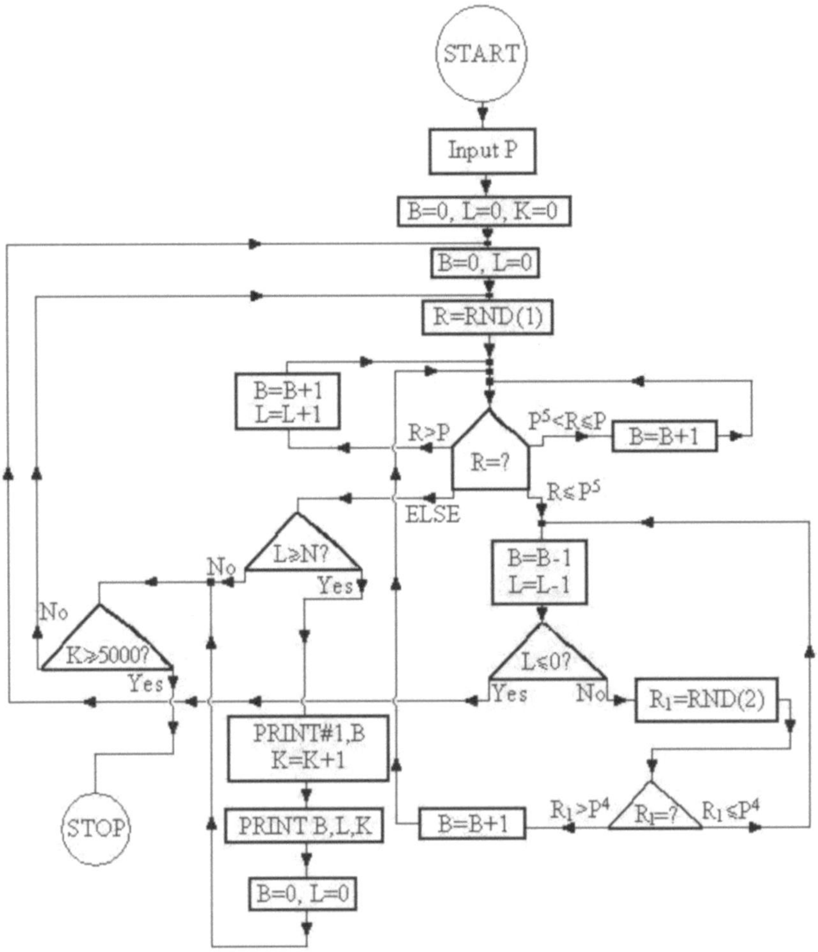

Fig. 9. Flowchart of Monte Carlo algorithm, for computation of the number of grains on the current path through the varistor.

As said above, for P=0, there is no non-conducting grains and all path lengths are the same, equal to L. With increasing fraction of non-conducting grain boundaries P, the conducting grains number B, augments substantially, which will increase the voltage per unit thickness of the ZnO element. P can also be augmented by increasing the amount of non-conducting inter-grain material, often as a by-product of attempting to reduce grain size. This non-conducting phase can be a spinel phase. Obviously increasing the number of non-conducting grain boundaries increases the current density in the remaining grain boundaries and results in greater grain boundary power dissipation and temperature rise.

By running the Monte Carlo program with different values of L, the number of ZnO grain layers across the varistor, and P, the probability of non-conducting grain boundaries in varistor, we obtain statistical sets of data for B, i.e. the number of grains crossed by the current.
As an example, a probability density histogram of B's data for the case of a very thin varistor with L=10 and P=0.3 is seen in Figure 10, which is related to a varistor of about 0.1 mm thick.

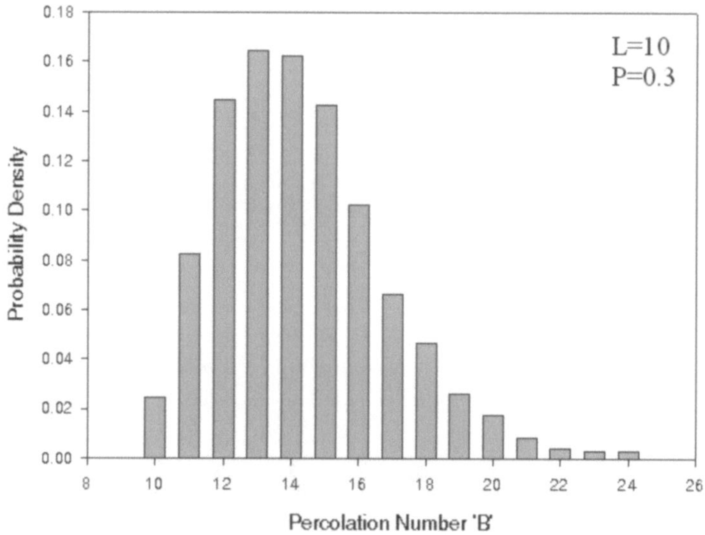

Fig. 10. A probability density histogram of the number of grains crossed by the current, obtained for special case of a very thin varistor of about 0.1 mm thick with probability of non-conducting grains equal to 30%.

Analyzing the statistical distribution of B by fitting different distribution curves on it, several distributions such as Normal, Lognormal, Weibull, Logistic, Loglogistic and Exponential were used for fitting our computational data. The best fitness was seen to be for the three distributions of Normal, Weibull and Lognormal (identically for Log_eNormal and Log_{10}Normal), comparing to the others.
In Figure 11 we can observe the fitted curves for these three distributions concerning the special case of Figure 10.
The Anderson-Darling statistic is a measure of how far the plot points fall from the fitted line in a probability plot. Using the Anderson-Darling measure to calculate the fit goodness of these distributions, we obtain the curves of Figure 12.
The statistic is a weighted squared distance from the plot points to the fitted line with larger weights in the tails of the distribution. In this method, a smaller Anderson-Darling (AD) measure indicates that the distribution fits the data better.
As can be observed in Figure 12, the LogNormal distribution has the best fit for the B data concerning the thin varistors of this study.
If the probability of a grain boundary to be non-conducting is P and as we supposed in our model that the grains are cubes, then it can be shown that to a first approximation, the mean number of active grain boundaries through which the current passes between electrodes is:

$$B = L\ (1 + \frac{P}{1\text{-}P}) \qquad (7)$$
$$0 < P < 1$$

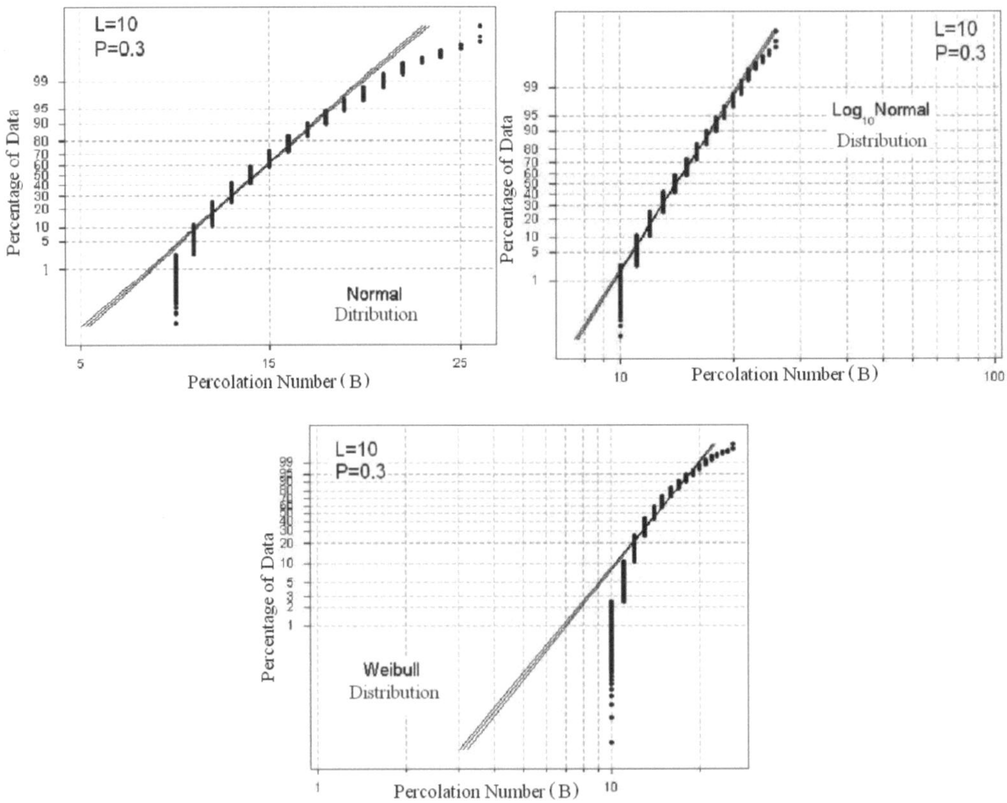

Fig. 11. Curve fitting of the number of grains crossed by the current, on three different distributions for the data of Figure 10.

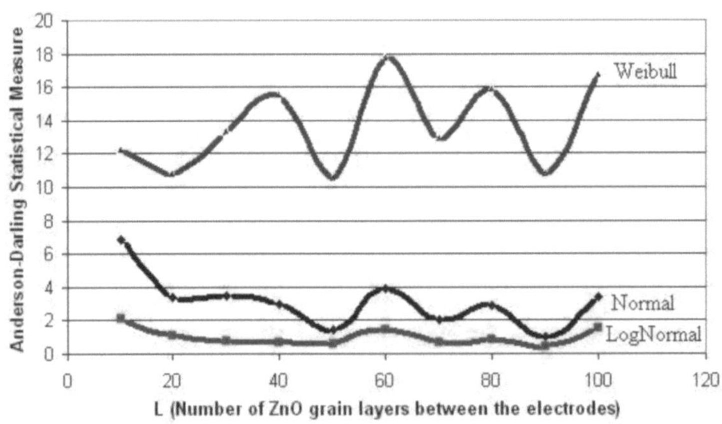

Fig. 12. Comparing the fitness of three different distributions on our data, concerning varistors of 0.1 to 1 mm thick.

And as we realized in our research work that the percolation number data for thin varistors obey the lognormal distribution, we deduce, using the Maple software, the relation (8) as an analytical formula for the standard deviation, s, of thin varistors data, having the lognormal distribution.

$$s := \sqrt{(\mathrm{e})^{\left(\frac{\ln\left(L\left(1-\frac{P}{P-1}\right)\right)-\frac{\ln\left(L\left(1-\frac{P}{P-1}\right)\right)^{2}}{L}}{L}\right)}\left((\mathrm{e})^{\left(\frac{\ln\left(L\left(1-\frac{P}{P-1}\right)\right)-\frac{\ln\left(L\left(1-\frac{P}{P-1}\right)\right)^{2}}{L}}{L}\right)}-1\right)} \quad (8)$$

By plotting this equation for different values of L and P in Maple software, we obtain Figure 13. As it is seen, the standard deviation is not high for amounts of P less than 0.5, while it is great for bigger P's in thinner varistor blocks.

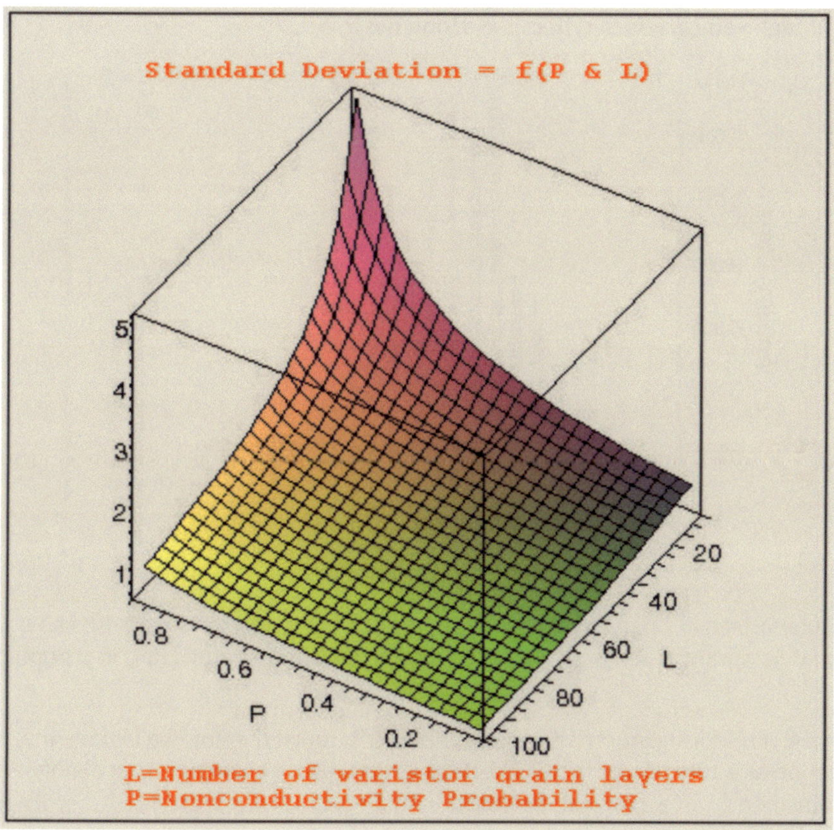

Fig. 13. Standard Deviation for percolation number B of thin varistors, having lognormal distribution.

We assumed that each current path is independent of every other path. In fact, at large P, the number of non-conducting grain boundaries would reduce the likelihood of interconnection of paths. But in P less than 0.5, especially in thin varistors even if two paths of differing length are near each other, the probability of their having substantially differing potentials is not great.

We realize from the form of the statistical distribution for B, that as P increases, the varistor conduction turn-on will be more rapid. This can be deduced from the low B tail of the statistical distribution. For small L, the average value of B increases with P, but the minimum value of B, which is L, remains the same. Thus the ratio of the mean to minimum possible value of B increases. The turn-on characteristics are determined mainly by the first few paths to conduct. Thus the number or fraction of completed paths for various P must be considered in addition to α.

Figure 14 compares lognormal and Normal distributions with the same mean (200) and with variances selected to give the same minimum value (~100) in a population of 600 random numbers. This figure indicates that the probability density of B increases much more rapidly at low values of B for the lognormal than for the Normal distribution. Thus conditions, which drive the statistical distribution for B toward the lognormal distribution, are likely to result in more rapid turn-on of the varistor element. The lognormal distribution also has a long tail at high values, which will cause a long tail in α.

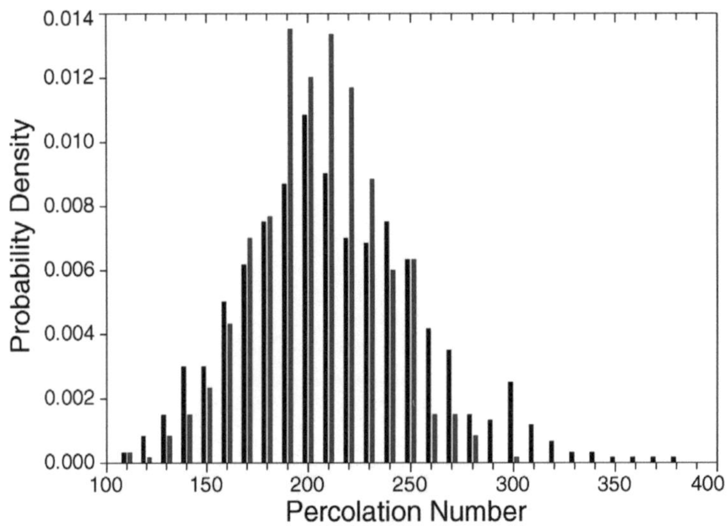

Fig. 14. Comparison of Lognormal (black) and Normal distribution data with same mean (200) and with variances set to give about the same minimum value (100) in a population of 600.

Based on the numerical computations and the distributions thereof, we believe that the more rapid turn-on as a function of increased P for large L (thick elements) is probably associated with a transition from a Normal distribution at P=0 toward a lognormal distribution with increasing P.

This transition can be rationalized from the probability density of B for a thin varistor with a reasonable probability of non-conducting grains, for which the distribution is clearly asymmetric with a rapid turn-on, when the shortest path across the arrester becomes conducting, followed by a rapid increase in the number of conducting paths with increasing voltage.

The Lognormal distribution increases more rapidly in the low end tail of the distribution which would result in a more rapid turn-on of an arrester element.

In Figure 15, the variation of α as a function of the applied voltage is seen for a non-conducting grain probability 0.5 for a thick varistor of 10 mm thickness (L=1000) and a thin varistor of 0.5 mm (L=50) thickness.

Both cases result in asymmetrical α characteristics, while the varistor thickness has an obvious influence on the shape of the curve. We accomplished the same analysis for varistors with different thickness (Number of grain layers L) and probability of non-conducting grains (P).

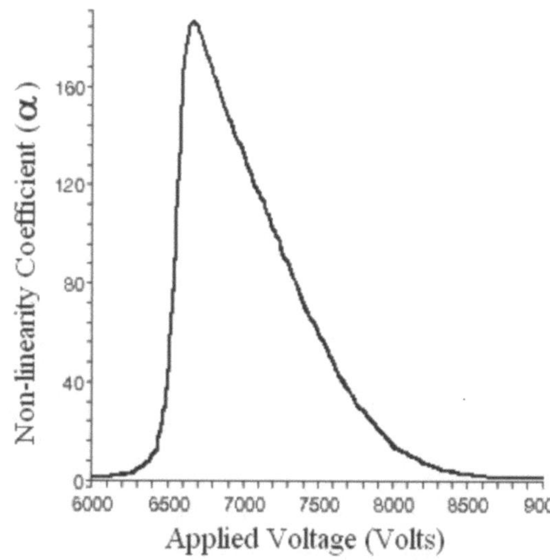

a) 10mm thick varistor (L=1000)

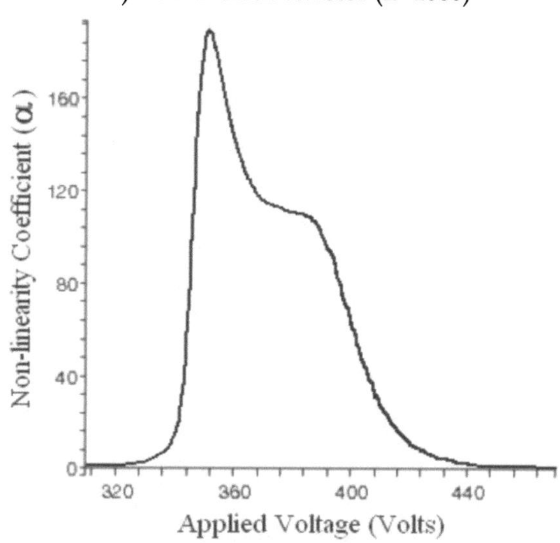

b) 0.5mm thick varistor (L=50)

Fig. 15. Varistor nonlinearity coefficient α as a function of the applied voltage for thin and thick varistors at a non-conducting grain probability of 0.5.

To provide a basis for comparison of the α(V) curves, we define the parameters FWHH and β as measures for broadness and rate of rise of the α(V) curve (Figure 16). We define FWHH as the Full Width at Half Height of the curve and β is defined as:

$$\beta = (V_{90\%} - V_{10\%}) / V_{10\%} \tag{9}$$

According to this definition for β, the small β means large slope of α(V) curve.
In Figure17, we see the variation of the (FWHH) of the α(V) curves as a function of L, which is linear as might be expected.

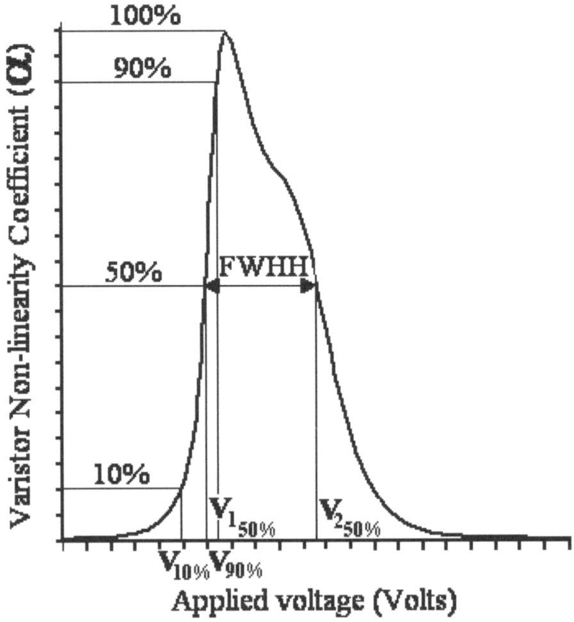

Fig. 16. Definition of parameters FWHH and β.

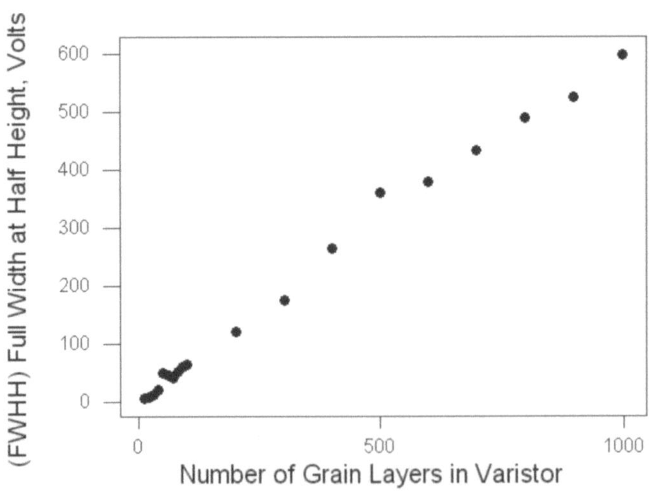

Fig. 17. Full Width at Half Height (FWHH) of the α(*V*) curves with P=0.5 as a function of L.

For P=0 and large L, the characteristics are just a multiple of the grain boundary characteristics which are modeled as symmetric. As the fraction of non-conducting grains increases, the mean percolation path increases and the probability of a short percolation path decreases.

However the minimum possible path remains the same (L) and above the minimum path, the number of conducting paths appears to increase rapidly which results in an asymmetric α with more rapid turn-on.

As both the threshold voltage and width of α are the sum of the contribution from each grain, i.e. a 10 mm thick varistor (L=1000) is equal to twenty 0.5 mm thick (L=50) varistors in series, so that the I(V) curve of the former will be sum of the I(V) curves of the latter. Thus FWHH(L) should increase linearly with L, and the result of Figure 16 can be taken as a verification of the computational methods.

Figure 18 shows the variation of β with non-conducting grain probability (P) for a thin (L=50) and a thick (L=1000) ZnO element. As we can see in this figure, the rate of rise of the α(V) curve depends on both the non-conducting grain probability and the element thickness.

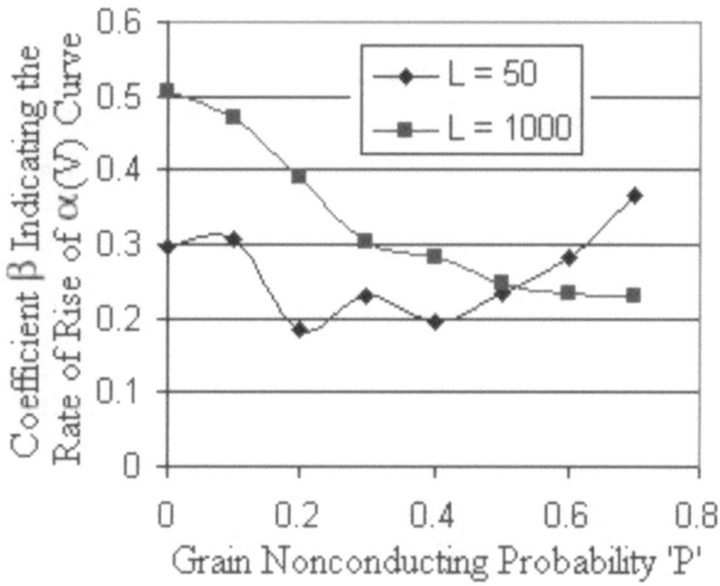

Fig. 18. Variation of β as a function of P, for L=50 (0.5 mm thick) and L=1000 (10 mm thick) varistors. Smaller β tends to indicate more rapid "turn-on" of the varistor with applied voltage.

For P>0.5 and large L (thick varistor), β remains constant, which means that increasing P has little effect on the varistor turn-on characteristics. This is probably a result of the fact that for large L, the standard deviation in B decreases as a fraction of L, so that the extreme value in B decreases relative to the mean.

With increasing P, the mean number of grain boundaries increases, as does the distribution of the number of grain boundaries through which the current passes from one electrode to the other. As the grain boundary characteristic is highly nonlinear, the conductivity of the

ZnO element rises rapidly once the first few current paths become conductive. This probably accounts for the increasing asymmetry in α for the whole varistor, and increasingly rapid current onset with increasing P, as the statistical distribution of path lengths broadens with P.

For large L, β decreases with increasing conducting grain probability, P, but for small L this is not the case. As well, the peak value of α increases with increased P for thick elements but not for thin elements. This must result from competition between the larger variance in B, the percolation path for small L, with the nature of the tail of the statistical distribution at low values of B, which determines the turn-on characteristics.

One would prefer the ZnO "turn on" (become substantially conductive) to be very rapidly so that the AC operating voltage can approach more closely the protection level of the varistor without causing excessive power dissipation. On the other hand, how the varistor approaches its ultimate conductivity with voltage is less important. As we observed, the probability, P, of non-conducting varistor grains has an influence on parameters such as the rate of rise of $\alpha(V)$ curve.

As a conclusion, we say that the characteristics of the thin ZnO varistors were statistically studied. The number of ZnO grains on each conducting path through a ZnO varistor, crossing by the current, is a statistical parameter (B).

It was shown that the nonlinearity of ZnO ceramics can be controlled, to some degree, by the fraction of non-conducting grains. Thus we can choose the best value for P to have the maximum rate of rise of the $\alpha(V)$ curve. This will result in a rapid "turn on" of the ZnO element, which allows the circuit being protected to operate more closely to the protection level without excessive power dissipation in the arrester element. This optimum value of P certainly depends L, which is related to the thickness of the varistor.

With increasing fraction of non-conducting grain boundaries P, the percolation number B, increases substantially, which will increase the voltage per unit thickness of the ZnO element. This can be exploited commercially in order to increase the percolation number.

These results can help us to have a better understanding of the behavior of these varistors, and the dependence of this behavior on their geometrical dimensions and the constituting materials. This will also enable us to have more realistic electric models for these ceramic elements.

5.3 Direct Binomial method

We propose also that the Binomial distribution can be used directly to explain the conduction phenomena in ZnO varistors. Here is a Maple program using the Binomial Distribution for computation of the current in the varistor and calculation of its α, to predict directly the turn-on characteristics. We use the Binomial Distribution formula to calculate the probability function of the L success in B trials (B=Percolation Number & L=Number of Layers).

For example we consider a varistor block with a diameter of 4cm (Dvaristor = 0.04).

Then we calculate the number of grains in the first layer, from which a current can be started, as follows:

$$\begin{gathered} S_{varistor} = \pi.(D_{varistor}^2) \; / \; 4 \\ D_{grain} = 0.000010 \\ S_{grain} = D_{grain}^2 \\ N_S = S_{varistor} \; / \; S_{grain} \end{gathered} \tag{10}$$

The number of the expected conducting grains just next to the upper electrode is:

$$N_P=N_S\,(1-P) \qquad (11)$$

Now we compute the probability for the current to advance L layers in crossing B grains, from one electrode to the other, for a given L & P (in the example here L=100 and P=0.3):

$$F = (B!) \,/\, ((L!).(B-L)!) \qquad (12)$$

$$\text{Probability}(L=100) = F\,.\,(((1-P)^{L}).(P^{(B-L)})); \qquad (13)$$

Now we can have the number of current paths in the varistor as a function of B, i.e., we know that how many paths there are for each B:

$$N_B = (\text{Probability}(L=100))\,.\,N_P \qquad (14)$$

Here we plot the number of the conduction paths, in a varistor with 100 layers and 30% of non-conducting grains, as a function of the percolation number B:

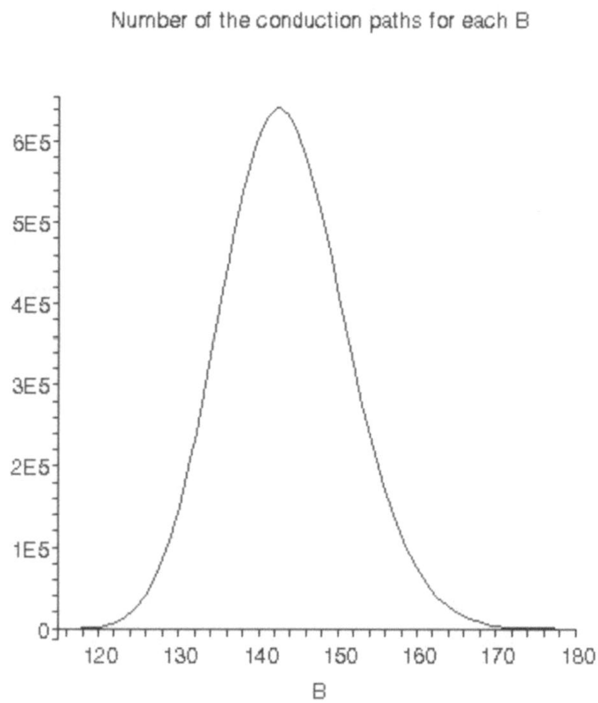

Fig. 19. Variation of P as a function of *B* for *L*=100 (~1 mm thick)

Now we compute the current per grain boundary, using the relation of $J(V_G)$ for one single grain boundary :

$$J = 10000\ 10^{[-4.5\tanh[-\frac{50\ \ln(VG)}{\ln(10)}+28]\ -5.5]}\ 10^{[\frac{10\ \ln(VG)}{9\ \ln(10)}+\frac{20}{9}]} \tag{15}$$

Then we substitute VG by V/B, as the total voltage V which is applied on the whole varistor is distributed on B grains en series :

$$V_G = V/B \tag{16}$$

In L trials, the mean number of cases without a non-conducting grain is L(1-P) and the mean number with a non-conducting grain is L.P. We can ignore the cases, which are "dead ends", as they do not count:

$$L.(1\text{-}P) = B_M \tag{17}$$

Since these are the cases which move us forward. Thus for a probability of a non-conducting grain, P, the mean percolation number is:

$$B_M = N/(1\text{-}P) \tag{18}$$

B_M is the number of active grain boundaries through which the current passes between electrodes.

Now we compute the current flowing per each single path, from which the total current through the varistor can be obtained as follows:

$$I_B = J.S_{grain} \tag{19}$$

$$I_V = \Sigma(I_B.N_B\,,\ B\text{=}L \text{ to } 10L) \tag{20}$$

This is because the total current will be the sum of the number of paths multiplied by the current in each path. Using this second formula for calculating the current in whole range of B, the α will be obtained. We substitute V by V+1 in IV to obtain the derivative and calculate the Alpha as follows:

$$I_{V1} = \text{subs}(V\text{=}V\text{+}1, I_V) \tag{21}$$

$$\alpha = (\log_{10}(10^{-2}\ I_{V1})\text{-}(\log_{10}(10^{-2}\ I_V)))\ /(\log_{10}(V+1)\text{-}\log_{10}(V)); \tag{22}$$

From Figure 20, it is obvious that how asymmetric the α(V) curve is. For large L, β decreases with increasing conducting grain probability, P, but for small L this is not the case. As well the peak value of α increases with increased P for thick elements but not for thin elements. This must result from competition between the larger variance in B, the percolation path for small L, with the nature of the tail of the statistical distribution at low values of B, which determines the turn-on characteristics. Previous works indicated the statistical distribution of the percolation number, B, for large L (thick varistors) is Gaussian. We used the Monte Carlo computations for thin varistors (L<100). For L<100, histograms of percolation number fit a Lognormal distribution better than a Normal distribution. We realize from the form of the statistical distribution for the percolation number, B, that as P increases, the turn-on will be more rapid. This can be deduced from the low B tail of the statistical distribution. For small L, the average value of B (percolation number) increases with P, but the minimum value of B, which is L, remains the same.

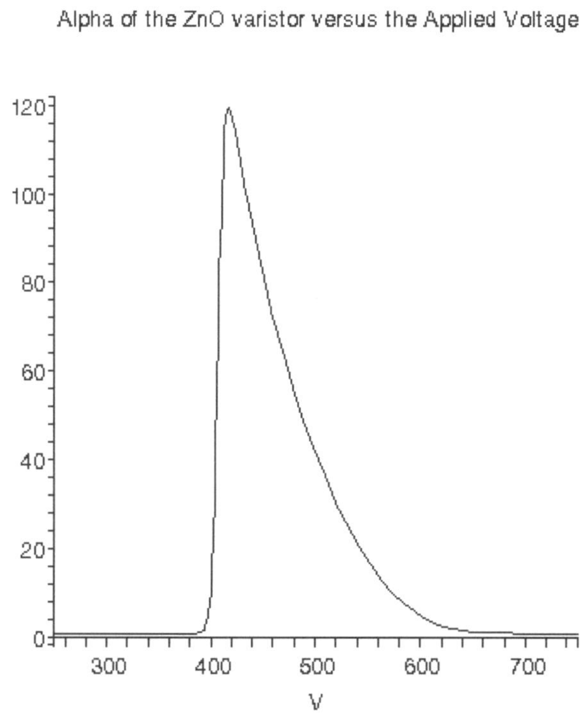

Fig. 20. Variation of α as a function of V for a L=100 (~1 mm thick) varistor.

6. Computer programs

6.1 Basic program for Monte Carlo modeling

```
4   OPEN "D:\Mohammad\B data\L200_1000. K5000. P 0.5\L200_K5000_P0.5.DAT" FOR APPEND
AS #1
'P = Probability of nonconducting grain
5   P = 0.5
'LMAX = Number of the layers across the varistor
6   LMAX = 200
'KMAX = Number of iterations (Current injection to the upper electrode)
7   KMAX = 5000
10  B = 0 : L = 0: K = 0
15  B = 0 : L = 0
20  R = RND(1)
30  IF R > P THEN GOSUB 100
40  IF P >= R AND R > P^5 THEN GOSUB 200
50  IF R <= P^5 THEN GOSUB 300
60  IF L >= LMAX THEN GOSUB 400
70  IF K >= KMAX THEN GOTO 450
80  GOTO 20
```

```
100 B = B + 1
110 L = L + 1
120 RETURN
200 B = B + 1
210 RETURN
300 B = B - 1
310 L = L - 1
320 IF L <= 0 THEN GOTO 15
330 R1 = RND(2)
340 IF R1 > P ^ 4 THEN B = B + 1
350 IF R1 <= P ^ 4 THEN GOTO 300
360 RETURN
400 PRINT #1, B
410 K = K + 1
420 PRINT B, L, K
430 B = 0: L = 0
440 RETURN
450 STOP
500 END
```

6.2 Maple program for varistor's conduction modeling

```
restart;
B:=L*(1+sum(P^n, n=1..infinity)); #  B=The mean number ofactive
grain boundaries through which the current passesbetween
electrodes.
s1:=(log(B) - (1/L)*(log(B))) / (sqrt(L));
s := (((exp(1))^(s1^2))*(((exp(1))^(s1^2))-1))^(1/2);
readlib(log10);
plot3d(s,P=0.1..0.9, L= 10..1000,axes=boxed, title="Standard
Deviation = f(P & L), P=Nonconductivity Probability, L=Number
of varistor grain layers");
J:=10^((tanh(50*log10(VG)-28))*4.5-
5.5)*10^((500*log10(VG)+1000)/450)*10^4;
plot([log10(VG),log10(J),VG=.1..10]);
J1:=subs(VG=VG+.01,J);
alpha_grain := ((log10(evalf(J1)*10^(-2))-log10(evalf(J)*10^(-
2)))/(log10(VG+.01)-log10(VG)),VG=2.8..4.5);
plot(alpha_grain, title="Alpha of a ZnO grain versus Grain
Boundary Voltage");VG:=(V/N);
JJLN:=sum(N*J*(sqrt(s*2*Pi))^(-1)*(exp(-1/2*((log(N)-
B)/sqrt(s))^2)),N=round(B-5*sqrt(s))..round(B+5*sqrt(s)));
L:=20; P:=0.1; JJLN;
JJ1LN:=subs(V=V+1,JJLN);
ALN201:=(log10(evalf(JJ1LN)*10^(-2))-log10(evalf(JJLN)*10^(-
2)))/(log10(V+1)-log10(V));
P; L; plot(ALN201,V=0..1000, title="Alpha LogNormal of the ZnO
varistor versus the Applied Voltage");
B:=L1*(1+sum(P1^n, n=1..infinity));
sn:=B*sum(P1^n,n=1..infinity);readlib(log10);
J:=10^((tanh(50*log10(VG)-28))*4.5-
5.5)*10^((500*log10(VG)+1000)/450)*10^4;
J1:=subs(VG=VG+.01,J);VG:=(V/N); J;
```

```
JJN:=sum(J*(sqrt(sn*2*Pi))^(-1)*(exp(-1/2*((N-
B)/sqrt(sn))^2)),N=round(B5*sqrt(sn))..round(B+5*sqrt(sn)));L1:
=20; P1:=0.1;JJN; JJN1:=subs(V=V+1,JJN);
AN201:=(log10(evalf(JJN1)*10^(-2))-log10(evalf(JJN)*10^(-
2)))/(log10(V+1)-log10(V));
P1; L1; plot(AN201,V=0..1000, title="Alpha of the Normal ZnO
varistor versus the Applied Voltage");
P;L;plot([ALN201,AN201], V=40..120, color=[red ,green ],
style=[line]);
```

7. References

[1] Meshkatoddini, M.R.; Boggs, S.; "Nonlinear Properties of ZnO Ceramics as a Function of Threshold Voltage and Fraction of Non-conducting Grains", Conference Record of ISEI-2006, The 2006 IEEE International Symposium on Electrical Insulation, 11-14 June 2006 Page(s):269 - 272

[2] Andoh H. et al, “Failure mechanisms and recent improvements in ZnO arrester elements,” IEEE Electrical Insulation Magazine, vol. 16, no. 1, pp. 25–31, Jan./Feb. 2000. Symposium on Electrical Insulating Materials, Toyohashi, Japan, Sept. 27-30, 1998

[3] Greuter F. & Blatter G., “Electric properties of grain boundaries in polycrystalline compound semiconductors”, Semiconductor Science and Technology, vol. 5, pp. 111–137, 1990

[4] Shengtao Li, Jianying Li, Fuyi Liu, M A Alim and G Chen, “The dimensional effect of breakdown field in ZnO varistors”, Journal of Applied Physics D, 35, pp.1884-1888, 2002

[5] Tao. M, Bui Ai, Dorlanne O. and Loubiere A., "Different Single Grain Junctions Within a ZnO Varistor", J. Appl. Phys. 61(4), 15 February 1987

[6] Shengtao Li, Feng Xie & Fuyi Liu, “Relation between Residual Voltage Ratio and Micro-structural Parameters of ZnO Varistors”, Proceedings of 1998 International Symposium on Electrical Insulating Materials, Toyohashi, Japan, Sept. 27-30, 1998

[7] Robert C.P. and Casella. G., "Monte Carlo Statistical Methods" (second edition). New York: Springer-Verlag, ISBN 0-387-21239-6, 2004

[8] Ueberhuber C.W., Numerical Computation 2: Methods, Software, and Analysis. Berlin: Springer-Verlag, pp. 124-125, 1997

[9] Steven Boggs & Hideyasu Andoh, “A Statistical Approach to Prediction of ZnO Arrester Element Characteristics”, IEEE Transactions on Power Delivery, VOL.16, NO. 4, October 2001

[10] H. Andoh, S. Nishiwaki, H. Suzuki, S. Boggs, and J. Kuang, “Failure mechanisms and recent improvements in ZnO arrester elements,” IEEE Electrical Insulation Magazine, vol. 16, no. 1, pp. 25–31, Jan./Feb. 2000.Symposium on Electrical Insulating Materials, Toyohashi, Japan, Sept. 27-30, 1998.

[11] M Bartkowiak & G.D.Mahan, “Nonlinear Currents in Voronoi Networks”, Physical Review B, Volume 51, Number 16, pp 10825-10831,15 April 1995.

[12] F. Greuter and G. Blatter, "Electric properties of grain boundaries in polycrystalline compound semiconductors", Semiconductor Science and Technology, vol. 5, pp. 111–137, 1990.

[13] Mohammad R. Meshkatoddini & Steven Boggs, " Investigation of the statistical behavior of thin ZnO-based varistors using a Monte Carlo Algorithm", ICEE2006, 14th Iranian Conference on Electrical Engineering, 16-18 May 2006, Tehran, Iran.

[14] M.R.Meshkatoddini et al, "New High Performance ZnO-Based Varistors with Rare-Earth Oxides, for Using in Surge Arresters", PSC 2003, 18th International Power System Conference, Tehran, Iran, 20-22 October 2003.

[15] M.R.Meshkatoddini, "Comparative Study of Different Models for ZnO Varistors", International Conference on Modelling, Simulation and Design of Dielectrics, Institute of Physics, Cambridge, UK, 6-8 April 2005.

[16] D.R.Clarke, "Varistor Ceramics", Journal of American Ceramics Society, 82, 485-502 (1999).

[17] M.Inada, "Effect of Heat-treatment on Crystal Phases, Microstructure and Electrical Properties of Non-ohmic Zinc Oxide Ceramics", Japanese Journal of Applied Physics, 18, 8, 1439-1446 (1979).

[18] J.Kim, T.Kimura, and T.Yamaguchi, "Microstructure Development in Sb2O3-doped ZnO", Journal of Materials Science, 24, 7, 2581-2586 (1989).

[19] T.Senda and R.C.Bradt, "Grain Growth of Zinc Oxide During the Sintering of Zinc Oxide-Antimony Oxide Ceramics", Journal of American Ceramics Society, 74, 6, 1296-1302 (1991).

[20] N.Daneu, A.Recnik, S.Bernik, and D.Kolar, "Microstructural Development in SnO-doped ZnO-Bi2O3 Ceramics", Journal of American Ceramics Society, 83, 12, 3165-3171 (2000).

[21] K.Eda, "Destruction Mechanism of ZnO Varistors Due to High Currents" Journal of Applied Physics, 56, 10, 2948-2955 (1984).

[22] T.K.Gupta, "Application of Zinc Oxide Varistors", Journal of American Ceramics Society, 73, 7, 1817-1840 (1990).

[23] A.Bui, H.T.Nguyen, and A.Loubiere, "High-field ZnO-Based Varistors", Journal of Applied Physics D, 28, 774-782 (1995).

[24] M. Inada, "Formation Mechanism of Non-ohmic Zinc Oxide Ceramics", Japanese Journal of Applied Physics.19, 3, 409-419 (1980).

[25] [25] J.Kim, T.Kimura and T.Yamaguchi, "Sintering of Zinc Oxide Doped with Antimony Oxide and Bismuth Oxide", Journal of American Ceramics Society, 72, 8, 1390-1395 (1989).

[26] S.G.Cho, H.Lee, and S.K.Kim, "Effect of Chromium on the Phase Evaluation and Microstructure of ZnO Doped with Bismuth and Antimony", Journal of Materials Science, 32, 4283-4284 (1997).

[27] M.R.Meshkatoddini, "Measurement of the Potential Barrier Height in ZnO-based Varistors", Proceedings of ICEE2001 Conference, Tehran, Iran, Spring 2001.

[28] M.R.Meshkatoddini, "Impedance Spectroscopy, A New Method for Quality Evaluation of Zinc Oxide Varistors", Proceedings of ICEE1998, Tehran, Iran, Spring 1998.

[29] Nguyen The Hung, "Contribution à l'étude des varistances ZnO à tension de seuil élevée", Thèse de l'Université Paul Sabatier, Toulouse, France, Mars 2002.

[30] Meshkatoddini, M. R., "Investigation of the Influence of the ZnO varistor preparation method on its characteristics", ISEI 2008, Conference Record of the 2008 IEEE International Symposium on Electrical Insulation,Volume , Issue , 9-12 June 2008, Page(s): 320-323, Digital Object Identifier 10.1109/ ELINSL.2008.4570338

[31] Tylor H.F.W., Cement Chemistry, Thomas Telford Publishing, London, pp: 12 - 28. 1997.

[32] Stutzman, P, Scanning Electron Microscopy Imaging Of Hydraulic Cement Microstructure, Cement & Concrete Composites 26: 957– 966. 2004.

[33] Idris, M.S., Kajian Pencirian Mikrostruktur Klinker Semasa Pengeluaran Simen, Unpublished MSc Thesis, School of Materials Engineering, Northern Malaysia University College of Engineering. 2006

[34] Hills, L.M, Clinker Formation and the Value of Microscopy, Proceeding of Twenty-Second International Conference on Cement Microscopy, Montreal, pp: 1 – 12. 2000.

[35] F. Fernandez & R. Diaz, "Metal-Oxide Surge Arrester Model for Fast transient Simulations", IPST-2001, Conference proceedings, 2001.

[36] H.J. Li, S. Birlasekaran & S.S. Choi, "A Parameter Identification Technique for Metal-Oxide Surge Arrester Models", IEEE Transactions on Power Delivery, Vol.17, No.3, July 2002.

[37] C.Talhi, A.Bui, "L'effet de l'oxyde de praséodyme Pr_6O_{11} sur l'augmentation du champ de seuil des varistances à base d'oxyde de zinc", Laboratoire de Génie Électrique de Toulouse 118, route de Narbonne 31062 Toulouse Cedex, ICSD 2004, Toulouse, France.

[38] Renée Legros, André Loubiére, "Etude du vieillissement par la méthode de l'impédance complexe des varistances réalisees par la voie de la chimie douce", LCMI, LGET, Toulouse, France, 1997.

[39] Ramón Puyané1, Isabelle Guy and Renaud Metz, "High Performance Varistor Discs Obtained from Chemically Synthesized Doped Zinc Oxide Powder", Journal of Sol-Gel Science and Technology, Volume 13, Numbers 1-3, pp. 575 - 578, January 1998.

[40] IEEE Working Group 3.4.11 - "Modeling of Metal Oxide Surge Arresters", IEEE Transactions on Power Delivery, Vol.7, No.1, pp. 302-309, January 1992.

[41] A.Bui, C.Talhi, A.Loubière, J.M.Laffargue, "Study of ageing ZnO-based varistor under electrical combined stresses", High Temp. Chem. Proc., 2, 1993, 371-379.

[42] M. Popov, L. van der Sluis, G.C. Paap, and P.H. Schavemaker, "On a hysteresis model for transient analysis," *IEEE Power Eng. Rev.*, vol. 20, pp. 53-54, May 2000.

[43] Daniel W. Durbak, "Surge Arrester Modeling", 0-7803-6672-7/01/$10,00 © IEEE 2001.

[44] R. Diaz, F. Fernandez & J. Silva, "Simulation and Test on Surge Arrester in High-Voltage Laboratory", IPST-2001, Conference proceedings, 2001.

[45] J.Kim, T.Kimura, and T.Yamaguchi, "Microstructure Development in Sb2O3-doped ZnO", Journal of Materials Science, 24, 7, 2581-2586 (1989).

[46] T.Senda and R.C.Bradt, "Grain Growth of Zinc Oxide During the Sintering of Zinc Oxide-Antimony Oxide Ceramics", Journal of American Ceramics Society, 74, 6, 1296-1302 (1991).
[47] N.Daneu, A.Recnik, S.Bernik, and D.Kolar, "Microstructural Development in SnO-doped ZnO-Bi2O3 Ceramics", Journal of American Ceramics Society, 83, 12, 3165-3171 (2000).

Part 2

Topics in Bioceramics

15

Biocompatibility

Waleed Elshahawy
College of Dentistry/Tanta University
Egypt

1. Introduction

During the past few years, the biocompatibility of biomaterials (non-vital material intended to interact with biological systems within or on the human body) has evolved into a comprehensive, complex, and independent discipline of biomaterials science. Consequently, a number of terms have been developed or were adopted from toxicology. Some of these terms may be familiar to patients and clinicians from daily life - for example, the term "safety". Safety in relation to the evaluation of biomaterials means freedom from unacceptable risks. Thus, safety does not stand for a complete lack of risks.

2. Biocompatibility

2.1 Definition of biocompatibility

Biocompatibility is a word that is extensively used within biomaterials science, but there still exists a great deal of uncertainty about what it actually means and about the mechanisms that are subsumed within the phenomena that collectively constitute biocompatibility. During the 2nd Consensus Conference in Liverpool, biocompatibility was defined as "the ability of a material to perform with an appropriate host response in a specific application" (Gatti & Knowles, 2002, as cited in 2nd Consensus Conference, 1991). A biocompatible material may not be completely "inert"; in fact, the appropriateness of the host response is decisive. Previously, the selection criteria for implantable biomaterials evolved as a list of events that had to be avoided, most of these originating from those events associated with the release of some products of corrosion or degradation, or additives to or contaminants of the main constituents of the biomaterial, and their subsequent biological activity, either locally or systemically. Materials were therefore selected, or occasionally developed, on the basis that they would be non-toxic, non-immunogenic, non-thrombogenic, non-carcinogenic, non-irritant and so on, such a list of negatives becoming, by default, the definition of biocompatibility. A re-evaluation of this position was initiated by two important factors. Firstly, an increasing number of applications required that the material should specifically react with the tissues rather than be ignored by them, as required in the case of an inert material. Secondly, and in a similar context, some applications required that the material should degrade over time in the body rather than remain indefinitely. It was therefore considered that the very basic edict that biocompatibility, which was equated with biological safety, meant that the material should do no harm to the patient, was no longer a sufficient pre-requisite. Accordingly, biocompatibility was redefined in 2008 as "the ability of a material to perform its desired function with respect to a medical therapy, without eliciting any undesirable local or systemic effects in the recipient or beneficiary of that therapy, but generating the most appropriate beneficial cellular or tissue response in that specific situation, and optimizing the clinically relevant performance of that therapy" (Williams, 2008).

2.2 Components of biocompatibility

In addition to the beneficial tissue response and the clinically relevant performance of a biomaterial, cytotoxicity, genotoxicity, mutagenicity, carcinogenicity and immunogenicity are considered to be the components which constitute "biocompatibility" (Table 1).

Beneficial tissue response and the clinically relevant performance
Cytotoxicity (systemic and local)
Genotoxicity
Mutagenicity
Carcinogenicity
Immunogenicity

Table 1. Components of biocompatibility

Toxicity of a material describes the ability to damage a biological system by chemical means. In higher organisms (animals, human beings), local toxicity – that is, adverse reactions emerging at the application site – is differentiated from systemic toxicity, in which adverse reaction appear in an area distant from the application site. Cytotoxicity refers to damage to individual cells, for example in cell cultures. Cells can die because of necrosis or apoptosis (programmed cell death).

Immunogenicity is referred to the ability of a substance to provoke an immune response or the degree to which it provokes a response. An allergic reaction to a substance can be triggered if the organism was previously sensitized to this substance. The concentrations that elicit a reaction in a previously sensitized person vary between subjects. The dose levels causing allergic reactions are generally significantly lower than those causing toxic reactions.

Genotoxicity describes an alteration of the basepair sequence of the genome DNA. Cells possess numerous mechanisms to repair genotoxic damages. Alternatively, a transfer of these genetic damages to subsequent generations of cells can be avoided by programmed cell death (apoptosis). Nonetheless, if theses genetic damages are passed on to the next generation, this effect is called mutagenicity. Mutagenicity and carcinogenicity are not the same. Carcinogenicity means that alterations in the DNA have caused a cell to grow and divide inappropriately; in other words, alterations of DNA promoted the generation of malignant tumors. Carcinogenicity results from several mutations. It is important to understand that not all mutagenic events lead to carcinogenesis. However, mutagenicity can be assessed as an indicator of "possible" carcinogenicity of substances that directly attack DNA.

The components of biocompatibility will be discussed in relation to bioceramics later in the current chapter.

3. Bioceramics

3.1 Definition of bioceramics

In practical sense, the term bioceramics can be referred to a group of ceramics, which are used in the field of biomedicine. These biomaterials are ceramics, which are manufactured or processed to be suitable for use in or as a medical device that comes into intimate contact with proteins, cells, tissues, organs, and organ systems.

3.2 Benefits and clinical performance of bioceramics

Bioceramics are used to restore normal activity of diseased or damaged parts of the body. As people age, progressive deterioration of tissues requires replacements in many critical applications. After successful researches, various bioceramic products are now commercially available in the medical market as substitutes for the original damaged body parts and for many other critical applications (Table 2).

Dentistry	Dental restorations
	Prosthodontic devices
	Orthodontic brackets
	Repair of periodontal disease
	Maxillofacial reconstruction
Orthopedics	Joint replacements
Cardiology	Prosthetic heart valves
Neurosurgery	Cranioplasty repair
Otolaryngology	Middle ear implants, Vocal cord paralysis
Miscellaneous	Magnetic treatment of bone tumors
	Drug delivery systems

Table 2. Benefits of ceramics in biomedicine.

Traditionally, ceramics have seen widescale use as restorative materials in dentistry. Dental ceramics are rigid materials that are shaped by sintering, casting, pressing, milling, or sonoerosion. Dental ceramics are also available as prefabricated inlays (inserts). Dental ceramic restorations include materials for denture teeth, fixed partial dentures, full crowns, veneers, inlays, onlays, and post - cores to restore missing tooth part, a tooth, or teeth. Restorative dental ceramics could be bonded to metal (Metal-Ceramics) or be metal free ceramics (All-Ceramics). High-performance ceramics yield excellent technical properties, which make them suitable to be used as copings or frameworks for crowns and bridges. To improve their aesthetics, they have to be veneered with other, mainly silicium oxide ceramics. Dental ceramics are further applied as implant materials, for example as coating for titanium implants, or as full ceramic implants. The most recent use for ceramics in dentistry is orthodontic brackets. The development and demand for these items has been driven solely by aesthetics. Also, ceramics are used for repair for periodontal diseases. They are also useful for maxillofacial reconstruction, augmentation and stabilization of the jaw bone because bioceramics may develop the clinical applications of bone substitutes. The physical, chemical and biological properties of bioceramics can be used for preparing advanced bone substitutes. Bioactive glass ceramics and calcium phosphate ceramics are the two ceramic types used as bone substitute or for bone healing process. Bioactive glass ceramics bonds to bone without an intervening fibrous connective tissue interface (Schepers et al., 1991). When granules of bioactive glass ceramics are inserted into bone defects, ions are released in body fluids and precipitate into a bone-like apatite on the surface, promoting the adhesion and proliferation of osteogenic cells (Neo et al., 1993). After long-term implantation, this biological apatite layer is partially replaced by bone (Neo et al., 1994). Bioactive glass with a macroporous structure has the properties of large surface areas, which are favorable for bone integration. The porosity provides a scaffold on which newly-formed bone can be deposited after vascular ingrowth and osteoblast differentiation. The porosity of bioactive glass ceramics is also beneficial for resorption and bioactivity (De Aza et al., 2003). Calcium phosphate polycrystalline ceramic materials can be produced by precipitation from aqueous solutions and by solid-state reactions. The rationale for using hydroxyapatite as a biomaterial is the advantage of using a material having similar composition and crystalline structure as natural calcified tissues. Hydroxyapatite and other calcium-based ceramic materials can actively encourage bone regeneration at the surface of an implant. It has been postulated that the use of calcium phosphate ceramic biomaterials might replace the use of bone grafts. The chemistry of these materials is reasonably well established (Nascimento et al., 2007) and significant animal experiments have shown these materials to be both biocompatible and bioactive.

However, bioceramics use in other fields of biomedicine has not been as extensive, compared to metals and polymers. For example, in orthopedics, ceramics such as alumina (aluminum oxide ceramics) and zirconia (zirconium oxide ceramics) are used for wear applications in joint replacements. Bioceramics can now be used for hips, knees, tendons and ligaments replacements. In cardiovascular or circulatory system (the heart and blood vessels involved in circulating blood throughout the body), problems can arise with heart valves and arteries. The heart valves suffer from structural changes that prevent the valve from either fully opening or fully closing, and the diseased valve can be replaced with a variety of substitutes. As with dental implants, ceramics may be used as pyrolytic carbon coatings for prosthetic heart valves (Sarkar & Banerjee, 2010). Less obvious examples of the use of ceramics as biomaterials are in neurosurgical cranioplasty repair of the skull bone defects, in hand arthroplasty of the metacarpophalangeal joint, in otolaryngology as implants in the middle ear, or the use of bioactive glass ceramics in the treatment of vocal cord paralysis. Bioactive glass ceramics containing magnetite can be used to kill bone tumors when a magnetic field is applied. Ceramics implants can also be used as drug delivery systems (Nascimento et al., 2007).

3.3 Classification of bioceramics

Based on their chemical reactivity with the physiological environment, bioceramics can be broadly categorized in three types (Fig. 1):

3.3.1 Bioinert ceramics

They are such as alumina, result in little or no physiological reaction in the human body and tend to exhibit inherently low levels of reactivity which peak in the order of hundreds of years. They are attached by compact morphological fixation.

3.3.2 Surface reactive or bioactive ceramics

They are such as bioactive glass ceramics (bioglass), react in a positive way with local cells, i.e. they directly attach by chemical bonds and have a substantially higher level of reactivity, peaking in the order of 100 days.

3.3.3 Resorbable bioceramics

They are porous or nonporous structures which are slowly and gradually replaced by bone such as tricalcium phosphate, have even higher levels of reactivity, peaking in the order of 10 days (Shackelford, 2005).

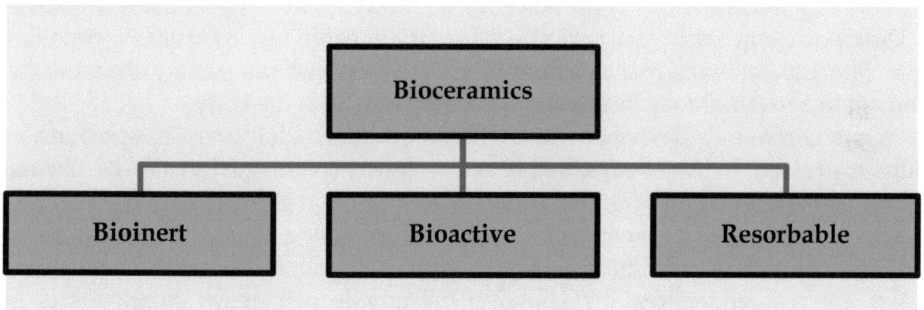

Fig. 1. Classification of bioceramics according to biocompatibility

4. Why is it significant to study biocompatibility of ceramics?

Biocompatibility of ceramics is a critical issue because of three different reasons. The first is that these materials are in intimate contact with human tissues for long terms and cannot be removed by the patient. Secondly, biocompatibility is an ongoing process and not a static one. For example, it is possible that a dental implant that is osseointegrated today may or may not be osseointegrated in the future. Thirdly, it has to be stressed that biocompatibility of fixed prosthodontic materials like ceramics is often overlooked because many practitioners assume that, if the material is on the market, its biocompatibility does not need to be questioned. For example, two systems are currently responsible for standards that can be used to document dental products quality: the American National Standard Institute / American Dental Association (ANSI/ADA) document No. 41 (1997) and addendum No. 41A (1982) and the International Standards Organization (ISO) 10993 document (1993). The ANSI/ADA and ISO do not require specific biologic tests to approve the quality of a new dental material. Rather, they place the responsibility on the manufacturer to present evidence for a compelling case for approval. So, it is up to the manufacturer to defend the substantial equivalence argument. The evidences used for approval of quality of a dental material consist of in vitro tests (cell-culture), in vivo tests (animal experiments), and clinical tests (clinical trials of the material). However, it is becoming increasingly impractical to test all new materials through all of these stages. The problems of time, expense, and ethics have limited the usefulness of this traditional biologic testing scheme. Therefore, companies market materials with little clinical experience, and may rely heavily on in vitro and animal experiments (Wataha, 2001).

Although most ceramic materials are generally regarded as being more or less inert, their possible effects of degradation products on biological systems must not be overlooked. The composition and physical properties of ceramic materials can affect the inertness. Safety cannot be inferred from measurements of one ceramic formulation to other compositions or conditions. Since bioceramics have been mainly used in dentistry, biocompatibility and its relevant properties for ceramics will be mainly discussed in relation to oral health.

5. Biologically relevant properties

5.1 Ceramics composition

It is believed that biologic reactions in general are mainly based on the interaction of a substance eluted from a material with a biologically relevant molecule. Thus, the composition of a material is of importance for its biocompatibility (Schmalz & Garhammer, 2002). Different elements in the periodic table of elements can be used in ceramics. The diversity of these ceramics makes understanding their biocompatibility difficult, because any element in a material may be released and may influence the body.

Ceramics are commonly described by their composition. However, composition can be generally expressed in two ways; either as weight percentage (wt %) of elements or percentage of the number of atoms of each element in the material (atomic percentage = at %). Weight percentage is the most common way of describing a material's composition, and is used by material manufacturers and by standard organizations. However, biologic properties are best understood by knowing the atomic percentage composition. Atomic percentage better predicts the number of atoms available to be released and affect the body. The wt% and at% of a material or an alloy may be substantially different from each another.

Ceramics could be oxide or non-oxide ceramics. Oxide ceramics in dentistry are primarily based on silicon oxide (SiO_2), aluminum oxide (Al_2O_3), and zirconium oxide (ZrO_2). Non-oxides, such as silicon carbide, silicon nitrite, and aluminum nitrite, are of minor importance in dentistry due to their black color. Some dental ceramics can be combined, such as an Al_2O_3-ceramic framework veneered with SiO_2 ceramic. Lanthanum glass is used as a coupling agent, which infiltrates the aluminum oxide framework. Lanthanum glass consists of 39% lithium oxide. Additives (such as leucite) are intended to improve the mechanical properties of the ceramics, in particular to limit crack propagation. Further additives in dental ceramics are fluxing agents and coloring pigments, such as metal oxides, as well as fluorescents such as oxides of cesium and samarium. Some calcium phosphate materials are regarded as ceramics, too. These substances represent a very heterogeneous group of materials, including sintered hydroxyl apatite (HA) with a very low solubility and tricalcium phosphate (TCP) ceramics with varying resorption behaviors. Calcium phosphate ceramics usually consists of 100% of the respective mineral phase (TCP or HA).

5.2 Biodegradation and corrosion

Biological systems may have harmful or destructive effects on bio- materials, classified as biodegradation. In the oral environment, this includes not only the process of destruction and dissolution in saliva but also chemical/physical destruction, wear and erosion caused by food, chewing and bacterial activity. Therefore, it is important to evaluate the material

reactivity in the oral cavity, which is governed by thermo-dynamic principles and electrochemical reaction kinetics. This means that when a material is placed in the oral cavity, the material-saliva system will be driven toward a state of thermo-dynamic equilibrium. At equilibrium, the material either will remain stable in its elemental form or oxidize into its ionic form (corrosion). Thus, the initially uncharged elements inside the material lose electrons and become positively charged ions as they are released into solution. Corrosion is a chemical property that has consequences on other material properties, such as esthetics, strength, and biocompatibility. From a biocompatibility standpoint, the corrosion of a material indicates that some of the elements are available to affect the tissues around it.
The chemical durability of dental ceramics is basically good. They are commonly regarded as insoluble or only very slightly soluble at best. However, the degradation of dental ceramics can generally occur because of mechanical forces (wear) or chemical attack (solubility in an acidic, neutral, or alkaline environment), or a combination of the two. Some calcium phosphate ceramics are internationally engineered for a gradual resorption (TCP). The release of substances can generate unwanted effects (biological and mechanical) on one hand, or it may promote biocompatibility on the other hand, such as in terms of improved bone apposition (bioactivity). The multiphase microstructure of many dental ceramic materials results in complicated corrosion modes, as each phase is likely to react individually to the corrosive medium. Besides, chemical durability of ceramic materials may be influenced by many other factors, such as the chemical character of the corrosive medium, the exposure time, and the temperature. For glass ceramics, the initial surface reaction is mainly an acid-base reaction in which leaching ions are replaced by H^+ ions, the result of which will be an alkali-ion-depleted leach layer overlying a permeable gel layer. Beneath the alkali-depleted layer, the corrosion process will produce a silica-rich layer, offering some protection to the bulk material. However, because of differences in composition, microstructure, and local corrosion conditions, the corrosion process is far more complicated and may also lead to the partial breakdown of the silicate structure at the surface. In addition, glasses high in K_2O have been less chemically durable than glasses made with soda (Na_2O) as an added flux material, whereas the presence of zirconia and alumina has shown to improve the chemical durability of glasses. When exposed to hydrolysis testing, ultra-low-temperature sintering ceramics displayed higher solubility than traditional high-temperature sintering ceramics. However, in repeated hydrolysis tests, high- and low-sintering ceramic materials did not react in the predicted manner. Alumina, which is regarded as a very stable material, may also undergo compositional changes when exposed to a corrosive environment (Milleding et al., 2002).

5.3 Ion release

Corrosion, as mentioned before, is always accompanied by a release of elements and a flow of current. The release of substances from dental materials is considered to be gradual and to occur in small amounts. Evaluation of mass release from dental ceramics is not common in the literature, although there are some studies that have demonstrated such mass release. The leakage of inorganic ions from ceramics has been found to take place in aqueous media and vary with the glass composition and environmental conditions. Under more severe conditions (as the concentration of alkali ions increases), the Si-O-Si bonds may be broken, and the entire glass structure may be impaired. The reduction in chemical durability is of importance, since an increased susceptibility to chemical attack may release ions of the elements (K2O.Al2O3.4SiO2), which in certain circumstances, could be considered undesirable from a biocompatibility perspective (Milleding et al., 2002).

Two dominant mechanisms could be responsible for the aqueous corrosion of alkali-silicate glasses: (1) the selective leaching of alkali ions and (2) the dissolution of the glass network. At a pH of 9 or less, selective leaching of alkali ions could be the dominant mechanism. This mechanism can be controlled by the diffusion of H^+ or H_3O^+ ions from an aqueous solution into the glass and the loss of alkali ions from the glass surface. In general, alkali metal ions from glass are much less stable in the glass phase than in the crystalline phase and thus could be leached more rapidly.

In contact with saliva or other organic fluids, biomaterials are instantly covered with organic films, the composition and properties of which undoubtedly influence the surface corrosion process and subsequent bio-reactions. It has been assumed that organic films on ceramic surfaces reduce the surface degradation by building up concentration gradients and reducing the diffusion of ionic elements through the surface films. In addition, it has been found that leaching of inorganic ions can be influenced by pH of the corrosion solutions and the ions potential for the complex binding of dissolved glass constituents, resulting in more extensive corrosion than indicated by the pH value alone (Milleding et al., 2002).

Sjögren et al. (2000) tested the release of elements from different dental ceramics (low-fusing, conventional veneering, press-casting ceramics) into a cell culture medium by inductively coupled plasma optical emission spectrophotometry. They found multiple released elements such as aluminum (Al), silicon (Si), sodium (Na), potassium (K), magnesium (Mg), and calcium (Ca). Also, Milleding et al. (2002) studied the in vitro ion dissolution from glass-phase ceramics, with or without crystalline inclusions, and from all-crystalline ceramics using the inductively coupled plasma optical emission spectroscopy. A large number of inorganic elements leached out from the previous dental ceramics. The major leaching elements were sodium and potassium. There were also magnesium, silicon, and aluminum. The various glass-phase ceramics displayed significant differences in ion release and significantly higher release values than all-crystalline alumina and zirconia ceramics. No significant difference in dissolution was found between high and low-sintering glass-phase ceramics or between glass-phase ceramics with high volume fractions of crystallites in the glass phase in comparison with those with lower crystalline content.

Logically, it has to be noted that the type of released elements depends on the composition of ceramic material itself. From silicon oxide ceramics, silicon, sodium, potassium, boron, and aluminum are released into various diluents at different pH values; silicon, sodium, and potassium are leached in higher amounts than are aluminum and boron. Aluminum oxide ceramics leach only minimal amounts of ions under physiological conditions. Calcium phosphate ceramics release calcium and phosphate into adjacent tissues. Overall, hydroxyapatite and fluorine apatite ceramics are less soluble than tricalcium phosphate ceramics (Lacefield, 1999).

6. Types of biocompatibility tests

Biomaterials are developed in order to evaluate, treat, augment or replace human tissue, organ or function. Biocompatibility is the main prerequisite for their safe use as medical devices. In order to assess the biocompatibility of a material, it is necessary to do a battery of tests, depending on the intended use, location and duration the material is to come in contact with the tissues. The evaluation of biocompatibility is dependent not only on the tested biomaterial but also on the test method used. So clinicians need to be familiar with these methods. Biocompatibility is measured with 3 types of biologic tests: in vitro tests, animal experiments and clinical tests (Fig. 2).

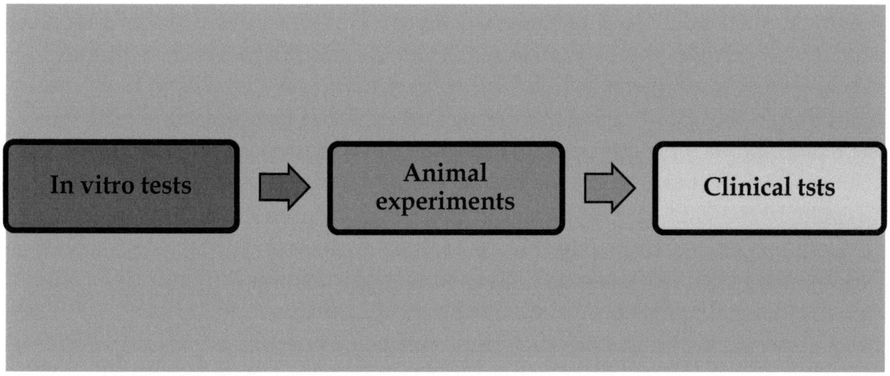

Fig. 2. Plan of biocompatibility tests in order.

The common approach when testing the biological behavior of materials is to start with simple in vitro tests. If these experiments and investigations of a material's efficiency deliver promising findings, then more comprehensive studies on experimental animals (in vivo evaluation) will be performed. Clinical trials (usage tests) are the final step of this evaluation process.

6.1 In vitro tests

In vitro biocompatibility tests are less expensive ways to survey newly developed materials. They simulate biological reactions to materials when they are placed on or into tissues of the body. These tests are performed in a test tube, cell-culture dish, or otherwise outside of a living organism in which cells or bacteria are generally placed in contact with a material. For example, a strain of bacteria may be used to assess the ability of a material to cause mutations (the Ames test). The advantages of in vitro biocompatibility tests are, being experimentally controllable, repeatable, fast, relatively inexpensive and relatively simple. Another major advantage is that these tests generally avoid the ethical and legal issues that surround the use of animals and humans for testing. The primary disadvantage of in vitro biocompatibility tests is their questionable clinical relevance.

6.2 Animal experiments

In animal experiments, the material is placed into an animal, usually a mammal. For example, the material may be implanted into a mouse or placed into the tooth of a rat, dog, cat, sheep, goat or monkey. Animal models allow the evaluation of materials over long time durations and in different tissue qualities (e.g. normal healthy or osteopenic bone) and ages. Not only can the tissues in the immediate vicinity be assessed, but, tissues in remote locations of the implanted material can also be studied, which is particularly relevant to the study of wear particle debris. However, questions arise about the appropriateness of an animal species to represent the human response and that they are time-consuming and expensive. In animal experiments, ethical concerns and animal welfare issues are very important.

6.3 Clinical tests

The clinical test is, by definition, the most relevant biocompatibility test. These tests are essentially clinical trials of a material in which the material is placed into a human volunteer in its final intended use. In a controlled clinical study, test and control materials are examined at the same time. Controlled clinical studies possess a higher level of significance/evidence compared with studies in which only one material is investigated. Biocompatibility data from clinical studies are naturally of special interest for the clinician, since the examination was done on the target group of this material (patients). But this should not conceal the fact that clinical studies reveal limitations, too. An uncritical transfer of such results to patients in daily practice may result in problems, for instance, if data are not based on a blinded study. Therefore, at least treatment and subsequent assessment should be done by different persons. Many unwanted reactions appear only after chronic exposure. But clinical studies - in particular those with new materials - are frequently limited to comparatively short periods of time (some are only 6 months). In addition, only a small and often strictly selected group of patients is included in the study, for instance in a university hospital. The clinical studies are also expensive, time-consuming, extraordinarily difficult to control its variables, difficult to interpret and may be legally and ethically complex. Clinical tests are done only if satisfactory results are obtained in the in vitro and animal experiments.

7. Systemic toxicity

7.1 Means of systemic toxicity testing

Experimental animals are usually used to determine systemic toxicity. Previously, the acute lethal dose 50% (acute LD_{50}) was determined as routine. Acute LD_{50} is the dose required to kill half the members of a tested population after specified test duration. Today, other methods that are more sparing of animals are used, such as the so-called limit test (administration of a fixed dose, e.g., 2,000 mg/kg body weight). The chronic systemic toxicity will be determined by administering the material or extract over several months. Tests are sometimes extended over the lifetimes of the experimental animals. At the end of these studies, survival rates of the animals and patho-histological alterations of the main organs will be determined. Further information regarding chronic toxicity is obtained from accidents (high exposure level) and based on observations of occupationally exposed subjects (e.g., dental personnel) who are often in contact with the "active" unset material.

7.2 Systemic toxicity related to ceramics

One fundamental concern about the safety of ceramics used as fixed prosthodontic materials is their ability to cause systemic toxicity in the body. A stress must be applied on several key concepts that affect this concern. For example, in dentistry, the following should be concerned (Wataha, 2000):

7.2.1 Presence inside the body

Elements released from a dental fixed prosthodontic material into the oral cavity are not inside the body because these elements may gain access to the inside of the body through absorption in the gastrointestinal tract, in the oral mucosa, from the skin, or in the respiratory system. The mechanism for this absorption depends on the nature of the chemical properties of the released elements - whether they exist as ions, as hydrophilic and lipophilic compounds, as volatile substances, or as particles. In contrast, elements that are released from dental implants into the bony tissues around the implant are, by definition, inside the body. Therefore, elemental release from ceramic implants is thought to be more critical systemically than elemental release from dental ceramics used for prosthetic restorations.

7.2.2 Route of access to the body

The route by which an element gains access inside the body is critical to its biological effect. Some elements become more toxic when administered intravenously into mice than when administered orally.

7.2.3 Distribution in the Body

Any biomaterial, once inside the body, can release ions which can be distributed to many tissues by diffusion through tissues, the lymphatic system, or the blood stream. Released metallic particles (0.5 to 10.0 um) may also be ingested by cells such as macrophages. Almost all dental materials release substances into the oral cavity, from where they may enter the human body through different routes, including swallowing of saliva and inhalation, with subsequent passage of the epithelial barriers in the gastrointestinal tract or the lungs. These substances may, via the blood circulation, be transported to different organs. The oxidation state and chemical form of the metallic ions will significantly influence its absorption, distribution, retention half-life, and excretion. Ultimately, the body generally eliminates the released ions through the urine, feces, or lungs. The application site may thus be in a different location from the effect. At the location of the effect, there may be interference with the function of the specific organ if the concentration is sufficiently high (systemic toxicity). According to the time frame, acute (up to an exposure period of 24 h), subacute (up to 3 months), and chronic toxicity are differentiated.

In general, the systemic toxicity of ceramics is considered to be extremely low (Aldini et al., 2002). In dentistry, only dental laboratory technicians might be exposed to an inhalation of ceramic dust due to processing and finishing of dental ceramics that may cause silicosis (fibrotic pneumoconiosis). These lung diseases have been observed in workers in the ceramic industry who were exposed to ceramic dust for an extended period of time. The risk to a dental laboratory technician of developing silicosis due to ceramic dust is currently unknown. The patient's silicosis risk is considered "very minimal" (Mackert, 1992) if commonly accepted safety measures, such as dust removal, are followed. On the other hand, there is evidence that released metallic ions from fixed prosthodontic materials can and do gain access to the body, and these metallic ions may be widely distributed (Wataha, 2000). Person-Sjögren & Sjögren (2002) found a statistically significant increase in levels of insulin release from the Langerhans cells after exposure to lithium-containing ceramic (Empress ceramics). The danger lies in overseeing the possibility that minimal amounts of ions eluted due to chemical or mechanical wear might adversely affect the pancreas, or other organs or tissues.

8. Local toxicity

8.1 Means of local toxicity testing

Current knowledge about biomaterials-tissue interactions has been gained through bioassays in vitro and in vivo. Taking into account biocompatibility tests available in the general field, cytotoxicity assays are of special concern. In vitro studies are mainly performed to evaluate the cytotoxicity. A vast number of different in vitro test methods exists which include both quantitative and qualitative methods of cytotoxic effect, i.e. cell damage or lysis caused by membrane leakage. However, each test method basically consists of three components: (a) the biological system, (b) the cell/material contact, and (c) the biological endpoint and corresponding recording system. The biological system used in in vitro cytotoxicity tests may be (i) organ cultures, (ii) cells in culture or (iii) cell organelles. The cell-material contact may be direct; the cells grow next to, or even on the test material. In in vitro tests, direct cell/material contact methods simulate the in vivo situation in certain instances. In indirect contact, materials and cells are separated by a barrier. Eluates derived from a dental material by storing it for a specific period of time in a liquid, such as the nutrient medium, may be used for toxicity testing instead of the material itself. Besides the description of cell morphology, different biological endpoints can be used as indicators for cell damage: membrane effects, cell activity and proliferation rate. The cell reaction can be described morphologically as is done with the lysis index in the agar overlay test. However, this method is considered to be only qualitative, or at most, semi-quantitative in nature. Furthermore, some dental filling materials contain or produce considerable amounts of ingredients, which if applied to cells in culture; the morphology of the cells will appear to be normal, indicating no cell damage even though the cells are no longer vital (Schmalz & Netuschil, 1985). The use of membrane effects, cell activity and proliferation rate have no such drawbacks. Membrane effects can be demonstrated by dye exclusion (trypan blue). The trypan blue exclusion assay can be used to indicate cytotoxicity, where the dead cells take up the blue stain of trypan blue, and the live cells have yellow nuclei. Direct cell counting is easy to perform and can be combined with a vital stain in order to exclude dead cells.

8.2 Local toxicity related to ceramics

Different researches have been performed to study local cytotoxicity of dental ceramics. Cobb et al. (1988) investigated the in vitro biocompatibility of porous air-fired opaque porcelain with human gingival fibroblasts. Their results indicated that porous air-fired opaque porcelain is biocompatible. Then, Josset et al. (1999) studied the reaction of human osteoblasts cultured with zirconia and alumina by investigating cellular functions, and found that no cytotoxic effect was observed because neither material altered cell growth rate in accordance with the absence of any inducing effect on DNA synthesis or proliferation. Also, Sjögren et al. (2000) evaluated the cytotoxicity of different types of feldspathic porcelain ceramics by using cells from a mouse fibroblast cell line and the agar overlay test, Millipore filter test, and MTT (3-(4, 5-dimethylthiazol-2-y1)2, 5-diphenyl tetrazolium bromide)-based calorimetric assay. All the ceramics studied were rated "non-cytotoxic". Consistent with the former study, Uo et al. (2003) tested the cytotoxicity of different feldspathic, leucite-reinforced glass, and lithium-containing ceramics against human gingival fibroblasts that were cultured using extraction solutions of ceramics, with the aid of almar blue assay. They found that no ceramic extractions showed any evidence of significant cytotoxicity.

Different implantation studies have been also performed for different types of ceramics in different tissues. Silicon oxide ceramic did not cause inflammation after implantation in muscle (G. Schmalz & C. Schmalz, 1981). Bioglasses based on silicon oxide were osteoconductive and osteoinductive when implanted in bone (Chan et al., 2002). Aluminum oxide ceramic, before and after infiltration with lanthanum glass, was found to cause a significantly thicker connective tissue encapsulation and an increased number of inflammatory cells 12 weeks after subcutaneous implantation, compared with Teflon and titanium (Limberger & Lenz, 1991). On the other hand, aluminum oxide ceramic resulted in osseointegration in other studies and thus revealed a good compatibility with surrounding bone (Piatelli et al., 1996). There are obviously differences between the compatibility of various ceramics, and these may be correlated to different indications and applications and different contact with tissue (for example, core ceramic versus implant ceramics). Zirconium oxide ceramic showed good osseointegration when implanted in guinea pigs (Aldini et al., 2002). Calcium phosphate ceramics have been implanted in various animal models. Results were heterogeneous according to the materials tested and depended mainly on the following parameters: calcium (Ca)/phosphate (P) ratio, chemical purity, removal of organic compounds from raw materials, sintering technique, crystal structure (monophase or polyphase), and size and type of pores. Numerous macrophages and foreign body giant cells were observed histologically during the first weeks after implantation of absorbable TCP ceramics. The integration of non-soluble hydroxyl-apatite ceramic in bone without any cellular interface (osseointegration) indicates good biocompatibility (Lacefield, 1999).

Various degrees of ceramic toxicity have been stated. Messer et al. (2003) studied the cytotoxicity of feldspathic porcelains, lithium-disilicate ceramics, and leucite-based glass ceramics by testing their ability to alter cellular mitochondrial dehydrogenase activity (SDH activity) using tetrazolium assay. Their results revealed that dental ceramics are not equivalent in their in vitro biologic effect, even with the same class of material and most ceramics caused only mild in vitro suppression of cell function to levels that would be acceptable on the basis of standards used to evaluate alloys and composites (< 25% suppression of SDH activity). However, the lithium-containing ceramics exhibited cytotoxicity that would not be deemed biologically acceptable on the basis of prevailing empirical standards for dental alloys. Additionally, Pera et al. (2005) investigated the in vitro cytotoxicity of different ceramic materials (lithium-containing, aluminous, zirconium, and feldspathic ceramics) with the use of MTT testing on mouse fibroblasts. Their results revealed that not all tested materials were free from cytotoxicity. Other confirmatory studies have been reported by Elias et al. (2002); Yamamoto et al. (2004) who revealed a varying ability to induce inhibition of cell proliferation, cytotoxicity (as measured by colony forming efficiency) of silica, and alumina components in ceramic materials used for orthopedic prostheses.

It has to be noted that the biocompatibility has been mainly studied for traditional feldspathic porcelains. Most newer ceramic materials, such as those for computer aided design - computer aided manufacture (CAD-CAM) all-ceramic systems, have not been tested for biologic response with the same scrutiny as has been applied to dental casting alloys or even traditional ceramics. In vitro studies have reported different mass loss and cytotoxicity of some newer formulations of all-ceramic materials. An in vitro study done by the author of the current chapter (Elshahawy et al., 2009a) investigated the ion release from CAD-CAM leucite-reinforced glass ceramic material into both sodium chloride and lactic acid immersing solutions using inductively coupled plasma mass spectroscopy and showed that transient exposure of tested material to an acidic environment for one week is likely to significantly increase elemental release from it (e.g. aluminum and potassium ions). However, the amounts of these released elements (ions) were shown by the author of the

current chapter to be not enough to show high evidence of toxicity against cultured fibroblasts using the trypan blue assay (Elshahawy et al., 2009b).
Whatever is the dental material used for fixed prosthodontic appliance, it is nevertheless difficult to predict the clinical behavior of a material from in vitro studies, since oral factors such as changes in the quantity and quality of saliva, diet, oral hygiene, polishing of the material surface, amount and distribution of occlusal forces, or brushing with toothpaste, can all influence corrosion to varying degrees. From a biocompatibility standpoint, the corrosion of a material indicates that some of the elements are available to affect the tissues around it. Therefore, a study was performed by the author of the current chapter (Elshahawy et al., 2010) which quantitatively assess the element release from CAD-CAM fabricated leucite-reinforced glass ceramic crowns into saliva of fixed prosthodontic patients. They revealed the release of silicon and aluminum ions from them after three months in service. These released amounts were not enough to produce pronounced cytotoxic effects against fibroblasts.

9. Genotoxicity, mutagenicity, and carcinogenicity

9.1 Means of testing

The Ames assay is used worldwide as an initial screen to determine the mutagenic potential of new chemicals and drugs. It is perhaps the most rapid, simple, sensitive and economical screening test for mutagenicity and has an extensive database and good correlation with carcinogenicity.
The comet assay is a quick, simple, sensitive, reliable and fairly inexpensive genotoxicity test which is widely used to evaluate the genotoxic potential of chemical and physical substances. Ostling & Johanson (1984) first demonstrated "comets" and described the tails in terms of DNA with relaxed supercoiling through a process of electrophoresis (pH 9.5) of cells embedded and lysed in agarose on a microscope slide. Later, Singh et al. (1988) used alkaline electrophoresis to analyze DNA damage from treatments with X-rays or hydrogen peroxide (H_2O_2). Since then, the worldwide acceptance of Comet assay makes it a good assay for detecting DNA damage.

9.2 Genotoxicity, mutagenicity, and carcinogenicity related to ceramics

The mutagenicity (genotoxicity) of dental ceramics is not clear due to the lack of research focusing on this aspect. Takami et al. (1997) tested the mutagenicity of aluminous (Al_2O_3) ceramic by using Ames assay and tester Salmonella typhimurium strains TA98, TA100 and TA1535. Mutagenicity was not induced by extracted samples of the Al_2O_3 ceramic with and without metabolic activation in Salmonella typhimurium strains TA98 and TA1535. Another study by Covacci et al. (1999) in which zirconia ceramic stabilized by yttria (Y-TZP) was evaluated for mutagenic and carcinogenic potential in the form of discs did not show any mutagenic or oncogenic effects in vitro. A study by Noushad et al. (2009) found that dental ceramics did not induce any DNA damage after using tester Salmonella strains TA98 and TA1537 to detect frameshift mutations whereas using tester strains TA100 and TA1535 to detect base-pair substitution mutations. From previous studies, it is noted that some biomaterials are mutagenic to one tester strain while it is not mutagenic to another. Even though many investigators have sometimes used just 2 strains to determine the mutagenic potential of biomaterials, it is felt that the use of at least four tester strains as recommended by Mortelmans & Zeiger (2000) gives a more definite result.

Other studies tested the mutagenicity and carcinogenicity of the different components of ceramics separately. For example, lithium is a component of certain ceramics. Leonard et al. (1995) reviewed the information available on the mutagenicity, carcinogenicity and teratogenicity of lithium. It was concluded that lithium is unlikely to be carcinogenic. Weiner et al. (1990) studied the effects of lithium hypochlorite in a series of tests including five strains of Salmonella. Lithium was not found to be genotoxic in any of the test systems with the exception of an equivocal response in the Chinese hamster ovary/hypoxanthine-guanine phosphor-ribosyl transferase assay, which was not replicable in a subsequent experiment.

Silica and alumina are main components of ceramics on which genotoxic studies have been reported. In one Comet assay, Zhong et al. (1997) indicated that silica and glass fibers can induce DNA damage in mammalian cells and that crystalline silica has a higher DNA-damaging activity than amorphous silica. Simon et al. (2007) assessed the genotoxicity of alumina and titanium oxide (TiO_2) using the Comet assay and showed that DNA damage was limited to single-strand breaks and/or alkali-labile sites and that genotoxicity was weak. From previous studies, the genotoxicity of some of the components of dental ceramics remains controversial.

Three decades ago, uranium salts were previously added at a concentration of 1,000 ppm to dental ceramics for simulating the natural luminescence of teeth. Because of the radioactivity of uranium salts, alternatives are now applied, such as oxides of rare earths. Today, radiation of dental ceramics is only due to natural radionuclides (mainly α and γ emitters) and much below the materials dated back to the times, when uranium salts had been added. Feldspathic ceramic specimens showed an activity concentration (Uranium/Thorium chains) that is in the same order of magnitude as for the human body. No radiation related adverse effects of dental ceramics have been documented in the literature (Veronese et al., 2006). Raw materials used for zirconium oxide ceramics (e.g., Zirkon, $ZrSiO_4$) may contain contaminants such as thorium and uranium. These contaminants generate α-, β-, and γ-radiation. However, the effective activity of zirconium oxide ceramic was far below the mean value of the annual exposure to natural radiation (Piconi & Macauro, 1999).

So far, no clinical reports have been published that document a carcinogenic effect of certain dental ceramic materials in the oral cavity. The long exposure time that is necessary for the emergence of a malignant tumor is a very aggravating factor for clinical assessment of potential carcinogenic properties. Therefore, it is only possible to draw indirect conclusions from other areas (e.g., occupational exposure to chemicals) to a possible carcinogenic effect.

10. Immunogenicity

As mentioned before, the term immunogenicity is referred to the ability of a substance to provoke an immune response or the degree to which it provokes a response. Sun et al. (2009) studied the clinical effects and security of nanometer ceramics artificial bone transplantation to treat the bone defect. After follow-up period for 24 months, the artificial bone has no immunogenicity, no rejection, does not affect the blood calcium and

phosphorus content, and has higher osteogenic activity. According to our knowledge, there is no documentation about sensitivity to ceramics.
Ceramics are rigid materials and therefore generally need to be luted to human hard tissues like teeth. There can be allergies/sensitivities to the cements/bonding agents that are necessary for the attachment of ceramic fixed prosthodontic restorations. Postoperative sensitivities have been observed in a few cases after the (adhesive) luting of ceramics (inlays, crowns) (Pallesen & Dijken, 2000; Studer et al., 1996). Also, one thing that may be an issue is that if the ceramic fixed prosthodontic restoration is impinging on a vital tissue, e.g. if margins of dental ceramic veneers are impinging on what is called the biologic width of gingiva (the amount of space under the gum where nothing can be placed) then a chronic state of inflammation will ensue.

11. Conclusion

1. Substances are released from ceramics into the surrounding tissues; mainly silicon, aluminum, potassium.
2. Systemic toxicity of ceramics is unlikely to occur due to the relatively low amounts of released elements such as lithium and lead.
3. Few ceramics have shown to be cytotoxic in vitro. The clinical relevance of these findings remains unclear.
4. Generally, local toxicity of ceramics is considered as low. However, more cytotoxicity researches are needed due to possible exceptions.
5. Overall, there is no evidence that ceramics cause or contribute to neoplasia in the body.
6. Ceramics are generally considered as biocompatible materials, although relatively little data are available.
7. Future biocompatibility studies should be performed to study more measurements dealing with cells functions such as protein fabrication (e.g. collagen synthesis), respiratory and digestive cell functions in a response to elements released from ceramics.
8. Future biocompatibility studies should be also performed to test the combinations of the elemental salts released from ceramic materials for the detection of synergistic, antagonistic, or additive effects caused by different mixtures of cations.

12. References

Aldini, N.N., Fini, M., Giavaresi, G., Torricelli, P., Martini, L., Giardino, R., Ravagliolo, A., Krajewski, A.M., Mazzocchi, M., Dubini, B., Ponzi- Bossi, M.G., Rustichelli, F. & Stanic, V. (2002). Improvement in zirconia osseointegration by means of a biological glass coating: an in vitro and in vivo investigation. *Journal of Biomedical Materials Research*, Vol.61, No.2, (October 2001), pp. 282–289, ISSN 1097-4636.

Chan, C., Thompson, I., Robinson, P., Wilson, J. & Hench, L. (2002). Evaluation of Bioglass/dextran composite as a bone graft substitute. *International Journal of Oral and Maxillofacial Surgery*, Vol.31, No.1, pp. 73–77, ISSN 0001-6357.

Cobb, C.M., White, C.L., Gillahan, R.D. & Tira, D.E. (1988). In vitro biocompatibility of air-fired opaque porcelain with human gingival fibroblasts. *Journal of Prosthetic Dentistry*, Vol.59, No.2, pp. 187-194, ISSN 0022-3913.

Covacci, V., Bruzzese, N., Maccauro, G., Andreassi, C., Ricci, G.A., Piconi, C., Marmo, E., Burger, W. & Cittadini, A. (1999). In vitro evaluation of the mutagenic and carcinogenic power of high purity zirconia ceramic. *Biomaterials,* Vol.20, No.4, (August 1998), pp. 371-876, ISSN 0142-9612.

De Aza, P., Luklinska, Z., Santos, C., Guitian, F. & De Aza, S. (2003). Mechanism of bone-like formation on a bioactive implant in vivo. *Biomaterials,* Vol.24, No.8, (October 2002), pp. 1437-1445, ISSN 0142-9612.

Elias, Z., Poirot, O., Danière, M.C., Terzetti, F. & Binet, S. (2002). Surface reactivity, cytotoxicity and transforming potency of iron-covered compared to untreated refractory ceramic fibers. *Journal of Toxicology and Environmental Health. Part A,* Vol.65, No.23, (September 2001), pp. 2007-2027, ISSN 1528-7394.

Elshahawy, W., Ajlouni, R., Watanabe, I., James, W. & Abdellatif, H. (2010). Clinical ion release from type IV gold and CAD-CAM fabricated ceramic crowns, proceedings of 88th International Association of Dental Research, Barcelona, Spain, July 14-17, 2010.

Elshahawy, W., Watanabe, I. & Koike, M. (2009a). Elemental ion release from four different fixed prosthodontic materials. *Dental Materials,* Vol.25, No.8, (February 2009), pp. 976-981, ISSN 0109-5641.

Elshahawy, W., Watanabe, I. & Kramer, P. (2009b). In vitro cytotoxicity evaluation of elemental ions released from different prosthodontic materials. *Dental Materials,* Vol.25, No.12, (July 2009), pp. 1551-1555, ISSN 0109-5641.

Gatti, A. & Knowles, J. (2002). Biocompatibility and biological tests, In: *Integrated Biomaterials Science,* R. Barbucci, 793-813, Kluwer Academic/Plenum Publishers, ISBN 0-306-46678-3, New York, USA.

Josset, Y., Oum'Hamed, Z., Zarrinpour, A., Lorenzato, M., Adent, J.J & Laurent-Maquin, D. (1999). In vitro reactions of human osteoblasts in culture with zirconia and alumina ceramics. *Journal of Biomedical Materials Research,* Vol.47, No.4, (April 1999), pp. 481-493, ISSN 1097-4636.

Lacefield, W.R. (1999). Materials characteristics of uncoated/ceramic coated implant materials. *Journal of Advanced Dental Research,* Vol.13, pp. 21–26, ISSN 2229-4112.

Leonard, A., Hantson, P. & Gerber, G.B. (1995). Mutagenicity, carcinogenicity and teratogenicity of lithium compounds. *Mutation Research/DNA Repair,* Vol.339, No.3, pp. 131-137, ISSN 0921-8777.

Limberger, F. & Lenz, E. (1991). Biological tests on In-Ceram ceramics compared with cobalt alloys and the metals titanium, tantal, and niobium in animal experimentation. *Deutsche Stomatologie,* Vol.41, No.11, pp. 407–410, ISSN 00120790.

Mackert, J.R. (1992). Side-effects of dental ceramics. *Advances in Dental Research,* Vol.6, No. 9, pp. 90–93, ISSN 1685-7667.

Messer, R.L.W., Lockwood, P.E., Wataha, J.C., Lewis, J.B., Norris, S. & Bouillaguet, S. (2003). In vitro cytotoxicity of traditional versus contemporary dental ceramics. *Journal of Prosthetic Dentistry,* Vol.90, No.5, pp. 452-458, ISSN 0022-3913.

Milleding, P., Haraldsson, C. & Karlsson, S. (2002). Ion leaching from dental ceramics during static in vitro corrosion testing. *Journal of Biomedical Materials Research,* Vol.61, No.4, (October 2001), pp. 541-550, ISSN 1097-4636.

Mortelmans, K. & Zeiger, E. (2000). The Ames Salmonella/microsome mutagenicity assay. *Mutation Research/DNA Repair*, Vol.455, No.1-2, pp. 29-60, ISSN 0921-8777.

Nascimento, C., Issa, J.P.M., Oliviera, R.R., Iyomasa, M.M., Siéssere, S. & Regalo, S.C.H. (2007). Biomaterials applied to the bone healing process. *International Journal of Morphology*, Vol.25, No.4, (August 2007), pp. 839-846, ISSN 0717-9367.

Neo, M., Nakamura, T., Ohtsuki, C., Kasai, R., Kokubo, T. & Yamamuro, T. (1994). Ultrastructural study of the A-W GC-bone interface after long-term implantation in rat and human bone. *Journal of Biomedical Materials Research*, Vol.28, No.3, (October 1993), pp. 365-372, ISSN 1097-4636.

Neo, M., Nakamura, T., Ohtsuki, C., Kokubo, T. & Yamamuro, T. (1993). Apatite formation on three kinds of bioactive material at an early stage in vivo: a comparative study by transmission electron microscopy. *Journal of Biomedical Materials Research*, Vol.27, No.8, (February 1993), pp. 999-1006, ISSN 1097-4636.

Noushad, M., Kannan, T.P., Husein, A., Abdullah, H. & Ismail, A.R. (2009). Genotoxicity evaluation of locally produced dental porcelain - An in vitro study using the Ames and Comet assays. *Toxicology in Vitro*, Vol.23, No.6, (May 2009), pp. 1145-1150, ISSN 0887-2333.

Ostling, O. & Johanson, K.J. (1984). Microelectrophoretic study of radiation induced DNA damage in individual mammalian cells. *Biochemical and Biophysical Research Communications*, Vol.123, No.1, pp. 291-298, ISSN 0006-291X.

Pallesen, U., van Dijken, J.W.V. (2000). An 8-year evaluation of sintered ceramic and glass ceramic inlays processed by the Cerec CAD/CAM system. *European Journal of Oral Science*, Vol.108, No.3, (February 2000), pp. 239–246, ISSN 0909-8836.

Pera, P., Conserva, E., Pin, D., Acquaviva, A., Riboldi, A., Mariottini, G.L. & Pane, L. (2005). Cytotoxicity in vitro analysis of ceramic materials for metal free prosthetic substructures. *Minerva Stomatologica*, Vol.54, No.6, pp. 363-371, ISSN 0026-4970.

Person-Sjögren, S. & Sjögren, G. (2002). Effects of dental materials on insulin release from isolated islets of Langerhans. *Dental Materials*, Vol.18, No.1, (January 2001), pp. 20-25, ISSN 0109-5641.

Piatelli, A., Podda, G. & Scarano, A. (1996). Histological evaluation of bone reactions to aluminum oxide dental implants in man: a case report. *Biomaterials*, Vol.17, No.7, (April 1995), pp. 711–714, ISSN 0142-9612.

Piconi, C. & Macauro, G. (1999). Zirconia as a ceramic biomaterial. *Biomaterials*, Vol.20, No.1, (December 1997), pp. 1–25, ISSN 0142-9612.

Sarkar, R. & Banerjee, G. (2010). Ceramic based bio-medical implants. *International Ceramic Review*, Vol.59, No.2, (January 2010), pp. 98-102, ISSN 0020-5214.

Schepers, E., de Clercq, M., Ducheyne, P. & Kempeneers, R. (1991). Bioactive glass particulate material as a filler for bone lesions. *Journal of Oral Rehabilitation*, Vol.18, No.5, pp. 439-452, ISSN 0305-182X.

Schmalz, G. & Garhammer, P. (2002). Biological interactions of dental cast alloys with oral tissues. *Dental Materials*, Vol.18, No.5, (February 2001), pp. 396-406, ISSN 0109-5641.

Schmalz, G. & Netuschil, L. (1985). A modification of the cell culture agar diffusion test using fluoresceindiacetate staining. *Journal of Biomedical Materials Research*, Vol.19, No.6, (January 1985), pp. 653-661, ISSN 1097-4636.

Schmalz, G. & Schmalz, C. (1981). Toxicity tests on dental filling materials. *International Dental Journal*, Vol.31, No.3, ( ), pp. 185–192, ISSN 0020-6539.

Shackelford, J.F. (2005). *Bioceramics - Advanced ceramics series* (vol.1), Gordon & Breach Science Publishers, ISBN *90-5699-612-6*, New York, USA.

Simon, A., Gouget, B., Mayne, M., Herlin, N., Reynaud, C., Degrouard, J. & Carriere, M. (2007). In vitro investigation of TiO_2, Al_2O_3, Au nanoparticles and multi-walled carbon nanotubes cyto- and genotoxicity on lung, kidney cells and hepatocytes. *Toxicology Letters*, Vol.172, No.1, pp. 172S, S36, ISSN .

Singh, N.P., McCoy, M.T., Tice, R.R. & Schneider, E.L. (1988). A simple technique for quantitation of low levels of DNA damage in individual cells. *Experimental Cell Research*, Vol.175, No.1, pp. 184-191, ISSN 0014-4827.

Sjögren, G., Sletten, G. & Dahl, J.E. (2000). Cytotoxicity of dental alloys, metals and ceramics assessed by Millipore filter, agar overlay and MTT test. *Journal of Prosthetic Dentistry*, Vol.84, No.2, pp. 229-236, ISSN 0022-3913.

Studer, S., Lehner, C., Brodbeck, U., Schekrer, P. (1996). Short-term results of IPS-Empress inlays and onlays. *Journal of Prosthodontics*, Vol.5, No.4, pp. 277–287, ISSN 1059-941X.

Sun, Y., Xiao, J.D., Xiong, J.Y. & Liu, J.Q. (2009). Clinical study of nanometer calcium phosphate ceramic artificial bone. *Zhongguo Gu Shang*, Vol.22, No.11, pp. 819-821.

Takami, Y., Nakazawa, T., Makinouchi, K., Glueck, J. & Nose, Y. (1997). Biocompatibility of alumina ceramic and polyethylene as materials for pivot bearings of a centrifugal blood pump. *Journal of Biomedical Materials Research*, Vol.36, No.3, (August 1996), pp. 381-386, ISSN 1097-4636.

Uo, M., Sjören, G., Sundh, A., Watari, F., Bergman, M. & Lerner, U. (2003). Cytotoxicity and bonding property of dental ceramics. *Dental Materials*, Vol.19, No.6, (July 2002), pp. 487-492, ISSN 0109-5641.

Veronese, I., Guzzi, G., Giussani, A., Cantone, M.C. & Ripamonti, D. (2006). Determination of dose rates from natural radionuclides in dental materials. *Journal of Environmental Radioactivity*, Vol.91, No.1-2, (August 2006), pp. 15–26, ISSN 0265-931X.

Wataha, J.C. (2001). Principles of biocompatibility for dental practitioners. *Journal of Prosthetic Dentistry*, Vol.86, No.2, pp. 203-209, ISSN 0022-3913.

Weiner, M.L., Batt, K.J., Putman, D.L., Curren, R.D. & Yang, L.L. (1990). Genotoxicity evaluation of lithium hypochlorite. *Toxicology*, Vol.65, No.1-2, pp. 167-175, ISSN 0300-483X.

Williams, D.F. (2008). On the mechanisms of biocompatibility. *Biomaterials*, Vol.29, No.20, (April 2008), pp. 2941-2953, ISSN 0142-9612 .

Yamamoto, A., Honma, R., Sumita, M. & Hanawa, T. (2004). Cytotoxicity evaluation of ceramic particles of different sizes and shapes. Journal of Biomedical Materials Research, Vol. 68A, No.2, (June 2003), pp. 244-256, ISSN .

Zhong, B.Z., Whong, W.Z. & Ong, T. (1997). Detection of mineral-dust-induced DNA damage in two mammalian cell lines using the alkaline single cell gel/Comet assay. *Mutation Research*, Vol.393, No.3, pp. 181-187, ISSN 0921-8777.

16

The Role of Aluminium Ceramics in Total Hip Arthroplasty

Wedemeyer Christian and Kauther Max Daniel
Department of Orthopaedic Surgery, Department of Trauma Surgery, University of Duisburg-Essen Germany

1. Introduction

The investigation into the reaction of tissues on organic and inorganic particles is the topic of a multitude of medical disciplines. In the area of hygienics, industrial medicine and pulmonology for instance, the effects of fine dust particles in the lungs are examined (Wottrich et al. 2004). In the field of oncology, now and then the possible influence of nano foreign-body particles on carcinogenesis is being investigated (Jacobs et al. 1998). From the viewpoint of the immunologist, the different attitudes of cells and variable receptor activities during the phagocytosis of foreign bodies are of interest (Sun et al. 2003).
In the region of surgical disciplines like orthopaedics, trauma surgery or maxillary surgery, particle studies that investigate the influence of the abrasion of different prosthetic materials on the peri-prosthetic tissue have been undertaken in order to get evidence on their respective biocompatibility. In orthopaedics, this question is of importance mainly in the field of the total endoprosthetic hip replacement. Hip joint prostheses rank as most frequently performed implantations of biomaterials in orthopaedics and are particularly affected with regard to the formation of wear debris and loosening because of their exceptional mechanical use.

2. The development of the artificial hip joint replacement

At about 1.3 million implantations of biomaterials worldwide per year, the artificial replacement of the hip joint counts among the most frequent operations in the frame of orthopaedic surgery (Harris et al. 1995). In Europe, 500,000 hip joint prostheses are implanted annually. In Germany, there are about 200,000 every year including up to ten percent replacement of an already existing prosthesis (Özkir et al. 2007). In the USA, 600,000 total hip arthroplasties (THA) are implanted yearly at present. Up to the year of 2030 the annual requirements of primary THA will probably rise from 209,000 to 572,000, an increase by 174%. The future need by patients younger than 65 years will presumably come to 52% of the primary THA. The annual hospital expenses for primary THA will rise to 17.4 billion US$ up to 2015. According to estimations the number of THA revisions will increase from 40,800 (year 2005) to 96,700 (2030) in United States. The expected rise of THA revisions will be at 137% (Kurtz et al. 2007).
The idea of an artificial hip joint can be traced back to design drawings made by Gluck at the end of the 19th century. The first translation of this idea into action took place by Hey-

Groves in 1922, who used an implant from ivory as a femur head prosthesis. The first metallic total hip arthroplasty - consisting of an acetabulum and a femur head fastened to the neck of the femur by a bolt - was implanted by Wiles in 1938.
Today, the fixation occurs in the femur by an intra-medullary metallic stem. The first operation that was performed this way was done by Moore in 1940.
The results of the hip joint replacement, at first scanty concerning durability and load capacity, could be improved in 1959 by the introduction of bone cement (polymethyl methacrylate, PMMA) by Sir John Charnley (Herren et al. 1987).
Today, the most frequently used materials in hip arthroplasty are titanium alloys (e.g. titanium-aluminium-vanadium, TiAlV) or cobalt-chromium-compounds, polyethylene (e.g. ultra-high-molecular-weight-polyethylene, UHMWPE), and aluminium ceramics (Al_2O_3).

2.1 The significance of aluminium ceramics in hip arthroplasty

Bio-ceramics are an alternative to metal alloys for total hip replacement. Alumina is the most widely used oxide ceramic material for orthopaedic implants. Alumina ceramics are manufactured from powdered aluminium oxide which is mixed with an organic binder and pressed into a mold. The material is then dried to evaporate the water and calcinated to burn the organic binder. The quality and purity of the powdered raw material and the temperature of calcination determine the microstructure of ceramics (Hannoche et al. 2005).
Ceramics became of great interest in hip arthroplasty (Manley et al. 2008) as they promised a solution to the problem of aseptic loosening. Clinical and experimental studies showed that pairings of alumina ceramics in comparison to the widely used polyethylene/metal pairings lead to less debris. Wear particles can stimulate osteoclasts around the implant resulting in particle-induced osteolysis - the major cause of long time failure in hip arthroplasty (Sunfeldt et al. 2006).
Many authors concluded that the number of revision surgeries of ceramic/ ceramic pairings due to aseptic loosening is lower compared to polyethylene/ metal pairings (Clarke et al. 1994). Until 2001, more than 3 million ceramic prosthetic heads and more than 300,000 ceramic acetabular cups were successfully implanted worldwide (Thompsen et al. 2002). The combination of a ceramic head combined with a polyethylene cup reduces the particle wear by half compared to a metal/polyethylene pairing. (Hannoche et al. 2005) The combination of a ceramic head and a ceramic cup can further reduce the particle wear by ten times (Hannoche et al. 2005). Nowadays, the most widely used combination in total hip arthroplasty is that of a ceramic head with a polyethylene cup (Rössler and Rüther 2005).
The idea to use ceramic materials for hip replacement comes from the French surgeon Pierre Boutin. In 1970, he implanted a cemented ceramic cup combined with a ceramic head for the first time. In 1974, Peter Griss and Heinz Mittelmeier introduced the regular use of ceramic heads and cups in Germany. Meanwhile, most ceramic materials are based on alumina (Al_2O_3) (Hannoche et al. 2005). An advantage of using ceramics in hip joint replacement is that laboratory experiments showed ceramics to be bio-inert. The material does not corrode, it is well tolerated by the surrounding tissue and produces significantly less wear than current alternative materials of ultra-high molecular weighted polyethylene (UHMWPE) and titanium aluminium vanadium TiAlV (Manley et al. 2008, Warashina et al. 2003).
Ceramics possess advantageous tribological characteristics (Bierbaum et al. 2002). In-vivo experiments showed that the ceramic surface forms a lubricating protein-rich film after implantation that works as sliding layer (Christel 1992).

In the 70ties, so-called monolithic or bloc cups entirely made of ceramics were used at the beginning. They demonstrated good bio-inert characteristics and a high corrosion resistance but a poor osteo-integration. A fibrous tissue was frequently found between implant and bone leading to migration and loosening of the prosthesis due to change of position of the components and resulting in increased abrasion (Willmann1998). Therefore, the results in the early era of ceramic hip implants have been poor. Many studies from this period reported revision surgeries in 30 % of the patients after 5 years compared to significantly better results of metal / polyethylene combinations (Mahoney et al. 1990). Furthermore, the initial poor performance of ceramic hip joints is attributed to the mechanical characteristics of ceramics. The brittle, inelastic material can hardly compensate forces by deformation. Therefore, fractures of ceramics, mostly found in the femoral part, were more frequent than in prostheses out of polyethylene or titanium alloys (Manley et al. 2008). Ceramic implants demand an experienced surgeon as the positioning of the implant e.g. inclination, anteversion, offset and neck length, is of great importance for the long-term results. The correct positioning of cup and stem reduces asymmetric strain on the implant, avoids an impingement and reduces the material abrasion and fracture. Hamadouche et al. published excellent long-term results in correctly positioned ceramic prosthesis (Hamadouche et al 2002). Savarino et al. recently published a study showing that trauma, infection, mechanical instability and incorrect positioning mostly lead to implant failure of alumina ceramic on alumina ceramic hip prosthesis. They pointed out that particle-induced osteolysis was no longer the number one reason of long-term failure (Savarino et al. 2009).

Another reason for the improvement of durability of alumina ceramics was a constant development of the material. First generation alumina ceramic materials consisted of large (7.2 micron particle size), less strong (400 MPa) and less dense (3, 94 g / cm ^{3}) crystals and showed impurities of up to 5 percent by volume. Currently, third generation alumina ceramics consist of crystals with a much finer grain size (1.8 microns). They have a higher density (3.98 g / cm ^{3}), higher stability (580 MPa) and impurities of less than 0.5 percent by volume. (Hannoche et al. 2005)

In addition to the improving surgical technique and the higher quality of the raw material, the geometric shape of the femoral head and acetabular cup were optimized leading to a significant increase in the longevity of ceramic prosthesis (Thompsen et al. 2002). Clinical outcome studies confirmed that a total hip prosthesis made of alumina ceramic has good durability and low particle wear. Rousseau et al. (2004) reported of an average survival rate of 62.8% after 11 years of cemented alumina ceramic on alumina ceramic hip replacements. Lusty et al. (2007) found a survival rate of 99% after a mean of 7 years in cementless alumina-on-alumina ceramic hip replacements.

2.2 The aseptic loosening of the hip prosthesis

While in the early times of hip arthroplasty secondary complications were most frequently caused by material fatigue, inaccurate implantation techniques, and infections, today - following improvements in the methods of operation, sterilisation, and material development - the aseptic loosening of the prosthesis is the most common reason for orthopaedic revision operations (Sundfeldt et al. 2006).

During the aseptic loosening of a prosthesis, there is a progressive periprosthetic osteolysis in the course of time leading to a growing instability and increasing wear triggered by the advancing mobility of the prosthesis. In the long term, this results in the loosening and

dislocation of the prosthesis. A reason for the limited durability of the endoprosthesis is therefore caused by the composite of the implant and the surrounding bone tissue. In addition, the aspect of material wear plays a role. For the patient the loosening of the prosthesis appears in the form of pain following exercise or on movement (Rössler and Rüther 2005). As things develop revision operations are often inevitable. They weigh heavily upon the patients and are susceptible to risks because of peri- and postoperative complications (e.g. embolism or infections). Moreover, the secondary implant requires a greater intramedullary volume at each replacement of a hip prosthesis. Thus, the durability of the replacement prosthesis is significantly reduced in comparison to the first implant and the number of possible replacements is limited (Callaghan et al. 1985, Hanssen et al. 1988).
The real cause for the formation of periprosthetic osteolysis followed by aseptic loosening is still disputed. There is a multitude of theories:

2.2.1 Bone resorption induced by wear debris

Many authors in the contemporary technical literature assume that the wear debris generated by the abrasion of the prosthesis is phagocyted by cells of the surrounding tissue, thereafter activating these cells and giving rise to an aseptic reaction by release of different messenger substances (e.g. cytokines like interleukins or TNF-α), eventually causing the osteolysis of the surrounding bone. A significant role in this connection is assigned to the macrophages activating already present osteoclasts in the described manner or stimulating precursor cells to become osteoclasts or even further differentiating themselves into osteoclasts (Fujikawa et al. 2005). Furthermore, studies are mentioned in literature stating that other tissue cells, e.g. osteoblasts, are also able to phagocyte wear particles and to react to them (Lohmann et al. 2000). The cellular response caused by the wear debris and the connected bone resorption are dependent on different physical characteristics such as the material the particles are made from as well as their size, amount and morphology.
Schmalzried et al. (1992) consider that the aseptic loosening of the femur shaft is mainly due to mechanical reasons, while merely the loosening of the acetabulum, mostly consisting of polyethylene, is caused in the manner described above by bone resorption induced by wear debris.

2.2.2 The hydrostatic pressure

Schmalzried et al. (1992) and Aspenberg et al. (1998) regard an increased intracapsular pressure as an important factor in the origin of aseptic osteolysis. According to their theory the intra-articular cartilage in normal joints prevents the contact between bone and synovial fluid. As part of a disease or a trauma, a damage of the cartilage and consequently a loss of its osteoprotective properties is however possible. There will be an inflammation of the joint with the development of more intra-articular liquid. The result is an increased intracapsular pressure which reduces the blood circulation to the bone and leads to the ischaemic death of osteocytes consequentially leading to increased bone resorption. This theory is supported by studies of Robertsson et al. (1997), who discovered an increased intracapsular pressure in 18 implanted hip joints shortly before a revision operation because of aseptic loosening, compared to 34 clinically and radiologically stable hip joint implants. In addition, Aspenberg and van der Vis (1998) were able to show in a rabbit experiment that an oscillating intracapsular liquid pressure results in osteolysis and the death of osteocytes.

2.2.3 Endotoxic contamination of implants and particles

Hitchins and Merritt (1999) consider bacterial cell components attached to the hip joint implants (lipopolysaccharides, endotoxins) to be responsible for osteolysis. This theory is also supported by Akisue et al. (2002), who incubated macrophage-like THP-1 cells with titanium particles and endotoxins and observed almost no effect of endotoxin-free titanium particles on the expression of cytokines in their experiment.

2.2.4 Genetic predisposition

Matthews et al. (2000) pursued observations where certain patients showed no failure of their implant in spite of massive polyethylene abrasion, while in other patients with only small wear amount the implants failed with a distinct osteolysis. In a cell culture study, where macrophages from three different donors had been incubated by endotoxin-free polyethylene particles, they discovered a high variability of the macrophage cytokine expression induced by particle contact dependent on the donors. A similar discovery was made by Hatton et al. (2003), who incubated peripheral blood monocytes from six different donors with particles of aluminium ceramics and observed significant differences between the donor cells in the release of TNF-α. Genetic investigations on the correlation between genetic nucleotide polymorphisms and aseptic hip prosthesis loosening are currently under way (Wedemeyer et al. 2009, Bachmann et al. 2008).

2.3 In vivo and in vitro studies on wear particles

Until now, a lot of studies to clarify the morphological characteristics of wear particles have been carried out. It turned out that the peculiarity of the wear debris and hence the amount of wear particles in the periprosthetic tissue was dependent on the used material of the prosthesis.

Different methods have been used for the quantitative investigation of wear debris. These include calculations in the frame of radiographic studies and recently also calculation models with the help of prosthetic simulators. The latter have also been used to determine morphology and amount of the wear particles arising over time. At the same time there were clinical studies analysing the intra-operatively recovered particles afterwards, e.g. during a revision operation or the replacement of a loosened prosthesis.

The above mentioned investigation methods showed the highest rate of wear debris (6 to 140 mm^3 per year) for polyethylene in comparison to other materials (Livermore et al. 1990). On the other hand, for prostheses made from ceramics comparatively little wear debris is described. Lusty et al. (2007) mentioned a mean abrasion of 0.2 mm^3 per year. You et al. (2005) were not able to ascertain any wear debris in their radiological study, whereby in this case the assessment took place five years after the implantation of the prosthesis.

Concerning the size of the particles found in vivo, there is a dependence on the used prosthesis material, too, but like in the abrasion rate there are differences between authors also concerning the particles' size. Maloney et al. (1995) generally observed a mean particle size of just below 1μm for non-cemented polyethylene and metallic prostheses and an absolute number of 1.7 billion particles per gram tissue. Studies by Lee et al. (1992) on tissue recovered from cemented prostheses during revision operations yielded a similar particle size for metallic particles (TiAlV, CoCr, steel) between 0.3 and 1.8 μm and for polyethylene particles a magnitude between 2 and 13 μm.

During electron microscopic analysis of wear particles from Mittelmeier-prostheses (ceramics on ceramics) Yoon et al. (1998) found sizes between 0.13 and 7.20 µm (mean 0.7 µm). Hatton et al. (2003) on the other hand identified particles from ceramics in two magnitudes by a laser-based micro-dissection method (likewise on failing Mittelmeier-prostheses), one group with a mean diameter of 0.503 µm, while the other group showed a diameter on the nanometre scale (24 ± 19 nm).

The signalling pathways that have been investigated until now in cell-particle co-culture systems mainly contain cytokines (interleukins, TNF-α), prostaglandins and recently the RANK(L)/OPG-system, too, which so far, however, has only been tested on metallic and polyethylene particles (Baumann et al. 2004) and not as yet on aluminium ceramics. On this occasion, an increased expression resp. release of the corresponding messenger substances was connected with growing particle concentration and also in these cell culture studies polyethylene and different metallic particles induced a more distinct cellular response than particles from aluminium ceramics (Sterner et al. 2004). Concerning the influence of the particle size, however, different effects have been observed. For instance Sterner et al. (2004) described a proportional correlation between the size of aluminium ceramics particles and the release of TNF-α in macrophage-like cells, while Yagil-Kelmer et al. (2004) as the only authors so far concluded from their studies that smaller ceramics particles in single individuals might have a more significant biological answer than particles of the same material but with greater diameter.

Catelas et al. (1999) investigated the impact of aluminium ceramics particles on cell death mechanisms (apoptosis versus necrosis) in macrophage cultures and reported on a volume effect where cells showed apoptotic changes dependent from a combination of size and concentration of the deployed particles.

In a co-culture with a murine macrophage cell line, Petit et al. (2006) examined the influence of bisphosphonates on the release of TNF-α and the induction of apoptosis. It turned out that only the TNF-α-release induced by polyethylene (UHMWPE) particles could be inhibited, not the one induced by particles from aluminium ceramics and that bisphosphonates lead to an induction of apoptosis in the cells.

2.4 Monocytes and macrophages

Monocytes are mononuclear phagocytes circulating in the blood. They represent precursors of macrophages. In particular at inflammatory processes in the neighbouring tissue, monocytes are caused by chemotaxis to enter the tissue concerned where they differentiate to macrophages under the prevalent humoral stimuli and are able to survive there for a period from a few days up to several months. During this time the macrophages are performing most diverse functions: By secretion of messenger substances (cytokines, leukotrienes, prostaglandins, proteases, complement factors, etc.) for instance, these cells initiate inflammatory reactions as well as conversion and repair processes in the surrounding tissue. They phagocytose very different organic and inorganic materials, both foreign-bodies and own tissue, whereby the phagocytosis is being facilitated by previous opsonization of the materials being phagocyted by antibodies or complement proteins. The macrophages participate in immunological processes (immune defence) by antigen presentation at lymphocytes (Schiebler et al. 2002).

Within the framework of the particle-induced periprosthetic osteolysis the macrophages have a special significance in double regard: on the one hand, especially these cells react to the particles by introduction of a pro-inflammatory signalling cascade, which eventually leads to an increased activation and recruitment of osteoclasts. On the other hand, the wear debris is able to inhibit the effect of osteoprotective mechanisms, e.g. the release of the osteoprotective interferon-gamma, whereby a differentiation of macrophages into osteoclasts is being facilitated (Purdue et al. 2006).

2.4.1 The THP-1 cell line and cell-particle co-culture systems

Tsuchiya et al. (1982) established this cell line and were able to verify their numerous monocytic properties, especially the potentiality for phagocytosis of foreign-body particles. Furthermore, the authors demonstrated that a treatment of this cell line by the phorbol ester phorbol-12-myristate-13-acetate (PMA), also called 12-O-tetradecanoylphorbol-13-acetate (TPA), results in a further differentiation along the monocytic development line up to macrophage-like cells (Tsuchiya et al. 1982). This happens via a signalling pathway mediated by protein kinase C, where PMA functions as an analogue for diacylglycerole, the physiological activator of protein kinase C. Schwende et al. (1996) observed a significant amplification of characteristics typical for macrophages if THP-1 cells were stimulated by PMA in a concentration of 10 nmol over 72 hours: This includes an increased expression of the surface proteins CD14 and CD11b typical for monocytes and especially macrophages and an intensified potential for the secretion of oxygen radicals. Additionally, they concluded that the PMA treatment induces an increase of TNF-α-cytokine release from the cells, inducible by lipopolysaccharides. Further observations showed cells stimulated by PMA are able to phagocytose about ten times more latex particles than unstimulated cells, and the morphological development to macrophages reached by the PMA treatment is quasi accompanied by a termination of the very high cell division rate known from the untreated cells. Beside different humane and murine cell lines, fresh peripheral blood monocytes (PMBC) extracted from donors have often been used (Yagil-Kelmer et al. 2004). There are only occasional reports about PMA stimulated THP-1 cells used for the analysis of osteolysis inducing properties of biomaterials. More often particle studies have been carried out with THP-1 cells using other substances than PMA for the stimulation to macrophage-like cells. A research group from Würzburg, Germany, used THP-1 cells which had been pre-treated by a granulocyte macrophage colony-stimulating-factor (GM-CSF) and vitamin D_3 (Baumann et al. 2004).

2.5 TNF-α und das RANK(L)/OPG- system

2.5.1 Osteoprotegerin (OPG)

The identification and characterisation of the RANK(L)/OPG system began in 1997 by discovery of Osteoprotegerin (OPG). It just happened by accident at the AMGEN company where cDNA from rat intestinal tissue had been characterised in a study. The researchers were looking for TNF receptor-like molecules, hoping for their therapeutic benefit. Certain transgenic mice which overexpressed the gene for OPG attracted attention because of a distinct osteopetrosis (Simonet et al. 1997). It was discovered that this was due to a decrease of the number of osteoclasts in the mouse skeleton and it was concluded that OPG plays a key role in the regulation of the osteoclastogenesis.

Apart from that, a Japanese research group discovered osteoprotegerin almost at the same time. Here, they were searching specifically for an undiscovered factor produced by

osteoblasts and stromal cells which they awarded regulatory features concerning the osteoclastogenesis and hence the bone resorption (Yasuda et al. 1998).
In detailed analyses of OPG, it turned out that OPG is a peptide comprising 380 amino acids. OPG is secreted as a 401 amino acids-long peptide where a propeptide of 21 amino acids is being split off (Simonet et al. 1997, Yasuda et al. 1997). It represents the only protein of the TNF receptor super-family which does not possess transmembrane and cytoplasmatic domains and which is secreted as a soluble protein. Using OPG-knockout mice, it was found that OPG apparently plays an important role in other organ systems, too. This is because a suppression of the OPG in the mice not only culminated in strong osteoporosis but also in changes in the vascular system, e.g. calcifications of the large arteries, proliferations of the intima and media and aortic dissections. Meanwhile, it is known that OPG-mRNA is being expressed in almost all organs (Simonet et al.1997, Yasuda et al. 1997).

2.5.2 Receptor Activator of Nuclear Factor kappa B Ligand (RANKL)

Soon after the discovery of OPG it became apparent that a protein already known concerning its ability to stimulate dendritic cells, the RANKL, works as a ligand of OPG. For this reason it is also called osteoprotegerin ligand (OPGL). In the presence of a low M-CSF level, RANKL is necessary and sufficient for the complete differentiation of osteoclast precursor cells to mature osteoclasts (Lacey et al. 1998).
Again during knockout mice experiments it was demonstrated that the elimination of the RANKL gene entailed a strong osteopetrosis and a total missing of osteoclasts (Kong et al. 1999). RANKL has been identified as a peptide with a length of 317 amino acids. It exists in a membrane bound and also in a soluble mode. The latter emerges through separation of a peptide part at position 140 or 145 (Lacey et Al: 1998).

2.5.3 Receptor Activator of Nuclear Factor kappa B (RANK)

By the discovery of RANKL as a potential ligand for OPG, it was concluded that also RANK, an already known receptor for RANKL, should play a crucial role in osteoclastogenesis and bone resorption.
The absence of RANK in knockout mice induces a distinct osteopetrosis, attributable to the complete missing of osteoclasts (Kong et al. 1999). Detailed analyses of RANK showed it to be a peptide with a length of 616 amino acids. It is composed of a signal peptide with 28 amino acids, an N-terminal extracellular domain, a short transmembrane domain of 21 amino acids and a large C-terminal cytoplasmatic domain. It is mainly expressed on monocytic cells and macrophages and also on pre-osteoclasts, B- and T-cells, dendritic cells and fibroblasts (Anderson et al. 1997).
The signal transduction running via RANKL and RANK eventually ends with the regulation of the nuclear transcription factor NF-kB. In its inactive form, NF-kB is bound to different inhibitor proteins in the cytoplasm. In the case of activation, these are degrading and the NF-kB molecule is able to translocate to the nucleus where it controls the transcription of a multitude of genes and, besides the osteoclastogenesis, it also regulates significantly apoptotic, inflammatory and autoimmune modulatory processes (Holt et al. 2007).

2.5.4 Tumor Necrosis Factor alpha (TNF-α)

The cytokine TNF-α is being secreted mainly by activated macrophages and owns a multitude of functions. Through its capability to stimulate osteoclasts, it is a potent mediator of bone

resorption (König et al. 1988) and it also plays an important role in the frame of inflammatory diseases and post-menopausal osteoporosis (Komine et al. 2001). The role of TNF-α in bone resorption has been studied by means of a knockout survey: Merkel et al. (1999) observed a hardly detectable bone resorption after elimination of the TNF-α receptor in mice following subperiostal implantation of bone cement particles. Furthermore, it could be noticed that TNF-α not only shows a synergistic effect to RANKL, but it also initiates a signal cascade via its receptor leading again to the activation of the transcription factor NF-kB.
The stimulation of osteoclasts, however, is not due to TNF-α alone. The presence of M-CSF is needed, too. Both factors together ensure a differentiation of haematopoietic progenitors to dendritic cells and hence an enlargement of the pool of pre-osteoclastic cells (Udagawa et al. 1990).

2.5.5 The RANK(L)/OPG system and TNF-α in the particle-induced osteolysis

Because of the affiliation of RANK and OPG to the TNF receptor superfamily, the following synonyms exist in the current nomenclature: RANK is also called Tumor Necrosis Factor Receptor Superfamily, member 11A. For RANKL there is the synonym Tumor Necrosis Factor Receptor Superfamily, member 11 and OPG is also considered as Tumor Necrosis Factor Receptor Superfamily, member 11B.
Like a multitude of other pro-inflammatory proteins (e.g. interleukin-1), TNF-α is in the position to stimulate the osteoclastogenesis via an activation of NF-kB. However, these cytokines do not belong to the final step of the signal cascade. This is rather composed by the RANK(L)/OPG system (Holt et al. 2007).
Because of the observations made so far, RANK must be the only receptor for RANKL on the cell surface of osteoclast precursors. OPG, which may be bound by RANKL, but with an opposite effect in bone metabolism, hence turned out as a soluble "decoy receptor" catching RANKL in the intercellular space without bone resorption resulting from this binding.
The relation of RANK and RANKL on one hand and OPG on the other is decisive for the way (construction resp. resorption of bone) and the extent of bone metabolism because of the above mentioned observations. The relevance of this relation has been confirmed in the area of the particle-induced aseptic osteolysis. During analyses of periprosthetic tissue surrounding loosened hip joint prostheses, a strong increase of RANK and RANKL expression could be found and in comparison a low expression of OPG (Mandelin et al. 2003). A further conclusion that could be drawn from this study was the fact that an imbalance in favour of RANKL and to the detriment of OPG induces the enhancement of osteoclastogenesis and hence loosening of the prosthesis.

2.6 THP cells and ceramic particles

We recently performed a study to illuminate the effect of alumina ceramic particles with different diameters and concentrations on the mRNA expression of certain key regulators in particle-induced aseptic osteolysis (RANK, RANKL, OPG, and TNF-α) in THP-1 macrophage-like cells (Bylski et al. 2009). Titanium particles were used as a positive control. RNA was analyzed by quantitative RT-PCR. Our results demonstrate that alumina ceramic particles, regardless of particle size, caused only slight up-regulations of RANK, TNF-α, and OPG mRNA, whose levels were significantly lower in comparison to those of titanium particles ($p < 0.05$). The continuous increasing tendency to time-dependent and particle-dependent mRNA-expression of all the parameters stimulated by titanium particles was not found after stimulation with ceramic materials. Even after the concentration of ceramic particles was increased, only a mild up-regulation of mRNA-expression was found. Furthermore, we observed that the bio-inert characteristics of ceramic particles did not change much in diameters ranging from 0.5 to 1.5 μm. At most of the measuring time points, there was no significant difference between the reactions of the large and small particles in this range. Our results support the theory about the relative bio-inert characteristics of alumina ceramic particles.

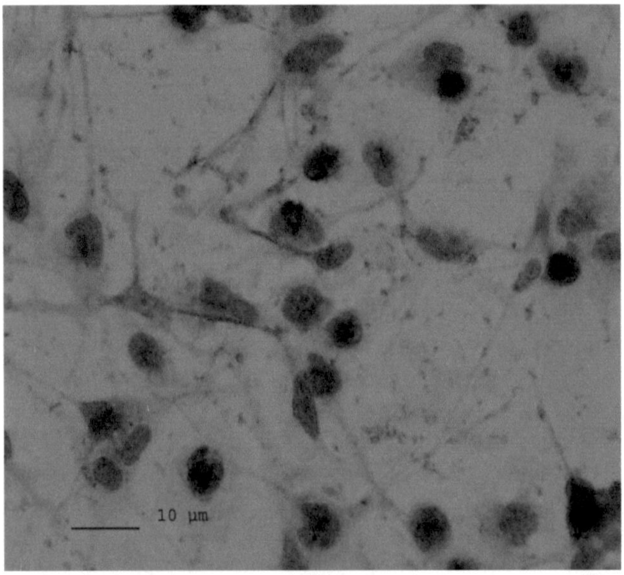

a)

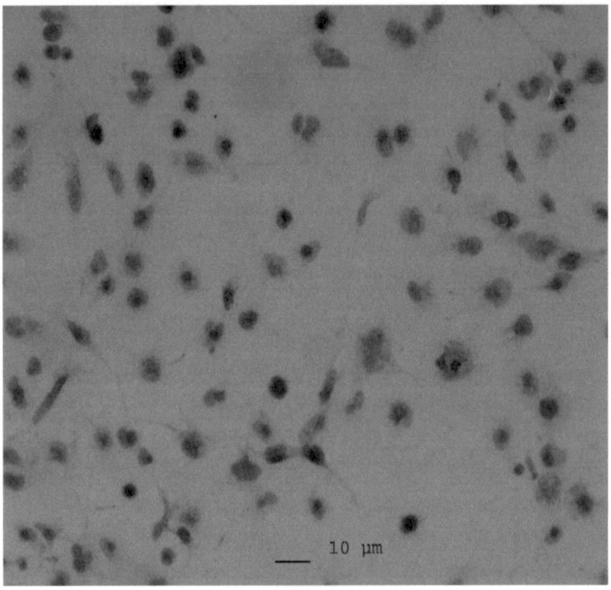

b)

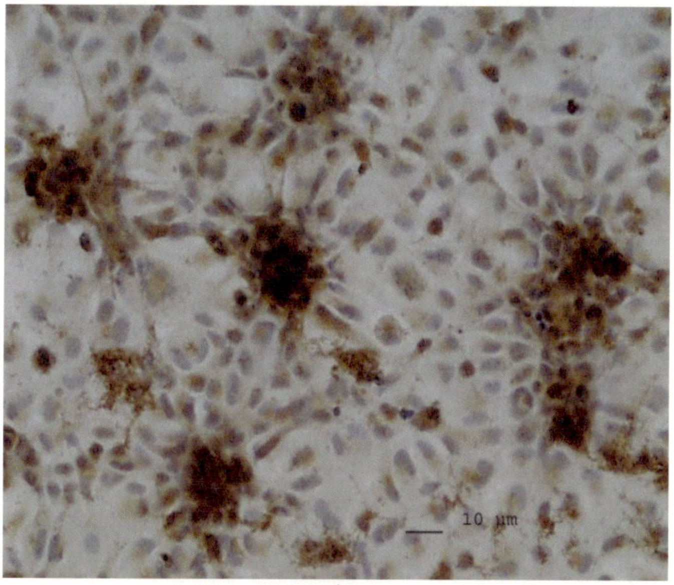

c)

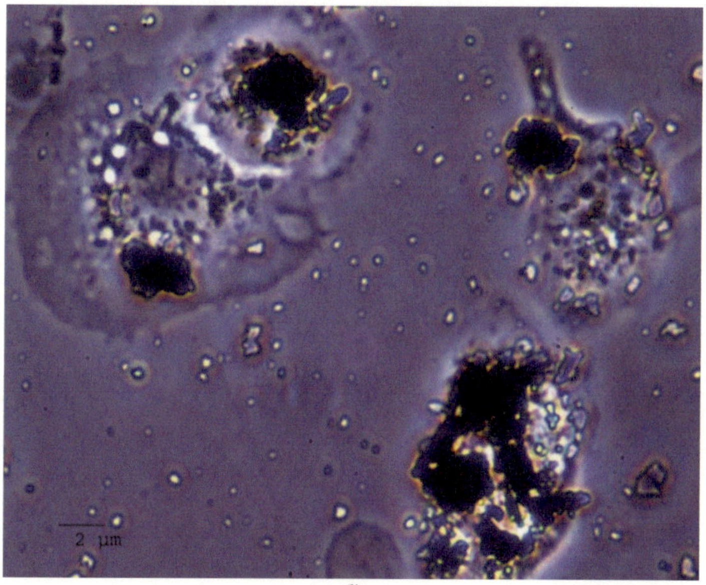

d)

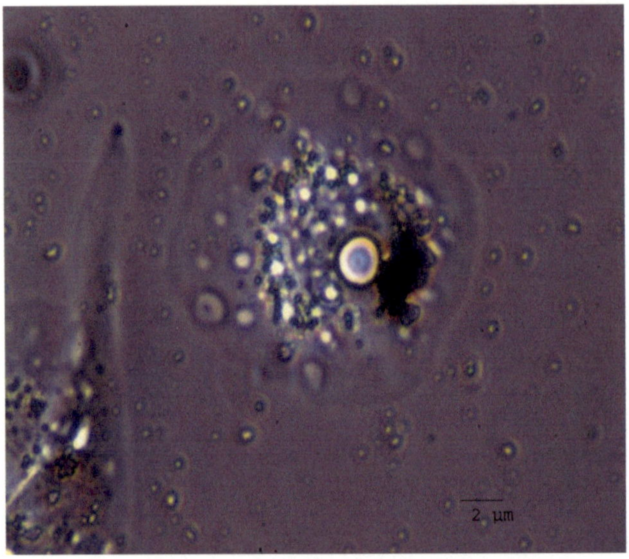

e)

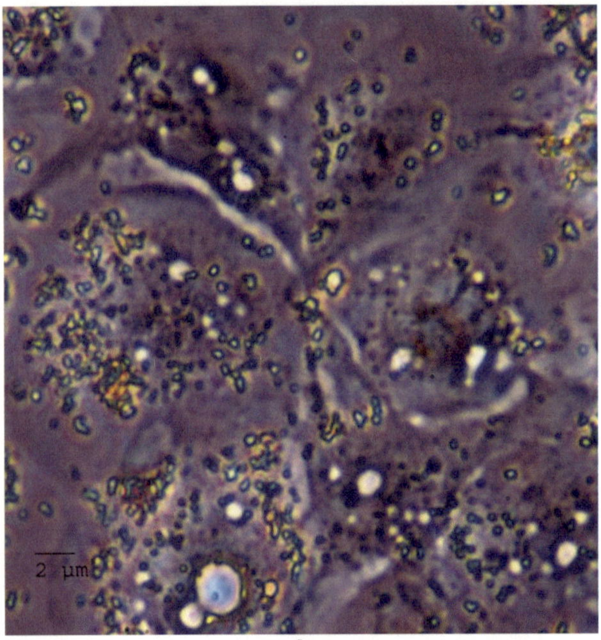

f)

Fig. 1. Light microscopic photographs of PMA-stimulated THP-1 cells after **immunohistochemical staining by CD68 antibodies (a-c) resp. at co-incubation with particles (d-f):** (a): 3 million THP-1 cells treated by 50 nmol PMA for 72 h, magnification 4x. (b): 1 million cells treated by 50 nmol PMA for 48 h, magnification 10x. (c): 5 million cells treated by 50 nmol for 72 h, magnification 10x. Cell aggregations are recognizable because of the high density. (d) + (e): THP-1 cells treated by PMA for 72 h phagocytosing titanium particles (for better overview at the rate of 1:10). After about 1 h already, a particle accumulation in the cytoplasmic area could be recognized by light microscope. Besides the clearly visible large titanium particles (deep black), the smaller particles appear more transparent. Magnification 100x. (f): Ceramic particles of the kind CT1200SG in the cytoplasm of THP-1 cells treated in the same way. In comparison to the titanium particles, the smaller size of the ceramic particles is distinct. Under the light microscope, they are only distinguishable from the roundish intracellular vacuoles by their more oval shape.

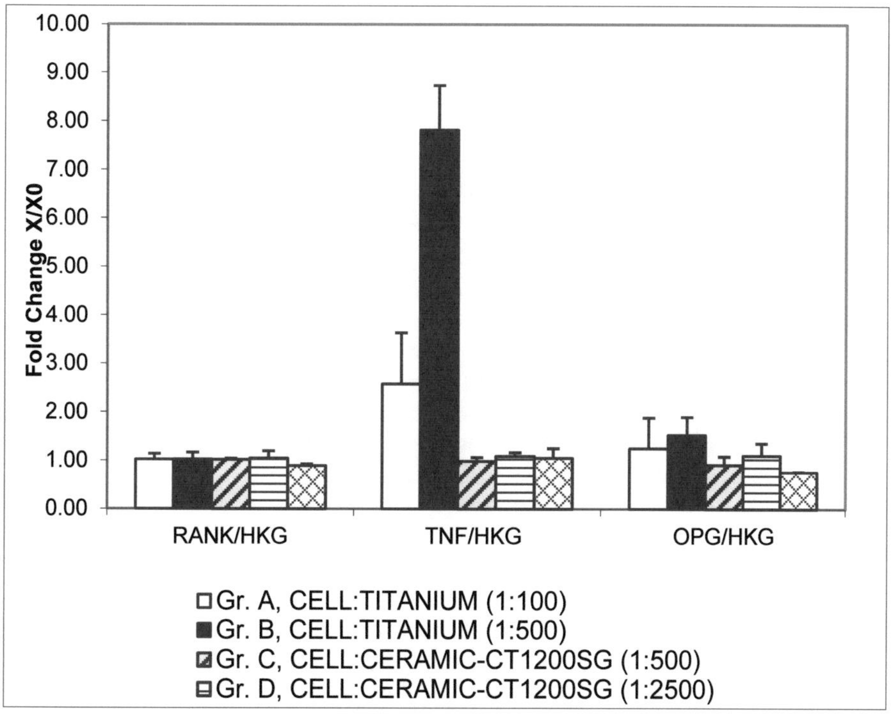

Fig. 2. Alteration of expression of RANK, TNF-α, and OPG caused by titanium and ceramic particles after an incubation time of 6 hours. Like before, the normalization took place by aid of the housekeeping-genes ß-actin and HPRT (here called HKG). A significant increase in expression could only be verified for TNF-α, which rose significantly in both titanium particle concentrations (group A and B, each $p<0.05$), while ceramic particles did not create a significant alteration of expression in any of the three genes. The group with smaller ceramic particles (group E) rather seemed to produce a decent but not significant suppression of RANK- and OPG-mRNA.

2.7 Possibilities and sense of prevention and intervention on particle-induced osteolysis

Currently, there exist several possibilities to prevent or to retard aseptic hip joint loosening whereas the different measures can start at different stages of anamnesis of patients with artificial hip joints.

In keeping with the philosophy of prevention, it could be targeted to avoid the necessity of hip implantation, for example by the reduction of adiposity as one of the most common causes for joint arthrosis, its early detection and conservative treatment.

If hip implantation is inevitable, its durability also depends on the surgeons experience and surgical technique. To avoid material abrasion or even fracture of the prosthesis, especially in case of ceramic prosthesis - described in chapter 2.1 - the exact positioning of the prosthesis components is particularly important.
To improve durability of prostheses, some decisive technical optimization were carried out referring to material properties as well as biocompatibility and durability. In the case of polyethylene, a smaller material abrasion has been reached by increasing the number of intermolecular binding (highly cross-linked UHMWPE), by high dosage gamma irradiation or by application of vitamin E for capturing the originated free radicals (Oral et al. 2007).
After the already undertaken further development of aluminium ceramics - e.g. by increasing purity level, strength and density - the research on this material continues to improve its tribological properties and especially its capacity (Masson et al. 2009). Using non-oxide ceramics, for instance silicon nitride or silicon carbide, Cappi et al. (2010) are aiming at applying ceramics for thin-walled implants (e.g. for resurfacing hip prostheses) which had not been possible previously because of the low mechanical strength of oxide ceramics.

2.8 Possible future problems of ceramics/ceramics sliding contact surfaces

The growing interest of orthopaedic surgeons in modern sliding contact surfaces like ceramics/ceramics is reflected by the more frequent clinical application to reduce wear debris and consequentially osteolysis in the young and active patient. Their wide range of use over the past decade, however, has led to complications, too, one being the developing of movement-related noises in hip-TEP patients. Most recent publications show that noises can occur in all sliding contact surfaces (Pokorny et al. 2010).

3. Conclusion

As ceramic bearing designs continue to improve with modified materials and manufacturing techniques, use will increase, especially in young and active patients and in concern to an earlier indication for total joint replacement and considering an eldery population. Clinical results for ceramic joint replacement , especially in young and active patients, show lower wear rates and a significant reduction of osteolysis. Furthermore our results support the thesis of bioinertness of ceramic components. New ceramic components are still under development, and the long-term results have to be evaluated over the next decades. Not every modern concept will likely stand the test of time, but some will be beneficial for patients undergoing total hip arthroplasty in the future.

4. Acknowledgements

The biomaterials research is supported by the German Research Foundation (DFG), Bonn, Germany (WE 3634/1-1).

5. References

Akisue, T, Bauer, TW, Farver, CF, Mochida, Y (2002). The effect of particle wear debris on NFkappaB activation and proinflammatory cytokine release in differentiated THP-1 cells. *J Biomed Mater Res,* 59 (3), pp. 507–515

Anderson, MA, Maraskovsky, E, Billingsley, WL, Dougall, WC, Tometsko, ME., Roux, ER, Teepe, MC, DuBose, RF, Cosman, D, Galibert, L (1997). A homologue of the TNF receptor and its ligand enhance T-cell growth and dendritic-cell function. *Nature*, 390, pp. 175–179

Aspenberg, P, Van der Vis, H (1998). Fluid pressure may cause peri-prosthetic osteolysis. Particles are not the only thing. Acta Orthop Scand, 69 (1), pp. 1–4

Bachmann HS, Hanenkamp S, Kornacki B, Frey UH, Bau M, Siffert W, Wedemeyer C (2008). Gender-dependent association of the GNAS1 T393C polymorphism with early aseptic loosening after total hip arthroplasty. *J Orthop Res*, 26(12), pp.1562-1568

Baumann, B, Rader, CP, Seufert, J, Noth, U, Rolf, O, Eulert, J, Jakob, F (2004). Effects of polyethylene and TiAlV wear particles on expression of RANK. RANKL and OPG mRNA. *Acta Orthop Scand*, 75, pp. 295–302

Bierbaum, BE, Nairus, J, Kuesis, D, Morrison, JC, Ward, D (2002). Ceramic-on-ceramic bearings in total hip arthroplasty. *Clin Orthop Relat Res*, 405, pp. 158–163

Bylski, D, Wedemeyer, C, Xu, J, Sterner, T, Löer, F, von Knoch, M (2009). Alumina ceramic particles, in comparison with titanium particles, hardly affect the expression of RANK-, TNF-alpha- and OPG-mRNA in the THP-1 human monocytic cell line. *J. Biomed. Mater. Res. A*, 89 (3),pp. 707–716

Callaghan, JJ, Salvati, EA, Pelicci, PM, Wilson, PD Jr, Ranawat, CS (1985). Results of revision for mechanical failure after cemented total hip replacement, 1979 to 1982: a two to five year follow up. *J Bone Joint Surg*, 67-A, pp. 1074–1085

Cappi, B, Neuss, S, Salber, J, Telle, R, Knüchel, R, Fischer H (2010). Cytocompatibility of high strength non-oxide ceramics. *J Biomed Mater Res A*, 93(1), pp. 67-76

Catelas, I, Petit, A, Marchand, R, Zukor, DJ, Yahia, LH, Huk, OL (1999). Cytotoxicity and macrophage cytokine release induced by ceramic and polyethylene particles in vitro. *J Bone Joint Surg*, 81-B, pp. 516–521

Christel, PS (1992). Biocompatibility of surgical-grade dense polycrystaline alumina. *Clin Orthop*, 282, pp. 10–18

Fujikawa, Y, Itonaga, I, Kudo, O, Hirayama, T, Taira, H (2005). Macrophages that have phagocytosed particles are capable of differentiating into functional osteoclasts. *Mod Rheumatol*, 15(5), pp. 346-351

Hamadouche, M, Boutin, P, Daussange, J, Bolander, ME, Sedel, L (2002). Alumina-on alumina total hip arthroplasty: A minimum 18.5-year follow-up study. *J Bone Joint Surg Am*, 84A, pp 69–77

Hannoche, D, Hamadouche, M, Nizard, R, Bizot, P, Meunier, A, Sedel L (2005). Ceramics in total hip replacement. *Clin Orthop Relat Res*, 430,pp. 62–71

Hanssen, AD, Rand, JA (1988). A comparison of primary and revision knee athroplasty using the kinematic stabilizer prosthesis. *J Bone Joint Surg*, 12-A, pp. 720–731

Harris, WH (1995).The problem is osteolysis. *Clin Orthop Relat Res*, (311), pp. 46-53

Hatton, A, Nevelos, JE, Matthews, JB, Fisher, J, Ingham, E (2003). Effects of clinically relevant alumina ceramic wear particles on TNF-alpha production by human peripheral blood mononuclear phagocytes. *Biomaterials*, 24(7), pp. 1193–1204

Herren, T, Remagen, W, Schenk, R (1987). Histology of the implant-bone interface in cemented and uncemented endoprostheses. *Orthopade*, 6(3), pp. 239-251

Hitchins, VM, Merritt, K (1999). Decontaminating particles exposed to bacterial endotoxin (LPS). *J Biomed Mater Res*, 46(3), pp. 434–437

Holt, G, Murnaghan, C, Reilly, J, Meek, RM (2007). The biology of aseptic osteolysis. *Clin Orthop Relat Res*, 460, pp. 240–252

Jacobs, JJ, Skipor, AK, Patterson, LM, Hallab, NJ, Paprosky, WG, Black, J, Galante, JO (1998). Metal release in patients who have had a primary total hip arthroplasty. A prospective, controlled, longitudinal study. *J Bone Joint Surg (Am)*, 80 (10), pp. 1447–1458

König, A, Mühlbauer, RC, Fleisch H (1988). Tumor necrosis factor alpha and interleukin-1 stimulate bone resorption in vivo as measured by urinary [3H]tetracycline excretion from prelabeled mice. *J Bone Miner Res*, 3(6), pp. 621-627

Komine, M, Kukita, A, Kukita, T, Ogata, Y, Hotokebuchi, T, Kohashi O (2001). Tumor necrosis factor-alpha cooperates with receptor activator of nuclear factor kappaB ligand in generation of osteoclasts in stromal cell-depleted rat bone marrow cell culture. *Bone*, 28(5), pp. 474-83

Kong, YY, Yoshida, H, Sarosi, I, Tan, HL, Timms, E, Capparelli, C, Morony, S, Oliveira-dos-Santos, AJ, Van, G, Itie, A, Khoo, W, Wakeham, A, Dunstan, CR, Lacey, DL, Mak, TW, Boyle, WJ, Penninger, JM (1999). OPGL is a key regulator of osteoclastogenesis, lymphocyte development and lymph-node organogenesis. *Nature*, 397, pp. 315–323

Kurtz, SM et al. (2007). Projections of Primary and Revision Hip and Knee Arthroplasty in the United States from 2005 to 2030. *J Bone Joint Surg Am*, 89 Suppl 3, pp. 780-785 Lacey, DL, Timms, E, Tan, HL, Kelly, MJ, Dunstan, CR, Burgess, T, Elliott, R, Colombero, A, Elliott, G, Scully, S, Hsu, H, Sullivan, J, Hawkins, N, Davy, F, Capparelli, C, Eli, A, Qian, YX, Kaufman, S, Sarosi, I, Shalboub, V, Senaldi, G, Guo, J, Delaney, J, Boyle, WJ (1998). Osteoprotegerin (OPG) ligand is a cytokine that regulates osteoclast differentiation and activation. *Cell*, 93, pp. 165–176

Lee, JM, Salvati, EA, Betts, F, DiCarlo, EF, Doty, SB, Bullough, PG (1992). Size of metallic and polyethylene debris particles in failed cemented total hip replacements. *J Bone Joint Surg (Br)*, 74 (3), pp. 380–384

Livermore, J, Ilstrup, D, Morrey, B (2000). Effect of femoral head size on wear of the polyethylene acetabular component. *J Bone Joint Surg (Am)*, 72 (4), pp. 518–528

Lohmann, CH, Schwartz, Z, Köster, G, Jahn, U, Buchhorn, GH, Mac Dougall, MJ, Casasola, D, Liu, Y, Sylvia, VL, Dean, DD, Boyan, BD (2000). Phagocytosis of wear debris by osteoblasts affects differentiation and local factor production in a manner dependent on particle composition. *Biomaterials*, 21, pp. 551–561

Lusty, PJ, Tai, CC, Sew-Hoy, RP, Walter, WL, Walter, WK, Zicat, BA (2007). Third-generation alumina-on-alumina ceramic bearings in cementless total hip arthroplasty. *J Bone Joint Surg (Am)*, 89 (12), pp. 2676–2683

Manley, MT, Sutton, K (2008). Bearings of the future for total hip arthroplasty. *J Arthroplasty* , .23(7 Suppl), pp. 47-50

Mahoney, OM, Dimon, JH (1990). Unsatisfactory results with a ceramic total hip prosthesis. *J Bone Joint Surg Am*, 72

Maloney, WJ, Smith, RL, Schmalzried, TP, Chiba J, Huene, D, Rubash, H (1995). Isolation and characterization of wear particles generated in patients who have had failure of a hip arthroplasty without cement. *J Bone Joint Surg (Am)*, 77 (9), pp. 1301–1310

Mandelin, J, Li, TF, Liljestrom, M, Kroon, ME, Hanemaaijer, R, Santavirta, S, Konttinen, YT (2003). Imbalance of RANKL/RANK/OPG system in interface tissue in loosening of total hip replacement. *J Bone Joint Surg Br*, 85, pp. 1196–1201

Matthews, JB, Besong, AA, Green, TR, Stone, MH, Wroblewski, BM, Fisher, J, Ingham, E (2000). Evaluation of the response of primary human peripheral blood mononuclear phagocytes to challenge with in vitro generated clinically relevant UHMWPE particles of known size and dose. *J Biomed Mater Res*, 52 (2), pp. 296–307

Masson, B (2009). Emergence of the alumina matrix composite in total hip arthroplasty. *Int Orthop*, 3(2), pp. 359-363.

Merkel, KD, Erdmann, JM, McHugh, KP, Abu-Amer, Y, Ross, FP, Teitelbaum, SL (1999). Tumor necrosis factor-alpha mediates orthopaedic implant osteolysis. *Am J Pathol*, 154 (1), pp. 203–210

Oral, E, Muratoglu, OK (2011). Vitamin E diffused, highly crosslinked UHMWPE: a review. *Int Orthop*, 35(2), pp. 215-23

Özkir, S, Droste, P, Echtermeyer, V (2007): Wann wird zementiert? *Trauma Berufskrankh*, 9 (Suppl 3), pp. 351–358,

Petit, A, Mwale, F, Antoniou, J, Zukor, DJ, Huk, OL (2006).Effect of bisphosphonates on the stimulation of macrophages by alumina ceramic particles: a comparison with ultra-high-molecular-weight polyethylene. *J Mater Sci Mater Med*, 17(7), pp. 667-673

Pokorny, A, Knahr, K (2010). The noisy hip. Is it only a ceramic issue? In: Cobb J. (ed), Modern Trends in THA Bearings. *Material and Clinical Performance*. Springer, pp. 85-90.

Purdue, PE, Koulouvaris, P, Nestor, BJ, Sculco, TP (2006). The central role of wear debris in periprosthetic osteolysis. HSSJ, 2 (2), pp. 102–113

Robertsson, O, Wingstrand, H, Kesteris, U, Jonsson, K, Önnerfält, R (1997). Intracapsular pressure and loosening of hip prostheses. Preoperative measurements in 18 hips. *Acta Orthop Scand*, 68 (3), pp. 231–234

Rössler H, Rüther W (2005): Orthopädie und Unfallchirurgie. München: Elsevier GmbH; s. bes. S. 50.

Rousseau, MA, Le Mouel, S, Goutallier, D, Van Driessche, S (2004). Long-term results of alumina-on-alumina total hip arthroplasty. *Rev Chir Orthop Reparatrice Appar Mot*, 90(8), pp. 741- 748

Savarino, L, Baldini, N, Ciapetti, G, Pellacani, A, Giunti, A (2009). Is wear debris responsible for failure in alumina-on-alumina implants? *Acta Orthop*, 80 (2), pp. 162–167

Schiebler, TH, Schmidt, W (2002). Anatomie. 2. Ed. Berlin, Springer-Verlag, Heidelberg, Germany.

Schmalzried, TP, Jasty, M, Harris, WH (1992). Periprosthetic bone loss in total hip arthroplasty. Polyethylene wear debris and the concept of the effective joint space. *J Bone Joint Surg (Am)*, 74 (6), pp. 849–863

Schwende, H, Fitzke, E, Ambs, P, Dieter, P (1996). Differences in the state of differentiation of THP-1 cells induced by phorbol ester and 1,25 dihydroxyvitamin D3. *J Leukoc Biol*, 59 (4), pp. 555–561

Simonet, WS, Lacey, DL, Dunstan, CR, Kelley, M, Chang, MS, Luthy, R, Nguyen, HQ, Wooden, S, Bennett, L, Boone, T, Shimamoto, G, DeRose, M, Elliott, R, Colombero, A, Tan, HL, Trail, G, Sullivan, J, Davey, E, Bucay, N, Renshaw-Gregg, L, Hughes, TM, Hill, D, Pattison, W, Campbell, P, Boyle, WJ (1997). Osteoprotegerin: a novel secreted protein involved in the regulation of bone density. *Cell*, 89, pp. 309-319

Sterner, T, Schütze, N, Saxler, G, Jakob, F, Rader, CP (2004). Effects of clinically relevant alumina ceramic, zirconia ceramic and titanium particles of different sizes and concentrations on TNF-alpha release in a human macrophage cell line. *Biomed Tech (Berl.),* 49 (12), pp. 340–344

Sun, DH, Trindade, MC, Nakashyma, Y, Maloney, WJ, Goodman, SB, Schurman, DJ, Smith, RL (2003). Human serum opsonization of orthopedic biomaterial particles: protein-binding and monocyte/macrophage activation in vitro. *J Biomed Mater Res A,* 65 (2), pp. 290–298

Sundfeldt, M, Carlsson, LV, Johansson, CB, Thomsen, P, Gretzer, C (2006). Aseptic loosening, not only a question of wear: A review of different theories. *Acta Orthop,* 77, pp. 177–197

Thompson, DP (2002). Materials science: cooking up tougher ceramics. *Nature,* 16, 417 (6886), p 237

Tsuchiya, S, Kabayashi, Y, Goto, Y, Okumura, H, Nakae, S, Konno, T, Tada, K (1982). Induction of maturation in cultured human monocytic cells by a phorbol diester. *Cancer Res,* 42, pp. 1530–1536

Udagawa, N, Takahashi, N, Akatsu, T, Tanaka, H, Sasaki, T, Nishihara, T, Koga, T, Martin, TJ, Suda, T (1990). Origin of osteoclasts: mature monocytes and macrophages are capable of differentiating into osteoclasts under a suitable microenvironment prepared by bone marrow-derived stromal cells. *Proc Natl Acad Sci USA,* 87, pp. 7260–4726

Warashina, H, Sakano, S, Kitamura, S, Yamauchi, KI, Yamaguchi, J, Ishiguro, N, Hasegawa, Y (2003). Biological reaction to alumina, zirconia, titanium and polyethylene particles implanted onto murine calvaria. *Biomaterials,* 24(21), pp. 3655-3661

Wedemeyer, C, Kauther, MD, Hanenkamp, S, Nückel, H, Bau, M, Siffert, W, Bachmann, HS (2009). BCL2-938C>A and CALCA-1786T>C polymorphisms in aseptic loosened total hip arthroplasty. *Eur J Med Res,* 14 (6), pp. 250–255

Willmann G (1998). Ceramics for total hip replacement--what a surgeon should know. *Orthopedics,* 21(2), 173- 177,

Wottrich, R, Diabaté, S, Krug HF (2004). Biological effects of ultrafine model particles in human macrophages and epithelial cells in mono- and co-culture. *Int J Hyg Environ Health,* 207(4), pp. 353-361

Yagil-Kelmer, E, Kazmier, P, Rahaman, MN, Bal, BS, Tessman, RK, Estes, DM (2004). Comparison of the response of primary human blood monocytes and the U937 human monocytic cell line to two different sizes of alumina ceramic particles. *J Orthop Res,* 22, pp. 832–838

Yasuda, H, Shima, N, Nakagawa, N, Mochizucki, SI, Yano, K, Fujise, N, Sato, Y, Goto, M, Yamaguchi, K, Kuriyama, M, Kanno, T, Murakami, A, Tsuda, E, Morinaga, T, Higashio, K (1998). Identity of osteoclastogenesis inhibitory factor (OCIF) and osteoprotegerin (OPG): a mechanism by which OPG/OCIF inhibits osteoclastogenesis in vitro. *Endocrinology,* 39, pp. 1329–1337

Yoon, TR, Rowe, SM, Jung, ST, Seon, KJ, Maloney, WJ (1998). Osteolysis in association with a total hip arthroplasty with ceramic bearing surfaces. *J Bone Joint Surg Am,* 80(10), pp. 1459-1468

17

Application of Zirconia in Dentistry: Biological, Mechanical and Optical Considerations

Cláudia Ângela Maziero Volpato[1], Luis Gustavo D´Altoé Garbelotto[1], Márcio Celso Fredel[2] and Federica Bondioli[3]
[1]*Department of Dentistry – Federal University of Santa Catarina*
[2]*Department of Mechanical Engineering – Federal University of Santa Catarina*
[3]*Department of Materials and Environmental Engineering – University of Modena and Reggio Emilia*
[1,2]*Brazil*
[3]*Italy*

1. Introduction

With the aim of replacing the infrastructure of metallic dental prostheses, structural ceramics have been improved and have become increasingly more popular in dentistry. Among the dental ceramics, zirconia has emerged as a versatile and promising material because of its biological, mechanical and optical properties, which has certainly accelerated its routine use in CAD/CAM technology for different types of prosthetic treatment.
Zirconia based ceramics are routinely used in structural applications in engineering, such as in the manufacture of cutting tools, gas sensors, refractories and structural opacifiers. (Rashad & Baioumy, 2008). To meet structural demands, zirconia is doped with stabilizers to achieve high strength and fracture toughness. (Kelly & Denry, 2008). The bioceramics that are currently used in medical and dental care are derived from structural materials used in aerospace and military armor, which were modified to suit the additional requirements of biocompatibility. (Thompson et al., 2007; Vagkopoulou et al., 2009).

2. Structural bioceramics based in zirconia

2.1 Zirconia

Zircon is a shiny gray-white metal, which may look blue-black when in powder form. Zirconia is an oxide which has a high tensile strength, high hardness and corrosion resistance. It is not found as a pure oxide in nature. The main sources of zirconium are zirconate (ZrO_2-SiO_2, $ZrSiO_4$) and baddelyite (ZrO_2), and most of the material used is chemically extracted from these two minerals. The zirconate is more abundant, but less pure, requiring significant processing to get zirconia. (Picone & Maccauro, 1999). Baddelyite already contains levels of zirconia ranging from 96.5% to 98.5%. As this mineral shows significant levels, it is known as a source of extreme purity in obtaining zirconium metal and its compounds. Zirconium dioxide (ZrO_2) resulting from baddelyite, which is also known as zirconia, is a course oxide that presents a monoclinic crystal structure at room temperature. However, the powder can be purified and processed synthetically at high temperatures, forming a cubic structure called cubic zirconia. The resulting material is hard, optically flawless and translucent, usually used for making precious stones or gas sensors, P. ex. (Koutayas et al., 2009).

2.2 Phases of zirconia (monoclic, tetragonal and cubic)

The spatial arrangement of the atoms in zirconia is characterized by distinct crystallographic structures, characterizing a property known as polymorphism. Its three phases, or crystal structures, are characterized by specific geometry and dimensional parameters: monoclinic, tetragonal and cubic. (Fig. 1a,b,c). Pure zirconia has a monoclinic structure at room temperature, which is stable up to 1170°C. Between this temperature and 2370°C, tetragonal zirconia is formed, while cubic zirconia is formed at temperatures above 2370°C. After processing, and depending on the cooling process, the tetragonal phase becomes monoclinic at about 970°C. Due to polymorphism, pure zirconia cannot be used at elevated temperatures due to a large volume change (3-5%) which occurs during cooling to the monoclinic phase. This change is sufficient to exceed the elastic and fracture limits, resulting in cracks and flaws in ceramics. (Denry & Kelly, 2008).

The transformation of the tetragonal to monoclinic phases can be employed to improve the mechanical properties of zirconia, especially its tenacity. The mechanism involved is known as a booster from transformation. This transformation is martensitic in nature, therefore, a process that occurs by shear without diffusion, ie the atomic position change occurs abruptly at a speed close to the speed of sound propagation in solids. The reverse transition, ie the monoclinic > tetragonal transformation and occurs at approximately 1170° C, while the tetragonal > monoclinic transformation, which occurs during cooling, is observed between 850 and 1000°C, depending on the strain energy. Therefore, the manufacturing of components of pure zirconia is not possible due to spontaneous failure. The addition of stabilizing oxides is important because it allows the maintenance of the tetragonal form at room temperature. (Hannink et al., 2000).

Different oxides, such as yttrium oxide (Y_2O_3), calcium oxide (CaO) or magnesium oxide (MgO), can be added to zirconia to stabilize it, allowing the tetragonal form to exist at room temperature after sintering. The addition of varying amounts of stabilizers allows the formation of partially or fully stabilized zirconia which, when combined with changes in processes, may result in ceramics with exceptional properties such as high flexural strength and fracture toughness, high hardness, excellent chemical resistance and good conductivity ions. A fully stabilized zirconia is obtained by adding sufficient amounts of stabilizing oxides, such as 16mol% magnesia (MgO), 16mol% of limestone (CaO) or 8 mol% yttria (Y_2O_3). Since the partial stabilization of zirconia is obtained with the same oxides, but in smaller amounts (eg 2 mol% to 3mol% yttria), a multiphase structure is created, which usually consists of tetragonal and cubic zirconia majority / monoclinic precipitated in small amounts. (Picone & Maccauro, 1999). The transformation of tetragonal zirconia into monoclinic is a phenomenon influenced by temperature, vapor, particle size, micro-and macrostructure of the material, and also by the concentration of stabilizing oxides. The critical particle size for the partially stabilized zirconia to be maintained in the tetragonal form at room temperature is 0.2µm to 1µm (for compositions ranging from 2% to 3 mol% yttria), because, under 0.2 micrometres, the transformation to the monoclinic phase is not possible. (Kelly & Denry, 2008).

2.2.1 Monoclinic zirconia

The natural form of zirconia, known as baddelyite, contains approximately 2% HfO_2 (hafnium oxide), which is very similar to zirconia in structure and chemical properties. ZR_4+ ions have a coordination number of seven for the oxygen ions occupying tetrahedral interstices, with the average distance between the zirconia ion and three of the seven oxygen ions is 2.07Å. Since the average distance between the zirconium ion and four oxygen ions is 2.21Å, in the structure, one of the angles (134.3°) differs significantly from the tetrahedral value (109.5°). Thus, the structure of the oxygen ion is not planar and a curve occurs in the plane of the four oxygens, and the plane of three oxygens is completely erratic. (Hannink et al., 2000)

2.2.2 Tetragonal zirconia

Zirconia in its tetragonal phase has the form of a straight prism with rectangular sides. Ions ZR_4+ have a coordination number of eight, where the shape once again appears distorted due to the fact that four oxygen ions are at a distance of 2.065Å in the form of a tetrahedron plan, and four others are at a distance of 2.455Å in a tetrahedron that is elongated and rotated 90°. (Vagkopoulou et al, 2009).

2.2.3 Cubic zirconia

The structure of cubic zirconia can be represented by a simple cubic lattice with eight oxygen ions, which are surrounded by a cubic arrangement of cations, known as fluorite, ie the oxygens occupy the tetrahedral interstices of a cubic lattice (CFC) of cations. (Vagkopoulou et al, 2009).

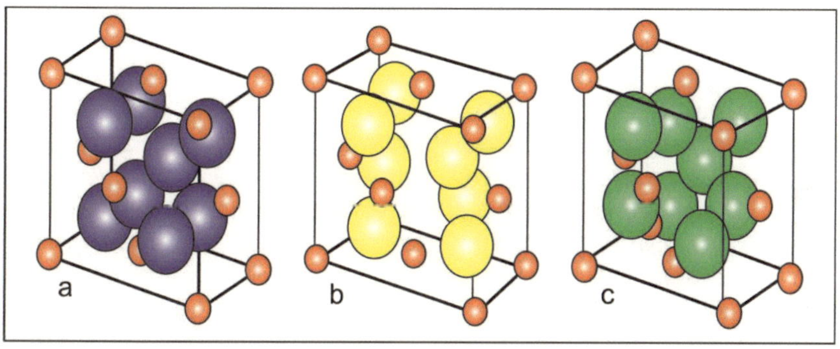

Fig. 1. Crystal structure of monoclic (a), tetragonal (b) and cubic zirconia(c).
Source: (Hannink et al., 2000)

2.4 Technology of powder (powder size and morphology)

The increasing use of ceramics in more advanced technological applications has resulted in a heightened demand for improvements in properties and reliability. (Lange, 1989). In the last decades is has been the realized that such improvements can be achieved only through careful attention not only to the chemical and mineralogical composition but also to the fabrication process. (Rahaman, 1995). The engineering properties of a polycrystalline ceramic are controlled by the microstructure, which depends on the processing method used to fabricate the body. Therefore the microstructures due to the fabrication processes govern the production of an object with the desired properties (Fig. 2).

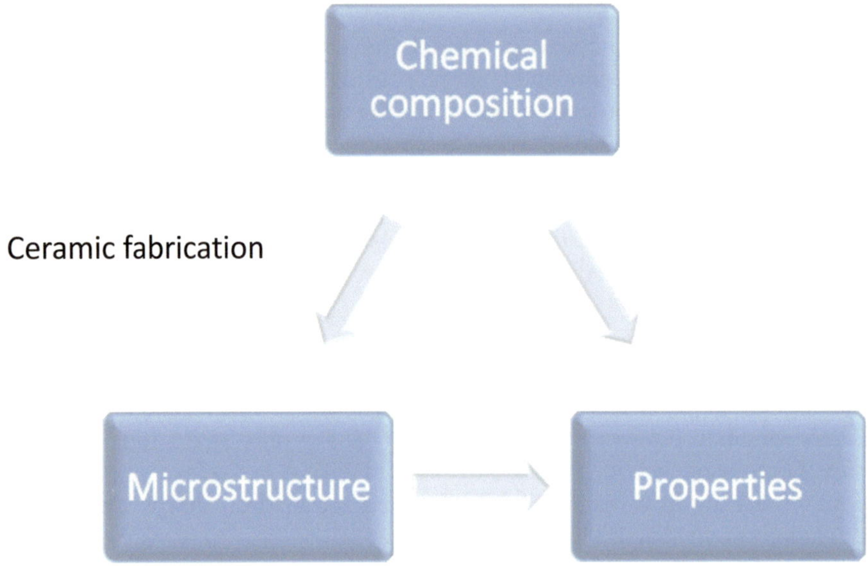

Fig. 2. Relationships in the ceramic fabrication process. Source: Rahaman, 1995

The production of polycrystalline ceramic from powders can be divided in two steps: processes prior to the firing of the green body and processes that occur during firing. The most useful approach requires that close attention be paid to each processing step because each step has the potential for producing undesirable microstructural flaws that can limit the properties and the reliability of the obtained body.

In particular during the last 20-25 years there was a large acceptance that utilization of powders with controlled purity, size and size distribution, shape and degree of agglomeration are required to obtain high quality ceramics. The properties of the starting powders are then determined by the method of their production and many methods are available for the preparation of ceramic powders. Different mechanical mixing (involving the comminution of a coarse, granular material), thermal decomposition, precipitation or hydrolysis and hydrothermal methods have been proposed to fulfill these requirements (Table 1). (Sõmiya & Akiba, 1999).

The chemical methods, although often more expensive than the mechanical methods, offer in general a strictly control of the powder characteristic (Fig. 3a,b) taking into account all the specific synthesis variables. (Tsukada et al., 1999; Piticescu et al., 2005). In practice, the choice of a powder preparation method will depend on the production cost and the capability of the method to achieve a certain set of desired characteristic. Normally a particle size of less than ≈1 µm allows the achievement of a high density material within a reasonable time (e.g., a few hours). Whereas a powder with a wide distribution of particle size may lead to higher packing density in the green body, this benefit is usually vastly outweighed by difficulties in the microstructure control during the consolidation step. The large grains coarsen rapidly at the expense of the smaller grains making the control of the grain size impossible. Homogeneous packing of a powder with a narrow size distribution, especially with spherical shape, generally improves the densification step and the microstructure control.

Processing route	Synthesis method
1. Thermal decomposition	a. Heating (evaporation) b. Spray drying c. Flame spraying d. Plasma spraying e. Vapor phase (CVD) f. Freeze drying g. Hot kerosene drying h. Hot petroleum drying
2. Precipitation or hydrolysis	a. Neutralization and precipitation b. Homogeneous precipitation c. Coprecipitation d. Salts solution e. Alkoxides f. Sol-gel
3. Hydrothermal	a. Precipitation (coprecipitation) b. Crystallization c. Decomposition d. Oxidation e. Synthesis f. Electrochemical g. Mechanochemical h. RESA (reactive submerged arc) i. Hydrothermal + microwave j. Hydrothermal + Ultrasonic
4. Melting and rapid quenching	

Table 1. Processing route and synthesis method used to obtain ZrO_2 powders
Source: Sōmiya & Akiba, 1999

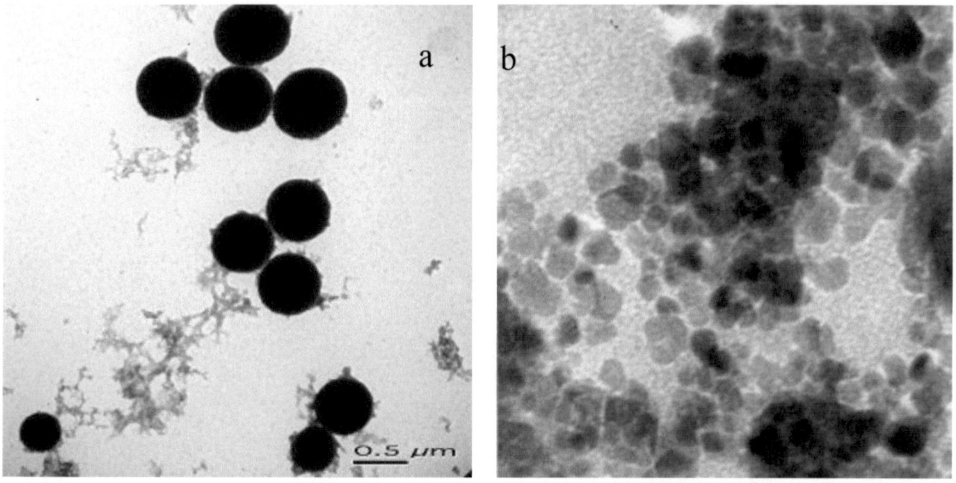

Fig. 3. a,b. An example of a ZrO_2 powder prepared by sol-gel method (a) and by microwave assisted hydrothermal method (b).

Moreover, chemical methods involving chemical reactions under carefully controlled conditions, allow obtaining ultrafine, nanometric powders (grain size below 100 nm). With this aim, the so-called wet methods are now the most popular: coprecipitation (Bondioli et al., 2001; Carter et al., 2009), sol-gel (Diaz-Parralejo et al., 2001; Heshmatpour & Aghakhanpour, 2011), hydrothermal synthesis also supported by microwave radiation (Bondioli et al., 2001; Kanade et al., 2008). The hydrothermal method (hydrolysis) to produce ZrO_2-based nanocrystalline powders was first used in 1990. (Sõmiya et al., 1991). Powders with primary particles 20–22 nm in size and specific surface area 20–95 m^2/g were obtained in the ZrO_2–Y_2O_3 (0, 5.2, and 13.9 wt.% Y_2O_3) system. The bending strength of materials sintered from these powders varied from 300 to 1000 MPa and critical fracture toughness coefficient from 2.5 to 6.0 MPa/m. Variation in temperature, pressure, synthesis time and chemical composition permits flexible control of the production of practically mono disperse powders with preset particle sizes and crystal structure.

Nanoceramics materials are in the early stages of development but already show many processing and property advantages over conventional coarse grained alternatives. Interest in nanosized ceramic powders for processing of ceramics is motivated by the promise of the improved sinterability at low temperature (below 0.5Tm), an increase in toughness due to a reduction in flaw size, and low-temperature superplastic deformation. Properties of nanostructured materials are directly related to their unusual microstructure that features extremely small grains and a large fraction (up to 40 vol% depending on grain size)of highly disordered interfaces. Grain size dependent properties have been studied intensively in nanostructured ceramics. For example,an increased surface area such that possessed by nanopowders should result in an enhanced densification rate at a given sintering temperature as expected by the Herring's law (Herring, 1950). Indeed it would suggest that full densification should be achievable at much lower sintering temperatures.

Figure 4 shows the relative shrinkage of nanometric ZrO_2 and commercial Y-stabilized tetragonal ZrO_2 measured by dilatometry. Shrinkage of nanometric ZrO_2 starts at 650°C and

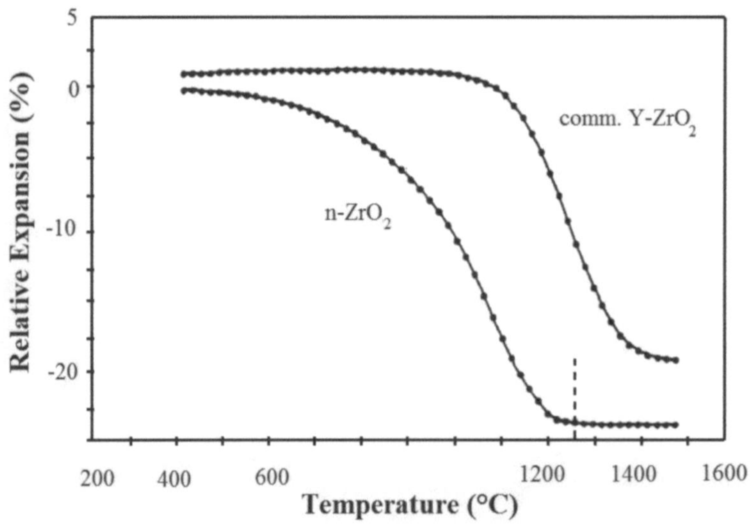

Fig. 4. Relative expansion as a function of sintering temperature for nanostructured and commercial Y-stabilized ZrO_2 measured in a dilatometer. The dashed vertical line represents 1/2 Tm.

the sample is completely dense at 1200°C. In the commercial material no densification is observed until 1100°C and full density is obtained only above 1400°C. The sintering temperatures are drastically reduced due to the short diffusion distances in the nanostructured ceramic (pressureless sintering). Experimental evidence shows, moreover, that faster densification rate allow a given density to be achieved at smaller grain size before serious grain growth takes place. (Chen & Wang, 2000; Mazaheri et al., 2008; Tartaj & Tartaj, 2009). Success in avoiding the latter is related to control of the competition between densification and grain growth, something that is extremely difficult because the driving forces for both are proportional to the reciprocal grain size and hence comparable in magnitude. A successful approach is the so-called "two steps sintering" (Lin et al., 1997) generally performed by heating the green compacts initially to a high temperature (e.g. 1250 or 1310°C) for a very short time and then cooling it down rapidly to a lower temperature (e.g. 1150°C) where they are held until complete densification is achieved. (Binner & Vaidhyanathan, 2008). Alternatively, high-density nanostructured ceramic systems including ZrO_2 have been achieved by means of pressure-assisted sintering. Applying some pressure during sintering can increase the densification rate and suppress the particle growth. (Madhav Reddy et al., 2010).
Regarding the mechanical properties, the empirical Hall-Petch equation predicts that as the grain size (d) decreases the yield strength increases as given by the following equation, here σ is the measured flexural strength, σ_0 is the flexural strength at an infinite grain size, k is the Hall–Petch constant and d is the average grain size.

$$\sigma = \sigma_0 + kd^{1/2} \tag{1}$$

Figures 5a,b shows that the flexural strength decreased monotonically with an increase in average grain sizes for the nanostructured ZrO_2 samples (Madhav Reddy et al., 2010) following a Hall–Petch type relationship. However it is obvious that this relationship cannot be extrapolated to arbitrarily small grain size and some form of lower limit to this behavior must exist. It has been observed that even if nanocrystalline materials are inherently stronger than their microcrystalline counterparts, the increment in strength falls below the estimated strength based on the Hall-Petch equation.

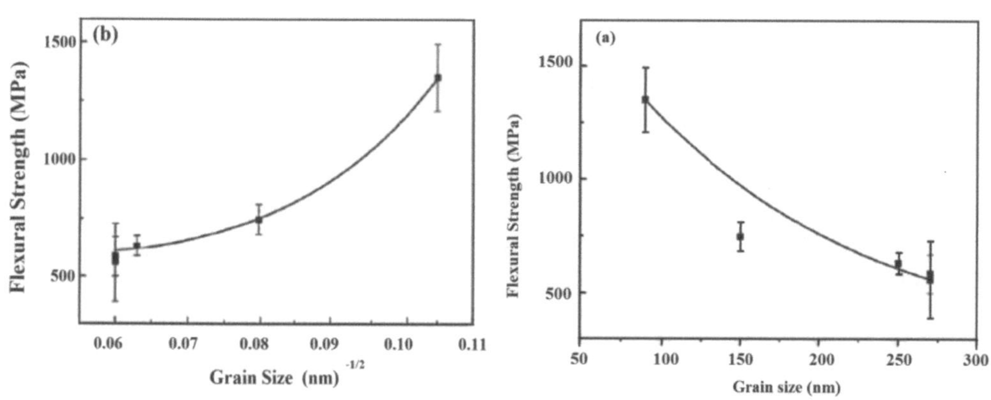

Fig. 5. a,b. Variations of flexural strength with (a) mean grain size and (b) inverse of square root of grain size of ZrO_2 samples. Source: Madhav Reddy et al., 2010

However is important to underline that the consolidation of ultrafine powder is very difficult because of its low apparent density, low flow rate, high contents of absorbed gases and admixtures, high surface area relative to its weight, and severe interparticle friction. Generally, as the size decrease below ≈0.5μm, the particles exhibit a greater tendency to interact, giving rise to the formation of agglomerates. One consequence of the presence of agglomerates is that the packing of the consolidated powders can be quite non uniform and, during the firing step, little benefit can be obtained over that of a coarse powder. To take advantage of the unique properties of bulk nanocrystalline materials, the nanometer range powders have to be densified with minimal microstructural coarsening and/or undesirable microstructural transformations. However the use of nanopowders requires proper control of the handling and consolidation procedure. The densification process for conventional powders is well known, both theoretically and practically.

3. Biological characteristics

3.1 Biocompatibily and bioinercia

In vitro and in vivo studies have confirmed a high biocompatibility of zirconia, especially when it is completely purified of its radioactive contents. (Gahlert et al., 2007; Andreiolli et al., 2009). Generally, ceramics are inert materials, which have no adverse local or general tissue reactions. As the ceramic prostheses are made with highly polished surface, they can contact the gum tissue and assist in the maintenance of gingival architecture. Depending on the smoothness, the ceramics prevent the buildup of plaque, creating a favorable surface for the gingival tissues. Zirconia based ceramics are chemically inert materials, allowing good cell adhesion, and while no adverse systemic reactions have been associated with it. (Ichikawa et al., 1992). However, particles from the degradation of zirconia at low temperature (LTD) or from the manufacturing process can be released, promoting an immune localized inflammatory reaction. (Chevalier, 2006).

3.2 Degree of toxicity

In vitro tests have shown that zirconia has a lower toxicity than titanium oxide and similar to alumina. Cytotoxicity, carcinogenicity, mutagenic or chromosomal alterations in fibroblasts or blood cells has not been observed. (Vagkopoulou et al., 2009).

3.3 Radioactivity

Zirconia is often accompanied by radioactive elements of long half-life, such as thorium (Th) and uranium (U). The separation of these elements is difficult and costly. Two types of radiation are correlated with zirconia, alpha and gamma. Significant amounts of alpha radiation have been observed in zirconia based ceramics used in the manufacture of surgical implants, because, due to their high ionization, the alpha particles destroy cells of hard and soft tissues. As for gamma radiation, the literature suggests that the radiation level is not worrisome in zirconia. (Vagkopoulou et al., 2009).

4. Mechanical characteristics

4.1 Flexural strength

Flexural strength is an important mechanical property that aids in predicting the performance of fragile materials. It can be defined as the final force required to cause fracture and is strongly affected by the size of flaws and defects on the surface of the material tested. (Mecholsky, 1995). Microcracks and defects that inherently grow during the thermal and mechanical processes can significantly influence the measurement of resistance. However, data alone cannot be extrapolated to predict the clinical performance of a material. Resistance values are significant when incorporated into context, through knowledge of the material microstructure, processing history, methodology, test environment and failure mechanisms. Structural failure probabilities are determined by additional failure, variables that describe the stress distribution and sizes of defects, which may be considered as single or multiple failure modes. An understanding of the current clinical failure modes is absolutely necessary before the results of in vitro resistance testing can be considered with clinical validity. (Kelly, 1995). Values of mechanical strength of fragile materials usually exhibit a large dispersion of values (above 50%), even for high-performance ceramics. (Absi & Glandus, 2002). This known phenomenon is based on the distribution of defects or failures.

The hypothesis that surface defects and microcracks in ceramic Y-TZP zirconia are made internally on the surface machined by CAD-CAM technique was confirmed by Luthardt et al. (2004). Milling may introduce residual surface compressive stresses that can significantly increase the resistance of zirconia ceramics. On the other hand, severe wear can make profound defects, which act as stress concentrating areas. Alternative methods, such as the partially sintered method of ceramics manufacturing, as well as wear-free procedures, should be developed to obtain crowns and bridges of the Y-TPZ system that increases strength and reliability. Another important fact is that the accumulation of microcracks resulting from loading in an aqueous environment (such as that found in the oral cavity), can cause surface defects that act as enhancers of tension in areas of local concentration, facilitating the initiation of fracture under low level applied stresses. (Lee et al., 2000).

4.2 Fracture toughness

Fracture toughness is defined as the level of critical stress at which a particular defect starts to grow. This property indicates the material's ability to resist rapid crack propagation and catastrophic fracture. (Scherrer et al., 1998). It also measures the ease of crack growth from an initial failure. The toughness is determined numerically as the critical value that causes crack extension in the "mode I", ie the opening mode by tension perpendicular to the crack, being designated as KIC. Steel and ductile metals show values above 50MPa m1/2. Ceramics cover a range of fracture toughnesses which rarely exceeds 5MPa m1/2. Not surprisingly, these low values affect their clinical performance. (Quinn et al., 2003).

In zirconia, the process of phase transformation induces compressive stress at the crack tip and shear stresses that act against the stress field generated in this region. The addition of an oxide stabilizes the system transformation of zirconia in the tetragonal phase and retains a layer of compressive stresses, resulting in the formation of a tougher stabilized tetragonal zirconia polycrystal. (Kelly & Denry, 2008).

4.3 Subcritical crack growth

The subcritical crack growth (SCG), which consists of a slow propagation of failures, is one of the major causes of damage to ceramics and usually occurs as a function of time. SCG under constant load is due to the corrosive action in the region under stress at the crack tip. Three distinct phases of spreading, designated I, II and III are observed at a velocity curve of

the crack versus intensity factor of the applied force (V-KI). Region I corresponds to low growth velocities, and gets the most attention of research because it allows the calculation of the survival of the ceramic piece. (Tsalouchou et al., 2008).
The amount of SCG is affected by different factors that add to the strain rates. The format, depth and width of the defects within the material affect the tension intensity factor. In ceramics, the cyclic loading also accelerates the crack propagation and decreases its threshold due to degradation of toughening mechanisms. (Lawson, 1995). A faster spread of the crack is observed in the presence of water, which can be attributed to a high concentration of water molecules around the crack. This environment increases the rate of crack growth because it facilitates the Zr-O-Zr union cleavage at the end of the crack. (Chevalier et al., 1999).

4.4 Toughening

The addition of alumina to zirconia Y-TZP provides a material with a higher elastic modulus and finer microstructure, often with greater toughness. The increase in KIC from pure zirconia to composites, for example containing 4% vol. Al_2O_3, is a process based on the toughness of the zirconia grain due to the presence of impurities of alumina at the edges and, consequently, provides an increase in transgranular fracture mode. (Kelly & Francis Rose, 2002).
In the case of the tetragonal phase, zirconia remains metastable at environmental conditions, meaning that it is theoretically unstable and may persist indefinitely. Thus, the stress field in front of a crack causes these particles to transform into a stable tetragonal monoclinic phase. Following this transformation, there is a slight increase in the volume of the particle and the result is that compressive stresses are set on the surface of the cracks in regions close to their terminus, which tend to throttle and close the cracks, thus preventing their growth. The result is an increase in the toughness of zirconia in the face of this transformation, preventing the propagation of cracks and improving the mechanical behavior of ceramics in the face of tension stresses. (Kelly & Denry, 2008).
Another toughening mechanism that happens in ceramic materials is crack deflection, which occurs when a crack changes its propagation direction after encountering a particle of the second phase, pore or grain boundaries. In zirconia, the pattern of crack propagation is usually transgranular. An increase of crystalline content present in the fully sintered zirconia equals an increase in the mechanical properties. However, in crystalline materials with the same content, the difference in strength and fracture toughness is related to porosity and effectiveness of each toughening mechanism. (Guazzato et al., 2004).

4.5 Martensitic transformation

The characteristic of phase transformations that zirconia suffers is of the martensite type. This phenomenon was named by Osmond, who described the microstructure of a hardened steel and introduced the term 'martensite' (Kelly & Francis Rose, 2002). In metallurgy, this process is characterized by a transformation that takes place without mass transfer, in a certain range of temperatures and changes the shape of the nucleus. Phase transformations are reversible, and involve the expansion of the core volume by approximately 3 to 4%.
With the stabilization of zirconia by oxides (CaO, MgO, Y_2O_3, CeO_2 or other rare earths), tetragonal and cubic phases may be partially or fully stabilized, allowing the determination of mechanical properties necessary for their use. (Tsalouchou et al., 2008). The

transformation is thermodynamically reversible to 1174°C and the change in atomic position shows a curve of thermal hysteresis between heating and cooling cycles. The intensity of the martensitic transformation is influenced by several parameters, such as size, shape and location of the particles of ZrO_2 (inter or intragranular), amount of oxide stabilizers, and difference in coefficient of thermal expansion, among others. (Quinn et al., 2003; Kelly & Francis Rose, 2002).

4.6 Aging

Aging or zirconia degradation at low temperature (LTD) is a progressive and spontaneous phenomenon that is exacerbated in the presence of water, steam or fluids. The consequences of the material aging process are many, including surface deterioration, microcracks and decreased resistance in medium and long term periods. Although this degradation has been demonstrated and indirectly associated with a number of flaws in femoral head ceramic prostheses, its relationship does not seem to be clear in predicting failure, due to the absence of rigorous follow up scientific studies to support it. (Denry & Kelly, 2007).

Aging occurs through a slow surface transformation to the monoclinic stable phase. This transformation begins in individual particles on the surface through a mechanism of stress corrosion. The initial transformation of specific particles can be related to a state of imbalance: greater particle size, lower yttria content, specific guidance from the surface, the presence of residual stress, or even the presence of a cubic phase. The transformation occurs through nucleation and growth processes. This phenomenon leads to a cascade of events occurring in neighboring particles, leading to an increase in volume that stresses the particles and results in subcritical crack growth (SCG), offering a way for water to penetrate inside the material. The stage of growth again depends on various microstructure patterns, such as: porosity, residual stresses, and particle size, among others (Figs. 6 to 8). (Chevalier, 2006).

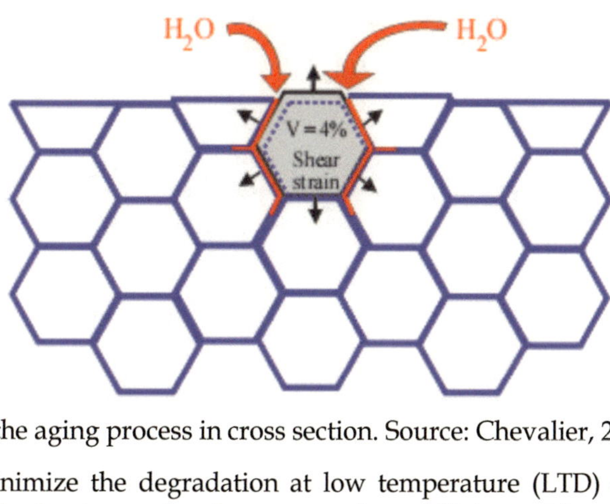

Fig. 6. Diagram of the aging process in cross section. Source: Chevalier, 2006

The attempt to minimize the degradation at low temperature (LTD) of 3Y-TZP includes reducing the particle size, increasing the content of a stabilizing oxide, or even the formation of composites with aluminum oxide (Al_2O_3). The addition of alumina particles prevents the relaxation of the network of tetragonal zirconia under stress during the aging process, since relaxation is responsible for degradation. (Lee et al., 2000).

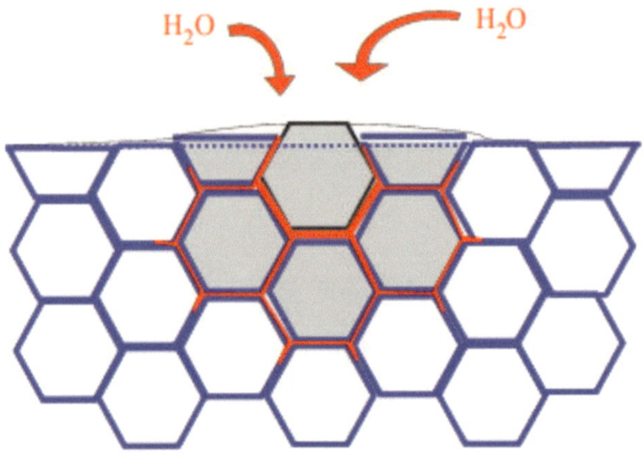

Fig. 7. Nucleation in a specific grain surface, leading to the formation of microcracks and stress on neighboring grains. Source: Chevalier, 2006

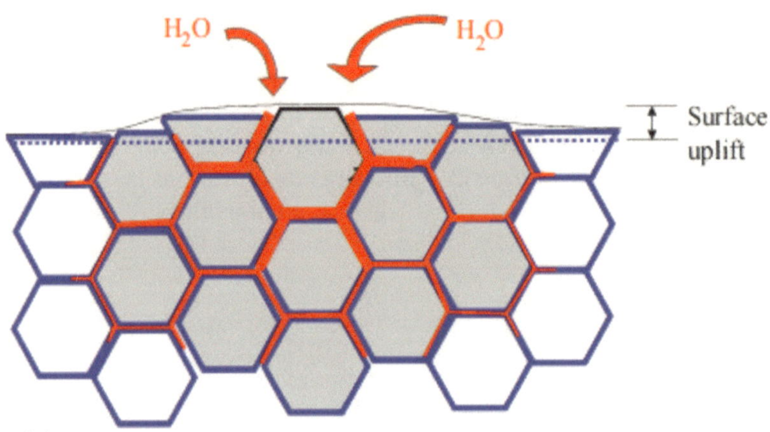

Fig. 8. Growth of the transformed zone, leading to the extensive formation of microcracks and surface roughening. The processed grains appear in gray. The paths in red represent the penetration of water due to the formation of microcracks around the processed grains. Source: Chevalier, 2006

In the literature, there is a major concern in evaluating the influence of superficial (Ardlin, 2002; Kim et al., 2005) and termic treatments (Tsalouchou et al., 2008) on the mechanical properties of partially stabilized zirconia. Fine polishing of the surface can reduce surface defects created in the finishing, improving the mechanical properties of the surface. However, prior to polishing, surface modifications such as adjustments and finishing, can introduce a compressive surface tension, initially increasing the flexural strength, but then changing the phase integrity of the material and increasing the susceptibility to aging. (Deville et al., 2006). Table 1 summarizes the potential effects of the different stages of processing on the microstructure of zirconia and, consequently, on aging. (Chevalier, 2006).

Processing stage	Potential effect to the microstructure
Starting powder	Yttria content and distribution, presence of additives (leading to secondary phases)
Training	Pore distribution of green compacts, and hence porosity of final components
Sintering temperature and duration	Density, particle size, amount of cubic phase
Cooling rate after sintering	Fractionation step, residual stresses
Hot isostatic pressure	Density, defined microstructure, increasing the concentration of oxygen gaps, residual stresses
Whitening	Decrease of oxygen gaps, changes in residual stresses
Wear and machining	Surface roughness, residual stresses, initial monoclinic content
Cleaning, sterilization	Initial content of monoclinic phase

Table 2. Potential effect of the different stages of processing on the microstructure of ceramics based on zirconia.

5. Optical characteristics

5.1 Degree of opacity and translucency

The ceramic systems used in dentistry must have adequate translucency to achieve good dental aesthetics while at the same time provide adequate strength during chewing. Considering the currently available ceramic materials, these two properties cannot be obtained by a single material, especially in the manufacturing of fixed prostheses. Thus, an oxide ceramic material should be used as an infrastructure, while a glass or feldspathic ceramic must be used as an esthetic coating material.

The infrastructures of zirconia provide good masking of darkened substrates due to an adequate level of opacity, and they also allow a controlled translucency after lamination, due to their homogeneity and high density (residual porosity <0.05%), even limited in thickness (0.5mm). Its opaque optical behavior can be attributed to the fact that the grain size is greater than the length of light, and also that zirconia has a high refractive index, low absorption coefficient and high opacity in the visible and infrared spectrum. (Heffernan et al., 2002a; Denry & Kelly, 2008). Therefore, zirconia cannot be used as a restorative material alone. Because of its opacity and the current processing technologies, zirconia must be covered with translucent ceramics which exhibit characteristics that may look like natural teeth. (Heffernan et al., 2002b).

The opacity of zirconia is very useful in clinical situations where polychromatic substrates need to be masked so as not to interfere with the aesthetic result. Blackened teeth, pins and metal cores can be adequately masked by zirconia infrastructures. Through a retrospective of the first zirconia introduced in dentistry (ZTA - alumina matrix strengthened / toughened by zirconia), the opacity of the final prosthetic piece was found to be either similar or higher than metaloceramic crowns. (Heffernan et al., 2002b). This fact, coupled with the difficulty of processing ceramics based on zirconia, boosted research focused on the microstructure and processing of this biomaterial. Changes in size and distribution of grains, methods of obtaining industrial blocks for machining, various additives, stabilizers and pigments have influenced the porosity of the material and therefore its optical behavior, improving its degree of final translucency. (Cho et al., 2009).

Pre-sintered ceramic blocks have been obtained by techniques of industrial forming (isostatic compaction) that have an impact on the final degree of homogeneity and density of the material. (Denry & Kelly, 2008). Thus, the use of these blocks in machining systems (CAD/CAM) has allowed the construction of a thinner infrastructure (0.5mm), due to the high density post-sintering zirconia block, creating more space for the application of a ceramic layer. (Cho et al., 2009; Miyazaki et al., 2009). Zirconia 3Y-TPZ blocks are close to pure white. The main advantage of these blocks is that the white ceramic substrate can be adequately masked. However, an exceedingly white background could compromise the aesthetic outcome of the prosthesis, creating a new substrate that is difficult to be masked by layers of feldpathic or glass ceramic. Solutions such as pre-soaking the sintered infrastructure in solutions based in nanosized pigments of iron oxide or lanthanum has helped in obtaining an infrastructure that is easier to be covered.

6. Zirconia in dentistry

The first study on the use of zirconia in dentistry was an assessment on the ceramic surface of metal implants, with the aim of improving their biocompatibility in the oral cavity. Histological examinations revealed fibrointegration of these implants, but with a higher clinical success of those coated with alumina. (Cranin et al., 1975). With the advancement in technology for stabilizing the tetragonal phase at room temperature, and confirming how it was possible to obtain a structurally more tenacious material, several studies were developed by leveraging the use of zirconia as a structural material in dentistry. The first work along these lines studied implants manufactured from the slipcasting process, followed by a superficial treatment with Nd:YAG laser. Thereafter, several studies involving the peri-implant bone remodeling (Miani et al., 1993), the quality of bone apposition directly on ceramic implants (Akagawa et al., 1993), the stability and contour of the mucosa around these implants (Akagawa et al., 1998), the use of a zirconium ceramic based pin in the interior of root canals (Asmussen et al., 1999), prosthetic systems obtained by machining CAD/CAM (Wang et al., 2008) and the construction of abutments on implants (Yildirim et al., 2000) were reported in the literature, reinforcing the viability of this material.

In Prosthetic Dentistry, the introduction of zirconia began with the InCeram Zirconia system (Vita Zanhfabrick, Germany). In this handmade ceramic system, the presence of zirconia resulted in an increase in structural strength of 30 to 40%. Later on, industrial ceramic systems, based on CAD/CAM, allowed the improvement of machining techniques for pre-sintered ceramic blocks to obtain infrastructure and prosthetic abutments.

6.1 Major types of zirconia used in dentistry

Three different materials were developed for use in dentistry. The first two ceramics are materials with at least two ZrO_2 t- phases as a minor phase (dispersed and precipitated, respectively), and the latter is essentially a t-ZrO_2 (single phase). The origin and details of the stabilization of the tetragonal phase differ among these three toughened microstructures. The three materials share the requirement of stabilizing the tetragonal phase and the toughening involves the martensitic transformation. (Denry & Kelly, 2008).

6.1.1 Zirconia toughned alumina (ZTA)

Ceramics based on zirconia are combined with a matrix of alumina (Al_2O_3), forming a structure known as ZTA (alumina reinforced with zirconia grains). The stability of the tetragonal phase at room temperature did not initially involve the use of doping, but instead is controlled by the size, morphology and particle localization (intra-or intergranular). In ZTA, particles above a critical size will attain monoclinic symmetry after cooling. Among the dental ceramics, the only commercial example of a toughened ceramic through dispersion is the In-Ceram Zirconia (Vita Zahnfabrik, Germany), which is an interpenetrating composite which was developed with this philosophy, with the addition of 33mol% zirconia stabilized with 12mol% ceria (12Ce-TPZ) to the precursor InCeram Alumina (70 to 80% aluminum oxide) to be used initially by the craft technique of infiltration slip casting (slipcasting). This technique has a small contraction with sintering; however, the amount of porosities incorporated during the handmade infrastructure has reduced the resistance of prosthetics made with this material. On the other hand, the industrial processing of pre-sintered blocks of the same material results in parts with higher mechanical properties, creating tougher prostheses, but with contractions around 25% (Wang et al., 2008).

6.1.2 Mg-PSZ (magnesia partially stabilized zirconia)

The microstructure of Mg-PSZ consists of an array of cubic zirconia partially stabilized by 8 to 10mol% of magnesium oxide. Due to difficulty in obtaining free silica Mg-PSZ precursors (SiO_2), magnesium silicates can form a low content of magnesia, favoring the transformation from tetragonal to monoclinic (t→m) and resulting in lower mechanical properties and stability of the material. Fully sintered blocks have been manufactured with this material, and require rigid and strong machining systems (Denry & Kelly, 2008).

6.1.3 Yttriafullstabilized tetragonal zirconia polycrystal (3Y-TZP)

The 3Y-TZP consists of an array of partially stabilized zirconia with a 2% 4mol yttria oxide. In 1977, it was reported that ZrO_2 fine grain (usually <0.5 mm) with small concentrations of Y_2O_3 stabilizers could contain up to 98% of the metastable tetragonal phase after sintering. The main feature of this microstructure is to be formed by tetragonal grains of uniform diameter in the order of nanometers, sometimes combined with a small fraction of the cubic phase. As explained above, the YSZ (yttrium oxide) is suitable for optical applications due to its high refractive index, low absorption coefficient and high opacity in the visible and infrared spectrum. 3Y-TZP was first applied in the medical field of orthopedics, with

significant success due to its good mechanical properties and biocompatibility (Piconi & Maccauro, 1999). In dental applications, it is fabricated with microstructures containing small grains (0.2 to 0.5 mm in diameter) depending on the sintering temperature, which avoids the phenomenon of structural deterioration or destabilization in the presence of saliva, slowing the growth of subcritical cracks (CST). (Kelly & Denry, 2008).
Prosthetic restorations with 3Y-TZP are obtained by milling pre-sintered blocks, followed by a sintering performed at high temperature, or by machining of sintered blocks completely (Fig 9a,b). The blocks are machined with the aid of CAD/CAM systems, and, in the case of pre-sintered blocks, prosthetic restorations are pre-shaped into a size 25 to 30% higher than desired (depending on the batch of material) to compensate for the sintering shrinkage. The final sintering temperature is between 1350 and 1550°C. (Miyazaki et al., 2002). This processing reduces the level of tension present and prevents the transformation from the tetragonal phase to the monoclinic phase (t→m), which leads to a final surface virtually free of the monoclinic phase. Procedures to adjust or sandblast this material should be eliminated to maintain surface integrity and prevent the transformation of t→m. Infrastructures obtained from these blocks are more stable, have a high crystalline content and a flexural strength around 900 to 1200MPa (Kelly & Denry, 2008). The blocks are processed fully sintered by isostatic pressure at a temperature between 1400 to 1500°C. This process causes the block to have a final density close to 99%, high hardness and low machinability. Thus, machining systems must be robust. (Miyazaki et al., 2002). Restorations produced by these blocks tend to exhibit a large amount of monoclinic zirconia as a function of compression generated by the machining process, which usually results in microcrack on the surface and a susceptibility to degradation at low temperature (CST). Recently, to produce ceramic blocks with greater durability and stability under high temperatures and humid environments, the industry has introduced small amounts of alumina to 3Y-TZP, constituting a variation, called TZP-A. However, a disadvantage of alumina addition is a reduction in the translucency of the block.

Fig. 8. a,b. Micrographs of different zirconia 3Y-TZP ceramic blocks.

6.2 Ceramic infrastructure for crowns and bridges

6.2.1 Technique of ceramic infiltration (*slipcasting*)

In 1988, Sadoun refined the slipcasting technique to produce a high-strength infrastructure that was named InCeram (Zahnfabrik Vita, Germany). This system has two interpenetrated phases: in the first phase, an alumina structure is obtained by the slipcasting process. Later, this structure is manually infiltrated by a thin layer of colored glass (lanthanum glass), granting an increased resistance (650MPa), and providing the ability to receive a ceramic coating. Three infiltrating systems were developed: strengthening with alumina, zirconia and magnesium. The system InCeram Zirconia is a ceramic body consisting of 67% aluminum oxide, and the remainder (33%) is composed of tetragonal zirconium oxide. This mixture provides an increase in tenacity and on the flexural strength of the material (750MPa). In this system, increasing the final strength is considerable, but a decrease in translucency can be observed, due to the presence of an overly white zirconia, which gives the mix an opacity similar to that of metal-ceramic crowns (Heffernan et al., 2002a,b)

6.2.2 Compaction technique – CAD/CAM

This ceramic system was first described in 1993 by Andersson and Oden, and was named All-Ceram Procera (Nobel Biocare, Sweden). With this system, it is possible to fabricate all-ceramic crowns from an infrastructure of densely sintered and non-porous (99.5%) pure aluminum oxide. The powder is compressed into a die refractory in a ceramic processing known as uniaxial vacuum, which supports a sintering of 1550°C for 1 h, producing a dense and crystalline infrastructure, with a particle size of 4μm and flexural strength of 600MPa. Afterwards, these infrastructures are tried on clinically, and covered by low-fusing ceramics. The infrastructure can also be obtained with a mixture of alumina and zirconia, through the same process of compression.

Clinical procedures basically consist of obtaining an impression of prepared areas and making models and stone dies. The die is positioned on the rotating platform of a scanner developed for this system. A probe with a spherical sapphire tip performs a global mapping of the die at 360°. The outline of the preparation is registered in three dimensions, reaching 25,000 to 50,000 points. The points obtained are sent to the computer where the operator works on the image generated by the scanner, starting with the identification of the cervical end. After delineating the end, the emergence profile and the thickness of the future infrastructure are defined. Upon completion of the digital design, it is sent via modem to a Production Station (Sandvik AB Estolcomo PROCERA or PROCERA Law Fair USA). With this information, an infrastructure can be made without the presence of the model. It is formed by a compaction process (uniaxial or isostatic) that consists of pressing the powder into a mold, creating a "compressed green" in the desired shape and sufficient strength for handling. It is necessary be careful with the compression to minimize the density gradient, due to the characteristics of the process and the state of powder agglomeration. The compacted infrastructure is then sintered in a vacuum furnace at high temperatures. (Oden et al. 1998).

6.2.3 Machining technique – CAD/CAM

Pre-sintered blocks have been introduced in the market to try to correct the deficiencies of the craft systems. (Vagkopoulou, 2009). During the industrial processing, advanced isostatic techniques decrease the concentration of defects within the block, reducing the number of catastrophic failures (Miyazaki et al., 2009). Through the obtained digital design (CAD), usually obtained by a scanner, the system starts turning a CAM ceramic block for 15 to 50 minutes. The resulting piece is crispy, but most of its outline and adaptation should be checked before being subjected to sintering. Upon being subjected to machining, the parts produced present a negative influence from this process. Systems employing pre-sintered blocks have been used because they seem less machine influenced, because the blocks are more porous. On the other hand, the fully sintered blocks present high hardness, requiring robust devices that will generate more power and thus greater compressive tension on the outer surface of the block, enabling the transformation of the tetragonal to monoclinic phase (t → m). The subsequent sintering of pre-sintered machined infrastructure increases the hardness and fracture toughness. However, repeated heat treatments, in which parts are submitted with an application of feldspathic or glass porcelain, seem to have a negative effect on fracture resistance of the material. (Guazzato et al., 2004).

6.3 Abutments for implant-supported prostheses

Intermediaries are devices used to make the connection between the implant and the prosthesis. They can be commercially purchased in a pre-manufactured form or fabricated by laboratory manual systems or CAD/CAM. The intermediaries are mostly used for custom cemented prostheses, since they present a gingival profile already in the intermediary stage, keeping the mucosal and peri-implantar gingival concave architecture. Custom ceramic abutments resulting from the machining of pre-sintered blocks or industrial compression techniques are widely used because they provide good results due to their structural stability. They have been made of alumina, alumina/zirconia or zirconia and are particularly suitable for implant supported restorations located in the anterior maxilla, especially in cases where the gingival margin is very thin, and a dark halo or a shadow under the gum tissue due to the metallic color of the intermediary can compromise the aesthetic outcome of the prosthesis. (Yildirim et al., 2000).

6.3.1 Degree of cervical adaptation

As the abutment is in direct contact with the implant platform, this region is a possible area to be colonized by bacteria. Therefore, a misfit between the abutment and the implant can result in bone loss around the implants. Studies have shown that pre-fabricated abutments have a higher degree of adaptation to those manufactured in a laboratory. (Byrne et al.,1998; Vigolo et al., 2006) Another important fact is that the union between metallic implants and ceramic abutments is performed by the interface of different materials. The system of connection between them (eg: a hexagonal metal and a ceramic inner hexagon) is not a mechanically efficient system. Metallic devices are being associated with ceramic abutments so that the connection between the parts is made of metal, with only the support between the implant platform and the abutment being ceramic.

6.3.2 Flexural strength

Abutments suffer tensile stresses when subjected to oral function. Metal parts have a higher mechanical strength and deformation capacity when compared with ceramic parts. In different studies, ceramic abutments based on zirconia showed average values of fracture 737.6N (Yildirim et al., 2003), 443.6N (Att et al., 2006), 649N (Sundh & Sjörgren, 2008) and 740N (Kerstein & Radke, 2008). These results demonstrated that zirconia has a higher resistance to established values for maximum incisal forces reported in the literature (90N to

370N). When noting the findings of a study by Att Wael (2006), ceramic abutments subjected to aging showed the lowest resistance when compared to abutments that were not aged, confirming that, despite the structural potential of zirconia, a probability of failure exists due to the aging and growth of subcritical cracks (SCG).

6.4 Ceramic implants and osseointegration

The first attempt to use ceramics in the manufacture of a dental implant was in 1975; however, the first study of implants on zirconia was recorded in 1993, when a group of researchers inserted experimental Y-TPZ implants in the mandible of dogs. (Cranin et al., 1975; Miani et al., 1993). With fluorochrome markers, the authors reported the direct apposition of new bone formation to implants after 120 days from the intervention. Confirming the early findings, several studies have shown that there were no adverse reactions, mutagenic or genetic effects on bone formation, pathologic or peri-implant soft tissue inflammatory states, or the mobility of the implant after installation of the prosthesis. A high proliferation of osteoblasts was also observed, presenting an excellent tissue response and good density of cortical bone newly formed around 97.5% of the implants after the period of osseointegration. However, despite the initial encouraging results when compared with longitudinal and multicenter studies made with titanium implants, the clinical and laboratory data are scarce for a wide and safe clinical application.

7. Conclusion

In dentistry, zirconia has been indicated and used for making crowns, bridges, abutments and implant prosthesis infrastructures. As is well known, dental ceramics are often associated with brittle fractures, due to their low resistance to fracture, which has delayed the routine clinical use of all-ceramic prostheses. The introduction of stabilized zirconia has created a real possibility and promise for the application of ceramics in dental reconstructions. However, the mechanical properties of zirconia are not still great and a suitable processing and clinical application protocol is still not fully known and controlled.

8. References

Absi, J.; Glandus, J.C. (2002). Numerical separation of bi-modal strength distribuitions. *J Eur Ceramic Soc,* Vol. 22, No. 5, (May, 2002) pp. 591-601, ISSN 0955-2219.

Akagawa, Y.; Hosokawa, R.; Sato, Y.; Kamayama, K. (1998). Comparasion between freestanding and tooth-connected partially stabilized implants after two years function in monkeys: a clinical and histologic study. *J Prosthet Dent,* Vol. 80, No. 5, (Nov, 1998) pp. 551-558, ISSN 0022-3913.

Akagawa, Y.; Ichikawa, Y.; Nikai, H.; Tsuru, H. (1993). Interface histology of unloaded and early loaded partially stabilized zirconia endosseous implant in initial bone healing. J Prosthet Dent, Vol. 69, No. 6, (Jun, 1993) pp. 599-604, ISSN 0022-3913.

Andreiolli, M.; Wenz, H.J.; Kohal, R.J. (2009). Are ceramic implants a viable alternative to titanium implants? A systematic literature review. *Clin Oral Implants Res,* Vol. 20, No. 4, (Sep, 2009) pp. 32-47, ISSN 0905-7161.

Ardlin, B.I. (2002). Transformation toughened zirconia for dental inlays, crowns and bridges: chemical stability and effect of low temperature aging on flexural strength and surface structure. *Dent Mater,* Vol. 18, No. 8, (Dec, 2002) pp. 590-595, ISSN 0109-5641.

Asmussen, E.; Peutzfeld, A.; Heitmann, T. (1999). Stiffness, elastic limit and strength of newer types of endodontic posts. *J Dent,* Vol. 27, No. 4, (May, 1999) pp. 75-78, ISSN 0300-5712.

Att, W.; Kurun, S.; Gerds, T.; Strub, JR. (2006). Fracture resistance of single-tooth implant-supported all-ceramic restorations: an in vitro study. *J Prosthet Dent,* Vol. 95, No. 2, (Feb, 2006) pp. 111-116, ISSN 0022-3913.

Binner, J.; Vaidhyanathan, B. (2008). Processing of bulk nanostructured ceramics. *J Eur Ceram Soc,* Vol. 28, No. 7, (2008) pp. 1329-1339, ISSN 0955-2219.

Bondioli, F.; Ferrari, A.M.; Corradi, A.B.; (2001). Role of praseodymiun on zirconia phases stabilization. *Chem Mater,* Vol. 13, No. 12, (Nov, 2001) pp. 4550-4554, ISSN 0897-4756.

Bondioli, F.; Ferrari, A.M.; Leonelli, C.; Siligardi, C.; Pellacani, G.C. (2001). Microwave-hydrothermal synthesis of nanocrystalline zirconia powders. *J Am Ceram Soc,* Vol. 84, No. 11 (Nov, 2001) pp. 2728-2730, ISSN 0002-7820.

Byrne, D.; Houston F.; Cleary, R.; Claffey, N. (1998). The fit of cast premachined implant abutments. *J Prosthet Dent,* Vol. 80, No. 2, (Ago, 1998) pp. 184-192, ISSN 0022-3913.

Carter, G.A.; Rowles, M.; Ogden, M.I.; Hart, R.D.; Buckley, C.E. (2009). Industrial precipitaion of zirconyl chloride: the effect of pH and solution concentration on calcination of zirconia. *Mater Chem Phys,* Vol. 116, No. 2-3, (Aug, 2009) pp. 607-614, ISSN 0254-0584.

Chen, I.W.; Wang, X.H. (2000). Sintering dense nanocrystalline ceramics without final-stage grain growth. *Nature,* Vol. 404, No. 9, (Mar, 2000) pp. 168-171, ISSN 0028-0836.

Chevalier, J. (2006). What future for zirconia as a biomaterial? *Biomaterials,* Vol. 27, No. 4, (Jan, 2006) pp. 535-543, ISSN 0142-9612.

Chevalier, J.; Olangnon, C.; Fantoz, G. (1999). Subcritical crack propagation in 3Y-TPZ ceramics: static and cyclic fatigue. *J Am Ceram Soc,* Vol. 82, No 11, (1999) pp. 3129-3138, ISSN 1551-2916.

Cho, M.S.; Yu, B.; Lee, Y.Y. (2009). Opalescence of all-ceramic core and venner materials. *Dent Mater,* Vol. 25, No. 6, (Jun, 2009) pp. 695-702, ISSN 0109-5641.

Cranin, A.N.; Schnitman, P.A.; Rabkin, S.M.; Onesto, E.J. (1975) Alumina and zirconia coated vitallium oral endosteal implants in beagles. *J Biomed Mater Res,* Vol. 9, No. 9, (Jul, 1975) pp. 257-262, ISSN 1549-3296.

Denry, I. & Kelly, J. R. (2008). State of the art of zirconia for dental applications. *Dent Mater,* Vol. 24, No. 3, (Mar, 2008) pp. 299-307, ISSN 0109-5641.

Deville, S.; Chevalier, J.; Gremillard, L. (2006). Influence of surface finish and residual stresses on the aging sensitivity of biomedical grade zirconia. *Biomaterials,* Vol. 27, No. 10, (Apr, 2006) pp. 2186-2192, ISSN 0142-9612.

Diaz-Parralejo, A.; Cuerda-Correa, E.M.; Macias-Garcia, A.; Diaz-Diez, M.A.; Sanchez-Gonzalez, J. (2011). Tailoring the properties of yttria-stabilized zirconia powders prepared by the sol-gel method for potential use in solid fuel cells. *Fuel Proces Tech,* Vol. 92, No. 2, (Fev, 2011) pp. 183-189, ISSN 0378-3820.

Gahlert, M.; Gudehus, T.; Eichhorn, S.; Steinhauser, E.; Kniha, H.; Erhardt, W. (2007). Biomechanical and histomorphometric comparasion between zirconia implants with varying surface texture and a titanium implant in the maxilla of miniature pigs. *Clin Oral Implants Res,* Vol. 18, No. 5, (Jun, 2007) pp. 662-668, ISSN 0905-7161.

Guazzato, M., Albakry, M.; Ringer, S.P.; Swain, M.V. (2004). Strength, fracture toughness and microstructure of a selection of all-ceramic materials. Part II. Zirconia based dental ceramics. *Dent Mater,* Vol. 20, No. 5, (June 2004) pp. 449-456, ISSN 0109-5641.

Hannink, R.H.J.; Kelly, P.M.; Muddle, B.C. (2000). Transformation toughening in zirconia – containing ceramics. *J Am Ceram Soc,* Vol. 83, No 3, (Mar, 2000) pp. 461-487, ISSN 1551-2916.

Heffernan, M.J.; Aquilino, S.A.; Diaz-Arnold, A.M.; Haselton, D.R.; Stanford, C.M.; Vargas, M.A. (2002). Relative translucency of six all-ceramic systems. Part I: Core materials. *J Prosthet Dent,* Vol. 88, No. 1, (Jul, 2002) pp. 4-9, ISSN 0022-3913. (a)

Heffernan, M.J.; Aquilino, S.A.; Diaz-Arnold, A.M.; Haselton, D.R.; Stanford, C.M.; Vargas, M.A. (2002). Relative translucency of six all-ceramic systems. Part II: Core and venner materials. *J Prosthet Dent,* Vol. 88, No. 1, (Jul, 2002) pp. 10-15, ISSN 0022-3913. (b)

Herring, C. (1950). Effect of change of scale on sintering phenomena. *J Appl Phys,* Vol. 21, (1950) pp. 301-303, ISSN 0021-8979.

Heshmatpour, F.; Aghakhanpour, R.B. (2011). Synthesis and characterization of nanocrystalline zirconia powder by simple sol-gel method with glucose and fructose as organic additives. *Powder Tech,* Vol. 205, No. 1-3, (Jan, 2011) pp. 193-200, ISSN 0032-5910.

Ichikawa, Y.; Akagawa, Y.; Nikai, H.; Tsuru, H. (1992). Tissue compatibility and stability of a new zirconia ceramic in vivo. *J Prosthet Dent,* Vol. 68, No. 2, (Aug, 1992) pp. 322-326, ISSN 0022-3913.

Kanade, K. G.; Baeg, J.O.; Apte, S.K.; Prakash, T.L. Kale, B.B. (2008). Synthesis and characterization of nanocrystallined zirconia by hydrothermal method. *Mater Res Bul,* Vol. 43, No. 3, (Mar, 2008) pp. 723-729, ISSN 0025-5408.

Kelly, J. R. & Denry, I. (2008). Stabilized zirconia as a structural ceramic: An overview. *Dent Mater,* Vol. 24, No. 3, (Mar, 2008) pp. 289-298, ISSN 0109-5641.

Kelly, J. R. (1995). Perspectives on strength. *Dent Mater,* Vol. 11, No. 2, (Mar, 1995) pp. 103-110, ISSN 0109-5641.

Kelly, P.M.; Francis Rose, L.R. (2002). The martensitic transformation in ceramics-its role in transformation toughening. *Prog Mater Sci,* Vol. 47, (Mar, 2002) pp. 463-557, ISSN 0079-6425.

Kerstein, R.; Radke, J. (2008). A comparasion of fabrication precision and mechanical reliability of 2 zirconia implant abutments. *Int J Oral Maxillofac Implants,* Vol. 23, No. 6, (Nov-Dec, 2008) pp. 1029-1036, ISSN 0882-2786.

Kim, B.K.; Bae, H.E.; Shim, J.S.; Lee, K.W. (2005). The influence of ceramic surface treatments on the tensile bond strength of composite resin to all-ceramic coping materials. *J Prosthet Dent,* Vol. 94, No. 4, (Oct, 2005) pp. 357-362, ISSN 0022-3913.

Koutayas, S.O.; Vagkopoulou, T.; Pelekanos, S.; Koidis, P. & Strub, J.R. (2009). Zirconia in Dentistry: Part 2. Evidence-based clinical breakthrough. *Eur J Esthet Dent,* Vol. 4, No. 4, (Winter, 2009) pp. 348-380, ISSN 1862-0612.

Lange, F.F. (1989). Powder processing science and technology for increased reliability. *J Am Ceram Soc,* Vol. 72, No. 1, (Jan, 1989) pp. 3-15, ISSN 0002-7820.

Lawson, S. (1995). Environmental Degradation of zirconia ceramics. *J Eur Ceram Soc,* Vol. 15, No 6, (1995) pp. 485-502, ISSN 0955-2219.

Lee, S.K.; Tandon, R.; Readey, M.J.; Lawn, B.R. (2000). Scratch damage on zirconia ceramics. *J Am Ceram Soc,* Vol. 83, No. 6, (June, 2000) pp. 1482-1432, ISSN 0002-7820.

Lin, F.J.T.; De Jonghe, L.C.; Rahaman, M.N. (1997). Microstructure refinement of sintered alumina by a two-step sintering technique. *J Am Ceram Soc,* Vol. 80, No. 9, (Sep, 1997) pp. 2269-2279, ISSN 0002-7820.

Luthardt, R. G.; Holzhüter, M.; Rudolph, H.; Herold,V. & Walter, M. (2004). CAD/CAM machining effects on Y-TZP zirconia. *Dent Mater,* Vol. 20, No. 7, (Sep, 2004) pp. 655-662, ISSN 0109-5641.

Madhav Reddy, K.; Mukhopadhyay, A.; Basu, B. (2010). Microstructure-mechanical-tribological property correlation of multi stages park plasma sintered tetragonal ZrO_2. *J Eur Ceram Soc,* Vol. 30, No. 16, (Dec, 2010) pp. 3363-3375, ISSN 0955-2219.

Mazaheri, M.; Simchi, A.; Golestani-Fard, F. (2008). Densification and graingrowth of nanocrystalline 3Y-TZP during two-step sintering. *J Eur Ceram Soc,* Vol. 28, No. 15, (Nov, 2008) pp. 2933-2939, ISSN 0955-2219.

Mecholsky Jr, J.J. (1995). Fracture mechanics principles. *Dent Mater,* Vol. 11, No. 2, (Mar, 1995) pp. 111-112, ISSN 0109-5641.

Miani, C.; Piconi, C.; Piselli, D.; Ponti, M. (1993). Prove sperimentali in vivo della zirconia in Implantologia Orale. *Rev Ital Osseoint,* Vol. 3, pp. 23-34, ISSN 2036-413X.

Miyazaki, T.; Hotta, Y.; Kunii, S.; Kuriyama, S. & Tamaki, Y. (2009). A review of dental CAD/CAM: current status and future perspectives from 20 years of experience. *Dent Mater,* Vol. 28, No. 1 (Jan, 2009) pp. 44-56, ISSN 0109-5641.

Oden, A.; Andersson, M.; Krystek-Ondracek, I.; Magnusson, D. (1998). Five year clinical evaluation of Procera AllCeram crowns. *J Prosthet Dent,* Vol. 80, No. 4, (Oct, 1998) pp. 450-456, ISSN 0022-3913.

Piconi, C.; Maccauro, G. (1999). Zirconia as a ceramic biomaterial. *Biomaterials,* Vol. 20, No. 1 (Jan, 1999) pp. 1-25, ISSN 0142-9612.

Piticescu, R.; Monty, C.; Millers, D. (2005). Hydrothermal synthesis of nanostructured zirconia materials: present state and future prospects. *Sensors and Actuators B,* Vol. 109, No. 1, (Aug, 2005) pp. 102-106, ISSN 0925-4005.

Quinn, J.B.; Sundar, V.; Lloyd, I.K. (2003). Influence of microstructure and chemistry on the fracture toughness of dental ceramics. *Dent Mater,* Vol. 19, No. 7, (Nov, 2003) pp. 603-611, ISSN 0109-5641.

Rahaman, M.N. (1995). *Ceramic Processing and Sintering* (2ed), Ed M. Becker, ISBN 9780824795733, New York.

Rashad, M.M.; Baioumy, H.M. (2008). Effect of thermal treatment on the crystal structure and morphology of zircon nanopowders produced three different routes. *J Mat Proc Tech,* Vol. 195, No. 1-3, (Jan, 2008) pp. 178-185, ISSN 0924-0136.

Scherrer, S.S.; Denry, I.L.; Wiskott, H.W. (1998). Comparasion of three fracture toughness testing techniques using a dental glass and a dental ceramic. *Dent Mater,* Vol. 14, No. 4, (Jul, 1998) pp. 246-255, ISSN 0109-5641.

Sõmiya, S.; Kumaki, T.; Hishinuma, K.; Nakai, Z.; Tokuji, A.; Suwa, Y. (1991). Hydrothermal precipitation of ZrO2 powder. *Prog Crystal Growth Charact,* Vol. 21, No. 1-4, (Oct, 1991) pp. 195-198, ISSN 0960-8974.

Sõmiya, T.; Akiba, T. (1999). Hydrothermal zirconia powders: a bibliography. *J Eur Ceram Soc,* Vol. 19, No. 1, (Apr, 1999) pp. 81-87, ISSN 0002-7820.

Sundh, A,; Sjörgren, G. (2008). A study of the bending resistance os implant-supported reinforced alumina and machined zirconia abutments and copies. *Dent Mater,* Vol. 24, No. 5, (Aug, 2008) pp. 611-617, ISSN 0109-5641.

Tartaj, J.; Tartaj, P. (2009). Two-stage sintering of nanosize pure zirconia. *J Am Ceram Soc,* Vol. 92, No. S1, (Jan, 2009) pp. 103-106, ISSN 0002-7820.

Thompson, J.Y.; Stoner, B.R.; Piascik, J.R. (2007). Ceramics for restorative dentistry: Critical aspects for fracture and fatigue resistance. *J Mat Science Eng,* Vol. 27, No. 3, (April, 2007) pp. 565-569, ISSN 0022-2461.

Tsalouchou, E.; Cattell, M.J.; Knowles, J.C.; Pittayachawan, P.; McDonald, A. (2008). Fatigue and fracture properties of yttria partially stabilized zirconia crown systems. *Dent Mater,* Vol. 24, No. 3, (Mar, 2008) pp. 308-318, ISSN 0109-5641.

Tsukada, T.; Venigalla, S.; Morrone, A.A.; Adair, J.H. (1999). Low temperature hydrothermal synthesis of yttrium-doped zirconia powders. *J Eur Ceram Soc,* Vol. 82, No. 5, (May, 1999) pp. 1169-1974, ISSN 0955-2219.

Vagkopoulou, T.; Koutayas, S.O.; Koidis, P. & Strub, J.R. (2009). Zirconia in Dentistry: Part 1. Discovering the nature of an upcoming bioceramic. *Eur J Esthet Dent,* Vol. 4, No. 2, (Summer, 2009) pp. 130-151, ISSN 1862-0612.

Vigolo, P.; Fonzi, F.; Majzoub, Z.; Cordioli, G. (2006). An in vitro evaluation of titanium, zirconia and alumina Procera abutments with hexagonal connection. *Int J Oral Maxillofac Implants,* Vol. 21, No. 4, (Jul-Ago, 2006) pp. 575-580, ISSN 0882-2786.

Wang, H.; Aboushelib, M.N.; Feilzer, A.J. (2008). Strength influencing variables on CAD/CAM zirconia frameworks. *Dent Mater,* Vol. 24, No. 5, (May, 2008) pp. 633-638, ISSN 0109-5641.

Yildirim, M.; Edelhoff, D.; Hanisch, O.; Spiekermann, H. (2000). Ceramic abutments. A new era in achieving optimal esthetics in impalnt dentistry. *Int J Periodontics Rest Dent,* Vol. 20, No. 1, (Feb, 2000) pp. 81-91, ISSN 0198-7569.

Yildirim, M.; Fischer, H.; Marx, R.; Edelhoff, D. (2003). In vivo fracture resistance of imaplnt-supported all-ceramic restorations. *J Prosthet Dent,* Vol. 80, No. 2, (Oct, 2003) pp. 325-331, ISSN 0022-3913.

18

The Rationale for Silicon Nitride Bearings in Orthopaedic Applications

B. Sonny Bal[1] and Mohamed Rahaman[2]
[1]*Department of Orthopaedic Surgery, School of Medicine, University of Missouri Columbia, Missouri*
[2]*Department of Materials Science and Technology Missouri University of Science and Technology, Rolla, Missouri USA*

1. Introduction

The history of silicon nitride (Si_3N_4) has been described previously; detailed analyses of particles of meteroritic rock have been shown to contain silicon nitride crystals, suggesting that this material exists naturally in the galaxy.[1] Synthetic Si_3N_4 was probably developed by Deville and Wöhler in 1859. Commercial interest in this material increased in the 1950s, as the material properties of silicon nitride were better understood, and its application in internal combustion engines was contemplated.

Today, industrial applications of silicon nitride ceramics and related composites are common; these include bearings, turbine blades, and glow plugs, related to the fact that silicon nitride has high fracture toughness, strength, and attractive wear properties.[2] Ceramic ball bearings made of silicon nitride, for example, have been used in technical applications, and their extreme strength has been validated using a number of techniques.[3] These material qualities have led many investigators to speculate that silicon nitride may also have applicability in biomedical fields, especially since it exhibits biocompatibility[4, 5] and is visible on plain radiographs as a partially radiolucent material. Because of the fortuitous combination of these properties, silicon nitride has been investigated for applications in skeletal repair, and for the bearings of prosthetic replacements of arthritic hip and knee joints.

Since clinical data are yet sparse, this review will attempt to give the reader an overview of the rationale for the use of silicon nitride in biomedical, specifically orthopaedic applications. As of the present time, surgical screws, plates, and bearings for use in prosthetic hip and knee joints have been developed and tested, using silicon nitride as the source material.[6-8] Cervical spacers and spinal fusion devices made of silicon nitride composites are presently in use, although clinical results have yet to be reported given the relatively short follow-up time. The goal of this chapter is to examine the scientific rationale to support the adoption of silicon nitride ceramics for use in biomedical, specifically, orthopaedic applications.

2. Clinical rationale for ceramics

Modern biomaterials, such as titanium alloys, polished cobalt-chromium, and high-density polyethylene have revolutionized orthopaedic surgery since the 1970's, such that at the present time, total hip replacement and total knee replacement for diseased joints are commonplace around the world, with very predictable and durable outcomes.[9, 10] One concern, over the long-term, is the host biological response to accumulated, microscopic wear debris particles in the periprosthetic joint space. Strategies to decrease bearing wear in the ball-and-socket joint of artificial hip joints, and the sliding-rolling articulation of artificial knee joints have been pursued vigorously by orthopaedic implant manufacturers and material scientists, with variable success.

The bearing surface credited with ushering in modern hip and knee replacement surgery is metal-on-polyethylene, specifically, a highly polished cobalt chromium surface articulating against a polyethylene surface. In total hip replacements, cobalt-chromium femoral heads are still used widely by surgeons, usually with prosthetic socket components made of ultrahigh-molecular-weight or cross-linked polyethylene. In total knees, the femoral components are made of cobalt-chromium that articulate with a polyethylene spacer, designed to reproduce, at least in part, the complex articular geometry of the human knee joint.

The cross-linking of polyethylene is a manufacturing strategy designed to reduce the incidence of bearing wear in prosthetic hip and knee joints.[11, 12] Low wear is desirable since particulate wear in hip and knee replacement ultimately results in inflammation, periprosthetic bone loss, and premature implant loosening, necessitating repeat surgery. These considerations apply even more acutely to the young and active patients who will place greater demands on the prosthetic joint. Alumina and zirconia ceramics were introduced to replace metal surfaces in joint replacement surgery; the goal was to offer a smooth, low-friction surface that could reduce wear more than the metal-polyethylene bearings. Evidence has shown reduced wear when ceramic surfaces are used in hip[13, 14] and knee[15] replacement surgery, instead of metal.

Despite their promising role in orthopaedic bearings, zirconia and alumina ceramics have had their drawbacks. Zirconia is an unstable material that can undergo phase transformation *in vivo*, leading to catastrophic failure; this material has been largely withdrawn from the medical market.[16] In comparison studies with cobalt-chromium, zirconia femoral heads have not shown superior wear reduction in total hip replacements.[17, 18] Alumina is the most widely used ceramic material in biomedical applications today; despite very attractive wear reductions, sporadic cases of catastrophic bearing failure requiring major repeat surgery continue to be reported.[19, 20] Clearly, there is a need for even tougher ceramic materials that are biocompatible, and can provide reliable, long-term service in orthopaedic bearings, with negligible wear rates.

2.1 Material Properties of Silicon Nitride

Silicon nitride has been developed as an industrial ceramic for more than 50 years, and during that time, its mechanical properties have been significantly improved by refining processing methods, and using additives to create composite structures. Of the different processing methods used to make silicon nitride, there are two typical processing routes, known as reaction-bonding and hot isostatic processing (HIP), respectively. Reaction-bonded silicon nitride processing is a method to produce ceramic material by incorporation and nitridation of silicon powders; this method was developed in the 1950s with the goal of developing internal combustion engines with hot-zone components made entirely from ceramics.[1] The resulting material has relatively low density, high porosity, and low strength. The HIP method was developed to address these concerns; it uses silicon nitride powder as the raw material and various sintering additives to produce bulk Si_3N_4 in confined graphite dies under a hot, nitrogen environment. Silicon nitride thus prepared has improved strength, albeit at a higher manufacturing cost.[21] A compromise is to combine the two technologies; thus, Si_3N_4 can be made by post-sintering the reaction-bonded silicon nitride in order to achieve a relatively high strength, at a fraction of the fabrication cost of HIPed Si_3N_4.

Polished test bars with a shape of 3 x 4 x 30 mm made of silicon nitride with 10% Y_2O_3 and Al_2O_3 as additives have shown an initial bending strength of approximately 600 MPa; ion implantation of the structural ceramic can increase this strength significantly, as shown by Shi et al.[22] These investigators found increases in specimen bending strength of 56%, 66%, and 35% by the implantation of Ti, Zr, and Cr ions, respectively.[22] Silicon nitride, like other ceramics, is a brittle material; typical material property tests have shown that silicon nitride has a Vickers hardness of 12-13 GPa; Young's modulus of 299 GPa, Poisson's ratio of 0.270, and a typical grain size of 0.59 μm.[23]

Composites of silicon nitride with 6 wt% yttrium oxide and 4 wt% alumina were fabricated to measure mechanical strength and related properties, according to ASTM C-1161 standards, using specimens with dimensions of 3 x 4 x 45 mm; results showed a near 100% theoretical material density (3.20 g/cm^3), average grain width of 1.5 μm, flexural strength of 923 ± 70 MPa, with a Weibull modulus of 19 and a fracture toughness of 10 ± 1 MPa.m$^{1/2}$.[24] Both material strength and toughness are at least an order of magnitude higher than typical values reported for alumina, the most common ceramic bearing material in orthopaedic bearings today. These data have been validated by other authors; using two less favorable compositions of silicon nitride doped with yttrium and aluminum, Guedes et al reported a fracture toughness of 5 MPa m$^{1/2}$ and Vickers hardness values of 13 GPa.[25]

The intrinsic material properties of silicon nitride make it suitable for articulation against bearing steel, which is a softer material than ceramic. Thus, silicon nitride has been used in rolling contact applications in automotive, turbomachinery, and power industries, where it has a significant advantage due to its low density (half that of bearing steel), low friction, corrosion resistance, and reliable performance under extreme conditions.[21] In modern aircraft and space vehicles, very demanding bearing operating conditions such as high vacuum ($<10^{-6}$ Torr), extreme temperatures (e.g. +230 to -150°C), long life (both wear and fatigue life, usually 10-15 years without maintenance) and low friction are common requirements.[26]

Fully densified Si_3N_4 has many advantages in such extreme applications; all-ceramic silicon nitride ball or roller bearings can operate against silicon nitride rolling elements and rings at temperatures up to 1000°C, at very high speeds. Hybrid ceramic-steel bearings perform just as well under these conditions; silicon nitride ceramic bearings in industry have met the requirements of higher efficiency, greater stiffness, higher speed, higher reliability, higher accuracy, lower friction, corrosion-resistance, and non-conductivity.[26] Thus, from a mechanical standpoint, silicon nitride ceramic should be adaptable to orthopaedic bearings, whether articulating against polyethylene, metal, or silicon nitride itself.[8] Practical barriers to widespread adoption of this technology are related to material and processing costs, and the need for reproducibility and reliability; these problems are common to all ceramic-powder blending and sintering processes.

3. Tribological properties of silicon nitride

The suitability of silicon nitride for hip and knee bearings has been debated in the literature, but most authors agree that in the absence of material oxidation *in vivo*, silicon nitride has the low friction necessary to articulate against itself, even when water is the only lubricant.[27] Published friction and wear data for different types of silicon nitride show a wide scatter due to different test conditions used in various studies.[28] Prototype total hip bearings have been fabricated using sintered Si_3N_4 composites (Amedica Corporation, Salt Lake City, Utah, USA); these have confirmed improved fracture toughness and strength over medical-grade alumina, and when tested in a hip simulator, both cobalt-chromium and silicon nitride femoral heads produced low wear rates against silicon nitride acetabular bearings that were comparable to Al_2O_3-Al_2O_3 wear rates, which are the lowest of any orthopaedic bearing.[24] Other work has validated the observation that water-lubricated silicon nitride has very low friction when sliding against itself.[29] The *in vivo* environment should be more favorable yet, since human synovial fluid is an excellent lubricant, regardless of the bearing material used.[30, 31]

Studies on the tribological behavior of ceramics have shown that the wear mechanisms depend on contact conditions during laboratory testing. In most structural ceramics such as silicon nitride, wear occurs through a small-scale surface fracture process if the contact load exceeds a threshold value specific to that material. Another mechanism whereby wear can be generated is surface oxidation of the material. In laboratory testing with pure silicon nitride, both mechanisms of wear have been confirmed.[28] Accordingly, for stable, long-term steady performance of silicon nitride orthopaedic bearings, a chemically-inert material composition that is impervious to surface oxidation is mandatory. Precise control of the raw powders and processing parameters is critical in manufacturing bearing components with predictable, reliable *in vivo* behavior that can last for the remaining decades of a patient's lifespan.

Laboratory investigations using finite element analyses support the use of silicon nitride in load-bearing hip resurfacing components; these differ from hip replacement in that the diseased femoral head is resurfaced with a prosthetic cap, rather than being cut out and replaced entirely.[32] Stress distributions in the proximal femur bone with implanted silicon nitride hip prostheses are similar to those of intact, healthy bone. Mazzocchi et al investigated silicon nitride ceramics for their potential use in orthopaedic implants, and validated several properties that are critical to biomedical applications, such as wetting behavior and wear performance that simulates conditions typical of a hip joint prosthesis.[33] In three different silicon nitride ceramic materials prepared, these investigators found a lower contact angle of water when compared to oxide ceramics such as alumina and zirconia. Also, very low friction coefficients were consistently measured with undetectable surface modifications and wear tracks in the silicon nitride materials tested, using a disc-on-ball model of wear detection.[33]

4. Material stability of silicon nitride

A key concern in evaluating the use of any new material for *in vivo* implantation is the long-term stability of that material, i.e., a lack of corrosion, oxidation, and other chemical alterations that can affect material properties after implantation. Silicon nitride in raw, unimplanted form, when exposed to flowing oxygen, rapidly forms oxides on the surface that are populated with cracks and pores. Investigations with additives of other materials can dramatically change this corrosive behavior, such that the resulting composites are suitable for long-term material stability, with imperviousness to oxidative degradation.

Surface modifications for different silicon nitride material preparations were investigated for a duration of 45 days, using liquid media water and isotonic physiological saline solution.[33] Both weight changes in the specimens, and scanning electron microscopic examination of the exposed surfaces were done to identify morphological and chemical modifications. These experiments showed a very limited surface modification related to exposure to oxygen-containing media; the newly formed phases were limited to the boundary zone, in the nano-scale.[33] Silicon nitride ceramics contain, besides silicon nitride grains, the grain boundary phases formed by sintering additives, and SiO_2 that usually exists on the surface of the starting silicon nitride powders. Independent of additives, during sintering, SiO_2 partially decomposes forming surface gradients, or even leads to metallic Si inclusions in the ceramic. The often-used additive Al_2O_3 partially dissolves into the Si_3N_4 grains by a chemical reaction; the resulting boundary phase has a decisive influence on the mechanical properties and oxidative behavior of the bulk ceramic. Different Y_2O_3/ Al_2O_3 containing silicon nitride ceramics with amorphous grain boundaries exhibit varying degrees of corrosion-resistance, even to acids.[34]

Aluminum implantation into raw silicon nitride is a valid strategy to control material oxidation that is mediated by sodium.[35] The beneficial role of aluminum is in surface modification of the ceramic, so that sodium-accelerated oxidation processes can be reversed. In addition, the surface morphology and phase characteristics of the oxides are enhanced, resulting in smooth and glassy oxide layers that may play a protective role during oxidation. Related work has identified the optimal concentrations of aluminum ion implantation necessary for the optimization of the oxidation resistance of Si_3N_4 ceramics.[36]

Accelerated aging *in vivo* was modeled using autoclaving of prototype Si_3N_4 bearings; the autoclave environment exposes the material to high temperatures and humidity. Despite exposure to autoclaving for 100 hours, composite silicon nitride doped with Al_2O_3 and Y_2O_3 (Amedica Corporation, Salt Lake City, UT) showed no phase changes on x-ray diffraction, and the material maintained its superior flexural strength.[8] Thus, while phase changes and material degradation are a concern with implantable materials, and the history of catastrophic failures of zirconia bearings in total hip replacement[37-39] sound a cautionary note in this regard, existing evidence concerning the properties of silicon nitride composites, and the successful use of silicon nitride in critical industrial environments attest to the stability of this material, and to its suitability for use in the *in vivo* environment. Finally, the material has been implanted in spinal applications in the United States for at least two years now, with no adverse reports concerning any implant.

5. Biocompatibility of silicon nitride

A requirement of any material used for *in vivo* applications is that it must be bioinert, i.e., the material must not demonstrate toxicity in bulk or particulate form. Oxide ceramics, such as zirconia and alumina were attractive as initial ceramic materials in orthopaedic applications because of their excellent biocompatibility, in addition to their wear resistance. Recent evidence with cytotoxicity assays shows that silicon nitride ceramics may have a similar, favorable biocompatibility profile.

Not only are silicon nitride ceramics non-toxic, but the material may encourage cell adhesion, normal proliferation, and differentiation. Neumann et al investigated silicon nitride ceramics of different surface properties, with titanium alloy as a reference; cytotoxicity testing, cell viability, and morphology assessment were performed applying the

L929-mice fibroblast cell culture model in a direct contact assay.[40] These investigators reported favorable results with all silicon nitride materials tested; cell growth, viability, and morphology were comparable to titanium, and polished silicon nitride surfaces appeared to promote cell growth. Further investigation compared industrial-grade silicon nitride using the L929-cell culture model, with alumina and titanium alloy as controls.[41] Again, silicon nitride ceramics showed no cytotoxicity and favorable physicochemical properties. Investigators concluded that silicon nitride ceramic should be considered for biomedical applications.

The biocompatibility of Si_3N_4 has also been assessed in an in vitro model using the human osteoblast-like MG-63 cell line.[42] Results showed that silicon nitride is a non-toxic, biocompatible ceramic surface for the propagation of functional human bone cells in vitro. Its high wear resistance and ability to support bone cell growth and metabolism make silicon nitride an attractive candidate for clinical application. Cappi et al performed mechanical investigations and cell culture tests with mouse fibroblast cells (L929) and human mesenchymal stem cells on silicon nitride; excellent cytocompatibility was demonstrated by live/dead staining for both types of cells.[43] Furthermore, the human mesenchymal stem cells were able to differentiate towards osteoblasts on all silicon-based ceramic materials tested. Guedes et al implanted silicon nitride ceramic constructs into rabbit tibias for 8 weeks, and showed no adverse reaction, with bone ingrowth occurring into and around the implants.[44] In a separate investigation, the authors also found that silicon nitride-based ceramics did not elicit any toxic response when tested with standard cell culture models.[25]

Howlett et al investigated the effect of silicon nitride on rabbit marrow stromal cells and their differentiation when grown in vitro and *in vivo*.[4] In vitro, marrow stromal cells attached to the ceramic discs; fresh marrow stromal cells formed cartilage, bone, and fibrous tissue when implanted with silicon nitride, or without, into the intraperitoneal cavity of rabbits. When inserted into living bone, silicon nitride promoted the formation of a cuff of bone, conferring osseous stability; the material itself remained unchanged during the animal's life, with morphologically normal tissue found adjacent to the implant upon autopsy.[4]

6. Orthopaedic applications of silicon nitride

Ceramic materials have been used in orthopaedic bearings for several decades; their advantages over cobalt-chrome metal in terms of low friction and improved wear qualities are well known, and have been reviewed extensively.[45] Silicon nitride ceramic materials are markedly different from the other, alumina-based ceramics presently used in orthopaedic surgery. While alumina, and oxidized zirconium are used presently in the bearings of total hip[46-49], and total knee replacements[50-53], one unique property of silicon nitride ceramics pertains to its ability to be formulated into a porous substrate as well as a hard glassy bearing surface. As a porous material, silicon nitride is capable of direct bone ingrowth. Of all ceramic materials used in biomedical applications therefore, silicon nitride is the only one that addresses the possibility of monolithic implants, capable of an articulating smooth surface on one side, with a porous ingrowth surface fabricated on the opposing side of the same implant. Thus, several skeletal applications of silicon nitride are feasible.

Total joint replacements, like prosthetic hip and knee arthroplasty, require materials with low wear rates and favorable frictional coefficients that remain stable *in vivo*, for several

decades of service life. Silicon nitride articulating against itself, or metal, or polyethylene seems to satisfy this requirement since under test conditions, the contact surface of silicon nitride becomes ultra-smooth due to tribochemical polishing, and the friction becomes very low at increasing sliding distances.[27] Thus, silicon nitride should be an excellent material for total hip bearings, especially in light of very favorable tribological properties when this material articulates against itself in water. In theory, if oxidation were a significant concern, non-oxide ceramics like silicon nitride should not be suitable for hip and knee bearings, but as studies have demonstrated, surface oxidation can be controlled or eliminated by doping raw silicon nitride with selected additives, thereby increasing its resistance to oxidation.[24, 54]
Osteofixation using plates and screws, such as in maxillofacial surgery is another potential application of silicon nitride. Unlike metal devices, silicon nitride is partially radiolucent, which means that both the implant and the underlying bone can be visualized on plain radiography. Such is not the case with metal implants. Reaction-bonded porous silicon nitride yields an implant material suitable for spinal surgery, particularly fusion of intervertebral bodies; in this application, silicon nitride has already been in clinical use in the United States for at least two years, with no reports of adverse effects.[55] Other potential biomedical applications of silicon nitride include drug-release devices, microelectro-mechanical systems (Bio-MEMS), and traumatic reconstructions of otorhinolaryngologic skeletal defects.[56-59]
A cancellous-structured porous silicon nitride composite ceramic has been developed and is in commercial use as a spinal fusion implant; cylindrical implants have shown bone ingrowth rates similar to those reported for porous titanium, indicating that porous silicon nitride is an excellent substrate for implants designed for direct, biological skeletal fixation.[60] New bone forms even in the cortical region of the rabbit tibia, and around silicon nitride implants, suggesting that the material is osteoconductive, and promotes stable osseous fixation.[61]

7. Conclusions

Ceramic materials have remarkable properties that have fueled speculation about their potential applications in the biomedical field, where the need for improved biocompatibility, strength, endurance, reliability, and related properties is especially acute. Oxide ceramics such as zirconium and alumina have been used in skeletal reconstruction; specifically as bearings in total hip and knee replacements. Today, alumina is the most common ceramic bearing used in orthopaedic surgery, and oxidized zirconium has replaced zirconia as a bearing surface.
As the world population increases, the demand for maintaining an active, healthy lifestyle has increased and will likely do so for the foreseeable future. Consistent with this demand, the need for artificial hip and knee replacements has continued a steady upward trend, especially in economically developed nations.[62] The limitations of the materials used in orthopaedic joint reconstructions are evident in the significant burden of repeat surgery, with attendant increases in costs and morbidity, associated with failed total hip and knee replacements.[63, 64] Improved materials, such as silicon nitride composites, when thoroughly investigated in terms of their mechanical properties and suitability for *in vivo* implantation, may play a role in the development of future orthopaedic implants that can relieve human suffering and dysfunction in the years to come. A half century after industrial silicon nitride ceramics were developed, this material may yet fulfill its promise in the biomedical field.

8. References

[1] Riley FL. Silicon Nitride and Related Materials. *Journal of the American Ceramic Society.* 2000;83(2):245-265.

[2] Brook RE. *Concise Encyclopedia of Advanced Ceramic Materials*: Pergamon Press: Oxford; 1991.

[3] Supancic P, Danze R, Harrer W, Wang Z, Witschnig S, Schöppl O. Strength tests on silicon nitride balls. Vol 4092009:193-200.

[4] Howlett CR, McCartney E, Ching W. The effect of silicon nitride ceramic on rabbit skeletal cells and tissue. An in vitro and *in vivo* investigation. *Clinical Orthopaedics and Related Research.* 1989(244):293-304.

[5] Neumann A, Jahnke K, Maier HR, Ragoß C. Biocompatibility of silicon nitride ceramic in vitro. A comparative fluorescence-microscopic and scanning electron-microscopic study. *Biokompatibilität von siliziumnitrit-keramik in der zellkultur. Eine vergleichende fluoreszenzmikroskopische und rasterelektronen-mikroskopische untersuchung.* 2004;83(12):845-851.

[6] Neumann A, Unkel C, Werry C, et al. Osteosynthesis in facial bones. Silicon nitride ceramic as material. *Osteosynthese im gesichtsschädelbereich. Siliziumnitritkeramik als werkstoff.* 2006;54(12):937-942.

[7] Neumann A, Unkel C, Werry C, et al. Prototype of a silicon nitride ceramic-based miniplate osteofixation system for the midface. *Otolaryngology - Head and Neck Surgery.* 2006;134(6):923-930.

[8] Bal BS, Khandkar A, Lakshminarayanan R, Clarke I, Hoffman AA, Rahaman MN. Testing of silicon nitride ceramic bearings for total hip arthroplasty. *Journal of Biomedical Materials Research - Part B Applied Biomaterials.* 2008;87(2):447-454.

[9] Kowalczewski JB, Milecki M, Marczak D. [What's new in total hip replacement?]. *Chirurgia Narzadow Ruchu i Ortopedia Polska.* 2005;70(6):401-405.

[10] Blumenfeld TJ, Scott RD. The role of the cemented all-polyethylene tibial component in total knee replacement: a 30-year patient follow-up and review of the literature. *Knee.* Dec 2010;17(6):412-416.

[11] Lachiewicz PF, Geyer MR. The use of highly cross-linked polyethylene in total knee arthroplasty. *Journal of the American Academy of Orthopaedic Surgeons.* Mar 2011;19(3):143-151.

[12] Capello WN, D'Antonio JA, Ramakrishnan R, Naughton M. Continued improved wear with an annealed highly cross-linked polyethylene. *Clinical Orthopaedics & Related Research.* Mar 2011;469(3):825-830.

[13] Bascarevic Z, Vukasinovic Z, Slavkovic N, et al. Alumina-on-alumina ceramic versus metal-on-highly cross-linked polyethylene bearings in total hip arthroplasty: a comparative study. *International Orthopaedics.* Dec 2010;34(8):1129-1135.

[14] Kim Y-H, Choi Y, Kim J-S. Cementless total hip arthroplasty with alumina-on-highly cross-linked polyethylene bearing in young patients with femoral head osteonecrosis. *Journal of Arthroplasty.* Feb 2011;26(2):218-223.

[15] Oonishi H, Kim SC, Kyomoto M, Masuda S, Asano T, Clarke IC. Change in UHMWPE properties of retrieved ceramic total knee prosthesis in clinical use for 23 years. *Journal of Biomedical Materials Research Part B, Applied Biomaterials.* Aug 2005;74(2):754-759.

[16] Clarke IC, Manaka M, Green DD, et al. Current status of zirconia used in total hip implants. *Journal of Bone & Joint Surgery - American Volume.* 2003;85-A Suppl 4:73-84.
[17] Fukui K, Kaneuji A, Sugimori T, Ichiseki T, Kitamura K, Matsumoto T. Wear comparison between a highly cross-linked polyethylene and conventional polyethylene against a zirconia femoral head: minimum 5-year follow-up. *Journal of Arthroplasty.* Jan 2011;26(1):45-49.
[18] Nakahara I, Nakamura N, Nishii T, Miki H, Sakai T, Sugano N. Minimum five-year follow-up wear measurement of longevity highly cross-linked polyethylene cup against cobalt-chromium or zirconia heads. *Journal of Arthroplasty.* Dec 2010;25(8):1182-1187.
[19] Iwakiri K, Iwaki H, Minoda Y, Ohashi H, Takaoka K. Alumina inlay failure in cemented polyethylene-backed total hip arthroplasty. *Clinical Orthopaedics & Related Research.* May 2008;466(5):1186-1192.
[20] Rhoads DP, Baker KC, Israel R, Greene PW. Fracture of an alumina femoral head used in ceramic-on-ceramic total hip arthroplasty. *Journal of Arthroplasty.* Dec 2008;23(8):1239.e1225-1230.
[21] Wang W, Hadfield M, Wereszczak AA. Surface strength of silicon nitride in relation to rolling contact performance. *Ceramics International.* 2009;35(8):3339-3346.
[22] Shi F, Miao H, Peng Z, Si W, Qi L, Li W. Bending strength of ceramics implanted by titanium, zirconium, and chromium ions with MEVVA source. Vol 280-2832005:1211-1212.
[23] Chen FC, Ardell AJ. Fracture toughness of ceramics and semi-brittle alloys using a miniaturized disk-bend test. *Materials Research Innovations.* 2000;3(5):250-262.
[24] Bal BS, Khandkar A, Lakshminarayanan R, Clarke I, Hoffman AA, Rahaman MN. Fabrication and Testing of Silicon Nitride Bearings in Total Hip Arthroplasty. Winner of the 2007 "HAP" PAUL Award. *Journal of Arthroplasty.* 2009;24(1):110-116.
[25] Guedes e Silva CC, Higa OZ, Bressiani JC. Cytotoxic evaluation of silicon nitride-based ceramics. *Materials Science and Engineering C.* 2004;24(5):643-646.
[26] Wang L, Snidle RW, Gu L. Rolling contact silicon nitride bearing technology: A review of recent research. *Wear.* 2000;246(1-2):159-173.
[27] Mazzocchi M, Bellosi A. On the possibility of silicon nitride as a ceramic for structural orthopaedic implants. Part I: Processing, microstructure, mechanical properties, cytotoxicity. *Journal of Materials Science: Materials in Medicine.* 2008;19(8):2881-2887.
[28] Jahanmir S. Wear transitions and tribochemical reactions in ceramics. *Proceedings of the Institution of Mechanical Engineers, Part J: Journal of Engineering Tribology.* 2002;216(6):371-385.
[29] Iliev C. On the wear behaviour of silicon nitride sliding against metals in water. *Industrial Lubrication and Tribology.* 2010;62(1):32-36.
[30] Blewis ME, Nugent-Derfus GE, Schmidt TA, Schumacher BL, Sah RL. A model of synovial fluid lubricant composition in normal and injured joints. *European Cells & Materials.* 2007;13:26-39.
[31] Mazzucco D, Spector M. The John Charnley Award Paper. The role of joint fluid in the tribology of total joint arthroplasty. *Clinical Orthopaedics & Related Research.* Dec 2004(429):17-32.

[32] Zhang W, Titze M, Cappi B, Wirtz DC, Telle R, Fischer H. Improved mechanical long-term reliability of hip resurfacing prostheses by using silicon nitride. *Journal of Materials Science: Materials in Medicine.* 2010;21(11):3049-3057.

[33] Mazzocchi M, Gardini D, Traverso PL, Faga MG, Bellosi A. On the possibility of silicon nitride as a ceramic for structural orthopaedic implants. Part II: Chemical stability and wear resistance in body environment. *Journal of Materials Science: Materials in Medicine.* 2008;19(8):2889-2901.

[34] Herrmann M, Schilm J, Hermel W, Michaelis A. Corrosion behaviour of silicon nitride ceramics in aqueous solutions. *Nippon Seramikkusu Kyokai Gakujutsu Ronbunshi/Journal of the Ceramic Society of Japan.* 2006;114(1335):1069-1075.

[35] Cheong YS, Mukundhan P, Du HH, Withrow SP. Improved oxidation resistance of silicon nitride by aluminum implantation: I. Kinetics and oxide characteristics. *Journal of the American Ceramic Society.* 2000;83(1):154-160.

[36] Cheong YS, Mukundhan P, Du HH, Withrow SP. Improved oxidation resistance of silicon nitride by aluminum implantation: II. Analysis and optimization. *Journal of the American Ceramic Society.* 2000;83(1):161-165.

[37] Chowdhury S, Vohra YK, Lemons JE, Ueno M, Ikeda J. Accelerating aging of zirconia femoral head implants: change of surface structure and mechanical properties. *Journal of Biomedical Materials Research Part B, Applied Biomaterials.* May 2007;81(2):486-492.

[38] Hayaishi Y, Miki H, Yoshikawa H, Sugano N. Phase transformation of a new generation yttria-stabilized zirconia femoral head after total hip arthroplasty. *Modern Rheumatology.* 2008;18(6):647-650.

[39] Masonis JL, Bourne RB, Ries MD, McCalden RW, Salehi A, Kelman DC. Zirconia femoral head fractures: a clinical and retrieval analysis. *Journal of Arthroplasty.* Oct 2004;19(7):898-905.

[40] Neumann A, Jahnke K, Maier HR, Ragoss C. [Biocompatibilty of silicon nitride ceramic in vitro. A comparative fluorescence-microscopic and scanning electron-microscopic study]. *Laryngo- Rhino- Otologie.* Dec 2004;83(12):845-851.

[41] Neumann A, Reske T, Held M, Jahnke K, Ragoss C, Maier HR. Comparative investigation of the biocompatibility of various silicon nitride ceramic qualities in vitro. *Journal of Materials Science-Materials in Medicine.* Oct 2004;15(10):1135-1140.

[42] Kue R, Sohrabi A, Nagle D, Frondoza C, Hungerford D. Enhanced proliferation and osteocalcin production by human osteoblast-like MG63 cells on silicon nitride ceramic discs. *Biomaterials.* Jul 1999;20(13):1195-1201.

[43] Cappi B, Neuss S, Salber J, Telle R, Knüchel R, Fischer H. Cytocompatibility of high strength non-oxide ceramics. *Journal of Biomedical Materials Research - Part A.* 2010;93(1):67-76.

[44] Guedes E Silva CC, König Jr B, Carbonari MJ, Yoshimoto M, Allegrini Jr S, Bressiani JC. Tissue response around silicon nitride implants in rabbits. *Journal of Biomedical Materials Research - Part A.* 2008;84(2):337-343.

[45] Bal BS, Garino J, Ries M, Rahaman MN. Ceramic Materials in Total Joint Arthroplasty. *Seminars in Arthroplasty.* 2006;17(3-4):94-101.

[46] Lewis PM, Al-Belooshi A, Olsen M, Schemitch EH, Waddell JP. Prospective randomized trial comparing alumina ceramic-on-ceramic with ceramic-on-conventional

polyethylene bearings in total hip arthroplasty. *Journal of Arthroplasty.* Apr 2010;25(3):392-397.

[47] Lombardi AV, Jr., Berend KR, Seng BE, Clarke IC, Adams JB. Delta ceramic-on-alumina ceramic articulation in primary THA: prospective, randomized FDA-IDE study and retrieval analysis. *Clinical Orthopaedics & Related Research.* Feb 2010;468(2):367-374.

[48] Garvin KL, Hartman CW, Mangla J, Murdoch N, Martell JM. Wear analysis in THA utilizing oxidized zirconium and crosslinked polyethylene. *Clinical Orthopaedics & Related Research.* Jan 2009;467(1):141-145.

[49] Good V, Ries M, Barrack RL, Widding K, Hunter G, Heuer D. Reduced wear with oxidized zirconium femoral heads. *Journal of Bone & Joint Surgery - American Volume.* 2003;85-A Suppl 4:105-110.

[50] Koshino T, Okamoto R, Takagi T, Yamamoto K, Saito T. Cemented ceramic YMCK total knee arthroplasty in patients with severe rheumatoid arthritis. *Journal of Arthroplasty.* Dec 2002;17(8):1009-1015.

[51] Vavrik P, Landor I, Denk F. [Clinical evaluation of the ceramic femoral component used for reconstruction of total knee replacement]. *Acta Chirurgiae Orthopaedicae et Traumatologiae Cechoslovaca.* Dec 2008;75(6):436-442.

[52] Innocenti M, Civinini R, Carulli C, Matassi F, Villano M. The 5-year results of an oxidized zirconium femoral component for TKA. *Clinical Orthopaedics & Related Research.* May 2010;468(5):1258-1263.

[53] Tsukamoto R, Chen S, Asano T, et al. Improved wear performance with crosslinked UHMWPE and zirconia implants in knee simulation. *Acta Orthopaedica.* Jun 2006;77(3):505-511.

[54] Luo M, Hou GY, Yang JF, et al. Manufacture of fibrous β-Si₃N₄-reinforced biomorphic SiC matrix composites for bioceramic scaffold applications. *Materials Science and Engineering C.* 2009;29(4):1422-1427.

[55] Sorrell CC, Hardcastle, P.H., Druitt, R.K.. McCartney, E.R. Paper presented at: Proceedings 5th Meeting and Seminar on: Implants for Spine, Ceramics, Cells and Tissues annaul conferences1999; Faenza, Italy.

[56] Kotzar G, Freas M, Abel P, et al. Evaluation of MEMS materials of construction for implantable medical devices. *Biomaterials.* Jul 2002;23(13):2737-2750.

[57] Kristensen BW, Noraberg J, Thiebaud P, Koudelka-Hep M, Zimmer J. Biocompatibility of silicon-based arrays of electrodes coupled to organotypic hippocampal brain slice cultures. *Brain Research.* Mar 30 2001;896(1-2):1-17.

[58] Davis DH, Giannoulis CS, Johnson RW, Desai TA. Immobilization of RGD to < 1 1 1 > silicon surfaces for enhanced cell adhesion and proliferation. *Biomaterials.* Oct 2002;23(19):4019-4027.

[59] Neumann A, Unkel C, Werry C, et al. [Osteosynthesis in facial bones: silicon nitride ceramic as material]. *HNO.* Dec 2006;54(12):937-942.

[60] Anderson MC, Olsen R. Bone ingrowth into porous silicon nitride. *Journal of Biomedical Materials Research - Part A.* 2010;92(4):1598-1605.

[61] Guedes e Silva CC, König Jr B, Carbonari MJ, Yoshimoto M, Allegrini Jr S, Bressiani JC. Bone growth around silicon nitride implants-An evaluation by scanning electron microscopy. *Materials Characterization.* 2008;59(9):1339-1341.

[62] Otten R, van Roermund PM, Picavet HSJ. [Trends in the number of knee and hip arthroplasties: considerably more knee and hip prostheses due to osteoarthritis in 2030]. *Nederlands Tijdschrift voor Geneeskunde.* 2010;154:A1534.

[63] Oduwole KO, Molony DC, Walls RJ, Bashir SP, Mulhall KJ. Increasing financial burden of revision total knee arthroplasty. *Knee Surgery, Sports Traumatology, Arthroscopy.* Jul 2010;18(7):945-948.

[64] Ong KL, Mowat FS, Chan N, Lau E, Halpern MT, Kurtz SM. Economic burden of revision hip and knee arthroplasty in Medicare enrollees. *Clinical Orthopaedics & Related Research.* May 2006;446:22-28.

19

Marine-Based Carbon and Silicon Carbide Scaffolds with Patterned Surface for Tissue Engineering Applications

Miriam López-Álvarez, Julia Serra,
Alejandro de Carlos and Pío González
University of Vigo
Spain

1. Introduction

This chapter deals with the development of bioinspired biomaterials based on carbon and silicon carbide (SiC) both derived from marine resources, specifically, from the sea rush *Juncus maritimus* Linnaeus. These three dimensional scaffolds with interconnected hierarchical porosity and a patterned surface, both preserved from the natural plant of origin, constitute a promising approach for tissue engineering.

In the last decade, a broad range of solutions have been developed for each particular function in tissue regeneration, as devices with high mechanical stability for large bone defects in load-bearing long bones or injectable materials for craniofacial surgery. The fabrication of patterned surfaces on biomaterials, especially focused on parallel channels, is another of the proposed solutions that is having a increasing interest in the last years. The need to promote an oriented growth of bone tissue, which presents in long bones, an aligned orientation of calcium phosphate crystals is one of the potential applications thought for these patterned surfaces. Thus, this oriented growth of osteoblasts will provide a bridge for bone regeneration in disrupted areas in order to correct defects in this type of bones and will improve the distribution of forces in load-bearing implants by guiding the growth of bone tissue in certain areas or in certain directions (Wiemann et al., 2007; Wise et al., 2008).

Different materials such as hydroxyapatite, silicon, polystyrene, polycaprolactone, etc. have been patterned for the same purpose using laser techniques, electrospinning or others, resulting in expensive processing protocols (Wise et al., 2008). Cells and extracellular matrix (ECM) fibrils in most natural tissues are not random, but exhibit well-defined patterns and specific spatial orientation. The variability in dimensions and distribution of channels on the patterned surfaces to mimic the natural biostructures that have, furthermore, a network of micro and nanoscale pores and channels, is required to optimize the flow of the cell growth medium, products of metabolism and waste. This is extremely difficult and very expensive to replicate artificially (Green et al., 2003; Huebsch & Mooney, 2009; Lakes, 1993) and has led to the inspiration in nature, where structures with a complex and hierarchical organization abound. Thus, the biodiversity that characterizes the marine environment represents an enormous potential for the acquisition of novel microstructures.

The use of biostructures derived from the marine environment for their application as biomaterials is very recent. For instance, several authors have proposed in the last years, the use of different marine species like coral skeletons, sea urchins and sponges as three-dimensional biomatrices (Abramovitch et al., 2006; Green et al., 2008, Cunningham et al., 2009). The results have confirmed that the three dimensional topography and the surface parameters of these materials influence positively in the cell differentiation. Topography and composition of the material have been proven to affect cellular functions, such as adhesion, growth, motility, secretion and apoptosis.

Apart from the patterning, scaffolds should have an internal structure designed with a predetermined density, pore shape, and size, with appropriate interconnection pathways. High porosity levels are necessary to support migration and proliferation of osteoblasts and mesenchymal cells, bone tissue ingrowth, vascular invasion, nutrient delivery, and matrix deposition in empty spaces. In fact, the main critical factor affecting bone formation is the presence of a combined macro- and microporosity, since macropores (size >100 μm) have a critical impact on osteogenic outcomes, promotion of vascularization, and mass transportation of nutrients and waste products (Wang et al., 2000), while micropores (size around 10 μm) favor capillary formation. Currently, it is commonly accepted that 3D scaffolds should also contain nanoporosity to allow diffusion of molecules for nutrition and signaling (Ratner, 2004). Pore interconnection also plays a key role in the overall biological system, since it provides the channel for cell distribution and migration, allowing efficient *in vivo* blood vessel formation. Furthermore, pore wall roughness contributes to increase the surface area, protein adsorption, and ion exchange (Yang et al., 2008; Bettinger et al., 2008).

Following these considerations, open structure–based scaffolds with appropriate pore size, interconnectivity, and total porosity should be modeled and developed in order to provide *in vivo* blood vessel invasion and neobone tissue ingrowth within the scaffolds. The proposed sea rush *Juncus maritimus* Linnaeus has a vascular system uniformly distributed throughout its section that provides a hierarchical interconnected porosity, with channels distributed along the entire plant. As an added value, this plant presents a surface patterning in the upper epidermal layers with channels aligned in the direction of the plant's growth (Dawes, 1998).

The particular motivation for this work was to engineer two different porous scaffolds of carbon and silicon carbide with a patterned surface. Both materials have been already proven as biocompatible. The use of carbon in the biomedical industry dates back to the late sixties when the unique blood compatibility of pyrolytic carbon as well as its physical and mechanical properties allowed, since 1968, its use in heart valve components (DeBakey-Surgitool). This material has also been applied as a coating on vascular grafts and recently, carbon has experienced an increasing interest in the form of nanotubes and nanofibers (Gómez et al., 2007; Harrison & Atala, 2007). The application of silicon carbide obtained from wood precursors in the biomaterials field derived from its potential to exceed the demand in orthopedic materials due to its inherent characteristic of being chemically inert and therefore resistant to aggressive chemical and mechanical environments, also with the interconnected and hierarchical porosity preserved from the plant vascular system (Borrajo, 2006). Several *in vitro* studies have demonstrated its biocompatibility with excellent results in proliferation and osteoblastic activity with different cell lines, as well as *in vivo* tests with implants in femur of rabbit where it has been shown penetration of bone tissue into the porous structure of the bio-inspired silicon carbide (de Carlos et al., 2006; López-Álvarez et al., 2010; González et al., 2009).

2. Scaffolds fabrication

Several techniques can be followed for the fabrication of ceramics. More specifically, for the silicon carbide obtaining, the sintering with or without applying pressure over silicon carbide powder and carbon, the chemical vapour deposition (CVD) by a gas containing carbon and silicon, or the compacted by reaction over carbon and silicon carbide particles, are the most commonly used. Other technique, the fabrication by pyrolysis and reactive infiltration with molten silicon from plant precursors, is the selected in this work. This one implies lower processing temperatures than in the sintering protocols, the substitution of silicon carbide powder by silicon powder and the plant precursor, both cheaper and higher velocity of fabrication than in the protocols with gas reactions as in chemical vapour deposition (Borrajo, 2006).

Thus, the bio-inspired carbon scaffolds obtained from the sea rush *Juncus maritimus* Linnaeus were fabricated by submitting the plant to air drying for several days and then introduced in a pyrolysis furnace inside an alumina melting pot. The pyrolysis process consisted in the thermal decomposition of the plant precursor (Maity et al., 2010; Singh et al., 2003; Varela-Feria et al., 2002) by a gradual increase in temperature of 2°C/min up to 500°C. The furnace was then maintained at that temperature for 10 minutes, followed by a gradual decrease of 20°C/min down to room temperature. This process was followed for the conversion of the structural carbon in the form of cellulose in the plant into solid carbon. Once obtained the carbon scaffolds, due to its marine origin, the removal of salts (sodium chloride and potassium chloride) was seen as necessary. Samples were subjected to ultrasonic baths of 60 minutes with warm milli-Q water.

According to the literature (Tang & Bacon, 1964) and under the hypothesis of low heating slopes (2°C/min in our case) this process is explained by four phases. The first one is the evaporation of absorbed water at 150°C. This water loss leads to the second phase, the breaking of the structure of cellulose between 150 and 240°C. The third one is the decomposition of organic polymers, mainly hemicellulose, due to the breaking of C-C and C-O bonds between 240 and 400°C, resulting in the release of carbon monoxide, carbon dioxide and water. Finally, the decomposition of cellulose and lignin at a high rate from 290°C until 400°C and more slowly up to the final 500°C, takes place. Therefore, it is at a temperature higher than 400°C that the induced thermal decomposition is accompanied by the restructuring of the carbon atoms, releasing residual hydrogen and eliminating defects, causing an increase in the degree of crystallinity (Tang & Bacon, 1964; Borrajo, 2006; Byrne & Nagle, 1997). However, Pappacena *et al.* (Pappacena et al., 2009) demonstrated recently that after subjecting wood to a one hour pyrolysis process with temperatures between 400 and 2500°C, the obtained carbon is always turbostratic and not graphite even if higher temperatures tend to slightly increase the degree of crystallinity.

To obtain the bio-inspired silicon carbide scaffolds, the carbon scaffolds were covered by the optimized amount of pure silicon powder and infiltrated, under vacuum conditions, at 1550°C during 30 minutes. Silicon melting point was exceeded (1410°C) and let it flow through the interconnected porosity reacting with carbon. That temperature was achieved by following a gradual heating ramp of 10°C/min up to 1200°C and of 5°C/min from 1200 to 1550°C. After the 30 minutes of permanence the furnace temperature was decreased by a well-controlled ramp of 10°C/min to room temperature and the silicon carbide (SiC) ceramics were obtained. It has been proven in the literature that the final structure of SiC depends on the temperature achieved, obtaining the form β-SiC at temperatures around

1550°C, while hexagonal polytypes (α-SiC) are obtained at 2200-2500°C (Dressler & Riedel, 1997).

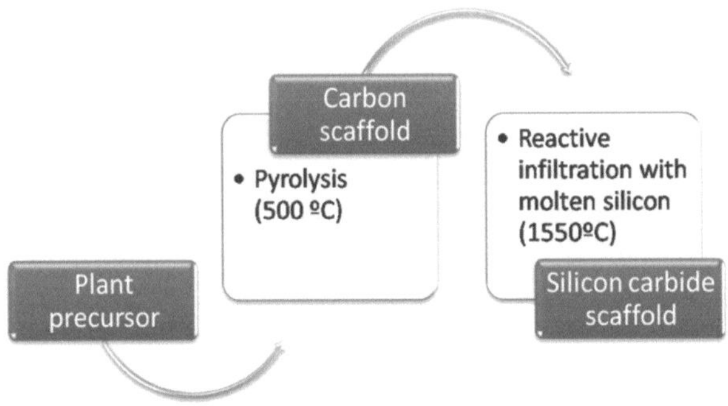

Fig. 1. Diagram of the fabrication process

Carbon graphite structure and β-SiC cubic poly-type are represented in Figure 2 (a) and (b) respectively.

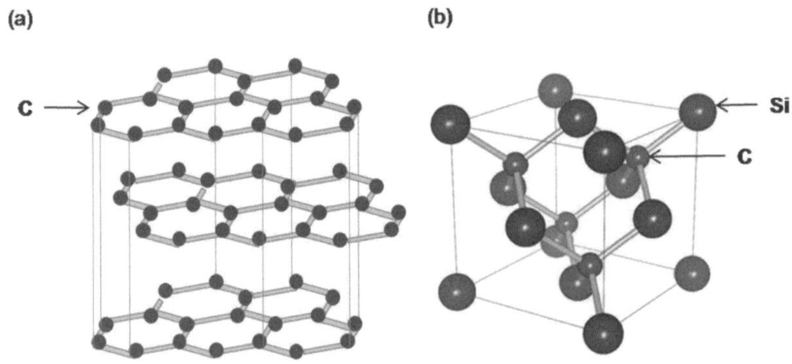

Fig. 2. Chemical structure of graphite (a) and β-SiC (b).

3. Physicochemical properties

Different techniques based on the interaction between a visible light beam (optical microscopy and Interferometric profilometry), high energy beam of electrons (SEM), X-ray beam (EDS/XRF/XRD) or a infrared radiation source (FTIR) and the atoms that constitute the material permit the evaluation of its morphology and chemical composition. At the same time the mercury´s strong surface tension prevents its penetration into a porous material by capillarity so that it is used to measure the pore distribution in the sample by applying pressure (mercury porosimetry) (Ishizaki et al., 1998).
The morphology of the marine plant *Juncus maritimus* presents in cross section the vascular bundles with pores of 20-25 μm that correspond to the xylem and of < 5 μm which are the phloem, and both of them distributed through the whole section of the plant communicating the central section with the most external layer of the epidermis of the plant (Figure 3 (a)). Therefore, abundant porosity with variable pore sized and homogeneously distributed and

interconnected with the epidermis and between them guaranteed by the vascular system constituents is presented in this marine plant. At the same time, the surface of this sea rush presents a macro-patterning with channels of around 100 μm in width oriented in the plant´s growth direction and, within each macrochannel, a micro-patterning with channels of around 7 μm oriented in the same direction (Figures 3 (b) to (d)). Arranged in rows on both sides of the macrochannels are the stomata, which correspond to oval-shaped pores used by the plant to regulate the gas exchange and water loss (Dawes, 1998), with a size of around 30 μm in diameter (Figure 3 (c)).

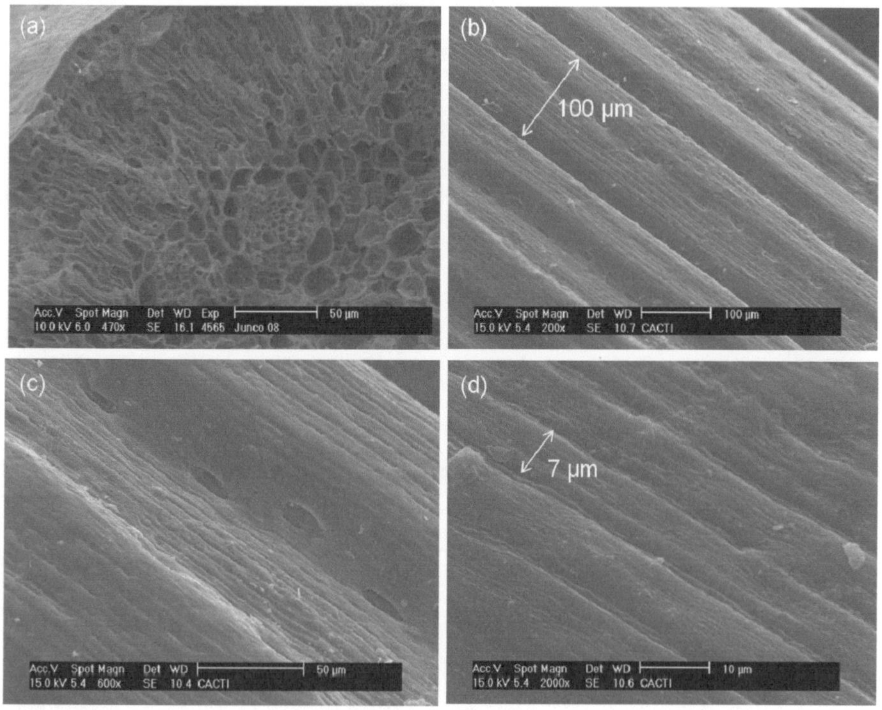

Fig. 3. SEM images of the sea rush *Juncus maritimus* in cross section (a) and surface patterning (b-d).

The carbon scaffold obtained from this sea rush preserved the double patterning with the macro and microchannels and the stomata observed previously in the natural plant as shown in Figure 4 (a). On the other hand, the silicon carbide scaffold showed the preservation of the natural microstructure after the ceramics processing, especially the macropatterning together with an increase at the surface roughness along the channels, mitigating the micro-channels topography, as consequence of the silicon carbide crystals formed on them (Figure 4 (b)). In both micrographs a detail at higher magnification (x4000) of each surface is shown, it is particularly interesting the β-SiC cubic poly-type crystal image in the Figure 4 (b). Three-dimensional images of the patterned surfaces of both materials obtained by interferometric profilometry showed by representing the variations in profile with blue for lower regions and red for the higher ones, the macro-pattern with parallel channels distributed along the surface in both materials (Figures 4 (c) and (d)). Profile variations along the channels in the silicon carbide surface as consequence of the SiC crystals can be observed in Figure 4 (d). When compared both surfaces in terms of the arithmetic average of the roughness profile (Ra), the values were 1.6 µm for the carbon and 3.5 µm for the silicon carbide pieces.

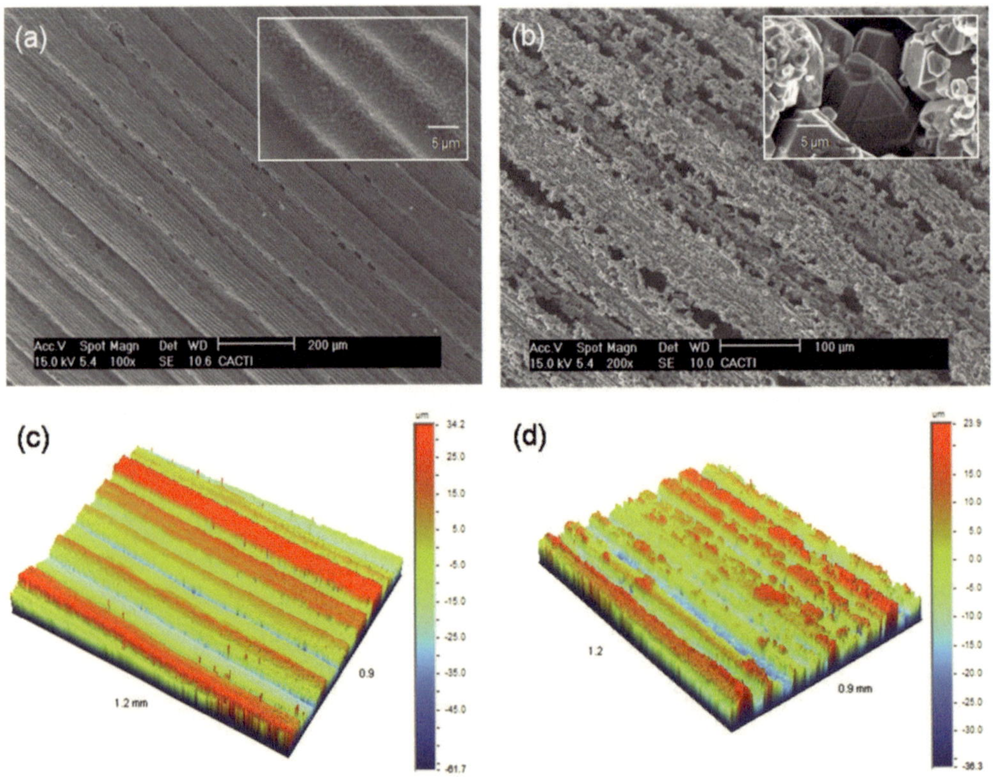

Fig. 4. SEM (a, b) and interferometric profilometry images (c, d) of a carbon scaffold (a, c) and a silicon carbide scaffold (b, d) both obtained from the sea rush *Juncus maritimus*.

The XRF analysis reveals that the composition of the C scaffold is of carbon in a percentage of 95.98% of the total mass (Table 1). Values of around 1% oxygen and sodium were detected, with close percentages of chlorine and potassium. Phosphorus and magnesium were also detected in concentrations of 0.241 and 0.150%, respectively. Calcium, sulfur, silicon and aluminum were detected in trace amounts (<0.01%). The SiC scaffold has a percentage in silicon of 69.64% and in carbon of 29.78%. In the remaining 0.58%, the most abundant element is oxygen and aluminum (1-0.1%) followed by, iron, calcium, chlorine, potassium, titanium, sulfur (0.1-0.01%) and, in trace concentrations (<0.01%), magnesium, nickel and zinc.

The presence of carbon in both materials comes from the natural plant, where the majority of the elemental composition is attributed to C, H and O, elements that can correspond to both structural compounds such as cellulose and lignin and other non structural organic matter (C-C, C-H, O, N, S, P, halogens…) in a total percentage of 93.958%. Chlorine, potassium and sodium are easily attributed to the most abundant sea salts, sodium chloride and potassium chloride, present in the natural plant in a percentage >1% for chlorine and

potassium and in the range 1-0.1% for sodium. Phosphates and silicates are nutrients found in the marine environment. Regarding the presence of zinc, this element is a micronutrient in the plant included in the structure of several enzymes what leads to attribute its presence to its own metabolism but, at the same time, the ability of this marsh plant to accumulate trace metals found in the sediment (bioaccumulation), such as cadmium, copper and zinc (Almeida et al., 2006), has been proven. The presence of aluminum is away from expected composition in this specie and can only be explained by its incorporation from the collection area. It should be noted the total absence of nitrogen that could be explained by one of the ecological requirements of this plant, a nitrogen-poor soil.
The comparison with the natural sample shows that the composition of the C scaffold has been greatly enriched in carbon with the removal of organic matter, while the presence of chlorine and potassium has been significantly decreased after subjecting the samples to several ultrasonic washes. Calcium, sulfur, silicon and aluminum were also considerably reduced and the presence of iron, zinc and nickel was no longer detected. In the case of SiC it was reduced the percentage of all elements, except silicon and it was detected the titanium for the first time. The presence of oxygen confirmed the formation of oxides with silicon or other elements. The composition of the SiC scaffold with oxides of different elements in smaller concentrations supposes a closer approximation to the living tissue composition, where the traces constituents are proven to benefit the cellular activity (Cousins, 1995).

(%)	C	H	O	Cl	K	Na	Si	Al	S	Ca	Fe	P	Mg	Ti	Ni	Zn
Natural	37.98	3.188	52.79	1.710	1.790	0.400	0.606	0.220	0.190	0.490	0.130	0.100	0.084		0.098	0.220
C scaf.	95.98		1.091	0.833	0.603	1.010	0.065	0.008	0.073	0.042		0.241	0.150			
SiC scaf.	29.78		0.216	0.038	0.015		69.64	0.130	0.014	0.052	0.091		0.007	0.016	0.003	0.003

Table 1. Elemental composition of the natural plant, carbon and silicon carbide scaffolds measured by XRF.

Both materials were analyzed by FT-IR (Figure 5). The main IR absorption bands can be attributed to the following functional groups (Smith, 1999; Dragnea et al., 2001; Qian & Jin, 2006): C-H groups exhibit sharp peaks located between 2800 and 2960 cm^{-1} associated to C-H_2 symmetric stretching (2860-2875 cm^{-1}), C-H_2 asymmetric stretching (2910-2930 cm^{-1}) and C-H_3 symmetric stretching (2950-2960 cm^{-1}). Moreover, well resolved peaks attributed to C-H_3 symmetric and asymmetric bending modes emerge at around 1380 and 1436 cm^{-1}, respectively. Carbonate groups can be identified by two absorption bands; the peak around 877 cm^{-1} associated to C-O bending vibration out-of-plane and 750 cm^{-1} related to C-O bending vibration in-plane. Other band at 1568 cm^{-1} can be associated to C=O stretching mode corresponding to carboxyle groups. The broad absorption band of absorbed water can be observed in the range 3200-3700 cm^{-1} (water stretching vibration). The hydroxyl group (-OH) has a stretching vibrational mode that appears at around 3571 cm^{-1} with a shoulder at 3550 cm^{-1}. In the case of the IR analysis of SiC scaffolds (Figure 5(b)) and besides of the previous described bands, new absorption peaks can be identified; SiC groups around 782 and 798 cm^{-1} associated to Si-C stretching vibration mode and Si-O groups which exhibit bands located around 1100 cm^{-1}, 800 cm^{-1} and 450 cm^{-1} identified as symmetric stretching, bending and rocking vibration modes respectively.

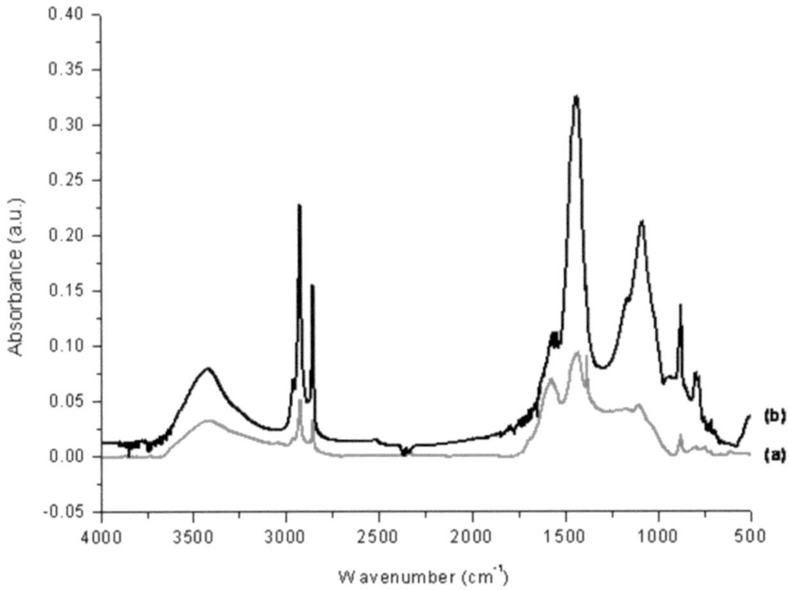

Fig. 5. FTIR analysis of the carbon (a) and silicon carbide scaffolds (b).

Two different three-dimensional scaffolds are presented in this chapter with different compositions but both made from the same natural plant precursor by following a relatively simple manufacturing process. A scaffold per se implies a porous matrix designed to guide the cells growth and to permit their ingrowth inside the matrix to favor the integration of the new tissue formed and also its vascularization. In both scaffolds the porosity of the original sea rush *Juncus maritimus* was preserved. However and due to the different chemical structure of materials, scaffolds obtained differ in size and total percentage of pores. The distribution of pores in the carbon scaffold (Figure 6) was measured, presenting a percentage of total porosity estimated in a 63% with nanopores (0.03-0.2 μm in diameter), mesopores (1-10 μm) and macropores (80-150 μm). Silicon carbide scaffolds presented a double size-scaled porosity with mesopores between 1 and 50 μm of diameter and macropores between 80 and 150 μm and a percentage of total porosity of 48%. After the infiltration with molten silicon in the ceramics processing, obviously, the nanoscaled pores were clogged.

When gone in depth into the porous structures found in the human tissues, a great variety was found even in the same type of tissue as, for instance, bone where the total percentage of pores per volume varies from 50 to 90% in femoral trabecular bone and between 3 to 12% in the cortical bone of the tibia. The same variety was observed in terms of pores sizes where two requirements have been stated as the necessity of pores larger than 100 μm to favor the formation of bone tissue with high vascularity and oxygenation, and also pores with diameters in the mesoscale (1-50 μm) or even nanoscale (<1μm) due to the recent results that indicate the higher adsorption of proteins for the anchorage of the cells when these small pores are present (Karageorgiou & Kaplan, 2005). Both scaffolds satisfy this hierarchical requirement, being a priori more beneficial for tissue engineering the carbon scaffold respect to the silicon carbide because of the nanopores presence.

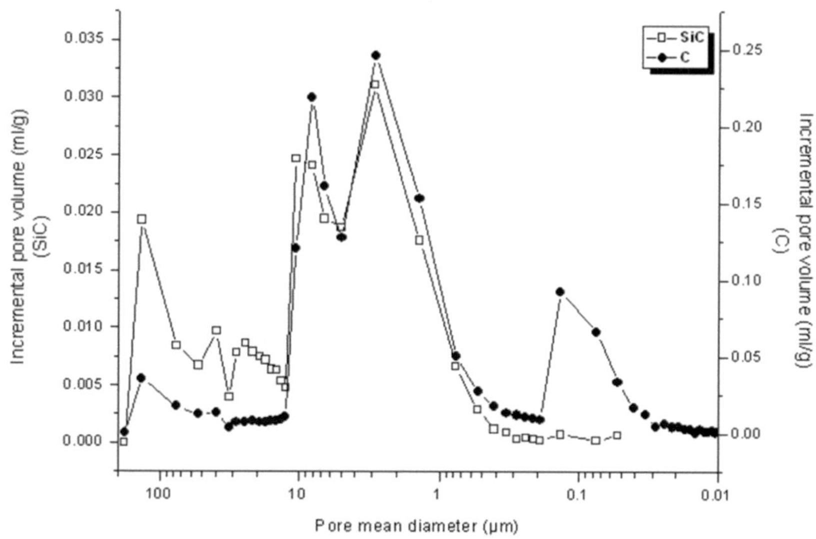

Fig. 6. Pore size distribution obtained by mercury porosimetry of a carbon scaffold (black point) and a silicon carbide scaffold (white square).

4. Evaluation of the cell morphology

The first step in the evaluation of a material as biocompatible is the cell morphology study. For testing the cell proliferation and alignment on the bio-inspired carbon and silicon carbide scaffolds proposed, pre-osteoblasts from the cell line MC3T3-E1 (ECACC, UK) were selected. This cell line was established from explants of mouse calvaria and it is considered as a standard model for *in vitro* testing of osteogenesis. In defined conditions of culture these cells differentiate to osteoblasts with the production of mature extracellular matrix and its mineralization (Wang et al., 1999; Raouf & Seth, 2000). The alignment of this cell line once differentiated to osteoblasts has been previously confirmed on different microgrooved substrata (Wang et al., 2000). Finally, and unlike other cell lines such as Saos-2 or MG-63, the selected cell line was not derived from osteosarcoma what implies a more adjusted to reality behavior. Cells alignment and proliferation on both carbon and silicon carbide scaffolds were analyzed by scanning electron microscopy.

As it is shown in the SEM micrographs of both surfaces (Figure 7), pre-osteoblasts grew aligned on both surfaces after 6 hours of incubation, by extending their filopodia along the plant-derived pattern. Cells clearly proliferated following the orientation marked by the channels, in addition, it can be appreciated how, after making contact with the surface of the material, the cells reoriented their filopodia in the marked direction (see white arrows Figure 7 (a)). Thus, the surface patterning promoted the alignment of pre-osteoblasts.

Once confirmed the cell alignment at short term, the seeded scaffolds were analyzed after longer periods of incubation with the pre-osteoblasts. Figure 8 presents both scaffolds after 7 days of incubation. On the patterned surfaces of both materials (a) and (c) it can be observed that cells have covered the whole macro-channeled surface and began to fill the spaces between. The cells present healthy appearance, with the characteristic flat morphology, adapting perfectly to the surface so that the micro-patterning underneath in the carbon scaffold (a) and the crystalline morphology underneath in the silicon carbide (c) can be still distinguished. As a porous and three-dimensional scaffold it was also important to evaluate the cell morphology in the cross section of the piece, where the porous structure is prevalent and the patterning is absent. As it can be observed in images (b) and (d) both sections have been also covered by the cells that grew appropriately inside the pores covering the cross sections, without a determined orientation due to the patterning absence.

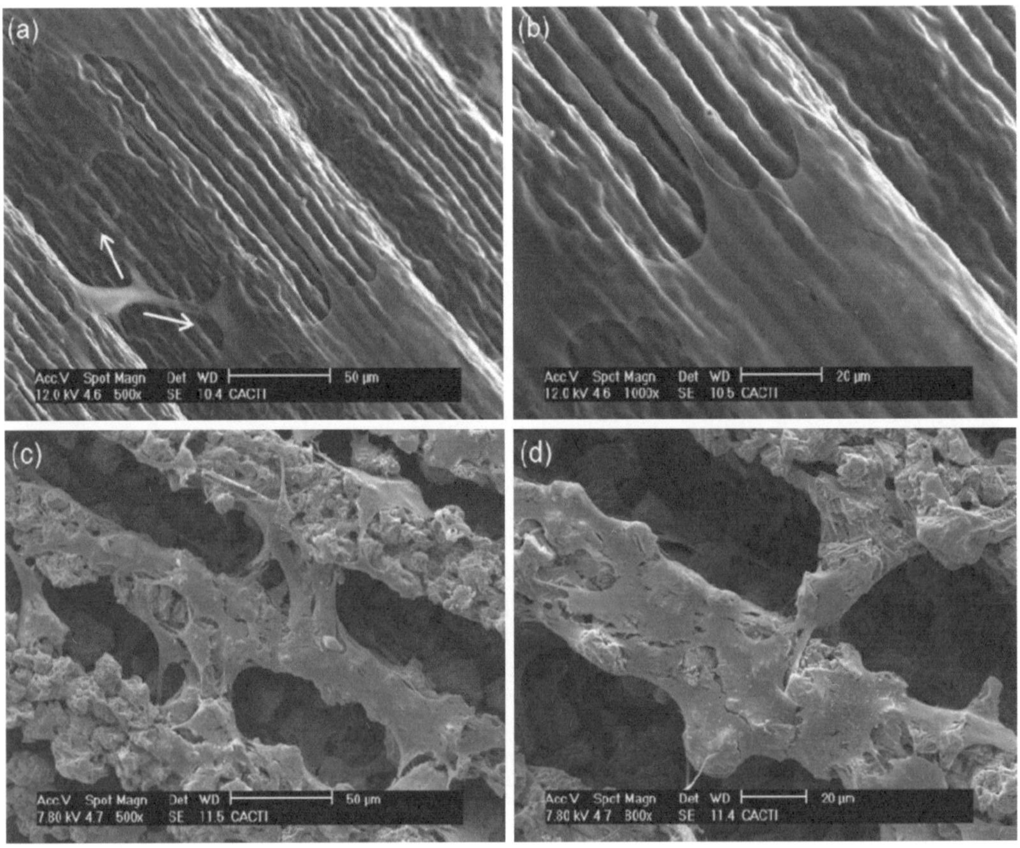

Fig. 7. SEM images showing the morphology of MC3T3-E1 cells on the patterned surface of a carbon scaffold (a, b) and a silicon carbide (c, d) after 6 hours of incubation.

The SEM micrographs of the carbon and SiC scaffolds after 28 days of incubation showed the surface completely covered by a thick layer of cells. However, in both cases the macro-patterning of the pieces was still clearly visible (a) and (c). Although after 28 days of incubation, the cells grow on a layer of cells and not directly on the surface structure of ceramics, it appears that they proliferate with a preferred orientation in the direction marked by the structure of both materials.

5. Evaluation of the cell differentiation

The evaluation of the differentiation of cells to osteoblasts is required as a proof of their functionality and of the suitability of the material. One of the most widely used tests consists on the measure of the activity of alkaline phosphatase. Alkaline phosphatase is an enzyme responsible for removing phosphate groups of various types of molecules in an alkaline environment. It is found in almost all body tissues, but its presence is particularly high in bone, and its expression, characteristic of early differentiation of osteoblast cells.

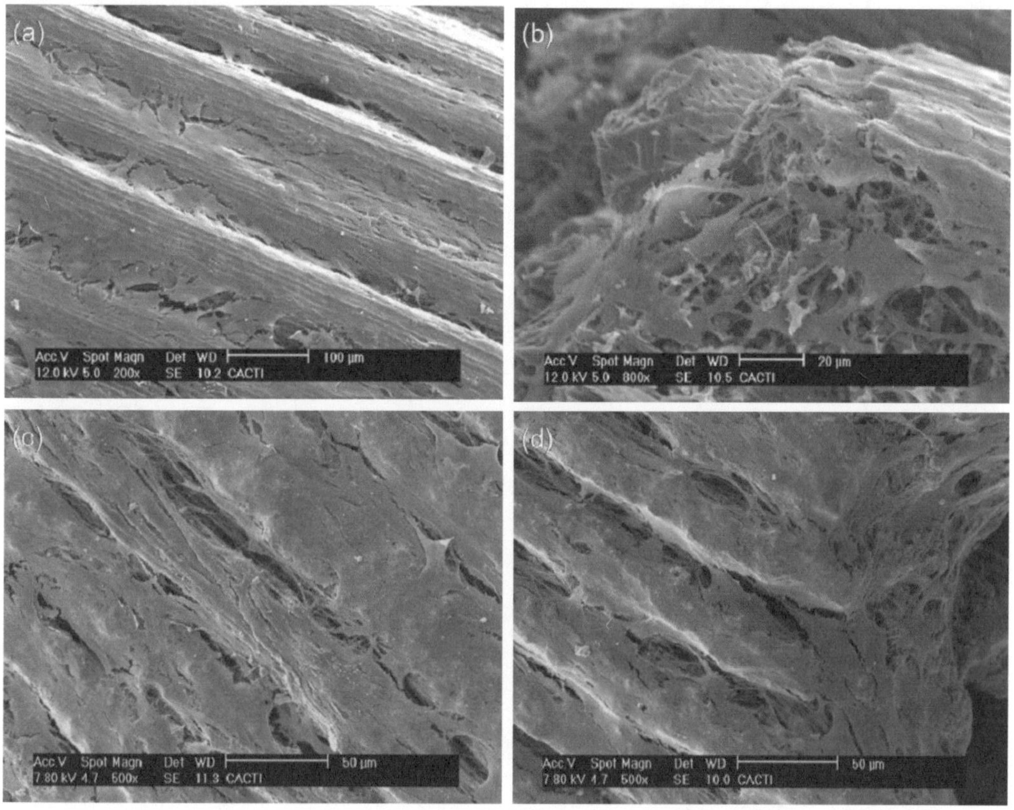

Fig. 8. SEM images showing the morphology of MC3T3-E1 cells on the carbon piece at patterned surface (a) and at the end of the scaffold in transversal section (b) and on the silicon carbide at the patterned surface (c) and at the end of the scaffold in transversal section (d) after 7 days of incubation.

To quantify osteoblastic activity and therefore evaluate the differentiation of MC3T3-E1 cells to osteoblasts enzyme activity of alkaline phosphatase (ALP) was measured. ALP activity on carbon and silicon carbide was evaluated over time (1, 7, 14, 21 and 28 days) using tissue culture polystyrene (TCP) as reference material (Figure 10). The enzyme activity estimated on carbon appeared with low values after 1 and 7 days of incubation followed by a dramatic increase on day 14th and continuing to increase until day 28 of culture. The sharp increase at day 14th suggested that the cells began to differentiate in that period. In the case of SiC the pattern followed is the same but with a much less marked increase on day 14th, to increase on day 21 and keeping up on day 28, but always in activity values lower than carbon (about 30% lower). The results for the reference material (TCP) followed the pattern found in the literature and validated the experiment. The statistical analysis of the data revealed a significant difference ($p < 0.05$) between alkaline phosphatase activity on C and TCP synthesized after 21 days of culture and SiC and TCP at 1 and 21 days of incubation ($p < 0.01$), and 28 days ($p < 0.05$).

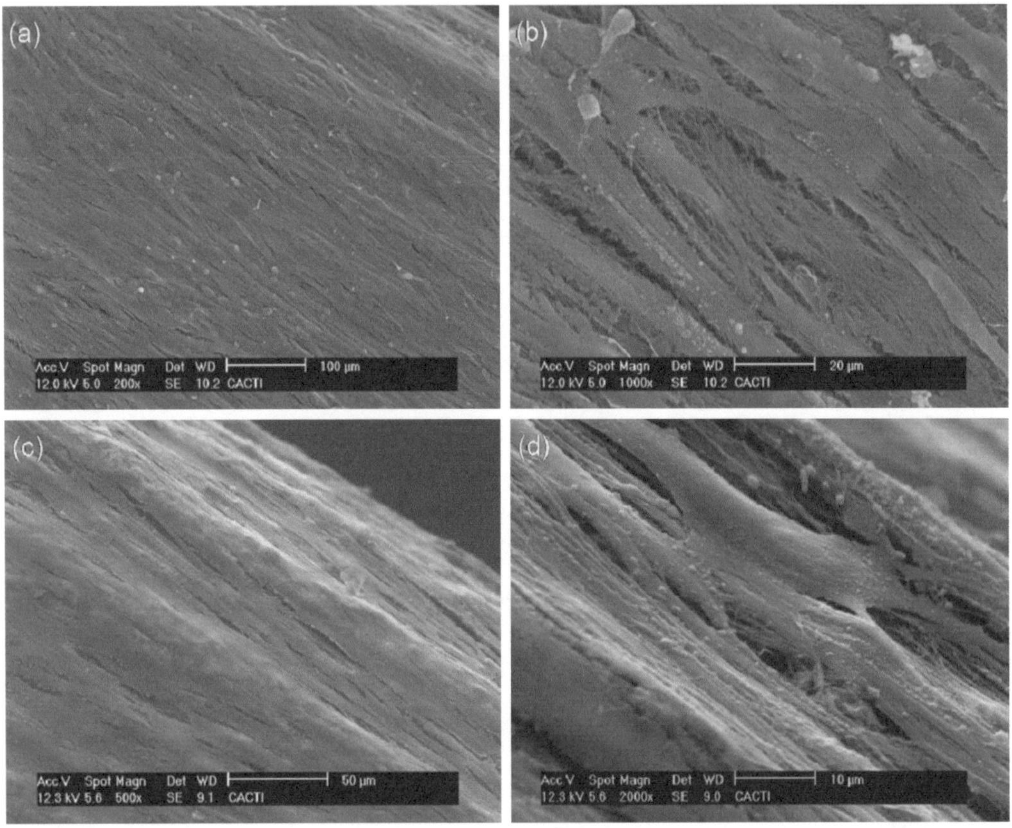

Fig. 9. SEM images showing the morphology of MC3T3-E1 cells on the patterned surface of a carbon scaffold (a, b) and of a silicon carbide scaffold (c, d), after 28 days of incubation.

Once differentiated to osteoblasts, cells appeared immersed in a complex network of fibers producing the extracellular matrix, oriented in all directions, as seen clearly in Figure 11 (a) corresponding to the surface layer of cells after 28 days in the carbon scaffold. These fibers can be appreciated in more detail in the micrograph (b). Taking into account the incubation period and the evident differentiation stage, these fibers could correspond to collagen. To confirm that, a striation across the fiber and through the whole one, should be distinguished. It can be detected a slightly striation along them (see arrow) however, this is mitigated by the gold coating of the SEM sample preparation. Regarding the silicon carbide surface after 28 days of incubation it can be observed how the cells start to secrete extracellular matrix including complex networks of filaments (c) and (d).

As it is well known bone tissue is constituted by the presence of an organic part, with the synthesis of extracellular matrix (collagen fibers) and a mineral part, production of calcium phosphate. EDS analysis can be used to reveal the presence of calcium phosphate deposits. Thus, the surface of the cell layer after 28 days of incubation on both C and SiC scaffolds was analyzed and the EDS spectra are represented in Figure 12. The presence of calcium was confirmed in several zones of the C scaffolds (Figure 12 (a, b)) and totally absent in the SiC scaffolds (Figure 12 (c)). The calcium presence can correspond to the mineralized extracellular matrix by the differentiated MC3T3-E1 pre-osteoblasts.

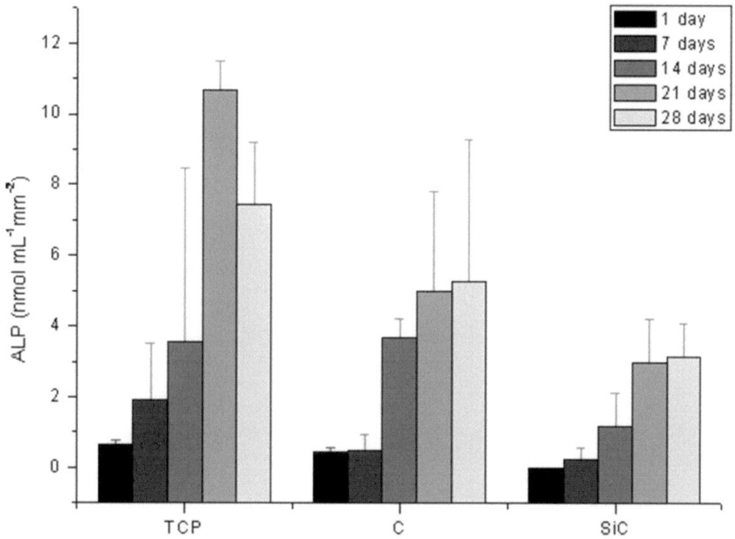

Fig. 10. ALP activity on C and SiC scaffolds after 1, 7, 14, 21 and 28 days of incubation.

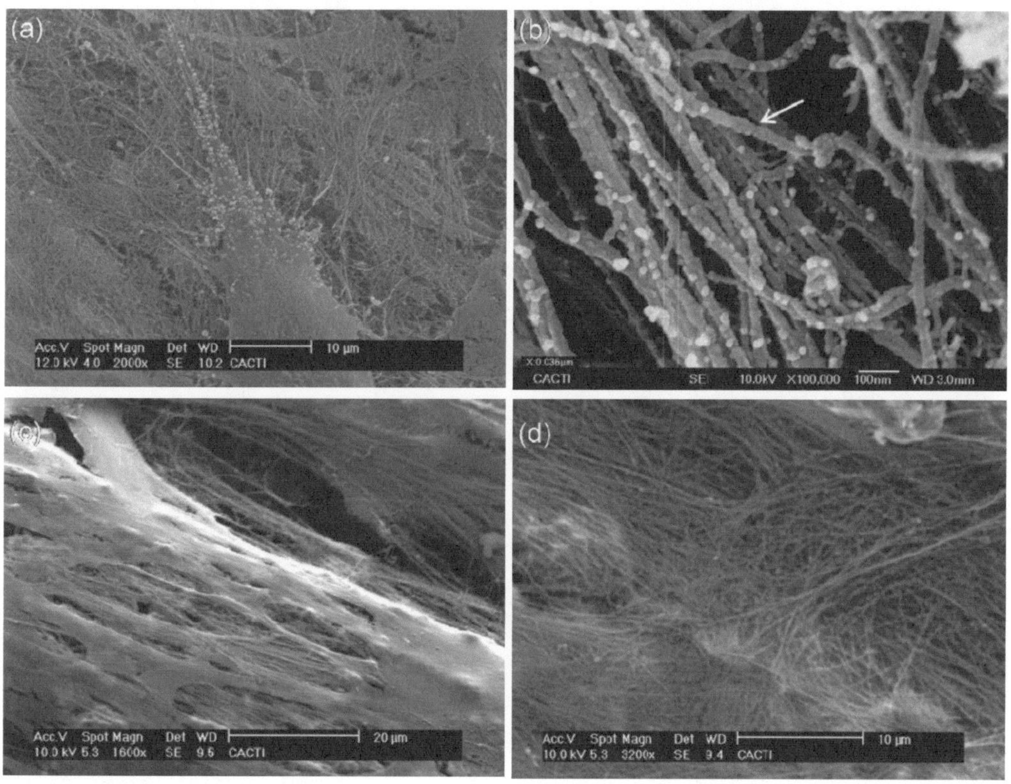

Fig. 11. SEM micrographs of the cell layer after 28 days on the surface of carbon (a, b) and silicon carbide (c, d).

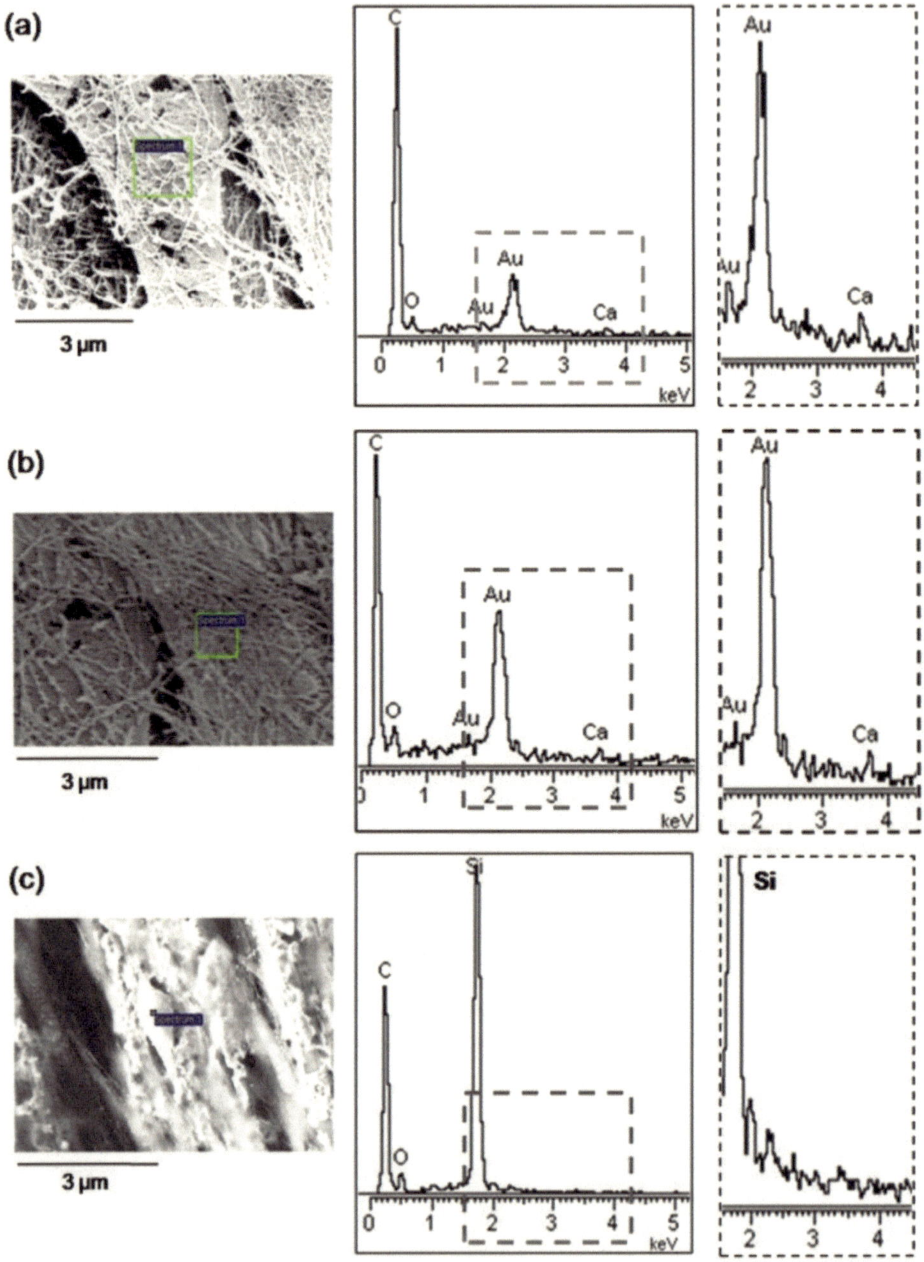

Fig. 12. EDS spectra of the carbon (a, b) and silicon carbide (c) cell layer after 28 days of incubation with MC3T3-E1. The site of interest analyzed, the general spectra and a magnification at the calcium peak area of 84 counts are represented for each one. The presence of gold comes from the SEM sample preparation

Apart from being attributed to the cells, the calcium presence at the carbon scaffolds could be assigned to the composition of the material itself (since that it could be possible that the electron beam has penetrated deeper in the monolayer of the cells achieving the material) or to the contributions of the culture medium of cells (MEM-alpha) that was renewed every two days during culture. After 28 days the quantity of calcium deposited could be higher enough to be detected by the EDS. Attending to the elemental composition (Table 1) calcium was detected in the carbon scaffold in a percentage of 0.042% and in the silicon carbide the percentage was similar (0.052%). The non-detection of calcium in the silicon carbide cell monolayer discarded the first possibility related to the material composition. Regarding with the second option, the calcium accumulation due to the culture medium contributions was also discarded since both materials were incubated under the same conditions. Therefore, taken into account the cells behaviour in both materials, the calcium detected on the monolayer of cells in the carbon scaffolds could have been produced by osteoblasts.

6. Conclusion

Recent market studies confirmed the global annual growth for the biomaterials sector, due to the aging of the population and to higher social welfare requirements. The pressing need to provide solutions for the replacement, repair and regeneration of tissues and organs brings new challenges. New disciplines such as tissue engineering and regenerative medicine require the convergence of biological sciences, materials engineering and medicine towards the design of scaffolds, with specific morphologies, which will be cultured with cells from the patient *in vitro* to obtain the required living tissue to be implanted. The variability required in the scaffolds properties comes from the specific tissue to engineer and increases the costs of their production. Natural bio-structures offer an enormous potential for the acquisition of microstructures and display the complex and hierarchical organization required in the human organism. The two porous three-dimensional bio-matrices detailed in this chapter aim to provide a promising potential solution for a specific application searched in the bone tissue requirements and focused in the alignment growth by a directional patterning.

7. Acknowledgment

This work was partially financed by POCTEP 0330IBEROMARE1P project, Xunta de Galicia 2010/83 and Ministerio de Ciencia e Innovación (MAT 2010-18281). The technical staff of CACTI (University of Vigo) and Dr. Mariana Landin (University of Santiago de Compostela) are gratefully acknowledged.

8. References

Abramovitch-Gottlib, L.; Geresh, S. & Vago, R. (2006). Biofabricated marine hydrozoans: A bioactive crystalline material promoting ossification of mesenchymal stem cells. *Tissue Engineering* Vol.12 [4], pp.729-739, ISSN 2152-4947

Almeida, C.M.R.; Mucha, A. P. & Vasconcelos M.T.S.D. (2006). Variability of metal contents in the sea rush *Juncus maritimus*-estuarine sediment system through one year of plant´s life. *Marine Environmental Research,* Vol.61, pp. 424-438, ISSN 0141-1136

Bettinger, C.J.; Borenstein, J.T. & Langer, R. (2008). Microfabrication techniques in scaffold development, In: *Nanotechnology and tissue engineering. The Scaffold,* C.T. Laurencin & L.S. Nair, (Ed.), 87-113, CRC Press, ISBN 978-1-4200-5182-7 Boca Raton, Florida

Borrajo, J.P. (2006). *Cerámicas biomórficas de carburo de silicio recubiertas con materiales bioactivos para aplicaciones biomédicas*. PhD thesis, University of Vigo.

Byrne, C.E. & Nagle, D.C. (1997). Carbonization of wood for advanced materials applications. *Carbon Journal*, Vol.35, pp. 259-266, ISSN 0008-6223

Cousins, R.J. (1995). Trace element micronutrients, In: *Molecular biology and biotechnology: a comprehensive desk reference*, R.A. Meyers (Ed.), 899-901, Wiley-Vch, ISBN 0-47118634-1 New York, US

Cunningham, E.; Dunne, N.; Walker, G.; Maggs, C.; Wilcox, R. & Buchanan, F. (2009). Hydroxyapatite bone substitutes developed via replication of natural marine sponges. *Journal of Materials Science: Materials in Medicine* Article in press, pp. 1-7, ISSN 0957-4530

Dawes, C.J. (1998). *Marine Botany*, 2nd edition, John Wiley & Sons, ISBN 0-471-19208-2 New York, US

de Carlos, A.; Borrajo, J.P.; Serra, J.; González, P. & León, B. (2006). Behaviour of MG-63 osteoblast-like cells on wood-based biomorphic SiC ceramics coated with bioactive glass. *Journal of Materials Science: Materials in Medicine*, Vol.17, pp. 523-529, ISSN 0957-4530

Dragnea, B.; Boulmer, J.; Débarre, D. & Bourguignon, B. (2001). Growth of a SiC layer on Si(100) from absorbed propene by laser melting. *Journal of Applied Physics*, Vol.90, pp. 449-455, ISSN 0021-8979

Dressler, W. & Riedel, R. (1997). Progress in silicon-based non-oxide structural ceramics. *International Journal of Refractory Metals and Hard Materials*, Vol.15, pp. 13-47, ISSN 0958-0611

Gómez de Salazar, J.M., Barrena, M.I.; Merino, C.; Plaete, O. & Morales, G. (2007). Preparación y estudio de materiales compuestos nanofibras de carbono/poliéster laminados con fibra de vidrio. *Anales de la Mecánica de Fractura*, Vol.1, pp. 234-238, ISSN 0213-3725

González, P.; Borrajo, J.P.; Serra, J.; Chiussi, S.; León, B.; Martínez-Fernández, J.; Varela-Feria, F.M.; de Arellano-López, A.R.; de Carlos, A.; Muñoz, F.M.; López, M. & Singh, M. (2009). A new generation of bio-derived ceramic materials for medical applications. *Journal of Biomedical Materials Research Part A*, Vol.88, No.3, pp. 807-813, ISSN 1549-3296

Green, D.; Howard, D.; Yang, X.; Kelly, M. & Oreffo, R.O.C. (2003). Natural marine sponge fiber skeleton: a biomimetic scaffold for human osteoprogenitor cell attachment, growth and differentiation. *Tissue Engineering* Vol.9, pp. 1159-1166, ISSN 2152-4947

Green, D.W. (2008). Tissue bionics: examples in biomimetic tissue engineering. *Biomedical Materials* Vol.3, pp.1-11, ISSN 1748-6041

Harrison, B.S. & Atala, A. (2007). Carbon nanotube applications for tissue engineering. *Biomaterials*, Vol.28, pp. 344-353, ISSN 0142-9612

Huebsch, N. & Mooney, D.J. (2009). Inspiration and application in the evolution of biomaterials. *Nature* Vol.462, pp. 426-432, ISSN 0028-0836

Ishizaki, K.; Komarneni, S. & Nanko, M. (1998). *Porous materials: Process technology and application*. Kluwer Academic Publishers, ISBN 0-412-71110-9, Dordrecht, The Netherlands

Karageorgiou, V. & Kaplan D. (2005). Porosity of 3D biomaterial scaffolds and osteogenesis. *Biomaterials*, Vol.26, pp. 5474-5491, ISSN 0142-9612

Lakes, R. (1993). Materials with structural hierarchy. *Nature* Vol.361, pp.511-515, ISSN 0028-0836

López-Álvarez, M.; de Carlos, A.; González, P.; Serra, J. & León, B. (2010). Cytocompatibility of bio-inspired silicon carbide ceramics. *Journal of Biomedical Materials Research Part B Applied Biomaterials,* Vol.95, No.1, pp. 177-183 ISSN 1552-4973

Maity, A.; Kalita, D.; Kayal, T.K.; Goswami, T.; Chakrabarti, O.; Maiti, H.S. & Rao, P.G. (2010). Synthesis of SiC ceramics from processed cellulosic bio-precursor. *Ceramics International Journal,* Vol.36, pp. 323-331, ISSN 0272-8842

Pappacena, K.E.; Gentry, S.P.; Wilkes, T.E.; Johnson, M.T.; Xie, S.; Davis, A. & Faber K.T. (2009). Effect of pyrolyzation temperature on wood-derived carbon and silicon carbide. *Journal of the European Ceramic Society,* Vol.29, pp.3069-3077, ISSN 0955-2219

Qian, J.M. & Jin, Z.H. (2006). Preparation and characterization of porous, biomorphic SiC ceramic with hybrid pore structure. *Journal of the European Ceramic Society,* Vol.26, pp. 1311-1316, ISSN 0955-2219

Raouf, A. & Seth, A. (2000). Ets transcription factors and targets in osteogenesis. *Oncogene,* Vol.19, pp. 6455-6463, ISSN 0950-9232

Ratner, B.D. (2004). A history of biomaterials, In: *Biomaterials science: an introduction to materials in medicine,* B.D. Ratner, A. S. Hoffmann and F. J. Schoen, (Ed.), 10-19, Elsevier Academic Press, ISBN 0-12582463-7, London, United Kingdom

Smith, B. (1999). *Infrared Spectral Interpretation A Systematic Approach,* CRC Press LLC, ISBN 0-84932463-7 Boca Raton, Florida, US

Singh, M.; Martínez-Fernández, J. & de Arellano-López A.R. (2003). Environmentally conscious ceramics (ecoceramics) from natural wood precursors. *Current Opinion in Solid State & Materials Science,* Vol.7, pp. 247-254, ISSN 1359-0286

Tang, M.M. & Bacon, R. (1964). Carbonization of cellulose fibers I. Low temperature pyrolysis. *Carbon Journal,* Vol.2, pp.211-214 ISSN 0008-6223

Varela-Feria, F.M.; de Arellano-López, A.R. & Martínez-Fernández, J. (2002). Fabricación y propiedades del carburo de silicio biomórfico: maderas cerámicas. *Boletín de la Sociedad Española de Cerámica y Vidrio,* Vol.41, pp. 377-384, ISSN 0366-3175

Wang, J.H.C.; Grood, E.S.; Florer, J. & Wenstrup, R. (2000). Alignment and proliferation of MC3T3-E1 osteoblasts in microgrooved silicone substrata subjected to cyclic stretching., *Journal of Biomechanics* Vol.33, pp. 729-735, ISSN 0021-9290

Wang, K.; Christensen, K.; Chawla, K.; Xiao, F.; Krebsbach, P.H. & Franceschi, R.T. (1999). Isolation and characterization of MC3T3-E1 preosteoblast subclones with distinct in vitro and in vivo differentiation/mineralization potential. Journal of Bone and Mineral Research, Vol.14, pp. 893-903, ISSN 08840431

Wiemann, M.; Bingmann, D.; Franzka, S.; Hartmann, N.; Urch, H. &, Epple, M. (2007). Oriented Growth of Osteoblast-like cells on two-dimensionally structured films of functionalized calcium phosphate nanoparticles on a silicon substrate. *Advanced Engineering Materials* Vol.9 [12], pp.1077-1081, ISSN 1438-1656

Wise, J.K.; Cho, M.; Zussman, E.; Megaridis, C.M. & Yarin, A.L. (2008). Electrospinning techniques to control deposition and structural alignment of nanofibrous scaffolds for celular orientation and cytoskeletal reorganization, In: *Nanotechnology and Tissue Engineering. The Scaffold,* C.T. Laurencin & L.S. Nair, (Ed.), 243-260, CRC Press, ISBN 978-1-4200-5182-7 Boca Raton, Florida

Yang, F.; Neeley, W.L.; Moore, M.J.; Darp, J.M.; Shukla, A. & and Langer, R. (2008). Tissue engineering: the therapeutic strategy of the twenty-first century, In: *Nanotechnology and tissue engineering. The Scaffold,* C.T. Laurencin & L.S. Nair, (Ed.), 3-33, CRC Press, ISBN 978-1-4200-5182-7 Boca Raton, Florida, US

20

ZrO_2-Bioglass Dental Ceramics: Processing, Structural and Mechanics Characterization

Luiz A. Bicalho[1], Carlos A. R. P. Baptista[2], Miguel J. R. Barboza[2], Claudinei dos Santos[3] and Renato C. Souza[4]
[1]*Universidade Federal Fluminense*
[2]*Universidade de São Paulo*
[3]*Centro Universitário de Volta Redonda*
[4]*Instituto Federal de Educação Ciência e Tecnologia*
Brazil

1. Introduction

The continuous evolution in the development and use of ceramics in various applications, which have hitherto not been considered, have been studied in order to reduce costs and increase the mechanical properties, promoting a longer life applications, with quality assurance.

When considering the use of ceramics in structural materials as materials for implants and implant components, it can be noted as acceptable to meet the demands from the work of mastication, bending fracture strength of about 250MPa and toughness fracture, fracture toughness, about $3MPa.m^{1/2}$. It is understood that results of these characterizations above indicators guarantees of reliability (ANUSAVICE, 2005).

The polycrystalline tetragonal zirconia is widely used as an agent for other toughened ceramics, because this material has a phase transformation induced by stress, a change of metastable tetragonal phase to monoclinic phase is accompanied by a volume expansion (3-6%), as specialized bibliographies. The transformation absorbs part of the energy required for crack propagation, with an increase in fracture toughness.

Bioglasses are bioactive materials, which are based on the following hypothesis: "The biocompatibility of an implant material is great if the material provides the formation of normal tissue on its surface and, additionally, if it establishes a seamless interface capable of withstanding the loads that normally occur at the site of implantation"(KOHN; DUCHEYNE; AWERBUCH, 1992).

The use of Bioglasses as sintering additives was studied by Amaral (AMARAL, 2002) in Si_3N_4 and Huang (HUANG, 2003) in ZrO_2. This practice reduced the final sintering temperature, without significantly affecting the properties of these materials for dental applications.

In the present work was used as sintering additive, a Bioglass system CaO-P_2O_5-SiO_2-MgO, for application as biomaterial. The use of this additive reduced the final sintering temperature, reducing the manufacturing cost of the product while maintaining the biocompatibility of the product. The bioactive, by having the thermal expansion coefficient

close to the materials used in coatings in cosmetic dental implants, improve the adhesion between implant components based on ZrO_2 and the crown (prosthetic) teeth. It is expected that the Bioglass intergranular additive occupies the interstices and gaps of zirconia and thereby minimize the internal porosity, increasing the mechanical strength and fracture toughness in sintered materials at low temperatures, since the interstices and voids represent the possibility of appearance of micro cracks.
The main objective of this study is to evaluate the microstructural aspects and the physical and mechanical properties of Y-TZP ceramics, $ZrO_2(Y_2O_3)$, sintered, and Bioglass system $3CaO-P_2O_5-SiO_2-MgO$ as an additive to liquid phase sintering.

2. Materials

The materials used in this work were commercially available:
Tetragonal yttria stabilized Zirconia (ZrO2) ceramic Y-TZP containing 3mol% Y2O3, with particle average size of 0.97 μm;Ca $(H_2PO_4)_2.H_2O$, high purity (99.99%); $CaCO_3$, high purity (99.99%); SiO_2, high purity (99.99%) and MgO, high purity (99.99%).

3. Experimental procedures

The necessary procedures of processing steps and characterization of materials used in this work are shown in Figure 1.

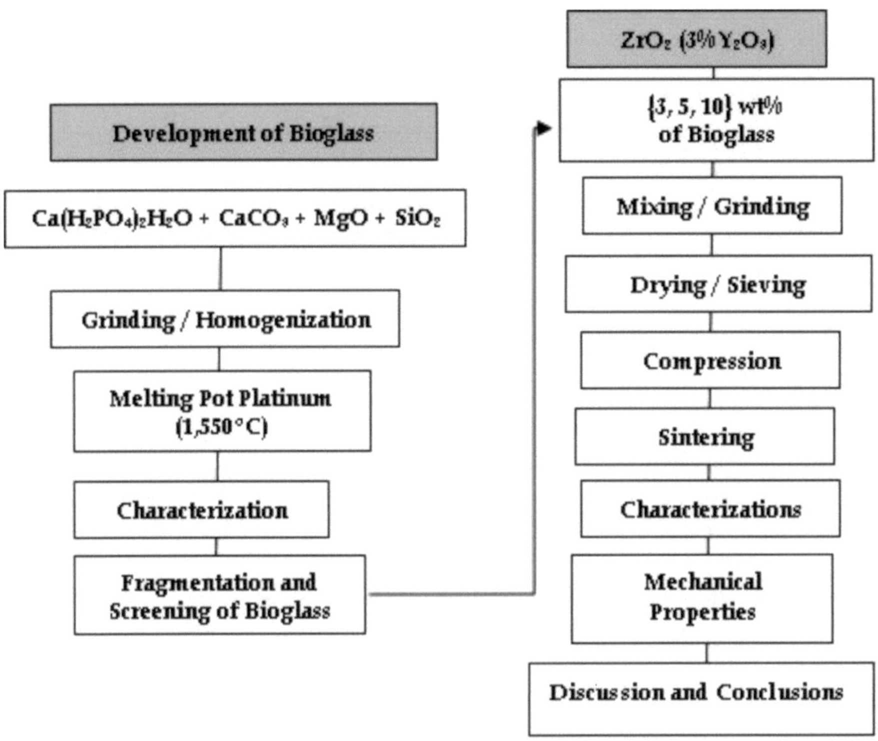

Fig. 1. Flowchart of activities

Was evaluated the content of Bioglass on the results of densification, flexural strength and fatigue, shrinkage during sintering, and the effects on the microstructural configuration. To determine density variations were measured and weighed the compressed green and after sintering.

3.1 Preparation of Bioglass

It was prepared a composition of Bioglass, based on 52.75wt% $CaOP_2O_5$, 30wt%SiO_2, 17.25wt% MgO. This composition was studied by Oliveira (OLIVEIRA et al., 1997) and presented biocompatibility with high bioactivity.

The powders were mixed in a rotary mill for 2 hours using pot stirring rod and polypropylene, in the midst of isopropyl alcohol with zirconia balls for sintered necessary homogenization.

Once mixed, the powders were dried in an oven (110°C) for 24 hours, sieved (sieve 63μm) and melted at a temperature of 1550°C, the air in a platinum crucible for 2 hours with a heating rate of 10°C/min. The cast (Bioglass) was then rapidly quenched in water at room temperature to obtain better fragmentation and amorphization.

The Bioglass was taken to the oven for drying, and subsequently fragmented with the use of an agate mortar, ground and passed through a sieve of 32μm. The powder, after screening, was subjected to characterization using the techniques of X-Ray Diffraction, Scanning Electron Microscopy (SEM) and Dilatometry.

3.2 Preparation of mixture of powders

Compositions were prepared from powder mixtures, adding distinct Bioglass content ranging from 3 to 5 and 10wt% by weight in the mixture with ZrO_2 (Y_2O_3). Higher values were inviable, since their low mechanical properties due to the small degree of densification, as the evaluations carried out previously by Habibe (HABIBE, 2007).

The raw materials were mixed in attrition mill amid isopropyl alcohol and stirred at 1000 rpm for 2 hours. For every 100g of powder mixtures during milling were used 180g of zirconia balls of sintered with an average diameter of 2mm. The stirring rod and grinding chamber, used herein, are made of polypropylene, to prevent contamination of powder mixtures for possible chafing with the surfaces in contact.

After milling, drying was performed for each mixture, using the vacuum absorption of excess fluid. The drying process was completed in an oven at a temperature of 100°C for 24 hours. The powders were then subjected to screens: 425, 125, 63 and 32μm. Again it was used agate mortar for each overflow of sieves.

3.3 Sample preparation

3.3.1 Characterization of powders

(i) Thermal analysis of ZrO_2-Bioglass

The coefficients of thermal expansion and glass transition temperature of the compositions were determined by dilatometry using dilatometer - BAHR Thermoanalyse GmbH DIL801L 2000 Model, furnace 7040 (1600°C). Samples of 3mm x 3mm, 10mm in length were prepared with standard measure based on Al_2O_3, and heated air, heating rate 25°C/min and cooling at 5°C/min.

(ii) Analysis of phases in mixing of powders

The phases present, both the powders of departure and the powder mixture were identified by X-Ray Diffraction, using diffractometer model XRD-6000 Shimadzu, which is a radiation 'Cu-Kα', with scanning between 20° and 80°, applying step angle of 0.05° and 3 seconds to scan for point counting. The peaks were identified by comparison with JCPDS standard file (JCPDS, 1988).

(iii) Morphology of powders

The powders were observed, the morphology of the particles by analysis by scanning electron microscopy - SEM, LEO 1450VP microscope using EDS and WDS engaged. In the analysis, the powders were coated with thin film of gold and observed using backscattered electron beam, allowing verification by the difference in tone, the phases and morphology of the particles. Were checked by X-Ray Diffraction, the results of possible chemical or crystallographic changes during the stages of milling and compacting the powder mixture. The study focuses on the compatibility analysis (green density) and concentration of monoclinic phase in the samples to be subjected to sintering.

3.3.2 Characterization sintered

Prior to tests of flexion characterizations were performed for density, hardness, fracture toughness, microstructure and surface phases.

(i) Determination of density

Density of the green bodies was determined by geometric method. The samples were measured in caliper with an accuracy of 0.01mm, and subsequently weighed on analytical balance (10^{-5}g). To a greater degree of accuracy, there were 15 measurements of each sample to obtain an average value reliably.

(ii) X-ray diffraction

The phases present in sintered samples were identified by X-Ray Diffraction using radiation "Cu-K α", scan from 10° to 80°, the step angle of 0.05° and speed of 3 sec/point count. The peaks were identified by comparison with JCPDS file.
Quantification of volume fraction of monoclinic phase (F_M) was calculated from the integrated intensities of monoclinic peaks $(\bar{1}11)_M$ e $(111)_M$ and also the tetragonal peak $(101)_T$.

$$F_M = \frac{1.311X_M}{1+0.311X_M} \tag{1}$$

For which:

$$X_M = \frac{(\bar{1}11)_M + (111)_M}{(\bar{1}11)_M + (111)_M + (101)_T} \tag{2}$$

where: $(\bar{1}11)_M$ 2θ=28.0°; $(111)_M$ 2θ=31.2°; $(101)_M$ 2θ=30.0° to represent the integrated intensity of diffracted peaks plans monoclinic $(\bar{1}11)_M$ and $(111)_M$ in the tetragonal $(101)_T$.

The calculation of the penetration depth of X-rays on the surface was analyzed based on the absorption of these rays by the material. The penetration depth of X-rays was given by equation (3) (KLUG, ALEXANDER, 1974):

$$h = -\frac{sen\theta}{2\left(\frac{\mu}{\rho}\right)\rho}\left[\ln\frac{I}{I_0}\right] \tag{3}$$

with

$$\left(\frac{\mu}{\rho}\right) = w_1\left(\frac{\mu}{\rho}\right)_1 + w_2\left(\frac{\mu}{\rho}\right)_2 + \ldots \tag{4}$$

where:
h = penetration depth [μm];
θ = diffraction angle,
I = intensity of X-ray beam diffracted
I_0 = intensity of X-ray beam focused,
μ = absorption coefficient;
w = weight fraction of component or element;
ρ = density [g/cm^3] (Zr = 6.511; O = 1.354; Y = 4.472; $ZrO_2.3\%Y_2O_3$ = 6.051).

(iii) Microstructural analysis

We performed observations of the sintered samples by scanning electron microscopy LEO 1450VP coupled with WDS. To observe the microstructure, the samples were ground and polished according to the procedure mentioned below. After mounting the samples in bakelite, thinning was performed in automatic grinding with diamond paste of particle size, in mesh, from 180 to 600, for the total removal of the inlay material and obtain a flat surface for analysis.

Then the samples were polished with diamond pastes, the sequence of 15, 9, 6, 3 and 1μm. To reveal the grain boundaries, surfaces polished attack suffered heat to air at 1,400°C for 15 min using a heating rate of 30°C/min to minimize the effects of temperature on grain size. The distribution of grain sizes were measured in order to study the influence of Bioglass on the content of final grain size of ZrO_2 with the purpose of these results are also correlated with the results of mechanical properties. The distribution of grain sizes were measured using image analyzer microscope, LEICA, aimed at studying the influence of Bioglass on the content of final grain size of ZrO_2. These results were also correlated with the results of mechanical properties.

4. Mechanical properties

4.1 Hardness Vickers (HV)

The methodology used to determine the hardness of the samples followed the ASTM C 1327-99, which provides the standard test method to obtain the Vickers hardness of advanced ceramics.

4.2 Fracture toughness (K_{IC})

The methodology for determining the values of fracture toughness by Vickers indentation of the samples followed the recommendations of ASTM C 1421-99. This is the pattern for obtaining the fracture toughness of advanced ceramics at room temperature.

4.3 Flexural strength - 4 points

For the analysis of flexural-bodies were used for proof-polished, dimensions (in mm) 45 x 4 x 3, as previously described. The flexural strength at room temperature, (σ_f) was evaluated by the collapse load of the body of evidence points determined by the method '4 points', following the specifications dictated by the standard DIN EN 843-1 (ASTM C 1161-90) with download speed of 0.5mm/min and with a spacing of 40mm and 20mm between the rollers of support and loading, (I_1 and I_2, respectively) as shown in Figure 2, using a universal machine mechanical testing kN MTS-250.

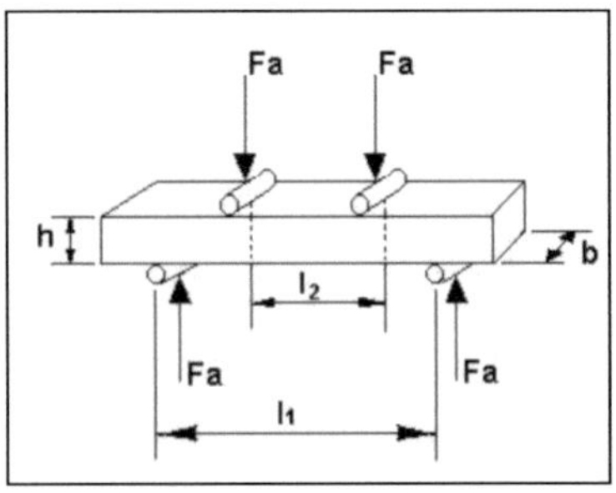

Fig. 2. Schematic representation of resistance to bending in four points. The polished face is turned down (KELLY, 2000).

The flexural strength of the specimens was calculated using Equation 5.

$$\sigma_f = \frac{3}{2} F_A \times \frac{(I_1 - I_2)}{b \times h^2} \tag{5}$$

where:

σ_f = resistance to bending (MPa);
F_A = breaking load (N);
b = measure the width of the samples (mm);
h = height measurement of samples (mm);
I_1 = wider spacing between the rollers loading (mm);
I_2 = smaller spacing between the rollers loading (mm).

5. Results and discussion

The identification and characterization procedures aimed to verify whether the characteristics of materials and products are expected in the present work, meet the conditions of sufficient quality when applying the final ceramic as dental materials.

5.1 Characterization of materials

The main objective focuses on the characterization of raw material, identifying the origin of each material used, assessing the crystallographic characteristics and morphological profile of the post in question. This goal aims to confirm whether such characteristics of powders, and mixtures of the powders were suitable for the final density of sintered.

5.1.1 Microstructural characterization

Powders of Zirconia – $ZrO_2(Y_2O_3)$ – and Bioglass were characterized by SEM and the results are shown in Figures 3 and 4. The zirconia powder used in this study was produced through a spray-drying, with additions of ligands, which promote the agglomeration of spherical shapes, as shown in Figure 3. These ligands are used to facilitate compaction of the samples.

Fig. 3. Particle morphology of $ZrO_2(Y_2O_3)$ as received.

In Figure 4 are performed by SEM micrographs of samples of Bioglass powder, sifted after. One observes the presence of acicular particles presenting larger dimensions than the sieve, the intrinsic characteristics of glassy materials.

Fig. 4. Morphology of the Bioglass after fragmentation and sieving.

The Bioglasses, after collection, were screened on meshes of up to 32μm in order to minimize the effect of their distribution in the zirconia matrix, increasing the compaction of the mixtures of powders, where the density on the green of the samples ranged from 48% to 42% due to the addition of Bioglass matrix $ZrO_2(Y_2O_3)$. Reducing the particle size of Bioglass facilitating the spreading of the fluid (liquid phase) during the sintering step.

5.1.2 Characterisation of compressed

The green relative density of the compacts are shown in Figure 5.

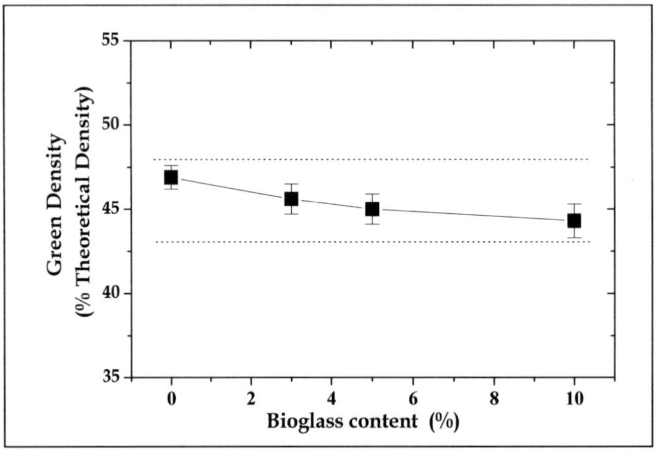

Fig. 5. Effect of the addition of Bioglass on the green density of compacts.

Note that there is a slight reduction in relative density due to the addition of Bioglass in its composition. This behavior occurs through the morphology of Bioglass, highly irregular (Figure 5), compared to the powder of $ZrO_2(Y_3O_2)$, used in these experiments. Be considered negligible differences in green density, sintering behavior observed in the samples with different amounts of Bioglass.

Figure 6 shows representative micrographs of samples sintered at each composition studied. For comparative analysis of dense material, some samples of $ZrO_2(Y_2O_3)$ were without Bioglass sintered at 1500°C/2 hours.

It is observed the presence of equiaxed grains of ZrO_2 in the whole area analyzed. Were obtained in all cases, with 3, 5 or 10wt% of Bioglass, microstructures were quite similar. There is also the presence of voids between grains resulting from the elimination of residual porosity and intergranular phase during thermal attack.

From the micrographs presented in Figure 6 and Table 4.4 lists the microstructural parameters, to verify whether the presence of liquid phase formed from the fusion of glass particles interfered with grain growth of ZrO_2.

Analyzing the results of Table 1 can be stated that the content of Bioglass little or almost nothing, interferes with the average grain size of ZrO_2 and density of grains per unit area, as shown in Figure 6.

These microstructural characteristics are a direct function of initial grain size and sintering temperature used. Dense ZrO_2 sintered solid phase is usually obtained at temperatures around 1500°C. In this temperature range, depending on the sintering time applied, the average grain size can vary from 0.5μm to 1μm, whose sizes are the result of higher levels with very long sintering, such as 1500°C/8h.

The use of relatively low sintering temperatures, as from 1200 to 1350°C, hinders the growth of the grains of the matrix, thus increasing the population of grain per unit area. In this study, the use of liquid phase has as one of several objectives to facilitate the densification at low temperatures, minimizing the grain growth, which could hamper the

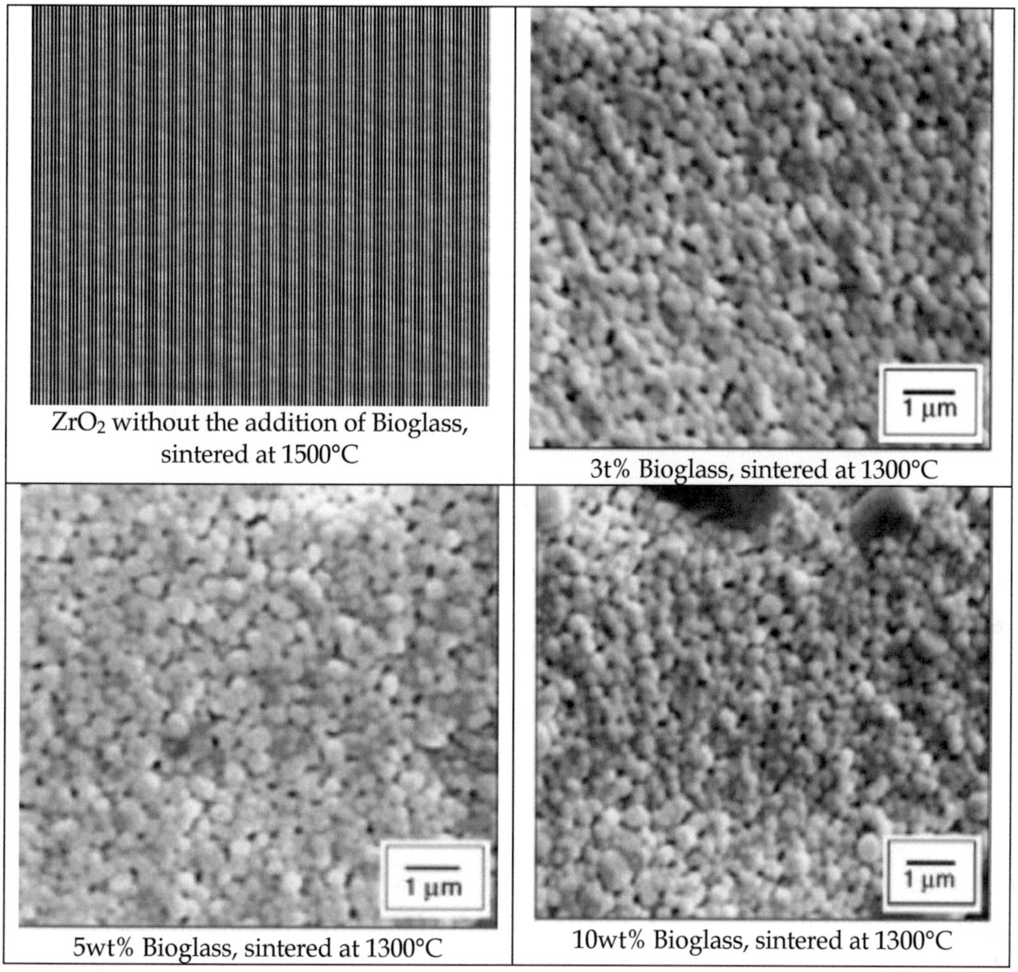

Fig. 6. Micrographs of ceramics $ZrO_2(Y_2O_3)$-Bioglass.

(ZrO_2: Bioglass) wt%	**Medium Size Grain (μm)**	**Density of grains (No. grains /μm^2)**
100:00 (1500°C)	0.803 ± 0.121	3.405
97:03	0.325 ± 0.065	9.982
95:05	0.329 ± 0.076	9.964
90:10	0.333 ± 0.070	9.939

Table 1. Parameters microstructure of $ZrO_2(Y_2O_3)$-Bioglass, sintered at 1300°C.

growth and propagation of cracks during the fracture of the material, knowing that the cracks propagate in this material so intergranular (following the grain boundaries), and has also the beneficial effect of martensitic transformation (tetragonal to monoclinic, as indicated by T→M), which occurs when the crack is tetragonal grain, and exerts compressive stress on them.

5.1.3 Dilatometry

These tests are carried out to verify the effect of the addition of Bioglass in the temperature of maximum shrinkage of the sintered body by the derivative of the shrinkage versus time. Thus have been able to verify the relationship between the percentage of Bioglass with the rates of dimensional variation of the material versus time and temperature.

The results of dilatometric analysis performed on samples previously consolidated raw material monolithic, had coefficients of thermal expansion ($\alpha_{200\text{-}1200°C}$) to 10.6x10^{-6}/°C tetragonal zirconia polycrystal (Y-TZP) and 10.2x10^{-6}/°C for Bioglass.

From the results, there is compatibility between the thermal expansion coefficients of the two phases (ZrO_2-Bioglass) for the formation of the composite ceramic-ceramic primary requirement for development of dual-phase ceramic materials (MEYERS, CHAWLA, 1998), due to reduction of residual stresses are generated between the phases of the composite and rigid after cooling.

Table 2 presents the results of calculations concerning the average values of the coefficients of thermal expansion, carried out for all compositions in this work, based on the weighting between the coefficients of thermal expansion and modulus of elasticity of the components of mixtures. These values are important in determining the residual stress generated between the phases in sintered.

	Bioglass		$ZrO_2(Y_2O_3)$[1]		Composite
Concentration of Bioglass (wt%)	Modulus of Elasticity E (GPa)[2]	Thermal Expansion Coefficient α (x 10^{-6}/°C)	Modulus of Elasticity E (GPa)	Thermal Expansion Coefficient α (x 10^{-6}/°C)	Thermal Expansion Coefficient α (x 10^{-6}/°C)
3					10.599
5	90	10.2	190	10.6	10.594
10					10.587

Table 2. General Physical Characteristics of the composites.

The curves of shrinkage and shrinkage rate as a function of temperature and hold time showed the following highlights:

(i) Part one

Can be analyzed at temperatures up to 600°C. The most significant variations in this region occur at temperatures around 450°C. At this temperature there is a smooth change of the shrinkage, which can be attributed to the volatilization of organic substances present in the

[1]The results of modulus of elasticity were obtained from the available literature (OLIVEIRA, 1997)
[2]Manufacturer's data.

compact. These are derived from organic raw material, $ZrO_2(Y_2O_3)$, which has a binder, and stearin used in compaction of powders.

(ii) Part two

The region represents the effect of temperature on the shrinkage of compacts, observed from 1,050°C. Observe that there is a characteristic temperature where the rate of shrinkage has a maximum.

Figure 7 shows the temperatures of shrinkage depending on the content of Bioglass. In this figure are also presented as the accumulated instantaneous values of shrinkage at these temperatures.

It is observed that the samples are reduced maximum temperature decrease with increasing amount of Bioglass added. This behavior implies that a larger amount of glass reduces the temperature, the greater formation of liquid phase, which in turn allows a greater shrinkage of the compact. Samples without the presence of additives, have different behavior, because it is sintered by solid phase, and therefore governed by other mechanisms of sintering.

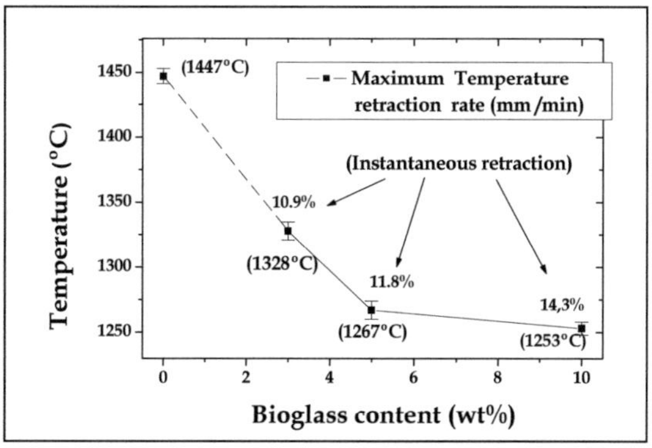

Fig. 7. Effect of the addition of Bioglass in the temperatures of greatest rate of shrinkage.

It was observed that for samples with 3wt% of Bioglass, at temperatures below 1300°C, was not reached maximum retraction of the derivative and thus, levels of incorporation of the sintering cycle time initially proposed.

The temperatures of maximum shrinkage determined for samples with 5 and 10wt% of Bioglass are respectively 1267°C and 1253°C.

(iii) Third party

In this third area of analysis is taken into account the hold time at 1300°C. From the analysis of Figure7, we observed that the samples with higher concentrations of Bioglass, achieved larger decreases until the maximum temperature testing (1300 ° C).

From there, observing Figure 8, note that there is an evolution of the shrinkage in the first minutes of landing, in all situations where Bioglass is used as an additive. Comparatively, the blocks of ZrO_2 sintered without the addition of Bioglass show continued growth as a function of hold time, because the weather influences the kinetics of densification of the sintered solid phase.

Observe that the first 20 minutes, in samples with Bioglass, retractions that occur faster when compared with the remaining time, always with a tendency to stabilize the rate of shrinkage (indicated by the rate of change of the curve). Both the rate of shrinkage and total shrinkage increases with increasing content of Bioglass. This is justified by the greater amount of liquid phase which facilitates the diffusion of the solid phase.

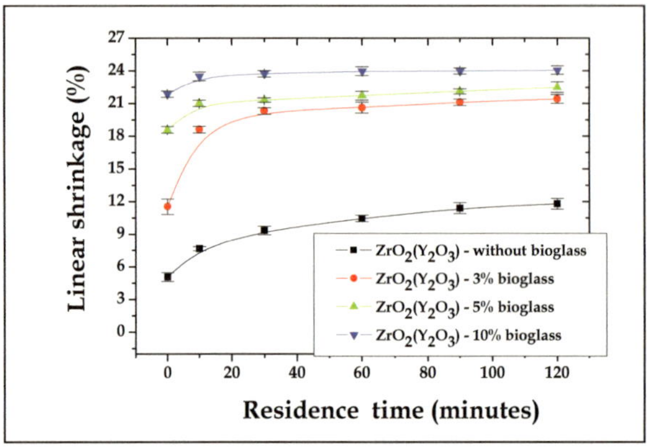

Fig. 8. Effect of isotherm plateau in shrinkage of the ceramics sintered at 1,300ºC.

The results seen above can be represented in percentage gains of shrinkage versus time of isotherm used. These results are shown in Figures 9 and 10.
An important detail presented by the geometrical behavior of the curves with respect to the correlation between the percentage increase in Bioglass and gain decrease with residence time at 1300°C. The three curves show asymptotic behavior, with the rate of linear shrinkage with time tends to proportionality with the difference between the instantaneous rate and a maximum rate for each composition in Bioglass. Larger amounts of Bioglass cause lower coefficients of thermal expansion, in agreement with Table 2. Add to it that lower the green density implies an increase of spaces to be filled, so it will have a higher rate of shrinkage.

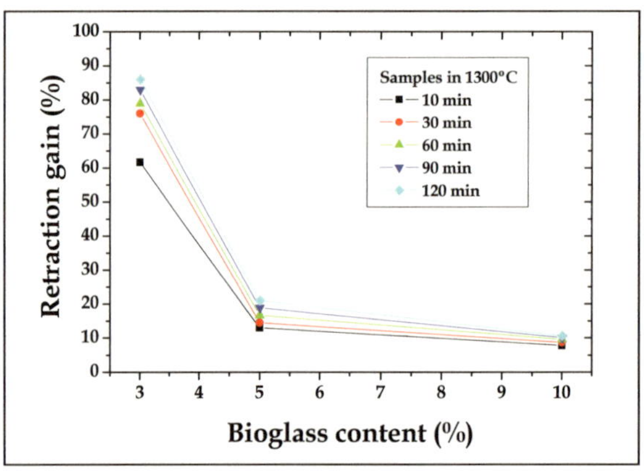

Fig. 9. Gains due to shrinkage of the content of Bioglass, for various treatment times.

It is evident that the ceramics with 3wt% of Bioglass have improved significantly with the use of level of sintering. At first, with 10 minutes of landing, there is a gain of 60%, indicating that this time was sufficient to reduce the viscosity of the glass, influencing their spread around the ZrO_2 grains. Times lead to higher cumulative gains exceeding 85%, as in 120 minutes. Samples with 5 and 10wt% are less influenced by the time of landing, for sintering at 1300ºC. Still, for the maximum times studied, 120 minutes, occur up to 20% gains, for ceramics with 5wt% Bioglass and 10% for ceramics with 10wt% of Bioglass. The level of decline observed in the samples will indicate that in all situations where Bioglass was used as an additive, and 1300°C, there is full densification of the ceramics studied. In the case of the ceramics of $ZrO_2(Y_2O_3)$, without addition of glass-forming liquid phase, the phenomenon does not occur.

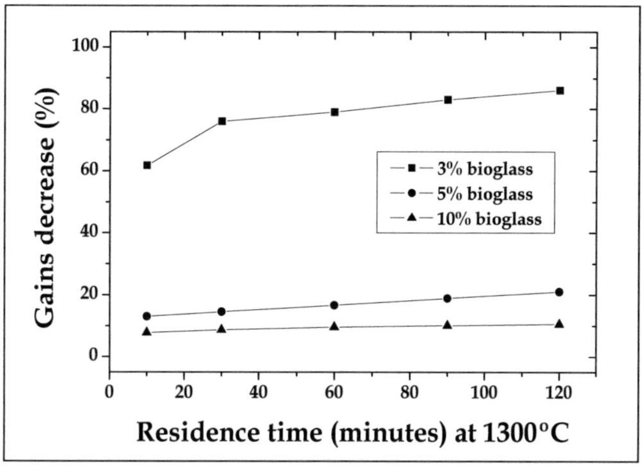

Fig. 10. Gains decrease as a function of residence time, for each composition of Bioglass.

5.2 Characterization of the sintered samples

It is intended to report the results of analysis of the crystalline phases present in sintered samples, and studies the influence of Bioglass content on densification. With this it will have the assurance that the sintering conditions (temperature, time, rate of heating and cooling) were adequate for obtaining ceramics with adequate strength for use in dental applications.
The research initiated by Habibe (HABIBE, 2007) showed that the addition of Bioglass in higher sintered $ZrO_2(Y_2O_3)$ to provide undesirable martensitic transformation (T ›M), which promote a volumetric expansion of the ceramic matrix generating increasing porosity. One concern has been established to optimize the densification, taking into account the interrelationship between low sintering temperature, sintering time and microstructural features.

5.2.1 Crystallographic characterization

In order to demonstrate the efficiency of the technique of X-Ray Diffraction in the characterization and measurement of the percentage of monoclinic and tetragonal phases of $ZrO_2(Y_2O_3)$ was proposed in this study, the use of determining the depth of penetration of radiation, based on parameters crystallographic theory.
Using the parameters presented the results obtained when applying these equations (3 and 4), was approximately 7.3μm. Grain sizes are below 0.5μm, so there is a layer thick enough to be detected by diffraction, thus allowing the identification of transformation of monoclinic phase in sub-surface levels with considerable degree of accuracy.
In pre-existing glasses of similar chemical composition to this study, obtained under the same conditions of melting and cooling, gave values of Vickers hardness near 6.2GPa, when

subjected to thermal treatment time exceeding 30 minutes. The values of fracture toughness and resistance to bending found were 0.93MPa.m$^{1/2}$ and 54MPa, respectively, for materials with rapid cooling (diamonds), and the values of 1.4MPa.m$^{1/2}$ and 115MPa, respectively, for materials with slow cooling (vitro-ceramic) (OLIVEIRA, 1997).
The percentage of transformed monoclinic phase after sintering, carried out according to Equations 1 and 2 are presented in Table 3 and illustrated in Figure 11.

Bioglass wt%	Monoclinic %
0	5.69 $\pm$ 0.02
3	6.71 $\pm$ 0.06
5	8.75 $\pm$ 0.03
10	14.37 $\pm$ 0.05

Table 3. Percentage of monoclinic phase in the sintered samples.

This behavior may be related to the gradient of contraction between the two phases (zirconia and Bioglass) after sintering, during cooling, since there is a difference between the thermal expansion coefficients between these materials (10.6x10^{-6}/°C for zirconia and the Bioglass 10.2x10^{-6}/°C). This difference promotes the generation of stress fields around the grains of $ZrO_2(Y_2O_3)$, which may exceed the maximum compressive stress needed to transform the tetragonal-monoclinic. Thus, the grains of $ZrO_2(Y_2O_3)$ tetragonal become monoclinic, with volume expansion of about 3 to 6% by volume (STEVENS, 1986), resulting in an overall structure microcracking, resulting in a reduction in density on the sample. In Figure 4.24 are the results related to this characterization.

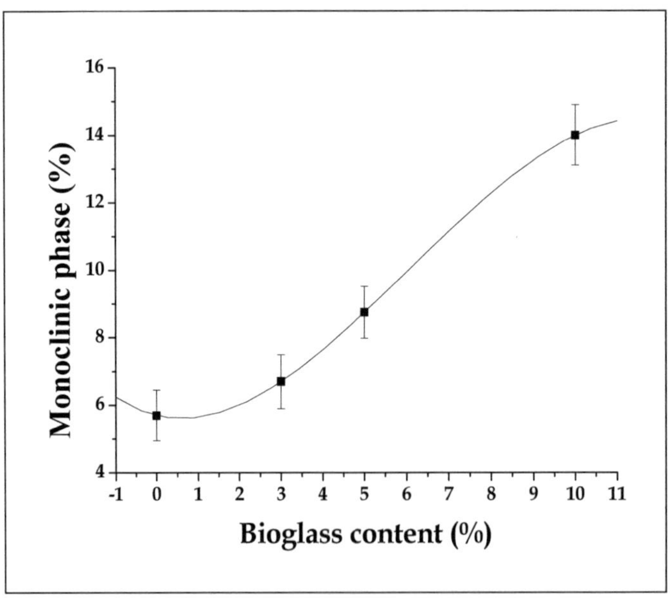

Fig. 11. Monoclinic phase concentration depending on the content of Bioglass in blocks sintered at 1300°C.

5.2.2 Relative density

As previously noted, there is elevation of monoclinic fraction considering increasing the amount of additive. Associated with this transformation occurred during cooling, which is due to thermal residual stress motivated by the difference in thermal expansion coefficient between the phases, may be the reason for the reduction of relative density versus the contents of Bioglass as shown in Figure 12.

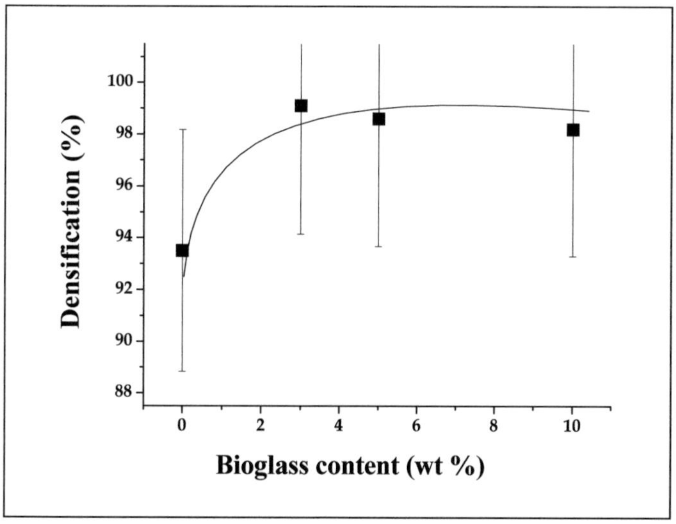

Fig. 12. Relative density as a function of the content of Bioglass, sintered blocks to 1,300°C.

The analysis of X-ray diffraction patterns indicated the presence of considerable fraction of the tetragonal phase and residual fraction of monoclinic phase. Intergranular crystalline phases were not detected in any of sintered body, indicating that the intergranular phase of Bioglass originally made can be fully amorphous or so, the fractions present in the sintered samples are not detected in the diffractometer. This can be considered in the light of that, in previous work (OLIVEIRA et al., 1997).

Bioglass cold considering controlled rates similar to those used in this study (10°C/min) showed the crystalline phases whitlockite and enstatite, and this last was not observed in this work. The possible crystallization of the Bioglass used in this work, and therefore the glass ceramics, may have contributed to the improvement of mechanical properties, with the increase in resistance (60 to 120MPa) and fracture toughness (1 to 1.5MPa.m$^{1/2}$) in the glassy phase (OLIVEIRA et al., 1997).

It is noteworthy that the material exhibits a tendency to decrease the densification with increasing amounts of Bioglass from 3wt%, and these results imply a direct function of increasing content of monoclinic phase transformed. Furthermore, the low density results on the composition submitted for "100-00", suitable only for comparison purposes, since this is accomplished by sintering the solid phase, since there have Bioglass in its composition, which could trigger mechanisms unique liquid phase sintering.

The results indicated that the dilatometry samples monolithic $ZrO_2(Y_2O_3)$ did not densify under these conditions because temperature and isothermal sintering times studied did not allow the efficient operation of the mechanisms for densification of the material. Moreover, samples with Bioglass showed maximum shrinkage temperature of about 1253, 1267 and

1328°C, for contents of Bioglass 10, 5 and 3wt% respectively. In all cases where it is applied Bioglass need for isothermal sintering at 1300°C for full densification is achieved. However, higher levels of Bioglass ($ZrO_2(Y_2O_3)$ containing 10wt% Bioglass) did not allow full densification, because during cooling, as proposed by previous study Habibe (HABIBE, 2007), there is generation of stress fields between the matrix and grain boundaries which promotes the phase transformation (T→M) that generates volume expansion and increased presence of pores and microcracks.

5.2.3 Critical failure size vs. surface roughness

The flexural strength, σ_f, the ceramic is directly proportional to the fracture toughness, K_{IC}, as predicted by linear elastic fracture mechanics (KIM et al., 2000):

$$\sigma = \frac{K_{IC}}{\sqrt{\pi c}} \tag{6}$$

The parameter "c" can be considered alternatively as the size of failure to initiate the fracture. Thus, the size of failure for start of fracture in samples of 3, 5 or 10wt% of Bioglass and sintered at 1300°C/2h, are valued between 80μm and 230μm.

The maximum surface roughness, assessed during the preparation of specimens for bending tests/fatigue, was less than 0.30μm. Whereas the roughness implies that a 'valley' is half of a crack, it is concluded that the roughness used did not affect the test results.

5.3 Mechanical properties

5.3.1 Vickers hardness and fracture toughness

Table 4 and Figure 13 present the results of Vickers hardness and fracture toughness, K_{IC}, the samples sintered at different temperatures and fractional percentages of Bioglass.

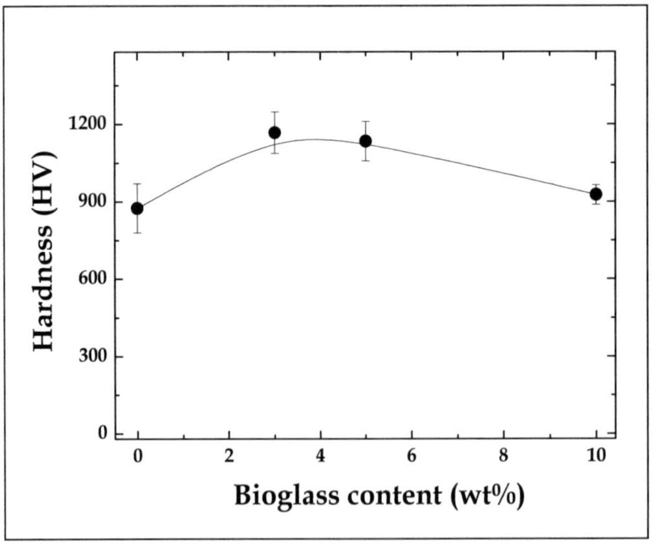

Fig. 13. Hardness of sintered samples as a function of sintering temperature and amount of Bioglass.

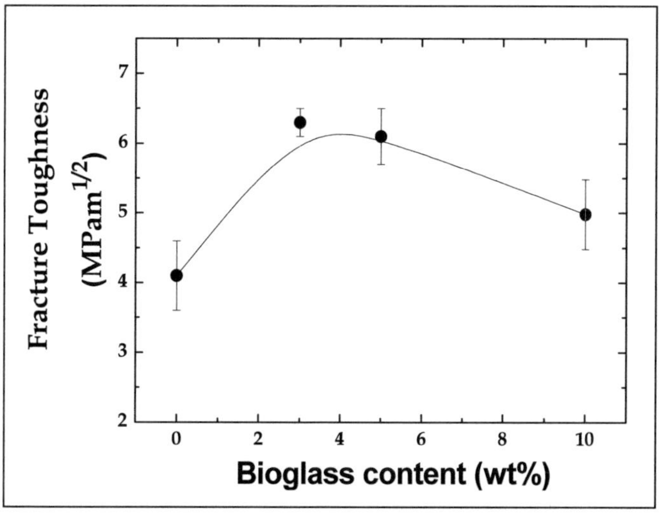

Fig. 14. Fracture toughness of sintered samples as a function of sintering temperature and amount of Bioglass.

The samples sintered at 1300ºC, containing 3wt% of Bioglass showed higher hardness and toughness, respectively, 1170HV and 6.3MPa.m$^{1/2}$. These results are related to indicators of relative density and low amount of martensitic transformation, shown in Figures 14 and 15. It is important to note that these samples show the best properties, possibly due to the high relative density, coming from the better spreading of the liquid formed during sintering and its penetration around the ZrO_2 particles. These phenomenons facilitates the elimination of pores and reduce accumulation of glass triple joints, minimizing the generation of stress fields during cooling and therefore points in favor of crack propagation.

5.3.2 Flexural strength

Samples of $ZrO_2(Y_2O_3)$, with addition of Bioglass, sintered at 1300ºC and without addition of Bioglass (sintered at 1500ºC) were tested for 4 point bending. The results are presented in Table 4 in Figure15.

Bioglass (wt %)	Vickers Hardness (HV)	K_{IC} (MPa.m$^{1/2}$)	Flexural Strength (MPa)
0 (1500ºC)	875 ± 95	4.1 ± 0.5	127.44 ± 57.15
3	1,167 ± 80	6.3 ± 0.2	453.28 ± 74.64
5	1,134 ± 76	6.1 ± 0.4	363.31 ±54.88
10	926 ± 38	5.0 ± 0.5	303.00 ± 77.40

Table 4. Vickers hardness, fracture toughness and Flexural strength of samples sintered.

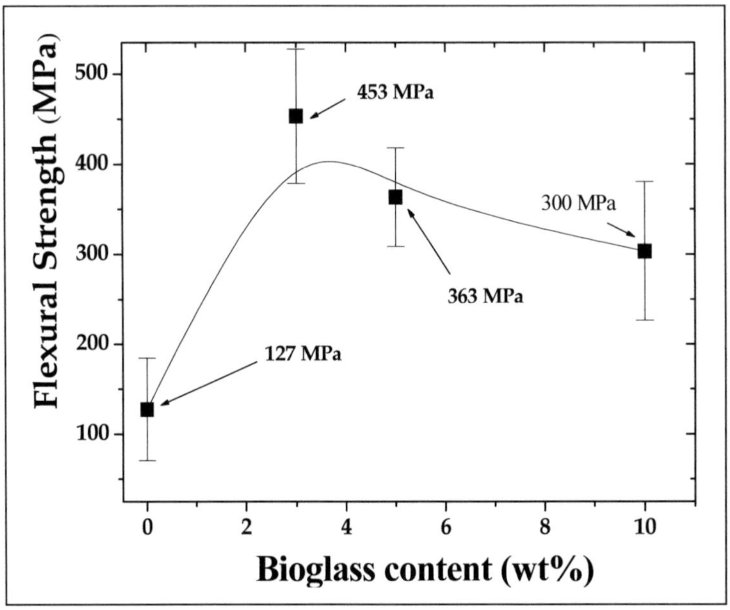

Fig. 15. Flexural strength of samples sintered at 1300°C, depending on the amount of Bioglass added to ZrO_2 matrix.

It is observed that the variations in the composition in the flexural strength is shown similar to those observed in the behavior of relative density (Figure 12), hardness (Figure 13) and fracture toughness (Figure 14), ie, an elevation in the range from zero to 3wt% Bioglass and reduction in the range of 3 to 5wt% Bioglass. Such behavior suggests that the 3wt% level gives better distribution in the zirconia matrix, leading to the conclusion that the higher values are dispersed causing islands to concentrate on Bioglass, causing some weakening and partial degradation of those characteristics. Such behavior is indicative of the concentration 3wt% Bioglass in $ZrO_2(Y_2O_3)$ would be the best choice among the four discussed compositions. This without taking into account other factors that could influence the choice.

5.3.3 Residual stresses

The calculation of the average thermal residual stress generated during cooling of the sintered samples was based on the consideration that there is homogeneous distribution of second phase in ceramic matrix ZrO_2, and is directly related to the difference in thermal expansion coefficients between the phases in the ZrO_2 matrix and intergranular glassy phase, composed of Bioglass (TAYA et al., 1990; SHI et al., 1998).
Was not taken into account the hypothesis of partial crystallization of the glass or the temperature range where there is softening of the glass present. This average residual thermal stress in the two phases can be calculated as a function of the percentage of intergranular phase (or second) that integrates the system, according to Equations 7 and 8, proposed by Shi (SHI et al., 2000).

$$\sigma_b = E_b\left(\langle\alpha\rangle - \alpha_b\right)\Delta T \quad (7)$$

$$\sigma_m = E_m\left(\langle\alpha\rangle - \alpha_m\right)\Delta T \tag{8}$$

Where σ_b and σ_m are the contours and residual stresses in the matrix, respectively. E_m and E_b indicate the modulus of elasticity of matrix and grain boundaries (intergranular phase), respectively, and α, α_m and α_b indicate the average thermal expansion coefficients, the matrix (index m) and the intergranular phase (index b), respectively. The average coefficient of thermal expansion of each composition varies, and is given by Equation 9:

$$\langle\alpha\rangle = \frac{\alpha_b C_b E_b + \alpha_m C_m E_m}{C_b E_b + C_m E_m} \tag{9}$$

Where $\langle\alpha\rangle$ is the coefficient of thermal expansion of the composition; α_b, C_b, E_b are, respectively, coefficient of thermal expansion, Young's modulus and fraction of Bioglass (grain boundary), α_m, C_m, E_m are respectively, the coefficient of thermal expansion, the fraction and the modulus of elasticity of ZrO_2 matrix.
By calculating the average coefficient of thermal expansion and residual stresses, it is found that when $\alpha_m > \alpha_b$ or $\sigma_b < 0$, the grain boundary is the transition between compression (intragranular) and tensile (matrix).
The residual stress in a multiphase composite is developed due to the discrepancy between the modulus of elasticity and Thermal Expansion Coefficient (TEC) between the constituent phases. Due to the lower TEC of Bioglass, α_b, compared to the array of ZrO_2, α_m, tensile residual stresses are developed in ZrO_2 matrix during cooling from the sintering temperature. (BASU, VLEUGELS, 2001).
The residual stress in zirconia matrix was calculated according to the model proposed by Taya (TAYA et al., 1990) and confirmed by Shi (SHI et al., 2000). In the calculations we used the modulus of elasticity (E) from 90GPa to 190GPa for the Bioglass and ZrO_2. The calculation results of compressive residual stress at grain boundaries and tensile stress in the grains of ZrO_2 matrix are shown in Figure 16, and provide a barrier to crack propagation, toughened materials.

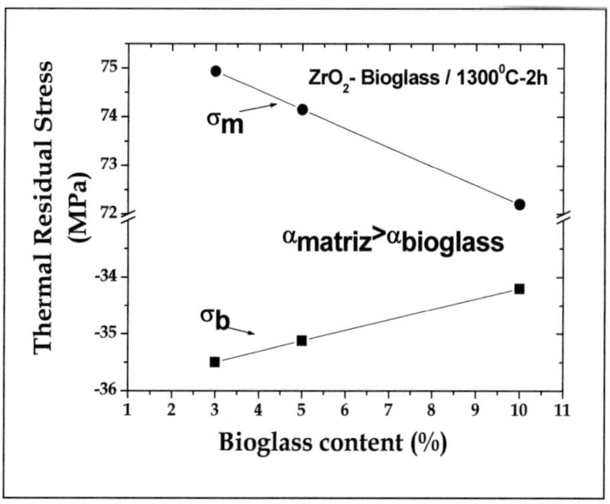

Fig. 16. Thermal residual stress due to the content of Bioglass (intergranular phase)

The toughening of ceramics developed in this work may be related to several phenomena, such as tetragonal-monoclinic transformation, crack deflection, stress-induced martensitic transformation thermal residual porosity of the sintered samples, or other possible causes unrelated. It may be noted that increasing the intergranular phase (Bioglass) leads to increased % of monoclinic phase and increased porosity associated with this phenomenon.
However, increasing the amount of Bioglass leads to a greater accumulation of concentrations of glass in triple junctions, with consequent formation of stress concentration, which permits growth and propagation of cracks. The thermal residual stresses in ZrO_2 matrix show a smaller and smaller effect as a function of the addition of Bioglass composition. However, there is a reduction in the contribution of residual stress on phase transformation (T→M), which can improve the toughness of ceramics.
Moreover, the presence of low amounts of Bioglass, facilitates the diffusional processes, reduce the possibility of transformation (T→M) to occur during cooling and increase the thermal residual stress between the phases, favoring the phase transformation during the emergence and growth of a crack, toughened material.
Previous studies have shown that propagation of intergranular cracks of the type prevalent in ZrO_2 based ceramic sintered by liquid phase (SHI et al., 1998, SUN et al., 2003, HUANG et al., 2003, SHI et al., 2000) due to the presence of glassy phase. The amount of intergranular phase in which the fracture toughness (K_{IC}), the maximum can be achieved, $C_{b,m}$, when $\alpha_b < \sigma_m$, is as follows:

$$C_{b,m} = \frac{\alpha_m - \alpha_b}{3\left(\alpha_m - \frac{E_b}{E_m}\alpha_b\right)} \tag{10}$$

Where α_b E_b are the Thermal Expansion Coefficient (TEC) and Modulus of Elasticity of Bioglass, respectively, and α_m, and E_m are, respectively, the Thermal Expansion Coefficient and Modulus of Elasticity of ZrO_2 matrix.
The calculated results show that a great theoretical value is achieved with 2.84wt% of Bioglass. This result is consistent with the composition of ZrO_2-Bioglass composite composed of 97wt% ZrO_2 and 3wt% Bioglass, which presents the best mechanical properties among the samples sintered at 1300°C/2h. Moreover, the results are consistent with previous work (SHI et al., 2000), which shows that only a small amount of intergranular glassy phase, an increase of fracture toughness can be obtained.

6. Conclusions

After the experiments, and based on these results, we can conclude that:
1. Samples of 3wt% Bioglass composition showed better densification compared to those of composition 0, 5 and 10wt% due to better spreading of the liquid phase between grains of $ZrO_2(Y_2O_3)$. These results are related to high density and low percentage of monoclinic ZrO_2 phase, present in the sintered samples.
2. The addition of higher concentrations of additives in an increase in 'islands' of Bioglass, the junctions between the grains of the matrix of $ZrO_2(Y_2O_3)$ causing residual stress fields, which led to greater amounts of martensitic transformations, after sintering, increasing the weakening of the material.

3. The results of mechanical characterization promote the use of Bioglass as sintering additive, instead of $ZrO_2(Y_2O_3)$ pure. Using the techniques of X-Ray Diffraction, high resolution, together with testing the hardness, fracture toughness, flexural strength at 4 points confirmed this statement.
4. The ceramic compositions suggested for studies, combined with the processing conditions (parameters of milling, pressing and sintering) were effective in obtaining the ceramic bodies of high relative density and with relatively fine grain. Regardless of the content of Bioglass added to zirconia, the average grain size of zirconia was in the order 0.30 to 0.35 μm.
5. The additions of 3 and 5% of Bioglass produced an increase in hardness in relation to zirconia with 3wt%Y_2O_3, and also in relation to the addition of 10wt% of Bioglass, with values of 1240 and 1210HV for 3Y-TZP composites-Bioglass (97-3) and 3Y-TZP-Bioglass (95-5), respectively. These results are due to higher densification of samples submitted for 3wt% of Bioglass.

7. Acknowledgment

The authors acknowledge to the FAPESP for financial support, under Grants no. 04/04386-1 and 05/52971-3. They also thank the Fundação Euclides da Cunha - Universidade Federal Fluminense.

8. References

Amaral, M.; Lopes, M.A.; Silva, R.F.; Santos, J.D. (2002). Densification route of Si_3N_4 - Bioglass biocomposites. *Biomaterials*, v.23, p.857-862.

Anusavice, K. J. (2005); *Phillips Materiais Dentários*. 11a Ed. Rio de Janeiro; Elsevier Editora. ISBN: 9788535215328.

Basu, D., Sarkar, B. K. Effect of zirconia addition on the fatigue behaviour of fine grained alumina. Bull. *Materials Science Forum*, v. 24, n. 2, p.101-104, Apr, 201. ISSN: 0255-5476

Bicalho, L. A.; Souza, R.C.; Santos; C.; Barboza, M. J. R.; Baptista, C.A.R.P. (2008). Fatigue of Zirconia - Bioglass Dental Ceramics. *Materials Science Forum*, v. 591-593, p.628-633. ISSN: 0255-5476

Bicalho, L. A.; Santos, C.; Souza, R. C.; Barboza, M. J. R.; Baptista, C.A.R.P. (2010). Mechanical behavior of ZrO_2-Bioglass dental ceramics under cyclic fatigue loading. *Materials Science Forum*, v. 636-37, p. 47-53. ISSN: 0255-5476

Bicalho , L. A. ; Santos, C. ; Habibe, A. F. ; Souza, R. C. ; Barboza, M. J. R. ; Baptista, C. A. R. P. (2007) Performance of ZrO_2-Bioglass Dental Ceramics under Cyclic Fatigue Loading. *Cadernos UniFOA*, v. 2, p. 30-33.

Bicalho, L. A.; Santos, C.; Barboza, M. J. R.; Moreira, M. E. S.; Strecker, K. (2010) . Investigation of the sinterability of $ZrO_2(Y_2O_3)$-Bioglass dental ceramics by dilatometry. In: *The 6th Latin American Congress of Artificial Organs and Biomaterials*, august-2010. Gramado-RS, Brazil. Proceedings of the 6th COLAOB, v. 1. p. 1-8.

Habibe, A. F. (2007). Desenvolvimento e caracterização de cerâmicas à base de ZrO_2 utilizando um biovidro do sistema $3CaO.P_2O_5$-SiO_2-MgO como aditivo de sinterização. 123f.; *MsC Dissertation in Materials Science.* Escola de Engenharia de Lorena - Universidade de São Paulo, Lorena, Brasil.

Huang, X.W.; Wang, S. W.; Huang, X. X. (2003). Micoestruture and mechanical properties of ZTA fabricated by liquid phase sintering. *Ceramics International,* v.29, p.765-769, 2003. ISSN: 0272-8842.

Huang, X.W.; Wang, S. W.; Huang, X. X. (2003). Micoestruture and mechanical properties of ZTA fabricated by liquid phase sintering. *Ceramics International,* v.29, p.765-769. ISSN: 0272-8842.

JCPDS - Powder Diffraction File Search Manual. Hanawalt Method Inorganic. (1998), USA. *Published by the International Center for Diffraction Data.*

Kelly, J. R.(1997). Ceramics in restorative and prosthetic dentistry. *Annuary of review in Material Science.,* v.27, p. 443–68, 1997.

Kim, D.-J., Lee, M.-H., Lee, D. Y., Han, J.-S. (2000). Mechanical Properties, Phase Stability, and Biocompatibility of (Y,Nb)-TZP/Al_2O_3 Composite Abutments for Dental Implant. *Journal of Biomedical Material Research,* v.53, n.4, p. 438-443. ISSN (printed): 1549-3296.

Klug, H. P.; Alexander, L. E. (1974). X-Ray Diffraction Procedures for Polycristllines and Amorphous materials. *John Wley and Sons,* New York.

kohn D. H.; Ducheyne P.; Awerbuch, J. (1992). Acoustic emission during fatigue of porous-coated Ti-6Al-4V implant alloy. *Journal of Biomedical Materials Research,* v. 26, n.1, p.19-38 January 1992. ISSN (printed): 1549-3296,

Kohn, D. H., Ducheyne P., Awerbuch, J. (1992). Sources of acoustic emission during fatigue of Ti-6Al-4V: effect of microstructure: *Journal of Materials Science.* v.27, n. 6, p.1633–1641. ISSN: 0022-2461,

Maeda, L. D. ; Habibe, A. F. ; Santos, C. ; Daguano, J. K. M. F. ; Bicalho, L. A. ; Souza, R. C. ; Barboza, M. J. R. (2007). Efeito da adição de Biovidro na resistência à fratura dos compósitos biocerâmicos ZrO_2-biovidro. In: 51° Congresso Brasileiro de Cerâmica, Salvador-BA, Brazil. Proceedings of 51° CBC-Congresso Brasileiro de Cerâmica, v. 1.

Meyers, M. A., Chawla, K. K. (1982). Princípios de Metalurgia Mecânica. Edgard Blücher. São Paulo.

Oliveira, A. P. A. (1997). Influência de fatores físico-químicos na produção de pós de zircônia. *Doctoral Thesis* . PUC-RJ.

Oliveira, J. M.; Fernandes, M. H.; Correia, R. N. (1997). Development of a New Glass Ceramic in the System MgO-3CaO.P_2O_5-SiO_2. *Bioceramics,* v.5, p. 7-14. ISSN (electronic): 2090-5025.

Santos, C.; Souza, R. C. ; Barboza, M. J. R. ; Bicalho, L. A. ; Baptista, C.A.R.P. ; Strecker, K. (2009). Fatigue Behavior of dental ceramics based on Zr_2O-Al_2O_3 composites. In: *64° Congresso anual da ABM,* Belo Horizonte. Anais do 64° Congresso ABM, lulho-2009. v. 1. p. 1-10.

Shi, J. L., Lu, Z. L., Guo, J. K. (2000). Model analysis of boundary residual stress and its effect on toughness in thin boundary layered yttria-stabilized tetragonal zirconia polycrystalline ceramics. *Journal of Materials Research,* v.15, n.3, p.727-732.

Shi, J. L.; Li, L. Guo, J .K. (1998). Boundary stress and its effect on toughness in thin boundary layered and particulate composites: model analysis and experimental test on T-TZP based ceramic composites. *Journal of the European Ceramic Society,* v.18, p.2035-2043, 1998.

Stevens, R. (1986). An introduction to zirconia: zirconia and zirconia ceramics. [S.L.]. 2nd Ed *Twickenham: Magnesium elektron.*

Sun, Y.-H. Zhanga, Y-F. Guo, J-K. (2003). Microstructure and bending strength of 3Y-TZP ceramics by liquid-phase sintering with CAS addition. *Ceramics International.* v.29, p.229-232. ISSN: 0272-8842.

Taya, M.; et al. (1990). Toughening of a particulate-reinforced ceramic-matrix composite by thermal residual stress. *Journal of the American Ceramic Society,* v.73, n.5, p.1382-1391.

21

Comparison of Apatite-Wollastonite Glass-Ceramic and β-tricalcium Phosphate used as Bone Graft Substitutes after Curettage of Bone Cysts

Pavel Sponer, Karel Urban and Tomas Kucera
Charles University in Prague, Faculty of Medicine and University Hospital in Hradec Kralove
Czech Republic

1. Introduction

The need to treat the bone defects arises throughout the whole spectrum of orthopaedic surgery. It is possible to produce materials with specific mechanical properties, architecture and biodegradability. The essential attributes for regeneration of new osseous tissue include biocompatibility, osteointegration, osteoconductivity, osteoinductivity and osteogenicity. Biocompatibility means that the immunogenic response and foreign body reactions to the implanted material are minimised or absent. Osteointegration is defined as a process leading to close bonding of the newly formed mineralised tissue with the implant material. Beyond these bone regeneration essentials, other matrix properties can influence the suitability of a material for its intended clinical applications, such as surface roughness, the mechanical integrity of the matrix and the matrix's porosity [1]. Several osteoconductive bone graft substitutes are available for clinical application, including coralline hydroxyapatite, collagen-based matrices, calcium phosphates, calcium sulphate and deproteinised bovine bone. These materials vary substantially in tems of their chemical composition, mechanical properties and biodegradability. Orthopaedic surgeons should understand the differences between the various bone graft substitutes to ensure they select a material that provides the desired properties for the intended clinical application.

An ideal synthetic bone graft substitute should be a porous matrix with interconnecting porosity that promotes rapid bone ingrowth, and at the same time, it should possess a sufficient strength to prevent its crushing under physiological loads during osteointegration and healing. Hydroxyapatite ($Ca_{10}(PO_4)_6(OH)_2$), β-tricalcium phosphate ($Ca_3(PO_4)_2$), their derivatives and combinations are the most commonly used ceramic materials in orthopaedics. While hydroxyapatite ceramic materials provide an osteoconductive matrix for bone ingrowth and ongrowth, slow in-vivo resorption profiles can potentially limit their clinical applications [2]. Although β-tricalcium phosphate ceramic has been studied substantially in animal models and its biocompatibility, osteoconductivity and resorbability have been reported, there have been only limited data regarding long-term outcome of its clinical use in surgery for bone tumours [3-12]. However, the informations about biological responses such as bone bonding and resorption of ceramics are very important in clinical applications [13-15].

The goal of this study was to analyze and compare the clinical and radiological outcome in patients after implantation of the nonresorbable oxyhydroxyapatite glass-ceramic and Beta-Tricalcium phosphate used to fill the defects of long bones after curettage of bone cysts.

2. Materials and methods

2.1 Patients

This retrospective study was approved by the local research ethics committee, and all participants gave their written, informed consent. We evaluated 39 consecutive patients who fulfilled the following inclusion criteria: (i) histologically confirmed unicameral bone cyst, (ii) treatment by curettage of the lesion and implantation of glass-ceramic or beta-tricalcium phosphate, (iii) follow-up after at least 24 months to confirm a static radiographic outcome without recurrence of benign bone lesion. There were 28 male and 11 female patients. The age of the 21 patients with the implanted nonresorbable glass-ceramic at the time of surgery was 4–44 years with an average age of 15 years. The age of the 18 patients with implanted beta-tricalcium phosphate at the time of surgery was 7–30 years with an average age of 14 years. The unicameral bone cysts were located in the humerus (16 patients), femur (13), fibula (4), tibia (2), calcaneus (2), ulna (1), and iliac bone (1). All patients had curettage of the cyst and filling of the resulting bone defect with glass-ceramic (15 male and 6 female) or tricalcium phosphate (13 male and 5 female). Cortical fenestration was carried out using an osteotome and the soft tissue membrane was removed by a curette. After the filling of the bone defect with glass-ceramic or tricalcium phosphate, the cortical window was replaced. Internal fixation was employed in four patients because a pathological fracture occurred in 3 cases and an impending fracture was in 1 patient. Splints or bandages were used postoperatively for patients judged to be at risk of pathological fracture. Full weight bearing was allowed after 8-12 weeks. The patients were scheduled for follow-up evaluations which included clinical and radiographic examinations at 4-6, 10-14 weeks, 6 and 12 months intervals. Thereafter they were seen yearly.

2.2 Filling

A bioactive glass-ceramic material with wollastonite, oxyhydroxyapatite, residual glass phase, and whitlockite at a ratio of 45–30–20–5% (BAS-0, Lasak Ltd., Prague, Czech Republic) was used. The apatite-wollastonite glass-ceramic was dense with a porosity of less than 0.5%. The sintering temperature was 1200°C, the bending strength 170 MPa and the compression strength 400MPa. Granules 2–4mm in diameter were implanted during operations. The amount of the glass-ceramic used ranged from 5 to 45 g.

β-tricalcium phosphate (Poresorb®, Lasak Ltd., Prague, Czech Republic) in granule form with particle size of 0,6 to 2 mm was implanted. The porosity of the interconnected β-tricalcium phosphate scaffold was 35±5 %, the average macropore size was 100 μm in diameter, the size of micropores was 1-5 μm, and the sintering temperature was 1180°C. The quantity of implanted β-tricalcium phosphate ranged from 3 to 66 g.

2.3 Clinical and radiographic follow-up

Clinical and radiographic examinations of all patients were carried out at 3–12 years (7 years on average) after implantation of the bioactive glass-ceramic and at 2–8 years (4 years on average) after implantation of tricalcium phosphate. The clinical follow-up included subjective complaints, objective findings focused on soft tissue status in the area of synthetic bone graft substitute filling, range of movement in adjacent joints and weight-bearing ability of the treated extremity. The radiographs were taken in standard projections and evaluated independently by three investigators. The results were then compared for inter-observer agreement and, in case of differences, the patient´s radiograph was reviewed by all three observers together. The radiographic integration of bioactive glass–ceramic was evaluated according to the criteria of Uchida et al. [16] to determine the presence of the markedly seen bridging trabeculae at the recipient bone–implant interface and the amount of incorporation of glass–ceramic. The radiographic integration of tricalcium phosphate was classiffied according to the criteria of Nigro and Grace [17]. Complete integration was associated with radiographical homogeneity of host bone and bone graft substitute, and trabecular remodelling was seen. Integration was judged to be partial when the formation of a partial gap between synthetic bone graft substitute and host bone was observed. Resorption was defined as gradually decreased radiopacity of the bone graft substitute until complete resorption.

3. Results

3.1 AW glass-ceramic

The clinical evaluation showed that 10 patients had no subjective complaints, 4 reported transient pain, 3 had pain at activity, 2 reported pain during weather changes, 1 woman observed pain during pregnancy, and 1 patient had pain at rest. No restriction in weightbearing of the limb treated was reported by any of the patients. Radiographs obtained immediately after surgery showed radiolucent zones between the implanted glass-ceramic and the surrounding bone. Periodic assessments revealed that, during the period of 3–4.5 months, the radiolucent zones faded and new bone developed. Continuing radiographic observations up to a maximum of 12 years after implantation have shown no evidence of biodegradation of the apatite-wollastonite glass-ceramic material. The distinct bridging bone trabeculae bound to the surface of glass-ceramic granules were observed in all patients (Fig. 1). In the metaphyseal region of long bones, the incorporated glass-ceramic granules still remained distinct and seemed to be connected to each other. However, in the diaphyseal region of long bones, there was little distinction between the glass-ceramic granules and adjacent bone. Increased bone density around the incorporated glass-ceramic material with remodeling of the cortex in the diaphyseal region of long bones was found in six of the nine patients. No degenerative changes were encountered in adjacent joints. No postoperative infections or fractures were observed in our patients. Recurrence of bone cysts was identified in one case. The patient had further curettage of recurrence located in the area surrounding an incorporated glass-ceramic material and the cavity was filled with allogeneic cancellous bone grafts; this patient had no pain. One male with a small residual defect required no further surgery.

3.2 Tricalcium phosphate

Neither postoperative infection nor adverse reaction due to the material were encountered. No patient complained of local pain af final examination and all patients were satisfied with their limb function. Radiographs obtained immediately after surgery demonstrated radiolucent zones between the implanted tricalcium phosphate and the surrounding bone. Over time, radiolucent zones faded and new bone developed in all 18 patients. The mean

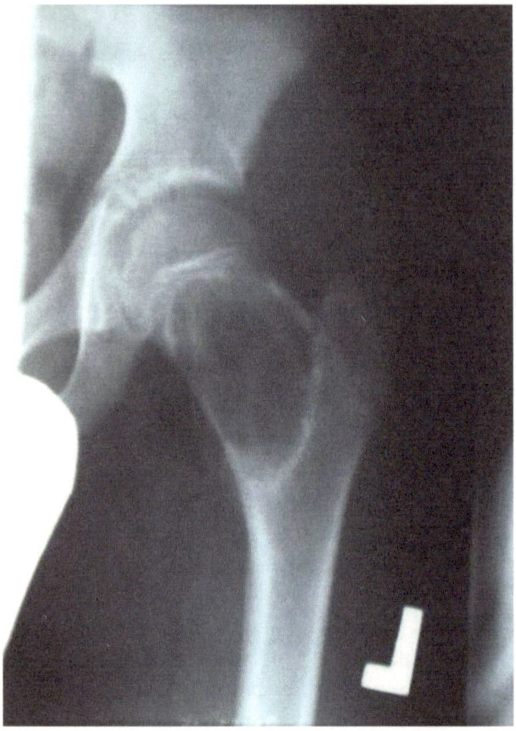

a)

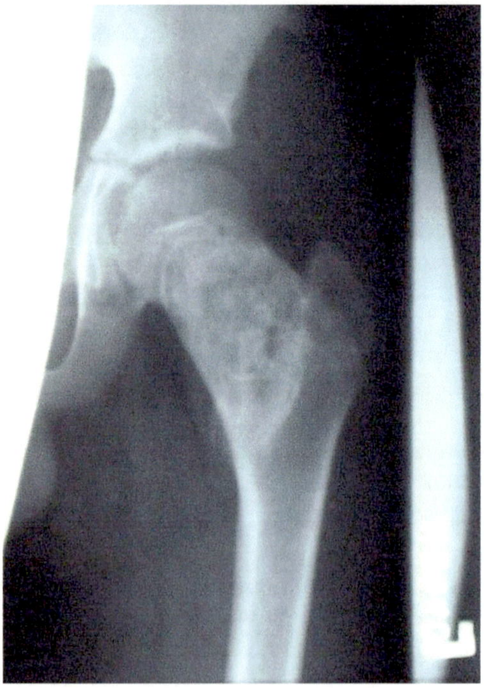

b)

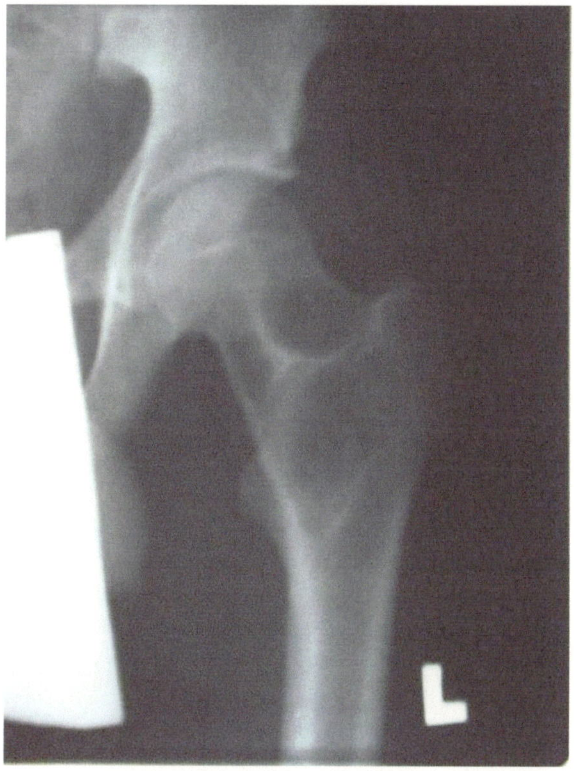

c)

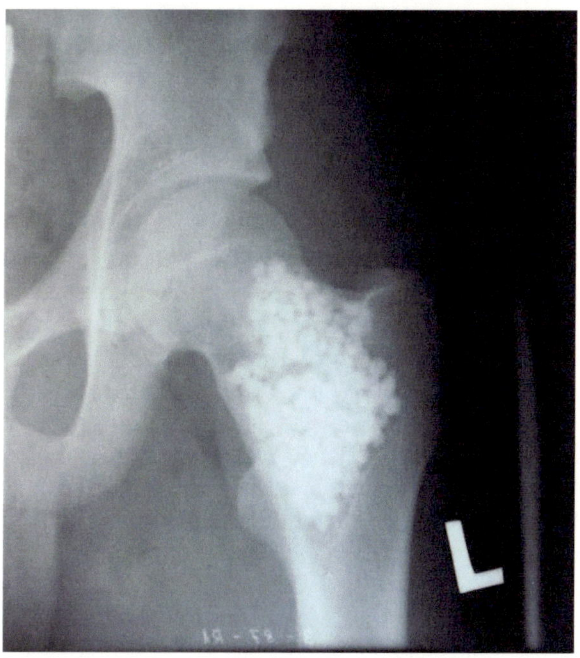

d)

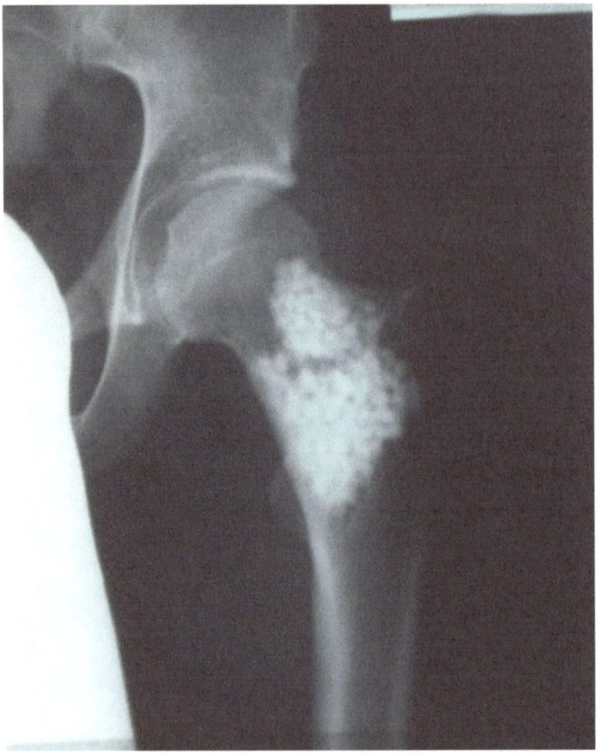

e)

Fig. 1. Anteroposterior radiograph of the left proximal femur showing a large unicameral bone cyst in a 13-year-old boy preoperatively (a); complete filling of the curetted cavity with autologous cancellous bone grafts 2 months after surgery (b); recurrence of the cyst 4 years later (c); complete filling of defect with apatite-wollastonite glass-ceramic granules 3 months after curettage of the recurrence (d); bone trabeculae bound to the glass-ceramic granules which still remained distinct 10 years after surgery (e)

period necessary for disappearance of these radiolucent zones was 9 weeks (range 5-13 weeks). Periodic radiographic assessments revealed decreased radiographic density of β-tricalcium phosphate and replacement of β-tricalcium phosphate granules by newly formed bone trabeculae. These processes apper to have started on the periphery of the synthetic filling and progressed centrally. Signs of the implanted β-tricalcium phosphate still remained radiographically at the final follow-up in all 18 cases, but the material was incorporated in the surrounding bone and gradually resorbed.

Postoperative fractures were recorded in two patients with a unicameral bone cyst in the humerus. One male fell 3 weeks after surgery; the fracture was treated conservatively. The other male patient has had a car accident 20 months after surgery; the displaced diaphyseal fracture was treated with open reduction and plate osteosynthesis. In two young patients, growth arrest or deformity were seen before curretage of the cyst and implantation of β-tricalcium phosphate. Premature closure of the physeal plate with resulting shortening of the upper extremity was found as a complications after pathological fracture of proximal humerus in one male with unicameral bone cyst. Recurrences of the unicameral bone cyst were seen only in 2 cases. Both patients had further curettage of recurrence in the area surrounding an incorporated tricalcium phosphate. In the skeletaly mature male patient, the cavity was filled with autologous cancellous bone grafts; in other patient, β-tricalcium phosphate was added to the cavity at the time of repeated curettage (Fig. 2).

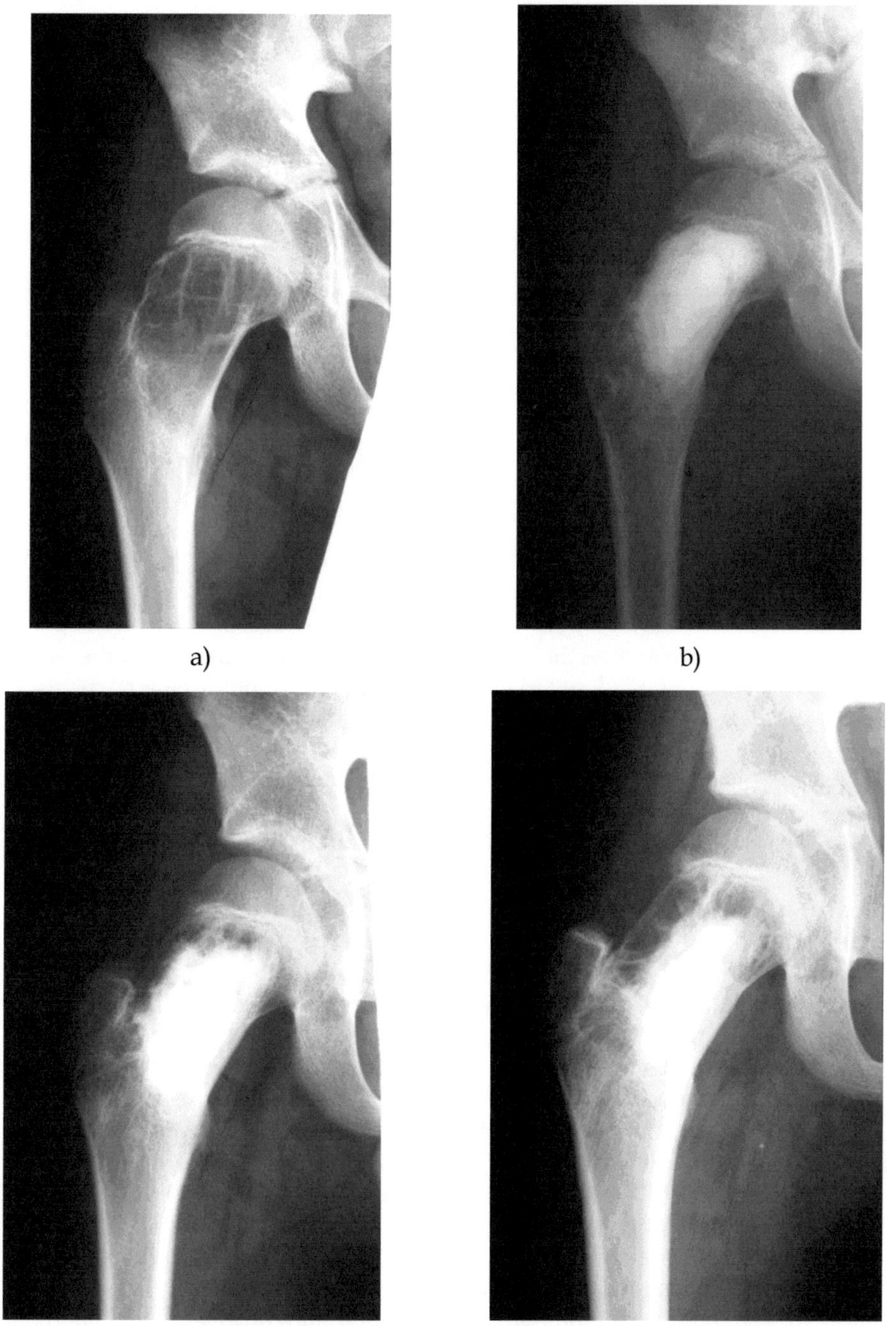

a) b)

c) d)

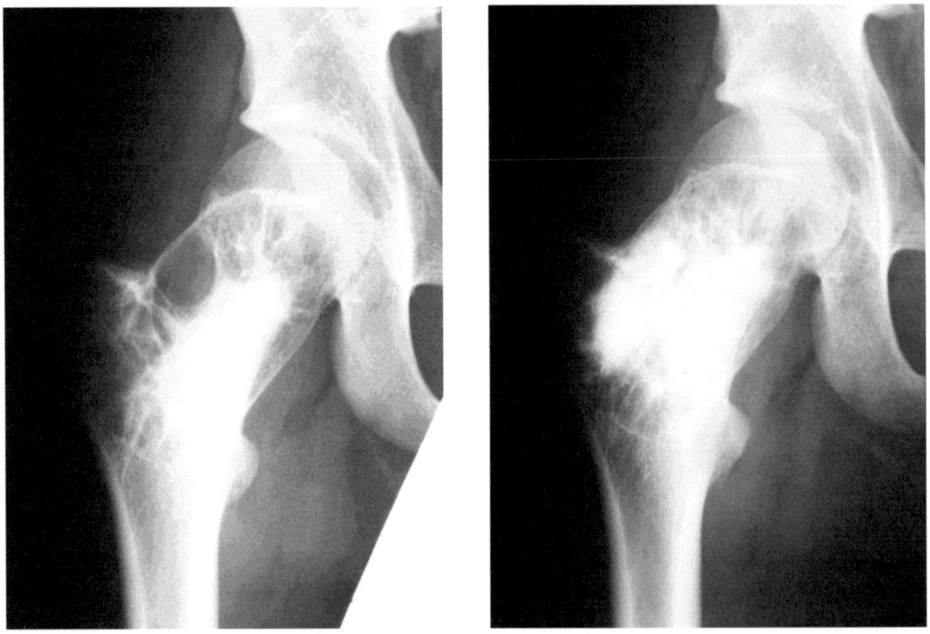

e) f)

Fig. 2. Anteroposterior radiograph of the right proximal femur showing a large unicameral bone cyst in a 11-year-old boy preoperatively (a); complete filling of the curetted cavity with β-tricalcium phosphate after surgery (b); decreased radiographic density of β-tricalcium phosphate apparent on the periphery of filling with recurrence of the cyst localized between the implanted material and growth plate 7 months later (c); further progression of the recurrence 13 months after surgery (d); large multicameral recurrence of unicameral bone cyst in 15 years of age (e); β-tricalcium phosphate added to the cavity at the time of repeated curettage (f)

4. Discussion

Conventional bone grafting with an autologous bone harvested from the iliac crest is the current gold standard because the autografts contain marrow cells including nondifferentiated mesenchymal stem cells, osteogenic bone cells, an osteoconductive matrix and osteoinductive proteins. However, the availability of autologous bone grafts is limited, their harvesting requires a further operation with prolonged operative time and is often associated with donor-site morbidity. The processed allograft bone is an attractive alternative to the autologous bone [18]. These allogeneic bone grafts are primarily osteoconductive, but they retain a variable number of osteoinductive proteins. The host immune response to freezedried allografts is less vigorous than the response to fresh or fresh-frozen allografts [19]. The transmission of infectious agents from donor to recipient with allogeneic bone grafts is their major risk and a loss of mechanical properties is their important disadvantage. The infectious risk is increased when fresh allografts are used and, therefore, serological testing has to be performed to reduce that risk.

In an effort to overcome limitations of autografts and allografts, the synthetic bone graft substitutes were developed from a variety of materials, including calcium phosphates, calcium sulphates, bioactive glasses and glass-ceramics which appear to be the ideal substances for use as matrices because the inorganic component of bone is composed of hydroxyapatite. Bioceramics can be divided into three categories: bioinert ceramics (alumina), surface-bioactive ceramics (sintered hydroxyapatite, bioactive glasses and apatite-wollastonite glass-ceramics), and resorbable bioactive ceramics (low-crystalline hydroxyapatite and tricalcium phosphate) [2].

The used dense bioactive glass-ceramic with main crystalic phases, oxyhydroxyapatite, and wollastonite, when compared with beta-tricalcium phosphate ceramic, has no microporosity and nanoporosity. After the filling of the bone cavity with 2–4-mm granules of nonresorbable material, new bone tissue ingrowth was seen in the interspaces between granules [20,22]. The nonresorbable bone graft substitutes can offer an excellent permanent function in certain conditions. In contrast, these synthetic materials can also affect tissue reparation and function. After osteointegration of the rigid nonresorbable material, the surrounding bone tissue is often mechanically influenced and protected (the stress-shielding phenomenon), thereby changing the local mechanical signals. This adversely affects the surrounding bone tissue. The concentration of load stress at the interface of the rigid implant–bone tissue can generally cause pain or mechanical failure [21, 22]. Equalizing to the mechanical properties of the surrounding bone is essential, because the mechanical signals are important mediators for the differentiation of connective tissue progenitor cells.

The ideal biodegradable bone graft substitutes should fulfill some requirements such as biocompatibility, adequate initial strength and stiffness, retention of mechanical properties throughout sufficient time to assure its biofunctionality and non-toxicity of the degradation by-products [23-25]. The continuous degradation of a resorbable material causes a gradual load transfer to the healing tissue, preventing stress-shielding phenomenon and stimulates the healing and remodeling of bones [22, 26]. The surgeon should be concerned with the mechanical and biological properties of the bone graft substitute as well as the handling and ability to assess healing of the grafted site.

Aside from chemical composition, the microstructure (the volume, density and size of pores and interconnections, the specific surface) plays a role in the bone ingrowth of porous biomaterials. An increase of porosity would make bone ingrowth inside synthetic materials easier but would decrease their biomechanical properties. The interconnections in a porous material act only as pathways for nutritional elements, vascularization and cells between the pores that are the sites for bone tissue growth proceeding from the outside to the inside [2]. Therefore, the size of pores should be larger than the size of interconnections [27]. Macroporosity (pore size >50 μm), microporosity (pore size <10 μm) and pore wall roughness play an essential role in new bone formation [12, 28-33]. The larger surface area can support a higher bone inducing protein adsorption, ion exchange and bone-like apatite formation by dissolution and reprecipitation while the surface roughness enhances attachment, proliferation and differentiation of bone forming cells [2, 32]. Thus, these data imply that more extensive dissolution and reprecipitation of low-crystalline calcium phosphates can cause more osteoconductive and cell-mediated degradation characteristics [34].

Resorption is an important characteristic of bone graft substitutes and can be divided in two mechanisms: solution-mediated dissolution processes and cell-mediated (phagocytic) processes. The degradation attributes of calcium phosphates are dependent on the chemical composition, crystal structure, crystal and grain size, microporosity, neck geometry, and neck dissolution rates of the materials [34]. The density of pores and interconnections which expresses the quantity of connections between pores of porous materials plays a more important role than their size that is modified during degradation of resorbable bioceramics, whereas the sizes and the densities are equally important in unresorbable biomaterials [2, 27]. Although tricalcium phosphate was incorporated in the surrounding bone and gradually resorbed, signs of the implanted β-tricalcium phosphate still remained radiographically in all cases. In our opinion, the implanted quantity of tricalcium phosphate was higher in our patients compared to other authors [5-7, 15]. More experimental and clinical studies will be required in order to resolve the healing problems of large bone defects combining scaffolds with osteoinductive factors and cell cultures. According to this study, interconnected β-tricalcium phosphate produced better clinical results than apatite-wollastonite glass-ceramic. Interconnected β-tricalcium phosphate is a successful and safe bone graft substitute for the treatment of benign bone tumours and tumour-like lesions because of its biocompatibility and bioresorbability.

5. References

[1] Khan Y, Yaszemski MJ, Mikos AG, Laurencin CT (2008) Tissue engineering of bone: material and matrix consideration. J Bone Joint Surg Am 90-A:S36–42

[2] Sponer P, Urban K, Kucera T, Kohout A, Brtkova J, Knizek J (2011) The use of interconnected β-tricalcium phosphate as bone substitute after curettage of benign bone tumours. Eur J Orthop Surg 21:235-241

[3] Merten HA, Wiltfang J, Grohmann U, Hoenig JF (2001) Intraindividual comparative animal study of α-and β-tricalcium phosphate degradation in conjunction with simultaneous insertion of dental implants. J Craniofacial Surg 12:59-67

[4] Wiltfang J, Merten HA, Schlegel KA, Schultze-Mosgau S, Kloss FR, Rupprecht S, Kessler P (2002) Degradation characteristics of α and β tri-calcium-phosphate (TCP) in minipigs J Biomed Mater Res (Appl Biomater) 63: 115–121

[5] Galois L, Mainard D, Delagoutte JP (2002) Beta-tricalcium phosphate ceramic as a bone substitute in orthopaedic surgery. Inter Orthop 26:109–115

[6] Ogose A, Hotta T, Kawashima H, Kondo N, Gu W, Kamura T, Endo N (2005) Comparison of hydroxyapatite and beta tricalcium phosphate as bone substitutes after excision of bone tumors J Biomed Mater Res Part B: Appl Biomater 72B:94–101

[7] Hirata M, Murata H, Takeshita H, Sakabe T, Tsuji Y, Kubo T (2006) Use of purified beta-tricalcium phosphate for filling defects after curettage of benign bone tumours. Inter Orthop 30: 510-513

[8] Brunner TJ, Grass RN, Bohner M, Stark WJ (2007) Effect of particle size, crystal phase and crystallinity on the reactivity of tricalcium phosphate cements for bone reconstruction. J Mater Chem 17:4072–4078

[9] Walsh WR, Vizesi F, Michael D, Auld J, Langdown A, Oliver R, Yu Y, Irie H , Bruce W (2008) β-TCP bone graft substitutes in a bilateral rabbit tibial defect model. Biomaterials 29:266–271

[10] Stubbs D, Deakin M, Chapman-Sheath P, Bruce W, Debes J, Gillies RM, et al. (2004) In vivo evaluation of resorbable bone graft substitutes in a rabbit tibial defect model. Biomaterials 25:5037–5044

[11] Gál P, Ondruš Š, Škvařil J, Straka M, Jochymek J, Plánka L (2009) Synthetic biocompatible degradable material for juvenilie bone cyst treatment. Acta Chir Orthop Traumatol Cech 76:495-500

[12] von Doernberg MC, von Rechenberg B, Bohner M, Grunenfelder S, van Lenthe GH, Muller R, et al. (2006) In vivo behavior of calcium phosphate scaffolds with four different pore sizes. Biomaterials 27:5186–5198
[13] Hibi A, Ishikawa T, Asano M, Ohsawa S, Tsuge K, Iyoda K. (1994) A study of failed implantation of hydroxyapatite for benign bone tumor. Orthoped Surg (Sekiekgeka) 45: 1423-1428
[14] Matsumine A, Myoui A, Kusazaki K, Araki N, Seto M, Yoshikawa H, Uchida A (2004) Calcium hydroxyapatite ceramic implants in bone tumour surgery: a long-term follow-up study. J Bone Joint Surg 86-B:719–725
[15] Nicholas RW, Lange TA (1994) Granular tricalcium phosphate grafting of cavitary lesions in human bone. Clin Orthop Rel Res 306:197-203
[16] Uchida A, Araki N, Shinto Y, Yoshikawa H, Kurisaki E, Ono K (1990) The use of calcium hydroxyapatite ceramic in bone tumour surgery. J Bone Joint Surg 72-B:298–302
[17] Nigro N, Grace D (1996) Radiographic evaluation of bone grafts.J Foot Ankle Surg 35:378–385
[18] Sponer P, Urban K, Urbanova E, Karpas K, Mathew PG (2009). Behaviour of nonresorbable bioactive glass-ceramic implanted into long bone defects: comparison with cancellous allografts. Arch Orthop Trauma Surg 129:1353–1360
[19] DeLong WG, Einhorn TA, Koval K, McKee M, Smith W, Sanders R, Watson T (2007) Bone grafts and bone graft substitutes in orthopaedic trauma surgery. J Bone Joint Surg 89-A:649–658.
[20] Strnad Z (1992) Role of the glass phase in bioactive glass-ceramics. Biomaterials 13:317–321.
[21] Muschler GF, Nakamoto C, Griffith LG (2004) Engineering principles of clinical cell-based tissue engineering. J Bone Joint Surg Am 86-A:1541–1558
[22] Sponer P, Urban K, Urbanova E, Karpas K, Mathew PG (2010) Behavior of bioactive glass-ceramic implanted into long bone defects: a scintigraphic study. J Pediatr Orthop B 19:102-107
[23] Botez P, Sirbu P, Simion L, Munteanu F, Antonia I (2009) Application of a biphasic macroporous synthetic bone substitutes CERAFORM®: clinical and histological results. Eur J Orthop Surg Traumatol 19:387–395
[24] Mano JF, Sousa RA, Boesel LF, Neves NM, Reis RL (2004) Bioinert, biodegradable and injectable polymeric matrix composites for hard tissue replacement: state of the art and recent developments. Compos Sci Technol 64:789- 817
[25] Naito K, Obayashi O, Mogami A, Itoi A, Kaneko K (2008) Fracture of the calcium phosphate bone cement which used to enchondroma of the hand: a case report. Eur J Orthop Surg Traumatol 18:405–408
[26] Dorozhkin SV (2009) Calcium orthophosphate-based biocomposites and hybrid biomaterials. J Mater Sci 44:2343–2387
[27] Lu JX, Flautre B, Anselme K, Hardouin P, Gallur A, Descamps M, et al. (1999) Role of interconnections in porous bioceramics on bone recolonization in vitro and in vivo. J Mater Sci Mater Med 10:111–120
[28] Bohner M, Baumgart F. (2004) Effects of geometrical factors on the resorption of calcium phosphate bone substitutes. Biomaterials 25:3569–3582

[29] Daculsi G, Passuti N. (1990) Effect of macroporosity for osseous substitution of calcium phosphate ceramics. Biomaterials 11:86–87

[30] Eggli PS, Muller W, Schenk RK. (1988) Porous hydroxyapatite and tricalcium phosphate cylinders with two different macropore size ranges implanted in the cancellous bone of rabbits. Clin Orthop 232:127–138

[31] Galois L, Mainard D. (2004) Bone ingrowth into two porous ceramics with different pore sizes: an experimental study. Acta Orthop Belg 70:598–603

[32] Karageorgiou V, Kaplan D. (2005) Porosity of 3D biomaterial scaffolds and osteogenesis. Biomaterials 26:5474–5491

[33] Yuan H, Kurashina K, de Bruijn JD, Li Y, de Groot K, Zhang X. (1999) A preliminary study on osteoinduction of two kinds of calcium phosphate ceramics. Biomaterials 20:1799–1806

[34] Oonishi H, Hench LL, Wilson J, Sugihara F, Tsuji E, Kushitani S,Iwaki H (1999) Comparative bone growth behavior in granules of bioceramic materials of various sizes. J Biomed Mater Res 44:31–43

Part 3

Topics in Ceramics and Environment

22

Development of Potassium Polytitanates Nanoadsorbents for the Removal of Lead Ions from Water - Dynamic Processes

Aguilar González Miguel Ángel
Centre for research and advanced studies of national polytechnic. Cinvestav Saltillo México

1. Introduction

Lead (Pb) is a toxic element that accumulates in the body by inhaled air or ingested in food or water. Lead has been used for centuries to make water pipes and cooking utensils. In fact, lead is a very rare element in the lithosphere. Lead is one of the heavy metal that has a great number of applications in industrial activities that are necessary to life and its contact effects have been studied by some of researchers. Lead is present in applications such as: manufacture of accumulators, centralized waste landfill, copper smelter, electric-electronic components, electrowinning, inorganic chemicals, cast iron and steel, metal finishing, fusion and metal forming, powder metallurgy ferrous and non-ferrous, non-ferrous metal casting, organic chemicals plastics and synthetic fibers, chemical pesticides, storage batteries, anti-knock gasoline, pigments porcelain enamel, plastics manufacturing, electricity generating by steam, transportation of cleaning equipment and combustion wastes. All these applications are due to its physicochemical properties than allows it to be compatible and shape easily to others materials with modified properties. It can be mixed with other metals to form useful materials. Most of these materials can be largely recovered, so that lead can be recycled. The environmental threat comes from the one used in chemical compounds.

Its asymptomatic presence is one of the most important aspects in different diseases and also of environment caused by its contact or consumption of waters with low contents of this element (Naicker et al, 2010, Albalak et al, 2003 & Noona et al, 2001). There is great concern in cities of underdeveloped countries on five continents by the fact that lead influence in the human body can develop: anemia, kidney disease, brain damage, impaired nervous system function peripheral, high blood pressure, reproductive abnormalities, developmental defects, abnormal metabolism of vitamins, low intelligence quotient and sometimes death, (Hrudey et al., 1995 USEPA, 1992). The greatest concern that its presence there affects mainly children under 6 years and unborn babies (Zhang et al., 2005, Naicker et al., 2010, & Riddell et al., 2007). On the other hand, lead plays an important role in the food chain and is of vital environmental importance because it alters the ecosystems and natural soils(Tong et al, 2000).

The lethal dose of lead absorbed is estimated at 0.5 mg. The accumulation and toxicity appear if was adsorbed more than 0.5 mg per day. The half life of lead in bone is 32 years and in the kidney of 7 years. The exposure limit for lead in air is 0.15 mg/m^3. The exposure

limit for lead in food is 2.56 mg/kg according to The U.S. Centers for Disease Control and Prevention (CDC) who has defined an elevated blood lead level (BLL) as 10 micrograms per deciliter (mg/dl).
Recently Potassium Polytitanates ($K_2Ti_nO_{2n+1}$) have been recognized like functional advanced materials and adsorbents of metallic ions(Tan et al., 2007), they can be present with different behaviors e.g. with relationships among n=4 to 6 they present adsorption and ion exchange properties (Wang et al, 2009) and with relationships between 6 to 8 they could be present excellent mechanical properties (Bavykin et al., 2006). Microstructure is influenced by this relationship for $K_2Ti_4O_9$ layer morphology regularly is presented, fibers, whiskers or cylindrical forms to $K_2O{\bullet}6TiO_2$ or $K_2Ti_6O_{13}$(potassium hexatitanate). Potassium polytitanates have been applied like raw materials to fabricate ceramics (filters, barriers, high temperature sensors (Gratzel, M., 2001; Fox M. A & Dulay M. T., 1993), and their excellent mechanical and frictional properties are useful for their application as:reinforcing additives, as well as, antifriction properties(Zhuang et al., 2007) in different materials. Low friction coefficient has shown that they could be excellent applications as solids lubricants and frictionant materials (Xie et al., 2010). Additionally its low thermal conductivity and its high reflectance at the ultraviolet range, near to infrared region favors its application in manufacture heat resistance materials, also as isolating of materials and coatings.
To counteract lead pollution of air, soil and water adsorbents have been developed based on activated coal (Ayyapan et al., 2005), biological mixture(TSezos., 1984), ion exchange resins(Matsumiya et al., 2003), (Plaza et al., 2009), and natural or synthetic zeolites(Barrier, 1978) which ones intended to diminish the concentration of lead in solution. Limitations such as high cost of synthesis reactors, processing equipment and maintenance, long times adsorbent synthesis, bogged down process, complications for storage the exhaust adsorbents all of these reasons generate the necessity to develop adsorbents that combine excellent adsorption properties, low cost, high efficiency, short and immediate processes and a post-treatment or confinement environmentally safe.
For this reason in this research has been synthesized PPN (Potassium Polytitanates Nanoadsorbents) as effective adsorbents of lead in contaminated water.
Among several chemical routes available for the synthesis of nanoadsorbents materials, molten salts technique has been proved to be very promising due to its uniqueness in being facile, cost effective, environmentally benign and crystalline regulated in order to adsorption applications (Afanasiev, 2006).
Below are described some investigations which have involved different techniques for obtaining nanomateriales with interesting adsorption properties.

1.1 Alkaline titanates synthesized by different techniques

1.1.1 Soft chemical route

One method to obtain layered alkali titanates in nanometric scale is the called soft chemical route, which is a type of chemical reactions used at room temperature in open systems, based on polymerization reactions in which a molecular precursor solution becomes a chemical reaction and subsequently dried and densified into a solid material with properties different from that of the starting materials. Riss et al., 2007 adjusted the photoelectronics properties of layered titanates nanostructures by changing the nature and bonding state of ions in the interlayer region studies. Also they have shown that a facile ion exchange behavior in titanates nanostructures control these properties by interactions in the interlayer region.

It is well known that cationic ions that exist in the structure of the titanates are weakly attached to each interlayer channel (Unal, et al, 2003); therefore the adsorption and ion exchange properties are due to that the distances more higher between alkali ions negatively charged and the oxygen ions than those distances of protons and octahedral oxygen ions. Zhang et al., 2005 proposed that the protonation of the TiO_6 block influences the electron density in the moiety of the Ti-O to the OH bonds leading to an increase in the length of adjacent Ti-O bond.

1.1.2 Hydrothermal synthesis route

The *hydrothermal synthesis* appeared in the decade of the 80´s and was proposed as a method of synthesis of crystalline ceramic materials with high purity features. Through this method can be obtained high-quality crystals without the need of routines of grinding or a further heat treatment of materials, using low surface area powders and high pressures at temperatures below 200° C. Besides exploits the reactivity of powders with low specific surface areas. Some types of zeolites and titanates as adsorbents for various applications have been synthesized (Hayashi, H. & Hakuta, Y. 2010).

Yang et al., 2010 prepared layered titanates nanofibers and sodium titanates and sodium titanates in acid form by hydrothermal synthesis, they studied these materials as adsorbents for removal of toxics radioactive (Sr^{2+} , Ba^{2+} as substitute of $^{226}Ra^{2+}$) and heavy metals ions from water (Pb^{2+}). This method consists in mixing hydroxides and titanium compounds under hydrothermal conditions. So two aqueous solutions containing salts are mixed y treated during 2 days in an autoclave at 245 °C (hydrothermal reaction) to yield titanates fibers. The solid after drying at 178° C during 6 h in the mixture can be recovered and washed with deionized water. They can convert to trititanate to a new phase $Na_{1.5}H_{0.5}Ti_3O_7$. The chemical equation for this transition is:

$$2Na_2Ti_3O_7+H_2O \rightarrow 2Na_{1.5}H_{0.5}Ti_3O_7+NaOH \qquad \text{at } 250° \text{ C}$$

1.1.3 Sol-gel synthesis route

Sol gel Route it is described as a structural synthesis from drawing upon particles in solution (Schmidt et al., 2010). It´s similar to the precipitation route but different due in this method is more stable the environment inside of the solution as result of chemical interactions. The conventional agglomerated sizes are presented around of 1-100 nm. The overall sol-gel process can be described in four stages: a) homogeneous precipitation (kinetic control), b) Initial nucleation where by decomposition of compounds it is release the necessary precursors to particle precipitation, c) metallic hydrolysis or deprotonation of hydrous metal oxides and d) control of parameters that influences in the hydrolysis as such as: control of the temperature, concentration, and pH of the solution. Finally hydrolysis process is promoted by compounds chemical interactions as organic liquids which ones OH^- ions in systematical form.

1.1.4 Calcinations synthesis route

In this technique analytical-grade reagents as KF and TiO_2 (anatase or amorphous gel) are used as starting materials. Afterward powders are dried at 720° C for 20 min before being mixed. The weight ratio of KF to TiO_2 in the mixtures can adjusted to various values, to investigate the influences of weight ratio on the formation of titanates whiskers. Powders are mixed homogeneously to form a precursor powder and then put into a furnace. The heat treatment is basically fixed at 720° C for 4 h(Li et al., 1999).

1.1.5 Molten salts synthesis route

The synthesis method of **molten salts**, is the fusion of salts to solubilize reagents that lead to better dissemination of materials and reduced the reaction temperature. The term "Molten Salt" is self-descriptive; it is melted salt(s). The molten salt fusion is usually performed in a temperature range from 200 to 600 ° C. When liquid is stable, has a heat capacity similar to water (by volume) and flows much like water does. In this method it loses the difference between "solvent" and "solute" in the sense that the flux must provide atoms or structural building units ending in a final crystal product or "quasi-crystalline". Moreover, the chemical properties of flux, as the acidity and redox potential, have a controlling influence on the nature of the products. Salts are simple, usually ionic (that is the chemical bonds are a simple ionic type) and stable compounds. The most common example of which is "table salt", or sodium chloride (NaCl). Both sodium and chlorine are notoriously reactive; sodium is one of the most electropositive substances (wants to lose an electron) and chlorine one of the most electronegative (wants to take an electron). These two opposite substances readily join to form stable sodium chloride via a strong ionic bond.

1.2 Potassium polytitanates nanoadsorbents obtained by molten salts route

Unique properties and behavior different among the oxides family, potassium polytitanates are a large class of semi-crystalline titanates materials, arrangements in their lamellar or stringy nanostructures, uniformly linked and with interlayer spaces between them. Also the exfoliation and intercalation could be properties present in this kind of materials (Unal et al., 2003).

The procedure for obtaining powders by molten salt consists as follow: TiO_2 powders were mixed with KNO_3 and KOH and then heated to a temperature range from 450 to 550 ° C and afterward treated for a period of time 1 hour. The synthesized powders should be washed (purified) with distilled water and separated with a filter paper. Finally the powders are treated inside furnace at a temperature of 90° C during 2 hours. Complete description for the potassium polytitanates is given in the experimental part of this chapter.

However, the fibrous powders have a bulky nature, poor fluidity, and inconvenience in handling. Thus, in spite of their attractive properties, fibrous potassium titanates are difficult to use to produce ceramic green bodies by conventional ceramic technologies and require very high pressures. Moreover, the fibrous powders tend to cause dusting, which represents a health risk in the working environment; the fibrous morphology of some potassium titanates has been associated with carcinogenicity. For these reason in this work potassium polytitanates nanoadsorbents were bonding with potato starch according previous studies (Gregorová et al, 2008). Potato starch was used to bind the nanoadsorbent and generate porous structure due that is an organic material with an ash content low to negligible (at least from the standpoint of materials science.) This is above the sintering can say that the end products of starch will always be CO_2 and H_2O, provided that during the burning is conducted in an oxidizing atmosphere. Of course, under reducing conditions may occur residual carbon. The potato starch breakdown begins to be detected below 300° C (Laurentin et al., 2003).

Molten salt through the following reaction is proposed:

$$KNO_3 + KOH + TiO_2 \xrightarrow{500} KTiO_3 + CO_2$$

In this work, we describe the preparation and characterization of nanoadsorbents of potassium polytitanates produced by the molten salt synthesis which ones contain both combined structures potassium titanate. Figure 1.

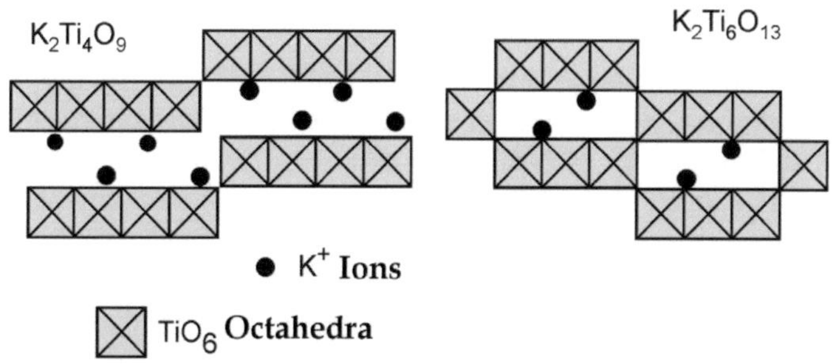

Fig. 1. Idealized structural arrangements from $K_2Ti_4O_9$ (potassium tetratitanate) and $K_2Ti_60_{13}$ (potassium hexatitanate).

2. Experimental part

Nanoadsorbents of potassium polytitanates were produced with KNO_3 (Merck brand with 98.9% of purity), KOH(brand Aldrich with flakes form and Powdered TiO_2 (anatase of purity of 99%, with average particle size of 3 mm brand Aldrich, Milwaukee, WI) in a relationship of 8.2:1.0:0.8 all of them were treated, using Steel Stainless crucibles(750 ml of capacity). First stage consists of treat in molten mixtures KNO_3 y KOH. The synthesis was carried out in an electric muffle furnace (Thermolyne 27000), 500° C for 1 h in air atmosphere compounds. Afterward having reached a temperature of 500 was added to the TiO_2 anhydrous. Preliminary tests showed that this time was sufficient to obtain products consisting mainly of potassium titanates with some of un-reacted TiO_2. The ideal time of heat treatment of this technique to produce pure potassium titanates is three hours, but this option was rejected for economic and energy saving reasons. After that molten material was separated, in this stage soluble compound were separated of the solid material. Nanoadsorbents fine powders were washed in distilled water to dilute any remaining of soluble nitrates and chlorides in the molten salt mixture, these powders were washer using 1000 ml of bidistilled water for each 250 g of powders and then were filtrated with paper Whatman No. 42 four times each set. The ultimate suspension was ultrasonically treated (Fisher Scientific, Model 550, USA) for 30 min (80.75 kW at 20 kHz) and filtrated. The powders thus obtained were finally dried for 2 h at 90° C.The dried powders Were used directly and also in granular form as adsorbent material in adsorption columns. Thermal events of nanoadsorbents powders and their dehydration processes were investigated by thermal analysis, (TGA/DTA, in Perkin Elmer, Pyris Diamond instrument).

Phase composition of nanoadsorbents were analyzed by X-ray diffractometry (XRD, Philips X'Pert using CuKa radiation with a nickel filter, with the continuous scan mode at 51-601 2y at a rate of 0.0212 y/s; the operation conditions were: 40 kV and 100 mA. The reflection positions and relative intensities were referred to the International Center for Diffraction Data (ICDD-2004). Quantitative XRD analysis was carried out to determine the TiO_2 contents in the PTP, using TiO_2-anatase phase as an internal standard.

The kinetics of adsorption of Pb (II) present in aqueous solutions was determined by an atomic emission spectrometer (ICP, Thermoelemental Thermojarrel Ash model Iris Intrepid II). The spectrometer was calibrated with NIST certified standards.
The specific surface area was determined on a sorptometer (Quantachrome Autosorb1C, Asic-xtcd6) using the principle of adsorption/desorption of a monolayer of nitrogen on the surface of the solid, based on the methods of Brunauer-Emmett-Teller (BET) and micro-pores analysis. The pore size distribution was investigated by mercury intrusion porosimetry (MIP Pore Master 60 GT). IUPAC (International Union of Pure and Applied Chemistry) classification was used to identify the porosity of the samples characterized. On the other hand, isotherms were adjusted according.(Gregg and Sing, 1952: Rouquerol et al., 1994)
The observations and energy dispersive X-ray spectroscopy (EDS) patterns of nanosorbents were obtained of the scanning electron microscopy (SEM) were carried out in a XL30-ESEM FEI Philips instrument, operating at 20 kV. Transmission electron microphotographs (TEM), high-resolution transmission electron microphotographs (HRTEM), selected-area electron diffraction (SAED) in a TEM Titan 80-300 FEG is a high-resolution analytical transmission electron microscope operating at 200 kV. The sample for TEM/HRTEM observations was prepared by deposition of a drop of the colloidal dispersion of $K_2Ti_6O_{13}$ onto 200 mesh Cu grids coated with a carbon layer.
Fact Sage software has been used to predict and of to calculate phases than could precipitate and phase transformation during adsorption processes in the aqueous system.
Figure 2 shows processing stages of manufacturing process of potassium polytitanates nanoadsorbents.

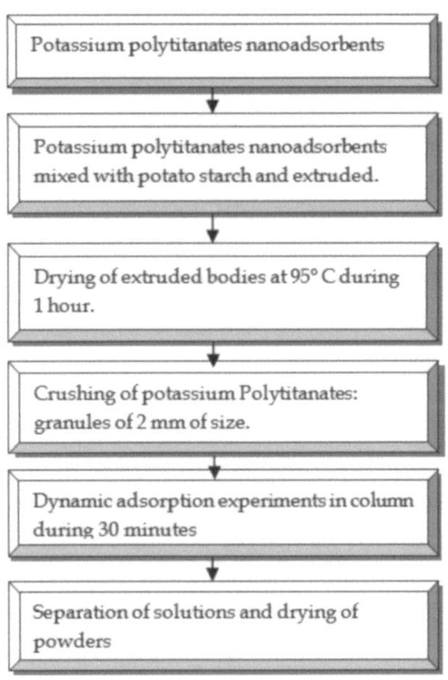

Fig. 2. Flow chart of manufacturing process of potassium polytitanates nanoadsorbents.

2.1 Materials

For the fixed-bed experiments, potassium polytitanates nanoadsorbents were prepared in the form of granules.

2.1.1 Solutions

The initial metal stock solution was prepared from the nitrate salt, $Pb(NO_3)_2 \cdot 6H_2O$. All chemicals were reagent grade and they were used without further purification. All solutions were prepared with deionized water and the glassware was cleaned by soaking in 10% HCl and rinsed with water. The background electrolyte was potassium nitrate.
pH, adsorbent dose and concentration of the solutions were selected considering that the pH of operation of polytitanate potassium in nanoasorbentes is 10.24 (stable after 24 hours) this was based on previous studies and preliminary evidences of PPN electrochemical studies than were performed.
According to these electrochemical studies that were carried out in the nanoadsorbents and research made by Dean et al, 1972, pH values for precipitation of lead in dilute solutions is located close to 6 (Table 1). These values were considered for the systematization and synchronization of the experiments, as well as taking the value of 3.5 which is a real industrial value in the treatment of lead wastewaters. In the table 1 are presented pH values of precipitation of different bivalent metallic ions.

Precipitation values in diluted solutions	pH
Fe^{3+}	2.0
Al^{3+}	4.1
Cr^{3+}	5.3
Cu^{2+}	5.3
Fe^{2+}	5.5
Pb^{2+}	6.0
Ni^{2+}	6.7
Cd^{2+}	6.7
Zn^{2+}	7.0
Hg^{2+}	7.3
Mn^{2+}	8.5
Co^{2+}	6.9

Table 1. pH values of precipitation in dilute solutions. (Dean et al 1972).

2.1.2 Determination of the columns features

Lead nitrate used in adsorption tests (Aldrich, 99% purity) was an aqueous solution with a concentration of 155 mg/dm3, using distilled water and pHo= 5.5 without initial pH adjustment. Then the solutions were stirred for a period of 30 minutes at 120 rpm, with the aid of a magnetic stirrer Corning brand. This solution, No. 1, was used in direct adsorption tests. Additionally, solution 2 was prepared by mixing 1 with nitric acid (1M) for a pH0= 3.5. Both solutions were used to investigate the influence of pH on the effectiveness and mechanism of removal of Pb2+ of acidic solutions, usually produced as industrial waste
Dean et al, 1972. pH measurements were made with a pH-Meter, Make plusmeter Orion Model 420, with a glass electrode Thermo brand.

2.2 Column adsorption conditions

All experiments were conducted at 25 ± 2° C. Pyrex ® glass column with the dimensions of 25 mm in diameter and volume Vc = 75 cm^3, which was filled with 165g of adsorbent granules (with sizes from 2 to 3.5mm), occupying a volume of Vo = 70 cm^3. Granules in turn were supported by a stainless steel mesh (standard mesh # 325) of 24mm of diameter, not to drag all this fine material produced during the elution process. The top of the column was covered with a lid that contains a connecting tube to remove air bubbles. Flow of metal solutions was controlled by a valve located at the bottom of the column. Others research have applied fiberglass to support the materials adsorbents. In this work, previous studies were conducted (with distilled water only) to determine the flow rate of solutions through the porous ceramic adsorbents. The optimum rate was determined at 70 ml/min, this was made trying to consider bibliographic information found for columns packed with activated carbon and porosity content in the adsorbents.

The corresponding aqueous solution was added with a burette to the adsorbent, to control the rate of delivery. The initial rate of elution corresponded to a value of 2Vo/h (140 cm^3/ h), this rate is within the typical values for industrial adsorption columns (Klein 1985) and samples were monitored and passed through the adsorbent at different times (0, 2, 5, 11, 20, 37 and 40 hours.). The aqueous solution was poured into to the column using a system of two attached vessels located over the column with a stable hydrostatic pressure of liquid (Figure. 2). To determine the elution rate, the solution was passed through the column and collected in samples of 100 cm^3, which determined the pH (Orion potentiometer 420) and the contents of K and Pb by ICP spectrometry (Thermoelemental Thermojarrel Ash, Iris Intrepid II).

The results of Pb and K in each period were evaluated by chemical analysis technique of atomic absorption spectrophotometry in order to study the mechanisms of reaction column and also pH of solutions were determined in each of the different samples. The amount added in this device was enough to supply each experiment by 40 hours continuously.

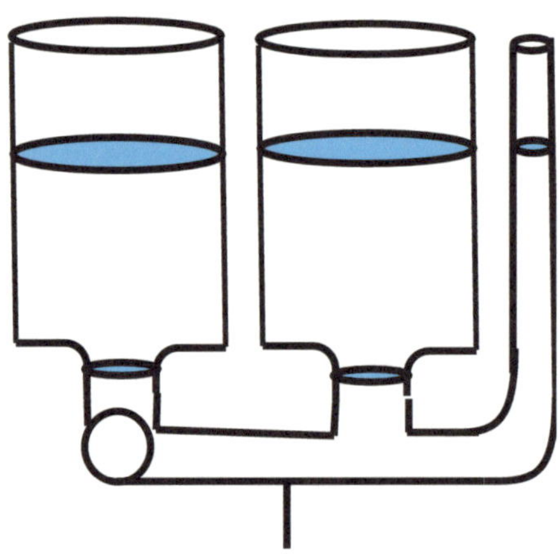

Fig. 3. Schematic representation of the attached vessels for dynamic adsorption experiments on lead removal by potassium polytitanates nanoadsorbents.

All experiments were performed independently for each condition of pH and concentration. The maximum time of pass of the lead nitrate solutions was 40 hours (adsorbents saturated), then the adsorbents were removed from the column, dried at a temperature of 60 ° C for one hour. These adsorbents were characterized by the techniques of SEM, XRD, thermal analysis, IR and specific surface area.

Operating conditions in the dynamic adsorption experiments in columns for aqueous solutions of lead are summarized in Table 2. This work has applied a method based on the originally described by Michaels et al., 1952 for ion exchange resins and subsequently by Weber, 1972 & Lukchis, 1973, for activated carbon beds.

Property/Feature	Inorganic Media
Agglomerate Size	0.25-3 mm.
Column Volume	175 ml.
Volume ocupaded by adsorbent	53 ml
Density	3.1
Weight of adsorbent	169.6 g
Volume of solution in the area occupied by the adsorbent.	35 ml.
Ion-Exchange class	cationic
pH	3.5, 5.5, 5.8
Elution Rate	2.33 ml/min.=140mL/H
Water content after draining	40-50%
Initial concentration of Pb^{2+} in solution	≈150, 330 mg/L

Table 2. Column and nanoadsorbents experimental conditions used in adsorption processes: Properties/features and Inorganic media.

3. Results

3.1 Reaction mechanism on adsorbent

The results showed that the introduction of KOH in molten potassium nitrate allows for different titanates as a function of the variation in the chemical composition of the melt. The probable mechanisms of the process that takes place in the investigated systems are represented in the figure 4.

In this work were considered two ways in the formation of simple titanates ions:

1. The dissolution of TiO_2 as a result of interaction with OH^- ions.
2. The acid-base interaction of the Lux-Flood of O_2-ions directly with TiO_2 particles

The interaction between TiO_2+O_2+K ions promotes the formation of ions, which ones may not participate in the formation of K_2TiO_3 as product by high solubility in the melt and hydrolysis by water during washing of the sediments.

Structural fragments of the TiO_2 network formed by interactions with OH^-(solution) can also be transformed into simple structural units titanate anion by the interaction of Lux-Flood (TiO_2+0^{2-}). This anion can participate in the formation of ions polytitanate ($K_2Ti_nO_{2n+1}$ +OH^-).

The formation of titanates (polytitanates) of potassium can occur in hot melt but the rate of hydrolysis and decomposition of them is high so that the formation of potassium titanate crystal in the melt is of low probability.

Kinetic data obtained showed that the amount of potassium titanate formed by the molten salts treatment after the end of induction period (1 hour) does not depend of temperature or KOH content in the melt.

This can consider this period of induction and the time needed to obtain the product system in thermodynamic equilibrium. The process of obtaining of this equilibrium system includes the processes indicated in Figure 4. Taking into account that the rate of these elementary processes is different, it is can say that the amount of product obtained is due to the ion content variation $O^{2-}(KNO_3)$ in the melt.
In the washing process, free positions of potassium cations in the crystal lattice may be occupied by cations of hydrogen during the washing process. Previous results indicate that the anionic structure of sintered polytitanate does not depend on potassium content. High rate of hydrolysis of simple potassium titanate by the washing allows obtain solids products in the form of potassium polytitanate formed as nanometric particles with molar relationships of $K_2TiO/TiO_2=5.1$, at 500° C of temperature.
In the cooling stage, it is possible that the formation of crystalline titanates includes both titanium compounds: oxides and titanates and then the generation of different forms of TiO_2 as: anatase, brookite, and rutile and one part of titanium oxide crystals was formed by interactions between anions of titanate.

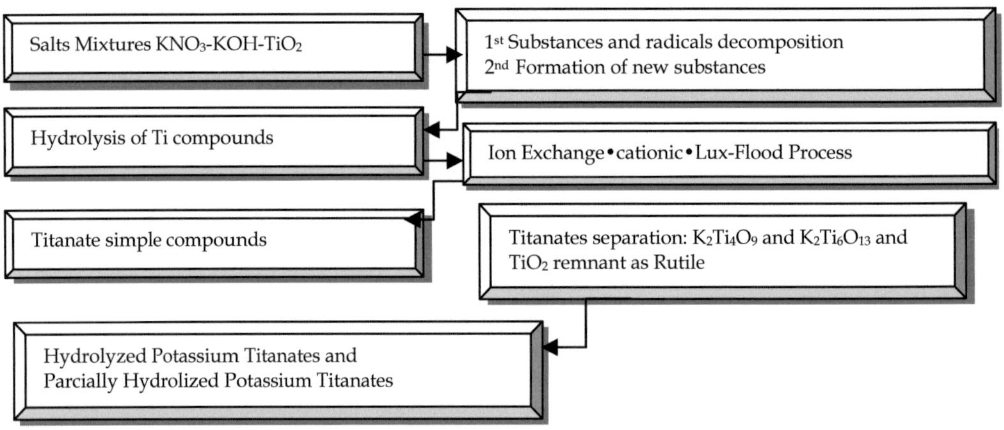

Fig. 4. Chemical processes scheme in systems treated by the KNO_3-KOH-TiO_2 molten salts processes.

3.1.1 Adsorption dynamic test

Before performing adsorption column, it is necessary to determine the coefficients of adsorption in static systems, for lead present in the different waters, with the aim of designing appropriate operating conditions.
In addition adsorption results have determined that it is possible to perform adsorption studies by column experiments under real conditions of flow (non-equilibrium situation).
Chemical analysis data (Figure 5.) showed than the adsorbent effectively removal Pb^{2+} ions in the solution with pH_0= 5.5. After passing the solution 1 through the adsorption column, the lead concentration decreased to 0.5-0.6 mg/dm^3, meeting the requirements of national rules for river water (1.0 mg/dm^3). The best lead removal of Pb^{2+} was observed after three hours of contact, and then the saturation time did not influence the effectiveness of the removal of lead is shown in Figure 4. However, for solution 2 (pH_0= 3.5), the remaining concentration of Pb^{2+} was relatively higher but decreased to 14-23 mg/dm^3.

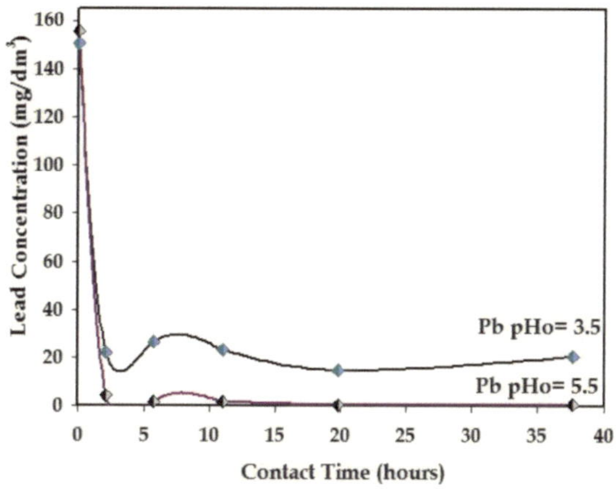

Fig. 5. Effect of contact time on the content of Pb in the passed aqueous solution prepared at pH of 3.5 and 5.5.

During adsorption process, both eluted (purified) solutions showed a high content of K, whose concentration changed during the time of saturation of the adsorbent (Figure 6.). During first ten hours, the content of K increased from 0.07 to 118 mg/dm^3 (solution 1) and 205 mg/dm^3 (solution 2) and then decreased between 86 and 93 mg / dm 3 (pH_0= 5.6) and between 133 and 143 mg/dm^3 (pH_0= 3.5).

Potassium concentration in the eluted solution was higher for the solution 1 than those in solution 2, which can be explained by the presence of two parallel processes of ion exchange:(1) $Pb^{2+} \leftrightarrow 2K$ + ($PbOH^+ \leftrightarrow K^+$) and (2) $H^+ \leftrightarrow K^+$, by reducing the pH, second process rate increases strongly, dominating the first. With increasing K content increased the pH value of the eluted solution to 10.6 (solution 1) to 11.3 (solution 2), (Figure 6). It is noteworthy that the maximum value of pH was observed in the first sample collected and gradually decreased (solution 1) or stabilized (solution 2). The elution rate retained its original value during 15 hours of contact under conditions of constant hydrostatic pressure, and then gradually decreased from 140 to 100 cm^3/h (38 contact hours) is shown in Figure 7.

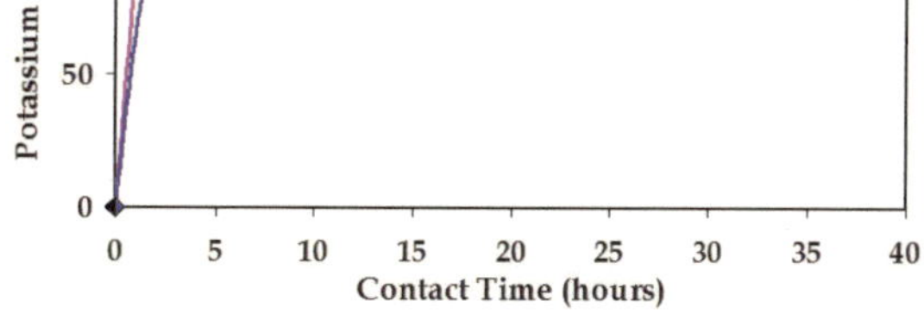

Fig. 6. Effect of contact time on the content of K in the passed aqueous solution prepared at pH of 3.5 and 5.5.

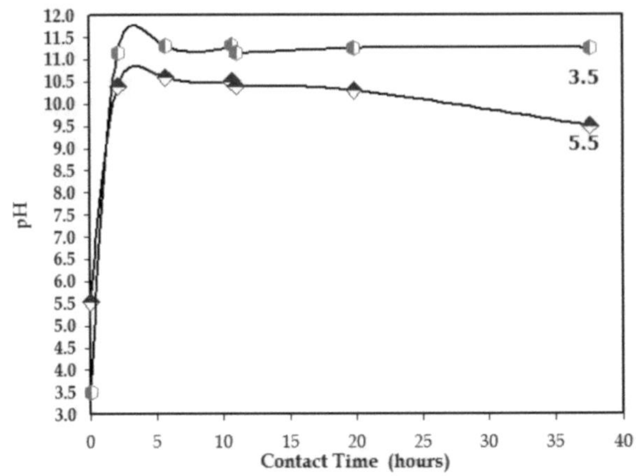

Fig. 7. Effect of contact time vs Ph in adsorption process with potassium polytitanates.

The behavior of pH in solutions with Pb, has shown that in solution with pH_0=3.5 were obtained higher values after the first hour of sampling compared with those in which ones carried out with pH_0 5.5. This may be due to adsorption capacity lower in solutions with lower pH_0 to 5.5 inducing the formation of more precipitation of hydroxide-like compounds in the initial stage of the experiment and therefore a lower activity between the adsorbent surface interactions and lead ions in solution.
Figure 9 shows a comparison in measuring the pH of the dynamic experiments carried out at this stage lead.

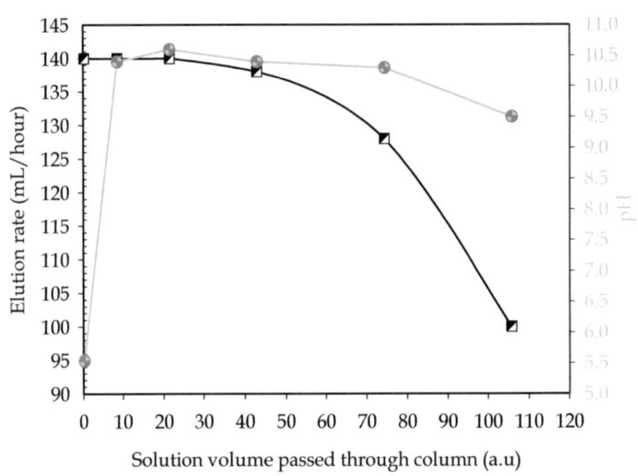

Fig. 8. Influence of solution volume passed through column versus elution rate of Pb solution and versus pH. The experiments were done on a pH_0=5.5. The initial concentration was: Pb = 150.55, K = .065 in mg/L.

Figure 8 shows a graph of the volume passed through the column vs. elution rate and versus pH. In the figure 9 it can be seen that the elution rate on dynamic adsorption experiments of Pb 150mg / L remains constant during the first stage almost as much as 50% of the passage of total solution volume, the rate decreases in the last quarter of the experiment, this may be due to saturation of available adsorption sites on the adsorbent.

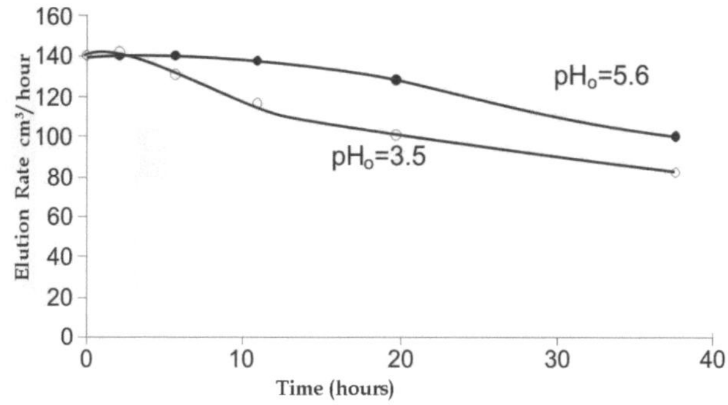

Fig. 9. Effect of contact time on the flow rate of aqueous solution.

3.2 XRD for nanoadsorbents synthesized

The goal of the XRD studies in this research was indentify crystalline phases, also chemical composition of materials applied in the different stages of processing. Studies were first realized in the start materials. The quantify of crystalline phases was made in previous studies by Li et al, 1999.

Figure 7 shows X-ray diffraction pattern corresponding to the obtained samples in the method of making of potassium polytitanates nanoadsorbents by means of precursors route in molten salts. Details of synthesis for each material is given and described in the experimental methodology of this research. Chart 74-0275 was identified as $K_2Ti_60_{13}$ of ICDD database.

Crystalline phases contents in selected composition as optimum have shown quasi-crystalline behavior for a N value (K_20/TiO_2) of 5.1. Whit this relationship the main XRD diffraction intensities obtained in 2θ degrees were: 11.2, 23.9, 29.8, 33.1, 34.7, 37.8, 43.1, 47.5, 47.7, 51.9, 55.1, 57.3, 58.7, 59.2, 61.9 y 66.4°(main reflexions in volume) these results are in agreement with Liu et al 2005.

Besides it is very important to note than potassium titanates pure compounds as: $K_2Ti_2O_5$ (potassium dititanate), $K_2Ti_4O_9$ (potassium tetratitanate) and $K_2Ti_6O_{13}$ (potassium hexatitanate) all of them includes closed reflections in 2 teta degrees scale. Figure 11 shows a graph corresponding to potassium titanates in form of pure compounds.

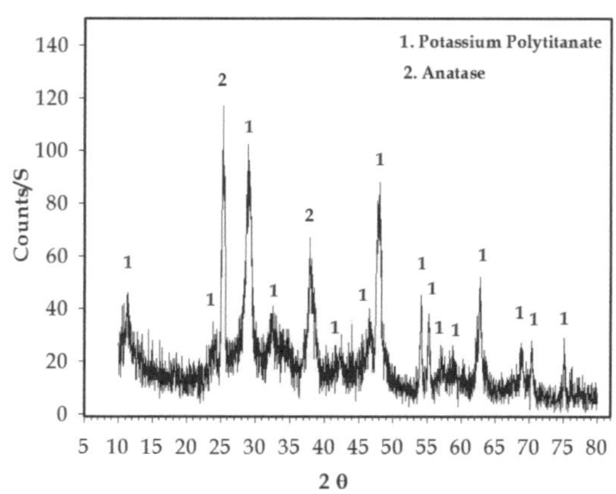

Fig. 10. X-ray diffraction (XRD) patterns of potassium polytitanates nanoadsorbents from samples prepared by molten salts synthesis, , with relationship of $TiO_2/K_2O = 5.1$.

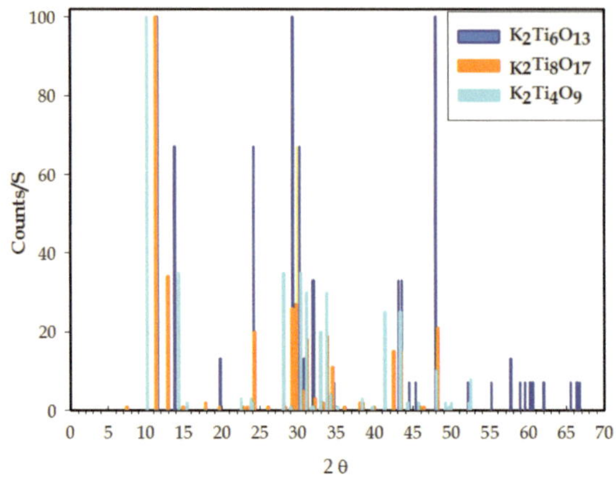

Fig. 11. X-ray diffraction (XRD) patterns from pure potassium titanate: $K_2Ti_4O_9$, $K_2Ti_6O_{13}$ y $K_2Ti_8O_{17}$.

According to EDX-S analysis results the crystalline phases of the material produced from mixtures of PTP containing oxides of titanium as $K_2Ti_60_{13}$.

Figure 12 shows atomic arrangements obtained from the data of X-ray diffraction for nanoadsorbents designed in powder cell program software.

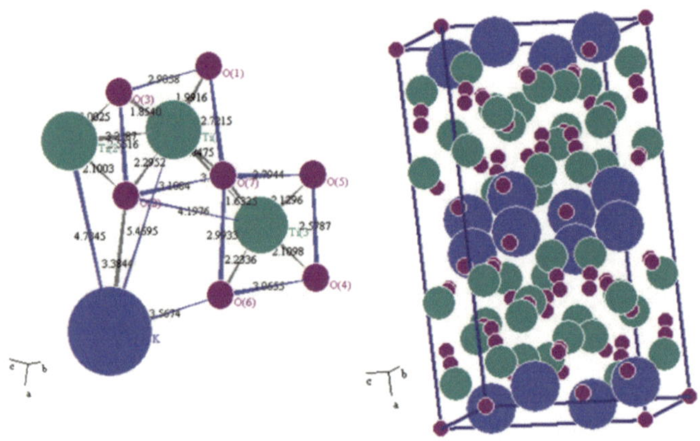

Fig. 12. Atomic arrangements obtained with nanosorbents data, nanostructure was modeled by Powder Cell Software and ICDD 2004 data base.

3.3 Surface area measurements

3.3.1 Textural properties

There exist three parameters used in the measurement of porosity: specific surface area, the specific pore volume (open porosity) and size and distribution pore. Results show values of surface area were similar than those obtained by Wallenberg et al.,1990 from potassium tetratitanate of $13.9m^2/g$, in potassium polytitanates nanoadsorbents was obtained $13.96m^2/g$ as result from without treatment samples

In order to know the surface characteristics of both external and internal, in the adsorbents, there was an isotherm nitrogen adsorption on a sample of granular adsorbent with BET technique, the method for sample preparation described previously in experimental part.

The nitrogen adsorption isotherm obtained of the based in potassium polytitanates, is presented in figure 13. BET measurements were realized for three sections of column: top, middle and bottom, also one sample without adsorption. Such plots are typical for microporous solids with a significant contribution to the meso and macroporous scale (Ruthven, 1884). The higher value was presented in the sample corresponding to top of column of 2.54m²/g. Figure 13

In the nanoadsorbent particles analyzed (sample without treatment), were found in the dimensions range from 100 to 150 microns. Results (shape of the isotherm) have indicated that the porosity detected, corresponding to type mesoporous adsorbents with regions of porosity of relative pressures (P/Po) 0 to 0.2, then from 0.2 to 0.32 on the scale corresponds to the inter-lamellar distances that exist in the structure of the potassium titanates. The results showed that there is a pore size distribution variation, containing a scale small fraction less than 50 A ° assessed by TEM.

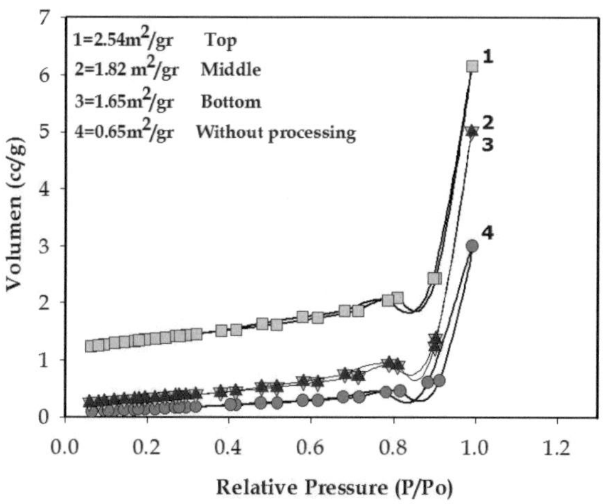

Fig. 13. Results of specific surface area measurement of dynamic experiments for samples with Pb: 1) granular adsorbents agglomerated with potato starch, without adsorption proceses, 2), 3) and 4) same adsorbents extracted from the adsorption column with solutions of Pb, on the top, middle and bottom, respectively

3.4 SEM Images

The morphologies of the start material samples were investigated by scanning electron microscopy. Figure 14 shows the SEM images of quasicristalline potassium polytititanate nanoadsorbent. From figure 2a, was found that the this material consisted of agglomerates of well-defined layers with sizes of 100 to 500 nm. The surfaces of these layers were relatively smooth. After crystallization, the sizes of these layers kept unchanged. However, their surfaces became rough (Figure 17). Obviously, the layers of quasicrystalline K2Ti4O9 were composed of small agglomerates primary particles of several nanometers in size.

Figure 15 shows a 10000X photomicrograph of a sample of adsorbent after the adsorption process of lead. Samples have remained for a contact time of one hour with the solution corresponding to the concentration of 155mg/L of $Pb(NO_3)_2$. They can be observed with the morphology of agglomerates of lead adsorbed on the surface of potassium polytitanates.

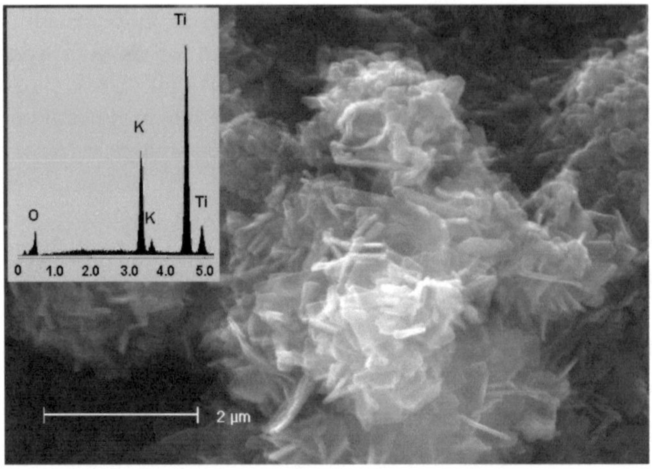

Fig. 14. SEM Photomicrography. Agglomerates of potassium polytitanates nanoadsorbents a) y b) spectrum obtained by SEM-EDXS Technique.

Fig. 15. Photomicrography of nanoadsorbents in lead adsorption. The concentration of solution was 155 mg/L of solution. The contact time corresponded to one hour.

3.5 Thermal analysis of nanoadsorbents of potassium polytitanates

Thermal analysis data of Figure 16, corresponding to PTP have indicated that the obtained material contained adsorbed water (removed from 350 to 550 ° C) and structural water (removed from 500 to 700 ° C). The thermogram ATG showed a partial loss of sample weight during their analysis of 8.8% (wt%) at 555 ° C, subsequently detected a loss of 4.7 (wt%) to 714 ° C generated by the decomposition of groups Ti -OH (Shimizu et al., 1981) (removal of structural water).

The DTA thermogram indicated the presence of an exothermic reaction to 703 ° C, which corresponds to the onset of crystallization of the PTP (Bergon et al., 1984). At a temperature of 1030 ° C, there is only the transformation of anatase and rutile TiO_2.

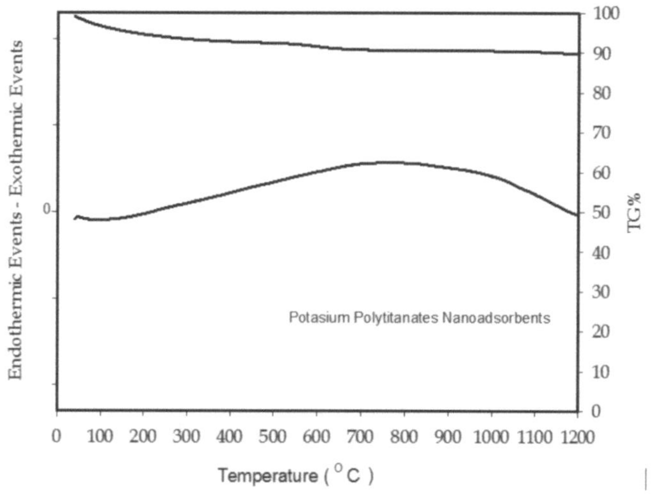

Fig. 16. Thermal analysis of Potassium polytitanates nanoadsorbents.

3.6 Transmission Electronic Microscopy (TEM)

Morphology of the nanaoadsorbents was further investigated by transmission electron microscopy. Figure 17 shows the TEM image of quasi-crystalline $K_2Ti_40_9$(layers) and $K_2Ti_6O_{13}$(fibers). It was found that some well-defined lamellar agglomerates of about 100 nm in sizes were connected to each other to form pores in the range of 30 to 50 nm. During the sonication for TEM measurement, some small primary particles were dropped from the fibers of $K_2Ti_6O_{13}$. These monodispersed primary particles are of about lower than 50 nm in size, in agreement with SEM observation. Most of the morphology of the particles in PPN was still preserved well after sonication. These secondary agglomerates are slightly larger than those of anatase phase. Mesopores formed among the small interlayer spaces in nanoadsorbents, while macropores or large mesopores produced among of PPN were also observed. This confirms the existence of a hierarchical porous structure in the nanoadsorbents. The TEM image of this is consistent with the XRD and SEM results obtained.

3.7 Infrared spectroscopy of samples after and before of the adsorption processes

The IR analysis confirms that there is no generation of new functional groups at the end of the adsorption process. When comparing the spectrums of figure 18, are very similar, there is only a small variation in the intensity of some small peaks located around 1440 and 1750 WLcm^{-1}, these peaks can be attributed to ion exchange occurred between ions of K^+ and Pb^{2+} during the adsorption process. Nakamoto et al, 1978.

The adsorption promotes a decrease in the intensity of the absorption bands of 1448 and 1613 cm.-1 (angular vibration groups Ti-OH and HOH). This indicates that the incorporation of Pb^{2+} in the surface structure of titanate decreases the concentration of Ti-OH groups and adsorption of H_2O in the air. While, there is a variety of characteristics of Ti-O bonds in the octahedral TiO_6, instead of an adsorption peak at 500 nm where two peaks at 500 and 650 nm.

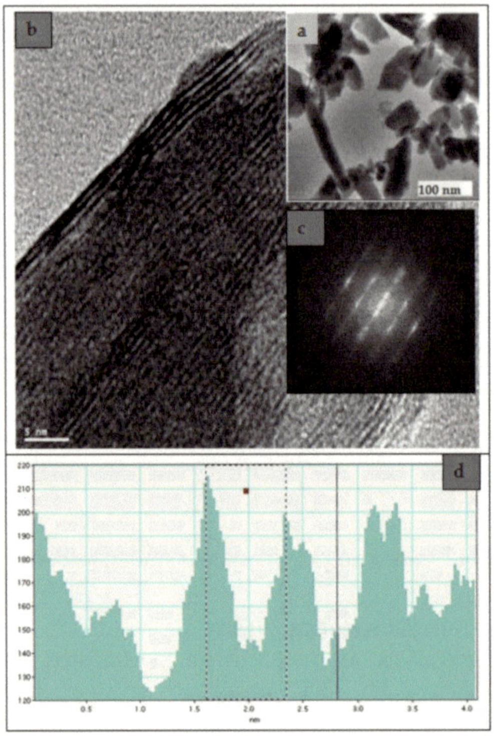

Fig. 17. TEM images of scratched nanoadsorbents: a) TEM image of a nanoadsorbents composed of several lamellae, b) scratched surface morphology (c) SAED pattern of this whisker showing the same crystallographic orientation of lamellae and d) interlayer spacing on potassium polytitanates structure.

3.8 Solubility in nanoadsorbents, saturated, press and heat treated

After the adsorption experiments samples were milled, mixed, heat treated (780 and 900° C) and afterward poured into glass flasks, in a 10% v/v of adsorbent with the rest of water. These solutions were heat treated of 96° C during a period of 5 hours (Japanese rule JISR-3520 and Russian rule GOST 10134-82). The finality of these processing was the assess of the final lead content and consider as free in solution and its total solubility. Results (Table 3) have shown, that final content in the samples were treated to 900°C has met with a lower level for water for human consume(Abalak et al, 2003).

Heat treatment Samples	Lead Concentration (mg/L)
780° C	**0.140**
900° C	**0.064**

Table 3. Results of chemical analysis by ICP technique for nanoadsorbents heat treated samples to 780 and 900° C.

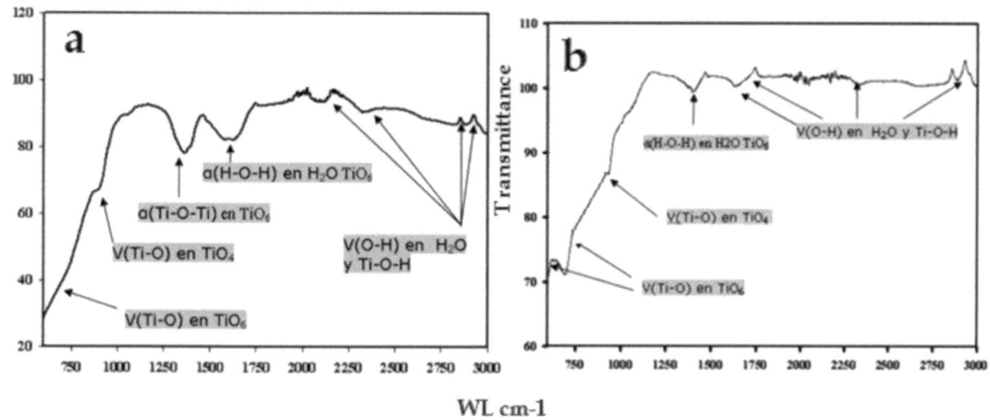

Fig. 18. IR spectrums for two samples a) without adsorption process and b) adsorbed with lead ions.

In this study was developed a new adsorbent for the removal of divalent cations on metal oxides, which allows continuity between surface reactions and precipitation. The model applied in Figura 19 fits the model of surface complexity, whereas the precipitation in the solid is described by the formation of a solid solution whose composition varies continuously between that of the original material and a pure precipitate cationic adsorption. It demonstrated the ability of the surface precipitation model to describe the adsorption equilibrium of metal cations in potassium hydroxide amorphous. The model can also be extended to describe competitive cationic and anionic adsorption.

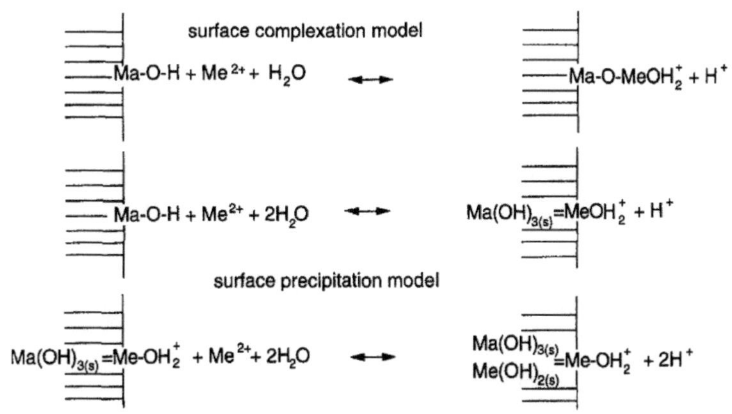

Fig. 19. Scheme of surface complexation/precipitation reactions. Modified from Lützenkirchen J. & Behra, Ph (1981) of the original model of Farley et al. (1996).

Farley et al, 1984 proposed surface precipitation model in order to cations adsorption in metallic oxides.

Our working group will shortly present a study on the transformation of potassium titanate about this nanoadsorbents kind being that it is possible carry out after the adsorption process a treatment to convert an orthorhombic structure(potassium titanate) in a new metallic titanate(metallic titanate) with new functional properties.

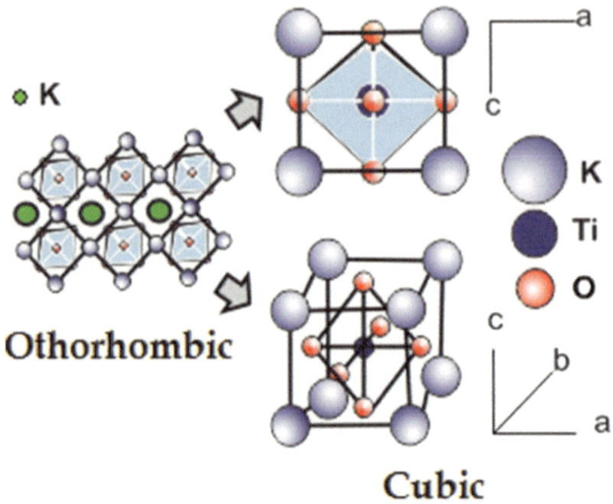

Fig. 20. Structural transformation of the nanoadsorbents with help of heat treatment.

The results confirmed that the nanoadsorbents based on potassium polytitanates promoted the removal of lead ions from aqueous solutions. The greatest efficiency of removal was obtained in the No.1 solution, slightly acidic (pH_0 5.6). To explain the action of the investigated nanoadsorbents, it is to consider the mechanism of interaction between potassium titanates and Pb-containing solutions with different pH. It is well known that the chemical speciation of Pb in aqueous solutions depends on the pH (Dean et al. 1972). In acidic solutions (pH <5) Pb^{2+} ions predominate, for pH values in the range from 5 and up to 6, the solutions have a high ion content $PbOH^+$, and also Pb^{2+}. In solutions with pH> 6 is favored precipitation of $Pb(OH)_2$. The investigated adsorbent has a hydrolysis high rate and becomes to its hydrated form (Zheng et al., 1998, Nunes et al., 2006), thus increasing the pH and the concentration of K^+ ions in aqueous solution. In the first stage of interaction of the adsorbent with the solution of lead salts, can be seen two parallel processes: the adsorption of Pb^{2+} and $PbOH^+$ ions on the inner surface of the potassium titanates crystals as well as the hydrolysis of titanate potassium. After adsorption of Pb^{2+} ions, they are involved in a process of ion exchange with K^+ cations, which are located between channels from the crystals of the adsorbent (Figure 1). Our results can be applied perfectly to Farley model et, al 1996.

The most important aspect in this work is that it is possible to obtain new materials from contaminated materials using a single heat treatment and without human health risks.

4. Conclusions

In this work were developed potassium polytitanates nanoadsorbents by molten salt synthesis with general formula ($K_2Ti_nO_{2n+1}$), where the relationship (K_2O/TiO_2)= 5.1, with a homogeneous structure and high functional properties. Nanoadsorbents obtained with a lamellar and fibrous structure, have shown a high intercalation of cations, ion exchange activity and structure layer, and fibrous, characterized by high adsorption properties.

Nanoadsorbents were obtained as ceramic materials whit porosity and quasi-crystalline phases regulated. They serve for the immobilization of Pb^{2+} and with an additional heat treatment can be obtained ceramics general interest.

The lead removal mechanisms in the adsorbent produced in granular form and applied to adsorption columns, was carried out by a combination of ion exchange-adsorption-coprecipitation of Pb^{2+}. During the continuous contact time, nanoadsorbents structure was deteriorated, slowing the elution rate, but this did not influence in the effectiveness of metallic removal.

The intra-particular Pb^{2+} diffusion into adsorbent, has occurred in 2 stages: 1) while the metal ions, rapidly diffuse between particles of the nanoadsorbents, at the beginning of adsorption and 2) when the intra-particular diffusion is reduced and then stabilized. The kinetics of adsorption depends of: metal ions concentration in solution, adsorbent dose, pH and contact time, and does not depend on the particle size of adsorbent.

In the sintering is transformed species generating titanate Pb metal. ($PbTiO_3$). One possible application can be as electronic sensors and piezoelectric devices.

Polytitanates potassium consisted of particles agglomerate like-flake shape with a nominal diameter of 50-300 nm and a thickness of 20-60 nm, which in turn have a lamellar structure, the space between the plates, which forms nanopores. When heating the particles to a temperature of potassium polytitanates 500 ° C, they crystallize to nanowiskers.

The chosen methodology has included in the study of the effect of synthesis parameters on the physico-chemical and adsorption properties of potassium polytitanates and optimization of the properties of nanomaterials for use as a highly effective adsorbent for bivalents cations.

Studies identified the optimal synthesis regimes of potassium polytitanates (weight ratio of raw materials TiO_2: KOH: KNO_3 = 1:1:32; at 500 °C of temperature, time 1 h) and its further processing (sequentially washed with water and 0.1 N aqueous solution of hydrochloric acid, drying, milling. In potassium polytitanates, always get ultrafine materials (the average size of agglomerates of nanoparticles of hydrated potassium polytitanates of 1.4 microns.

The structure and composition of Potassium Polytitanates studied by Xray diffraction analysis (Fig. 10), infrared spectroscopy and scanning electron microscopy (Fig. 18). IRS show slight differences in the adsorbed samples, data no revels differences among the obtained samples. According XRD data base ICDD 2004 shows that the composition of phase is variable and its depends of the reagents relationship, the selected technology and the synthesis temperature conditions . Its means that in the stage of synthesis is very important to have a PPN with low content of impurities as $K_2Ti_4O_9$ $K_2Ti_6O_{13}$, the product structure will depends largely of the synthesis temperature and their process of purification(washes).

5. Acknowledgment

Author thanks the financial support of Consejo Nacional de Ciencia y Tecnología (CONACYT, México), through the project SEP-CONACYT-2006 No 54883 and the scholarship given to M.A. Aguilar-González (No. 201784).

6. References

Afanasiev, P. (2006) *Molten Salts Synthesis of Alkali Metal Titanates.* J. Mater. Sci. Vol. 41 p.p 1187-1195.

Albalak, R. Noonan, G. Buchanan, S. Flanders, W.D. Gotway-Crawford, C. Kim, D. Jones, R.L. Sulaiman, R. Blumenthal, W, Tan, R. Curtis, G. & McGeehin, M.A. (2003). *Blood lead levels and risk factors for lead poisoning among children in Jakarta, Indonesia*. The Science of the Total Environment Vol.301 p.p 75–85

Anthony, R.M. & Breimburst, L.H. (1981).*Determining Maximum Influent Concentrations of Priority Pollutants for Treatment Plants*. Journal of the Water Pollution Control Federation. Vol. 53, No 10, p.p 1457-1468.

Ayyappan, R., Carmalin-Sophia, A., Swaminathanb, K., & Sandhya S. (2005). *Removal of Pb(II) from aqueous solution using carbon derived from agricultural wastes*. Proc. Biochem. Vol. 40, p.p 1293–1299.

Constantino, V.R.L. Barbosa, C.A.S. Bizeto, M.A. & Dias P.M (*2000). Intercalation Compounds involving Inorganic Layered Structures. Anais da Academia Brasileira de Ciências*. Vol. 72 (1).

Bansal, V. Poddar, P. Ahmad, A. & Sastry, M. (2006).Room-Temperature Biosynthesis of Ferroelectric Barium Titanate Nanoparticles, J. Am. Chem. Soc., Vol.128 No.36, pp 11958–11963.

Bavykin, D.V. Friedrich, J.M. & Walsh, F.C. (2006) Review: *Protonated Titanates and TiO_2 Nanostructured Materials: Synthesis, Properties, and Applications. Advanced Materials*, Vol. *18*. p.p 2807–2824. WILEY-VCH Verlag GmbH & Co. KGaA, Weinheim.

Bavykin, D.V. & Walsh, F.C. (2009). Elongated Titanate Nanostructures and Their Applications. Eur. J. Inorg. Chem., 977–997 © 2009 Wiley-VCH Verlag GmbH & Co. KGaA, Weinheim3.2.1. Fuel Cells and Batteries.

Barrier, R.M. (1978) Zeolites and Clay Minerals, Academic Press, London.

Bergon, C., y Risbud, S. (1984). *The One-Component System, The titania and zirconia systems In:Introduction to phase equilibrium in ceramic the Amerincan Ceramics Society*. Urbana, illinois. p.p 9-19.

Dean, J.G. Bosqui, F.L. & Lanouette, K.H. (1972). *Removal of heavy metals from waste-water*. Env. Sci. Tech. 6, 518-524.

Fox M. A. & Dulay M. T. (1993). Heterogeneous Photocatalysis. Chem. Rev., 93, 341–357.

Grätzel, M. (2001). *Photoelectrochemical cells*. Nature. Vol. 414, p.p 338–344.

Gregorová, E., y Pabst, W. 2007. *Porosity and Pore Size Control in Starch Consolidation Casting of Oxide Ceramics—Achievements and Problems*, J. Eur. Ceram. Soc. Vol. 27 p.p 669-672.

Hayashi, H. & Hakuta, Y. (2010) Review: *Hydrothermal Synthesis of Metal Oxide Nanoparticles in Supercritical Water* Materials, Open Access. Vol.*3*, p.p 3794-3817; doi:10.3390/ma3073794 ISSN 1996-1944 www.mdpi.com/journal/materials.

Hrudey, S.E. Chen. W & Rosseaux, C.G (1995) Bioavailability in environmental risk assessment. Lewis Publ., Boca Raton, FL.

Klein, M. (1985). Filtering device for treating contaminated aqueous solutions. US patent 4505823. 11 pp.

Ion, A.C. Ion, I. and & Culetu, A. (2010). *Lead adsorption onto exfoliated graphitic nanoplatelets in aqueous solutions*. Materials ScienceandEngineeringB. Article in press.

Li, Y. Gao, X.P. Li, G.R. & Pan, G.L. Yan, T.Y. & Zhu, H.Y.(2009). *Titanate Nanofiber Reactivity: Fabrication of $MTiO_3$ (M) Ca, Sr, and Ba) Perovskite Oxides. J*. Phys. Chem. C 2009, 113, 4386–4394.

Lanphear, B.P. Hornung, R. Ho, M. Howard, C.R. Eberly, S. & Knauf, K. *Environmental lead exposure during early childhood*. J Pediatr. 2002 Jan;140(1):40-7. General and

Community Pediatrics Department, Children's Hospital Medical Center, Cincinnati, OH 45229, USA.

Laurentin, A. Cardenas, M. Ruales, J. Pérez, E. & Tovar, J. (2003). *Preparation of indigestible pyrodextrins from diferent starch sources*, J. Agric. Food Chem, 51, 18, p.p 5510-5515.

Li, G.L. Wang, G.H. & Hong, J.M.(1999). *Synthesis of $K_2Ti_6O_{13}$ whiskers by the method of calcination of KF and TiO_2 mixtures.* Materials Research Bulletin, Vol. 34, Nos. 14/15, pp. 2341–2349.

Luckchis, G. M (1973) Chem. Eng. Vol. 80:p.p 111.

Lützerkitchen, J.&. Behra, Ph. (1996). On the surface precipitation model for cation sorption at the (Hydr)oxide Water Interface. Kluwer Academic Publishers. Printed in the Netherlands. Aquatic Geochemistry Vol.1: p.p 375-397

Matsumiya, H.), Masai, H.), Terazono, Y.), Iki, N.) & Miyano S. (2003). Chelating adsorbent for heavy metal ions prepared by loading anion-exchange resin with thiacalix[4]arenetetrasulfonate. Bulletin of the chemical society of Japan Vol. 76, No. 1 p.p.133-136

Naicker, N. Norris, S. Mathee, A. von Schirnding, Y. & Richter, L. (2010). *Prenatal and adolescent blood lead levels in South Africa: Child, maternal and household risk factors in the Birth to Twenty cohort.* Environmental Research Vol. 110 p.p 355–362. Nakamoto, K. (1978). Infrared and Raman Spectra of Inorganic and Coordination Compounds, John Wiley and Sons, New York.

Noonan, G. Albalak, R. Buchanan, S. Flanders, W.D. Gotway-Crawford, C. Kim, D. Jones R.L. Sulaiman, R. Blumenthal, W. Tan, R. Curtis, G. McGeehin, M.A. McElroy, R. Schurz-Rogers, R. Randolph, W. Rubin, C. & Jarrett J.(March 2001). *Blood lead levels and risk factors for lead poisoning among children, Torreon, Coahuila, Mexico.* Final Report. Lead Poisoning Prevention Branch/Health Studies Branch/Nutritional Biochemistry Brand, U.S. Centers for Disease Control and Prevention, U.S. Dept. of Health and Human Services, Atlanta, GA.
http://www.bvsde.paho.org/bvsea/fulltext/torreon.pdf.

Pfaff, G. (1993). *Sol–gel synthesis of strontium titanate powders of various compositions* J. Mater. Chem. Vol. 3, p.p 721-724.

Pfaff, G. (1994).*Synthesis of calcium titanate powders by the sol-gel process.* Chem. Mater. Vol. 6 No.1, p.p 58–62.

Plaza, M.G. Pevida, C. Arias, B. Fermoso, J. Casal, M.D. Martín, C.F. Rubiera, F & Pis J.J. (2009). *Development of low-cost biomass-based adsorbents for postcombustion CO_2 capture.* Fuel. Vol. 88, Issue 12 p.p 2442-2447.
7th European Conference on Coal Research and Its Applications.

Riddell, T. Solon, O. Quimbo, S.A. Tan, C.M, Butrick, E. & Peabody, J.W. (2007) *Elevated blood-lead levels among children living in the rural Philippines.* Bulletin of the World. Health Organization 85 (9) P.P 674-682.

Riss, A. Berger, T. Grothe, H. Bernardi, J. Diwald, O. & Knolzinger E. (2007). *Chemical Control of Photoexcited States in Titanate Nanostructures.* Nano Letters 2007 Vol 7. No.2 p.p 433-438.

Ruthven, D.M. (1984).*Principles of adsorption and adsorption processes.* Wiley Interscience, New York.(USA), 1984.

Schmidt, H.K. Geiter, E. Menning, M. Krug, H. Becker, C. &. Winkler, R. (2010).The sol- gel process for nano-technologies: *New nanocomposites with interesting optical and mechanical properties.*

http://scidok.sulb.uni-saarland.de/volltexte/2010/2895/pdf/sm199812.pdf.

Sperlich, A. Werner, A. Genz. A. Amy, G. Worch, E. & Jekel, M. *(2005).Breakthrough behavior of granular ferric hydroxide (GFH) fixed-bed adsorption filters: modeling and experimental approaches.* Water Research Vol.39 p.p 1190–1198.

Tsezos, M. (1984). Recovery of Uranium from Biological Adsorbents- Desorption Equilibrium, Biotechnology AND Bioengineering, Vol. 26, p.p 973-981.

Tong, S. von Schirnding, Y.E & Prapamontol, T. Environmental lead exposure: a public health problem of global dimensions.Bulletin of the Wordl Health Organization Vol.78 No.9 p.p 1068-1077.

Unal, U. Matsumoto, Y. Tanaka, N. Kimura, Y. & Tamoto, N.(2003). Electrostatic Self-Assembly Deposition of Titanate(IV) Layered oxides intercalated with transition metal complexes and their electrochemical properties. J. Phys. Chem. B. Vol. 107 (46), p.p 12680-12689.

Wallenberg, L., Sanati, M., & Andersson, A.(1990). On the transformation Mechanism of $K_2Ti_4O^9$ to TiO_2 and Formation of Microvoids, Microsc. Microanal Microstruct. 1:357-364.

Weber, W. (1972). *Physicochemical processes for water quality control "Kinetics of Adsorption in Columns of Fluidized Media,"* J. Water Pollut. Contr. Fed., 37, 4, 425. Weber, WJ, Jr.*SERIE: Environmental science and technology, New York Wiley-InterScience.*

Xie, G. Y, Zhong, Y. J, Sui, G. X, & Yang, R. (2010). *Mechanical properties and sliding wear behavior of potassium titanate whiskers-reinforced poly(ether ether ketone) composites under water-lubricated condition.* Journal of applied polymer science. Vol. 117 issue 1 p.p 186-193

Zhang, S. Dai, Y. Xie, X. Fan, Z. & Tan, Z. (2005). *Study on blood lead level and related risk factors among children aged 0-6 years in 15 cities in China.* Journal Zhonghua liu xing bing xue za zhi = Zhonghua liuxingbingxue zazhi Vol. 26 No. 9 p.p 651-654. http://alertaplomo.org

Zhao, M. & Duncan, J.R. *(1998) Bed-depth-service-time analysis on column removal of Zn^{2+} using Azolla filiculoides.* Biotechnology Letters, Vol 20, No 1, , pp. 37–39.

Zhang, S. Chen, Q. & Peng, L.-M. (2005). *Structure and formation of $H_2Ti_3O_7$ nanotubes in an alkali environment.* Phys. ReV. B, Condensed matter and materials physics Vol. 71, No. 1. pp. 014104.1-014104.11

Zhuang, G.S., Sui G.X., Meng H., Sun, Z.S. & R. Yang.(2007). *Mechanical properties of potassium titanate whiskers reinforced poly(ether ether ketone) composites using different compounding processes.* Composites Science and Technology Vol. 67, Issue 6, p.p 1172-1181..

Wang, Q. Guo, Z. & Chung, J. (2009) *Formation and structural characterization of potassium titanates and the potassium ion exchange property.* Materials Research Bulletin Vol. 44 p.p 1973–1977.

Yang, J. Shi, Y. & Yan, C. (2010). Selective Laser Sintering of Polyamide 12/Potassium Titanium Whisker Composites. Journal of Applied Polymer Science. Vol. 117 No.4 p.p 2196-2204.

Yeh, Y. C., Tseng, T.Y, & Chang D.A.(2005). Electrical Properties of TiO_2-$K_2Ti_6O_{13}$ Porous Ceramic Humidity Sensor J. Am. Ceram. Soc., Vol. 73 issue 7. p.p *1992-1998.*

23

Metal Stabilization Mechanisms in Recycling Metal-Bearing Waste Materials for Ceramic Products

Kaimin Shih and Xiuqing Lu
Department of Civil Engineering, University of Hong Kong Hong Kong SAR
China

1. Introduction

Ceramic materials are essential to a wide range of marketable products, and are of diverse compositions and characteristics. The effective thermal reaction capable of achieving mineral phase transformation is a unique and beneficial opportunity to convert many types of metals into more environment-ally friendly forms. This characteristic may be of particular value to the sustainable development in the 21st century, when more and more attention is focused on environment protection. It is well known that discharge of hazardous metals into natural environments, such as water bodies and soils, is detrimental to human health and the ecosystem. For example, nickel and copper enter the human body via food and water consumption. For human beings, continued inhalation of nickel and its compounds can cause lung cancer, while acute nickel exposure can lead to a variety of clinical symptoms, such as gastrointestinal disturbances, visual disturbance, headache and giddiness, and so on. For animals, prolonged exposure to nickel can lead to adverse effects on haematological parameters, decreased body weights and cancer and, therefore, nickel compounds are often treated as carcinogenic substances (Gang & Zhuang, 2007). Similarly, high accumulation of copper in human body is detrimental to liver and may even cause deadly cirrhosis (European Copper Institute [ECI], 2008). A large amount of waste containing hazardous metals is generated in a wide variety of industries, such as mining and ore processing, metallurgy, chemical industry, alloys industry, paint industry, glass industry, pulp and paper mills, leather tanning, textile dyeing and printing, chemical fertilizer, chloro-alkali industry, petroleum refining and coal burning (Agarwal, 2009). Some municipal solid wastes also contain hazardous metals, such as electrical and electronic equipments waste, barriers, paints and so on. In 2000, the total amount of hazardous waste in China was as much as 830 million tons (State Environmental Protection Administration of China, 2001). It's reported that hazardous waste of up to 963 million tons was generated in 2004, which was 116% of that in 2000 (State Environmental Protection Administration of China, 2005). In the United States, it has been reported that about 40% of hazardous wastes contain heavy metals (Hirschhorn & Oldenburg, 1991).

Traditional wastewater treatment methods use physiochemical processes, such as precipitation, coagulation, reduction, ion exchange, and membrane processes such as ultrafiltration, electrodialysis, and reverse osmosis to remove the pollutants (Park et al.,

2005). However, such physiochemical processes can also result in large quantities of hazardous metals getting into the resulting sludge which requires further treatment into a type of solid waste. Ashes and sludge containing these hazardous metals are usually more difficult to be treated because of their persistence for both biological and chemical degradation, compared to organic wastes and many other chemical pollutants. Moreover, metal concentration increases after degradation of organic materials. Therefore, sludge or its post-incineration ashes generated from municipal and industrial wastewater treatment processes have also become an increasingly serious problem for many regions in the world.

The use of low-cost sorbents has been investigated as an effective way to remove hazardous metals from water. Natural materials or waste products from certain industries having high capacity for accommodating hazardous metals can be employed at lower costs. These sorbents may include bark, chitosan, xanthate, zeolite, clay, peat moss, seaweed and dead biomass (Bailey et al., 1999). The mechanism of these cost-effective methods is to form a net negative charge to hold the hazardous cations; the large surface area of these materials also contributes to the absorption. Nevertheless, such a mechanism is without chemical bonding and may not be stable enough to resist acidic attacks under certain natural environments. In addition, the spent sorbents may also be subject to the corresponding treatment and disposal.

Although a number of processing strategies have the capability to recover metals from solids, metals commonly contained in sludge or spent sorbents are still difficult to be recycled. One common strategy currently adopted to dispose hazardous metal sludge or ashes of incinerated spent sorbent is landfill. However, due to the non-degradable property of metals, leachates of landfills may contain higher levels of hazardous metals, and thus may cause potential pollution to the surrounding land and groundwater resources (Alejandro, 2007a, 2007b; Bilgili 2006). Therefore, the U.S. Environmental Protection Agency (EPA) has proposed the Land Disposal Restriction (LDR) program to set up more strict standards for land disposal of hazardous sludge. LDR regulation requires that hazardous wastes must meet protective treatment standards before disposal in landfill. It also demands these wastes be stored in secure landfills with no hydraulic contact, restricted access, and continuous monitoring (Shih, 2005; Knecht, 2001). Since land resources have become more limited while the quantity of wastes is continuously increasing, the cost of landfill process will inevitably be higher in the future. Therefore, strong attention has been focused on the more economical and environmentally friendly alternatives to dispose hazardous-metal bearing sludge.

Portland cement is a type of hydraulic cement commonly used around the world. It is produced by pulverizing clinkers consisting essentially of hydraulic calcium silicates and contains one or more forms of calcium sulfate. Portland cement is usually added as a binder to stabilize/solidify heavy metals (Douglas & Brandstetr, 1990). After a short hydration period, the slurry mixture of heavy-metal wastes and Portland cement can be solidified. As a result, wastes are solidified in the mixture (Hou et al., 2006). With further mechanistic study of cement solidification, the main stabilization processes are found to be precipitation, chemisorption and encapsulation (Gougar et al., 1996; Andac & Glasser, 1999; Yousuf et al., 1995), rather than incorporating them into crystalline matrices, and such immobilization mechanisms may be reversible in many conditions. Experiments have showed that Portland cement binders may not be able to prevent heavy metals leaching in acidic environments (Cheeseman et al., 1993; Yousuf et al., 1995). For example, under the attack of dissolved CO_2 in natural environments, leaching of solidified metals may be highly affected by the decreased pH value (Lange et al., 1997; Andac & Glasser, 1999; Stegemann et al., 2000).

However, if hazardous metal ions are incorporated into crystalline matrices of minerals, they may usually achieve higher stability. Glass-bonded zeolite has been developed as a high-level waste form for nuclear waste generated during pyroprocessing of spent fuel from the Integral Fast Reactor (IFR) (Lewis et al., 1994; Sun et al., 1999). Treated radioactive wastes with remarkable leaching resistivity were then stored in geological repositories. Such a stabilization/solidification method for hazardous metals has been proven to be very promising for radioactive wastes. Nevertheless, if geological repositories or landfill facilities are still needed for disposal of treated wastes, such methods will still pose a serious problem because of paucity of land. If an environmental strategy is able to manage a closed-loop material flow by reusing and recycling hazardous waste to produce marketable products, it can help preserve the environment on a sustainable basis. Besides being a resolution of environmental problems, such a strategy can also create new resources and facilitate production of products. Addition of materials regenerated from waste may impact quality of products adversely but financial returns from sale of the new products can at least compensate the processing cost (André, 2010). Such a strategy will be a major step toward more sustainable development in the 21st century. It has drawn high interest from many researchers. Experiments have successfully demonstrated that hazardous metal waste can be sintered into ceramic raw materials to produce new ceramics with excellent leaching resistance (Wiebusch & Seyfried, 1997; Chen & Lin, 2009; Vieira et al., 1999).

Ceramic materials of diverse compositions and characteristics often have a crystalline or partially crystalline structure. They are brittle, hard and strong in compression and weak in shearing and tension, and are able to withstand chemical attack in an acidic or caustic environment. An effective thermal reaction capable of achieving mineral phase transformation could be a beneficial opportunity to convert many types of metals into their more environmentally friendly forms. The beneficial use of waste materials for ceramic sintering processes is a sustainable way of reducing the waste problem and it provides the new raw materials for the industry at the same time. Sintering clays with metal-bearing sludge has been employed to treat waste materials and to generate new bricks and tiles (Wiebusch & Seyfried, 1997; Reinosa et al., 2010; Zhang et al., 2007). By aiming at industrial application, such ceramic products were sometimes proven to be with even better mechanical performance than traditional bricks and tiles, in terms of properties such as hardness, transverse rupture strength, abrasion and erosion (Vieira et al., 1999). Hazardous metals immobilized in the new construction ceramic products also show excellent resistance against the leaching test (Zhang et al., 2007). To create a successful waste-to-resource strategy, the mechanism and efficiency of removing or deactivating pollutants in the waste need to be clearly identified and quantitatively evaluated for reliable control of product safety and quality. In the mechanistic study of thermal reaction, simulation of hazardous metals by their oxides may further illustrate the phase transformation process and provide the basic incorporation efficiency information.

Recent works involving mechanistic investigation of stabilizing of nickel and copper waste solids in alumina and iron-rich ceramics have had major breakthroughs in understanding of incorporation efficiencies and product leaching behavior (Hu et al. 2010; Shih, 2005, 2006a, 2006b, 2007; Tang et al., 2010). Such findings are crucial for the development of beneficial usage of metal bearing waste materials and for safely blending them into ceramic raw materials for manufacturing marketable products. Therefore, this chapter systematically introduces current developments in identified stabilization mechanisms, incorporation efficiencies, and observed metal leaching properties of product phases. Finally, the derived technical information is organized to suggest the feasibility of this waste-to-resource strategy.

2. Metal stabilization mechanisms

To facilitate the observation of metal stabilization mechanisms, the corresponding metal oxides were selected to simulate the solidification process and to simplify the system. Nickel and copper were selected as the target metals, and their oxide forms, NiO and CuO, were chosen to simulate metal-bearing sludge for sintering, as most of them exist as oxide forms at high temperatures. Kaolinite and alumina, both commonly found in ceramic raw materials, were used to incorporate nickel and copper. In addition, although iron is not a major element in ceramic raw materials, it often exists in ceramic raw materials as an impurity. Therefore, hematite, a common iron oxide phase, was also selected as a precursor for the incorporation test. As one of the most important processing parameters, sintering temperature was controlled in the range of 800 °C to 1480 °C to be corresponding to the temperature range currently being used in the ceramic industry. For a wider fit with industrial production processes, a short sintering time scheme of 3 h to 6 h was adopted.

2.1 Qualitative analysis by X-Ray Diffraction (XRD)

After its discovery by Wilhelm Conrad Röntgen, the X-ray technique has been applied to detect broken bones and metal cracks because of its penetrating ability (Lin, 1983). In 1919, Hull A. W. found that a crystalline substance gave a unique X-ray diffraction (XRD) pattern, just as a fingerprint. Since then, the XRD technique analysis has become one of the most powerful techniques for identifying crystalline materials and investigating crystal structures (Mittemeijer and Scardi, 2003).

When an X-ray beam impinges a crystalline sample whose atoms or molecules are arranged regularly, the scattered waves interact with each other and form new strengthening or weakening waves, also known as diffraction. The intensity and position of waves (called line profile or peak) in diffraction patterns are different. They can be used to identify the size and shape of the unit cell of the crystalline sample when comparing them with the standard patterns which come from a certain authorized database. In this study, Powder Diffraction Files (PDF) database of International Centre for Diffraction Data (ICDD) was selected as the standard patterns database. Considering that incorporating hazardous metals into ceramic materials leads to phase transformation, XRD technique was employed to investigate the reaction pathways and explore the metal stabilization mechanisms.

2.2 Ceramic raw materials

Kaolin, chiefly composed of kaolinite ($Al_2O_3{\cdot}2SiO_2{\cdot}2H_2O$), is considered to be one of the richest forms of clay in nature. After being preheated at 700°C for 12 hours, major elemental compositions of the kaolin material used in this study are expressed in their corresponding oxide forms (Fig. 1). Percentages of Si and Al elements expressed in SiO_2 and Al_2O_3 forms are 45.1% and 38.6%, respectively, in the sample, similar to theoretical mass percentages (46.5% and 38.6%) derived from its chemical formula. The XRD pattern of kaolin powder also shows kaolinite as the dominant phase, when matching with the Powder Diffraction Files (PDF) database of International Centre for Diffraction Data (ICDD) (Fig. 2 (a)).

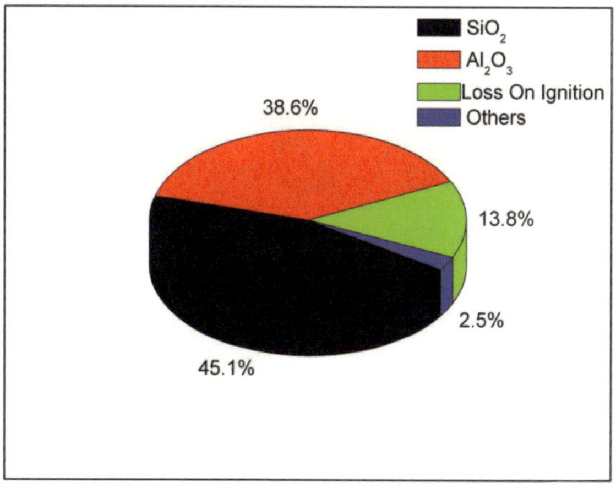

Fig. 1. Major elements (by weight, percentage) expressed in the forms of metal oxides for kaolin powder.

When heated, kaolinite transfers to the other Si-Al phases known as the kaolinite-mullite series. To investigate roles of different phases in incorporation of hazardous metals, phases within the kaolinite-mullite series were produced by heating kaolin raw materials at different temperatures. After heating at 600°C, kaolinite lost its physically bound water and was converted into amorphous substances (Fig. 2 (b)), which is believed to be metakaolin ($2Al_2O_3 \cdot 4SiO_2$), according to previous studies (Brindley and Nakahira, 1959). When the temperature was at or above 980°C, metakaolin changes into a poorly crystallized phase which may be one of the following reactions:

$$2Al_2O_3 \cdot 4SiO_2 \text{ (metakaolin)} \rightarrow 0.375Si_8[Al_{10.67}\square_{5.33}]O_{32} + SiO_2 \quad (1)$$

$$2Al_2O_3 \cdot 4SiO_2 \text{ (metakaolin)} \rightarrow 0.188Al_8[Al_{13.33}\square_{2.66}]O_{32} + 4SiO_2 \quad (2)$$

Where □ represents vacancy and the defect spinel $Al_8[Al_{13.33}\square_{2.66}]O_{32}$, is generally believed to be γ-Al_2O_3. However, metakaolin at 980°C is difficult to be identified by X-ray diffraction technology, due to its poorly crystallized nature. After sintering kaolin at 990°C for 3 hours (Fig. 2 (c)), the weak but still detectable peaks at around 2θ =37°, 46° and 67° are similar to characteristic peaks of γ-Al_2O_3 (Zhou and Snyde, 1990). Transformation is followed by mullite ($3Al_2O_3 \cdot 2SiO_2$) and cristobalite (SiO_2) formation at 1200°C (Fig. 2 (d)). When heating temperature reached 1480°C, the intensive peak of cristobalite at around 2θ=24° indicated nearly all amorphous silica had been crystallized to cristobalite (Fig. 2 (e)).

Since aluminium is suspected to be one of the major metals to react with nickel and copper, alumina (Al_2O_3) was also used as a raw material to simulate the processes of metal incorporation. Alumina has several polymorphs, including crystalline corundum (α-Al_2O_3) and metastable phases with defect crystal structures, such as γ-, η- and θ-alumina (Wolverton and Hass, 2000; Zhou and Snyder, 1990). γ-Al_2O_3, an important technological material with high surface area, can be converted from boehimite at 975°C and transformed into corundum by further calcination at above 1200°C (Shih and Leckie, 2007). Both γ-Al_2O_3 and α-Al_2O_3 were used as aluminium-rich precursors to immobilize hazardous nickel and copper metals in this study. As Fig. 3 shows, after heating HiQ®-7223 alumina powder (boehmite, AlOOH, ICCD PDF#72-0359) at 650°C for 6h and 1500°C for 3 hours, γ- and α-Al_2O_3 were formed.

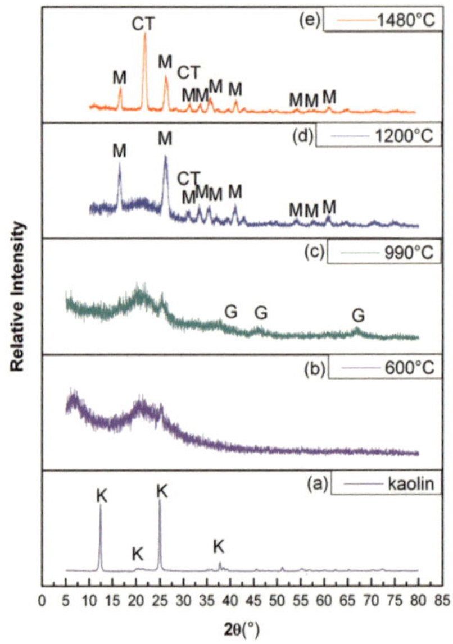

Fig. 2. X-ray diffraction (XRD) patterns of (a) USP grade acid-washed kaolin powder, and after being heated at (b) 600 °C/3 h, (c) 990 °C/3 h, (d) 1200 °C/3 h and (e) 1480 °C/3 h. "K" represents the peak positions of the referenced kaolinite (ICDD PDF#78-1996), "G" for γ-Al_2O_3, "CT" for cristobalite (SiO_2, ICDD PDF#76-0938) and "M" for mullite(ICDD PDF#79-1445).

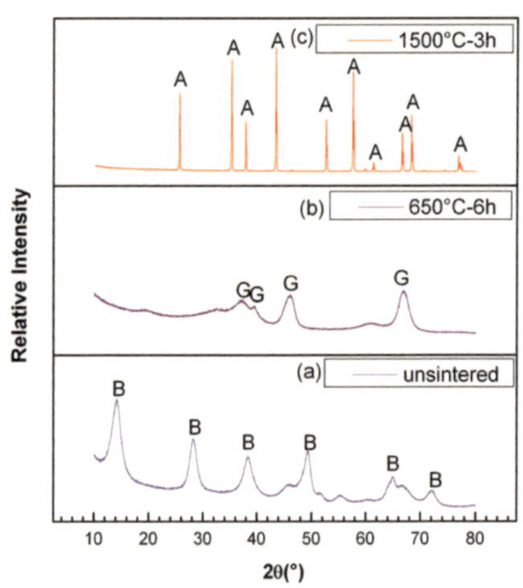

Fig. 3. XRD patterns of (a) HiQ®-7223 alumina powder, and sintered HiQ®-7223 alumina powder at (b) 650 °C/6 h, and (c) 1500 °C/3 h. "B" represents for beohmite (AlOOH, ICCD PDF#72-0359), "G" for γ-Al_2O_3, and "A" for corundum (α-Al_2O_3, ICCD PDF#76-0144)

2.3 Spinel formation

When sintering the mixture of nickel oxide and kaolinite, nickel spinel ($NiAl_2O_4$) was discovered from XRD patterns (Fig. 4(d)). In that experiment, an extended 6 hour sintering was designed to further facilitate the attainment of near equilibrium and observation of product phases. Molar ratio of Ni and Al was fixed at 1:2, corresponding with the molecular formula of nickel aluminate spinel. At 900°C, the sintered kaolinite was transferred into amorphous metakaolin (Fig. 4(a)), and no spinel was detected in the calcined mixture of kaolinite and nickel oxide (Fig. 4(c)). However, when sintering temperature was increased to 990°C, a poorly crystalline phase appeared, as shown in Fig. 4(b), at 2θ around 37°, 46° and 67°. Regardless of the true composition of this defect spinel phase, positions of its diffraction peaks were similar to those of γ-Al_2O_3 (Zhou and Snyde, 1990). When NiO was mixed in the same kaolinite precursor, formation of new nickel aluminate spinel phase in the product could be observed due to occurrence of diffraction peaks at different 2θ positions (Fig. 4(d)).

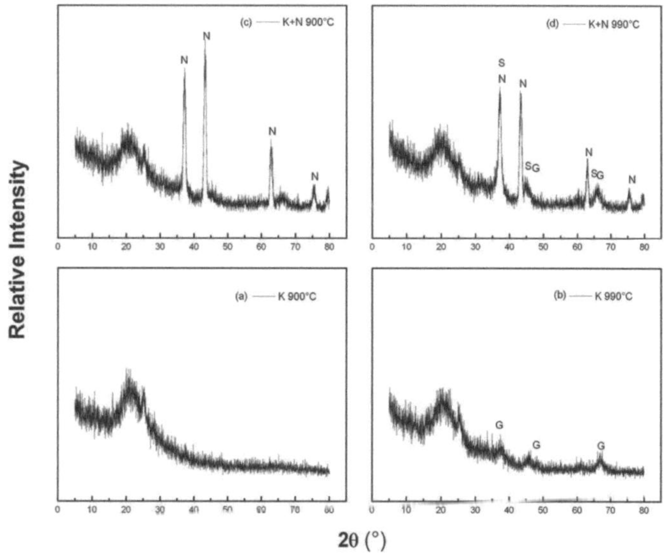

Fig. 4. XRD patterns of the 6 h sintered (a) kaolinite (K) at 900°C, (b) kaolinite 990°C, (c) kaolinite with NiO (N) at 990°C, and (d) kaolinite with NiO at 900°C. "G" represents for the peaks of γ-Al_2O_3, "N" for NiO (ICDD PDF#78-0429), and "S" for $NiAl_2O_4$ (ICDD PDF#78-0552).

Considering the defect spinel structure derived from kaolinite, as shown in Eqs. (1) and (2), the possible mechanism for the formation of nickel aluminate spinel at this sintering temperature may be:

$$NiO + \gamma\text{-}Al_2O_3 \rightarrow NiAl_2O_4 \quad (3)$$

$$5.33NiO + Si_8Al_{10.67}O_{32} \rightarrow 5.33NiAl_2O_4 + 8SiO_2 \quad (4)$$

However, due to very poor crystalline nature of these defect spinel structures, the XRD technique was not able to identify which reaction, i.e. Eq. (3) or Eq. (4), is the mechanism of forming nickel aluminate spinel in the 990°C and 3 h sintered kaolinite + NiO sample. The formation of $NiAl_2O_4$ from sintering the NiO and Al_2O_3 mixture has been widely studied, and a phase diagram of NiO-Al_2O_3 system at temperatures above 1350°C has been presented (Philips et al., 1963). However, very few literatures have referred to reactions at temperatures lower than 1350°C. To further confirm the possible formation mechanism of $NiAl_2O_4$, γ-Al_2O_3 was selected as a precursor to react with NiO at 990°C under 6 h sintering. The XRD result (Fig. 5) not only confirms the possible reaction of Eq. (3) at the lower (< 1350 °C) temperature, but also helps indicate the potential formation mechanism of $NiAl_2O_4$ on sintering kaolinite + NiO at 990°C.

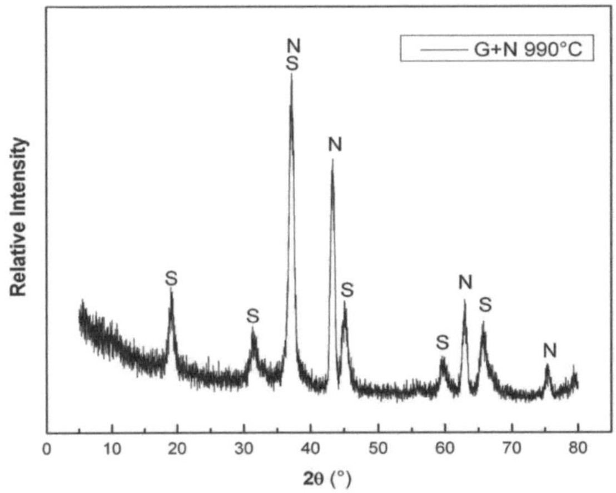

Fig. 5. XRD pattern of sintering the mixture of NiO (N) and γ-Al_2O_3 (G) at 990°C for 6 hours. "N" stands for NiO (ICDD PDF#78-0429), and "S" represents for $NiAl_2O_4$ (ICDD PDF#78-0552).

Although the presence of $NiAl_2O_4$ could be detected at 990°C on sintering NiO and kaolinite, NiO still showed strong XRD peaks in the result (Fig. 4). Therefore, higher temperatures, such as 1250°C and 1450°C, may be needed to further facilitate the mass transfer process. XRD patterns obtained from products of sintering kaolinite + NiO at 1250°C and 1450°C for 3 hours (Fig. 6(a) and Fig. 6(b)) reveal that spinel is a primary product phase, together with the crystalline cristobalite (SiO_2). The major diffraction peak of cristobalite (2θ=21.94°) at 1250°C sintered samples was weaker than the most intensive peak of $NiAl_2O_4$ (2θ=37.01°), but it turned to be stronger at 1450°C. At 1000°C, mullite starts to form with excess amorphous silica, and the silica crystallizes into cristobalite at a higher temperature. The $NiAl_2O_4$ is more likely to be formed from mullite at high temperatures due to the following reactions:

$$3[2Al_2O_3 \cdot 4SiO_2(\text{metakaolin})] \rightarrow 2[3Al_2O_3 \cdot 2SiO_2(\text{mullite})] + 8SiO_2(\text{cristobalite}) \quad (5)$$

$$3NiO + 3Al_2O_3 \cdot 2SiO_2(\text{mullite}) \rightarrow 3NiAl_2O_4 + 2SiO_2(\text{cristobalite}) \quad (6)$$

Eq. (6) was further confirmed by sintering the mixture of mullite, cristobalite and nickel oxide with a fixed molar ratio of Ni : Al : Si = 1 : 2 : 2, at 1250°C for 3 hours. The XRD patterns of the product are provided in Fig. 6(a), which shows that mullite together with nickel oxide could produce crystalline $NiAl_2O_4$, while cristobalite did not appear to have reacted with nickel oxide.

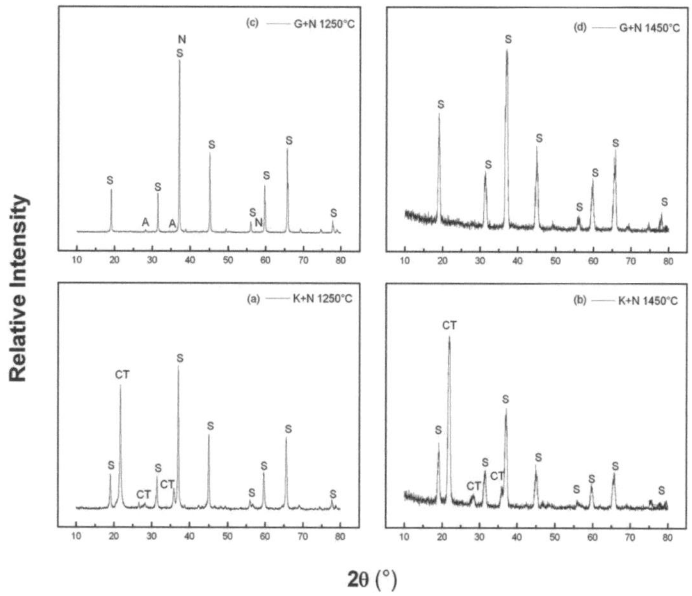

Fig. 6. XRD patterns of 3 h sintered products of (a) kaolinite (K) + NiO at 1250°C, (b) kaolinite + NiO at 1450°C, (c) γ-Al_2O_3 (G) + NiO at 1250°C, and (c) γ-Al_2O_3 (G) + NiO at 1450°C. "N" is for NiO (ICDD PDF#78-0429), "A" for corundum (α-Al2O3, ICDD PDF#83-2080), "S" for $NiAl_2O_4$ (ICDD PDF#78-0552), and "CT" for cristobalite (SiO_2, ICDD PDF#76-0938).

When being sintered with γ-Al_2O_3, nickel oxide disappeared in the system of 1450°C (Fig. 6(d)), while it was still slightly observable in the 1250°C system (Fig. 6(c)). Nickel aluminate spinel was the only detectable crystalline phase in 1450°C, while the residual NiO and corundum (α-Al_2O_3) existed in the 1250°C system, together with $NiAl_2O_4$. It has been reported that well-crystallized corundum (α-Al_2O_3) could be observed from calcining γ-Al_2O_3 at above 1200°C (Eq. (7); Shih and Leckie 2007). Therefore, besides the potential mechanism of Eq. (3) in the low temperature range, nickel incorporation may be proceeded with by two possible steps at higher temperatures:

$$\gamma\text{-}Al_2O_3 \rightarrow \alpha\text{-}Al_2O_3 \tag{7}$$

$$NiO + \alpha\text{-}Al_2O_3 \rightarrow NiAl_2O_4 \tag{8}$$

Feasibility of reaction represented in Eq. (8) was confirmed by the calcined product of α-Al_2O_3 and NiO with molar ratio of Ni : Al = 1:2 at 1250°C for 3 hours. The XRD pattern in Fig. 7(b) reveals that $NiAl_2O_4$ was the only phase in the product without detectable residual reactants. By comparing the XRD pattern of Fig. 6(c) to that of Fig. 7(b), it appears that at 1250°C, the spinel formation rate from sintering α-Al_2O_3 + NiO is higher than that from sintering NiO + γ-Al_2O_3.

When the aluminum-rich precursors were replaced by iron-rich precursors, similar spinel formation reaction was observed, as shown in Fig. 7(c). Nonoverlaid peaks, i.e. at 2θ around 18.4°, 30.2°, and 35.6°, clearly indicated the formation of nickel ferrite spinel ($NiFe_2O_4$, trevorite) from the sintered iron oxide and nickel oxide mixture under 1250°C for 3 h.

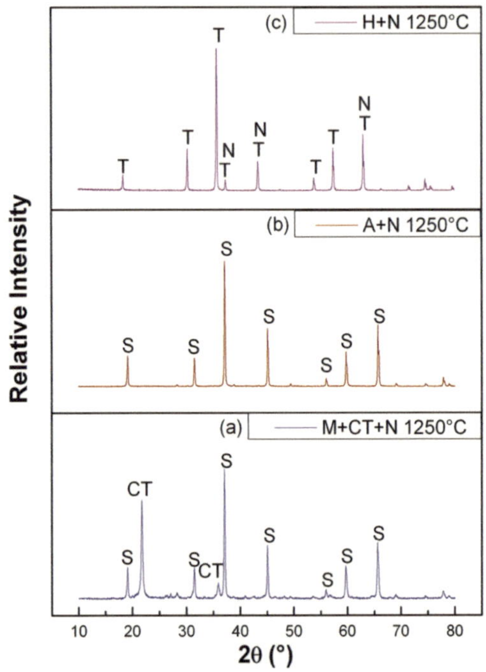

Fig. 7. XRD patterns of sintering (a) mullite(M) + cristobalite(CT) + NiO(N), (b) α-Al_2O_3(A) + NiO(N), (c) hematite(H) + NiO(N) at 1250°C for 3 hours. "S" stands for nickel aluminate spinel ($NiAl_2O_4$, ICDD PDF#78-0552), "CT" for cristobalite (SiO_2, ICDD PDF#76-0938), and "T" for nickel ferrite spinel ($NiFe_2O_4$, trevorite, ICDD PDF#86-2267).

Therefore, the thermal reaction between NiO and Fe_2O_3 as provided in the following equation could be a potential pathway for producing $NiFe_2O_4$:

$$NiO + Fe_2O_3\,(\text{hematite}) \rightarrow NiFe_2O_4 \quad (9)$$

When being sintered with kaolinite at 1000°C for 3 hours, substantial copper was incorporated into the copper aluminate spinel ($CuAl_2O_4$) structure, which clearly acted as a host to accommodate the hazardous copper in high temperature environments (Fig. 8(a)). The diffraction pattern of sintering the γ-Al_2O_3 + CuO system at 1000°C for 3 h indicates a large portion of copper in copper oxide has been successfully converted into spinel (Fig. 8(b)). The results indicate a feasible way to tackle the copper waste problem by sintering it with kaolinite or other aluminium-rich precursors for marketable ceramic products.

3. Incorporation efficiency

From the qualitative information provided in the previous section, metal incorporation is found to play an important role in determining composition of the final products. High temperature usually increases the reaction rate and leads to high incorporation efficiency. Sintering time is a parameter highly related to productivity of industrial process and energy cost. Therefore, the quantitative relationship between temperature and incorporation efficiency is of great importance for the design of reliable methods to effectively incorporate the metals into ceramic products. Metal incorporation efficiency at the temperature range from 600°C - 1480°C was investigated by a 3 h short sintering scheme to reveal such a relation in the metal stabilization process.

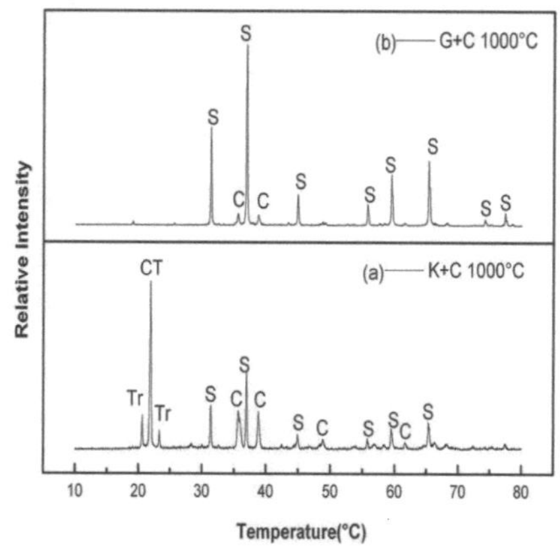

Fig. 8. XRD patterns of (a) kaolinite(K) + CuO(C) mixture sintered at 1000°C for 3 h, and (b) γ-Al_2O_3(G) + CuO mixture sintered at 1000°C for 3 h. "S" represents for $CuAl_2O_4$ (ICDD PDF#76-2295), "C" for CuO (ICDD PDF#48-1548), "CT" for cristobalite (SiO_2, ICDD PDF#76-0935), and "Tr" for tridymite (SiO_2, ICDD PDF#88-1535).

3.1 Quantitative XRD technique

Quantitative analysis of XRD data usually involves determination of the amounts in specific phases in a specimen, by modelling the observed XRD patterns. In this study, metal incorporation efficiencies for different precursors and temperatures were estimated by quantitative XRD analysis, using the Whole Pattern Fitting (WPF) strategy to model XRD patterns. The WPF was carried out by using the Pawley method integrated into a XRD data processing software, JADE (Material Data, Inc). Phase quantification by WPF was executed by matching the observed XRD patterns with the PDF database of ICDD. In this study, the weight percentage of each crystalline phase was generated together with the weighted (R) and expected (E) reliability values to indicate the quality of fitting for each refinement. Weighted and expected reliability values are calculated by the following equations:

$$R(\%) = 100 \times \sqrt{\frac{\sum[w(i) \times (I(o,i) - I(c,i))^2]}{\sum[w(i) \times (I(o,i)) - I(b,i))^2]}} \tag{10}$$

$$E(\%) = 100 \times \sqrt{\frac{(N-P)}{\sum I(o,i)}} \tag{11}$$

where I (o,i) and I (c,i) represent the observed intensity and calculated intensity of a fitted data point (i), respectively; I (b,i) is the background intensity of point (i); w (i) is the weight of the data point, as w(i)=1/I(o,i); N is the number of fitted data points and P is the number of refined parameters. The ratio of R/E, which is equal to 1 in an ideal model, stands for "goodness of fit", but due to the existence of background intensity, it's often greater than one.

3.2 Transformation ratio

To define metal incorporation efficiency, an index of transformation ratio (TR) was designed, as the following equation:

$$TR(\%) = \frac{\dfrac{wt\% \text{ of } MX_2O_4}{\text{MW of } MX_2O_4}}{\left(\dfrac{wt\% \text{ of } MX_2O_4}{\text{MW of } MX_2O_4} + \dfrac{wt\% \text{ of MO}}{\text{MW of MO}}\right)} \tag{12}$$

where M stands for metal ions (Ni or Cu in this study); X is Al or Fe; and MW means molecular weight. When TR = 0, no metal incorporates into the precursor, while all metal incorporate into precursors when TR = 100%.

Although the large amount of amorphous silica at temperature below 1150°C had made the quantitative analysis difficult, efficiencies of Ni incorporation by kaolinite and mullite precursors are shown in Fig. 9 (a), which demonstrates that both precursors can achieve a high Ni incorporation efficiency (more than 90%) at temperatures above 1160°C. In Fig. 9 (a), TR of kaolinite + NiO system is around 10% higher than that of mullite + cristobalite + NiO system when sintering at 1150°C for 3 hours.

Fig. 9 (b) shows nickel aluminate spinel generated from sintering the two NiO + Al_2O_3 mixtures at over 600°C. In terms of nickel incorporation efficiency, more than 90% of nickel in the α-Al_2O_3 precursor was transformed into $NiAl_2O_4$ structure at over 1250°C. It is also interesting to find a crossover point of these two alumina systems at around 1150°C in Fig. 9 (b), which demonstrates that reaction of NiO and γ-Al_2O_3 dominated at lower temperatures while corundum (α-Al_2O_3) precursor facilitated nickel incorporation at higher temperatures. Assigning k_1, k_2 and k_3 to stand for reaction rates of γ-Al_2O_3 reacting with NiO (Eq. 3), γ-Al_2O_3 transforming to corundum (Eq. 7), and NiO incorporating into corundum (Eq. 8), respectively, the higher free energy of γ-Al_2O_3 leads its reaction with nickel oxide to being more favourable energetically ($k_1 > k_3$) and then further obtains its higher incorporation efficiency at temperature below 1150°C. When temperature was increased, k_2 exceeded k_1, so reactions of both alumina precursors are likely identical with the case of having α-Al_2O_3 reacting with NiO.

$$NiO + \gamma\text{-}Al_2O_3 \xrightarrow{k_1} NiAl_2O_4 \tag{3}$$

$$\gamma\text{-}Al_2O_3 \xrightarrow{k_2} \alpha\text{-}Al_2O_3 \tag{7}$$

$$NiO + \alpha\text{-}Al_2O_3 \xrightarrow{k_3} NiAl_2O_4 \tag{8}$$

At high temperatures, silica content may be a potential flux to further facilitate mass transfer during the reaction, and thus the higher TR may occur in the system using kaolinite or mullite as precursors, compared to results of Al_2O_3 + NiO systems shown in Figs. 9 (a) and 9 (b). Compared to the interaction between nickel oxide and kaolinite under 900°C for 6 hours (Fig. 4 (c)), the transformation level of Ni in the NiO + hematite system was already much higher (~90%) even with sintering at 900°C for only 3 hours (Fig. 7). Furthermore, from TR curves of NiO + kaolinite and NiO + hematite (Fig. 9 (b) and Fig. 9 (c)), it appears that the ferrite spinel can be formed at a lower temperature range. This indicates that nickel bearing sludge may first react with iron impurities, before initiating reactions with kaolinite-based ceramic precursors during immobilization of hazardous nickel waste in ceramics.

For the case of copper incorporation, a low temperature range (650°C - 1000°C) was designed to investigate incorporation efficiency. As shown in Fig. 10, incorporation efficiency of CuO + γ-Al_2O_3 system is generally higher than that of CuO + kaolinite system.

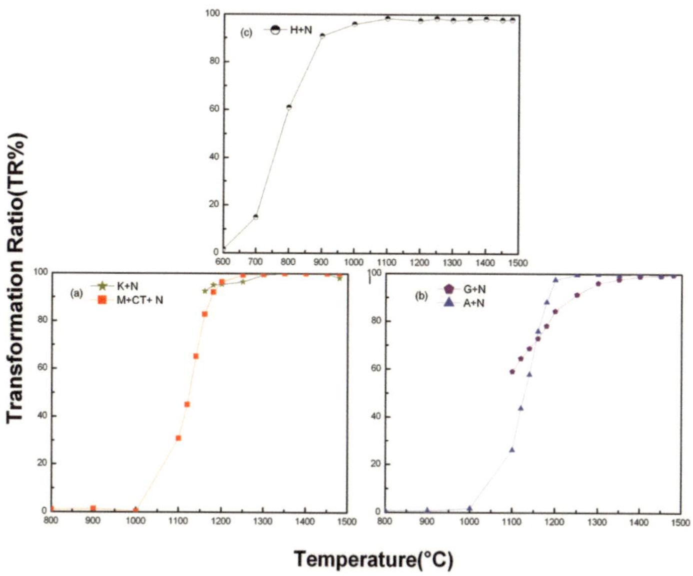

Fig. 9. Nickel incorporation efficiency when sintering NiO with different precursors for 3 hours: (a) kaolinite(K) from 1160°C to 1480°C and mullite(M) + cristobalite(CT) from 800°C to 1480°C; (b) γ-Al_2O_3(G) from 1100°C to 1480°C and α-Al_2O_3(G) from 800°C to 1480°C; and (c) hematite(H) from 600°C to 1480°C.

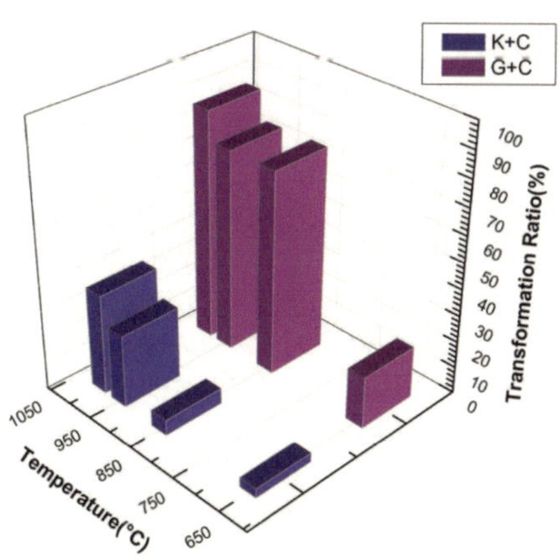

Fig. 10. Efficiency of copper incorporation by kaolinite and γ-Al_2O_3 precursors with 3 h sintering under 650°C, 850°C, 950°C and 1000°C. "K" represents kaolinite, "G" denotes γ-Al_2O_3, and "C" CuO.

The highest copper incorporation efficiency observed in CuO + γ-Al_2O_3 system was nearly 90% in TR, when the sample was sintered at 1000°C for 3 hours. Incorporation efficiency of CuO + kaolinite system was also found to increase with increase of sintering temperature, although incorporation efficiency was lower than that of the CuO + γ-Al_2O_3 system. A maximum of 33% copper incorporation was observed when sintering CuO with kaolinite precursor at 1000°C for 3 hours, and this result may still suggest the important contribution of kaolinite in stabilizing copper in copper-bearing sludge under ceramic sintering processes.

4. Metal leachability and leaching behavior

Since the major purpose of incorporating hazardous metal waste into ceramic products is to immobilize the waste and prevent contamination of the environment, leachability and leaching behavior of metals from the new product phases are of great importance to determine whether such products are more environmentally benign than their original forms.

4.1 Design of leaching experiment

The "Toxicity Characteristic Leaching Procedure (TCLP)" is a method designed by U.S. EPA to determine mobility of both organic and inorganic analytes present in liquid, solid and multiphasic wastes at a laboratory level. For solid phase, extraction fluid # 1 (pH = 4.88 ± 0.05 acetic + NaOH solution) or extraction fluid # 2 (pH = 2.88 ± 0.05 acetic solution) is used for the test, with the amount of extraction fluid equal to 20 times the weight of the solid phase (Environment Protection Agency [EPA], 1992). In this study, 10 ml of the more acidic # 2 extraction solution was applied to leach a 0.5 g solid sample. To further test the spinel leachability over a longer period and under harsher conditions, standard TCLP test was modified by grinding the samples into powders, to reach a larger surface area, as well as for prolonging leaching time to more than 20 days.

To further transform reactants into the designated spinels, the same oxide raw materials as were used in the previous spinel formation sections were mixed in molar ratio of 0.5 for Ni/Al, Ni/Fe, and Cu/Al systems to generate single phase $NiAl_2O_4$, $NiFe_2O_4$ and $CuAl_2O_4$ samples, respectively. Considering more than 98% TR under sintering at 1480°C for 3 hours (Fig. 9), extended treatment for 48 hours at 1480°C was used to convert NiO + γ-Al_2O_3 and NiO + hematite into $NiAl_2O_4$ and $NiFe_2O_4$, respectively. To form the single phase $CuAl_2O_4$ sample, CuO + γ-Al_2O_3 was under 990°C thermal treatment for 20 days. All products were ground into powders for XRD analysis to confirm the signal phase results in the products, as well as for BET analysis to evaluate their surface areas. The 0.5 g powder sample and 10 mL of # 2 extraction solution were filled into each leaching vial, which was then rotated end-over-end during the leaching time. At the designated sampling time, the collected leachate was filtered and diluted to measure ion concentrations by inductively coupled plasma atomic emission spectroscopy (ICP-AES). Four replications of samples collected at each time point were performed to reduce random errors.

4.2 Metal leaching results

After thermal processing at 1480°C for 48 hours, the NiO(N) + γ-Al_2O_3(G) mixture formed $NiAl_2O_4$; the NiO + kaolinite(K) mixture was converted into $NiAl_2O_4$ + cristbolite (SiO_2); and the NiO + hematite (H) mixture resulted in $NiFe_2O_4$. Changes of pH values in sample leachates are shown in Fig. 11. In the first few days, pH values of NiO (a) and $NiAl_2O_4$ (b) leachates increased rapidly, while pH of $NiFe_2O_4$ (c) leachate and sintered NiO + kaolinite (d) had only slight increases. After the first few days, pH values of NiO leachate gradually increased, but pH of leachates from $NiAl_2O_4$, sintered NiO + kaolinite, and $NiFe_2O_4$ samples maintained more stable levels till the end of leaching period (more than 25 days). As a higher pH in leachate stands for fewer protons in a solution, a larger change of pH value indicates that more protons have been consumed. A potential reason for the change of pH is the exchange of metal ions in solids with protons in the solution, which leads to decrease of proton concentrations in the leachate. Therefore, such change of pH value may imply higher long term acid resistance of sintered products, and it also illustrates that the leaching time of standard TCLP (18 h) may not be able to reflect long term leachability of the low soluble materials.

Because nickel leachability in spinel determines the quality of immobilization of hazardous metals in ceramics, concentrations of nickel in leachates were measured. From BET analysis, surface areas of different sample powders were determined to be 3.6 ± 0.5 m^2/g for NiO, 1.1 ± 0.1 m^2/g for $NiAl_2O_4$, 0.73 ± 0.12 m^2/g for sintered NiO + kaolinite ($NiAl_2O_4$ + cristobalite), and 1.7 ± 0.2 m^2/g for $NiFe_2O_4$. In addition to surface area, nickel content in each nickel containing compound, i.e. NiO, $NiAl_2O_4$, $NiAl_2O_4$ + cristobalite, and $NiFe_2O_4$, should also be normalized for comparison.

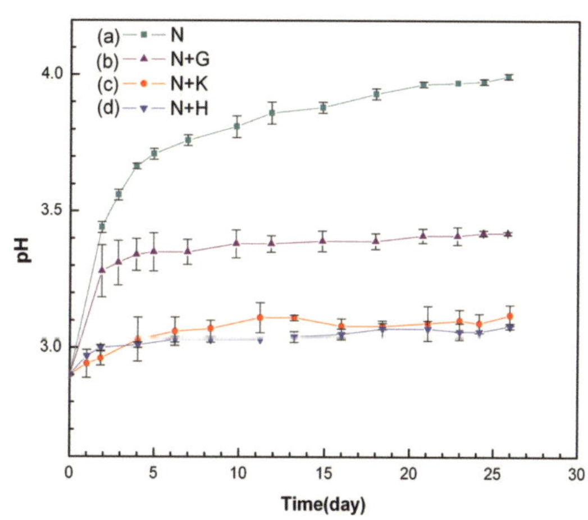

Fig. 11. The pH values of leachates from (a) NiO(N), and the 1480°C/48 h sintered (b) NiO + γ-Al_2O_3(G), (c) NiO + kaolinite(K), and (d) NiO + hematite(H). The product of (b) is $NiAl_2O_4$; of (c) is $NiAl_2O_4$ + cristobalite(SiO_2); and of (d) is trevorite($NiFe_2O_4$). The 0.5 g solid sample powder was leached by 10 ml # 2 extraction solution (acetic acid, pH = 2.9) for a period ranging from 0.75 day to 26 days. The mixture of powder sample and extraction solution was rotated end-over-end during the leaching period.

In Fig. 12, surface-area and metal-content normalized nickel concentrations in leachates illustrate the result of comparing the intrinsic leachability of different nickel bearing phases. Nickel in $NiAl_2O_4$ + cristobalite sample was found to be more invulnerable to acid attack, which may be due to encapsulation of cristobalite matrix for nickel aluminate spinel.

$NiFe_2O_4$ showed a higher leachability than NiO in the first two days, but the trend changed to a much slower process than that of NiO in the prolonged leaching period. Potential reason for the higher leachability of $NiFe_2O_4$ at the initial stage was likely due to incomplete formation of $NiFe_2O_4$ or acid attack on its weaker grain boundary. Although leaching curves of $NiAl_2O_4$ and $NiFe_2O_4$ were quite similar, $NiFe_2O_4$ shows a much smaller change of pH value than $NiAl_2O_4$ in Fig. 11. If the leaching mechanisms of these two spinels are the same, changes of their pH values should be similar. Therefore, further investigation was conducted to compare the results of the other ions leached out of these two solids.

For congruent dissolution, molar ratio of Ni/Al and Ni/Fe in leachates should be equal to 0.5 for both $NiAl_2O_4$ and $NiFe_2O_4$ phases. Fig. 13 (a) summarizes the Ni/Al ratio in the leachate and the result of Ni/Al ~ 0.5 confirmed the behavior of congruent dissolution for $NiAl_2O_4$. However, similar comparison found that Ni/Fe ratio in leachate of $NiFe_2O_4$ was much higher than 0.5, as shown in Fig. 13 (b). One possible explanation for the high Ni/Fe ratios is incongruent dissolution of Ni and Fe from the $NiFe_2O_4$ structure, which is different from the case of leaching $NiAl_2O_4$. Another possible explanation is that after the congruent dissolution of Ni and Fe from $NiAl_2O_4$, the Fe re-precipitated on the surface of particles.

At pH of around 3.0, the dissolved Fe may exist in the forms of Fe^{3+}, $FeOH^{2+}$, $Fe(OH)_2^+$ and $Fe(OH)_4^-$. From a pC-pH diagram of $Fe(OH)_{3\bullet Am}$, the total dissolved Fe concentration approximates to 0.88 ppm, which is considerably close to Fe concentrations detected in this study (0.5 ppm - 1.0 ppm). Therefore, regardless of the leaching mechanism of $NiFe_2O_4$, composition of its leachate was mainly controlled by reprecipitation of amorphous $Fe(OH)_3$ solid.

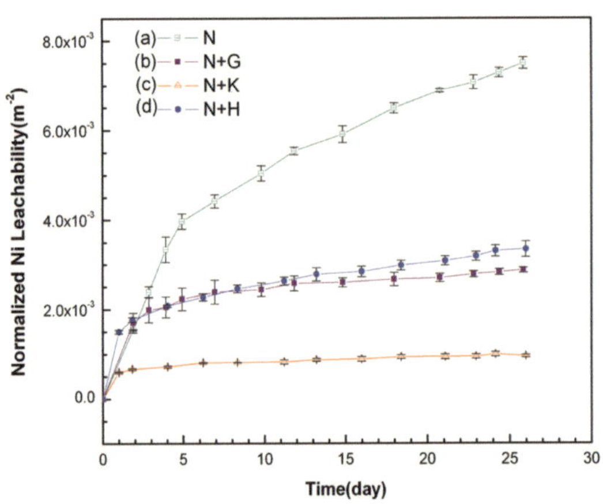

Fig. 12. Normalized Ni concentrations in leachates of (a) NiO(N), and the 1480°C/48 h sintered (b) NiO + γ-Al_2O_3(G), (c) NiO + kaolinite(K), and (d) NiO + hematite(H). The product of (b) is $NiAl_2O_4$; of (c) is $NiAl_2O_4$ + cristobalite(SiO_2); and of (d) is trevorite ($NiFe_2O_4$). The 0.5 g powder samples were leached by 10 ml # 2 extraction solution (acetic acid, pH = 2.9) over a period of time ranging from 0.75 day to 26 days. The mixture of sample and solution was rotated end-over-end during the leaching period. The Ni concentrations in leachates were normalized by the powder sample surface areas and nickel contents.

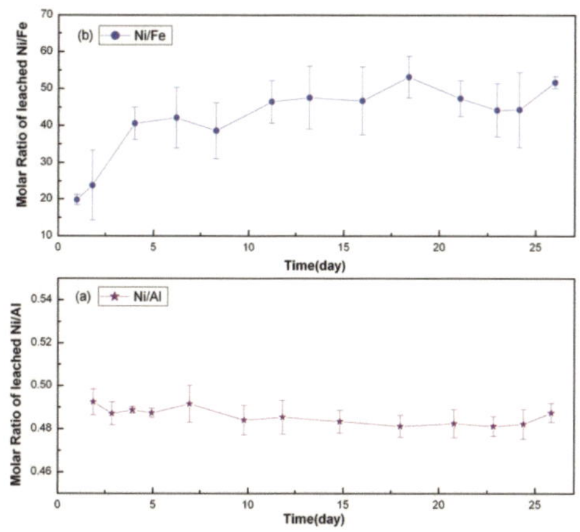

Fig. 13. Molar ratios of leached [Ni]/[Al] in (a) $NiAl_2O_4$ leachates, and [Ni]/[Fe] in (b) $NiFe_2O_4$ leachates. The leaching test was conducted using 10 mL TCLP # 2 extraction solution (acetic acid, pH = 2.9) to leach 0.5 g powder for 0.75 day ~ 26 days.

In the 22 days leaching period for copper-bearing phases, similar acid resisting capacity was found for copper aluminate spinel ($CuAl_2O_4$). As Fig. 14 shows, pH values of CuO leachates increased rapidly in the first two days, and then stabilized at around 4.9. However, pH values of $CuAl_2O_4$ leachates (b) were maintained at around 3.1, which indicates that much fewer protons had been consumed from the leaching solution, compared to CuO leachates during the leaching period. The normalized Cu leachability also clearly indicates the higher acid resistance of copper aluminate spinel (Fig. 15).

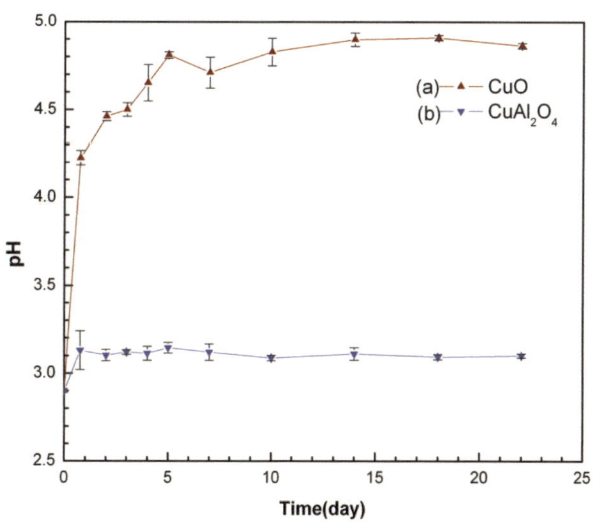

Fig. 14. The change of pH values of (a) CuO, and (b) $CuAl_2O_4$. The $CuAl_2O_4$ was prepared by sintering CuO+ γ-Al_2O_3 at 990°C for 20 days. The 0.5 g powder sample was added into 10 ml TCLP No. 2 extraction solution (pH ~ 2.9) and the vial was rotated end-over-end for 0.75 day~22 days.

Very little copper was leached from the $CuAl_2O_4$ solid sample; more copper was dissolved from CuO solid after normalization of sample surface area and copper content. The considerable difference between $CuAl_2O_4$ and CuO not only implies the superior acid resistivity of $CuAl_2O_4$, but also indicates that the waste containing copper oxide needs further stabilization before disposal.

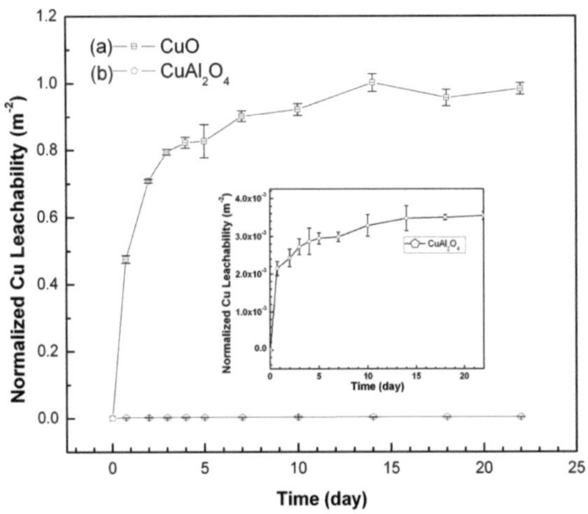

Fig. 15. Normalized Cu leachablity of (a) CuO; and (b) $CuAl_2O_4$. The $CuAl_2O_4$ was prepared by sintering CuO + γ-Al_2O_3 at 990°C for 20 days. Each leaching vial was filled with 10 ml TCLP extraction solution # 2 (acetic acid with a pH = 2.9) and 0.5 g powder sample. These vials were then rotated for 0.75 day and 22 days. Copper concentrations in leachates were normalized by different surface areas and Cu contents in the solid samples.

5. Conclusions

To investigate mechanisms of immobilization of hazardous waste metals by ceramic materials, nickel oxide and copper oxide were used to simulate the metal bearing sludge and sintered with a variety of ceramic precursors. The incorporation mechanisms together with reaction efficiencies were also identified to facilitate optimization of metal incorporation. To confirm the environmentally benign property of product phases, a modified leaching test was carried out to evaluate the amounts of leached hazardous metals. The changes of pH values were observed to reflect the degree of metal leaching, and the measured metal concentrations in leachates were normalized by solid surface area and metal content to reflect the intrinsic properties of the product phases. Overall, the current progress of recycling metal bearing waste for ceramic products can be organized as follows.

For results of incorporating nickel:

- Nickel oxide can react with kaolinite, mullite, γ-Al_2O_3 and α-Al_2O_3 under thermal conditions to form $NiAl_2O_4$; and with Fe_2O_3 to produce $NiFe_2O_4$.
- NiO can react with the defect spinel structure derived from kaolinite precursor at low temperature to form spinel, and it can also react with mullite to form $NiAl_2O_4$ at higher temperature.

- When reacting with aluminium-rich and iron-rich precursors, more than 90% of nickel oxide can be converted into spinel phase at temperatures above 1250°C.
- When sintering NiO with kaolinite and mullite precursors at high temperatures, SiO_2 content may act as a flux to facilitate the mass transfer.
- The formation temperature of $NiFe_2O_4$ is about 200°C lower than that of $NiAl_2O_4$, and, therefore, nickel may first react with Fe_2O_3 to form ferrite in ceramic products.
- Spinel-containing products show remarkable resistance to protonic attacks by pH 2.9 acetic acid solution.
- Leaching of $NiFe_2O_4$ may be accompanied by reprecipitation of amorphous $Fe(OH)_3$ from the leachate.

For the results of incorporating copper:

- Calcination of CuO with γ-Al_2O_3 or kaolinite precursor could produce copper aluminate spinel ($CuAl_2O_4$).
- When the sintering temperature is lower than 1000°C, transformation ratio of CuO + γ-Al_2O_3 system will be much higher than that of CuO + kaolinite system.
- The normalized copper leachability of CuO was found much higher than that of $CuAl_2O_4$ and this indicates that the CuO is quite vulnerable to acid attack, while $CuAl_2O_4$ shows much stronger intrinsic acid resistance.

As significant amount of nickel oxide was still found as the residual in alumina-rich systems when sintering temperature was below 1140°C, one needs to be cautious in attempts to incorporate nickel-bearing waste at such low temperatures in ceramic products, such as manufacturing of bricks or low-grade porous construction ceramics. However, similar work has revealed that copper bearing waste may be further transformed into copper spinel at temperatures below 1000°C, considering the 87% copper transformation ratio when sintering CuO and γ-Al_2O_3. Therefore, the current mechanistic development of incorporating hazardous metals into ceramic products is of great importance as it provides key information on applicability of incorporating different types of metals into marketable ceramic products as a waste-to-resource strategy.

6. References

Agarwal, S. K. (2009). *Heavy metal pollution,* A P H Publishing Corporation, Delhi, India.

Alejandro, S. & Jacint, N. (2007). Bioaccumulation of Metals and Effects of Landfill Pollution in Small Mammals. Part I. The Greater White-toothed Shrew, Crocidura Russula. *Chemosphere,* Vol. 68, pp. 703-711.

Alejandro, S. & Jacint, N. (2007). Bioaccumulation of Metals and Effects of a Landfill in Small Mammals. Part II. The Wood Mouse, Apodemu Sylvaticus. *Chemosphere,* Vol. 70, pp. 101-109.

Andac, M. & Glasser, F. P. (1999). Long-term Leaching Mechanisms of Portland Cement-Stabilized Municipal Solid Waste Fly Ash in Carbonated Water. *Cement and Concrete Research,* Vol. 29, No. 2, pp. 179-186.

André, Z. (2010). Ceramic Products from Waste. In: *Ceramic Materials,* InTech, ISBN 978-953-307-145-9, Vienna, Austria.

Bailey, S. E.; Olin, R. J.; Brickea, M. & Adrian D. A. (1999). A Review of Potentially Low-Cost Sorbents for Heavy Metals. *Water Research,* Vol. 33, No. 11, pp. 2469-2479.

Bilgili, M. S.; Demir, A.; Ince, M. & Ozkaya, B. (2006). Metal Concentrations of Simulated Aerobic and Anaerobic Pilot Scale Landfill Reactors. *Journal of Hazardous Materials*, Vol. 145, pp. 186-194.

Brindley, G. W. & Nakahira, M. (1959). Aolinite-Mullite Reaction Series: II, Metakaolin. *Journal of the American Ceramic Society*, Vol. 42, No. 7, pp. 14-318.

Cheeseman, C. R.; Butcher, E. J.; Sollars, C. J. & Perry, R. (1993). Heavy Metal Leaching from Hydroxide, Sulphide and Silicate Stabilized/Solidified Wastes. *Waste Management*, Vol. 13, No. 8, pp. 545-552.

Chen, L. & Lin, D. F. (2009). Applications of Sewage Sludge Ash and Nano-SiO_2 to Manufacture Tile as Construction Material. *Construction and Building Materials*, Vol. 23, No. 11, pp. 3312-3320.

Douglas, E. & Brandstetr, J. (1990). A Preliminary Study on the Alkali Activation of Ground Granulated Blast-Furnace Slag, *Cement and Concrete Research*, Vol. 20, No. 5, pp. 746-756, ISSN: 00088846.

Environment Protection Agent (1992). Toxic Characteristic Leaching Procedure, (accessed in February 2011), available from:
http://www.epa.gov/osw/hazard/testmethods/sw846/pdfs/1311.pdf

European Copper Institute (2008). European Union Risk Assessment Report, In: European Chemical Agent, 10.01.2011, (accessed in February 2011), available from:
http://echa.europa.eu/chem_data/transit_measures/vrar_en.asp

Gang, B. & Zhuang, Z. (2000). Nickel Toxicology Research Progress in China. *Healthy Toxicology*, Vol. 14, No. 3, pp. 129-135.

Gougar, M. L. D.; Scheetz, B. E. & Roy, D. M. (1996). Ettingite and C-S-H Protland Cement Phases for Waste Ion Immobilization: A Review. *Waste Management*, Vol. 16, No. 4, pp. 295-330.

Hirschhorn, J. & Oldenburg, K. (1991). *Prosperity Without Pollution : the Prevention Strategy for Industry and Consumers*, Van Nostrand Reinhold, New York.

Hou, H.; He X.; Zhu, S. & Zhang, D. (2006). The Cement Solidification of Municipal Solid Waste Incineration Fly Ash. *Journal of Wuhan University of Technology*, Vol. 21, No. 4, pp. 137-140.

Hu, C. Y.; Shih, K. & Leckie J. O. (2010). Formation of Copper Aluminate Spinel and Cuprous Aluminate Delafossite to Thermally Stabilize Simulated Copper-Laden Sludge. *Journal of Hazardous Materials*, Vol. 181, pp. 399-404.

Knecht, M. A. (2001). Overview of U.S. Federal Laws and Regulations Affecting Mixed Waste Treatment. In : *Hazardous and Radioactive Waste Treatment Technologies Handbook*, CRP Press, ISBN 978-08493-9586-4.

Lange, L.; Hills C. D. & Poole, A. B. (1997). Effect of Carbonation on Properties of Blended and Non-blended Cement Solidified Waste Forms. *Journal of Hazardous Materials*, Vol. 52, No. 2-3, pp. 171-191.

Lewis, M. A.; Fischer, D. & Murphy, C. (1994). Properties of Glass-bonded Zeolite Monoliths. In: *American Ceramic Society*. Indianapolis, Indiana.

Lin, L. (1983). Principles and Application of X-ray Diffraction. *Industry Materials*, Vol. 86, pp. 100-109.

Mittemeijer, E. J. & Scardi, P. (2003). *Diffraction Analysis of the Microstructure of Materials.* ISBN: 3-540-40519-4, Springer, Verlag Berlin Heidelberg New York.

Park, D.; Lee, D. S.; Park, J. M.; Chun, H. D.; Park, S. K.; Jistuhara, I.; Miki, O. & Kato, T. (2005). Metal Recovery from Electroplating Wastewater Usign Scidophilic Iron Oxidizing Bacteria: Pilot-Scale Feasibility Test. *Ind. Eng. Cjem. Res.*, Vol. 44, pp. 1854-1859.

Philips, B.; Hutta, J. J. & Warshaw, I. (1963). Phase Equilibria in the System NiO-Al_2O_3-SiO. *Journal of the American Ceramic Society*, Vol. 46, No. 12, pp. 579-583.

Reinosa, J. J.; Silva A. C.; Rubio-Marcos, F.; Mello-Castanho, S. R. H.; Moya, J. S. & Fernandez, J. F. (2010). High Chemical Stability of Stoneware Tiles Containing Waste Metals . *Journal of the European Ceramic Society*, Vol. 30, No. 14, pp. 2997-3004.

Shih, K. & Leckie J. O. (2007). Nickel Aluminate Spinel Formation during Sintering of Simulated Ni-Laden Sludge and Kaolinite. *Journal of the European Ceramic Society* Vol. 27, pp. 91-99.

Shih, K.; White, T. J. & Leckie J. O. (2006a). Spinel Formation for Stabilizing Simulated Ni-Laden Sludge with Aluminum-Rich Ceramic Precursors. *Environmental Science and Technology.* Vol. 40, No. 16, pp. 5077-5083.

Shih, K.; White, T. J. &Leckie J. O. (2006b). Nickel Stabilization Efficiency of Aluminate and Ferrite Spinels and Their Leaching Behavior. *Environmental Science and Technology* Vol. 40, No. 17, pp. 5520-5526.

Shih, K. (2005). Stabilization of Nickel by Aluminum- and Iron-rich Ceramic Materials: Reaction Pathways and Product Leaching Behavior. In: *Dissertation Abstracts International*, Vol. 66-08, Section B, pp. 4423-4527, ISBN: 978-054-2286-93-3.

State Environmental Protection Administration of China (2001). The State of Environment in China in 2000. *Environmental Protection*, Vol. 07.

State Environmental Protection Administration of China (2005). The State of Environment in China in 2004. *Environmental Protection*, Vol. 06.

Stegemann, J.; Roy A.; Caldwell, R. & Paul, J. S. (2000). Understanding Environmental Leachability of Electric Arc Furnace Dust. *Journal of Environmental Engineering*, Vol. 126, No. 2, pp. 112-120.

Sun, D.; Wronkiewicz, D. J. & Simpson, L. J. (1999). A Study of Alteration Phases on Glass-Bonded Zeolite and Sodalite Using the Vapor Hydration Test. *In Proceedings of the Material Research Society Symposium*, Vol. 556, pp. 189–196.

Tang, Y.; Shih, K. & Chan K. (2010). Copper Aluminate Spinel in the Stabilization and Detoxification of Simulated Copper-Laden Sludge. *Chemosphere*, Vol. 80, pp. 275-380.

Vieira, M. T.; Catarino, L.; Oliveira, M.; Sousa, J.; Torralba, J. M. & Cambronero, L. E. G. (1999). Optimization of the Sintering Process of Rawmaterial Wastes. *Journal of Materials Processing Technology*, Vol. 92-93, pp. 97–101.

Wiebusch, B. & Seyfried, C. F. (1997). Utilization of Sewage Sludge Ashes in the Brick and Tile Industry. *Water Science and Technology*, Vol. 36, No. 11, pp. 251-258.

Wolverton, C. & Hass, K. C. (2000). Phase Stability and Structure of Spinel-based Transition Aluminas. *Physical Review B*, Vol. 63 (024102), pp. 1-10.

Yousuf, M.; Mollah, A.; Vempati, R. K.; Lin, T. C. & Cocke, D. L. (1995). The Interfacial Chemistry of Solidification/Stabilization of Metals in Cement and Pozzolanic Material Systems. *Waste Management*, Vol. 15, No. 2, pp. 137-148.

Zhang, H.; Zhao, Y. & Qi, J. (2007). Study on Use of MSWI Fly Ash in Ceramic Tile. *Journal of Hazardous Materials*, Vol. 141, No. 1, pp. 106-114.

Zhou, R. S. & Snyder, R. L. (1990). Structure and Transformation Mechnaisms of the η,γ and θ Transition Alulminas. *Acta Cryst*, B47, pp. 617-630.

24

Using Ceramic Materials in Ecoefficient Concrete and Precast Concrete Products

César Medina[1], M. I. Sánchez de Rojas[2], Moisés Frías[2] and Andrés Juan[1]
[1]*Escuela Superior y Técnica de Ingeniería Agraria. U. of León (León)*
[2]*Instituto de Ciencias de la Construcción Eduardo Torroja (CSIC) (Madrid)*
Spain

1. Introduction

The industrial and economic growth witnessed in recent decades has brought with it an increase in the generation of different types of waste (urban, industrial, construction, etc.) despite the waste management policies which have been adopted nationally and internationally.

The practice of dumping and/or the inadequate management of waste from the various manufacturing sectors have had a notable impact on the receiving environment, leading to water, soil, air and noise pollution, amongst other complications, and adding to existing environmental problems. At the same time, these practices represent an economic cost. However, if waste is managed correctly it can be converted into a resource which contributes to savings in raw materials, conservation of natural resources and the climate, and promotes sustainable development, all of which complies with strategies for sustainable development within the European Union and Spain.

Spain occupies an important position in the ceramics industry world market, second only to China as a producer of wall and floor tiles and, according to data from the Spanish Ministry for Industry, Tourism and Commerce (Subdirección General de Estudios y Planes de Actuación, 2009) , the world leader in the ceramic sanitary ware industry. Within the European Union, Spain is the leading manufacturer of ceramics: 26.11 % of all ceramics manufacturing facilities in the European Union are located in Spanish territory (Fraunhofer Institute for Systems and Innovation Research Öko-Institut, November 2009).

The Spanish ceramics industry includes the following sectors: ceramic flooring and wall coverings (ceramic floor and wall tiles, respectively), ceramic sanitary ware, bricks and roofing tiles, refractory materials, ceramics for technological applications (insulators, etc.), and ceramic objects for domestic and decorative purposes (tableware and ornaments).

This sector has witnessed a fall in production as a direct result of the continuing world economic crisis. Nevertheless, production figures for the main ceramics subsectors in Spain for 2008 were as follows: the brick and roofing tile subsector, 20 million tons (according to data from HISPALYT); the tiles and flooring subsector, 495.2 million m^2 (according to ASCER); and the ceramic sanitary ware subsector, 7 million items, providing an indication of the volume of waste involved. The percentage of items rejected for sale and thus discarded depends on the type of industry in question, on product requirements and on other technical considerations. In total, rejects account for 5-8 % of final production according to data provided by manufacturers.

The manufacturing process inevitably generates a percentage of products deemed unsuitable for sale and thus discarded, regardless of any improvements made to the manufacturing process. The two principal reasons for rejection are breakage and deformation, defects which do not affect the intrinsic properties of the ceramic material, or firing defects as a result of too little or too much heat, which do affect the material's physical and chemical properties.

Waste from the ceramics industry is classified as non-hazardous industrial waste (NHIW). According to the Integrated National Plan on Waste, 2008-2015 (España. Ministerio de Medio Ambiente y Medio Rural y Marino, 26/02/2009), NHIW encompasses all waste generated in the course of industrial production which is not classified as hazardous in Order MAM /304/2002 (España. Ministerio de Medio Ambiente, 19/02/2002), of the 8th of February, establishing the classification of waste in accordance with the European Waste Catalogue (EWC) using the following codes:

- 10 Waste from thermal processes
- 10 12 Waste from the manufacture of ceramic goods, bricks, roofing tiles and construction materials
- 10 12 08 Waste ceramics, bricks, roofing tiles and construction materials (after thermal processing)

Ceramic goods are produced from natural materials containing a high proportion of clay minerals. Following controlled thermal processing, these acquire the characteristic properties of fired clay.

These properties include durability, stable behaviour throughout its service life, chemical inertia, heat- and fire-resistance and good resistance to electricity. As regards chemical composition these materials present a highly acid nature, with a predominance of silica, aluminium and even iron oxide, in addition to other compounds present in lesser proportions (Commission, August 2007).

This chapter will address and focus on the extent to the use of ceramic waste and ceramic and sanitary ware rubble in the applications which have been developed (structural and precast ecoefficient concretes) comply with current standards. The scientific aspects of this research have been described elsewhere (Juan et al., 2010).

2. State of the art

Much research worldwide has been conducted on the recycling of these kinds of wastes, and many researchers are involved in the study of this subject. The reason for this interest can be located in the characteristics presented by these materials, which make them extremely versatile and provide them with great potential as regards being used as raw materials in various construction sector applications; they can be used as an active additive thanks to their pozzolanic properties, or as recycled aggregate in the manufacture of mortars and concretes.

Concerning the study of ceramic waste as an active additive, research into the feasibility of reusing roofing tiles as a partial substitute for cement is of particular note. Among the studies which have been conducted in this field, those reported by Ay and Unal (Ay and Unal, 2000, Ay and Unal, 2001), Bensted y Munn (Bensted and Munn, 2001) and Lavat et al. (Lavat et al., 2009) are especially relevant. These authors found that substitution percentages of below 30% had no negative effects on the mechanical behaviour of Portland cement, thus demonstrating the viability of reusing ceramic roofing tile waste in the production of pozzolanic cements.

In Spain, the authors of this chapter, all members of the Eduardo Torroja Institute of Construction Sciences - CSIC (Instituto de Ciencias de la Construcción Eduardo Torroja - CSIC), have conducted intensive research into the possibility of reusing ceramic rubble as an additive in the production of cement and the manufacture of concrete roofing tiles, (de Rojas et al., 2001a, de Rojas et al., 2001b, Sánchez de Rojas et al., 2003, de Rojas et al., 2006, De Rojas et al., 2007) (Marín Andrés and Sánchez de Rojas, 2004) (Frías et al., 2008) (Sánchez et al., 2008), the results of which will be described below.

In addition to this waste product, the reuse of waste from clay blocks has also been amply analysed, examining the viability of partial substitution of cement in the production of mortars. Results indicate that such substitution yields improved mechanical properties and durability. Examples of recently published research in this field would include studies by Silva et al. (Silva et al., 2008, Silva et al., 2009) and Naceri and Hamina (Naceri and Hamina, 2009).

The use of waste from the ceramics industry as aggregate in the production of concretes has received less international research attention than research into other types of construction and demolition waste (CDW). Nevertheless, the studies mentioned below are worthy of note.

Koyuncu et al. (Koyuncu et al., 2004), Topcu (Topcu and Guncan, 1995), de Brito et al. (de Brito et al., 2005, de Brito, 2010), Correia et al. (de Brito et al., 2005) and Bakri (Bakri et al., 2006) all concluded that it was possible to use recycled ceramic aggregate in the production of non-structural concretes, finding that these subsequently presented good abrasion resistance and tensile strength, and in paving slabs, since recycled concrete also presented increased durability. Portella et al. (Portella et al., 2006) analysed the viability of incorporating ceramic waste from electrical insulator porcelain in concrete structures, finding that reuse was possible but necessitated the use of sulphate-resistant cements due to the negative effects of certain by-products which generated an alkali-aggregate reaction. Gomes et al. (Gomes and de Brito, 2007, Gomes and de Brito, 2009) studied the viability of incorporating coarse aggregate from concrete waste and ceramic block waste in the production of new concretes, and found that structural concrete made using recycled aggregates presented satisfactory durability, but that the 4-32 mm fraction of natural aggregates cannot be totally substituted.

Lastly, Guerra et al. (Guerra et al., 2009), López et al. (Lopez et al., 2007) and Juan et al. (Juan et al., 2007, Valdes et al., 2010) studied the mechanical and physical properties of concretes in which conventional coarse aggregate had been partially substituted by coarse ceramic aggregate obtained by crushing ceramic sanitary ware, and natural fine aggregate had been substituted by powdered ceramic material, obtaining satisfactory results in both cases.

3. Ceramic sanitary ware waste as recycled coarse aggregate in ecoefficient concretes

3.1 Introduction

In this section, the viability of using waste from the ceramic sanitary ware industry as coarse aggregate in the production of structural ecoefficient concretes will be analysed. Firstly, the aggregate will be characterised, examining physical, chemical and mechanical properties. Secondly, the compliance of concretes incorporating this kind of waste with the specifications established in current standards will be analysed.

3.2 Experimental studies

3.2.1 Ceramic sanitary ware waste as coarse aggregate

The waste used for this study came from ceramic sanitary ware industry rejects, which were crushed and sieved to obtain the 12.5/4 mm granulometric fraction.
The chemical composition of this waste was strongly acidic in nature, with a predominance of silica (66.57 %) and aluminium (21.60 %). Lesser quantities of other oxides were also present, principally iron oxide (1.41 %), calcium oxide (2.41 %), sodium oxide (1.41 %), potassium oxide (2.79 %) and zirconium oxide (1.48 %). The waste did not contain chlorides, soluble sulphates or total sulphur compounds.
The aggregate employed in the production of concrete, whether natural or recycled, must present physical, chemical and mechanical properties which comply with the specifications established in the standards EN 12620 (AENOR, 2009c) and EHE-08 (Comisión Permanente del Hormigón, 2008), article 28, Appendix 15.
The properties determined in the ceramic aggregate, together with the methodology employed, are given in Table 1.

Tests	Standards	Specification
Particle size grading	EN 933-1 / EN 933-2	Maximiun and minimun size
Maximum fines content	EN 933-1 / EN 933-2	$\leq$ 1.5 % maximium percentage which passes through a 0.063 mm sieve
Water absorption	EN 1097-6	< 4.5 % of the total weight of the sample
Resistance to wear	EN 1097-2	$\leq$ 40 % of the total weight of the sample
Flakiness index	EN 933-3	< 35 % of the total weight of the sample
Organic material	EN 1744-1	No presence (point 28.7.3 - EHE-08)
Alkali-aggregate reactivity	EN 932-3	No potential reactivity

Table 1. Specifications for coarse aggregate employed in concrete production according to the standards EN 12620 and EHE-08

3.2.2 Incorporation of ceramic sanitary ware waste in concrete production

Once it had been confirmed that the new recycled aggregate was apt for concrete production, substitution percentages of 20 and 25 % of natural coarse aggregate volume were established.
The next step was to calculate the mix design of the various concretes, using the de la Peña method (Arredondo, 1968), to obtain the mix proportions for the different components: sand, gravel, ceramics, cement (CEM I 52.5 R) and water. To this end, a characteristic strength of 30 N/mm^2 and a soft consistency were established, in compliance with the recommendations given in Section 31.5 of the EHE-08.

The final step was to mix, prepare and cure the concrete test specimens following standard EN 12390-2 (AENOR, 2009a), before conducting assays to assess compliance with the specifications established by the EHE-08 for structural concretes (see Table 2).

Test	Standard	Specification
Minimum cement content	-	250 kg/m^3
Maximum water/cement ratio	-	0.65
Compressive strength (28 days)	EN 12390-3	>25 N/mm^2
Resistance to water penetration	EN 12390-8	See table 6
Total chloride content	UNE 83987	0.4 and 0.2 % of the cement weight for passive and active reinforcements, respectively

Table 2. Specifications for structural concretes, EHE-08

3.3 Results

3.3.1 Ceramic sanitary ware waste as coarse aggregate

It can be seen from Figure 1 that the particle size grading obtained for recycled ceramic aggregate is located within the granulometric zone established in Point 4.3.2 of the standard EN 12620, although it presents a lower percentage than that established by this zone for the 4/6 fraction. Furthermore, it can be observed that recycled coarse ceramic aggregate complies with the maximum content of fine aggregate (< 0.063 µm) acceptable in coarse aggregate.

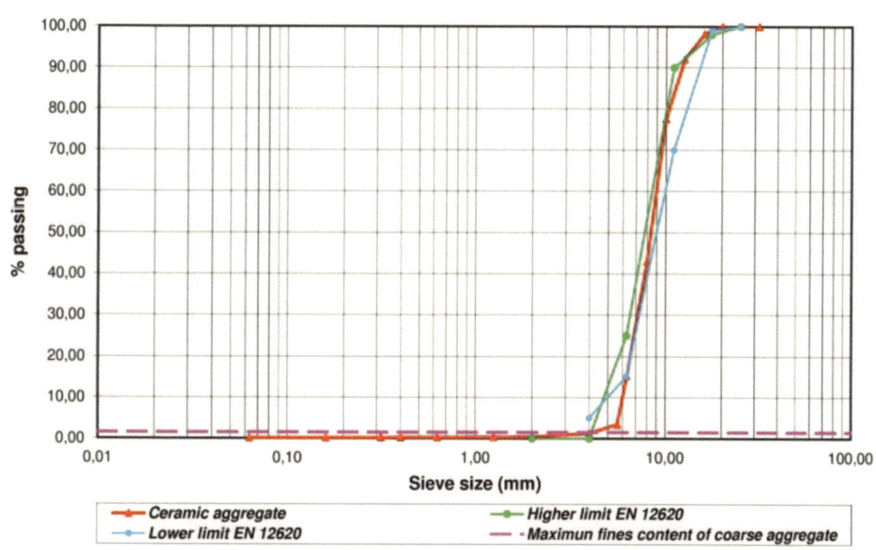

Fig. 1. Particle size grading of ceramic aggregate

The physical and mechanical properties of recycled ceramic aggregate were determined, and the results are presented in Table 3. In all cases, the properties complied with the specifications given in Table 1.

Properties	Value (of the total weight of the sample)
Water absorption	0.55
Resistance to wear	20
Flakiness index	23

Table 3. Physical and mechanical properties of ceramic aggregate

The water absorption values shown in Table 3 are lower than those found in other studies (Cachim, 2007) (Topcu and Canbaz, 2007), explained by the fact that these authors studied a different kind of ceramic waste, namely, blocks and roofing tiles, respectively.
Furthermore, it should be mentioned that this aggregate presented significant resistance to wear, with a lower value for the Los Angeles coefficient than that found for other recycled aggregates (Sánchez and Alaejos, 2006) (Martín-Morales et al., 2011), whilst also presenting a flaky morphology due both to the initial form of the waste and the subsequent crushing process.
As regards the presence of organic material which could affect time taken to harden and compressive strength, none were detected in the ceramic aggregate, which thus complied with the specifications established in the EHE-08 (see Table 1).
Lastly, with respect to the alkali-aggregate reaction, none of the minerals listed in Table 28.7.6 of the EHE-08 as being susceptible to reacting in an alkaline medium were found in the ceramic aggregate. Consequently, it can be concluded that this new recycled aggregate does not present an alkali-silica reactivity.

3.3.2 Incorporation of ceramic sanitary ware waste in concrete production

Results of concrete mix design for conventional concrete (CC) and recycled concretes incorporating 20 or 25 % of recycled aggregate (RC-20 and RC-25, respectively) are given in Table 4.

Type concrete	Material (kg/m³ concrete)					a/c
	Sand	Gravel	Ceramic	Cement	Water	
CC	548.63	1231.39	0.00	389.93	205.00	0.53
RC-20	575.65	971.64	219.96	379.10	205.00	0.54
RC-25	570.21	916.82	276.74	376.58	205.00	0.54

Table 4. Mix proportions of concretes

In the Table above, it can be seen that cement content and the water/cement ratio (w/c) comply with the specifications given in Table 1, thus ensuring satisfactory durability of the various concretes throughout their service lives and resistance to damage by chemical, physical and biological agents.
Concretes which incorporated ceramic aggregate presented greater compressive strength than conventional concrete, as shown in Figure 2. In the same figure, it can also be seen that compressive strength increased as the proportion of ceramic aggregate rose.

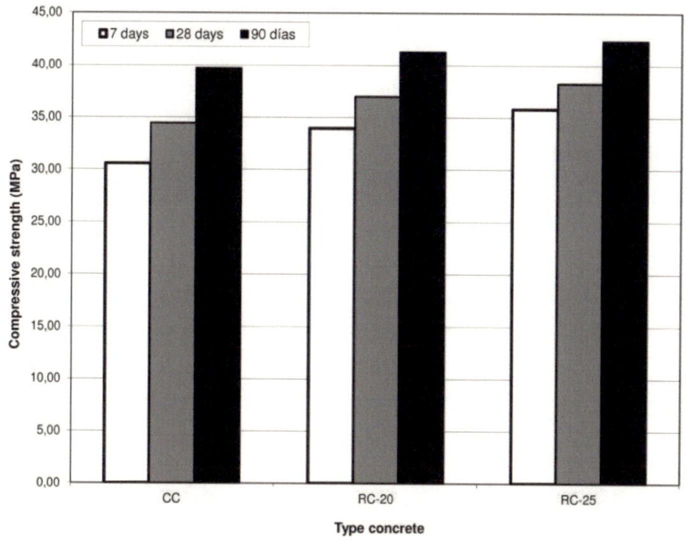

Fig. 2. Compressive strength of the concretes

The improved behaviour of the recycled concretes is primarily due to the morphology of ceramic aggregate, which provides better adhesion to the paste than natural aggregate. These results coincide with those obtained by Pacheco y Jalali (Pacheco-Torgal and Jalali, 2010) and Cachim (Cachim, 2009), who also found that recycled concretes presented improved mechanical properties.

In Figure 2 it can be seen that for all the test specimen ages assayed, the strength obtained for all concretes was better than the minimum characteristic strength required for reinforced concrete (see Table 2), as established in Point 31.4 of the EHE-08.

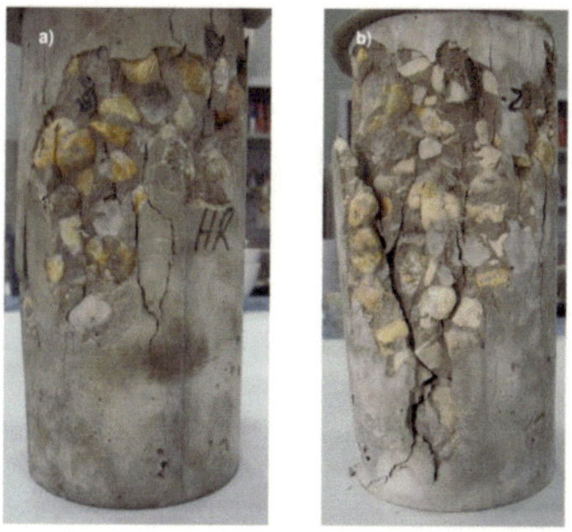

Fig. 3. Types of compressive fracture: a) Concrete CC. b) Recycled concrete RC-25

As regards the kinds of fractures observed in the various test specimens following fracture assays, these were all within the acceptable limits established in Point 8 of the standard EN 12390-3 (AENOR, 2009b). As is clearly depicted in Figure 3, the fractures present in both concretes (CC and RC-25) were similar, confirming that the incorporation of ceramic aggregate had no negative effects on the behaviour of concrete in this respect.

The results for impermeability of the recycled concretes are given in Table 5, showing maximum and average depth of water penetration under pressure. It can be seen from this table that the incorporation of ceramic aggregate did not lead to differences in maximum depth, which remained constant, whereas average depth increased slightly as the percentage of substitution rose.

Type concrete	Maximun depth (mm)	Average depth (mm)
CC	27	11
RC-20	26	12
RC-25	27	15

Table 5. Results of maximum and average depth of water penetration for concretes

Point 37.3.3 of the EHE-08 establishes that a concrete may be considered sufficiently impermeable to water if it complies with the specifications described in Table 6:

Environmental exposure class	Specification for the maximum depth (mm)	Specification for the average depth (mm)
IIIa, IIIb, IV, Qa, E, H, F, Qb (in the case of mass concrete or reinforced concrete elements)	50	30
IIIc, Qc, Qb (only in the case of pre-stressed concrete elements)	30	20

Table 6. Specifications for maximum and average depth of water penetration in concretes, as established in Point 37.3.3 of the EHE-08

From a comparative analysis of Tables 5 and 6, it can be concluded that the recycled concretes present a porous structure which renders them impermeable to water irrespective of the environment in which they are exposed, confirming their satisfactory behaviour when exposed to aggressive environmental.

Lastly, the results of the analysis to determine total chloride content in the concretes indicated the absence of chlorides at a depth of 25 mm in conventional concrete (CC), 26 mm in recycled concrete with 20% ceramic aggregate (RC-20) and 27 mm in recycled concrete with 25 % ceramic aggregate (RC-25).

According to Point 37.2.5 of the EHE-08, nominal thickness of concrete will vary depending on type of exposure, characteristic strength of the concrete, type of cement used and level of control. Consequently, assuming a service life of 100 years, the nominal thickness corresponding to the concretes produced in the present research would be 30 mm, it can be seen that the concretes incorporating ceramic aggregate comply with this durability specification.

3.4 Conclusions

Waste from the ceramic sanitary ware industry presents physical, chemical and mechanical characteristics which comply with the specifications established by current standards, confirming that they are apt for use as coarse aggregate in the production of concrete.

The recycled concretes complied with all requirements established in the EHE-08 for structural concretes, and perhaps most importantly, showed an increase in compressive strength.

The durability of the recycled concretes was similar to that of conventional concrete, confirming that they are apt for use as structural concretes in relatively aggressive environments.

To summarise, this ceramic waste can be used in the production of structural concretes, thus avoiding the use of new raw materials, reducing the generation of waste and making maximum use of the embodied energy contained in this waste, with all the consequent advantages this implies.

4. Use of ceramic rubble in the manufacture of concrete roofing tiles

4.1 Introduction

Ceramic goods are produced from natural materials containing a high proportion of clay minerals. Following a process of dehydration and controlled firing at temperatures between 700°C and 1000°C, they acquire the characteristic properties of fired clay. Thus, the manufacture of ceramic products involves high firing temperatures and these may lead to activation of the clay minerals, which then acquire pozzolanic properties. It is well-known that one of the first materials used for their pozzolanic properties were thermally treated clays, a material which has much in common with fired clay products.

Activation of the clays is achieved firstly through a process of dehydration, which starts to occur at temperatures of around 500°C, accompanied by the separation of amorphous aluminium. This latter is extremely active, and its maximum concentration is achieved at different temperatures depending on the type of mineral. Clay minerals such as kaolinite or montmorillonite, or a combination of both, acquire pozzolanic properties through controlled calcination at temperatures of between 540°C and 980°C. On the other hand, illite type clays require higher temperatures for activation, as do the schist clays containing a high proportion of vermiculite, chlorite and mica.

This section will focus on a study of ceramic rejects, investigating their application as substitute for aggregate and cement in the production of precast concrete and in particular, concrete roofing tiles. The initial phase was conducted in the laboratory, and subsequently industrial trials of concrete roofing tiles were carried out.

4.2 Experimental studies

4.2.1 Ceramic rubble as pozzolanic material: laboratory tests

The ceramic rubble used in this study was crushed and sieved to two different grades of fineness in order to provide two different kinds of ceramic material, CC1 (5000 cm^2/g) and CC2 (3500 cm^2/g).

The chemical composition of this ceramic rubble was similar to that of other pozzolanic materials, namely, it presented a strongly acidic nature where silica, aluminium and iron oxides represented around 70% of the total content in most cases.

Assessment of pozzolanic activity was carried out using an accelerated method (40ºC), studying the reaction of the material to a lime-saturated solution over time. The percentage of lime fixed by the sample was obtained by calculating the difference between the concentration of the initial lime-saturated solution and the concentration of CaO present in the solution following contact with the sample for a determined period of time.

4.2.2 Ceramic rubble in the manufacture of concrete roofing tiles: factory tests

The industrial trials were conducted at two different concrete roofing tile factories, and consisted of incorporating ceramic rubble as a component in the manufacture of concrete roofing tiles. Each factory used a different type of cement, which is indicated in each case.
Concrete roofing tiles are generally shaped using pressure or vibration of a mortar presenting suitable granulometry, composed mainly of mineral grains, pigments, water and the later incorporation of additives.

4.2.2.1 Manufacturing process

The manufacturing process comprises the following two stages:

- *Preparation of the concrete*: The mix design of components (aggregate, cement, pigments and additives) refers to the precise weight required to produce the mixture. Water is the only component added in volume rather than weight, where volume required depends on the moisture content of the aggregate.
  Generally, high intensity mixers are employed, capable of producing a homogeneous mix and obtaining excellent dispersion of pigments, to ensure the even colour required in the finished product.
  Once the mixing process is complete and the concrete presents the required consistency, the mixture is taken via a series of mechanical transport systems to the tileworks.
- *Chaping the tiles*: The tiles are shaped by automatic roofing tile machines, which compact and press the paste on moulds. The machine comprises a continuous conveyor belt of moulds, previously oiled with a mould release agent. As these pass through the machine they receive the paste, which is shaped into a tile by a mobile roller element, compacting the paste against a sloping fixed element.
  When the moulds emerge from the machine with their tiles, they are separated by a cutting system synchronised with the speed of the conveyor belt of moulds which shapes the front and back ends of the tiles. Once separated but still in its mould, each tile receives a top coating, when appropriate; if the texture required is smooth, the top coating will consist of various coats of coloured cement (slurries); if the texture required is grainy, coloured grains will also be added.
- *Curing the tiles*: Still in their moulds the tiles are deposited in special containers which are then placed into long tunnels or chambers. Here, curing is accelerated under controlled conditions of temperature and relative humidity until the tiles have hardened enough for them to be removed from the moulds and handled.
  Once curing is complete, the tiles are separated from their moulds, which are returned to the tile making machine, and enter the final stage of manufacture.
- *Selection, packing and loading*: Once separated from the moulds, those tiles presenting manufacturing defects are rejected, whilst the satisfactory tiles are given a polymer resin coating to prevent efflorescence. When identification is used, it is at this point that the tiles are stamped. The cured tiles are then transported to the packing section.

The most common packing method is to group the tiles vertically in packs of approximately 1 metre long. These packs are strapped to ensure stability.

The packs are then loaded three at a time onto wooden pallets, in a single or double layer, and they are sometimes wrapped in stretch or retractable plastic film.

Concrete roofing tiles must comply with a series of standard specifications included in standard EN 490 (AENOR, 2004a), using the assays described in standard EN 491 (AENOR, 2004b), and presented here in Table 7 and Figure 4.

Tests	**Specification**
Dimensions	± 4 mm; cover width ± 5 mm.
Mass	± 10 %
Mechanical strength	Minimum flexural strength 2.000 N
Water-tightness	Under water, no drops shall fall before 20 hours
Durability	Conformity to water-tightness and mechanical strength requirements after 25 freeze-thaw cycles +20 °C to -20 °C.
Nib support	1 min in vertical
Fire resistance	Class A1 for fire reaction; B_{roof} for external fire resistance

Table 7. EN 490 specifications for the concrete tiles

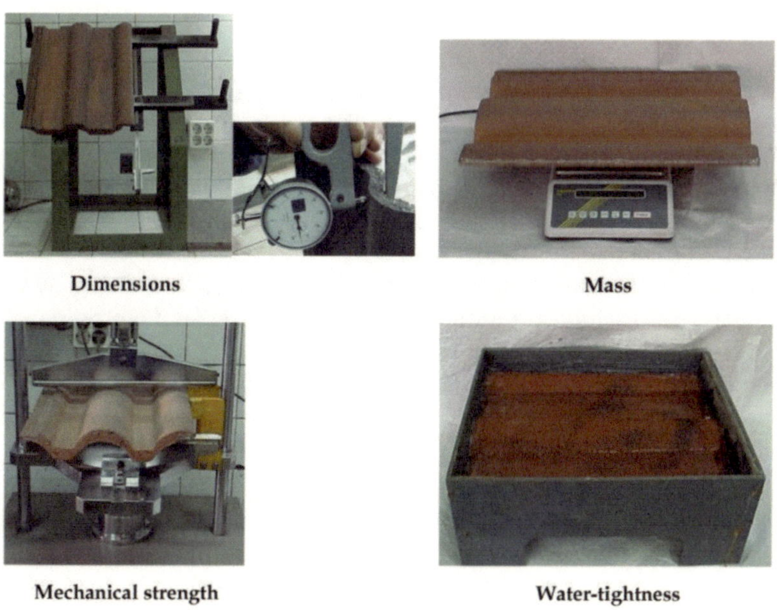

Fig. 4. Assays according to EN 491

4.2.2.2 Incorporation of ceramic rubble

In order to conduct these tests in the factory, the ceramic materials were combined into a single product (C) for incorporation as a natural aggregate substitute. To this end, the ceramic rubble was crushed and sieved so that the mixture (natural aggregate and ceramic material) corresponded to the granulometry used in the concrete roofing tile factory (Figure 5).

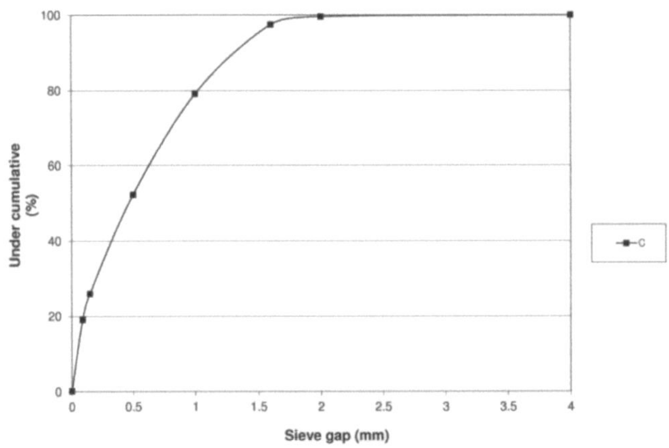

Fig. 5. Particle size grading of ceramic rubble used as aggregate

In order to be used as a pozzolanic material, it was necessary to crush the ceramic rubble until a suitable grade of fineness was achieved. For this study, the samples were crushed to two different grades of fineness, expressed as Blaine's specific surface:

- Ceramic material C1: 3500 cm^2/g
- Ceramic material C2: 3200 cm^2/g

4.3 Results

4.3.1 Ceramic rubble as pozzolanic material: laboratory tests

The results presented in Figure 6 show that the waste ceramic material presented an acceptable level of pozzolanic activity, since the percentage of lime fixed at one day in the CC1 sample was 46% of total available lime. In the case of the CC2 sample, the amount of lime fixed at one day was less (7%), due to the lower specific surface value, since grade of fineness plays a decisive role in the early stages.

At 7 days, the activity of both samples was more evenly matched, fixing over 50% of the lime. This activity was maintained until the end of the test at 90 days, by which time over 80% of the lime had been fixed. These results indicate that the firing temperature of these ceramic materials (around 900° C) is sufficient to activate the clays and for these to acquire pozzolanic properties (Murat, 1983, Sayanam et al., 1989, Johansson and Andersen, 1990, He et al., 1995, de Rojas et al., 2006).

4.3.2 Ceramic rubble in the manufacture of concrete roofing tiles: factory tests

The transverse tensile strength test established in the specifications of standard EN 490 were conducted on the concrete roofing tiles thus produced using the method described in standard EN 491. Results for median strength compared with the control tiles are given in Figure 7.

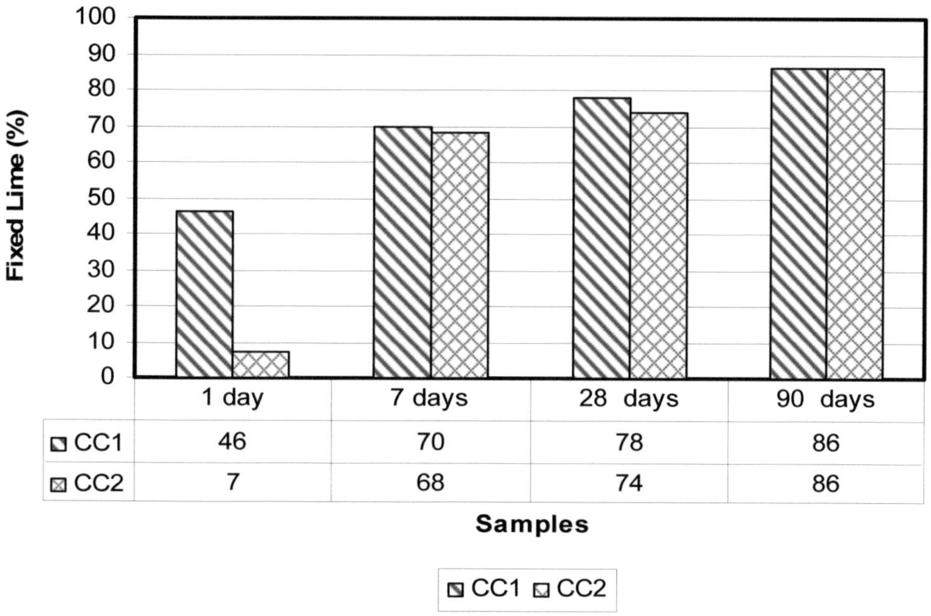

Fig. 6. Pozzolanic activity of the two types of ceramic rubble

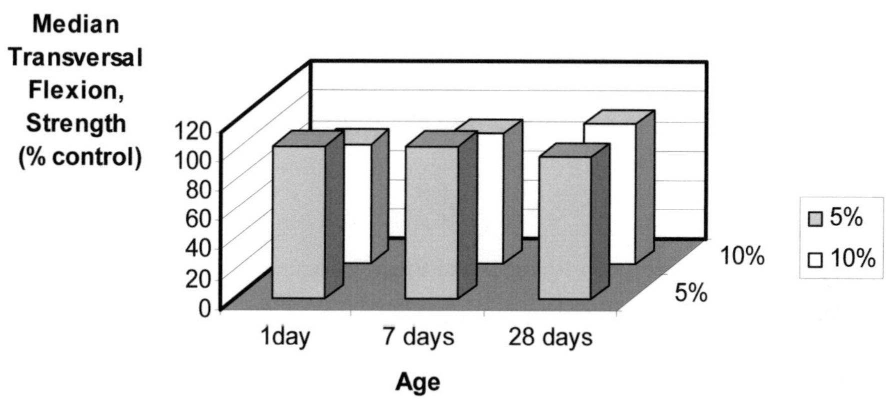

Fig. 7. Transversal flexion strength (% compared to control mortar): Ceramic rubble as aggregate

These tests of ceramic rubble as aggregate were conducted on products manufactured with a 5% and 10% substitution of natural siliceous sand by ceramic material, using the granulometries defined earlier. This factory used CEM I 42.5R cement, as established by current standards (AENOR, 2000).

The mean transverse tensile strength of the tiles manufactured with 5% of sand substituted by ceramic aggregate was very similar to that corresponding to concrete tiles without ceramic material, although at 28 days these values were slightly lower (3% lower).

When 10% of sand was substituted by ceramic material, mean transverse tensile strength was lower at 24 hours (19% lower). However, over time, the results became more equivalent to those for traditional concrete tiles (12% lower at 7 days and 5% lower at 28 days).

The tiles incorporating ceramic material performed well in the impermeability and frost resistance tests described in standard EN 490, conducted following the test methodology given in standard EN 491. However, after prolonged cycles of frost resistance tests (>400 cycles) which far exceeded the number of cycles (25) specified in the standard, it was observed that resulting strength was lower than that of the control tiles.

The factory tests of ceramic rubble as a pozzolanic material were conducted using materials with two different Blaine's specific surface values, C1 (3500 cm^2/g) and C2 (3200 cm^2/g), and three percentages of cement substitution: 5, 10 and 15%. The cement used by the factory was CEM II/A-V 42.5R.

Figure 8 presents the characteristic transverse tensile strength for tiles incorporating ceramic material at 9 hours, which is when the tiles were removed from their moulds, and at 7 and 28 days.

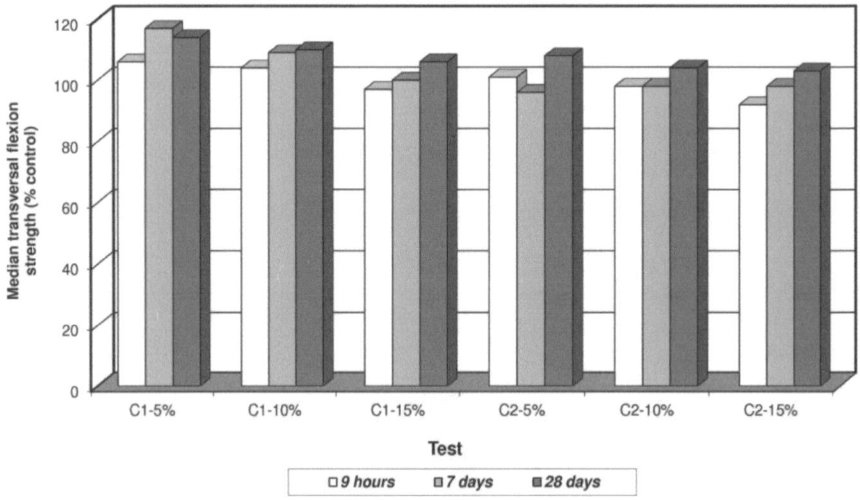

Fig. 8. Transverse tensile strength (% compared to control mortar): Ceramic rubble as pozzolanic material

As can be seen in the Figure 8, the tiles incorporating ceramic material presented transverse tensile strength values that were similar to or even higher than those for normal tiles (control tiles). The best results were obtained with cement substitution percentages of 5% and 10% using the C1 ceramic material.

These results demonstrate that the ceramic material employed acts as a pozzolanic material and, consequently, contributes to the development of mechanical strength. The use of such material thus enables the strength values required to be obtained whilst using less cement.

The results of impermeability tests conducted on roofing tiles incorporating ceramic material were satisfactory, complying with the specifications established in standards EN 490 and 491.

4.4 Conclusion

The ceramic material used (rejects, or rubble), presents acceptable pozzolanic characteristics, since the firing temperatures involved in the manufacture of these products is ideal for activating the clays contained, which thus acquire pozzolanic properties.

In the early stages, the pozzolanic activity depends more on the grade of fineness of the rubble than on the firing process it has previously undergone, as long as firing temperatures are within the normal industrial range (whether or not this is adequate for commercial ceramic products).

Ceramic rubble can be used in the production of precast concrete, whether as aggregate, in substitution of a percentage of natural aggregate, or as an active additive with pozzolanic characteristics. The latter application gave best technical results and also offers the most economic benefit. Additives endow the cements with positive characteristics, by contributing to an increase in mechanical strength in the medium and long term, and also enhance the chemical resistance of concrete to aggressive agents, which has a positive impact on the material's service life (Calleja, 1983).

As regards economic benefits, these are derived from energy savings in the production of cement. The substitution of a material which requires expensive thermal processing, such as clinkerization (Soria Santamaria, 1983), by a less expensive material in energy terms (such as an industrial reject, usually accumulated in dumps), represents a reduction in energy consumption even when it requires prior crushing, and a positive contribution to conservation of the environment.

5. Acknowledgements

This research has been made possible through funding from the Spanish Ministry for Science and Innovation's national project ref. BIA2010-21194-C03-(01-02), "Recovery of ceramic CDW as a pozzolanic material and as recycled aggregate. Influence of pollutants. Optimisation in recycling plants.", and through funding from the University of León project, ref. ULE2009-5, "Recycled ecoefficient concretes using the stony fraction of ceramics from construction and demolition waste".

6. References

AENOR 2000. UNE-EN 197-1:2000. Cement - Part 1: Composition, specifications and conformity criteria for common cements. Madrid: Asociación Española de Normalización y Certificación.

AENOR 2004a. EN 490:2004. Concrete roofing tiles and fittings for roof covering and wall cladding - Product specifications. Madrid: Asociación Española de Normalización y de Certificación.

AENOR 2004b. EN 491:2004. Concrete roofing tiles and fittings for roof covering and wall cladding - Test methods. Madrid: Asociación Española de Normalización y de Certificación.

AENOR 2009a. UNE-EN 12390-2:2009. Testing hardened concrete - Part 2: Making and curing specimens for strength tests. Madrid: Asociación Española de Normalización y Certificación.

AENOR 2009b. UNE-EN 12390-3:2009. Testing hardened concrete - Part 3: Compressive strength of test specimens. Madrid: Asociación Española de Normalización y Certificación.

AENOR 2009c. UNE-EN 12620:2003+A1:2009. Aggregates for concrete. Madrid: Asociación Española de Normalización y Certificación.

ARREDONDO, F. 1968. *Dosificación de hormigones,* Madrid, Instituto de Ciencias de la Construcción Eduardo Torroja (CSIC).

AY, N. & UNAL, M. 2000. The use of waste ceramic tile in cement production. *Cement and Concrete Research,* 30, 497-499.

AY, N. & UNAL, M. 2001. Reply to the discussion by J. Bensted and J. Munn of the paper "The use of waste ceramic tile in cement production". *Cement and Concrete Research,* 31, 163-163.

BAKRI, M. M., HUSSIN, K., MOHD, C., BAHARIN, S., RAMLY, R. & KHAIRIATUN, N. 2006. Concrete Ceramic Waste Slab (CCWS). *Journal of Engineering Research & Education,* 3, 139-145.

BENSTED, J. & MUNN, J. 2001. A discussion of the paper "The use of waste ceramic tile in cement production" by N. Ay and M. Unal. *Cement and Concrete Research,* 31, 161-162.

CACHIM, P. B. 2007. Concrete produced using crushed bricks as aggregate. *Portugal Sb07 - Sustainable Construction, Materials and Practices: Challenge of the Industry for the New Millennium, Pts 1 and 2,* 950-956.

CACHIM, P. B. 2009. Mechanical properties of brick aggregate concrete. *Construction and Building Materials,* 23, 1292-1297.

CALLEJA, J. 1983. Adiciones y cementos con adiciones. *Materiales de Construcción,* 25-52.

COMISIÓN PERMANENTE DEL HORMIGÓN 2008. *Instrucción Hormigón Estructural. EHE-08,* Madrid, Ministerio de Fomento. Centro de Publicaciones.

COMMISSION, E. August 2007. *Reference Document on Best Available Techniques in the Ceramic Manufacturing Industry [en línea]* [Online]. European Commission.

DE BRITO, J. 2010. Abrasion resistance of concrete made with recycled aggregates. *International Journal of Sustainable Engineering,* 3, 58-64.

DE BRITO, J., PEREIRA, A. S. & CORREIA, J. R. 2005. Mechanical behaviour of non-structural concrete made with recycled ceramic aggregates. *Cement & Concrete Composites,* 27, 429-433.

DE ROJAS, M. I. S., FRIAS, M., RIVERA, J., ESCORIHUELA, M. J. & MARIN, F. P. 2001a. Research about the pozzolanic activity of waste materials from calcined clay. *Materiales De Construccion,* 51, 45-52.

DE ROJAS, M. I. S., MARIN, F., RIVERA, J. & FRIAS, M. 2006. Morphology and properties in blended cements with ceramic wastes as a pozzolanic material. *Journal of the American Ceramic Society,* 89, 3701-3705.

DE ROJAS, M. I. S., MARIN, F. P., FRIAS, M. & RIVERA, J. 2001b. Viability of utilization of waste materials from ceramic products in precast concretes. *Materiales De Construccion,* 51, 149-161.

DE ROJAS, M. I. S., MARIN, F. P., FRIAS, M. & RIVERA, J. 2007. Properties and performances of concrete tiles containing waste fired clay materials. *Journal of the American Ceramic Society,* 90, 3559-3565.

ESPAÑA. MINISTERIO DE MEDIO AMBIENTE 19/02/2002. Orden MAM/304/2002, de 8 de febrero, por la que se publican las operaciones de valorización y eliminación de residuos y la lista europea de residuos. Boletín Oficial del Estado.

ESPAÑA. MINISTERIO DE MEDIO AMBIENTE Y MEDIO RURAL Y MARINO 26/02/2009. Resolución de 20 de enero de 2009, de la Secretaría de Estado de Cambio Climático, por la que se publica el Acuerdo del Consejo de Ministros por el que se aprueba el Plan Nacional Integrado de Residuos para el período 2008-2015. Boletín Oficial del Estado.

FRAUNHOFER INSTITUTE FOR SYSTEMS AND INNOVATION RESEARCH ÖKO-INSTITUT November 2009. *Methodology for the free allocation of emission allowances in the EU ETS post 2012. Sector Report for the Ceramics Industry*, European Commission.

FRÍAS, M., SÁNCHEZ DE ROJAS, M. I. & RODRÍGUEZ, O. Novedades en el reciclado de materiales en el sector de la construcción: adiciones puzolánicas. II Jornadas de Investigación en Construcción, 2008 Madrid. 1-10.

GOMES, M. & DE BRITO, J. 2007. Structural Concrete with Incorporation of Coarse Recycled Concrete and Ceramic Aggregates. *Portugal Sb07 - Sustainable Construction, Materials and Practices: Challenge of the Industry for the New Millennium, Pts 1 and 2*, 887-894.

GOMES, M. & DE BRITO, J. 2009. Structural concrete with incorporation of coarse recycled concrete and ceramic aggregates: durability performance. *Materials and Structures*, 42, 663-675.

GUERRA, I., VIVAR, I., LLAMAS, B., JUAN, A. & MORAN, J. 2009. Eco-efficient concretes: The effects of using recycled ceramic material from sanitary installations on the mechanical properties of concrete. *Waste Management*, 29, 643-646.

HE, C. L., OSBAECK, B. & MAKOVICKY, E. 1995. POZZOLANIC REACTIONS OF 6 PRINCIPAL CLAY-MINERALS - ACTIVATION, REACTIVITY ASSESSMENTS AND TECHNOLOGICAL EFFECTS. *Cement and Concrete Research*, 25, 1691-1702.

JOHANSSON, S. & ANDERSEN, P. J. 1990. POZZOLANIC ACTIVITY OF CALCINED MOLER CLAY. *Cement and Concrete Research*, 20, 447-452.

JUAN, A., LÓPEZ, V. M., MORÁN, J. M. & GUERRA, M. I. 2007. Reutilización de restos de cerámica blanca como áridos para la elaboración de hormigones. *Ingeniería Civil*, 1-5.

JUAN, A., MEDINA, C., GUERRA, M. I., MORÁN, J. M., AGUADO, P. J., SÁNCHEZ, M. I., FRÍAS, M. & RODRÍGUEZ, O. 2010. Re-use of ceramic wastes in construction. *In:* WUNDERLICH, W. (ed.) *Ceramic Materials*. Rijeka, Croatia: Sciyo.

KOYUNCU, H., GUNEY, Y., YILMAZ, G., KOYUNCU, S. & BAKIS, R. 2004. Utilization of ceramic wastes in the construction sector. *Euro Ceramics Viii, Pts 1-3*, 264-268, 2509-2512.

LAVAT, A. E., TREZZA, M. A. & POGGI, M. 2009. Characterization of ceramic roof tile wastes as pozzolanic admixture. *Waste Management*, 29, 1666-1674.

LOPEZ, V., LLAMAS, B., JUAN, A., MORAN, J. M. & GUERRA, I. 2007. Eco-efficient concretes: Impact of the use of white ceramic powder on the mechanical properties of concrete. *Biosystems Engineering*, 96, 559-564.

MARTÍN-MORALES, M., ZAMORANO, M., RUIZ-MOYANO, A. & VALVERDE-ESPINOSA, I. 2011. Characterization of recycled aggregates construction and demolition waste for concrete production following the Spanish Structural Concrete Code EHE-08. *Construction and Building Materials*, 25, 742-748.

MARÍN ANDRÉS, F. P. & SÁNCHEZ DE ROJAS, M. I. Sustitución parcial de materias primas por cascote cerámico en la fabricación de tejas de hormigón. *In:* (IETCC), I. D. C. D. L. C. E. T., ed. Reciclado de Materiales en Construcción. IV Semana de la Ciencia, 2004 Madrid. 11-23.

MURAT, M. 1983. Hydration reaction and hardening of calcined clays and related minerals. I. Preliminary investigation on metakaolinite. *Cement and Concrete Research,* 13, 259-266.

NACERI, A. & HAMINA, M. C. 2009. Use of waste brick as a partial replacement of cement in mortar. *Waste Management,* 29, 2378-2384.

PACHECO-TORGAL, F. & JALALI, S. 2010. Reusing ceramic wastes in concrete. *Construction and Building Materials,* 24, 832-838.

PORTELLA, K. F., JOUKOSKI, A., FRANCK, R. & DERKSEN, R. 2006. Secondary recycling of electrical insulator porcelain waste in Portland concrete structures: determination of the performance under accelerated aging. *Cerâmica,* 155-167.

SAYANAM, R. A., KALSOTRA, A. K., MEHTA, S. K., SINGH, R. S. & MANDAL, G. 1989. STUDIES ON THERMAL TRANSFORMATIONS AND POZZOLANIC ACTIVITIES OF CLAY FROM JAMMU REGION (INDIA). *Journal of Thermal Analysis,* 35, 99-106.

SILVA, J., DE BRITO, J. & VEIGA, R. 2008. Fine ceramics replacing cement in mortars Partial replacement of cement with fine ceramics in rendering mortars. *Materials and Structures,* 41, 1333-1344.

SILVA, J., DE BRITO, J. & VEIGA, R. 2009. Incorporation of fine ceramics in mortars. *Construction and Building Materials,* 23, 556-564.

SORIA SANTAMARIA, F. 1983. Las puzolanas y el ahorro energético en los materiales de construcción. *Materiales de Construcción,* 69-84.

SUBDIRECCIÓN GENERAL DE ESTUDIOS Y PLANES DE ACTUACIÓN. 2009. *Fichas Sectoriales 2009 [en línea]* [Online]. Ministerio de Industria, Turismo y Comercio.

SÁNCHEZ DE ROJAS, M. I., FRÍAS, M., RIVERA, J. & MARÍN, F. P. Waste products from prefabricated ceramic materials as pozzolanic addition. 11th International Congress on the Chemistry of Cement, 2003 Durban, South África.

SÁNCHEZ, M. & ALAEJOS, P. 2006. Influencia del árido reciclado en las propiedades del hormigón estructural. *Cemento - Hormigón,* 54-61.

SÁNCHEZ, M. I., FRÍAS, M., RIVERA, J. & MARÍN, F. P. 2008. Aprovechamiento del cascote cerámico como material puzolánico. *Cemento-Hormigón,* 32-41.

TOPCU, I. B. & CANBAZ, M. 2007. Utilization of crushed tile as aggregate in concrete. *Iranian Journal of Science and Technology Transaction B-Engineering,* 31, 561-565.

TOPCU, I. B. & GUNCAN, N. F. 1995. USING WASTE CONCRETE AS AGGREGATE. *Cement and Concrete Research,* 25, 1385-1390.

VALDES, A. J., MARTINEZ, C. M., ROMERO, M. I. G., GARCIA, B. L., DEL POZO, J. M. M. & VEGAS, A. T. 2010. Re-use of construction and demolition residues and industrial wastes for the elaboration or recycled eco-efficient concretes. *Spanish Journal of Agricultural Research,* 8, 25-34.